The Oxford Handbook of Molecular Psychology

OXFORD LIBRARY OF PSYCHOLOGY

Editor in Chief PETER E. NATHAN

The Oxford Handbook of Molecular Psychology

Edited by

Turhan Canli

OXFORD
UNIVERSITY PRESS

OXFORD
UNIVERSITY PRESS

Oxford University Press is a department of the University of
Oxford. It furthers the University's objective of excellence in research,
scholarship, and education by publishing worldwide.

Oxford New York
Auckland Cape Town Dar es Salaam Hong Kong Karachi
Kuala Lumpur Madrid Melbourne Mexico City Nairobi
New Delhi Shanghai Taipei Toronto

With offices in
Argentina Austria Brazil Chile Czech Republic France Greece
Guatemala Hungary Italy Japan Poland Portugal Singapore
South Korea Switzerland Thailand Turkey Ukraine Vietnam

Published in the United States of America by
Oxford University Press
198 Madison Avenue, New York, NY 10016

Library of Congress Cataloging-in-Publication Data
The Oxford handbook of molecular psychology / edited by Turhan Canli.
pages cm.—(Oxford library of psychology)
Includes bibliographical references and index.
ISBN 978–0–19–975388–8
1. Neuropsychology. 2. Neurochemistry. I. Canli, Turhan, editor.
QP360.O94 2015
612.8—dc23
2014025919

9 8 7 6 5 4 3 2 1
Printed in the United States of America
on acid-free paper

To Anne Moyer, Ian Moyer Canli, and all of my family.

SHORT CONTENTS

OXFORD LIBRARY OF PSYCHOLOGY

The *Oxford Library of Psychology*, a landmark series of handbooks, is published by Oxford University Press, one of the world's oldest and most highly respected publishers, with a tradition of publishing significant books in psychology. The ambitious goal of the *Oxford Library of Psychology* is nothing less than to span a vibrant, wide-ranging field and, in so doing, to fill a clear market need.

Encompassing a comprehensive set of handbooks, organized hierarchically, the *Library* incorporates volumes at different levels, each designed to meet a distinct need. At one level are a set of handbooks designed broadly to survey the major subfields of psychology; at another are numerous handbooks that cover important current focal research and scholarly areas of psychology in depth and detail. Planned as a reflection of the dynamism of psychology, the *Library* will grow and expand as psychology itself develops, thereby highlighting significant new research that will impact on the field. Adding to its accessibility and ease of use, the *Library* will be published in print and, later on, electronically.

The *Library* surveys psychology's principal subfields with a set of handbooks that capture the current status and future prospects of those major subdisciplines. The initial set includes handbooks of social and personality psychology, clinical psychology, counseling psychology, school psychology, educational psychology, industrial and organizational psychology, cognitive psychology, cognitive neuroscience, methods and measurements, history, neuropsychology, personality assessment, developmental psychology, and more. Each handbook undertakes to review one of psychology's major subdisciplines with breadth, comprehensiveness, and exemplary scholarship. In addition to these broadly-conceived volumes, the *Library* also includes a large number of handbooks designed to explore in depth more specialized areas of scholarship and research, such as stress, health and coping, anxiety and related disorders, cognitive development, or child and adolescent assessment. In contrast to the broad coverage of the subfield handbooks, each of these latter volumes focuses on an especially productive, more highly focused line of scholarship and research. Whether at the broadest or most specific level, however, all of the *Library* handbooks offer synthetic coverage that reviews and evaluates the relevant past and present research and anticipates research in the future. Each handbook in the *Library* includes introductory and concluding chapters written by its editor to provide a roadmap to the handbook's table of contents and to offer informed anticipations of significant future developments in that field.

An undertaking of this scope calls for handbook editors and chapter authors who are established scholars in the areas about which they write. Many of the nation's and world's most productive and best-respected psychologists have agreed to edit *Library* handbooks or write authoritative chapters in their areas of expertise.

For whom has the *Oxford Library of Psychology* been written? Because of its breadth, depth, and accessibility, the *Library* serves a diverse audience, including graduate students in psychology and their faculty mentors, scholars, researchers, and practitioners in psychology and related fields. Each will find in the *Library* the information they seek on the subfield or focal area of psychology in which they work or are interested.

Befitting its commitment to accessibility, each handbook includes a comprehensive index, as well as extensive references to help guide research. And because the *Library* was designed from its inception as an online as well as print resource, its structure and contents will be readily and rationally searchable online. Further, once the *Library* is released online, the handbooks will be regularly and thoroughly updated.

In summary, the *Oxford Library of Psychology* will grow organically to provide a thoroughly informed perspective on the field of psychology, one that reflects both psychology's dynamism and its increasing interdisciplinarity. Once published electronically, the *Library* is also destined to become a uniquely valuable interactive tool, with extended search and browsing capabilities. As you begin to consult this handbook, we sincerely hope you will share our enthusiasm for the more than 500-year tradition of Oxford University Press for excellence, innovation, and quality, as exemplified by the *Oxford Library of Psychology*.

Peter E. Nathan
Editor-in-Chief
Oxford Library of Psychology

ABOUT THE EDITOR

Turhan Canli is Associate Professor of psychology and radiology at Stony Brook University, where he is the founder and director of the *SCAN* (Social, Cognitive, and Affective Neuroscience) Center and past director of the graduate program in genetics. Dr. Canli's research focuses on individual differences in emotion, personality, and social behavior and integrates the tools of psychology, neuroscience, and molecular biology.

CONTRIBUTORS

Christina S. Barr
Chief, Section of Comparative Behavioral Genomics
NIH/NIAAA/Laboratory of Neurogenetics
Bethesda, MD

Catherine E. Barrett
Postdoctoral Fellow
Division of Behavioral Neuroscience and Psychiatric Disorders
Yerkes National Primate Research Center
Emory University
Atlanta, GA

Alessandro Bartolomucci
Department of Integrative Biology and Physiology
University of Minnesota
Minneapolis, MN

Ryan Bogdan
Director, BRAIN Lab
Assistant Professor
Department of Psychology, Division of Biology and Biomedical Sciences
Washington University
St. Louis, MO

Joshua W. Buckholtz
Assistant Professor of Psychology
Director, Systems Neuroscience of Psychopathology Lab
Center for Brain Science
Harvard University
Boston, MA

Turhan Canli
Associate Professor, Departments of Psychology and Radiology, and Programs in Genetics and in Neuroscience
Director, *SCAN* (Social, Cognitive, and Affective Neuroscience) Center
Past Director, Graduate Program in Genetics
Senior Fellow, Center for Medical Humanities, Compassionate Care, and Bioethics
Stony Brook, NY

Valeria Carola
Department of Experimental Neurosciences
IRCSS Fondazione Santa Lucia
Rome, Italy

Robin Chark
Department of Marketing
National University of Singapore
Singapore

Michael C. Chen
Postdoctoral Research Fellow
Beth Israel Deaconess Medical Center
Harvard Medical School, Department of Neurology
Boston, MA

Soo Hong Chew
Professor
Department of Economics
National University of Singapore
Singapore

Eco de Geus
Professor, Department of Biological Psychology
Director, EMGO+ Institute, VU University
Amsterdam, The Netherlands

Richard P. Ebstein
Professor, Department of Psychology
National University of Singapore
Singapore

Russell D. Fernald
Benjamin Scott Crocker Professor
Department of Biology
Stanford University
Stanford, CA

Joel Gelernter
Foundations Fund Professor of Psychiatry
Professor of Genetics and Neurobiology
Yale University School of Medicine & VA CT Healthcare Center
West Haven, CT

Ian H. Gotlib
David Starr Jordan Professor and Chair
Department of Psychology
Stanford University
Stanford, CA

Ahmad R. Hariri
Professor of Psychology & Neuroscience
Duke University
Durham, NC

Luke W. Hyde
Director, Michigan Neurogenetics and Developmental Psychopathology (MiND) Laboratory
Assistant Professor, Department of Psychology
Research Assistant Professor, Center for Human Growth and Development
Research Associate, Survey Research Center at the Institute for Social Research
University of Michigan
Ann Arbor, MI

Sheena A. Josselyn
Senior Scientist
Program in Neurosciences & Mental Health
Hospital for Sick Children
Associate Professor in the Departments of Psychology, Physiology and the Institute of Medical Sciences
University of Toronto
Toronto, Canada

Aniko Korosi
Assistant Professor
Swammerdam Institute for Life Sciences
Center for Neuroscience
University of Amsterdam
Amsterdam, the Netherlands

Poh San Lai
Research Associate Professor
Department of Paediatrics
Yong Loo Lin School of Medicine
National University of Singapore
Singapore

Paul J. Lucassen
Professor, Swammerdam Institute for Life Sciences
Center for Neuroscience
University of Amsterdam
Amsterdam, the Netherlands

Karen P. Maruska
Assistant Professor
Department of Biological Sciences
Louisiana State University
Baton Rouge, LA

Peter McGuffin
Emeritus Professor of Psychiatric Genetics
MRC SGDP Centre
Institute of Psychiatry, King's College London
London, UK

Andreas Meyer-Lindenberg
Director, Central Institute of Mental Health, Mannheim, Germany
Chair, Department of Psychiatry and Psychotherapy
University of Heidelberg
Heidelberg, Germany

Christian Montag
Department of Psychology
University of Ulm
Ulm, Germany

Stephen J. Morse
Ferdinand Wakeman Hubbell Professor of Law
Professor of Psychology and Law in Psychiatry
Associate Director, Center for Neuroscience & Society
University of Pennsylvania Law School
Philadelphia, PA

Eva Naninck
Ph.D. Student, Swammerdam Institute for Life Sciences
Center for Neuroscience
University of Amsterdam
Amsterdam, the Netherlands

Melanie Neijts
Ph.D. Student, Department of Biological Psychology
EMGO+ Institute, VU University
Amsterdam, the Netherlands

Maria Razzoli
Department of Integrative Biology and Physiology
University of Minnesota
Minneapolis, MN

Martin Reuter
Department of Psychology
Center for Economics and Neuroscience
University of Bonn
Bonn, Germany

James J. H. Rucker
Clinical Lecturer in Psychiatric Genetics
MRC SGDP Centre
Institute of Psychiatry, Kings College London
London, UK

Derya Sargin
Postdoctoral Fellow
Program in Neurosciences & Mental Health
Hospital for Sick Children
Toronto, Canada

Stephen J. Simpson
Professor, ARC Laureate Fellow
The Charles Perkins Centre and School of Biological Sciences
D17 - The Hub, Charles Perkins Centre
University of Sydney
NSW, Australia

Paul A. Stevenson
Professor
University of Leipzig
Faculty for Biosciences, Pharmacy & Psychology
Institute of Biology
Leipzig, Germany

Rene van Lien
Ph.D. Student, Department of Biological Psychology
EMGO+ Institute, VU University
Amsterdam, the Netherlands

Hasse Walum
Department of Medical Epidemiology and Biostatistics
Karolinska Institute
Stockholm, Sweden

Lars Westberg
Institute of Neuroscience and Physiology, Section for Pharmacology
University of Gothenburg
Gothenburg, Sweden

Gonneke Willemsen
Associate Professor, Department of Biological Psychology
EMGO+ Institute, VU University
Amsterdam, the Netherlands

Chen Yan
Ph.D. Candidate
Program in Neurosciences & Mental Health
Hospital for Sick Children
Department of Physiology
Institute of Medical Sciences
University of Toronto
Toronto, Canada

Larry J. Young
William P. Timmie Professor, Department of Psychiatry and Behavioral Sciences
Director, Center for Translational Social Neuroscience
Chief, Division of Behavioral Neuroscience and Psychiatric Disorders
Yerkes National Primate Research Center
Emory University
Atlanta, GA

Songfa Zhong
Assistant Professor
Department of Economics
National University of Singapore
Singapore

TABLE OF CONTENTS

PART 1

Introduction and Fundamentals

CHAPTER

1 Molecular Psychology: A Brief History and Introduction

Turhan Canli

Abstract

Molecular psychology is the study of behavior and its underlying brain systems using the tools of molecular biology. Although a variety of biological tools have been used to discover the relation of brain and behavior since the beginning of psychology, these have largely been confined to nonhuman animal models or brain-damaged patients. Technical advances in the past two decades in molecular genetics and in noninvasive brain imaging have transformed the field. These advances made it possible to conduct detailed examinations—across all species and subjects including healthy human volunteers—of the relation among genes, brains, and behavior; gene-by-environment (G×E) interactions; and the underlying regulatory mechanisms of gene expression through life experience and other environmental variables. This chapter provides a brief history of the major milestones of this field, along with updates from the contemporary literature, and introduces the reader to the contents of this volume.

Key Words: history, genetics, neuroscience, behavior, polymorphism, gene-environment interactions, brain, individual differences

The origins of psychology are firmly rooted in biology. The first psychology laboratory in the world was established in Leibzig, Germany, in 1879, by Wilhelm Wundt, a physiologist. William James's 1890 book *Principles of Psychology* begins with a review of brain function. Ivan Pavlov, best known among psychologists and lay people for his work on classical conditioning, received the 1904 Nobel Prize for his work on the physiology of digestion. Although many branches of psychology formed since those early years, the biological study of behavior never ceased to be conducted. Accordingly, researchers used whatever cutting-edge tools were available at the time. These tools were primarily applied to animal models of behavior, with a particular focus on brain structure and function.

Investigators with an interest in the biological basis of human behavior focused on genetics and the heritability of traits, or they worked with brain-damaged patients to understand human brain function. These research fields were transformed by the development of molecular genetics and noninvasive neuroimaging, respectively. These two technologies were the catalysts for launching molecular psychology because they made it possible to apply the tools of molecular biology and genetics to the study of behavior and its underlying brain systems across all species, including healthy human subjects.

In this chapter, I provide a brief history of some of these developments by sketching out a timeline along major milestones in the field and updates from the current literature in these domains: (1) the discovery of specific gene polymorphisms associated with individual differences in behavior, (2) the discovery of associated brain systems, (3) gene-by-environment (G×E) interactions in behavior, (4) G×E interactions in the brain, (5) their underlying epigenetic regulatory mechanisms, and (6) the development and future of optogenetics. I then conclude the chapter with an overview of the readings in this volume.

I. From Heritability to Candidate Gene Polymorphisms

Although twin studies have long illustrated the heritability of personality traits (Defries, McClearn, McGuffin, & Plomin, 2000; Dilalla & Gottesman, 2004), specific genes associated with any of these traits were unknown. This changed in 1996 with the first study to report an association between a complex behavioral trait, neuroticism, and a specific gene polymorphism of the serotonin (5-hydroxytryptamine, or 5-HT) transporter gene (*5-HTT*), which was conducted by Lesch and colleagues (Lesch et al., 1996). The gene contains a common deletion of a 44-base pair segment in its promoter region, which is the so-called "short" (as opposed to the "long") allele, and this polymorphism is known as the serotonin transporter-linked polymorphic region, 5-HTTLPR. Participants completed self-report questionnaires on five personality traits (neuroticism, extraversion, openness to experience, agreeableness, and conscientiousness) and were genotyped for their 5-HTTLPR genotype. Individuals who carried one or two copies of the short allele had significantly higher neuroticism scores (and significantly lower agreeableness scores) than did homozygous carriers of the long allele. This study has been "seminal" in the truest sense of the word, having been cited more than 2,800 times as of Spring 2014.

The work inspired numerous replication attempts, which produced both positive and negative results, as well as extension studies linking this polymorphism to other behaviors. Possible reasons for nonreplication have been identified, such as selection of assessment tools or factors related to the sample, such as small sample size, ethnic stratification, or selection of extreme scorers (Ball et al., 1997; Dreary et al., 1999; Reif & Lesch, 2003; Sirota, Greenberg, Murphy, & Hamer, 1999). A number of meta-analyses have been conducted that mostly confirmed an association between 5-HTTLPR genotype and trait neuroticism (Munafo, Clark, & Flint, 2005; Schinka, Busch, & Robichaux-Keene, 2004; Sen, Burmeister, & Ghosh, 2004). However, the association is sensitive to the selection of the trait measurement instrument: a meta-analysis of the existing literature, along with analysis of primary data from a large sample of Finnish subjects (N = 3,872) found no significant association between 5-HTTLPR genotype and trait neuroticism when measured using the Eysenck Personality Questionnaire (EPQ), but did find a significant association when measured using the NEO personality inventory (Munafo et al., 2009b).

The year 1996 also saw the publication of two other studies linking a specific gene polymorphism to a behavioral trait. The reported link was between the dopamine D_4 receptor gene polymorphism, which features a repeat sequence, and novelty seeking (Benjamin et al., 1996; Ebstein et al., 1996). These studies reported that carriers of the seven-repeat allele scored significantly higher in novelty seeking than did noncarriers. Later work and several meta-analyses suggested that the association with novelty seeking was not confirmed (Kluger, Siegfried, & Ebstein, 2002) but instead reflected an association with attention deficit hyperactivity disorder (Li, Sham, Owen, & He, 2006; Maher, Marazita, Ferrell, & Vanyukov, 2002; Thapar, O'Donovan, & Owen, 2005; Wohl, Purper-Ouakil, Mouren, Ades, & Gorwood, 2005).

II. Imaging Genetics

One reason that replication studies of polymorphic genes have been inconsistent may be that the behavioral phenotype under study is distant from the biological function of the purported associated gene. For example, the trait of neuroticism is comprised of a complex set of behaviors that may even differ from one theoretical conceptualization of the trait to another. Indeed, as discussed earlier, the association between 5-HTTLPR genotype and neuroticism depends, in part, on the selection of the self-report instrument (Munafo et al., 2009b). Yet the 5-HTTLPR is functional at the level of 5-HTT expression, whose association with these high-level behavioral patterns is mediated through multiple levels from serotonergic cells to brain regions and larger scale brain functional networks. Each intervening level introduces additional complexities that obscure the association between the gene and its behavioral phenotype.

One critical step in reducing the complexity of linking gene function to behavioral phenotypes has been the introduction of intermediate or "endo"-phenotypes (Baud, 2005; Castellanos & Tannock, 2002; Congdon & Canli, 2005; Congdon, Lesch, & Canli, 2007; de Geus, Wright, Martin, & Boomsma, 2001; Gottesman & Gould, 2003; Gottesman & Shields, 1973; Hasler, Drevets, Gould, Gottesman, & Manji, 2006; Hasler, Drevets, Manji, & Charney, 2004). These endophenotypes lie closer to the underlying biology of the behavior of interest and may thus map better onto its associated underlying genetic architecture.

The 5-HTTLPR endophenotype approach, directed at brain function, was initially based on recordings of event-related potentials using a response-inhibition (Go/No-Go) task showing electrophysiological differences between homozygous long and short allele carriers (Fallgatter, Jatzke, Bartsch, Hamelbeck, & Lesch, 1999). The study was important in reporting a neurophysiological correlate of genetic variation, but was limited in isolating specific brain regions or in linking these regions to emotional behavior.

The year 2002 saw the first magnetic resonance imaging (MRI)-based study on 5-HTTLPR function (Hariri et al., 2002). In this study, participants performed a matching task involving either fearful and angry facial expressions or horizontal and vertical oval shapes. Short-allele carriers, relative to long-allele carriers, exhibited higher levels of amygdala activation to the emotional facial expressions relative to the oval shapes. These and other study results available at the time (Bertolino et al., 2005; Furmark et al., 2004; Hariri et al., 2005; Heinz et al., 2005; Pezawas et al., 2005) were interpreted as showing that 5-HTTLPR short allele carriers are characterized by increased responsiveness to negative emotional stimuli, consistent with the observation that this allele is also associated with elevated levels of trait neuroticism, in what we called the "standard" or "phasic activation" model (Canli & Lesch, 2007).

In contrast to the standard or phasic model of activation, we proposed the "tonic" model, which stipulates that short-allele 5-HTTLPR carriers are characterized by elevated amygdala activation at rest (Canli & Lesch, 2007). We proposed this model on the basis of functional MRI (fMRI) data that included a resting baseline condition (Canli et al., 2005), allowing us to dissociate the negative and neutral control conditions from each other and on the basis of blood flow data at rest obtained through arterial spin labeling (ASL; Canli et al., 2006). Importantly, the results of the fMRI baseline analysis were independently confirmed and replicated by another group (Heinz et al., 2007), as were the results of the ASL study (Rao et al., 2007), and these were extended to depressed patients (Brockmann et al., 2011). However, one large-sample (N = 183) study of healthy individuals failed to replicate the elevated resting state finding using ASL (Viviani et al., 2010), suggesting that the variables that may affect resting brain blood flow as a function of 5-HTTLPR genotype continue to be incompletely understood.

III. Gene-Environment Interactions

A. Clinical and Behavioral Studies

The next milestone in molecular psychology, published in the same year, in the same journal as the Hariri et al. fMRI study, was a study of G×E interactions conducted by Caspi and colleagues (Caspi et al., 2002). This first G×E study focused on a polymorphism in the monoamine oxidase A (MAOA) gene, which resulted in a high-activity and a low-activity allele. Analyzing data from a cohort of 442 males over a period of more than 20 years, these researchers found an association between childhood maltreatment and later antisocial behavior as a function of *MAOA-LPR* genotype: presence of both risk factors—the low-activity variant of *MAOA* gene and a history of childhood maltreatment—doubled the risk for conduct disorder, antisocial personality disorder symptoms, and other measures of aggression. This G×E interaction was subjected to numerous replication studies that have produced both positive (Fergusson, Boden, Horwood, Miller, & Kennedy, 2011; Foley et al., 2004; Frazzetto et al., 2007; Nilsson et al., 2006; Weder et al., 2009; Widom & Brzustowicz, 2006) and negative (Derringer, Krueger, Irons, & Iacono, 2010; Haberstick et al., 2014; Huizinga et al., 2006; Prichard, Mackinnon, Jorm, & Easteal, 2008; Young et al., 2006) results. Several meta-analyses have confirmed a G×E interaction (Kim-Cohen et al., 2006; Taylor & Kim-Cohen, 2007). The most recent meta-analysis was based on 27 published studies and again confirmed that the low-activity MAOA genotype moderated the impact of early adversity on later antisocial behavior (Byrd & Manuck, 2014). Furthermore, this analysis revealed that this interaction was limited to males and specific to early adversity in the form of childhood maltreatment.

A follow-up study by the same group in the following year focused on the 5-HTTLPR polymorphism (Caspi et al., 2003). In that study, a significant G×E interaction was reported between the 5-HTTLPR short allele and depressive symptoms as a function of stressful life experiences. This interaction was presented as a "dosage" effect, which suggested that individuals with four or more stressful life experiences exhibited increasingly higher levels of depressive symptoms as the number of short alleles increased from zero to two. Later replication studies suggest that this

G×E interaction may be subject to additional moderating variables, such as participants' sex (Eley et al., 2004) and the availability of social support (Kaufman et al., 2004). Nonreplications in older subjects (Gillespie, Whitfield, Williams, Heath, & Martin, 2005; Surtees et al., 2005) suggest that age may also be an important moderating variable.

Two meta-analyses in 2009 stirred up controversy. Munafo and colleagues obtained data from five published studies out of a set of 14 eligible studies that met their criteria (out of a set of 33 publications) on the risk for depression as a function of the interaction between 5-HTTLPR and stressful life events (SLEs) and concluded that the results were at chance levels (Munafo, Durrant, Lewis, & Flint, 2009a). Risch and colleagues conducted a meta-analysis of 13 published studies and also concluded that there was no significant association between 5-HTTLPR genotype and SLEs conferring risk for depression (Risch et al., 2009).

A rebuttal was offered by Kaufman and colleagues who criticized both meta-analyses for using nonrepresentative datasets (Kaufman, Gelernter, Kaffman, Caspi, & Moffitt, 2010). They pointed out that of the 14 eligible studies in Munafo's meta-analysis, 3 out of 4 (75 percent) negative studies were included, but only 2 out of 10 (20 percent) positive studies were included. Similarly, of the 26 eligible studies in Risch's meta-analysis, 7 out of 9 (78 percent) negative studies were included, but only 6 out of 17 (35 percent) positive studies were included. Furthermore, Kaufman and colleagues noted that nine studies were excluded from the Risch meta-analysis because they were published after the dataset was closed, and that 7 out of these 9 studies were full or partial replications.

A more recent meta-analysis included 54 studies and found a highly significant association between 5-HTTLPR genotype and stress conferring risk for depression (Karg, Burmeister, Shedden, & Sen, 2011). Interestingly, these investigators took an additional step to analyze this relationship as a function of the type of stressor. They found that the interaction of 5-HTTLPR genotype with stressful life events, although still significant ($p = .03$), was far less robust a predictor for depression than were other factors such as stress ($p = .00002$), childhood maltreatment ($p = .00007$), or a medical condition ($p = .0004$). This broad perspective on G×E interactions has been criticized by others (Flint & Kendler, 2014) who concluded that "current attempts to subdivide MD [major depression] on the basis of interactions with environmental effects using candidate genes are unlikely to yield quick insights into the origins of the disease" (p. 497). And so the debate continues.

In addition to clinical studies, several behavioral studies have reported 5-HTTLPR G×E interactions. For example, we found that 5-HTTLPR genotype moderated the correlation between SLEs and self-reported rumination (Canli et al., 2006), such that short-allele carriers reported higher levels of rumination with increased numbers of SLEs and homozygous long-allele carriers reported lower levels of rumination. Elevated levels of rumination as a function of the 5-HTTLPR short allele, either by itself (Clasen, Wells, Knopik, McGeary, & Beevers, 2011) or in conjunction with the BDNF Val66Met genotype (Stone, McGeary, Palmer, & Gibb, 2013), have been replicated, although these studies did not measure SLEs.

In another study, Mueller and colleagues used a standardized lab stressor task, the Trier Social Stress Test (TSST), and found a similar interaction, such that short-allele carriers with increased numbers of SLEs produced higher levels of cortisol in response to the TSST and homozygous long-allele carriers produced lower levels of cortisol (Mueller et al., 2011). However, this interaction was only observed in younger adults (aged 18–31 years), but not in children (aged 8–12 years) or in older adults (aged 54–68 years), and only when SLEs had occurred in the first 5 years of life. Nonetheless, the interaction in younger adults (average age 24 years) was replicated in another study using a public speaking task, in which homozygous carriers of the 5-HTTLPR short allele with high levels of SLEs showed the highest cortisol response to the stressor, whereas homozygous carriers of the 5-HTTLPR long allele with high levels of SLEs showed the lowest cortisol response to the stressor (Alexander et al., 2009).

Taken together, these studies suggest that G×E interactions affect susceptibility for psychopathology as well as individual differences in behavior but that these interactions are likely moderated by additional variables that are currently poorly understood.

B. Brain G×E

We addressed the question of G×E in the brain using both fMRI and ASL (Canli et al., 2006). Environmental measures included self-reported life stress history, such as prior medical, financial, or legal problems or emotional or social adversity. We found that carriers of the 5-HTTLPR short allele showed positive correlations in amygdala resting activation as a function of life stress history, whereas homozygous carriers of the long allele showed negative correlations. This was an intriguing discovery, as life stress appeared to affect everyone, but in opposite directions, depending on 5-HTTLPR genotype.

In a study of healthy males, participants genotyped for the 5-HTTLPR and with high or low levels of SLEs were scanned as they viewed emotional or neutral facial expressions (Alexander et al., 2012). Homozygous carriers of the 5-HTTLPR short allele with high levels of SLEs exhibited elevated amygdala activation to fearful emotional facial expressions compared to other participants and also showed increased functional coupling between amygdala and hypothalamus. However, another fMRI study assessing brain response to emotional facial expressions found no evidence for such a G×E interaction (Walsh et al., 2012). This second study used both sexes, and participants were younger than in the Alexander study (age 18.6 vs. 26.8 years). It is possible that gender and age may be additional moderating factors affecting these inconsistent results.

Two imaging studies evaluated the interaction of 5-HTTLPR genotype and stressful life events on fear learning and extinction. In one, 74 healthy individuals were genotyped for the 5-HTTLPR and TPH2 G(-703)T polymorphisms and subjected to a differential fear conditioning paradigm (Hermann et al., 2012). Carriers of the 5-HTTLPR short allele with high numbers of traumatic events exhibited stronger electrodermal conditioning responses and also showed reduced activation in the left amygdala during extinction training than did noncarriers. Another study also used a differential fear condition paradigm and reported that homozygous 5-HTTLPR short-allele carriers showed elevated activity in the amygdala and elsewhere in response to the CS+ and also exhibited a G×E interaction with SLEs in the occipital cortex and insula, with a trend toward significance in the amygdala (Klucken et al., 2013).

IV. Mechanisms of Gene Regulation

A. Animal Studies

Following these studies demonstrating G×E interactions, the next milestone was a study in rodents that identified CpG gene methylation as a possible molecular mechanism for this interaction (Weaver et al., 2004). The starting point of the study was individual differences in rats' stress responses as a function of maternal care: rats that received poor maternal care as pups, such as little licking and grooming, exhibit exaggerated stress responses in adulthood. Weaver et al. (2004) showed that methylation of the glucocorticoid receptor gene in the hippocampus differed between animals as a function of the maternal care they had received as pups. The methylation differences had functional consequences because they were shown to affect glucocorticoid gene expression. Furthermore, by using a cross-fostering control condition, these investigators showed that differential methylation was associated with maternal care per se rather than maternal genotype. Later work demonstrated that these epigenetic changes were not limited to the promoter region of the glucocorticoid receptor but extended to the protocadherin-α, -β, and -γ (Pcdh) gene families (which are involved in synaptogenesis), as well as to nonpromoter regions of the genome (McGowan et al., 2011). These studies demonstrated that life experiences can activate epigenetic mechanisms that regulate gene expression in the brain, and in neurochemical stress and neurodevelopmental systems.

B. Human Studies

The potential epigenetic consequences of stressful early life experiences have also been studied in humans. Oberlander and colleagues studied the association between maternal mood during pregnancy and neonatal gene methylation and infant stress reactivity (Oberlander et al., 2008).

They studied mothers who were depressed and treated with serotonin reuptake inhibitors (SRIs; $N = 33$), depressed nontreated mothers ($N = 13$), and healthy controls ($N = 36$). Maternal self-report of depressive symptoms, controlling for SRI treatment, was associated with increased methylation of the glucocorticoid receptor (GR) gene (NR3C1) in the corresponding region previously reported by Weaver et al. (2004). Furthermore, these neonates later exhibited elevated stress responses at 3 months of age, as measured by salivary cortisol response to an infant habituation-information processing task. These data translate the findings by Weaver and colleagues (2004) to humans and provide a potential epigenetic mechanism by which maternal mood can program stress reactivity in the offspring.

The first genome-wide human methylation study of maternal transmission, using a probe array technology (27 K Illumina Infinium Methylation BeadChip) containing 27,578 CpG site probes, was conducted by Schroeder and colleagues (2012). The study was based on a larger sample (N = 201) of mother–infant pairs, in which all of the mothers had been diagnosed with a mood disorder and half were on antidepressant treatment. Despite the sample size, this study failed to discover any association between maternal depressive symptoms and neonatal DNA methylation in any of the CpG methylation sites. However, two genes, *TNFRSF21* and *CHRNA2*, were differentially methylated as a function of maternal antidepressant treatment (Schroeder et al., 2012).

The next genome-wide methylation study used a more advanced array (Illumina Infinium Human Methylation450 BeadChip) containing 485,755 CpG site probes (Non, Binder, Kubzansky, & Michels, 2014). This study compared genome-wide methylation levels in neonates born to depressed mothers on selective SRI treatment (N = 22), nonmedicated mothers with reported depression or anxiety (N = 13), or healthy controls (N = 23). Interestingly, there were no methylation differences between neonates born to selective SRI (SSRI)-treated mothers and normal controls, but there were 42 CpG sites that were differentially methylated in neonates born to nonmedicated depressed or anxious mothers. A gene ontology analysis showed that the most significant genes were known to play a role in gene transcription and translation, as well as in cell division processes. No significant methylation differences between groups were observed for the two genes previously reported by Schroeder and colleagues (2012). The authors noted that 36 of the 42 significant CpG sites could not be discovered by Schroeder and colleagues because they were not included in the earlier 27 K Illumina chip. Given the relatively small sample size and the small differences in methylation, as well as other limitations, it will be interesting to await the results of additional genome-wide epigenetic studies of this kind.

Another set of longitudinal studies has begun to examine the relationship between early childhood aggression and gene methylation in young adulthood. Based on earlier work showing lower levels of plasma cytokines in aggressive individuals, Provencal and colleagues (2013) examined gene methylation in T cells and in monocytes of selected cytokine candidate genes and associated transcription factors (interleukin [IL]-1α, IL-6, IL-4, IL-10 and IL-8, and NFκB1, NFAT5, and STAT6). These analyses were conducted in adult males who had exhibited either high levels of chronic physical aggression (CPA) between the ages of 6 to 15 years (N = 8), or not (N = 25). This study discovered 48 regions across all of the candidate genes that were differentially methylated as a function of CPA. A follow-up study used a microarray platform (custom designed 244 K promoter tiling arrays, Agilent technologies) to map the promoter methylation state of approximately 20,000 genes and 400 microRNAs to discover 448 differentially methylated genes as a function of CPA (Provencal et al., 2014), and a similar pattern of differentially methylated genes was observed in adult women with high CPA (N = 5) compared to adult women controls (N = 14; Guillemin et al., 2014). Taken together, these results are consistent with the idea that stressful childhood experiences in the form of CPA may modulate gene methylation. However, this conclusion cannot be reached without available comparison samples from the same individuals during the childhood period. Furthermore, the sample sizes of these studies were small, so that future replication studies with large longitudinal samples and multiple gene methylation measuring points would be helpful in further elucidating the mechanisms linking early childhood behavior to later adulthood gene methylation.

C. Postmortem Studies

Postmortem studies of human brain samples have significantly expanded our understanding of the interplay of life experience and epigenetic programming in neural circuits. The first such study was published in 2008 by McGowan and colleagues (2008). They analyzed postmortem hippocampal samples from suicide victims with or without a history of childhood neglect/abuse. Suicide victims with a history of childhood neglect/abuse were characterized by hypermethylation of ribosomal RNA genes, which was specific to the hippocampus (no hypermethylation was observed in a control region, the cerebellum) and was not due to genome-wide nonspecific methylation effects. A follow-up study by this group focused on methylation of the neuron-specific glucocorticoid receptor (NR3C1) promoter region (McGowan et al., 2009). Suicide victims with a history of childhood neglect/abuse, relative to those without such history or nonsuicide controls, showed increased methylation of an NR3C1 promoter (glucocorticoid receptor, GR, exon 1f), as well as decreased levels of corresponding gene expression. Importantly, the methylated region in the human gene corresponded to the region identified by Weaver et al. (2004). Thus, the earlier observations by Weaver and colleagues on the impact of early life adversity on epigenetic regulation of the stress response system translated successfully to humans. Furthermore, the earlier animal data by McGowan et al. (2011) showing differential methylation of protocadherin gene families as a function of early maternal care in rats was replicated in human postmortem samples, suggesting a high degree of conservation of epigenetic mechanisms of early life experience across species (Suderman et al., 2012).

V. Transgenerational Epigenetics

If life experiences can leave epigenetic fingerprints on the genome of an affected individual, can these fingerprints also be found on the genome of this individual's offspring? The next milestone study addressed this question in mice subjected to maternal separation (Franklin et al., 2010). Litters of pups (F1) were subjected to a combination of maternal separation and uncontrollable stress (MSUS) from postnatal day 1 to 14. F1 MSUS males and nonstressed controls were then mated with wild-type females, and their offspring (F2) were raised under normal rearing conditions. This process was carried through to the next (F3) generation, which again was raised under normal rearing conditions.

F1 MSUS males (but not male controls or F1 MSUS females) did show depressionlike behavior, as operationalized using forced-swim and tail suspension tests. Surprisingly, depressionlike behavior was also observed in F2 offspring, but in this generation it was observed in females not in males. Depressionlike behavior in F2 females was reversed by administration of the antidepressant desipramine. Perhaps most surprising, depressionlike behavior was still observed in F3 offspring, but now again in males.

Methylation analyses in sperm cells of F1 males were conducted for several candidate genes known to be involved in emotional behavior or the epigenetic regulation of gene expression. F1 MSUS sperm cells showed increased methylation within regions of the methyl CpG-binding protein 2 (*MeCP2*) and cannabinoid receptor-1 (*CB1*) genes, decreased methylation within a region of the corticotrophin releasing factor receptor 2 (*CRFR2*) gene, and no differential methylation as a function of MSUS in the serotonin receptor 1_A (*5-HT*$_{1A}$) or monoamine oxidase A (*MAOA*) genes. The critical question of epigenetic transmission was addressed with the analysis of gene methylation in the F2 generation. In F2, depressionlike behavior was observed in females (see previous paragraph), and analysis of samples from F2 female cortical brain regions again showed hypermethylation of *MeCP2* and *CB1* and hypomethylation of *CRFR2*. For all three genes, these methylation levels were functionally significant because they corresponded to reduced mRNA gene expression relative to controls (which was surprising in the case of *CRFR2* because methylation usually inhibits gene expression). A similar analysis in F2 male sperm cells revealed again hypermethylation of *MeCP2* (but not *CB1*) and hypomethylation of *CRFR2*. No analyses of F3 gene methylation were presented. Taken together, these were the first results demonstrating that adverse early life experiences can alter epigenetic marks on selected genes and that these epigenetic marks can be transmitted across generations into brain and sperm cells, although the transmission of corresponding behavior seems sex-specific and poorly understood.

In another study, the possible epigenetic transmission of male adverse experience was studied using a social defeat paradigm (Dietz et al., 2011). The F1 offspring of socially defeated F0 male mice, compared to F1 offspring from controls, showed elevated levels of social avoidance. It is possible that these behavioral effects may not represent some form of epigenetic inheritance based on experience but rather reflect standard genetic transmission of some preexisting trait. This possibility was elegantly addressed in a second experiment, in which Dietz and colleagues had F0 males mate both before ("predefeat") and after ("postdefeat") being subjected to a social defeat paradigm, and in which the investigators then studied the offspring conceived during those two periods. F1-postdefeat offspring, relative to F1-predefeat offspring, exhibited elevated anxietylike behavior, based on time spent in the open arm of an elevated plus maze, and depressionlike behavior, based on immobility latency in a forced-swim test. Analysis of plasma levels of the stress hormone corticosterone revealed significantly higher levels in F1-postdefeat relative to F1-predefeat offspring. This effect was not seen in F1 female offspring and was also not seen in response to an acute stressor.

One particularly critical aspect of the study by Dietz and colleagues is that it addressed the possibility that social defeat-related information could be transmitted through social channels rather than gametes. To dissociate these two possible routes, the authors used in vitro fertilization (IVF). IVF-offspring from socially defeated F0 mice and controls were then tested using the same behavioral battery as before. Under these conditions, the behavioral differences between social-defeat offspring and controls were much reduced: there were no differences in the social avoidance and elevated plus maze tasks, although there remained statistically significant but small differences in the forced-swim task. These data suggest that transmission of both anxiety- and depression-related behaviors from socially defeated F0 males to the F1 generation does occur and that the mode of transmission is in a larger part social, but also in a smaller part epigenetic.

A third study investigated the transgenerational epigenetic transmission of olfactory-based conditioned fear (Dias & Ressler, 2014). The study took advantage of the fact that the odorant acetophenone binds to the M71 receptor that is encoded by one gene, *Olfr151*, so that the focus of epigenetic marking was kept in sharp focus. F0 male mice were trained to associate acetophenone with a foot shock. The subsequent enhanced startle response in the presence of the trained odor (odor potentiated startle, OPS) relative to a control odor, was interpreted as a measure of enhanced sensitivity to acetophenone. F1 offspring exhibited OPS despite not having been subjected to any fear conditioning. In one control condition, the F1 offspring of F0 mice that had been fear conditioned with propanol instead of acetophenone showed OPS to propanol rather than acetophenone. Thus, inheritance of odor sensitivity was specific to whichever fear-associated stimulus the F0 generation had experienced. Indeed, behavioral testing of F2 animals demonstrated that the inheritance of odor sensitivity persisted into the third generation.

Dias and Ressler also used IVF to address the possibility that F0-trained odor information may be transmitted through social channels rather than gametes. For IVF, they used sperm from F0 fear-conditioned animals that was collected 10 days after the last conditioning trial. According to the authors, animal quarantine issues prevented behavioral assessment of the F1 IVF-generated animals, but neuroanatomical analysis of the olfactory bulb showed significantly larger dorsal and medial M71 glomeruli in offspring of acetophenone-trained F0 animals compared to offspring of propanol-trained F0 animals. These data suggest that paternal transmission of odor sensitivity did indeed pass through the germ line rather than through social channels from F0 male to F0 female.

The related possibility of maternal transmission mechanisms was addressed in a cross-fostering experiment in which F0 females were exposed to odor fear conditioning. Cross-fostering with nontrained F0 females demonstrated that the transmission of odor sensitivity to F1 offspring occurred through biological rather than social channels. Thus, either parent had the capacity to transmit experiential information.

Finally, the authors addressed the putative epigenetic mechanism of transmission. Given the specificity of acetophenone binding to the M71 receptor, which is encoded by *Olfr151*, and the increased size of M71 glomeruli in F1 offspring, Dias and Ressler analyzed methylation of *Olfr151* in F0 sperm cells. They found reduced methylation of *Olfr151*, but not a control gene, in sperm from acetophenone-trained F0 males relative to propanol-trained F0 males. Taken together, this study demonstrated transgenerational inheritance of a behavioral phenotype through methylation of relevant genes in parental gametes.

VI. Optogenetics

The development of optogenetics (Boyden, Zhang, Bamberg, Nagel, & Deisseroth, 2005) has transformed neuroscience, and it has offered tools to study genetically defined cell populations with unprecedented spatial and temporal precision. The method uses viral vectors to insert light-sensitive opsin molecules into cellular membranes of genetically defined cell types so that light can be used to excite or inhibit these cells (Bernstein & Boyden, 2011; Kramer, Mourot, & Adesnik, 2013; Packer, Roska, & Hausser, 2013; Tye & Deisseroth, 2012; Williams & Deisseroth, 2013; Zhang, Aravanis, Adamantidis, de Lecea, & Deisseroth, 2007).

Applications have been numerous and include studies of anxiety (Kim et al., 2013; Nieh, Kim, Namburi, & Tye, 2013; Tye et al., 2011), depression (Tye et al., 2013), memory (Huff, Miller, Deisseroth, Moorman, & LaLumiere, 2013; Liu et al., 2012; Ramirez et al., 2013), social behavior (Felix-Ortiz & Tye, 2014), addiction (Chen et al., 2013; Suska, Lee, Huang, Dong, & Schluter, 2013), and schizophrenia (Maher & LoTurco, 2012) among others.

Some recent technical advances have included the combination with neuroimaging techniques. For example, optogenetic fMRI uses optogenetics to activate selected sets of neurons and then uses fMRI to image system-wide brain responses (Abe et al., 2012; Fang & Lee, 2013; Lee, 2011; 2012). A related application has used micro positron emission tomography (μPET) in rodents: Thanos and colleagues optostimulated the nucleus accumbens (NAc) using the excitatory channelrhodopsin-2 and measured subsequent changes in regional brain glucose metabolism (BGluM) with [(18)F]2-fluoro-2-deoxy-D-glucose PET imaging (Thanos et al., 2013). Optogenetic stimulation was associated with increased BGluM in NAc but also produced increases in a larger network of regions including striatum, periaqueductal gray, somatosensory cortex, amygdala, hippocampus, and other regions. The study was limited in that the optostimulation paradigm activated all cells in NAc, rather than a more narrowly defined subset of cells, and in that animals had to be anesthetized during the PET-imaging procedure. Future studies using more refined optogenetic protocols and mobile PET technology with awake, behaving animals (Schulz et al., 2011) could produce exciting and rigorous studies of optogenetically activated brain circuits across a wide range of behavioral paradigms.

Another advance is the application of optogenetic regulation of gene expression. Konermann and colleagues developed light-inducible transcriptional effectors (LITEs) that combine optogenetics with genome editing using engineered nucleases (Konermann et al., 2013). The technology allows for light-controlled gene expression regulation that can be reversible and cell-specific. Konermann and colleagues demonstrated the power of this technology in primary mouse neurons, in the brains of awake mice, and with targeted epigenetic chromatin modifications. This technology will undoubtedly be instrumental in better understanding many of the processes discussed in the preceding sections because it allows for causal testing of both genetic and epigenetic mechanisms in behaving animals.

VII. Readings

The readings commissioned for this volume represent authoritative reviews and perspectives by many of the leading figures in the field, and they represent a cross-section of the approaches discussed in the preceding sections. Following this introductory chapter, Gelernter provides the reader with a review of the methods as a technical foundation for the remainder of the book.

The main section then introduces the reader to different applications of the molecular psychology approach, grouping chapters into three broad themes: Social Behavior, Clinical Topics, and Emotion, Cognition, and Stress.

The chapters on social behavior are widely stratified and cover social regulation of genes in insects (Simpson and Stevenson), the African cichlid fish *Astatotilapia burtoni* (Maruska and Fernald), voles (Barrett and Young), and nonhuman primates (Barr). In humans, social behaviors discussed include aggression and violence (Buckholtz and Meyer-Lindenberg), affiliative behavior (Westberg and Walum), and a wide range of economic and social decision-making processes (Ebstein, Zhong, Chark, Lai, and Chew).

The chapters on clinical topics discuss G×E mouse models for mood disorders (Razzoli, Bartolomucci, and Carola), and then move to humans to discuss imaging genetics approaches to understand risk for psychopathology (Hyde, Bogdan, and Hariri), the study of copy number variations in neuropsychiatric disorders (Rucker and McGuffin), and a molecular perspective on Major Depressive Disorder (Chen and Gotlib). I then round out the section with a new hypothesis, in which I entertain the question of whether depression is an infectious disease.

The chapters on Emotion, Cognition, and Stress discuss a possible role for nutrition and epigenetic mechanisms on cognition and emotion (Naninck, Lucassen, and Korosi) and the use of genetic tools for the erasure of emotional memories (Sargin, Yan, and Josselyn). Stress processes are discussed with regard to the genetics of autonomic nervous system activity (de Geus, van Lien, Neijts, and Willemsen). The section ends on a positive note, with a chapter on the genetic basis of positive emotionality (Reuter and Montag).

This volume concludes with two chapters that step outside the lab to take a broader perspective on the role of molecular psychology in science and society. Morse first discusses the role of genetics in clinical justice. I then conclude with my perspective on the ethical, legal, and social implications of molecular psychology. I argue that the intersection of neuroscience and molecular genetics/genomics provides a fertile ground for new sets of ethics considerations. Borrowing from the constructs of "genethics" and "neuroethics," I refer to this emerging field as "neurogenethics." The chapter spells out a research agenda for future ethics studies, examining the implications of laboratory-based discoveries in molecular psychology for the real world.

I hope that the reader will find the science on display in this volume exciting, surprising, and inspiring. I certainly had that experience editing this volume.

References

Abe, Y., Sekino, M., Terazono, Y., Ohsaki, H., Fukazawa, Y., Sakai, S., . . . Hisatsune, T. (2012). Opto-fMRI analysis for exploring the neuronal connectivity of the hippocampal formation in rats. *Neuroscience Research, 74*(3–4), 248–255. doi: 10.1016/j.neures.2012.08.007

Alexander, N., Klucken, T., Koppe, G., Osinsky, R., Walter, B., Vaitl, D.,... Hennig, J. (2012). Interaction of the serotonin transporter-linked polymorphic region and environmental adversity: Increased amygdala-hypothalamus connectivity as a potential mechanism linking neural and endocrine hyperreactivity. *Biological Psychiatry, 72*(1), 49–56. doi: 10.1016/j.biopsych.2012.01.030

Alexander, N., Kuepper, Y., Schmitz, A., Osinsky, R., Kozyra, E., & Hennig, J. (2009). Gene-environment interactions predict cortisol responses after acute stress: Implications for the etiology of depression. *Psychoneuroendocrinology, 34*(9), 1294–1303. doi: 10.1016/j.psyneuen.2009.03.017

Ball, D., Hill, L., Freeman, B., Eley, T. C., Strelau, J., Riemann, R., ... Plomin, R. (1997). The serotonin transporter gene and peer-rated neuroticism. *Neuroreport, 8*(5), 1301–1304.

Baud, P. (2005). Personality traits as intermediary phenotypes in suicidal behavior: Genetic issues. *American Journal of Medical Genetics Part C: Seminars in Medical Genetics, 133*(1), 34–42.

Benjamin, J., Li, L., Patterson, C., Greenberg, B. D., Murphy, D. L., & Hamer, D. H. (1996). Population and familial association between the D4 dopamine receptor gene and measures of novelty seeking. *Nature Genetics, 12*, 81–84.

Bernstein, J. G., & Boyden, E. S. (2011). Optogenetic tools for analyzing the neural circuits of behavior. *Trends in Cognitive Sciences, 15*(12), 592–600. doi: 10.1016/j.tics.2011.10.003

Bertolino, A., Arciero, G., Rubino, V., Latorre, V., De Candia, M., Mazzola, V.,... Scarabino, T. (2005). Variation of human amygdala response during threatening stimuli as a function of 5'HTTLPR genotype and personality style. *Biological Psychiatry, 57*(12), 1517–1525.

Boyden, E. S., Zhang, F., Bamberg, E., Nagel, G., & Deisseroth, K. (2005). Millisecond-timescale, genetically targeted optical control of neural activity. *Nature Neuroscience, 8*(9), 1263–1268. doi: 10.1038/nn1525

Brockmann, H., Zobel, A., Schuhmacher, A., Daamen, M., Joe, A., Biermann, K.,... Boecker, H. (2011). Influence of 5-HTTLPR polymorphism on resting state perfusion in patients with major depression. *Journal of Psychiatric Research, 45*(4), 442–451. doi: S0022-3956(10)00260-8 [pii] 10.1016/j.jpsychires.2010.08.016

Byrd, A. L., & Manuck, S. B. (2014). MAOA, childhood maltreatment, and antisocial behavior: Meta-analysis of a gene-environment interaction. *Biological Psychiatry, 75*(1), 9–17. doi: 10.1016/j.biopsych.2013.05.004

Canli, T., & Lesch, K. -P. (2007). Long story short: The serotonin transporter in emotion regulation and social cognition. *Nature Neuroscience, 10*(9), 1103–1109.

Canli, T., Omura, K., Haas, B. W., Fallgatter, A., Constable, R. T., & Lesch, K. P. (2005). Beyond affect: A role for genetic variation of the serotonin transporter in neural activation during a cognitive attention task. *Proceedings of the National Academy of Sciences of the United States of America, 102*(34), 12224–12229.

Canli, T., Qiu, M., Omura, K., Congdon, E., Haas, B. W., Amin, Z.,... Lesch, K. P. (2006). Neural correlates of epigenesis. *Proceedings of the National Academy of Sciences of the United States of America, 103*(43), 16033–16038.

Caspi, A., McClay, J., Moffitt, T. E., Mill, J., Martin, J., Craig, I. W.,... Poulton, R. (2002). Role of genotype in the cycle of violence in maltreated children. *Science, 297*(5582), 851–854.

Caspi, A., Sugden, K., Moffitt, T. E., Taylor, A., Craig, I. W., Harrington, H.,... Poulton, R. (2003). Influence of life stress on depression: Moderation by a polymorphism in the 5-HTT gene. *Science, 301*, 386–389.

Castellanos, F. X., & Tannock, R. (2002). Neuroscience of attention-deficit/hyperactivity disorder: The search for endophenotypes. *Nature Reviews Neuroscience, 3*(8), 617–628.

Chen, B. T., Yau, H. J., Hatch, C., Kusumoto-Yoshida, I., Cho, S. L., Hopf, F. W., & Bonci, A. (2013). Rescuing cocaine-induced prefrontal cortex hypoactivity prevents compulsive cocaine seeking. *Nature, 496*(7445), 359–362. doi: 10.1038/nature12024

Clasen, P. C., Wells, T. T., Knopik, V. S., McGeary, J. E., & Beevers, C. G. (2011). 5-HTTLPR and BDNF Val66Met polymorphisms moderate effects of stress on rumination. *Genes Brain and Behavior, 10*(7), 740–746. doi: 10.1111/j.1601-183X.2011.00715.x

Congdon, E., & Canli, T. (2005). The endophenotype of impulsivity: Reaching consilience through behavioral, genetic, and neuroimaging approaches. *Behavioral and Cognitive Neuroscience Reviews, 4*(4), 262–281.

Congdon, E., Lesch, K. P., & Canli, T. (2007). Analysis of DRD4 and DAT polymorphisms and behavioral inhibition in healthy adults: Implications for impulsivity. *American Journal of Medical Genetics Part B: Neuropsychiatric Genetics.*

Defries, J. C., McClearn, G. E., McGuffin, P., & Plomin, R. (Eds.). (2000). *Behavioral genetics* (4th ed.). New York: Worth.

de Geus, E. J., Wright, M. J., Martin, N. G., & Boomsma, D. I. (2001). Genetics of brain function and cognition. *Behavioral Genetics, 31*(6), 489–495.

Derringer, J., Krueger, R. F., Irons, D. E., & Iacono, W. G. (2010). Harsh discipline, childhood sexual assault, and MAOA genotype: An investigation of main and interactive effects on diverse clinical externalizing outcomes. *Behavioral Genetics, 40*(5), 639–648. doi: 10.1007/s10519-010-9358-9

Dias, B. G., & Ressler, K. J. (2014). Parental olfactory experience influences behavior and neural structure in subsequent generations. *Nature Neuroscience, 17*(1), 89–96. doi: 10.1038/nn.3594

Dietz, D. M., Laplant, Q., Watts, E. L., Hodes, G. E., Russo, S. J., Feng, J.,... Nestler, E. J. (2011). Paternal transmission of stress-induced pathologies. *Biological Psychiatry, 70*(5), 408–414. doi: 10.1016/j.biopsych.2011.05.005

Dilalla, L. F., & Gottesman, I. I. (Eds.). (2004). *Behavior genetics principles: Perspectives in development, personality, and psychopathology.* American Psychological Association.

Dreary, I. J., Battersby, S., Whiteman, M. C., Connor, J. M., Fowkes, F. G., & Harmar, A. (1999). Neuroticism and polymorphisms in the serotonin transporter gene. *Psychological Medicine, 29*(3), 735–739.

Ebstein, R. P., Novick, O., Umansky, R., Priel, B., Osher, Y., Blaine, D.,... Belmaker, R. H. (1996). Dopamine D4 receptor (D4DR) exon III polymorphism associated with the human personality trait of novelty seeking. *Nature Genetics, 12*, 78–80.

Eley, T. C., Sugden, K., Corsico, A., Gregory, A. M., Sham, P., McGuffin, P.,... Craig, I. W. (2004). Gene-environment interaction analysis of serotonin system markers with adolescent depression. *Molecular Psychiatry, 9*(10), 908–915.

Fallgatter, A., Jatzke, S., Bartsch, A., Hamelbeck, B., & Lesch, K. (1999). Serotonin transporter promoter polymorphism influences topography of inhibitory motor control. *International Journal of Neuropsychopharmacology, 2*, 115–120.

Fang, Z., & Lee, J. H. (2013). High-throughput optogenetic functional magnetic resonance imaging with parallel computations. *Journal of Neuroscience Methods*, *218*(2), 184–195. doi: 10.1016/j.jneumeth.2013.04.015

Felix-Ortiz, A. C., & Tye, K. M. (2014). Amygdala inputs to the ventral hippocampus bidirectionally modulate social behavior. *Journal of Neuroscience*, *34*(2), 586–595. doi: 10.1523/JNEUROSCI.4257-13.2014

Fergusson, D. M., Boden, J. M., Horwood, L. J., Miller, A. L., & Kennedy, M. A. (2011). MAOA, abuse exposure and antisocial behaviour: 30-year longitudinal study. *British Journal of Psychiatry*, *198*(6), 457–463. doi: 10.1192/bjp.bp.110.086991

Flint, J., & Kendler, K. S. (2014). The genetics of major depression. *Neuron*, *81*(3), 484–503. doi: 10.1016/j.neuron.2014.01.027

Foley, D. L., Eaves, L. J., Wormley, B., Silberg, J. L., Maes, H. H., Kuhn, J., & Riley, B. (2004). Childhood adversity, monoamine oxidase a genotype, and risk for conduct disorder. *Archives of General Psychiatry*, *61*(7), 738–744. doi: 10.1001/archpsyc.61.7.738

Franklin, T. B., Russig, H., Weiss, I. C., Graff, J., Linder, N., Michalon, A.,... Mansuy, I. M. (2010). Epigenetic transmission of the impact of early stress across generations. *Biological Psychiatry*, *68*(5), 408–415. doi: 10.1016/j.biopsych.2010.05.036

Frazzetto, G., Di Lorenzo, G., Carola, V., Proietti, L., Sokolowska, E., Siracusano, A.,... Troisi, A. (2007). Early trauma and increased risk for physical aggression during adulthood: The moderating role of MAOA genotype. *PLoS ONE*, *2*(5), e486.

Furmark, T., Tillfors, M., Garpenstrand, H., Marteinsdottir, I., Langstrom, B., Oreland, L., & Fredrikson, M. (2004). Serotonin transporter polymorphism related to amygdala excitability and symptom severity in patients with social phobia. *Neuroscience Letters*, *362*(3), 189–192.

Gillespie, N. A., Whitfield, J. B., Williams, B., Heath, A. C., & Martin, N. G. (2005). The relationship between stressful life events, the serotonin transporter (5-HTTLPR) genotype and major depression. *Psychological Medicine*, *35*(1), 101–111.

Gottesman, I. I., & Gould, T. D. (2003). The endophenotype concept in psychiatry: Etymology and strategic intentions. *American Journal of Psychiatry*, *160*(4), 636–645.

Gottesman, I. I., & Shields, J. (1973). Genetic theorizing and schizophrenia. *British Journal of Psychiatry*, *122*(566), 15–30.

Guillemin, C., Provencal, N., Suderman, M., Cote, S. M., Vitaro, F., Hallett, M.,... Szyf, M. (2014). DNA methylation signature of childhood chronic physical aggression in T cells of both men and women. *PLoS One*, *9*(1), e86822. doi: 10.1371/journal.pone.0086822

Haberstick, B. C., Lessem, J. M., Hewitt, J. K., Smolen, A., Hopfer, C. J., Halpern, C. T.,... Mullan Harris, K. (2014). MAOA genotype, childhood maltreatment, and their interaction in the etiology of adult antisocial behaviors. *Biological Psychiatry*, *75*(1), 25–30. doi: 10.1016/j.biopsych.2013.03.028

Hariri, A. R., Drabant, E. M., Munoz, K. E., Kolachana, B. S., Mattay, V. S., Egan, M. F., & Weinberger, D. R. (2005). A susceptibility gene for affective disorders and the response of the human amygdala. *Archives of General Psychiatry*, *62*(2), 146–152.

Hariri, A. R., Mattay, V. S., Tessitore, A., Kolachana, B., Fera, F., Goldman, D.,... Weinberger, D. R. (2002). Serotonin transporter genetic variation and the response of the human amygdala. *Science*, *297*(5580), 400–403.

Hasler, G., Drevets, W. C., Gould, T. D., Gottesman, I. I., & Manji, H. K. (2006). Toward constructing an endophenotype strategy for bipolar disorders. *Biological Psychiatry*, *60*(2), 93–105.

Hasler, G., Drevets, W. C., Manji, H. K., & Charney, D. S. (2004). Discovering endophenotypes for major depression. *Neuropsychopharmacology*, *29*(10), 1765–1781.

Heinz, A., Braus, D. F., Smolka, M. N., Wrase, J., Puls, I., Hermann, D.,... Buchel, C. (2005). Amygdala-prefrontal coupling depends on a genetic variation of the serotonin transporter. *Nature Neuroscience*, *8*(1), 20–21.

Heinz, A., Smolka, M. N., Braus, D. F., Wrase, J., Beck, A., Flor, H.,... Weinberger, D. R. (2007). Serotonin Transporter Genotype (5-HTTLPR): Effects of neutral and undefined conditions on amygdala activation. *Biological Psychiatry*, *61*(8), 1011–1014.

Hermann, A., Kupper, Y., Schmitz, A., Walter, B., Vaitl, D., Hennig, J.,... Tabbert, K. (2012). Functional gene polymorphisms in the serotonin system and traumatic life events modulate the neural basis of fear acquisition and extinction. *PLoS One*, *7*(9), e44352. doi: 10.1371/journal.pone.0044352

Huff, M. L., Miller, R. L., Deisseroth, K., Moorman, D. E., & LaLumiere, R. T. (2013). Posttraining optogenetic manipulations of basolateral amygdala activity modulate consolidation of inhibitory avoidance memory in rats. *Proceedings of the National Academy of Sciences USA*, *110*(9), 3597–3602. doi: 10.1073/pnas.1219593110

Huizinga, D., Haberstick, B. C., Smolen, A., Menard, S., Young, S. E., Corley, R. P.,... Hewitt, J. K. (2006). Childhood maltreatment, subsequent antisocial behavior, and the role of monoamine oxidase A genotype. *Biological Psychiatry*, *60*(7), 677–683. doi: 10.1016/j.biopsych.2005.12.022

Karg, K., Burmeister, M., Shedden, K., & Sen, S. (2011). The serotonin transporter promoter variant (5-HTTLPR), stress, and depression meta-analysis revisited: Evidence of genetic moderation. *Archives of General Psychiatry*, *68*(5), 444–454.

Kaufman, J., Gelernter, J., Kaffman, A., Caspi, A., & Moffitt, T. (2010). Arguable assumptions, debatable conclusions. *Biological Psychiatry*, *67*(4), e19–e20; author reply e21–e23. doi: 10.1016/j.biopsych.2009.07.041

Kaufman, J., Yang, B. Z., Douglas-Palumberi, H., Houshyar, S., Lipschitz, D., Krystal, J. H., & Gelernter, J. (2004). Social supports and serotonin transporter gene moderate depression in maltreated children. *Proceedings of the National Academy of Sciences USA*, *101*(49), 17316–17321. Epub 12004 Nov 17324.

Kim, S. Y., Adhikari, A., Lee, S. Y., Marshel, J. H., Kim, C. K., Mallory, C. S.,... Deisseroth, K. (2013). Diverging neural pathways assemble a behavioural state from separable features in anxiety. *Nature*, *496*(7444), 219–223. doi: 10.1038/nature12018

Kim-Cohen, J., Caspi, A., Taylor, A., Williams, B., Newcombe, R., Craig, I. W., & Moffitt, T. E. (2006). MAOA, maltreatment, and gene–environment interaction predicting children's mental health: New evidence and a meta-analysis. *Molecular Psychiatry*, *11*, 903–913.

Klucken, T., Alexander, N., Schweckendiek, J., Merz, C. J., Kagerer, S., Osinsky, R.,... Stark, R. (2013). Individual differences in neural correlates of fear conditioning as a function of 5-HTTLPR and stressful life events. *Social Cognitive and Affective Neuroscience, 8*(3), 318–325. doi: 10.1093/scan/nss005

Kluger, A. N., Siegfried, Z., & Ebstein, R. P. (2002). A meta-analysis of the association between DRD4 polymorphism and novelty seeking. *Molecular Psychiatry, 7*(7), 712–717.

Konermann, S., Brigham, M. D., Trevino, A. E., Hsu, P. D., Heidenreich, M., Cong, L.,...Zhang, F. (2013). Optical control of mammalian endogenous transcription and epigenetic states. *Nature, 500*(7463), 472–476. doi: 10.1038/nature12466

Kramer, R. H., Mourot, A., & Adesnik, H. (2013). Optogenetic pharmacology for control of native neuronal signaling proteins. *Nature Neuroscience, 16*(7), 816–823. doi: 10.1038/nn.3424

Lee, J. H. (2011). Tracing activity across the whole brain neural network with optogenetic functional magnetic resonance imaging. *Front Neuroinform, 5*, 21. doi: 10.3389/fninf.2011.00021

Lee, J. H. (2012). Informing brain connectivity with optogenetic functional magnetic resonance imaging. *Neuroimage, 62*(4), 2244–2249. doi: 10.1016/j.neuroimage.2012.01.116

Lesch, K. P., Bengel, D., Heils, A., Sabol, S. Z., Greenberg, B. D., Petri, S.,...Murphy, D. L. (1996). Association of anxiety-related traits with a polymorphism in the serotonin transporter gene regulatory region. *Science, 274*(5292), 1527–1531.

Li, D., Sham, P. C., Owen, M. J., & He, L. (2006). Meta-analysis shows significant association between dopamine system genes and attention deficit hyperactivity disorder (ADHD). *Human Molecular Genetics, 15*(14), 2276–2284.

Liu, X., Ramirez, S., Pang, P. T., Puryear, C. B., Govindarajan, A., Deisseroth, K., & Tonegawa, S. (2012). Optogenetic stimulation of a hippocampal engram activates fear memory recall. *Nature, 484*(7394), 381–385. doi: Nature11028 [pii] 10.1038/nature11028

Maher, B. S., Marazita, M. L., Ferrell, R. E., & Vanyukov, M. M. (2002). Dopamine system genes and attention deficit hyperactivity disorder: A meta-analysis. *Psychiatric Genetics, 12*(4), 207–215.

Maher, B. J., & LoTurco, J. J. (2012). Disrupted-in-schizophrenia (DISC1) functions presynaptically at glutamatergic synapses. *PLoS One, 7*(3), e34053. doi: 10.1371/journal.pone.0034053

McGowan, P. O., Sasaki, A., D'Alessio, A. C., Dymov, S., Labonte, B., Szyf, M.,...Meaney, M. J. (2009). Epigenetic regulation of the glucocorticoid receptor in human brain associates with childhood abuse. *Nature Neuroscience, 12*(3), 342–348.

McGowan, P. O., Sasaki, A., Huang, T. C., Unterberger, A., Suderman, M., Ernst, C.,...Szyf, M. (2008). Promoter-wide hypermethylation of the ribosomal RNA gene promoter in the suicide brain. *PLoS ONE, 3*(5), e2085.

McGowan, P. O., Suderman, M., Sasaki, A., Huang, T. C., Hallett, M., Meaney, M. J., & Szyf, M. (2011). Broad epigenetic signature of maternal care in the brain of adult rats. *PLoS One, 6*(2), e14739. doi: 10.1371/journal.pone.0014739

Mueller, A., Armbruster, D., Moser, D. A., Canli, T., Lesch, K. -P., Brocke, B., & Kirschbaum, C. (2011). Interaction of serotonin transporter gene-linked polymorphic region and stressful life events predicts cortisol stress response. *Neuropsychopharmacology, 36*(7), 1332–1339.

Munafo, M. R., Clark, T. G., & Flint, J. (2005). Promise and pitfalls in the meta-analysis of genetic association studies: A response to Sen and Shinka. *Molecular Psychiatry, 10*, 895–897.

Munafo, M. R., Durrant, C., Lewis, G., & Flint, J. (2009a). Gene X environment interactions at the serotonin transporter locus. *Biological Psychiatry, 65*(3), 211–219.

Munafo, M. R., Freimer, N. B., Ng, W., Ophoff, R., Veijola, J., Miettunen, J.,...Flint, J. (2009b). 5-HTTLPR genotype and anxiety-related personality traits: A meta-analysis and new data. *American Journal of Medical Genetics Part B: Neuropsychiatric Genetics, 150B*(2), 271–281. doi: 10.1002/ajmg.b.30808

Nieh, E. H., Kim, S. Y., Namburi, P., & Tye, K. M. (2013). Optogenetic dissection of neural circuits underlying emotional valence and motivated behaviors. *Brain Research, 1511*, 73–92. doi: 10.1016/j.brainres.2012.11.001

Nilsson, K. W., Sjoberg, R. L., Damberg, M., Leppert, J., Ohrvik, J., Alm, P. O.,...Oreland, L. (2006). Role of monoamine oxidase A genotype and psychosocial factors in male adolescent criminal activity. *Biological Psychiatry, 59*(2), 121–127. doi: 10.1016/j.biopsych.2005.06.024

Non, A. L., Binder, A. M., Kubzansky, L. D., & Michels, K. B. (2014). Genome-wide DNA methylation in neonates exposed to maternal depression, anxiety, or SSRI medication during pregnancy. *Epigenetics, 9*(7).

Oberlander, T. F., Weinberg, J., Papsdorf, M., Grunau, R., Misri, S., & Devlin, A. M. (2008). Prenatal exposure to maternal depression, neonatal methylation of human glucocorticoid receptor gene (NR3C1) and infant cortisol stress responses. *Epigenetics, 3*(2), 97–106.

Packer, A. M., Roska, B., & Hausser, M. (2013). Targeting neurons and photons for optogenetics. *Nature Neuroscience, 16*(7), 805–815. doi: 10.1038/nn.3427

Pezawas, L., Meyer-Lindenberg, A., Drabant, E. M., Verchinski, B. A., Munoz, K. E., Kolachana, B. S.,...Weinberger, D. R. (2005). 5-HTTLPR polymorphism impacts human cingulate-amygdala interactions: A genetic susceptibility mechanism for depression. *Nature Neuroscience, 8*(6), 828–834.

Prichard, Z., Mackinnon, A., Jorm, A. F., & Easteal, S. (2008). No evidence for interaction between MAOA and childhood adversity for antisocial behavior. *American Journal of Medical Genetics Part B: Neuropsychiatric Genetics, 147*, 228–232.

Provencal, N., Suderman, M. J., Caramaschi, D., Wang, D., Hallett, M., Vitaro, F.,...Szyf, M. (2013). Differential DNA methylation regions in cytokine and transcription factor genomic loci associate with childhood physical aggression. *PLoS One, 8*(8), e71691. doi: 10.1371/journal.pone.0071691

Provencal, N., Suderman, M. J., Guillemin, C., Vitaro, F., Cote, S. M., Hallett, M.,...Szyf, M. (2014). Association of childhood chronic physical aggression with a DNA methylation signature in adult human T cells. *PLoS One, 9*(4), e89839. doi: 10.1371/journal.pone.0089839

Ramirez, S., Liu, X., Lin, P. A., Suh, J., Pignatelli, M., Redondo, R. L.,...Tonegawa, S. (2013). Creating a false memory in the hippocampus. *Science, 341*(6144), 387–391. doi: 341/6144/387 [pii] 10.1126/science.1239073

Rao, H., Gillihan, S. J., Wang, J., Korczykowski, M., Sankoorikal, G. M., Kaercher, K. A.,...Farah, M. J. (2007). Genetic variation in serotonin transporter alters resting brain function in healthy individuals. *Biological Psychiatry, 62*(6), 600–606.

Reif, A., & Lesch, K. P. (2003). Toward a molecular architecture of personality. *Behavioral Brain Research*, *139*(1–2), 1–20.

Risch, N., Herrell, R., Lehner, T., Liang, K. Y., Eaves, L., Hoh, J., ... Merikangas, K. R. (2009). Interaction between the serotonin transporter gene (5-HTTLPR), stressful life events, and risk of depression: A meta-analysis. *Journal of the American Medical Association*, *301*(23), 2462–2471. doi: 301/23/2462 [pii] 10.1001/jama.2009.878

Schinka, J. A., Busch, R. M., & Robichaux-Keene, N. (2004). A meta-analysis of the association between the serotonin transporter gene polymorphism (5-HTTLPR) and trait anxiety. *Molecular Psychiatry*, *9*(2), 197–202.

Schroeder, J. W., Smith, A. K., Brennan, P. A., Conneely, K. N., Kilaru, V., Knight, B. T., ... Stowe, Z. N. (2012). DNA methylation in neonates born to women receiving psychiatric care. *Epigenetics*, *7*(4), 409–414. doi: 10.4161/epi.19551

Schulz, D., Southekal, S., Junnarkar, S. S., Pratte, J. F., Purschke, M. L., Stoll, S. P., ... Vaska, P. (2011). Simultaneous assessment of rodent behavior and neurochemistry using a miniature positron emission tomograph. *Nature Methods*, *8*(4), 347–352. doi: 10.1038/nmeth.1582

Sen, S., Burmeister, M., & Ghosh, D. (2004). Meta-analysis of the association between a serotonin transporter promoter polymorphism (5-HTTLPR) and anxiety-related personality traits. *American Journal of Medical Genetics*, *127B*(1), 85–89.

Sirota, L. A., Greenberg, B. D., Murphy, D. L., & Hamer, D. H. (1999). Non-linear association between the serotonin transporter promoter polymorphism and neuroticism: A caution against using extreme samples to identify quantitative trait loci. *Psychiatric Genetics*, *9*(1), 35–38.

Stone, L. B., McGeary, J. E., Palmer, R. H., & Gibb, B. E. (2013). Identifying genetic predictors of depression risk: 5-HTTLPR and BDNF Val66Met polymorphisms are associated with rumination and co-rumination in adolescents. *Frontiers in Genetics*, *4*, 246. doi: 10.3389/fgene.2013.00246

Suderman, M., McGowan, P. O., Sasaki, A., Huang, T. C., Hallett, M. T., Meaney, M. J., ... Szyf, M. (2012). Conserved epigenetic sensitivity to early life experience in the rat and human hippocampus. *Proceedings of the National Academy of Sciences USA*, *109*(Suppl. 2), 17266–17272. doi: 10.1073/pnas.1121260109

Surtees, P. G., Wainwright, N. W., Willis-Owen, S. A., Luben, R., Day, N. E., & Flint, J. (2005). Social adversity, the serotonin transporter (5-HTTLPR) polymorphism and major depressive disorder. *Biological Psychiatry*, *59*, 224–229.

Suska, A., Lee, B. R., Huang, Y. H., Dong, Y., & Schluter, O. M. (2013). Selective presynaptic enhancement of the prefrontal cortex to nucleus accumbens pathway by cocaine. *Proceedings of the National Academy of Sciences USA*, *110*(2), 713–718. doi: 10.1073/pnas.1206287110

Taylor, A., & Kim-Cohen, J. (2007). Meta-analysis of gene-environment interactions in developmental psychopathology: Meta-analysis. *Developmental Psychopathology*, *19*(4), 1029–1037. doi: 10.1017/S095457940700051X

Thanos, P. K., Robison, L., Nestler, E. J., Kim, R., Michaelides, M., Lobo, M. K., & Volkow, N. D. (2013). Mapping brain metabolic connectivity in awake rats with muPET and optogenetic stimulation. *Journal of Neuroscience*, *33*(15), 6343–6349. doi: 10.1523/JNEUROSCI.4997-12.2013

Thapar, A., O'Donovan, M., & Owen, M. J. (2005). The genetics of attention deficit hyperactivity disorder. *Human Molecular Genetics*, *14*(Spec No. 2), R275–R282.

Tye, K. M., & Deisseroth, K. (2012). Optogenetic investigation of neural circuits underlying brain disease in animal models. *Nature Reviews Neuroscience*, *13*(4), 251–266. doi: Nrn3171 [pii] 10.1038/nrn3171

Tye, K. M., Mirzabekov, J. J., Warden, M. R., Ferenczi, E. A., Tsai, H. C., Finkelstein, J., ... Deisseroth, K. (2013). Dopamine neurons modulate neural encoding and expression of depression-related behaviour. *Nature*, *493*(7433), 537–541. doi: 10.1038/nature11740

Tye, K. M., Prakash, R., Kim, S. Y., Fenno, L. E., Grosenick, L., Zarabi, H., ... Deisseroth, K. (2011). Amygdala circuitry mediating reversible and bidirectional control of anxiety. *Nature*, *471*(7338), 358–362. doi: 10.1038/nature09820

Viviani, R., Sim, E. J., Lo, H., Beschoner, P., Osterfeld, N., Maier, C., ... Kirchheiner, J. (2010). Baseline brain perfusion and the serotonin transporter promoter polymorphism. *Biological Psychiatry*, *67*(4), 317–322. doi: 10.1016/j.biopsych.2009.08.035

Walsh, N. D., Dalgleish, T., Dunn, V. J., Abbott, R., St. Clair, M. C., Owens, M., ... Goodyer, I. M. (2012). 5-HTTLPR-environment interplay and its effects on neural reactivity in adolescents. *Neuroimage*, *63*(3), 1670–1680. doi: 10.1016/j.neuroimage.2012.07.067

Weaver, I. C., Cervoni, N., Champagne, F. A., D'Alessio, A. C., Sharma, S., Seckl, J. R., ... Meaney, M. J. (2004). Epigenetic programming by maternal behavior. *Nature Neuroscience*, *7*(8), 847–854. Epub 2004 Jun 2027.

Weder, N., Yang, B. Z., Douglas-Palumberi, H., Massey, J., Krystal, J. H., Gelernter, J., & Kaufman, J. (2009). MAOA genotype, maltreatment, and aggressive behavior: The changing impact of genotype at varying levels of trauma. *Biological Psychiatry*, *65*(5), 417–424.

Widom, C. S., & Brzustowicz, L. M. (2006). MAOA and the "cycle of violence:" Childhood abuse and neglect, MAOA genotype, and risk for violent and antisocial behavior. *Biological Psychiatry*, *60*(7), 684–689. doi: 10.1016/j.biopsych.2006.03.039

Williams, S. C., & Deisseroth, K. (2013). Optogenetics. *Proceedings of the National Academy of Sciences USA*, *110*(41), 16287. doi: 10.1073/pnas.1317033110

Wohl, M., Purper-Ouakil, D., Mouren, M. C., Ades, J., & Gorwood, P. (2005). [Meta-analysis of candidate genes in attention-deficit hyperactivity disorder]. *Encephale*, *31*(4 Pt. 1), 437–447.

Young, S. E., Smolen, A., Hewitt, J. K., Haberstick, B. C., Stallings, M. C., Corley, R. P., & Crowley, T. J. (2006). Interaction between MAO-A genotype and maltreatment in the risk for conduct disorder: Failure to confirm in adolescent patients. *American Journal of Psychiatry*, *163*(6), 1019–1025. doi: 10.1176/appi.ajp.163.6.1019

Zhang, F., Aravanis, A. M., Adamantidis, A., de Lecea, L., & Deisseroth, K. (2007). Circuit-breakers: Optical technologies for probing neural signals and systems. *Nature Reviews Neuroscience*, *8*(8), 577–581. doi: nrn2192 [pii] 10.1038/nrn2192 tissue. It is unclear how this problem will be solved. Interesting findings have emerged from peripheral tissues: for example, childhood maltreatment has been shown to impact methylation throughout the genome (Yang et al., 2013), and this is something that could have a multitude of effects on gene regulation and risk for illness later in life. Other new methods that make use of genetics tools also show great promise for increasing understanding of neurobiology—for example, optogenetics (e.g., Lee et al., 2010), which allows direct observation of the actions, or excitation or inhibition, of specific populations of neurons.

This is an exciting time to study genetics of behavioral traits; to venture a rough guess, the amount of genotype and sequence data extant has been doubling every 6 months to a year for several years now. It is fair to predict that with the continued expansion of available data will come better understanding of the genetic underpinnings of behavior and probably lots of surprises.

Acknowledgments

Supported in part by National Institutes of Health grants R01 DA12690, R01 DA12849, R01 AA11330, R01 AA017535, and the VA Connecticut MIRECC.

References

Blackwood, D. H., Fordyce, A., Walker, M. T., St. Clair, D. M., Porteous, D. J., & Muir, W. J. (2001). Schizophrenia and affective disorders—cosegregation with a translocation at chromosome 1q42 that directly disrupts brain-expressed genes: Clinical and P300 findings in a family. *American Journal of Human Genetics*, 69, 428–433.

Caspi, A., Sugden, K., Moffitt, T. E., Taylor, A., Craig, I. W., Harrington, H., McClay, J., Mill, J., Martin, J., Braithwaite, A., & Poulton, R. (2003). Influence of life stress on depression: Moderation by a polymorphism in the 5-HTT gene. *Science, 301*(5631), 386–389.

Choi, M., Scholl, U. I., Ji, W., Liu, T., Tikhonova, I. R., Zumbo, P., Nayir, A., Bakkaloğlu, A., Ozen, S., Sanjad, S., Nelson-Williams, C., Farhi, A., Mane, S., & Lifton, R. P. (2009). Genetic diagnosis by whole exome capture and massively parallel DNA sequencing. *Proceedings of the National Academy of Science of the United States of America, 106*(45), 19096–19101.

Clark, V. E., Erson-Omay, E. Z., Serin, A., Yin, J., Cotney, J., Ozduman, K., Avsar T., Li, J., Murray, P. B., Henegariu, O., Yilmaz, S., Günel, J. M., Carrión-Grant, G., Yilmaz, B., Grady, C., Tanrikulu, B., Bakircioglu, M., Kaymakçalan, H., Caglayan, A. O., Sencar, L., Ceyhun, E., Atik, A. F., Bayri, Y., Bai, H., Kolb, L. E., Hebert, R., Omay, S. B., Mishra-Gorur, K., Choi, M., Overton, J. D., Holland, E. C., Mane, S., State, M. W., Bilgüvar, K., Baehring, J. M., Gutin, P. H., Piepmeier, J. M., Vortmeyer, A., Brennan, C. W., Pamir, M. N., Kiliç, T., Lifton, R. P., Noonan, J. P., Yasuno, K., & Günel, M. (2013). Genomic analysis of non-NF2 meningiomas reveals mutations in TRAF7, KLF4, AKT1, and SMO. *Science, 339*(6123): 1077–80.

de Jong, S., van Eijk, K. R., Zeegers, D. W., Strengman, E., Janson, E., Veldink, J. H., van den Berg, L. H., Cahn, W., Kahn, R. S., Boks, M. P., & Ophoff, R. A. (2012). PGC Schizophrenia (GWAS) Consortium: Expression QTL analysis of top loci from GWAS meta-analysis highlights additional schizophrenia candidate genes. *European Journal of Human Genetics, 9*, 1004–1008.

Edenberg, H. J. (2007). The genetics of alcohol metabolism: Role of alcohol dehydrogenase and aldehyde dehydrogenase variants. *Alcohol Research & Health, 30*(1), 5–13.

Edenberg, H. J., Dick, D. M., Xuei, X., Tian, H., Almasy, L., Bauer, L. O., Crowe, R. R., Goate, A., Hesselbrock, V., Jones, K., Kwon, J., Li, T.-K., Nurnberger, J. I., O'Connor, S. J., Reich, T., Rice, J., Schuckit, M. A., Porjesz, B., Foroud, T., & Begleiter, H. (2004). Variations in GABRA2, encoding the α2 subunit of the GABAA receptor, are associated with alcohol dependence and with brain oscillations. *American Journal of Human Genetics, 74*(4), 705–714.

ENCODE Project Consortium. (2012). An integrated encyclopedia of DNA elements in the human genome. *Nature, 489*, 57–74.

Glahn, D. C., Winkler, A. M., Kochunov, P., Almasy, L., Duggirala, R., Carless, M. A., Curran, J. C., Olvera, R. L., Laird, A. R., Smith, S. M., Beckmann, C. F., Fox, P. T., & Blangero J. (2010). Genetic control over the resting brain. *Proceedings of the National Academy of Science of the United States of America, 107*(3), 1223–1228.

Gudmundsson, J., Sulem, P., Gudbjartsson, D. F., Masson, G., Agnarsson, B. A., Benediktsdottir, K. R., Sigurdsson, A., Magnusson, O. T., Gudjonsson, S. A., Magnusdottir, D. N., Johannsdottir, H., Helgadottir, H. T., Stacey, S. N., Jonasdottir, A., Olafsdottir, S. B., Thorleifsson, G., Jonasson, J. G., Tryggvadottir, L., Navarrete, S., Fuertes, F., Helfand, B. T., Hu, Q., Csiki, I. E., Mates, I. N., Jinga, V., Aben, K. K., van Oort, I. M., Vermeulen, S. H., Donovan, J. L., Hamdy, F. C., Ng, C. F., Chiu, P. K., Lau, K. M., Ng, M. C., Gulcher, J. R., Kong, A., Catalona, W. J., Mayordomo, J. I., Einarsson, G. V., Barkardottir, R. B., Jonsson, E., Mates, D., Neal, D. E., Kiemeney, L. A., Thorsteinsdottir, U., Rafnar, T., & Stefansson, K. (2012). A study based on whole-genome sequencing yields a rare variant at 8q24 associated with prostate cancer. *Nature Genetics, 44*(12), 1326–1329.

Gusella, J. F., Wexler, N. S., Conneally, P. M., Naylor, S. L., Anderson, M. A., Tanzi, R. E., Watkins, P. C., Ottina, K., Wallace, M. R., Sakaguchi, A. Y., Young, A. B., Shoulson, I., Bonilla, E., & Martin, J. B. (1983). A polymorphic DNA marker genetically linked to Huntington's disease. *Nature, 306*(5940), 234–238.

Hariri, A. R., Mattay, V. S., Tessitore, A., Kolachana, B., Fera, F., Goldman, D., Egan, M. F., & Weinberger, D. R. (2002). Serotonin transporter genetic variation and the response of the human amygdala. *Science, 297*(5580), 400–40.

International Schizophrenia Consortium. (2009). Common polygenic variation contributes to risk of schizophrenia and bipolar disorder. *Nature, 406*, 748–752.

Kaufman, J., Yang, B. -Z., Douglas-Palumberi, H., Houshyar, S., Lipschitz, D., Krystal, J. H., & Gelernter, J. (2004). Social supports and serotonin transporter gene moderate depression in maltreated children. *Proceedings of the National Academy of Science of the United States of America, 101*, 17316–17321.

Kendler, K. S. (2010). The stress of internship and interactions with stress. *Archives of General Psychiatry, 67*(6), 566–567.
Kendler, K. S., Heath, A. C., Neale, M. C., Kessler, R., & Eaves, L. J. (1992*a*). A population-based twin study of alcoholism in women. *Journal of the American Medical Association, 268*, 1877–1882.
Kendler, K. S., Neale, M. C., Kessler, R. C., Heath, A. C., & Eaves, L. J. (1992*b*). Major depression and generalized anxiety disorder. Same genes (partly) different environments? *Archives of General Psychiatry, 49*(9), 716–722.
Klein, R. J., Zeiss, C., Chew, E. Y., Tsai, J. Y., Sackler, R. S., Haynes, C., Henning, A. K., SanGiovanni, J. P., Mane, S. M., Mayne, S. T., Bracken, M. B., Ferris, F. L., Ott, J., Barnstable, C., & Hoh, J. (2005). Complement factor H polymorphism in age-related macular degeneration. *Science, 308*(5720), 385–389.
Klengel, T., Mehta, D., Anacker, C., Rex-Haffner, M., Pruessner, J. C., Pariante, C. M., Pace, T. W., Mercer, K. B., Mayberg, H. S., Bradley, B., Nemeroff, C. B., Holsboer, F., Heim, C. M., Ressler, K. J., Rein, T., & Binder, E. B. (2013). Allele-specific FKBP5 DNA demethylation mediates gene-childhood trauma interactions. *Nature Neuroscience*, (1), 33–41.
Korbel, J. O., Urban, A. E., Affourtit, J. P., Godwin, B., Grubert, F., Simons, J. F., Kim, P. M., Palejev, D., Carriero, N. J., Du, L., Taillon, B. E., Chen, Z., Tanzer, A., Saunders, A. C., Chi, J., Yang, F., Carter, N. P., Hurles, M. E., Weissman, S. M., Harkins, T. T., Gerstein, M. B., Egholm, M., & Snyder, M. (2007). Paired-end mapping reveals extensive structural variation in the human genome. *Science, 318*(5849), 420–426.
Lee, J. H., Durand, R., Gradinaru, V., Zhang, F., Goshen, I., Kim, D. S., Fenno, L. E., Ramakrishnan, C., & Deisseroth, K. (2010). Global and local fMRI signals driven by neurons defined optogenetically by type and wiring. *Nature, 465*, 788–792.
Li, D., Zhao, H., & Gelernter, J. (2011). Strong association of the alcohol dehydrogenase 1B gene (*ADH1B*) with alcohol dependence and alcohol-induced medical diseases. *Biological Psychiatry, 70*(6), 504–512.
Lowing, P. A., Mirsky, A. F., & Pereira, R. (1983). The inheritance of schizophrenia spectrum disorders: A reanalysis of the Danish adoptee study data. *American Journal of Psychiatry, 140*(9), 1167–1171.
Malhotra, D., McCarthy, S., Michaelson, J. J., Vacic, V., Burdick, K. E., Yoon, S., Cichon, S., Corvin, A., Gary, S., Gershon, E. S., Gill, M., Karayiorgou, M., Kelsoe, J. R., Krastoshevsky, O., Krause, V., Leibenluft, E., Levy, D. L., Makarov, V., Bhandari, A., Malhotra, A. K., McMahon, F. J., Nöthen, M. M., Potash, J. B., Rietschel, M., Schulze, T. G., & Sebat, J. (2011). High frequencies of de novo CNVs in bipolar disorder and schizophrenia. *Neuron, 72*(6), 951–963.
Need, A. C., Shashi, V., Hitomi, Y., Schoch, K., Shianna, K. V., McDonald, M. T., Meisler, M. H., & Goldstein, D. B. (2012a). Clinical application of exome sequencing in undiagnosed genetic conditions. *Journal of Medical Genetics*, (6), 353–361.
Need, A. C., McEvoy, J. P., Gennarelli, M., Heinzen, E. L., Ge, D., Maia, J. M., Shianna, K. V., He, M., Cirulli, E. T., Gumbs, C. E., Zhao, Q., Campbell, C. R., Hong, L., Rosenquist, P., Putkonen, A., Hallikainen, T., Repo-Tiihonen, E., Tiihonen, J., Levy, D. L., Meltzer, H. Y., & Goldstein, D. B. (2012b). Exome sequencing followed by large-scale genotyping suggests a limited role for moderately rare risk factors of strong effect in schizophrenia. *American Journal of Human Genetics, 91*(2), 303–312.
Oslin, D. W., Berrettini, W., Kranzler, H. R., Pettinati, H., Gelernter, J., Volpicelli, J. R., & O'Brien, C. P. (2003). A functional polymorphism of the μ-opioid receptor gene is associated with naltrexone response in alcohol-dependent patients. *Neuropsychopharmacology, 8*, 1546–1552.
Richards, A. L., Jones, L., Moskvina, V., Kirov, G., Gejman, P. V., Levinson, D. F., & Sanders, A. R. Molecular Genetics of Schizophrenia Collaboration (MGS); International Schizophrenia Consortium (ISC), Purcell, S., Visscher, P. M., Craddock, N., Owen, M. J., Holmans, P., & O'Donovan, M. C. (2012). Schizophrenia susceptibility alleles are enriched for alleles that affect gene expression in adult human brain. *Molecular Psychiatry*, (2), 193–201.
Riordan, J. R., Rommens, J. M., Kerem, B., Alon, N., Rozmahel, R., Grzelczak, Z., Zielenski, J., Lok, S., Plavsic, N., & Chou, J. L. (1989). Identification of the cystic fibrosis gene: Cloning and characterization of complementary DNA. *Science, 245*(4922), 1066–1073.
Sanders, S. J., Murtha, M. T., Gupta, A. R., Murdoch, J. D., Raubeson, M. J., Willsey, A. J., Ercan-Sencicek, A. G., DiLullo, N. M., Parikshak, N. N., Stein, J. L., Walker, M. F., Ober, G. T., Teran, N. A., Song, Y., El-Fishawy, P., Murtha, R. C., Choi, M., Overton, J. D., Bjornson, R. D., Carriero, N. J., Meyer, K. A., Bilguvar, K., Mane, S. M., Sestan, N., Lifton, R. P., Günel, M., Roeder, K., Geschwind, D. H., Devlin, B., & State, M. W. (2012). De novo mutations revealed by whole-exome sequencing are strongly associated with autism. *Nature*, 485, 237–241.
Schizophrenia Psychiatric Genome-Wide Association Study (GWAS) Consortium. (2011). Genome-wide association study identifies five new schizophrenia loci. *Nature Genetics, 43*(10), 969–976.
Thomasson, H. R., Edenberg, H. J., Crabb, D. W., Mai, X. L., Jerome, R. E., Li, T. K., Wang, S. P., Lin, Y. T., Lu, R. B., & Yin, S. J. (1991). Alcohol and aldehyde dehydrogenase genotypes and alcoholism in Chinese men. *American Journal of Human Genetics, 48*(4), 677–681.
Thorgeirsson, T. E., Gudbjartsson, D. F., Surakka, I., Vink, J. M., Amin, N., Geller, F., Sulem, P., Rafnar, T., Esko, T., Walter, S., Gieger, C., Rawal, R., Mangino, M., Prokopenko, I., Mägi, R., Keskitalo, K., Gudjonsdottir, I. H., Gretarsdottir, S., Stefansson, H., Thompson, J. R., Aulchenko, Y. S., Nelis, M., Aben, K. K., den Heijer, M., Dirksen, A., Ashraf, H., Soranzo, N., Valdes, A. M., Steves, C., Uitterlinden, A. G., Hofman, A., Tönjes, A., Kovacs, P., Hottenga, J. J., Willemsen, G., Vogelzangs, N., Döring, A., Dahmen, N., Nitz, B., Pergadia, M. L., Saez, B., De Diego, V., Lezcano, V., Garcia-Prats, M. D., Ripatti, S., Perola, M., Kettunen, J., Hartikainen, A. L., Pouta, A., Laitinen, J., Isohanni, M., Huei-Yi, S., Allen, M., Krestyaninova, M., Hall, A. S., Jones, G. T., van Rij, A. M., Mueller, T., Dieplinger, B., Haltmayer, M., Jonsson, S., Matthiasson, S. E., Oskarsson, H., Tyrfingsson, T., Kiemeney, L. A., Mayordomo, J. I., Lindholt, J. S., Pedersen, J. H., Franklin, W. A., Wolf, H., Montgomery, G. W., Heath, A. C., Martin, N. G., Madden, P. A., Giegling, I., Rujescu, D., Järvelin, M. R., Salomaa, V., Stumvoll, M., Spector, T. D., Wichmann, H. E., Metspalu, A., Samani, N. J., Penninx, B. W., Oostra, B. A., Boomsma, D. I., Tiemeier, H., van Duijn, C. M., Kaprio, J., Gulcher, J. R.; ENGAGE Consortium, McCarthy, M. I., Peltonen, L., Thorsteinsdottir, U., & Stefansson, K. (2010). Sequence

CHAPTER

3 Neuromodulation of Social Behavior in Insects

Stephen J. Simpson *and* Paul A. Stevenson

Abstract

This review begins by introducing transformational concepts in neuromodulation and circuit function that have arisen from the study of insects and other invertebrates. These provide essential background not only to understanding behavior in insects (social or otherwise), but also for the functioning of nervous systems in all animals. The chapter details three (among many possible) examples in which neuromodulation of social behavior has been studied in depth. The first concerns the control of aggression and social dominance in crickets, the second involves swarming in locusts, and the third details the division of labor in honeybee societies. The authors conclude that insects have something profound to offer the study of psychology—not by providing direct analogues of human behavior, but rather by illustrating the power of neuromodulation to generate behavioral complexity in simple systems at the level of individuals and societies.

Key Words: neuromodulator, neuromodulation, octopamine, serotonin, biogenic amine, aggression, swarming, division of labor, cricket, locust, honeybee

The insects provide some of the most extraordinary examples of social behavior in the animal kingdom. The apotheosis of social organization among the insects is seen in the complex and highly ordered societies of ants, bees, wasps, and termites, but many other examples exist—from the dominance hierarchies that are common among the males of many insect species, to abundant examples of maternal care and the mass aggregations that typify species such as locusts. As well as being intrinsically fascinating from a scientific perspective, many of these examples have direct and major impacts on human societies as fundamental components in the functioning of the world's ecosystems; as providers of key services to agriculture, such as pollination and soil health; or as devastating pests. Insects also offer substantial advantages to the study of social behavior more generally.

The insects also offer a series of practical and conceptual advantages in the study of the neurochemical bases of social behavior. First, not only do insects show robust, often stereotyped behavioral responses, they are also far from being simple, hard-wired automata. In fact, plasticity and complex learning abilities are hallmarks of insect behavior (Dukas, 2008). Hence, the insects are not trivial or simplistic models for understanding complex social behavior. Second, insects possess accessible, comparatively "simple" nervous systems with individually identifiable neurons, thus allowing cells and neural circuits to be analyzed in detail. The central nervous system (CNS) of a typical insect contains several hundreds of thousands of neurons, rather than hundreds of billions, as in humans. These neurons are organized within defined nuclei, tracts, and commissures in a developmental plan comprising a series of segmental ganglia, beginning with the brain and a chain of ganglia linked by ventral nerve cords (Burrows, 1996). Third, the neurons of insects share the same basic genetic, molecular, and cellular properties of those in "higher" animals; for example, the same families of voltage-gated ion channels, transmitters, neuromodulators, neurohormones, and second-messenger systems. In fact, as we explain at the outset of this review, some of the foundational understanding concerning the modes of action of neuromodulators on cells and circuits has come from the study of insects, their relative the Crustacea, and other invertebrates.

Finally, we would argue that a major advantage in studying insects is that it frees the experimenter from complicating and obscuring concepts such as cognition, self-awareness, motivation, and personality. This advantage was noted by two of the most eminent practitioners of the study of insect behavior, Vincent Dethier (1976) and John Kennedy (1992)—although regrettably, in some quarters there has since been slippage back to unwarranted anthropomorphic and teleological thinking.

I. Polymorphic Networks Versus Dedicated Circuits

The traditional view of the control of behavior by the nervous system was that each behavior is produced by a discrete circuit of interneurons and that these dedicated circuits compete among each other for the attentions of the motor neurons and muscles in what David McFarland (1975) called the "behavioral final common path." However, research during the past three decades has shown that many circuits are not solely dedicated to the production of single behaviors. Instead, the same physical circuit may be functionally reconfigured to produce multiple outputs. Getting and Dekin (1985) coined the term "polymorphic network" when they found that the same circuit of interneurons in the marine snail *Tritonia* produced both escape swimming and also defensive withdrawal. Similarly, the same circuit in the lamprey produces crawling, burrowing, and swimming, whereas a single network produces hatching from the egg then walking in baby chickens (Harris-Warrick & Marder, 1991).

One of the most remarkable and intensively studied examples is the stomatogastric nervous system of the foregut in decapod Crustacea (lobsters, crayfish, crabs, shrimps, and their close relatives) (Harris-Warrick, Marder, Selverston, & Moulins, 1992; Selverston & Ayers, 2006). Indeed, some of the pioneering studies mapping the neuroanatomy of the lobster foregut nervous system were undertaken by Sigmund Freud. The central pattern generator circuits controlling movements of the decapod foregut are coupled to produce the various configurations of motor output necessary to coordinate food processing and swallowing. The most striking feature of the stomatogastric nervous system is the extent to which its networks (which contain only a few dozen neurons) are functionally reconfigured by the actions of chemical signaling molecules. The system is modulated by at least 20 compounds that are released into the circuits by neurons from higher neural centers, from peripheral sensory neurons, and from blood-borne compounds that are produced by glands elsewhere in the body. The net result of the actions of these modulatory chemicals is that the circuits controlling rhythmical movements of the esophagus, cardiac sac, gastric mill, and pylorus are highly dynamic. Neurons can switch between circuits, and networks can merge and be reconfigured—all without the physical circuit changing (Dickinson, 1995; Harris-Warrick & Marder, 1991).

II. Neuromodulators: A Fundamental Dimension in Neural Integration

The lobster foregut is not atypical—it has become increasingly appreciated that neuromodulation is a fundamental agent of neural control and integration not only in invertebrates but also in the more complex vertebrate nervous system. There are dozens of neuromodulatory chemicals, including peptides, biogenic amines, amino acids, and highly diffusive small molecules such as nitric oxide (NO) and carbon monoxide (see the section Signaling Molecules and Receptors in the Insect Nervous System). They are secreted by neurons that are either outside of the circuit in question (extrinsic) or part of the circuit itself (intrinsic) (Katz, 1995). As such, neuromodulators form part of the continuum of chemical communication within an animal, which spans classical neurotransmitters released across chemical synapses to hormones circulating around the entire blood system. The same compounds may serve multiple roles within this continuum, so that the distinctions between classical neurotransmission, neuromodulation, and hormones are often blurred.

A. *Mechanisms of Neuromodulation*

Neuromodulators have their effects by selectively activating and deactivating ion channels on other neurons. They bind to receptors on target cells and initiate a cascade of second messengers that leads to the opening or closing of ion channels. The result is that the strength of synapses is changed or else the intrinsic properties of neurons are altered.

A.1 ALTERING SYNAPTIC STRENGTH

Changing the strength of synapses can have profound effects on the behavior of a neural circuit, as can be shown to dramatic effect in artificial neural network simulations (Churchland, 1995). A change in synaptic strength may occur in response to patterns of activity within a circuit and involve intrinsic neuromodulation. Such activity-dependent effects are the basis for learning and memory (Pittenger & Kandel, 2003). Synaptic weights may also be changed in a more "hard-wired" manner, particularly in response to extrinsic neuromodulation, with modulators being released into circuits in a particular sensory context and flipping synaptic weights into behaviorally appropriate states.

But how do neuromodulators cause synaptic strength to alter? The efficacy of a synapse is ultimately determined by the number and type of ion channels opened on the postsynaptic neuron as a result of activity in a presynaptic cell. It follows that neuromodulation may alter synaptic strength either by changing the postsynaptic response to neurotransmitter molecules diffusing from the presynaptic cell or by causing the release of more or less neurotransmitter for each invading action potential in the presynaptic cell. Because the calcium ion (Ca^{++}) triggers transmitter release, it follows that one key to altering transmitter release is to change the amount of Ca^{++} entering the presynaptic terminal. Calcium ions enter the terminal of a neuron through depolarization-sensitive channels that are activated by the invading action potential. Neuromodulators have been shown to act both directly on these depolarization-sensitive Ca^{++} channels and indirectly by changing the shape of the action potential through their action on depolarization-sensitive potassium ion (K^+) channels. For example, enkephalins decrease the depolarization-dependent influx of Ca^{++} in mammalian primary pain fibers, which causes a decrease in the release of substance P and, ultimately, a reduced sensation of pain (Mudge, Leeman, & Fischbach, 1979). Serotonin (5-HT) released from modulatory interneurons in the marine mollusc *Aplysia* causes a strengthening of the synapse between gill/siphon sensory receptors and gill motor neurons (Pittenger & Kandel, 2003). This results from 5-HT triggering deactivation of depolarization-sensitive K^+ channels on the presynaptic terminal, which means that incoming action potentials do not repolarize as quickly as usual. Since the presynaptic terminal remains depolarized longer, more Ca^{++} flows into the terminal and thus more neurotransmitter (glutamate) is released. As a result, the sensory-motor synapse is strengthened and gill withdrawal sensitized to stimulation of the gill and siphon.

A.2 ALTERING INTRINSIC PROPERTIES OF NERVE CELLS

Neuromodulators can change membrane conductances by influencing the activity and kinetics of particular ion channels and, as a result, switch a neuron from one stable pattern of activity to another. For example, neuromodulation may result in a neuron changing from a passive to an active state in which it acts as a pacemaker, generates postinhibitory rebound when removed from inhibition, or produces a prolonged plateau potential. These changes will ramify throughout neural circuitry.

A.3 PACEMAKER ACTIVITY

Certain neurons (and muscle and gland cells) can produce rhythmical bursts of action potentials in the absence of synaptic input from other cells. These bursts sit on top of a baseline membrane potential that rises and falls cyclically. Such cells are called "burster" or pacemaker cells. They are widespread throughout both vertebrate and invertebrate nervous systems and play an important role in producing repetitive behaviors like chewing and walking, as well as controlling other cyclical pieces of physiology, including heart beat and gut contractions.(Marder & Bucher, 2001). Some nerve cells in both vertebrate and invertebrates act like "normal" neurons unless a neuromodulator is present, when they switch to pacemaker mode—they are what are known as "conditional bursters." For example, the neuron denoted AB in the stomatogastric circuit of lobsters produces rhythmical bursts of action potentials in the presence of dopamine, 5-HT, octopamine, and proctolin, with the precise pattern of bursting depending on the modulator.

A.4 PLATEAU POTENTIALS

A related phenomenon to pacemaker activity is the generation of plateau potentials. Instead of responding relatively linearly to its input, the neuron responds by switching between two states: it either sits at resting potential or, when it receives sufficient depolarization, flips into a sustained state of depolarization, the amplitude of which is virtually independent of the size of the input and which lasts far longer than the duration of the input. Neurons demonstrating plateau potentials are found throughout the nervous systems of vertebrates and invertebrates. Some cells only exhibit plateau potential activity when in the presence of neuromodulators (Jones & Thompson, 2001).

A.5 POSTINHIBITORY REBOUND

A third active property that is important within neural networks is postinhibitory rebound. This is where a neuron responds after being released from a hyperpolarizing input not by returning to its resting potential but by rebounding and often generating a prolonged burst of action potentials. Release from inhibition causes excitation. Various ionic mechanisms could produce postinhibitory rebound, including hyperpolarization-sensitive channels in conjunction with other low-voltage conductances (Jones & Thompson, 2001). Neuromodulators could block or induce the response in similar ways to those discussed for pacemakers and plateau potentials.

B. A Multitude of Network Possibilities

There are dozens of neuromodulatory compounds. It happens that there are also multiple types of receptor molecule for many or all of these. Different receptor types and subtypes may be localized within specific regions of a single cell: where they are, and in what numbers, will influence how a neuromodulator affects the cell's behavior. The same type of receptor molecule for a neuromodulator may be linked to a very large number of different types of ion channels and, due the actions of interlinking second messengers, the same neuromodulator acting on the same receptor molecule may have opposite effects on the same ion channels, opening or closing them. Add to these the facts that the same cell may co-release two or more transmitters and neuromodulators, in combinations determined by the nature of its own inputs, and that cells releasing neuromodulators may themselves be modulated ("metamodulation"), and the integrative possibilities for even simple neural circuits are almost infinite (Marder & Bucher, 2001; Nusbaum, Blitz, Swensen, Wood, & Marder, 2001).

III. Signaling Molecules and Receptors in the Insect Nervous System

Insects employ essentially the same neurotransmitter substances as mammals and possess evolutionary and pharmacologically related receptors (Hauser, Cazzmali, Williamson, Blenau, & Grimmelikhuijzen, 2006; Homberg, 2002). Acetylcholine is the most common neurotransmitter in the insect CNS and is employed by mechanoreceptors but not motor terminals (Homberg, 2002). As in vertebrates, the amino acids glutamate and γ-amino-butyric acid (GABA) act in insects mainly as excitatory and inhibitory neurotransmitters, respectively. As opposed to mammals, glutamate is the excitatory neurotransmitter at the motor terminals of insect striated muscle, where GABA can also directly inhibit muscular contraction. Insects additionally employ a large number of peptides as synaptic signaling molecules that have a wide spectrum of functions (Nässel & Homberg, 2006), as well as unconventional signaling molecules, such as the gas NO (cf. Ott, Delago, & Elphick, 2004). Regarding the modulation of social behavior, however, most information is available on biogenic amines, and we will focus mainly on this category of signaling molecules.

Biogenic amines can act in insects as dedicated neurotransmitters, as neuromodulators, and as neurohormones released into the hemolymph. As in mammals, they include the catecholamine dopamine, indolamine 5-HT, and histamine, although not noradrenaline and adrenaline, which are lacking in insects and other protostomes (Pflüger & Stevenson, 2005). Instead, insects convert the catecholamine substrate l-tyrosine first to tyramine and then to octopamine, which are known only as "trace amines" in the mammalian brain (Evans, 1985).

Octopamine is generally referred to as the invertebrate analogue of adrenaline/noradrenaline, underlying the fight-or-flight response in insects. It thereby fulfills numerous functions, primarily as a neuromodulator that influences sensory neurons, contractions of skeletal and visceral muscle, and synaptic transmission in the CNS. It is implicated in controlling many aspects of behavior, including the generation of motor patterns for locomotion, escape responses, appetitive learning and memory, aggression, and division of labor in social insects (Scheiner et al., 2006; Verlinden et al., 2010). Its precursor tyramine is becoming recognized as a bona fide neuromodulator, with actions often opposing those of octopamine (Roeder, 2005). The effects of octopamine on behavior are also often antagonized by serotonin (Erber, Kloppenburg, & Scheidler, 1993), which in insects acts as a messenger substance involved in olfaction, circadian rhythms, feeding behavior, aggression, behavioral gregarization in locusts (see section V, Phase Change in Locusts: From Neuromodulation to Mass Migration), phototactic behavior, and learning and memory (for references, see Anstey, Rogers, Ott, Burrows, & Simpson, 2009; Thamm, Balfanz, Scheiner, Baumann, & Blenau, 2010).

In insects, dopamine is reputedly involved in regulating rest and activity ("arousal," Kume, Kume, Park, Hirsh, & Jackson, 2005), aversive learning (Heisenberg et al., 2003) and courtship behavior (Liu, Wei, Yuan, & Guo, 2008). Comparatively little is known about the behavioral role of histamine. It is the neurotransmitter used by photoreceptors and some mechanoreceptors, and it modulates the circadian system and temperature preferences (Haas, Sergeeva, & Selbach, 2008; Nässel, 1999).

As in vertebrates, the actions of biogenic amines in insects are mediated by multiple receptor subtypes, which mostly belong to the super family of G-protein coupled receptors that cause changes in the intracellular concentration of the second messengers cAMP or calcium (Blenau & Baumann, 2001). The locations of biogenic amines and, in some cases, their receptors have been mapped by immunocytochemistry in the brains of several model insects (e.g., Homberg, 2002; Sinakevitch, Mustard, & Smith, 2011).

Aminergic neurons are comparatively few in number, but their projections and receptors are widespread. Only a few neuron types have been physiologically characterized. Most is known about the octopaminergic dorsal and ventral unpaired median (DUM/VUM-) neurons, which arise from a single median neuroblast (Bräunig & Pflüger, 2001). Neurons of this type, located in the thoracic abdominal system, are primarily associated with the control of muscular activity (Evans, 1985; Mentel, Duch, Stypa, Wegener, Müller & Pflüger, 2003), whereas a subpopulation of around 10 nerve cell bodies located in the subesophageal ganglion project to the brain, where they invade all the major neuropils (e.g., Schröter, Malun, & Menzel, 2007) and are implicated in controlling higher integrative functions including learning and memory (Hammer & Menzel, 1995) and social behaviors (Certel et al., 2010; Zhou, Rao, & Rao, 2008). The picture now emerging from numerous studies is that insect aminergic neurons can have selective influences on behavior by recruiting functionally distinct cell types that have specific terminal projections and activate functionally distinct amine receptor subclasses.

IV. Modulation of Aggression in Crickets

A. *On Aggression*

All animals and animal societies must cope with a simple biological fact—conspecifics are normally their own greatest competitors—rivals for the same territories, shelters, food stuffs, and sexual partners. It is thus no surprise that intraspecific aggression is common throughout the animal kingdom, from the lowest multicellular organisms (Brace & Purvey, 1978) to humans (Albert, Walsh, & Jonik, 1993). Aggression is an inherently dangerous behavior that aims to secure a limited resource at minimal risk of injury or cost to the individual (Archer, 1988). Game theory predicts that aggressive behavior between conspecifics is optimized in evolutionary stable strategies (Maynard Smith & Price, 1973), which are typically stereotyped contests that enable combatants to decide when it would be more opportune to persist in fighting or to flee. The ritualized exchange of agonistic signals during these escalating contests is thought to convey increasingly more accurate information for the *assessment* of the contestants' ability to maintain possession of resources. The latter depends not only on size, strength, and weaponry, but also on aggressive *motivation*, a factor determined by experiences such as winning, losing, and the possession resources and their value. These largely theoretical (and it must be said, somewhat teleological if used inappropriately) considerations (summarized in Hurd, 2006) provide a convenient functional framework to explain most behavioral observations. But what are the proximate mechanisms? How does experience control aggressive *motivation*, and how is it encoded in the CNS? How do animals assess agonistic signals, and how do they make the *decision* to fight or flee?

Research on the role of biogenic amines in aggression has been largely dominated by investigations into the role of serotonin, mainly because it is reputed to suppress aggression in mammals and humans (Nelson & Trainor, 2007). Although the adrenergic–noradrenergic system is traditionally viewed as preparing an animal for fight or flight, no consistent relationship with aggression has been found in vertebrates, although most recent data points toward promoting effects (Haden & Scarpa, 2007; Nelson & Trainor, 2007). These generalized actions of amines on aggression were thought to be reversed in invertebrates because, in arthropods such as lobsters and crabs, serotonin enhances aggressiveness, whereas the invertebrate adrenergic analogue, octopamine, promotes submissiveness (Huber, & Delago, 1998; Huber, Smith, Delago,

Isaksson, & Kravitz, 1997; for review see Kravitz & Huber, 2003). In the following sections, we review evidence, mainly from field crickets, suggesting that the aminergic control of aggression in insects is more similar to that in vertebrates.

B. Stereotyped Fighting and Its Initiation in Crickets

When two crickets meet, they are faced with the choice to court, fight, or flee, and the outcome is largely controlled by information exchanged during antennal contact (Adamo & Hoy, 1994; Hofmann & Schildberger, 2001; see also Fernandez et al., 2010, on *Drosophila*). Species and sex is perceived by the pheromonal signature (Iwasaki & Katagiri, 2008), which induces males to court conspecific females. Antennal contact between males, which takes the form of "fencing," is also a key releasing stimulus for cricket aggression and involves both mechanical and olfactory signaling (Hofmann & Schildberger, 2001). Agonistic responses, such as mandible spreading can be evoked by simply lashing the antennae with a bristle (Alexander, 1961) or by highly volatile male odors (Iwasaki & Katagiri 2008), which have been identified in fruit flies (Wang & Anderson, 2010). The antennal sensory neuropils of crickets include the antennal lobe, an olfactory center equivalent to the mammalian olfactory bulb (Ache & Young, 2005), as well a characteristic "ventral area of flagella afferents" (vfa), which is unknown in most other insects (Staudacher, Gebhardt, & Dürr, 2005). Mechanical antennal stimulation directly activates a small set of fast conducting giant interneurons (Schöneich, Schildberger, & Stevenson, 2011) that trigger directed turning responses and walking (Baba, Tsukada, & Comer, 2010; Zorovic & Hedwig, 2013), but their role in aggression is unknown.

Aggressive interactions between male crickets follow a stereotyped sequence of levels to which a fight escalates before one opponent retreats (Hofmann & Stevenson, 2000). Antennal contact is always followed by mandible spreading, then mandible engagement and pushing, which culminates in a grappling contest. Most fights involve physical contact without injury and last only several seconds, but can go on for minutes and result in the loss of an antenna or leg. The end of a fight is marked by the loser retreating, upon which the winner frequently starts to sing a characteristic rivalry song and produce erratic jerking body movements. Winners repeatedly attack losers, whereas losers actively avoid contact with other males for hours after defeat. There seems to be no individual recognition, since losers will also retreat from unfamiliar opponents (Hofmann & Stevenson, 2000), although in *Drosophila* losers fight differently against familiar and unfamiliar opponents (Yurkovic, Wang, Basu, & Kravitz, 2006).

Female crickets rarely interact but fight vigorously in the presence of a courting male or his courtship song, and winning females have a greater probability of receiving the male's spermatophore (Rillich, Buhl, Schildberger, & Stevenson, 2009). In contrast, *Drosophila* males and females adopt different fighting strategies, whereby the males but not the females establish dominance relationships (Nilsen, Chan, Huber, & Kravitz, 2004). The sexually dimorphic fighting patterns in fruit flies are specified by sex-specific splicing of the *fruitless* gene (Vrontou, Nilsen, Demir, Kravitz, & Dickson, 2006) and controlled by specific subsets of neurons expressing the male form of *fruitless* proteins (Chan & Kravitz, 2007).

When male crickets' antennae are ablated, they no longer fight but they frequently court one other (Hofmann & Schildberger, 2001; see also Fernandez et al., 2010 on *Drosophila*). This is claimed to result from reduced levels of serotonin accompanying antennal ablation (Murakami & Itoh, 2003); however, crickets that lack serotonin due to treatment with the synthesis inhibitor α-methyl-tryptophan show normal aggressive behavior (Stevenson, Hofmann, Schoch, & Schildberger, 2000). In contrast, crickets depleted of octopamine and dopamine by treatment with α-methyl-tyrosine rarely interact if left to their own devices, although they can be coaxed into fighting (Stevenson et al., 2000). Accordingly, amines are not essential for initiating aggression, although octopamine and possibly dopamine, rather than serotonin, promote its expression. This corresponds to the proposed modulatory role of octopamine in initiating flight behavior (Brembs, Christiansen, Pflüger, & Duch, 2007; Buhl, Schildberger, & Stevenson, 2008).

C. Experience-Dependent Modulation of Aggression

As in other animals, aggression in crickets is influenced markedly by a variety of social and other experiences. Male crickets tend to be submissive toward other males when raised under crowded conditions or after just having lost a fight. Conversely, social isolation, winning, and the possession of resources such as food (Nosil, 2002), a female (Killian & Allen, 2008), shelter (Rillich, Schildberger, & Stevenson, 2011), and even flying (Hofmann & Stevenson, 2000; Stevenson, Dyakonova, Rillich, & Schildberger, 2005; Stevenson et al., 2000), all promote a high level of aggressiveness. Here, we outline the role of neuromodulators in experience-dependent plasticity of aggressive behavior.

C.1 OCTOPAMINE AND THE FLIGHT EFFECT

Cricket fighting has been a popular pastime in China for centuries, and there is a wealth of folklore regarding how best to rear and promote the best fighters (Hofmann, 1996). For example, Chinese aficionados recommend "punishing" poor fighters by shaking and throwing them several times. This does in fact significantly increase a cricket's aggression, but it proved to be far more effective to make the animals fly for a minute or so (Hofmann & Stevenson, 2000). After flying, crickets escalate to the highest level of aggression and fight 2–3 times longer than usual. Moreover, although losers usually remain submissive for several hours after a fight, the losers of fights between the flown crickets regained their aggressiveness within only 15 minutes. These effects occurred without enhancing general excitability, as evaluated from the animals' startle responses, and clearly depended on the execution of flight motor activity but not on wind stimulation alone. This highlights the impact that motor activity can have on the operation of seemingly unrelated command centers.

Subsequent studies showed that the flight effect is mediated by the amine octopamine. Flying, as well as other physical activities, leads to activation of selected sets of octopaminergic neurons (Duch, Mentel, & Pflüger, 1999) and to a considerable increase in the hemolymph titer of octopamine (Adamo, Linn, & Hoy, 1995). The latter is probably spillover from the CNS and will itself not influence aggression since the concentration is too low to pass the insect "blood–brain" barrier (cf. Stevenson et al., 2005). The flight effect can, however, be mimicked by injecting the tissue permeable octopamine agonist chlordimeform (CDM) into the hemocoel, and it is abolished in crickets treated with epinastine, a selective octopamine receptor antagonist (cf. Roeder, Degen, & Gewecke, 1998), or α-methyl-p-tyrosine, which depletes octopamine and dopamine from the CNS (Stevenson et al., 2000; 2005). Notably, serotonin depletion induces hyperactivity and enhances startle responses, but without obviously affecting aggression, which indicates that octopamine's influence on aggression is not a simple "arousal" effect.

Work on *Drosophila* has confirmed that octopamine promotes aggression in insects. Baier et al. (2002) observed that male tyramine-β-hydroxylase mutant flies, which cannot synthesize octopamine and have 10-fold elevated tyramine levels in their brains, have deficits in aggressive behavior, whereas Certel et al. (2007) reported that these males tend to court rather than fight each other. Hoyer et al. (2008) later verified that tyramine-β-hydroxylase mutants as well as tyrosine decarboxylase mutants, which lack both octopamine and tyramine, display almost no aggression and that the defect could be rescued partially by octopamine treatment or substituting gene function. This was later confirmed by Zhou et al. (2008), who also identified a subset of octopaminergic neurons important for aggression (see also section C.3 Octopamine, Resources and the Residency Effect) and showed that enhancing octopaminergic signaling enhanced aggressiveness. Altogether the data suggest that octopamine increases aggression by promoting the tendency to fight (Stevenson et al., 2005) and to exhibit agonistic behaviors such as lunging (Hoyer at al., 2008; Zhou et al., 2008) and mandible spreading (Rillich & Stevenson, 2011). In this respect, octopamine can be considered as representing the motivational component of aggression.

C.2 OCTOPAMINE AND THE WINNER EFFECT

Various hypotheses have been put forward to explain the widespread occurrence in the animal kingdom of the winner effect, in which the experience of winning a conflict makes an individual more likely to win a subsequent encounter, but comparatively little is known of the proximate causes (reviewed in Hsu, Earley, & Wolf, 2006; Hsu, Lee, & Lu, 2009). Recent data for vertebrates implicate androgens as physiological mediators (Fuxjager & Marler, 2010), whereas in crickets the amine octopamine is clearly involved (Rillich & Stevenson, 2011).

Winning increases the probability of a cricket subsequently defeating an inexperienced opponent (Khazraie & Campan, 1999), and when winners are matched against each other, the fights are more severe and longer (Rillich & Stevenson, 2011). This winner effect is transient and persists for only 15–20 minutes after winning, which is far shorter than in rodents, where it can last for days. Changes in social status in crickets are thus not necessarily associated with (long-term) learning and memory, as suggested for fruit flies (Yurkovic et al., 2006). As is the case for the flight effect, the winner effect in crickets depends on the amine octopamine. It is prohibited by treatment with the octopamine receptor blocker epinastine, but remains unaffected by antagonists for other amine receptors (Rillich & Stevenson, 2011).

Rillich and Stevenson (2011) also demonstrated that when fights between two crickets are interrupted before either wins, both contestants become hyperaggressive in subsequent encounters. Hence, a winner effect, without actually winning, can result simply from the physical exertion of fighting. This is feasible considering that octopaminergic neurons are activated by a variety of mechanosensory signals (Morris, Duch, & Stevenson, 1999) and that fighting itself leads to an almost five-fold increase in hemolymph levels of octopamine (Adamo et al., 1995). However, a winner effect also develops in crickets that win against submissive opponents that retreated prior to any physical interaction. Thus, aggression is also enhanced by some aspect of the experience of winning. In humans, watching a previous victory elevates levels of androgens (Carré & Putnam, 2010), and, in other mammals, aggressive encounters lead to increased activity in dopaminergic pathways and increased androgen receptor expression in brain regions that mediate motivation and reward (Barron, Søvik, & Cornish, 2010; Couppis, & Kennedy, 2008; O'Connell & Hofmann, 2011). The possibility that fighting and winning may in some way represent a positive reinforcing or rewarding experience is discussed in the next section (C3).

C.3 OCTOPAMINE, RESOURCES, AND THE RESIDENCY EFFECT

Irrespective of species, animals in possession of a key resource are more likely to win disputes against contenders, but it is hotly debated how this is controlled (see reviews by Hsu et al., 2006, 2009; Kemp & Wiklund, 2004). For male field crickets, burrows are valuable assets offering shelter from predators and an aid in attracting females, which mate preferentially with burrow owners, and these burrow-owning males zealously fight off any intruding male (Alexander, 1961; Simmons, 1986).

Under laboratory conditions, initially submissive crickets become highly aggressive after occupying an artificial shelter and frequently win against an aggressive intruder (Rillich et al., 2011). As for winning, this residency effect is transient and has a similar time course. It first becomes evident after the cricket has experienced 2 minutes of residency, is maximal after 15 minutes, and declines within 15 minutes after the shelter is taken away from the cricket. The residency effect is also octopamine-dependent. It is not evident in crickets depleted of octopamine and dopamine, whereas it remains unaffected by serotonin depletion and is selectively blocked by treatment with octopamine antagonists (Rillich et al., 2011).

It is not yet clear what the experience of residency constitutes for a cricket and how it leads to activation of the octopaminergic system. An intriguing idea is that experiences such as residency, winning, and perhaps even physical activity, are evaluated as being in some way rewarding (Rillich et al., 2011; Stevenson & Rillich, 2011). Evidence in honeybees suggests that the value of food sources could be encoded by octopamine modulating associative reward pathways (Barron et al., 2010). Octopamine is known to convey reward signals in appetitive learning paradigms in honeybees (Hammer & Menzel, 1995), fruit flies (Schwärzel et al., 2003), and crickets (Mizunami et al., 2009). In the honeybee, the activity of even a single identified octopaminergic neuron can substitute for the sucrose reward in an associative learning paradigm (Hammer, 1993). This neuron, VUMmx1, has its soma in the subesophageal ganglion and projects to the brain, giving rise to extensive, bilaterally symmetrical dendritic branches in the antennal lobes and mushroom bodies, a neuropil evolutionarily related to the mammalian cerebral cortex (cf. Tomer, Denes, Tessmar-Raible, & Arendt, 2010). Insect mushroom bodies process olfactory and, to a lesser extent, mechanosensory inputs from the antenna (see review in Galizia & Rössler, 2010). They are noted for their role in olfactory learning and memory (Menzel, 1996) but also seem to be involved in decision-like processes (Heisenberg, 2003). The VUMmx1 neuron is one of a group of 15 related octopaminergic neurons occurring in the subesophageal ganglion of honeybees (Schröter et al., 2007) and other insects (for a review, see Bräunig & Pflüger, 2001). In *Drosophila*, a distinct subset of these octopaminergic neurons was found to be functionally important for expressing aggression (Zhou et al., 2008). Another subset express the sex determining factor *fruitless* and appear to be involved in mediating the choice between courtship and aggression (Certel et al., 2007; 2010). The function of these neuron types in crickets remains to be elucidated.

C.4 INFLUENCES OF GROUPING AND ISOLATION

Social isolation results in dramatic behavioral and physiological changes in many animal species.

With respect to aggression, numerous studies have noted that individuals reared in isolation have higher aggression, whereas aggression is depressed in individuals reared together (Hsu et al., 2006).

In *Drosophila*, evidence suggests that olfactory signaling mediates the reduction of aggression in socially grouped males. Crowding increases the expression of a gene (*Cyp6a20*) encoding a cytochrome P450 in non-neuronal support cells associated with olfactory sensillae that detect the pheromone 11-cis-vaccenyl acetate (cVA; Wang & Anderson 2010). Although acute exposure to cVA induces aggression, chronic exposure reduces it, whereby the two effects are mediated by different olfactory receptors on the flies antennae (Or67d and Or67a, respectively; Lui et al., 2011). Furthermore, although inhibition of Or67d neurons selectively attenuated acute aggression, blockade of Or65a neurons abolished the reduction in aggression due to chronic cVA or social exposure.

Aggression is also reduced in male crickets maintained in groups (e.g., Adamo & Hoy 1995; Alexander, 1961), but this does not appear to depend on olfactory signaling (Stevenson & Rillich, 2013). For example, male crickets that were tightly grouped, but each confined in individual wire mesh cages that permitted visual, olfactory, and tactile antennal contact, were found to be equally aggressive as males kept in total social isolation. In free-grouped males, some individuals frequently attacked other males, which invariably retreated without escalating. When randomly matched, grouped crickets had a lower average aggressive score than naïve isolated crickets but attained the same level after 2–3 hours of isolation, which is the same time span required by male crickets to regain their aggressiveness after losing (Stevenson & Rillich, 2013). The depressing effect of crowding on aggression in male crickets is thus a consequence of social subjugation while crowded.

C.5 SOCIAL DEFEAT AND MECHANISMS FOR SUPPRESSING AGGRESSION

Social defeat leads to subdued aggressive tendency in many animal species (Hsu et al. 2006; 2009), but the mechanisms leading to the actual retreat of an individual and subsequent suppression of aggression are not known. Most theories (outlined in Rillich, Schildberger, & Stevenson, 2007) agree that information gathered from agonist signals exchanged during fighting is somehow assessed to determine when it would be more opportune to flee rather than fight, but disagree in essence with respect to which individual evaluates these cues (sender, receiver, or both) and their net impact.

Work on crickets revealed that agonistic signals act to suppress aggressiveness of the receiver, but are probably not evaluated by the sender (Rillich et al., 2007). For example, pairs of crickets with lamed mouthparts, which are otherwise formidable weapons, still fight and even escalate higher and fight longer than sham treated crickets. In facts, fights became progressively longer the more the animals were handicapped, lasting minutes rather than seconds, for example, between opponents that both had lamed mandibles, clipped hind-leg claws to limit pushing, and blackened eyes. Furthermore, whereas "blinded" crickets or crickets with lamed mouthparts fought nonhandicapped crickets with almost unaltered win chances, blinded crickets practically always (98 percent) defeated crickets with lamed mouthparts (Rillich et al., 2007). Although not specifically designed to test it, the findings of these experiments conform to the cumulative assessment model postulated by Payne (1998), which predicts that a contestant will persist in fighting until the sum of the perceived adversary's actions accumulated during fighting surpasses some threshold to flee. Accordingly, the blinded cricket persists because it receives no visual and only limited physical input from an opponent with lamed mandibles, whereas the latter accumulates the full brunt of his adversary's actions and hence becomes the first to flee.

It is not known how the sensory information of various modalities conveyed by an opponent's agonistic signals is summated in the nervous system and leads to submission. Retreat and depressed aggressiveness are unlikely to result from depressed octopaminergic signaling. First, the levels of this amine are similar in winners and losers (Adamo et al., 1995). Second, although the potent octopamine agonist chlordimeform (CDM), which binds almost irreversibly to octopamine receptors (cf. Evans, 1985), restores aggression in losers, it cannot protect them from actually losing and subsequently behaving submissively (Stevenson et al., 2005). Thus, there must be some opposing control mechanism.

The actions of octopamine in arthropods are functionally antagonized by serotonin in a variety of behaviors including antennal scanning (Erber et al., 1993), escape running (Goldstein & Camhi, 1991), and aggression, at least in crustaceans (Kravitz & Huber, 2003). Serotonin is also implicated in mediating behavioral changes following social interactions (Edwards & Kravitz,

1987). In desert locusts, sensory inputs derived from conspecifics during social encounters promote the release of serotonin, which is sufficient and necessary for switching their behavior from active avoidance to social tolerance and aggregation (Anstey et al., 2009 see also section V). In crickets, evidence suggests that serotonergic signaling may depress escape responses in aggressive individuals (Dyakonova, Schürmann, & Sakharov, 1999), whereas losers are claimed to show enhanced escape behavior due to lower brain levels of serotonin after defeat (Murakami & Itoh, 2001). This phenomenon is better understood in crayfish, for which the effects of serotonin on escape and body posture change with social status due to a shift in the relative expression of different serotonin receptor subtypes to a pattern more appropriate for the new status (Cattaert, Delbecque, Edwards, & Issa, 2010).

The role of serotonin in insect aggression is, however, unclear. Corroborating findings in crickets (Stevenson et al., 2000; 2005), Baier et al. (2002) found that aggression in *Drosophila* is unaffected when serotonin synthesis is either disrupted or its level elevated by treatment with the precursor of 5-HT (5-HTP). Contrasting this, Dierick and Greenspan (2007) found that 5-HTP treatment promotes aggression in fruit flies. Similarly, Alekseyenko et al. (2010) report that selective disruption of serotonergic neurotransmission using molecular genetic techniques yielded flies that fought with reduced ability to escalate fights, whereas acute activation of serotonergic neurons resulted in flies that escalated faster and fought at higher intensities. These conflicting findings may be due to differences in behavioral protocols together with difficulties in dissecting out differential effects of serotonin operating via different receptor subtypes. Johnson et al. (2009), for example, could show that pharmacological activation of 5-HT_2-type receptors reduced total aggression in *Drosophila* and, conversely, that activating 5-HT_{1A}-type receptors increased it. Thus, different serotonin receptor subtypes are likely to influence different aspects of the total aggressive behavioral repertoire, as previously shown in mammals (de Boer & Koolhaas, 2005). This and other findings in mammals challenge the dogmatic view of serotonin acting simply to suppress aggression. Currently, serotonin is thought to limit impulsivity (for review, see Nelson & Trainor, 2007) or promote the drive to withdraw (Tops, Russo, Boksem, & Tucker, 2009), rather than suppress aggression per se. An analogous scenario is conceivable in insects.

The expression of aggression in insects is probably also limited by the action of a variety of other neuromodulators. In crickets, treatment with the opiate antagonist naloxone elevates aggressiveness in losers without affecting winners or socially naïve animals (Dyakonova, Schürmann, & Sakharov, 2002) and in *Drosophila* aggression is increased following genetic silencing of circuitry employing neuropeptide F, the invertebrate homologue of neuropeptide Y (Dierick & Greenspan, 2007). A further promising candidate is the gaseous modulator NO, which suppresses aggression in mammals, at least partly by influencing serotonergic signaling (Nelson & Trainor, 2007). Its role in insect aggression needs further clarification. Dyakonova and Krushinskii (2006) report that treatment with an NO synthesis inhibitor prohibits the aggression-promoting effects of flying in crickets, indicating that NO enhances aggressiveness. Iwasaki et al. (2007), in contrast, found that NO synthesis inhibitors relieve the depression of aggressiveness in losers but do not affect aggression in socially naïve crickets. Ongoing work supports the notion that NO promotes the tendency of both socially naïve and loser crickets to flee (Stevenson & Rillich, 2014 submitted).

D. A Threshold Model for Controlling the Fight-or-Flee Response

To conclude this section, what is generally referred to as the decision to fight or flee could be accounted for in crickets by simply modulating relative behavioral thresholds (Figure 2.1; Stevenson & Rillich, 2012; Stevenson & Schildberger, 2013). Experiences evaluated as being in some way rewarding (winning, resource possession) promote the tendency to fight to a level determined by the modulatory action of octopamine, which can be considered as representing the motivational component of aggression. Opposing this, aversive experiences (i.e., agonistic signals gathered from the opponent during ritualized fighting) will accumulate and eventually promote the tendency to flee upon surpassing the set level, possibly via the actions of serotonin, NO, and selected peptides. This idea is in line with the roles proposed for noradrenaline, serotonin, and NO in mammals (Tops et al., 2009), suggesting that some basic mechanisms of aggressive modulation may be shared.

V. Phase Change in Locusts: From Neuromodulation to Mass Migration

Locusts are a group of 20 or so otherwise typical grasshopper species that are defined by one

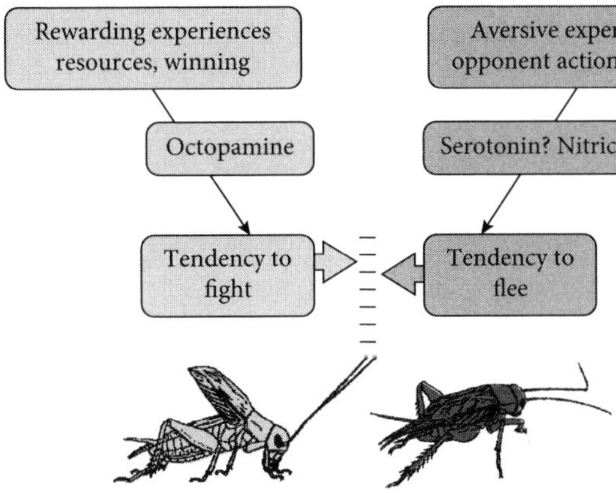

Fig. 2.1 Chart illustrating how the decision to fight or flee in crickets could be implemented by modulating relative behavioral thresholds. Rewarding experiences (winning, resource possession) lead to activation of the octopaminergic system, which promotes the tendency to fight. In this respect, octopamine represents the motivational component of aggression. Aversive experiences accumulated during fighting (opponent's agonist signals, defeat), in contrast, promote the tendency to flee, most likely via actions of serotonin, nitric oxide, and selected peptides. Hence, a cricket will attack and persist in fighting for as long as the sum of rewarding experiences (motivation) exceeds the sum of accumulated aversive experiences, but retreat the moment the latter is greater.

Figure originally published in Stevenson, P.A., & Rillich J. (2012). The decision to fight or flee—insights into underlying mechanism in crickets. *Frontiers in Neuroscience. 6*(118): doi:10.3389/fnins.2012.00118.

extraordinary feature of their biology: they are two animals packed within the same genome. When reared at low population densities, they develop into the cryptic "solitarious" phenotype (or phase); however, when crowded, they switch into the swarm-forming "gregarious" phase (Pener & Simpson, 2009; Simpson & Sword, 2008). Depending on the species of locust, this transition involves a suite of continuously varying traits—including color, shape, metabolic and hormonal physiology, brain structure, immune function, and reproductive and other life history traits. These do not all share the same underlying mechanisms or follow the same time course, but they are effectively coupled together by one of the most labile traits—social behavior (Burrows, Rogers, & Ott, 2011; Pener & Simpson, 2009; Simpson & Sword, 2009).

Solitarious phase locusts avoid one another except when seeking mates, whereas gregarious locusts are much more active and form aggregations. The transition from solitarious to gregarious behavior in response to crowding occurs within a few hours. The reverse behavioral transition from the fully gregarious to the solitarious state may either be slower, as seen in the desert locust, *Schistocerca gregaria* (Simpson, McCaffrey, & Hägele, 1999), or equally rapid, as in the Australian plague locust, *Chortoicetes terminifera* (Gray, Sword, Anstey, Clissold, & Simpson, 2009). In the desert locust, there is transfer of behavioral state across generations via a chemically mediated maternal epigenetic effect, which has been traced to an L-DOPA analogue produced in the female's accessory reproductive glands (Miller, Islam, Claridge, Dodgson, & Simpson, 2008). Epigenetic inheritance of other traits, such as color and size, also occurs in *S. gregaria*, but via different mechanisms (Maeno & Tanaka, 2011; Tanaka & Maeno, 2010), which illustrates that density-dependent phenotypic plasticity in locusts is a set of functionally coupled, but mechanistically distinct traits.

A. The Sensory Cues Triggering Behavioral Phase Change

In nymphs of *S. gregaria*, the combination of the sight and smell of other locusts or tactile stimulation alone is sufficient to induce behavioral gregarization within a few hours (Roessingh, Bouaïchi, & Simpson, 1998), with the hind legs being the principal sites for detecting tactile stimulation (Simpson, Despland, Hägele, & Dodgson, 2001). Intriguingly, in another species, the Australian plague locust, *C. terminifera*, touching of the antennae rather than the hind legs induces behavioral gregarization (Cullen, Sword, Dodgson, & Simpson, 2010). That different locust species have similar behavioral responses to crowding, but that these are produced by different mechanisms, reflects the fact that locust-like traits have evolved independently on several occasions within the grasshopper lineage (Pener & Simpson, 2009).

B. Modulation of Neural Pathways for Gregarization

The neural pathways carrying information about crowding from mechanoreceptors on the hind legs to the CNS have been identified in *S. gregaria* (Rogers et al., 2003). Because the shift in behavior in response to crowding is too rapid to involve physical remodeling of the neural circuitry, it must follow that neuromodulators are involved. Indeed, neuromodulation of CNS circuits controlling behavior is critical for the initial stages of gregarization. A survey of changes in putative modulators within the CNS (Rogers et al., 2004) and subsequent pharmacological interventions revealed that a transient pulse of serotonin in the metathoracic ganglion, induced either by stimulation of the hind legs or a combination of sight and odor cues from other locusts, is both necessary and sufficient to induce behavioral gregarization in *S. gregaria* (Anstey et al., 2009; Figure 2.2). Experiments on the migratory locust, *Locusta migratoria*, involving genome-wide gene expression profiling, RNA interference (RNAi), and pharmacological interventions have implicated both dopaminergic and serotonergic pathways in behavioral phase change in this species (Ma, Guo, Guo, Wang, & Kang, 2011). Dopamine has also been found to differ with phase in *S. gregaria*, but with a longer time course than that for the initial stages of behavioral change (Rogers et al., 2004). Whether these locust species differ in the neuromodulatory mechanisms of phase change is yet to be determined, as is the nature and time course of interactions between the various catecholamine pathways.

Ott et al. (2012) studied the involvement of two protein kinase systems in the early stages of behavioral gregarization in desert locusts: a cGMP-dependent protein kinase (PKG) that plays a role in group-related and social behaviors in various insects (Ament, Wang, & Robinson, 2010; see section entitled "Neuromodulators and the Division of Labor in Honeybee Society") and has been implicated in behavioral gregarization in locusts (Lucas et al., 2010), and cAMP-dependent protein kinase (PKA), which has been shown to play a key role in mediating the effects of serotonin on learning and memory (Pittenger & Kandel, 2003). Injecting the PKA inhibitor KT5720 before crowding prevented behavioral gregarization in solitarious nymphs, but injecting KT5823, an inhibitor of PKG, did not block crowding-induced behavioral change. Ott et al. (2012) also blocked *foraging* gene expression by RNA-interference (RNAi) without affecting gregarization on crowding. By contrast, RNAi-induced reduction in expression of PKA catalytic subunit C1 reduced behavioral gregarization after solitarious nymphs were crowded, whereas RNAi against the inhibitory R1 subunit exaggerated gregarization. These results are consistent with serotonin being a key initiator of behavioral gregarization (Anstey, Rogers, Ott, Burrows, & Simpson (2009).

C. Physical Restructuring of the Nervous System and Genomic Mechanisms

Longer term behavioral changes that are set in train by neuromodulators presumably involve physical changes in neural circuitry, perhaps involving gene expression changes akin to the shift from short- to long-term memory (Pittenger & Kandel, 2003). Indeed, fully solitarious and gregarious locusts differ in several respects, from the size and proportions of different brain regions (Ott & Rogers, 2010) to the electrophysiological responses of motor (Blackburn, Ott, Matheson, Burrows, & Rogers, 2010) and sensory pathways (Matheson, Rogers, & Krapp, 2004; Rogers, Krapp, Burrows, & Matheson, 2007; Rogers et al., 2010). Collectively, these changes reflect the different social circumstances and environmental conditions experienced by solitarious and gregarious locusts (Burrows et al., 2011)—differences that are also reflected in other aspects of their biology, including nutritional regulatory responses (Pener & Simpson, 2009; Simpson, Raubenheimer, Behmer, Whitworth, & Wright, 2002).

Surveys of gene expression differences between gregarious and solitarious locusts have implicated numerous compounds as being involved in phase change, including hexamerins, hemocyanins, juvenile hormone binding proteins, and heat shock proteins (Badisco et al., 2011; Wang et al., 2007), as well as differences in methylation profiles (Boerjan et al., 2011) and the small RNA transcriptome (Wei, Chen, Yang, Ma, & Kang, 2009). Next-generation de novo sequencing of *Locusta* transcriptomes has pinpointed 242 transcripts as being phase-related (Chen et al., 2010). The challenge remains to attribute functions to these candidates and to discover their roles in the pathway from initiation to maintenance of phase change. For example, Guo et al. (2011) reported that *CSP* (chemosensory protein) genes and a gene called *takeout* are differentially expressed in antennae of gregarious and solitarious nymphs of the migratory locust, *L. migratoria*. They combined the use of

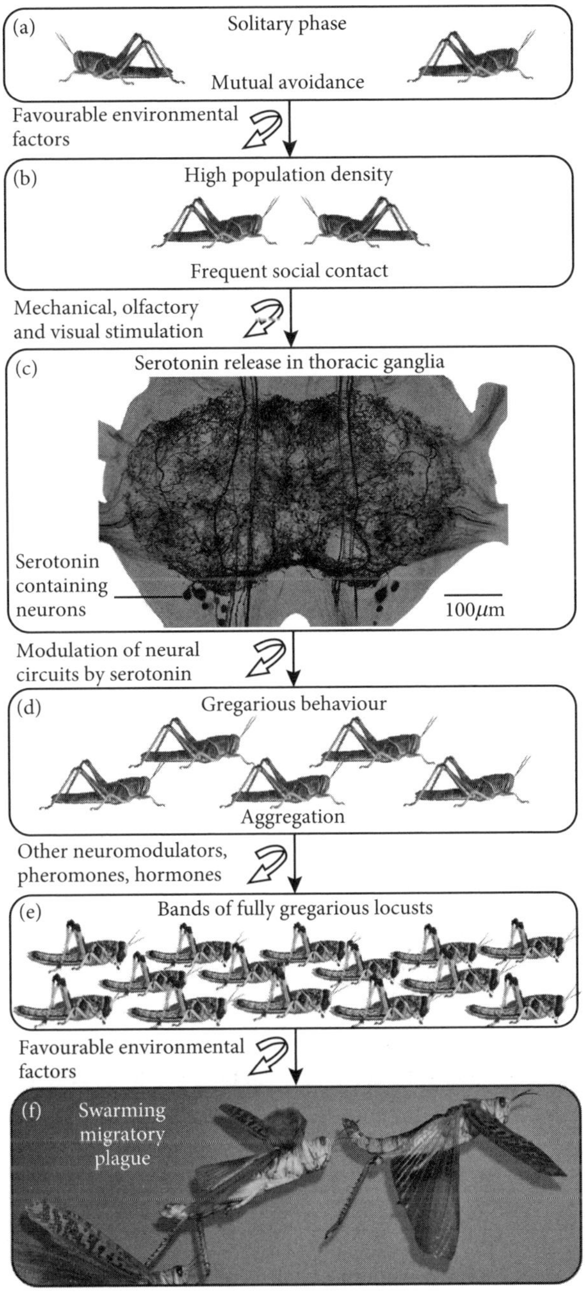

Fig. 2.2 Chart illustrating the role of serotonin in controlling phase change and swarming in the desert locust. (a) Solitary locusts are scarce and avoid each other. (b) Under favorable climatic conditions, population density and hence social contact increases. (c) The resulting mechanical, olfactory, and visual stimuli induce serotonin release, possibly from neurons in the thoracic nervous system (*arrow*). (d) Serotonin promotes gregarious behavior and aggregation, presumably by selectively modulating specific neuronal circuits. (e) Crowding recruits additional chemical messengers, leading to acquisition of all gregarious traits, including color change from green to black and yellow. (f) Environmental factors, such as drought and food shortage, promote swarming and migration of adult gregarious locusts.

(Figure adapted from Stevenson, 2009, and based on data from Anstey et al., 2009.)

RNA interference with olfactory behavioral experiments to implicate these genes in the shift from olfactory repulsion to attraction between individuals during behavioral gregarization.

D. From Individuals to Mass Migration

The shift from solitarious to gregarious behavior is a process occurring within the nervous systems of individual locusts, but its effects scale up to the behavior of groups and populations and may ultimately produce mass swarms. The patchiness of food and other resources at local spatial scales within the habitat determines the likelihood that solitarious locusts will be brought together—although they would otherwise avoid one another—and hence gregarize (Collett, Despland, Simpson, & Krakauer, 1998; Despland, Collett, & Simpson, 2000). Once a local aggregation of juvenile locusts (hoppers) reaches a critical local density, they suddenly commence marching en masse. Such marching bands can comprise millions of individuals and travel hundreds of meters each day; once they become adults, the bands turn into vast winged swarms that can travel hundreds of kilometers per day, wreaking havoc on agriculture. The sudden switch to collective mass movement within a band emerges as a result of local interactions between individuals, with locusts aligning with their moving neighbors (Buhl et al., 2006; 2011). This alignment response has been shown to be driven in part by the risk of cannibalism from others in the band (Bazazi et al., 2008; 2011; Simpson, Sword, Lorch, & Couzin, 2006).

E. The Function of Phase Change

The mechanisms of locust density-dependent phase polyphenism are now understood at levels spanning genetic and molecular events within the nervous systems of individuals through to continental-scale mass migration. Efforts have also been made to understand the function (adaptive significance) of phase polyphenism in the biology of locusts. Experiments and models have implicated roles for predation (including cannibalism), disease pressure, and migration (Hansen, Buhl, Bazazi, Simpson, & Sword, 2011; Reynolds, Sword, Simpson, & Reynolds, 2009; Simpson & Sword, 2009; Sword, Simpson, El Hadi, & Wilps, 2000; Wilson et al., 2002). Other avenues of research include mapping the genetic structure of locust populations across different regions (Chapuis et al., 2009; 2011) and tracking the origins of phase change within the grasshopper family tree (Song, 2005; Song & Wenzel, 2007).

VI. Neuromodulators and the Division of Labor in Honeybee Society

Insect societies are exemplars of cooperative group living (Wilson, 1971). Their success is mainly attributable to an efficient division of labor, in which only a small number of individuals reproduce and all others belong to a functionally sterile worker caste that specializes in different tasks important for colony growth and development. Honeybees (*Apis mellifera*) exhibit an advanced level of eusociality and are probably the most intensely studied obligate social insect. In this section, we outline recent insights into neuromodulatory mechanisms that control key aspects of the division of labor in these insects (for broader reviews on selected aspects see, e.g., Johnson, 2010; Page, Scheiner, Erber, & Amdam, 2006; Robinson, 1992; Sasaki, 2010; Zayed & Robinson, 2012).

Caste determination in honeybees is mainly determined by environmental factors (Schwander, Lo, Beekman, Oldroyd, & Keller, 2010). Each colony comprises a single, morphologically and physiologically differentiated queen, specialized for reproduction; a sparse number of males (drones), with which she mates; and tens of thousands of female workers that are sterile and perform all other tasks required to maintain the colony. Communication between colony members is achieved primarily using more than 50 pheromones (Le Conte & Hefetz, 2008) that sculpt social interactions by their effects on the behavior of individual colony members. Research over the last decade has revealed that some of the key actions of pheromones are mediated via interactions with neuromodulators, primarily biogenic amines, and that neuromodulators can tune a bee's response to a given pheromone.

A. The Queen, Her Pheromone, and Dopaminergic Signaling

Reproductive hegemony of queens is maintained by the action of the queen's mandibular gland pheromone (QMP; Slessor, Winston, & Le Conte, 2005), a chemical blend of five components transmitted throughout the colony by contact and by sharing regurgitated food (trophallaxis). QMP is an atypical pheromone in that it functions both as a releaser of rapid behavioral responses and as a primer for long-term changes in behavior and physiology (Le Conte & Hefetz, 2008). As a releaser pheromone, QMP stimulates the *retinue response*, in which workers are attracted to the queen to feed and groom her (Slessor et al., 2005). As a primer pheromone, QMP inhibits development of ovaries in young bees (Hoover, Keeling, Winston, & Slessor, 2003) and slows the age-dependent switch in tasks from brood care (nursing) to foraging (Pankiw, Huang, Winston, & Robinson, 1998).

Multiple lines of evidence suggest that key actions of QMP are mediated by influences on dopaminergic signaling pathways (Beggs et al., 2007; Beggs & Mercer 2009; Mercer et al. 2007; Vergoz et al., 2009; Vergoz, Lim, & Oldroyd, 2011) and that dopamine is important for determining behavioral profiles and regulating shifts in the expression of different behaviors in social bees (Mustard, Pham, & Smith, 2010). In the absence of a queen, dopamine levels increase in the brains of young workers but diminish in her presence (Harris & Woodring, 1995; Sasaki & Nagao, 2001). Complementing this, young workers exposed to QMP exhibit lower dopamine levels in the brain, altered responsiveness of tissue to dopamine, and changes in the expression levels of genes encoding different dopamine receptor subtypes (Beggs et al., 2007; Vergoz et al., 2009; 2011). Intriguingly, these effects of QMP can be reproduced by one of its major components, homovanillyl alcohol (HVA), a metabolite of dopamine that can itself selectively activate a specific dopamine receptor subtype (*Am*DOP$_3$, Beggs & Mercer, 2009; for details see section V1, B Queen Pheromone and Worker Sterility). Honeybees are equipped with three dopamine receptor subtypes, *Am*DOP$_1$, *Am*DOP$_2$, and *Am*DOP$_3$, that are coupled to different G proteins (Scheiner, Baumann, & Blenau, 2006). *Am*DOP$_1$ and *Am*DOP$_2$ are related to vertebrate D$_1$ dopamine receptors and mediate activation of the intracellular second messenger cAMP, whereas *Am*DOP$_3$ is related to vertebrate D$_2$-like receptors and lowers cAMP production via inhibition of adenylate cyclase. Changes in the relative activation of different subtypes can produce opposing effects on cells targeted by dopamine, and this is thought to be the means by which QMP suppresses reproductive development of workers and promotes their attendance in the queen's retinue.

B. Queen Pheromone and Worker Sterility

Dopaminergic signaling plays a key role in maintaining worker sterility and enabling fertility in the absence of a queen. In her presence, or following exposure to QMP, dopamine levels are significantly decreased in the brain of workers, and this correlates with low levels of ovarian activation (Beggs et al., 2007; Harris & Woodring, 1995). Conversely, feeding queenless workers with dopamine promotes ovarian development (Dombroski, Simões, & Bitonid, 2003). Furthermore, young workers respond to QMP, and particularly HVA, by expressing enhanced levels of *Am*DOP$_3$ receptor in their ovaries (Vergoz et al., 2011). It is suggested that this suppresses ovarian development by inhibiting the cAMP signaling cascade, possibly by the direct action of HVA, which can itself selectively activate *Am*DOP$_3$ receptors in vitro (Beggs & Mercer, 2009). With the loss of the queen from a colony, the levels of dopamine increase in the brains of young workers (Harris & Woodring, 1995) while, in the ovaries, the expression of *Am*DOP$_3$ gene declines and ovarian development is facilitated (Vergoz et al., 2011). Without a queen, workers begin to feed a few young larvae with royal jelly, of which one component, a 57-kDa protein designated as royalactin, induces the differentiation of honeybee larvae into queens. By acting on an epidermal growth factor receptor, which in turn activates a number of effectors to increase body size, royalactin decreases developmental time and promotes ovary development in honeybees and, surprisingly, evokes similar effects in the fruit fly (Kamakura, 2011).

C. Promoting Attendance of Workers to the Queen

The exposure of young bees to QMP early in life is important for ensuring their attraction to this pheromone, and this is achieved by actions both on brain tissue and olfactory receptors of the antennae. Exposing young bees for the first time to QMP causes a significant decrease in the expression of *Am*DOP$_1$ receptor genes relative to those encoding *Am*DOP$_3$ in the brain (Beggs et al., 2007). Similarly in the antennae, QMP also reduces expression of the *Am*DOP$_1$ gene, whereas the expression levels of the genes encoding the *Am*DOP$_3$ receptor and an octopamine receptor (*Am*OA$_1$) are significantly greater in bees highly attracted to QMP compared to bees less attracted to it (Vergoz et al., 2009). These changes cause young workers to subsequently respond to QMP or HVA with a decreased production of cAMP in the brain (Beggs et al., 2007). In effect, inhibited accumulation of cAMP via the action of QMP on *Am*DOP$_3$ receptors in young adults will curtail signaling via the D$_1$-like receptors (*Am*DOP$_1$ and *Am*DOP$_2$) that promote cAMP production.

Modulation of amine receptor profiles by QMP is thought to mediate retinue behavior by influencing associative perception of this pheromone. In honeybees and other insects, the dopaminergic system is instrumental in mediating aversive memory, and exposure to QMP or HVA will block aversive learning in young workers (Vergoz et al., 2007). The queen pheromone could thus prevent young workers from forming negative associations with her odors (Beggs & Mercer, 2009), which can be repellent (Moritz, Crewe, & Hepburn, 2002).

Furthermore, increased expression of octopamine receptors in antennae of young workers exposed to QMP (Vergoz et al., 2009) could confer an attractive quality to her odor signature. Octopamine is known to enhance the sensitivity to pheromones by modulating insect olfactory receptors (Pophof, 2000) and conveying the positive reinforcing properties of appetitive stimuli in honeybee associative learning (Hammer, 1993; Hammer & Menzel, 1995).

Octopamine, acting via cAMP and PKA, has been shown to increase the opening times of potassium-activated sodium channels in neurons (Kenyon cells) isolated from the mushroom bodies of crickets, whereas dopamine, acting via cGMP/PKG, shortens their opening times (Aoki, Kosakai, & Yoshino, 2008). These effects of octopamine and dopamine lead to changes in the width of action potentials, but the consequences for associative memory formation are not known.

D. Switching Tasks in Workers

In addition to attending the queen, young worker bees are engaged in other activities in the hive, including nursing the brood, food processing, and hive maintenance, but switch tasks after 2–3 weeks to foraging for food outside the hive. Although stereotyped, this "temporal polyethism" is flexible, and bees can respond to changing colony needs by accelerating, delaying, or even reversing their pattern of behavioral development (Robinson, 1992). When change occurs, it is accompanied by dramatic changes in metabolism, circulating hormones, endocrine glands, exocrine glands, brain structure, and gene expression (Amdam & Page, 2010; Denison & Raymond-Delpech, 2008; Schulz, Barron, & Robinson, 2002). Some of these changes, such as enlargement of the mushroom bodies, appear to be a direct result of experiencing the outside world and involve cholinergic signaling (Ismail, Robinson, & Fahrbach, 2006) and other changes that occur in response to environmental conditions, particularly nutritional status of the colony (Ament et al., 2010).

E. The Second Messenger cGMP and Priming Workers for Foraging

With aging, workers become progressively less attracted to QMP and even repelled by it at foraging age (Vergoz et al., 2009). Since this coincides with a progressive decline in *Am*DOP$_3$ gene expression in the antennae (Vergoz et al., 2009) and an increase in *Am*DOP$_1$ and *Am*DOP$_3$ receptor genes in the brain (Beggs et al., 2007), dopaminergic signaling may be involved. However, since the retinue response in workers is unaffected by dopamine (Fussnecker, McKenzie, & Grozinger, 2011; see also Schulz et al. [2002] and Barron et al. [2002] on octopamine), other regulatory mechanisms must control QMP's declining attractiveness.

Recent studies point to the involvement of cGMP-dependent PKG as a regulator of foraging, food acquisition, and energy balance in social insects (Ament et al., 2010). The gene encoding PKG is known as the *foraging* gene (*Amfor*) and is associated with food-related locomotion in many organisms (Ben-Shahar, 2005). It is expressed more highly in the brains of honeybee foragers than in nurses, and the timing of the increase marks the onset of foraging (Heylen et al., 2008). Treating young workers with a tissue-permeable analogue of the neuronal second messenger cGMP (8-Br-cGMP) accelerates the transition from nursing to foraging behavior (Ben-Shahar et al., 2003) and significantly reduces attraction of workers to QMP (Fussnecker et al., 2011). cGMP is likely to promote foraging by increasing the attraction of workers to light (Ben-Shahar et al., 2003) and modulating the expression of genes associated with this behavioral state (Fussnecker et al., 2011; Nelson, Ihle, Fondrk, Page, & Amdam, 2007; Whitfield et al., 2006). cGMP can thus be considered a major modulator of division of labor in honeybee society.

Insect brains contain a large number of neurons that can synthesize cGMP (Ott et al., 2004) under the control of the enzyme-soluble guanylyl cyclase, which is the only known receptor for the gas NO, a major neuromodulator in insects. However, to our knowledge, there is no evidence that NO directly influences the behavioral switch from nursing to foraging in honeybee workers.

F. Octopamine and Foraging Behavior

Work mainly by Andrew Barron, Gene Robinson, and David Schulz (Schulz et al., 2002) has shown that the biogenic amine octopamine plays a crucial role in mediating the phenotypic changes underlying the transition of nurses to workers. Independent of their age, brain levels of octopamine are high in foragers, but low in nurses (Wagener-Hulme, Kuehn, Schulz, & Robinson, 1999). This difference is particularly marked in the olfactory center of the insect brain, the antennal lobe (Schulz & Robinson, 1999), and is correlated with increased expression of the gene encoding tyramine-β-hydroxylase, an enzyme vital for octopamine synthesis (Lehman et al., 2006). In contrast to octopamine, levels of serotonin and dopamine do not show changes that would implicate their involvement in controlling foraging behavior (Schulz, Elekonich, & Robinson, 2003).

Treatment with octopamine, but not its precursor tyramine or serotonin, causes a significant dose-dependent increase in the numbers of new foragers in a colony, provided they are old enough to forage in the first place (Schulz & Robinson, 2001; Schulz et al., 2003). This is likely to result from modulatory effects of octopamine on foraging-related stimuli. For example, octopamine decreases social inhibition of foraging by the presence of older bees (Barron et al., 2002) and increases the response of workers to brood pheromone, a mixture of cuticular waxes produced by larvae that stimulates foraging (cf. Le Conte & Hefetz, 2008). Although octopamine is known to promote numerous other social behaviors in bees, such as aggression toward non-nest mates (Robinson et al., 1999) and hygienic behavior in nurses (Spivak, Masterman, Ross, & Mesce, 2003), it is clearly not acting simply as a general "arousal" factor. For example, octopamine does not enhance all flight-related tasks (e.g., corpse removal from the hive) or the responsiveness to all olfactory stimuli (e.g., QMP; Barron & Robinson, 2005). This neatly illustrates that octopamine selectively and differentially modulates different aspects of behavior to specifically promote foraging.

F.1 FOOD SHARING, AMINES, AND SOCIAL ADHESION

On returning to the hive, foragers communicate the distance, direction, and value of resources in their dance displays. Octopamine treatment increases the reported resource value (Barron, Maleszka, Vander Meer, & Robinson, 2007), and low doses of cocaine, which blocks the reuptake of neuronally released biogenic amines, have a similar effect (Barron, Maleszka, Helliwell, & Robinson, 2009). These drugged foragers danced more frequently and more vigorously for the same quality of food but never at the wrong place or time, and treated nonforagers do not start dancing. In effect, cocaine appears to cause exaggerated reporting of food value by modulating associated reward pathways (Barron et al., 2007; 2009; 2010) in which the octopaminergic system may be involved (Hammer, 1993; Hammer & Menzel, 1995).

Interestingly, and in contrast to solitary insects (e.g., Long & Murdock, 1983), honeybees do not respond to octopamine by feeding more, but will accept a lower food quality and report its whereabouts so that others can share. This has led to the intriguing idea that altruistic behavior in social insects may be promoted by associating it with a positive, rewarding experience. Indeed, in many social insects, the exchange between nest mates of regurgitated food (trophallaxis), and with it pheromones, clearly serves not only as a channel for altruistic food distribution, but also as a means to maintain social adhesion and possibly as a mechanism for establishing and maintaining hierarchies without agonistic conflicts (Contrera, Imperatriz-Fonseca, & Koedam, 2010).

VII. Conclusions and Future Directions

Research over the past decade has revealed that social behavior in insects is orchestrated by the actions of a variety of neuromodulators, primarily by biogenic amines and often via their interactions with pheromones. Biogenic amines act at the interface between the animal's social environment and central brain circuits. They mediate the influences of previous and momentary experiences on social behavior, often seemingly by signaling rewarding and aversive qualities. The neuronal representation of such qualities may be the key to mechanisms controlling conspecific aggression, repulsion, or attraction and establishing harmonious social groups. Perhaps even altruistic behavior, the cement of society, may have been promoted in evolution by associating it with a positive rewarding experience.

Amines do not seem to function as releasers of selected behaviors; rather, they promote their expression, allowing them to become overt under appropriate conditions. As such, they act as true neuromodulators. Their actions are by no means restricted to modifiers of general excitability or "arousal"—on the contrary, amines act selectively to differentially bias the expression of specific aspects of behavior. Simply by modulating behavioral thresholds, adaptive responses—"decisions"—are produced when in social conflict (Figure 2.2), and sophisticated collective decisions are made at the collective level (Robinson et al., 2011).

The specificity of action of biogenic amines is achieved by context-dependent recruitment of selected aminergic neurons, which in insects are comparatively few, each with unique projections to defined targets. The impact of amines released in the vicinity of these targets will depend on the proximity, locality (presynaptic, postsynaptic), and distribution of receptors for different amines, the receptor subtypes present, and the various

Bullerjahn, A., Mentel, T., Pflüger, H. J., & Stevenson, P. A. (2006). Nitric oxide: A co-modulator of efferent peptidergic neurosecretory cells including a unique octopaminergic neurone innervating locust heart. *Cell and Tissue Research, 325*, 345–360.

Burrows, M. (1996). *The neurobiology of an insect brain*. Oxford, UK: Oxford University Press.

Burrows, M., Rogers, S. M., & Ott, S. R. (2011). Epigenetic remodelling of brain, body and behaviour during phase change in locusts. *Neural Systems & Circuits, 1*, 11.

Carré, J. M., & Putnam, S. K. (2010). Watching a previous victory produces an increase in testosterone among elite hockey players. *Psychoneuroendocrinology, 35*, 475–479.

Cattaert, D., Delbecque, J. P., Edwards, D. H., & Issa, F. A. (2010). Social interactions determine postural network sensitivity to 5-HT. *Journal of Neuroscience, 30*, 5603–5616.

Certel, S. J., Leung, A., Lin, C. Y., Perez, P., Chiang, A. S., & Kravitz, E. A. (2010). Octopamine neuromodulatory effects on a social behavior decision-making network in *Drosophila* males. *PLoS ONE, 5*, e13248.

Certel, S. J., Savella, M. G., Schlegel, D. C., & Kravitz, E. A. (2007). Modulation of *Drosophila* male behavioral choice. *Proceedings of the National Academy of Sciences, USA, 104*, 4706–4711.

Chan, Y. B., & Kravitz, E. A. (2007). Specific subgroups of FruM neurons control sexually dimorphic patterns of aggression in *Drosophila melanogaster. Proceedings of the National Academy of Sciences, USA, 104*, 9577–9582.

Chapuis, M.-P., Berthier, K., Popple, J.-A., Simpson, S. J., Deveson, T., Spurgin, P., & Sword, G. A. (2011). Challenges to assessing connectivity between massive populations of the Australian plague locust. *Proceedings of the Royal Society B, 278*, 3152–3160.

Chapuis, M.-P., Loiseau, A., Michalakis, Y., Lecoq, M., Franc, A., & Estoup, A. (2009). Outbreaks, gene flow and effective population size in the migratory locust, *Locusta migratoria*, a regional-scale comparative survey. *Molecular Ecology, 18*, 792–800.

Chen, S., Yang, P., Jiang, F., Wei, Y., Ma, Z., & Kang, L. (2010). *De novo* analysis of transcriptome dynamics in the migratory locust during the development of phase traits. *PLoS ONE, 5*, e15633.

Churchland, P. M. (1995). *The engine of reason, the seat of the soul, a philosophical journey into the brain*. Cambridge, MA: MIT Press.

Collett, M., Despland, E., Simpson, S. J., & Krakauer, D. C. (1998). The spatial scales of locust gregarisation. *Proceedings of the National Academy of Sciences, USA, 95*, 13052–13055.

Contrera, F. A. L., Imperatriz-Fonseca, V. L., & Koedam, D. (2010). Trophallaxis and reproductive conflicts in social bees. *Insectes Sociaux, 57*, 125–132.

Couppis, M. H., & Kennedy, C. H. (2008). The rewarding effect of aggression is reduced by nucleus accumbens dopamine receptor antagonism in mice. *Psychopharmacology, 197*, 449–456.

Cullen, D., Sword, G. A., Dodgson, T., & Simpson, S. J. (2010). Behavioural phase change in the Australian plague locust, *Chortoicetes terminifera*, is triggered by tactile stimulation of the antennae. *Journal of Insect Physiology, 56*, 937–942.

Darwin, C. (1859). *On the origin of species by means of natural selection*. London: Jon Murray.

de Boer, S. F., & Koolhaas, J. M. (2005). 5-HT1A and 5-HT1B receptor agonists and aggression: A pharmacological challenge of the serotonin deficiency hypothesis. *European Journal of Pharmacology, 526*, 125–139.

Denison, R., & Raymond-Delpech, V. (2008). Insights into the molecular basis of social behaviour from studies on the honeybee, *Apis mellifera. Invertebrate Neuroscience, 8*, 1–9.

Despland, E., Collett, M., & Simpson, S. J. (2000). Small-scale processes in desert locust swarm formation, how vegetation patterns influence gregarization. *Oikos, 88*, 652–662.

Dethier, V. G. (1976). *The hungry fly. A physiological study of the behavior associated with feeding*. Cambridge, MA: Harvard University Press.

Dickinson, P. S. (1995). Interactions among neural networks for behavior. *Current Opinion in Neurobiology, 5*, 792–798.

Dierick, H. A., & Greenspan, R. J. (2007). Serotonin and neuropeptide F have opposite modulatory effects on fly aggression. *Nature Genetics, 39*, 678–682.

Dombroski, T. C. D., Simões, Z. L. P., & Bitondi, M. M. G. (2003). Dietary dopamine causes ovary activation in queenless *Apis mellifera* workers. *Apidologie, 34*, 281–289.

Duch, C., Mentel, T., & Pflüger, H. J. (1999). Distribution and activation of different types of octopaminergic DUM neurons in the locust. *Journal of Comparative Neurology, 403*, 119–134.

Dukas, R. (2008). Evolutionary biology of insect learning. *Annual Review of Entomology, 53*, 145–160.

Dyakonova, V. E., & Krushinkskii, A. L. (2006). Effects of an NO synthase inhibitor on aggressive and sexual behavior in male crickets. *Neuroscience & Behavioral Physiology, 36*, 565–571.

Dyakonova, V. E., Schürmann, F.-W., & Sakharov, D. A. (1999). Effects of serotonergic and opioidergic drugs on escape behaviors and social status of male crickets. *Naturwissenschaften, 86*, 435–437.

Dyakonova, V., Schürmann, F.-W., & Sakharov, D. A. (2002). Effects of opiate ligands on intraspecific aggression in crickets. *Peptides, 23*, 835–841.

Edwards, A. C., Zwarts, L., Yamamoto, A., Callaerts, P., & Mackay, T. F. (2009). Mutations in many genes affect aggressive behavior in *Drosophila melanogaster. BMC Biology, 7*, 29.

Edwards, D. H., & Kravitz, E. A. (1997). Serotonin, social status and aggression. *Current Opinions in Neurobiology, 7*, 812–819.

Erber, J., Kloppenburg, P., & Scheidler, A. (1993). Neuromodulation by serotonin and octopamine in the honeybee: Behaviour, neuroanatomy and electrophysiology. *Experientia, 49*, 1073–1083.

Evans, P. D. (1985). Octopamine. In G. A. Kerkut & L. I. Gilbert (Eds.), *Comprehensive insect physiology biochemistry and pharmacology* (pp. 499–530). Oxford: Pergamon.

Fenno, L., Yizhar, O., & Deisseroth, K. (2011). The development and application of optogenetics. *Annual Review of Neuroscience, 34*, 389–412.

Fernandez, M. P., Chan, Y. B., Yew, J. Y., Billeter, J. C., Dreisewerd, K., Levine, J. D., & Kravitz, E. A. (2010). Pheromonal and behavioral cues trigger male-to-female aggression in *Drosophila. PLoS Biology, 8*, e1000541.

Fussnecker, B. L., McKenzie, A. M., & Grozinger, C. M. (2011). cGMP modulates responses to queen mandibular pheromone in worker honey bees. *Journal of Comparative Physiology A, 197*, 939–948.

Fuxjager, M. J., & Marler, C. A. (2010). How and why the winner effect forms: Influences of contest environment and species differences. *Behavioral Ecology, 21*, 37–45.

Galizia, C. G., & Rössler, W. (2010). Parallel olfactory systems in insects: Anatomy and function. *Annual Review of Entomology, 55*, 399–420.

Getting, P. A., & Dekin, M. A. (1985). Tritonia swimming. A model system for integration within rhythmic motor systems. In A. I. Selverston (Ed.), *Model networks and behavior* (pp. 3–20). New York: Plenum Press.

Goldstein, R. S., & Camhi, J. M. (1991). Different effects of the biogenic amines dopamine, serotonin and octopamine on the thoracic and abdominal portions of the escape circuit in the cockroach. *Journal of Comparative Physiology A, 168*, 103–112.

Gray, L. J., Sword, G. A., Anstey, M. L., Clissold, F. J., & Simpson, S. J. (2009). Behavioural phase polyphenism in the Australian plague locust (*Chortoicetes terminifera*). *Biology Letters, 5*, 306–309.

Guo, W., Wang, X., Ma, Z., Xue, L., Han, J., Yu, D., & Kang, L. (2011). *CSP* and *takeout* genes modulate the switch between attraction and repulsion during behavioral phase change in the migratory locust. *PLoS Genetics, 7*, e1001291.

Haas, H. L., Sergeeva, O. A., & Selbach, O. (2008). Histamine in the nervous system. *Physiological Reviews, 88*, 1183–1241.

Haden, S. C., & Scarpa, A. (2007). The noradrenergic system and its involvement in aggressive behaviors. *Aggression & Violent Behavior, 12*, 1–15.

Hammer, M. (1993). An identified neurone mediates the unconditioned stimulus in associative olfactory learning in honeybees. *Nature, 366*, 59–63.

Hammer, M., & Menzel, R. (1995). Learning and memory in the honey bee. *Journal of Neuroscience, 15*, 1617–1630.

Hansen, M. J., Buhl, J., Bazazi, S., Simpson, S. J., & Sword, G. A. (2011). Cannibalism in the lifeboat—collective movement in Australian plague locusts. *Behavioral Ecology & Sociobiology, 65*, 1715–1720.

Harris, J. W., & Woodring, J. (1995). Elevated brain dopamine levels associated with ovarian development in queenless worker honey bees. *Comparative Biochemistry & Physiology C, 111*, 271–279.

Harris-Warrick, R. M., & Marder, E. (1991). Modulation of neural networks for behavior. *Annual Review of Neuroscience, 14*, 39–57.

Harris-Warrick, R. M., Marder, E., Selverston, A. I., & Moulins, M. (Eds.). (1992). *Dynamic biological networks. The stomatogastric nervous system*. Cambridge, MA: MIT Press.

Hauser, F., Cazzamali, G., Williamson, M., Blenau, W., & Grimmelikhuijzen, C. J. P. (2006). A review of neurohormone GPCRs present in the fruitfly *Drosophila melanogaster* and the honey bee *Apis mellifera*. *Progress in Neurobiology, 80*, 1–19.

Heisenberg, M. (2003). Mushroom body memoir: From maps to models. *Nature Reviews Neuroscience, 4*, 266–275.

Heisenberg, M., Schwaerzel, M., Monastirioti, M., Scholz, H., Friggi-Grelin, F., & Birman, S. (2003). Dopamine and octopamine differentiate between aversive and appetitive olfactory memories in *Drosophila*. *Journal of Neuroscience, 23*, 10495–10502.

Heylen, K., Gobin, B., Billen, J., Hu, T. T., Arckens, L., & Huybrechts, R. (2008). Amfor expression in the honeybee brain: A trigger mechanism for nurse-forager transition. *Journal of Insect Physiology, 54*, 1400–1403.

Hofmann, H. (1996). The cultural history of Chinese fighting crickets: A contribution not only to the history of biology. *Biologisches Zentralblatt, 115*, 206–213.

Hofmann, H., & Schildberger, K. (2001). Assessment of strength and willingness to fight during aggressive encounters in crickets. *Animal Behaviour, 62*, 337–348.

Hofmann, H. A., & Stevenson, P. A. (2000). Flight restores fight in crickets. *Nature, 403*, 613–613.

Homberg, U. (2002). Neurotransmitters and neuropeptides in the brain of the locust. *Microscopy Research & Technique, 56*, 189–209.

Hoover, S. E. R., Keeling, C. I., Winston, M. L., & Slessor, K. N. (2003). The effect of queen pheromones on worker honey bee ovary development. *Naturwissenschaften, 90*, 477–480.

Hoyer, S. C., Eckart, A., Herrel, A., Zars, T., Fischer, S. A., Hardie, S. L., & Heisenberg, M. (2008). Octopamine in male aggression of *Drosophila*. *Current Biology, 18*, 159–167.

Hsu, Y., Earley, R. L., & Wolf, L. L. (2006). Modulation of aggressive behaviour by fighting experience: Mechanisms and contest outcomes. *Biological Reviews of the Cambridge Philosophical Society, 81*, 33–74.

Hsu, Y. Y., Lee, I. H., & Lu, C. K. (2009). Prior contest information: Mechanisms underlying winner and loser effects. *Behavioral Ecology & Sociobiology, 63*, 1247–1257.

Huber, R., & Delago, A. (1998). Serotonin alters decisions to withdraw in fighting crayfish, *Astacus astacus*—he Motivational concept revisited. *Journal of Comparative Physiology A, 182*, 573–583.

Huber, R., Smith, K., Delago, A., Isaksson, K., & Kravitz, E. A. (1997). Serotonin and aggressive motivation in crustaceans: Altering the decision to retreat. *Proceedings of the National Academy of Sciences, USA, 94*, 5939–5942.

Hurd, P. L. (2006). Resource holding potential, subjective resource value, and game theoretical models of aggressiveness signalling. *Journal of Theoretical Biology, 241*, 639–648.

Ismail, N., Robinson, G. E., & Fahrbach, S. E. (2006). Stimulation of muscarinic receptors mimics experience-dependent plasticity in the honey bee brain. *Proceedings of the National Academy of Sciences, USA, 103*, 207–211.

Iwasaki, M., & Katagiri, C. (2008). Cuticular lipids and odors induce sex-specific behaviors in the male cricket *Gryllus bimaculatus*. *Comparative Biochemistry & Physiology A, 149*, 306–313.

Iwasaki, M., Nishino, H., Delago, A., & Aonuma, H. (2007). Effects of NO/cGMP signaling on behavioral changes in subordinate male crickets, *Gryllus bimaculatus*. *Zoological Science, 24*, 860–868.

Johnson, B. R. (2010). Division of labor in honeybees, form, function, and proximate mechanisms. *Behavioral Ecology & Sociobiology, 64*, 305–316.

Johnson, B. R., & Tsutsui, N. D. (2011). Taxonomically restricted genes are associated with the evolution of sociality in the honey bee. *BMC Genomics, 12*, 164.

Johnson, O., Becnel, J., & Nichols, C. D. (2009). Serotonin 5-HT2 and 5-HT1A-like receptors differentially modulate aggressive behaviors in *Drosophila melanogaster*. *Neuroscience, 158*, 1292–1300.

Jones, B. R., & Thompson, S. H. (2001). Mechanism of postinhibitory rebound in molluscan neurons. *American Zoologist, 41*, 1036–1048.

Kamakura, M. (2011). Royalactin induces queen differentiation in honeybees. *Nature, 473*, 478–483.

Katz, P. S. (1995). Intrinsic and extrinsic neuromodulation of motor circuits. *Current Biology, 5*, 799–808.

Kemp, D. J., & Wiklund, C. (2004). Residency effects in animal contests. *Proceedings of the Royal Society B, 271*, 1707–1711.

Kennedy, J. S. (1992). *The new anthropomorphism*. Cambridge: Cambridge University Press.

Khazraie, K., & Campan, M. (1999). The role of prior agonistic experience in dominance relationships in male crickets *Gryllus bimaculatus* (Orthoptera: Gryllidae). *Behavioural Processes, 44*, 341–348.

Killian, K. A., & Allen, J. R. (2008). Mating resets male cricket aggression. *Journal of Insect Behavior, 21*, 535–548.

Kravitz, E., & Huber, R. (2003). Aggression in invertebrates. *Current Opinion in Neurobiology, 13*, 736–743.

Kume, K., Kume, S., Park, S. K., Hirsh, J., & Jackson, F. R. (2005). Dopamine is a regulator of arousal in the fruit fly. *Journal of Neuroscience, 25*, 7377–7384.

Le Conte, Y., & Hefetz, A. (2008). Primer pheromones in social hymenoptera. *Annual Review of Entomology, 53*, 523–542.

Lehman, H. K., Schulz, D. J., Barron, A. B., Wraight, L., Hardison, C., Whitney, S., et al. (2006). Division of labor in the honey bee (*Apis mellifera*), the role of tyramine beta-hydroxylase. *Journal of Experimental Biology, 209*, 2774–2784.

Liu, T., Wei, H. P., Yuan, C. Y., & Guo, A. K. (2006). Disturbance of dopamine level triggers male-male courtship behavior in *Drosophila*. *Neuroscience Research, 55*, S192.

Long, T. F., & Murdock, L. L. (1983). Stimulation of blowfly feeding-behavior by octopaminergic drugs. *Proceedings of the National Academy of Sciences, USA, 80*, 4159–4163.

Lucas, C., Kornfein, R., Chakaborty-Chatterjee, M., Schonfeld, J., Geva, N., Sokolowski, M. B., & Ayali, A. (2010). The locust *foraging* gene. *Archives of Insect Biochemistry and Physiology, 74*, 52–66.

Lui, W., Liang, X., Gong, J., Yang, Z., Zhang, Y. H., Zhang, J. X., & Rao, Y. (2011). Social regulation of aggression by pheromonal activation of Or65a olfactory neurons in *Drosophila*. *Nature Neuroscience, 14*, 896–902.

Ma, Z. Y., Guo, W., Guo, X., Wang, X., & Kang, L. (2011). Modulation of behavioral phase changes of the migratory locust by the catecholamine metabolic pathway. *Proceedings of the National Academy of Sciences, USA, 108*, 3882–3887.

Maeno, K., & Tanaka, S. (2011). Adult female desert locusts require contact chemicals and light for progeny gregarization. *Physiological Entomology, 37*, 109–118. doi: 10.1111/j.1365-3032.2011.00812.x

Marder, E., & Bucher, D. (2001). Central pattern generators and the control of rhythmic movements. *Current Biology, 11*, R986–R996.

Matheson, T., Rogers, S. M., & Krapp, H. G. (2004). Plasticity in the visual system is correlated with a change in lifestyle of solitarious and gregarious locusts. *Journal of Neurophysiology, 91*, 1–12.

Maynard Smith, J., & Price, G. R. (1973). The logic of animal conflict. *Nature, 246*, 15–18.

McFarland, D. J., & Sibly, R. M. (1975). The behavioural final common path. *Philosophical Transactions of the Royal Society B, 270*, 265–293.

Mentel, T., Duch, C., Stypa, H., Wegener, G., Müller, U., & Pflüger H.-J. (2003). Central modulatory neurons control fuel selection in flight muscle of migratory locust. *Journal of Neuroscience, 23*, 1109–1113.

Menzel, R. (1996). Learning and memory in honeybees: From behavior to neural substrates. *Annual Review of Neuroscience, 19*, 379–404.

Miller, G. A., Islam, M. S., Claridge, D. W., Dodgson, T., & Simpson, S. J. (2008). Swarm formation in the desert locust (Schistocerca gregaria), isolation and NMR analysis of the primary maternal gregarizing agent. *Journal of Experimental Biology, 211*, 370–376.

Mizunami, M., Unoki, S., Mori, Y., Hirashima, D., Hatano, A., & Matsumoto, Y. (2009). Roles of octopaminergic and dopaminergic neurons in appetitive and aversive memory recall in an insect. *BMC Biology, 7*, 46.

Moritz, R. F. A., Crewe, R. M., & Hepburn, H. R. (2002). Queen avoidance and mandibular gland secretion of honey bee workers (*Apis mellifera* L.) *Insectes Sociaux, 49*, 86–91.

Morris, O. T., Duch, C., & Stevenson, P. A. (1999). Differential activation of octopaminergic (DUM) neurones via proprioceptors responding to flight muscle contractions in the locust. *Journal of Experimental Biology, 202*, 3555–3564.

Mudge, A. W., Leeman, S. E., & Fischbach, G. D. (1979). Enkephalin inhibits release of substance P from sensory neurons in culture and decreases action potential duration. *Proceedings of the National Academy of Sciences, USA, 76*, 526–530.

Murakami, S., & Itoh, M. T. (2001). Effects of aggression and wing removal on brain serotonin levels in male crickets, *Gryllus bimaculatus*. *Journal of Insect Physiology, 47*, 1309–1312.

Murakami, S., & Itoh, M. T. (2003). Removal of both antennae influences the courtship and aggressive behaviors in male crickets. *Journal of Neurobiology, 57*, 110–118.

Mustard, J. A., Pham, P. M., & Smith, B. H. (2010). Modulation of motor behavior by dopamine and the D1-like dopamine receptor AmDOP2 in the honey bee. *Journal of Insect Physiology, 56*, 422–430.

Nässel, D. R. (1999). Histamine in the brain of insects: A review. *Microscopy Research and Technique, 44*, 121–136.

Nässel, D. R., & Homberg, U. (2006). Neuropeptides in interneurons of the insect brain. *Cell and Tissue Research, 326*, 1–24.

Nelson, C. M., Ihle, K. E., Fondrk, M. K, Page, R. E., & Amdam, G. V. (2007). The gene *vitellogenin* has multiple coordinating effects on social organization. *PLoS Biology, 5*, 673–677.

Nelson, R. J., & Trainor, B. C. (2007). Neural mechanisms of aggression. *Nature Reviews Neuroscience, 8*, 536–546.

Nilsen, S. P., Chan, Y. B., Huber, R., & Kravitz, E. A. (2004). Gender-selective patterns of aggressive behavior in *Drosophila melanogaster*. *Proceedings of the National Academy of Sciences, USA, 101*, 12342–12347.

Nosil, P. (2002). Food fights in house crickets, *Acheta domesticus*, and the effects of body size and hunger level. *Canadian Journal of Zoology, 80*, 409–417.

Nusbaum, M. P., Blitz, D. M., Swensen, A. M., Wood, D., & Marder, E. (2001). The roles of co-transmission in neural network modulation. *Trends in Neurosciences, 24*, 146–154.

O'Connell, L. A., & Hofmann, H. A. (2011). The vertebrate mesolimbic reward system and social behavior network: A comparative synthesis. *Journal of Comparative Neurology, 519*, 3599–3639.

Ott, S. R., Delago, A., & Elphick, M. R. (2004). An evolutionarily conserved mechanism for sensitization of soluble guanylyl cyclase reveals extensive nitric oxide-mediated upregulation of cyclic GMP in insect brain. *European Journal of Neuroscience, 20*, 1231–1244.

Ott, S. R., & Rogers, S. M. (2010). Gregarious desert locusts have substantially larger brains with altered proportions compared to the solitarious phase. *Proceedings of the Royal Society B, 277*, 3087–3096.

Ott. S. R., Verlinden, H., Rogers, S. M., Brighton, C. H., Quah, P. S., Vleugels, R. K., Verdonk, R. & Vanden Broecl, K. (2012). Critical role for protein kinase A in the acquisition of gregarious behavior in the desert locust. *Proceedings of the National Academy of Sciences of the United States of America, 109*, E381–7.

Page, R. E., Jr., Scheiner, R., Erber, J., & Amdam, G. V. (2006). The development and evolution of division of labor and foraging specialization in a social insect (*Apis mellifera* L.). *Current Topics in Developmental Biology, 74*, 253–286.

Pain, S. P. (2009). Signs of anger: Representation of agonistic behaviour in invertebrate cognition. *Biosemiotics, 2*, 181–191.

Pankiw, T., Huang, Z. Y., Winston, M. L., & Robinson, G. E. (1998). Queen mandibular gland pheromone influences worker honey bee (*Apis mellifera* L.) foraging ontogeny and juvenile hormone titers. *Journal of Insect Physiology, 44*, 685–692.

Payne, R. J. H. (1998). Gradually escalating fights and displays: The cumulative assessment model. *Animal Behaviour, 56*, 651–662.

Pener, M. P., & Simpson, S. J. (2009). Locust phase polyphenism, an update. *Advances in Insect Physiology, 36*, 1–286.

Pflüger H. J., & Stevenson, P. (2005). Evolutionary aspects of octopaminergic systems with emphasis on arthropods. *Arthropod Structure & Development, 34*, 379–396.

Pittenger, E. R., & Kandel, E. R. (2003). In search of general mechanisms for long-lasting plasticity: *Aplysia* and the hippocampus. *Philosophical Transactions of the Royal Society, B, 358*, 757–763.

Pophof, B. (2000). Octopamine modulates the sensitivity of silkmoth pheromone receptor neurons. *Journal of Comparative Physiology A, 186*, 307–313.

Reynolds, A. M., Sword, G. A., Simpson, S. J., & Reynolds, D. R. (2009). Predator percolation, insect outbreaks and phase polyphenism. *Current Biology, 19*, 20–24.

Rillich, J., Buhl, E., Schildberger, K., & Stevenson, P. A. (2009). Female crickets are driven to fight by the male courting and calling songs. *Animal Behaviour, 77*, 737–742.

Rillich, J., Schildberger, K., & Stevenson, P. A. (2007). Assessment strategy of fighting crickets revealed by manipulating information exchange. *Animal Behaviour, 74*, 823–836.

Rillich, J., Schildberger, K., & Stevenson, P. A. (2011). Octopamine and occupancy - an aminergic mechanism for intruder-resident aggression in crickets. *Proceedings of the Royal Society B, 278*, 1873–1880.

Rillich, J., & Stevenson, P.A. (2011). Winning fights induces hyperaggression via the action of the biogenic amine octopamine in crickets. *PLoS ONE, 6*, e28891.

Robinson, E. J. H., Franks, N. R., Ellis, S., Okuda, S., & Marshall, J. A. R. (2011). A simple threshold rule is sufficient to explain sophisticated collective decision-making. *PLoS ONE, 6*, e19981.

Robinson, G. E. (1992). Regulation of division of labour in insect societies. *Annual Review of Entomology, 37*, 637–665.

Roeder, T. (2005). Tyramine and octopamine, Ruling behavior and metabolism. *Annual Review of Entomology, 50*, 447–477.

Roeder, T., Degen, J., & Gewecke, M. (1998). Epinastine, a highly specific antagonist of insect neuronal octopamine receptors. *European Journal of Pharmacology, 349*, 171–177.

Roessingh, P., Bouaïchi, A., & Simpson, S. J. (1998). Effects of sensory stimuli on the behavioural phase state of the desert locust, *Schistocerca gregaria. Journal of Insect Physiology, 44*, 883–893.

Rogers, S. M., Harston, G. W., Kilburn-Toppin, F., Matheson, T., Burrows, M., Gabbiani, F., & Krapp, H. G. (2010). Spatiotemporal receptive field properties of a looming-sensitive neuron in solitarious and gregarious phases of the desert locust. *Journal of Neurophysiology, 103*, 779–792.

Rogers, S. M., Krapp, H. G., Burrows, M., & Matheson, T. (2007). Compensatory plasticity at an identified synapse tunes a visuomotor pathway. *Journal of Neuroscience, 27*, 4621–4633.

Rogers, S. M., Matheson, T., Despland, E., Dodgson, T., Burrows, M., & Simpson, S. J. (2003). Mechanosensory-induced behavioural gregarization in the desert locust *Schistocerca gregaria. Journal of Experimental Biology, 206*, 3991–4002.

Rogers, S. M., Matheson, T., Sasaki, K., Kendrick, K., Simpson, S. J., & Burrows, M. (2004). Substantial changes in central nervous system neurotransmitters and neuromodulators accompany phase change in the locust. *Journal of Experimental Biology, 207*, 3603–3617.

Sasaki, K. (2010). Biogenic amines and division of reproduction in social insects. In E. M. Stuart (Ed.), *Social Insects: Structure, Function and Behaviour* (pp. 199–223). Hapauge, NY: Nova Science Publishers.

Sasaki, K., & Nagao, T. (2001). Distribution and levels of dopamine and its metabolites in brains of reproductive workers in honeybees. *Journal of Insect Physiology, 47*, 1205–1216.

Scheiner, R., Baumann, A., & Blenau, W. (2006). Aminergic control and modulation of honeybee behaviour. *Current Neuropharmacology, 4*, 259–276.

Schöneich, S., Schildberger, K., & Stevenson, P. A. (2011). Neuronal organization of a fast-mediating cephalothoracic pathway for antennal-tactile information in the cricket (*Gryllus bimaculatus* DeGeer). *Journal of Comparative Neurology, 519*, 1677–1690.

Schröter, U., Malun, D., & Menzel, R. (2007). Innervation pattern of suboesophageal ventral unpaired median neurones in the honeybee brain. *Cell and Tissue Research, 327*, 647–667.

Schulz, D. J., Barron, A. B., & Robinson, G. E. (2002). A role for octopamine in honey bee division of labor. *Brain Behavior & Evolution, 60*, 350–359.

Schulz, D. J., Elekonich, M. M., & Robinson, G. E. (2003). Biogenic amines in the antennal lobes and the initiation and maintenance of foraging behavior in honey bees. *Journal of Neurobiology, 54*, 406–416.

Schulz, D. J., & Robinson, G. E. (1999). Biogenic amines and division of labor in honey bee colonies, behaviorally related changes in the antennal lobes and age-related changes in the mushroom bodies. *Journal of Comparative Physiology A, 184*, 481–488.

Schulz, D. J., & Robinson, G. E. (2001). Octopamine influences division of labor in honey bee colonies. *Journal of Comparative Physiology A, 187*, 53–61.

Schwander, T., Lo, N., Beekman, M., Oldroyd, B. P., & Keller, L. (2010). Nature versus nurture in social insect caste differentiation. *Trends in Ecology & Evolution, 25*, 275–282.

Schwärzel, M., Monastirioti, M., Scholz, H., Friggi-Grelin, F., Birman, S., & Heisenberg, M. (2003). Dopamine and octopamine differentiate between aversive and appetitive olfactory memories in *Drosophila*. *Journal of Neuroscience*, *23*, 10495–10502.

Selverston, A. I., & Ayers, J. (2006). Oscillations and oscillatory behaviour in small neural circuits. *Biological Cybernetics*, *95*, 537–554.

Simmons, L. W. (1986). Inter-male competition and mating success in the field cricket, *Gryllus bimaculatus* (de Geer). *Animal Behaviour*, *34*, 567–579.

Simpson, S. J., & Sword, G. A. (2008). Locusts. *Current Biology*, *18*, R364–R366.

Simpson, S. J., & Sword, G. A. (2009). Phase polyphenism in locusts, mechanisms, population, consequences, adaptive significance and evolution. In D. W. Whitman & T. N. Ananthakrishnan (Eds.), *Phenotypic plasticity of insects, mechanisms and consequences* (pp. 147–189). Enfield: Science Publishers.

Simpson, S. J., Sword, G. A., Lorch, P. D., & Couzin, I. D. (2006). Cannibal crickets on a forced march for protein and salt. *Proceedings of the National Academy of Sciences USA*, *103*, 4152–4156.

Simpson, S. J., Despland, E., Hägele, B. F., & Dodgson, T. (2001). Gregarious behaviour in desert locusts is evoked by touching their back legs. *Proceedings of the National Academy of Sciences, USA*, *98*, 3895–3897.

Simpson, S. J., McCaffery, A. R., & Hägele, B. (1999). A behavioural analysis of phase change in the desert locust. *Biological Reviews*, *74*, 461–480.

Simpson, S. J., Raubenheimer, D., Behmer, S. T., Whitworth, A., & Wright, G. A. (2002). A comparison of nutritional regulation in solitarious and gregarious phase nymphs of the desert locust, *Schistocerca gregaria*. *Journal of Experimental Biology*, *205*, 121–129.

Sinakevitch, I., Mustard, J. A., & Smith, B. H. (2011). Distribution of the octopamine receptor AmOA1 in the honey bee brain. *PLoS ONE*, *6*, e14536.

Slessor, K. N., Winston, M. L., & Le Conte, Y. (2005). Pheromone communication in the honey bee (*Apis mellifera* L.). *Journal of Chemical Ecology*, *31*, 2731–2745.

Song, H. (2005). Phylogenetic perspectives on the evolution of locust phase polyphenism. *Journal of Orthoptera Research*, *14*, 235–245.

Song, H., & Wenzel, J. W. (2007). Phylogeny of bird-grasshopper subfamily Cyrtacanthacridinae (Orthoptera, Acrididae) and the evolution of locust phase polyphenism. *Cladistics*, *23*, 1–28.

Spivak, M., Masterman, R., Ross, R., & Mesce, K. A. (2003). Hygienic behavior in the honey bee (*Apis mellifera* L.) and the modulatory role of octopamine. *Journal of Neurobiology*, *55*, 341–354.

Staudacher, E., Gebhardt, M., & Dürr, V. (2005). Antennal movements and mechanoreception: Neurobiology of active tactile sensors. *Advances in Insect Physiology*, *32*, 49–205.

Stevenson, P. A. (2009). The key to Pandora's box. *Science*, *323*, 594–595.

Stevenson, P. A. (2011). Neurochemical control of the decision to fight or flee in crickets. *Proceedings of the* 9th *Göttingen Meeting of the German Neuroscience Society*, March 23–27 (S20–S26).

Stevenson, P. A., Dyakonova, V., Rillich, J., & Schildberger, K. (2005). Octopamine and experience-dependent modulation of aggression in crickets. *Journal of Neuroscience*, *25*, 1431–1441.

Stevenson, P. A., Hofmann, H. A., Schoch, K., & Schildberger, K. (2000). The fight and flight responses of crickets depleted of biogenic amines. *Journal of Neurobiology*, *43*, 107–120.

Stevenson P. A., & Rillich, J. (2012). The decision to fight or flee - insights into underlying mechanism in crickets. *Frontiers in Neuroscience*, *6*, 118. doi:10.3389/fnins.2012.00118

Stevenson, P. A., & Rillich, J. (2013). Isolation associated aggression – A consequence of recovery from defeat in a territorial animal *PLoS ONE,,6 e74965*.

Stevenson, P. A., & Rillich, J. (2014). *Adding up the odds - NO/cGMP mediates opponent assessment for the decision to flee in fighting crickets*. Manuscript submitted for publication.

Stevenson, P. A., & Schildberger, K. (2013). Mechanisms of experience dependent control of aggression in crickets. *Current Opinion in Neurobiology 23*:1-6

Sword, G. A., Simpson, S. J., El Hadi, O. T. M., & Wilps, H. (2000). Density dependent aposematism in the desert locust. *Proceedings of the Royal Society B*, *267*, 63–68.

Tanaka, S., & Maeno, K. (2010). A review of maternal and embryonic control of phase-dependent progeny characteristics in the desert locust. *Journal of Insect Physiology*, *56*, 911–918.

Thamm, M., Balfanz, S., Scheiner, R., Baumann, A., & Blenau, W. (2010). Characterization of the 5-HT1A receptor of the honeybee (*Apis mellifera*) and involvement of serotonin in phototactic behavior. *Cellular and Molecular Life Sciences*, *67*, 2467–2479.

Tomer, R., Denes, A. S., Tessmar-Raible, K., & Arendt, D. (2010). Profiling by image registration reveals common origin of annelid mushroom bodies and vertebrate pallium. *Cell*, *142*, 800–809.

Tops, M., Russo, S., Boksem, M. A., & Tucker, D. M. (2009). Serotonin: Modulator of a drive to withdraw. *Brain & Cognition*, *71*, 427–436.

Vergoz, V., Lim, J., & Oldroyd, B. P. (2011). Biogenic amine receptor gene expression in the ovarian tissue of the honey bee *Apis mellifera*. *Insect Molecular Biology*, *21*, 21–29.

Vergoz, V., McQuillan, H. J., Geddes, L. H., Pullar, K., Nicholson, B. J., Paulin, M. G., & Mercer, A. R. (2009). Peripheral modulation of worker bee responses to queen mandibular pheromone. *Proceedings of the National Academy of Sciences, USA*, *106*, 20930–20935.

Vergoz, V., Schreurs, H. A., & Mercer, A. R. (2007). Queen pheromone blocks aversive learning in young worker bees. *Science*, *317*, 384–386.

Verlinden, H., Vleugels, R., Marchal, E., Badisco, L., Pflüger H.-J., Blenau, W., & Broeck, J. V. (2010). The role of octopamine in locusts and other arthropods. *Journal of Insect Physiology*, *56*, 854–867.

Vrontou, E., Nilsen, S. P., Demir, E., Kravitz, E. A., & Dickson, B. J. (2006). Fruitless regulates aggression and dominance in *Drosophila*. *Nature Neuroscience*, *9*, 1469–1471.

Wagener-Hulme, C., Kuehn, J. C., Schulz, D. J., & Robinson, G. E. (1999). Biogenic amines and division of labor in honey bee colonies. *Journal of Comparative Physiology A*, *184*, 471–479.

Wang, H.-S., Wang, X.-H., Zhou, C.-S., Huang, L.-H., Zhang, S.-F., Guo, W., & Kang, L. (2007). cDNA cloning of heat shock proteins and their expression in two phases of the migratory locust. *Insect Molecular Biology*, *16*, 207–219.

Wang, L., & Anderson, D. J. (2010). Identification of an aggression-promoting pheromone and its receptor neurons in *Drosophila*. *Nature*, *463*, 227–231.

Wei, Y., Chen, S., Yang, P., Ma, Z., & Kang, L. (2009). Characterization and comparative profiling of the small RNA transcriptomes in two phases of locust. *Genome Biology, 10*, R6.

Whitfield, C. W., Ben-Shahar, Y., Brillet, C., Leoncini, I., Crauser, D., LeConte, Y., Rodriguez–Zas, S., & Robinson, G. E. (2006). Genomic dissection of behavioral maturation in the honey bee. *Proceedings of the National Academy of Sciences USA, 103*, 16068–16075.

Wilson, E. O. (1971). *The insect societies.* Cambridge, MA: Belknap Press of Harvard University Press.

Wilson, K., Thomas, M. B., Blanford, S., Doggett, M., Simpson, S. J., & Moore, S. L. (2002). Coping with crowds, density-dependent disease resistance in desert locusts. *Proceedings of the National Academy of Sciences USA, 99*, 5471–5475.

Yurkovic, A., Wang, O., Basu, A. C., & Kravitz, E. A. (2006). Learning and memory associated with aggression in *Drosophila melanogaster. Proceedings of the National Academy of Sciences USA, 103*, 17519–17524.

Zayed, A., & Robinson, G. E. (2012). Understanding the relationship between brain gene expression and social behavior: Lessons from the honey bee. *Annual Review of Genetics, 46*, 591–615.

Zhou, C., Rao, Y., & Rao, Y. (2008). A subset of octopaminergic neurons are important for *Drosophila* aggression. *Nature Neuroscience, 11*, 1059–1067.

Zorovic, M., & Hedwig, B. (2013). Descending brain neurons in the cricket *Gryllus bimaculatus* (de Geer): Auditory responses and impact on walking. *Journal of Comparative Physiology, A,199*, 25–34.

CHAPTER 4

Social Regulation of Gene Expression in the African Cichlid Fish *Astatotilapia burtoni*

Karen P. Maruska *and* Russell D. Fernald

Abstract

How does an animal's social environment shape its behavior and physiology, and what underlying molecular and genetic mechanisms lead to phenotypic changes? To address this question, the authors used a model system that exhibits socially regulated plastic phenotypes, behavioral complexity, molecular level access, and genomic resources. The African cichlid fish *Astatotilapia burtoni*, in which male status and reproductive physiology are under social control, has become an important model for studying the mechanisms that regulate complex social behaviors. This chapter reviews what is known about how information from the social environment produces changes in behavior, physiology, and gene expression profiles in the brain and reproductive axis of *A. burtoni*. Understanding the mechanisms responsible for translating perception of social cues into molecular change in a model vertebrate is important for identifying selective pressures and evolutionary mechanisms that shape the brain and ultimately result in diverse and complex social behaviors.

Key Words: behavior, brain, dominant, plasticity, reproduction, social behavior network, social status, subordinate, teleost, transcription

Animals must interact with others and their environment to survive and reproduce. Thus, a fundamental challenge in biology is to understand how animals acquire, evaluate, and then translate information from their environment into adaptive physiological and behavioral changes. Social interactions in particular can profoundly influence an individual's behavior and physiology, which is often mediated by a diverse array of cellular and molecular mechanisms. These links between social information and molecular plasticity ultimately shape the evolution of a species. It is well established that the brain controls the expression of behaviors, but how might an animal's behavior or perception of its social and physical environment sculpt its brain? To address this and related questions requires model systems that allow controlled manipulation of the social environment in naturalistic or seminaturalistic settings. Furthermore, with recent advances in molecular biological techniques, it is also advantageous to conduct studies of social behavior in a model system in which the genome sequence is known. Genomic resources allow investigators to use a combination of approaches, including candidate gene studies and large-scale technologies (e.g., transcriptomics, proteomics, metabolomics, epigenomics), to identify suites of genes that may be associated with particular behavioral patterns, thus providing insight into the molecular basis and evolution of social behaviors.

What is social behavior? For the purpose of this chapter, we define social behavior as interactions among members of the same species that influence immediate or future behaviors (Robinson, Fernald, & Clayton, 2008). This includes, but is not limited to, behaviors such as aggression, courtship, and mating. These interactions involve the production, reception, and interpretation of communicative signals that influence individual behaviors in a context-dependent manner. Plasticity in social behavior can arise due to either temporal or spatial variation in gene expression, and it is likely that the evolution of social behaviors involves both the evolution of new genes with novel functions and genes that are regulated in new ways (Robinson & Ben-Shahar, 2002). But how does an animal's social environment shape its brain function to alter behavior as it interacts with conspecifics in different contexts? Social information is complex and can influence brain function and physiology on many biological levels (e.g., behavioral, hormonal, cellular, molecular) (Fernald & Maruska, 2012) as an animal integrates this information with other external and internal cues (Figure 4.1). Our goal for this chapter is to review examples from an established social behavior model system, the African cichlid fish *Astatotilapia* (formerly *Haplochromis*) *burtoni*, to focus specifically on how social interactions in different contexts are associated with changes in gene expression in the brain and hypothalamic-pituitary-gonadal (HPG) axis (for a discussion of socially mediated changes in gene expression in insects, see the chapter by Simpson and Stevenson, elsewhere in this handbook).

Through the course of this chapter, we discuss many aspects of the social life of *A. burtoni* and how different behavioral contexts and position in a dominance hierarchy are associated with behavioral, physiological, and gene expression changes from the brain to the reproductive organs. Although we use examples from both males and females to illustrate how information is communicated between sexes, and, within a sex, to coordinate behaviors and cause molecular change, the majority of presented information focuses on males that are transitioning between subordinate and dominant social status. This subordinate to dominant transition highlights how changes in an animal's social environment and position in a dominance hierarchy can initiate rapid changes in gene expression. For example, imagine a male *A. burtoni* of low-ranking social status who, as a result, has little chance to pass on his genes because he lacks a territory to attract females for spawning, has a suppressed reproductive system, and is constantly chased and attacked by aggressive higher ranking males. If one of these high-ranking males

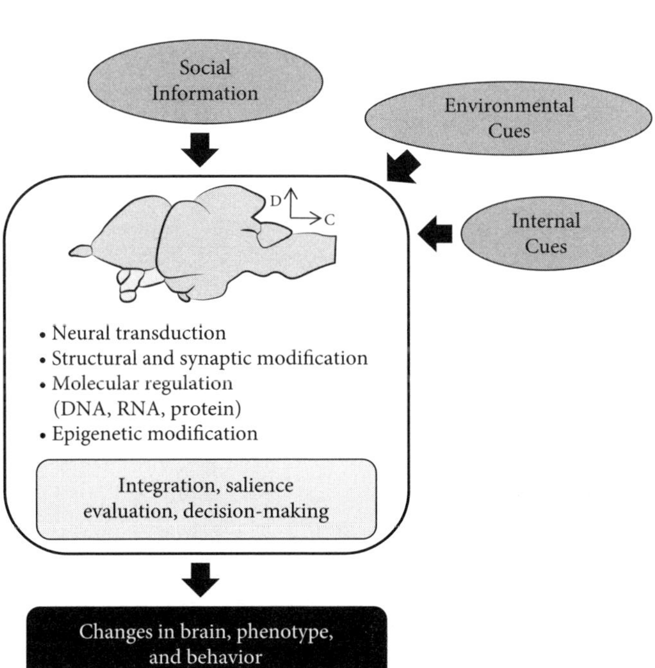

Fig. 4.1 Schematic diagram of the complex relationships between social information, the brain, and adaptive behavioral outputs. The brain constantly receives and evaluates the salience of inputs from external social and environmental cues, as well as from internal physiological cues such as hormonal and nutritional status. This information results in brain responses via mechanisms that include neural transduction, structural and synaptic modifications, molecular regulation including gene expression changes, and epigenetic modifications that ultimately lead to alterations in brain function, phenotypic variation, and adaptive behaviors.

Illustration of *Astatotilapia burtoni* brain is shown. D, dorsal; C, caudal.

is removed by a predator, however, and a vacated territory becomes available, the low-ranking male must now quickly be able to detect his absence, seize the opportunity to acquire this limited valuable resource, and then initiate a dramatic transformation that spans from whole-organism behavior and coloration changes, to hormonal, cellular, and transcriptional-level changes throughout the body. In fact, within a matter of minutes, his appearance and physiology has changed radically as he prepares for a new lifestyle as a dominant and reproductively active territory holder. How does the subordinate male accomplish this? In the following sections, we review what is known about how these males undergo this social transition, as well as highlight which details remain enigmatic. We begin by describing the natural history and behavioral repertoire of *A. burtoni* and introduce it as a suitable model system for studying social regulation of gene expression. We then illustrate how social ascent (subordinate to dominant transition) in males causes rapid changes in the brain and along the reproductive axis, including how sex steroids are linked to behavior. Next, we discuss how changes in gene expression in the brain are influenced not only by social ascent, but also by stable social status and other contexts, such as spatial learning tasks. Last, we review how females respond to social information and how changes in gene expression are related to their reproductive cycle. Because *A. burtoni* allows controlled studies in a seminaturalistic environment, its genome has recently been sequenced, and a wealth of information on social behaviors and transcriptional plasticity is currently available, we propose that this species will continue to be an important behavioral and genomic vertebrate model to expand our current understanding of the molecular underpinnings of complex social behaviors.

I. The African Cichlid Fish *Astatotilapia burtoni* as a Model for Social Regulation of Gene Expression

To understand how contextual social information is translated into changes in gene expression, it is crucial to have a model organism with well-described stereotypical social behaviors for which the social environment can be easily manipulated and quantified. In addition, such a system would have the reception of social information linked to specific phenotypic changes. The African cichlid fish *A. burtoni* is just such a model and has become an important vertebrate system for investigating how social information regulates brain function, behavior, and the reproductive axis.

A. *Natural History and Social Behavior*

Astatotilapia burtoni is endemic to Lake Tanganyika in the rift valley system in Eastern Africa where it lives in shallow shore pools and river estuaries (Fernald & Hirata, 1977). This species also has the remarkable benefit of the adult males existing in two distinct, but reversible, phenotypes. These two male phenotypes and their HPG axis activity—and hence reproductive capacity—are tightly coupled to their social status (Fernald, 2009) (see the Reproductive Physiology section). Dominant (also called territorial) males represent a small percentage of the population (10–30 percent) and are brightly colored (blue or yellow), with a black stripe through the eye (eye-bar), an opercular black spot at the caudal tip of the gill cover, prominent egg-spots on the anal fin, and a red humeral patch on the side of the body (Figure 4.2). These dominant males hold territories that they defend vigorously from rival males, and they spend significant amounts of time actively courting and eventually spawning with females (Fernald, 1977; Fernald & Hirata, 1977). In contrast, subordinate (also called nonterritorial) males make up the majority of the male population (70–90 percent), are duller in coloration (lacking eye-bar and humeral patch), do not hold territories or typically reproduce, school with females and other subordinate males, and flee from the aggressive attacks of dominant males. *A. burtoni* lives in a lek-like social system in which dominant males defend clustered territories to guard food, shelter, and spawning substrates from rival males. These dominant males perform 19 distinct behavioral patterns during social interactions that are associated with territoriality and reproduction (Fernald, 1977). For example, dominant males establish a spawning area by digging a pit in their territory, engage in agonistic threat displays and border disputes with neighboring dominant males, chase subordinate males away from their territory, and perform courtship quivers toward passing females in an attempt to lead them into their territory to spawn. Since defensible territory substrate for spawning and feeding is often limited, and females are less likely to mate outside the protection of a spawning shelter, there is fierce competition for this resource, and, as a result, only a minority of males at any one time will defend a territory and mate. Once a receptive (gravid, "ripe with eggs") female follows a dominant male into his territory and is

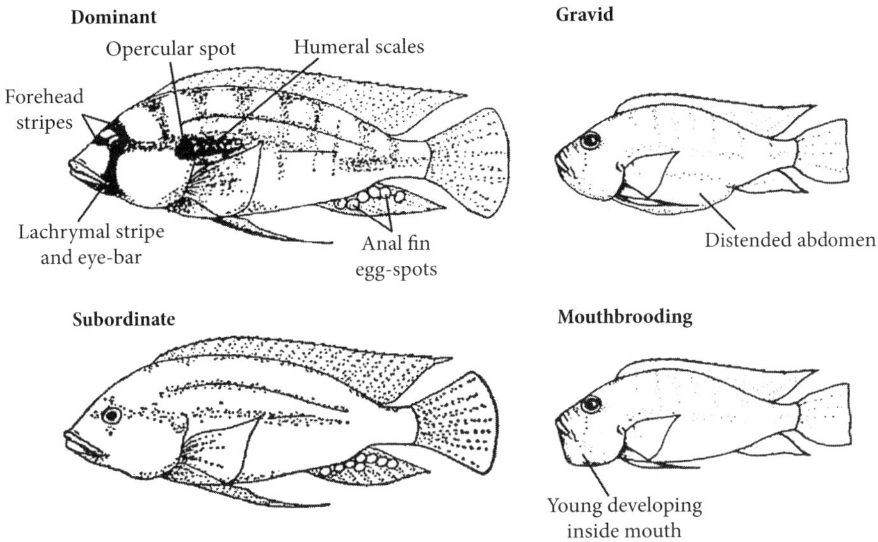

Fig. 4.2 Illustrations of the body patterns for typical dominant (territorial) and subordinate (nonterritorial) *Astatotilapia burtoni* males, and sexually receptive gravid and mouthbrooding parental females. Dominant males are brightly colored (yellow or blue), have distinct yellow-orange egg-spots on their anal fins, dark forehead stripes, a dark opercular spot on the caudal edge of the gill cover, a dark lachrymal stripe or eye-bar extending from the eye to the lower jaw, and a bright orange-red patch on the humeral scales. Subordinate males lack the robust markings of their dominant counterparts and are more similar in coloration to females. Females cycle between a gravid receptive phase in which they develop distended abdomens from growing oocytes as they get closer to spawning and a mouthbrooding phase in which their jaws protrude forward to accommodate the developing young inside the mouth.

Modified in part from Fernald (1977).

appropriately stimulated, she will deposit eggs on the substrate and then immediately collect them in her mouth. The male then displays his anal fin egg-spots on the substrate in front of her, and while she attempts to collect the egg-spots on his fin, the male releases sperm near her mouth to fertilize the eggs (Figure 4.3). There are often several bouts of egg-laying and fertilization that may be briefly interrupted as the dominant male leaves to chase away intruders or interact with neighboring males. During these spawning bouts, subordinate males can also attempt to interrupt the pair and "sneak" fertilization attempts; although these subordinate males do not defend territories, they do maintain sperm production and retain viable sperm in their testes during social suppression (Kustan, Maruska, & Fernald, 2012; Maruska & Fernald, 2011*a*). When spawning and fertilization are complete, the female leaves the territory to brood the young in her mouth (mouthbrooding) for approximately 2 weeks until releasing them as fully developed fry, while the dominant male resumes his territorial defense and continues to court other receptive females.

B. Social Communication

Communication is a fundamental component of social behaviors because these behaviors involve the production, reception, and interpretation

Fig. 4.3 Spawning behavior in the African cichlid fish *Astatotilapia burtoni*. The female lays a small batch of eggs in the territory of a male (egg-laying) and then immediately takes up the eggs into her mouth (egg-uptake). The male then presents the conspicuous egg-spots on his anal fin, which the female tries to pick up, thereby bringing her mouth close to the male's genital opening, through which sperm is released to fertilize the eggs (fertilization).

Modified from Salzburger, Braasch, & Meyer (2007).

of context-specific signals. Social communication in *A. burtoni* relies heavily on visual signals (Clement, Parikh, Schrumpf, & Fernald, 2005; Fernald, 1977; 1984; Fernald & Hirata, 1977), but studies have also demonstrated the importance of multimodal signaling, including chemosensory and acoustic communication (Crapon de Caprona, 1974; 1980; Maruska & Fernald, 2010*b*, 2010*c*, 2012; Maruska, Ung, & Fernald, 2012; Nelissen, 1977; Robison, Fernald, & Stacey, 1998). In addition to the distinct behavioral displays just mentioned, presentation of the black eye-bar increases aggression or readiness to attack, whereas presentation of the red humeral patch decreases attack readiness and may play a larger role in reproduction (Heiligenberg & Kramer, 1972; Heiligenberg, Kramer, & Schulz, 1972; Leong, 1969). The number of anal fin egg-spots may also be important for male–male communication because in aggression trials males with fewer egg-spots received more aggressive attacks than males with more egg-spots (Theis, Salzburger, & Egger, 2012). Chemosensory signaling is also important during social interactions in *A. burtoni* (Crapon de Caprona, 1980; Maruska & Fernald, 2012; Robison et al., 1998). As in other vertebrates, chemical signals in fish urine (or compounds released via the gills, skin, feces) are thought to influence both territorial and reproductive behaviors in receivers (Almeida et al., 2005; Arakawa, Blanchard, Arakawa, Dunlap, & Blanchard, 2008). Indeed, dominant male *A. burtoni* regulate release of these chemosensory signals in the form of urine pulses during territorial interactions with males and in reproductive contexts with receptive females (Maruska & Fernald, 2012). Dominant males also produce sounds during their courtship body quiver displays, and gravid females prefer males associated with sound production, suggesting that they gain some valuable information on male quality or motivation from his sounds (Maruska & Fernald, 2010*c*; Maruska, Ung, et al., 2012; Nelissen, 1977). Thus, this species exists in a complex social environment in which their interactions with conspecifics occur in multiple sensory modalities and different behavioral contexts, which ultimately shapes their phenotypes, survival, and reproductive fitness.

Lake Tanganyika is the oldest, deepest, and most morphologically and behaviorally diverse of the African rift lakes and may have originated the cichlid radiation that gave rise to species flocks in the other rift lakes (Sturmbauer, Husemann, & Danley, 2011). Thus, understanding the evolution of communication and social behaviors in this species from Lake Tanganyika is essential to fully appreciate the driving forces, mechanisms, and pathways of diversification that led to the incredible variety of cichlid fishes. Previous studies have suggested that single traits, such as those involved in visual signaling, are often insufficient to explain phenotypic diversity in cichlids and that species richness is a function of the number of traits involved in diversification (i.e., the "multifarious selection" hypothesis) (Blais et al., 2009; Nosil & Harmon, 2009; Nosil, Harmon, & Seehausen, 2009). Thus, the use of multiple communication systems for social interaction provides more traits on which sexual selection can act, allowing for a greater number of taxa potentially resulting in the high diversity of cichlid fishes (Blais et al., 2009; Sturmbauer et al., 2011). Although accompanying gene expression data are not yet available for many of the multimodal communicative aspects of social behavior in *A. burtoni*, the above-mentioned studies all represent an excellent starting point for future work on the molecular mechanisms underlying sensory processing and social communication, sexual selection, and speciation.

C. Reproductive Physiology

The disparity in behavior and appearance between male phenotypes described earlier is also associated with important and dramatic reproductive physiological differences, such that dominant males have an active and up-regulated HPG axis compared to subordinate males. For example, along the HPG axis, dominant males have larger gonadotropin-releasing hormone (GnRH1) neurons in the preoptic area of the brain (Davis & Fernald, 1990) that have distinct membrane properties (e.g., higher membrane capacitance, lower input resistance, shorter action potential duration; Greenwood & Fernald, 2004) and greater dendritic complexity (Scanlon, Greenwood, & Fernald, 2003) compared to subordinate males. Dominant males also have higher mRNA levels of *GnRH1*, kisspeptin receptor (*gpr54-2b;* previously *kiss1r*), and some sex-steroid receptor subtypes in the brain (Burmeister, Kailasanath, & Fernald, 2007; Grone, Maruska, Korzan, & Fernald, 2010; White, Nguyen, & Fernald, 2002), as well as higher cell proliferation rates in the preoptic area and other socially relevant regions of the brain (Maruska, Carpenter, & Fernald, 2012).

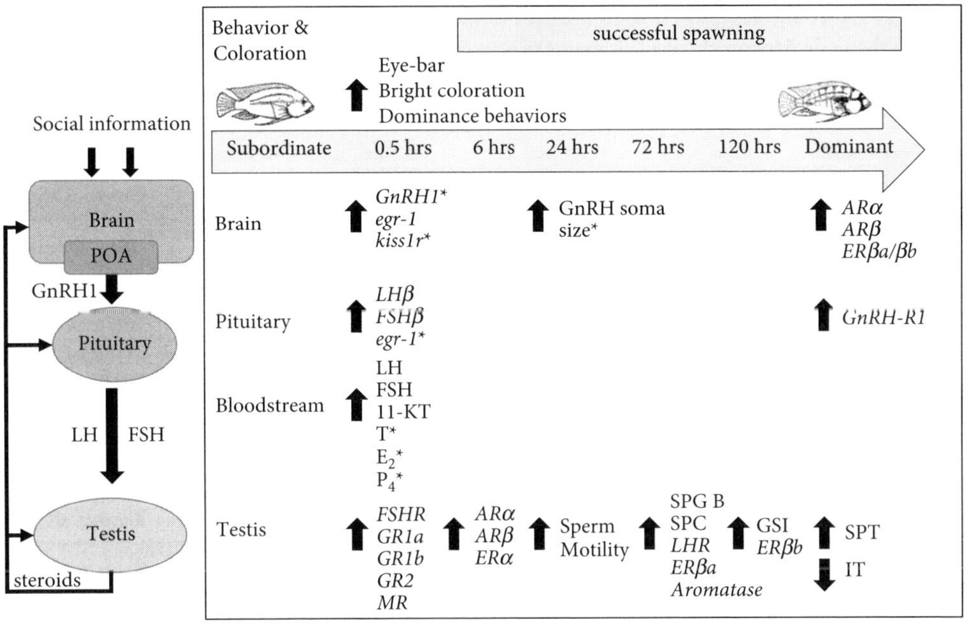

Fig. 4.4 Temporal summary of behavioral, morphological, and physiological changes in the hypothalamic-pituitary-gonadal axis during social ascent in the male cichlid fish *Astatotilapia burtoni*. Arrows indicate the time point after social opportunity at which the first significant increase (*up arrows*) or decrease (*down arrows*) from stable subordinate male values was observed. mRNA changes (as determined by in situ hybridization or quantitative polymerase chain reaction [qPCR]) are indicated in italics. Data were compiled from studies by Au et al., 2006; Burmeister et al., 2005; Burmeister et al., 2007; Kustan et al., 2012; Maruska & Fernald, 2010*a*, 2011*a*; Maruska et al., 2011; Maruska & Fernald, 2013; Maruska, Zhang, Neboori, & Fernald, 2013, as well as from some unpublished data from our lab (*). Note also that not all measures were quantified at each time point, and several are only known for subordinate and dominant states.

ARα/β, androgen receptor subtypes α and β; E_2, 17β-estradiol; egr-1, early growth response factor-1; ERα/βa/βb, estrogen receptor subtypes α, βa and βb; FSH, follicle stimulating hormone; FSHβ, β-subunit of FSH; FSR, FSH receptor; GnRH1, gonadotropin-releasing hormone 1; GR1a/1b/2, glucocorticoid receptor subtypes 1a, 1b and 2; GSI, gonadosomatic index; IT, interstitial tissue; kiss1r, kisspeptin receptor 1; LH, luteinizing hormone; LHβ, β-subunit of LH; LHR, LH receptor; MR, mineralocorticoid receptor; P_4, progesterone; POA, preoptic area; SPG B, B-type spermatogonia; SPC, spermatocytes; SPT, spermatids; T, testosterone; 11-KT, 11-ketotestosterone.Modified from Maruska & Fernald (2011*a*, 2011*b*).

At the level of the pituitary, dominant males have higher GnRH receptor type I (*GnRH-R1*), *LHβ*, and *FSHβ* mRNA levels than do subordinate males (Au, Greenwood, & Fernald, 2006; Maruska, Levavi-Sivan, Biran, & Fernald, 2011). In the bloodstream, dominant males also have higher circulating levels of androgens (testosterone and 11-ketotestosterone), 17β-estradiol, luteinizing hormone (LH), and follicle-stimulating hormone (FSH) (Maruska & Fernald, 2010*a*, 2010*c*; Maruska et al., 2011; Parikh, Clement, & Fernald, 2006*b*). Dominant males also have higher mRNA levels of LH receptor (*LHR*), FSH receptor (*FSHR*), and multiple steroid receptor subtypes (androgen, estrogen, glucocorticoid) in the testes (Maruska & Fernald, 2011*a*), as well as larger testes with a higher density of luminal sperm and spermatogenic potential (Fraley & Fernald, 1982; Kustan et al., 2012; Maruska & Fernald, 2011*a*) compared to subordinate males (Figure 4.4; Table 4.1). Thus, the greater reproductive capacity of dominant males is evident at every level of the reproductive axis (e.g., brain, pituitary, circulation, testes), as well as at multiple levels of biological organization (e.g., behavioral, morphological, hormonal, cellular, molecular).

Importantly, these above-mentioned behavioral, morphological, and physiological features of each male phenotype are reversible and under social control such that when a territory is vacated, a subordinate male will quickly rise in social rank and take it over. In nature, the social and physical environment fluctuates often, providing frequent opportunities for this phenotypic switching (Fernald & Hirata, 1977). Further, this transition between subordinate and dominant states can be experimentally controlled in the laboratory by manipulating the composition of the social environment (Burmeister, Jarvis, & Fernald, 2005; Maruska & Fernald, 2010*a*), which allows us to examine the

will facilitate interpretation of transcriptome studies that might use technologies such as RNA-seq to examine gene expression profiles in animals exposed to different social situations or to compare profiles among different tissues or cell types. The genome is also valuable for comparative genomic studies to identify conserved molecules or regulatory pathways involved in social behaviors and to determine how they differ among species and how they have been shaped through evolution. Genomic sequences can also be scanned for duplications and mutations, selective sweeps (beneficial mutations and linked alleles that rise in frequency in a population), binding sites for transcription factors, and to identify promoters and other upstream or downstream regulatory elements to help develop genetic manipulation tools. This newly sequenced genome of *A. burtoni* makes this species a powerful model to address specific questions on how the social environment is translated into transcriptional change on multiple levels and temporal scales, as well as on the evolution of complex social behaviors. Since utilization of the *A. burtoni* genome is still in its infancy, our main focus in the following sections is to review what we currently know about how tightly and rapidly the social environment can be linked to gene expression changes in the brain and at multiple levels of the reproductive axis.

III. Social Ascent Paradigm: Providing an Opportunity to Gain a Territory and Rise in Social Rank

How and why do we manipulate the social status of male *A. burtoni*? The natural phenotypic plasticity in male *A. burtoni* provides us with an excellent opportunity to examine physiological changes, including plasticity in gene transcription, that are induced by changes in dominance status. To provide an opportunity for social ascent, we use an experimental paradigm originally designed by Burmeister et al. (2005) and later modified to more closely mimic the natural territory tenure of 4–5 weeks (Fernald & Hirata, 1977; Hofmann, Benson, & Fernald, 1999; Maruska & Fernald, 2010*a*, 2011*a*; Maruska et al., 2011). This paradigm, described briefly here, allows us to control precisely the subordinate to dominance transition and to examine the associated changes in behavior, gene expression, and other physiological measures at specific time points following social ascent (Figure 4.5).

To create socially and reproductively suppressed males for the ascension paradigm, dominant subject males from community tanks are placed into aquaria for 4–5 weeks (a period sufficient to significantly suppress their entire reproductive axis) with several larger dominant suppressor males, females, and subordinate males. At the end of the suppression period, subjects are moved into the central compartment of an experimental tank that contains one large resident dominant male and several females. This central compartment is separated from mixed-sex community tanks on either side, with transparent barriers so that fish can interact visually but not physically. On the day of ascent, the resident suppressor male is removed with a net 1 hour prior to light onset by researchers wearing infrared night vision goggles, which minimizes disturbance in the tank and ensures that visual absence of the suppressor occurs consistently only at light onset for all tested individuals. Fish presented with a social opportunity in this paradigm intensify their body coloration, turn on their eye-bar, and start to behave like dominant males within just a few minutes of light onset (Figure 4.6). Stable dominant and stable subordinate males are also used as control comparisons to males ascending in social status. Ascending males and stable dominant males are then sacrificed at different time points after they display dominance behaviors (combined territorial and reproductive) at a rate of three behaviors per minute, whereas stable subordinate males are sacrificed at equivalent times after light onset to match those of ascending and stable dominant males. Thus, any gene expression changes elicited in the ascending males likely result from the recognition of the social opportunity because ascending males and stable dominant males perform behaviors at similar rates (Burmeister et al., 2005; Maruska & Fernald, 2010*a*) (Figure 4.6). By sampling these males at multiple time points after ascent, we can discover *what* changes occur at the transcriptional level, as well as *how quickly* they occur during the social transition (see the following sections).

IV. Rapid Social Regulation of Gene Expression in the HPG Axis of Males

What types of reproductive physiological and molecular changes occur when subordinate males become dominant? There is a direct relationship between social dominance and reproductive physiology in male *A. burtoni*, which makes it an excellent model to examine how behaviors associated with social status, or rank, can influence gene expression related to reproductive function. GnRH1 neurons in the hypothalamic-preoptic

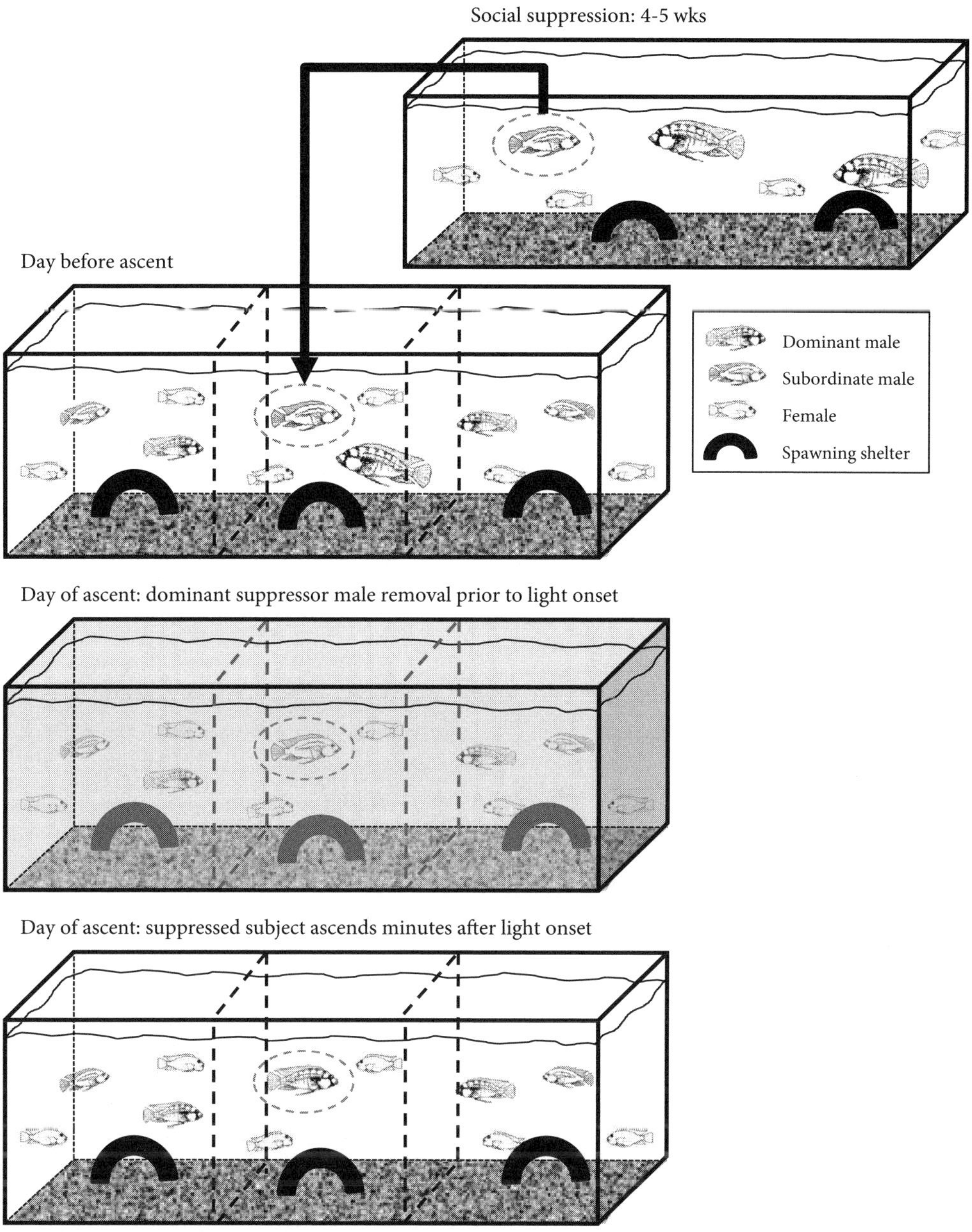

Fig. 4.5 Schematic representation of the ascent experimental paradigm used to provide socially suppressed subordinate male *Astatotilapia burtoni* with a social opportunity to gain a territory and become dominant. Previously dominant subject males (*circled*) are placed into community suppression tanks for 4–5 weeks that contain several large dominant males, subordinate males, and females. Subject males are then transferred to a central compartment of an experimental tank for 2 days prior to social opportunity. The central compartment contains a large resident dominant male and 3–4 females and is separated with transparent acrylic barriers (*dashed lines*) on either side from mixed-sex community tanks. On the day of ascent, the resident suppressor male is removed from the central compartment 1 hour before light onset. At light onset, the subject male recognizes the vacant territory, performs dominance behaviors within minutes, and ascends in social status.

Modified from Maruska and Fernald (2010*a*).

area of the brain sit at the apex of the HPG axis and are thus the master regulators of reproduction across vertebrates (Bliss, Navratil, Xie, & Roberson, 2010; Dellovade, Schwanzel-Fukuda, Gordan, & Pfaff, 1998). In male *A. burtoni*, the size of these GnRH1 neurons is under social control, such that they grow and shrink as an animal transitions between dominant and subordinate status, potentially many times in its lifetime (Francis, Soma, & Fernald, 1993). Dominant

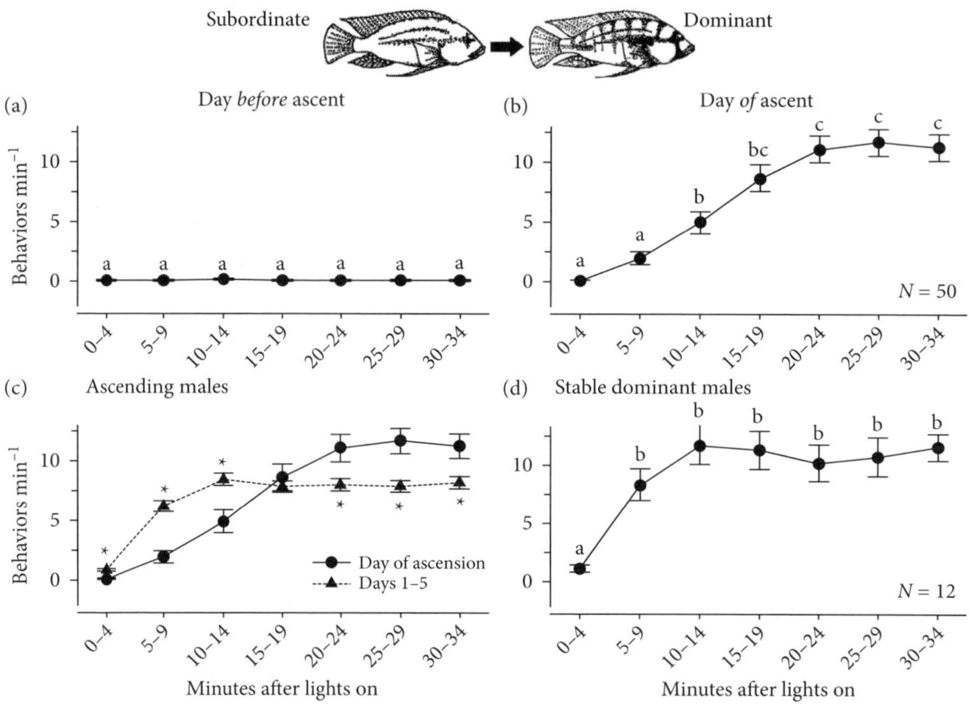

Fig. 4.6 Ascending male *Astatotilapia burtoni* display dominance behaviors within minutes of the perception of a social opportunity. (A) Subordinate males show virtually no territorial or reproductive behaviors on the day before ascent while they are still in the presence of the large suppressor male. (B) Suppressed subject males then rapidly increase dominance behaviors within minutes after light onset when presented with an opportunity to ascend in social status by removal of the suppressor resident male. (C) Ascending males show reduced rates of dominance behaviors on the day of ascent compared to subsequent days (days 1–5) for the first 15 minutes after light onset, but higher rates from 20–35 minutes. (D) Stable dominant control males also showed a rapid increase in dominance behaviors by 5–9 minutes after light onset, which is then maintained throughout the sampling period. Data are plotted as mean dominance behaviors (reproductive and territorial behaviors combined) min^{-1} ± SE compiled into 5-minute bins. Time points with different letters in A, B, and D are statistically different, and asterisks in C indicate differences between behavior rates on the day of ascent compared to days 1–5 after ascent at each time period. Note that some error bars are obscured by symbols.

Modified from Maruska and Fernald (2010*a*).

males have larger GnRH1 neurons with increased dendritic complexity compared to subordinate males (Davis & Fernald, 1990; Scanlon et al., 2003). When a subordinate male gets a chance to ascend in status and become dominant, however, there are several rapid changes that occur in the GnRH1 neurons. First, the social opportunity is associated with a rapid (20–30 minute) induction of the immediate early gene (IEG) *egr-1* (a transcription factor-encoding gene; also called *zenk, zif-268, ngfi-a, krox-24, tis8*) in the preoptic area and in GnRH1 neurons (Burmeister et al., 2005; Maruska, Zhang, Neboori, & Fernald, 2013) (Figure 4.7). This molecular response is likely due to the recognition of the social opportunity because it is not elicited in males who are already dominant and performing similar behaviors. This type of molecular response to an opportunity may be conserved across vertebrates because socially relevant reproductive stimuli are also known to induce IEG expression within GnRH1 neurons from fishes (Burmeister et al., 2005) to mammals (Gelez & Fabre-Nys, 2006; Meredith & Fewell, 2001; Pfaus, Jakob, Kleopoulos, Gibbs, & Pfaff, 1994). Second, there is an increase in GnRH1 mRNA levels in the brain at 30 minutes after ascent (Maruska & Fernald, 2013), suggesting that the reproductive axis is quickly stimulated. Third, an increase in GnRH1 soma size is detected in as little as 1 day after ascent (the earliest time point measured) (Maruska & Fernald, 2013), and these neurons reach dominant male sizes within 5–7 days (White et al., 2002). The significance of this change in GnRH1 soma size is unknown, but it may function to accommodate changes in synaptic inputs or variations in the cellular and molecular demands of the cell. Collectively, these studies suggest that suppressed males are well adapted to swiftly recognize and take advantage of an opportunity to gain a territory, become dominant, and reproduce.

The primary targets of the GnRH1 neurons are the gonadotropin-producing cells in the anterior pituitary gland. The released GnRH1 peptide, delivered to the pituitary via direct neuronal projections in fishes, binds to GnRH receptors (members of the G-protein coupled receptor superfamily) on the gonadotrope cells to induce release and synthesis of the two gonadotropin hormones, LH and FSH, which then target the gonads (testes or ovaries) to stimulate steroid production and gamete development. Multiple forms of GnRH receptors (i.e. types I, II, III) are found in mammals (Millar, 2005), amphibians (Wang et al., 2001), and fishes (Flanagan et al., 2007; Lethimonier, Madigou, Munoz-Cueto, Lareyre, & Kah, 2004; Moncaut, Somoza, Power, & Canario, 2005; Robison et al., 2001), and they often show differential spatial and temporal expression patterns (e.g., across tissue and cell types, and across season, reproductive stage, development, or dominance status) and varying responses to regulation by steroids, GnRH, and monoamines, all of which suggest functional specializations (Au et al., 2006; Chen & Fernald, 2006; Crowley et al., 1998; Levavi-Sivan, Safarian, Rosenfeld, Elizur, & Avitan, 2004; Lin et al., 2010). In male *A. burtoni*, pituitary mRNA levels of *GnRH-R1*, but not *GnRH-R2*, are socially regulated such that stable dominant males have higher levels compared to subordinate males (Au et al., 2006; Maruska et al., 2011). The increase in *GnRH-R1* during the social transition appears to occur more slowly (days), however, than changes in mRNA levels of other genes that occur within minutes to hours (Maruska et al., 2011). For example, pituitary mRNA levels of the IEG *egr-1, LHβ*, and *FSHβ* are more rapidly increased at just 30 minutes after social ascent, suggesting that GnRH1 release from its axon terminals has quickly activated the pituitary gland (Maruska et al., 2011;

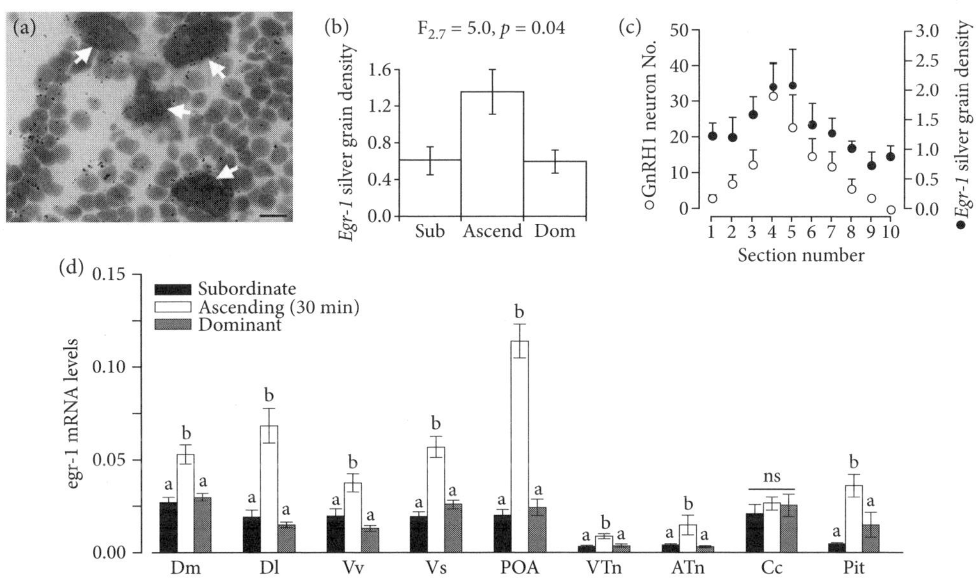

Fig. 4.7 Rapid increase in mRNA expression of the immediate early gene *egr-1* in GnRH1 neurons and throughout the nuclei of the social behavior network in males transitioning from subordinate to dominant status in *Astatotilapia burtoni.* (a) Photomicrograph of *egr-1* silver grains (*small black dots*; in situ hybridization) on GnRH1 neurons (*arrows*) in the anterior parvocellular preoptic nucleus (aPPn) of a male *A. burtoni*. Scale bar, 10 μm. (b) *Egr-1* silver grain density (mean ± SE) of the entire aPPn in subordinate (Sub), ascending (Ascend), and dominant (Dom) males shows greater *egr-1* staining at 20 minutes after social opportunity in ascending males compared to the stable social states. (c) GnRH1 neuron number (*open circles*) and *egr-1* silver grain density (*closed circles*) (mean ± SE) within adjacent sections of the aPPn to show the greater *egr-1* staining in sections that have more GnRH1 neurons. (d) Social opportunity rapidly increases *egr-1* mRNA levels (measured by quantitative polymerase chain reaction [PCR]) in microdissected nuclei of the social behavior network. Relative mRNA levels (normalized to the geometric mean of the reference genes *18s* and *g3pdh*) were higher in ascending males compared to the stable subordinate and dominant states. Different letters indicate significant differences among social groups at $p < 0.05$. ns, not significant.

ATn, anterior tuberal nucleus; Ce, cerebellum; Dm, medial part of the dorsal telencephalon; Dl, lateral part of the dorsal telencephalon; Pit, pituitary; POA, preoptic area; Vs, supracommissural nucleus of the ventral telencephalon; VTn, ventral tuberal nucleus; Vv, ventral nucleus of the ventral telencephalon.

A–C, modified from Burmeister et al. (2005), and D from Maruska, Zhang, et al. (2013).

Maruska, Zhang, et al., 2013) (Figures 4.4 and 4.7). Furthermore, circulating levels of LH and FSH protein are also higher by 30 minutes after ascent, suggesting that GnRH1 activation of the pituitary stimulates both the release *and* synthesis of gonadotropins (Maruska et al., 2011). Thus, within minutes of a social opportunity, the brain-pituitary portion of the HPG axis has already been stimulated.

In addition to small GnRH1 neurons and low HPG axis activity, subordinate males also have small testes. However, despite their reduced size, the testes continue to produce sperm during the suppression period, and these males may also retain viable sperm from when they were last dominant (Kustan et al., 2012; Maruska & Fernald, 2011*a*). This is significant, and evolutionarily adaptive, because it allows reproductively suppressed males to immediately spawn with females upon social ascent, without having to wait 5–7 days for the testes to grow or the 10–11 days required for new sperm production (Kustan et al., 2012; Maruska & Fernald, 2011*a*). In fact, behavioral experiments showed that these suppressed subordinate males can successfully spawn and fertilize eggs within hours of social ascent (Kustan et al., 2012). During the subordinate to dominant male social transition, the morphological and structural changes in testicular cell composition and relative testes size takes several days, whereas many molecular changes in the testes are detected more quickly (Huffman, Mitchell, O'Connell, & Hofmann, 2012; Maruska & Fernald, 2011*a*). In the testes, social opportunity triggers rapid (minutes to hours) changes in mRNA levels of some receptor types (FSHR, androgen receptors, corticosteroid receptors), as well as slower (days) changes in other receptor types (LHR, estrogen receptors, and aromatase, the enzyme that converts testosterone to estradiol) (Maruska & Fernald, 2011*a*) (Figure 4.4). This rapid transcriptional response in the most distal component of the HPG axis highlights the sensitivity and plasticity of the entire reproductive system to social information. Thus, there are measurable changes in transcriptional activity from the brain to the testes, all within minutes of a social opportunity, which is much more rapid than previously realized. Furthermore, the swift molecular changes in the testes raise the alternate possibility that there could be other signaling pathways that perhaps bypass the inferred linear cascade from brain GnRH1 release to pituitary LH/FSH release to testicular gonadotropin receptor activation, but this hypothesis requires future testing.

V. Social Regulation of Circulating Steroid Hormones and Steroid Receptor Gene Expression

The product of the up-regulated HPG axis in dominant males is mature testes that, in addition to producing sperm, also produce and release sex steroid hormones, such as androgens, progestins, and estradiol, that can modulate behaviors. Circulating sex steroids, as well as stress hormones, play vital roles in translating social and physiological cues into behavioral responses by acting via both membrane-bound receptors and nuclear steroid receptors that function as transcription factors to modulate gene expression (Sakamoto et al., 2012). Thus, steroids can directly influence behavioral circuits through either a rapid nongenomic mechanism or by modulating the expression levels of many downstream genes that can have important consequences for physiological and phenotypic change. As in other vertebrates, sex steroid and corticosteroid receptors are widespread throughout the brain of *A. burtoni* and therefore have the potential to integrate hormonal state with external social inputs in numerous neural circuits (Greenwood et al., 2003; Harbott, Burmeister, White, Vagell, & Fernald, 2007; Munchrath & Hofmann, 2010). Feedback of sex steroids on the GnRH1 system is also important for regulation of the HPG axis in male *A. burtoni*. For example, androgen receptors are expressed in GnRH1 neurons (Harbott et al., 2007), and androgens, but not estrogens, were shown to regulate GnRH1 cell size (Soma, Francis, Wingfield, & Fernald, 1996). Castrated *A. burtoni* males have hypertrophied GnRH1 neurons (Francis, Jacobson, Wingfield, & Fernald, 1992; Francis et al., 1993; Soma et al., 1996), and the set point for GnRH1 cell size appears to be determined by social cues and then maintained by negative feedback from gonadal androgens (Soma et al., 1996). Future studies are needed, however, to directly test this "social set point hypothesis" and to determine what role individual sex steroid receptor subtypes might have on GnRH1 neuron morphology, plasticity, and function.

Levels of circulating steroids often rapidly respond to social interactions in all vertebrates as part of a physiological response to challenges (i.e., the *challenge hypothesis*), which may also prepare the animal for current and future aggressive and reproductive interactions (Dijkstra, Schaafsma, Hofmann, & Groothuis, 2012; Hirschenhauser & Oliveira, 2006; Oliveira, Hirschenhauser, Carneiro, & Canario, 2002; Wingfield, Hegner, Dufty Jr, &

Ball, 1990). When subordinate male *A. burtoni* are given an opportunity to rise in rank, there is a robust increase in circulating levels of androgens (testosterone and 11-ketotestosterone), 17β-estradiol, and cortisol within just 30 minutes after ascent (Maruska & Fernald, 2010*a*; Maruska, Zhang, et al., 2013). This steroid response may function to activate and maintain aggressive behaviors to allow establishment of his newly acquired status and territory and to prepare for reproduction. There is also evidence for rapid changes in sex steroid receptor mRNA levels in distinct brain regions on this same time scale, suggesting that localized changes in steroid sensitivity may be important during and after social transition (Maruska, Zhang, et al., 2013). Dominant males also maintain higher levels of circulating androgens and 17β-estradiol compared to subordinate males (Maruska & Fernald, 2010*a*, 2010*c*; Parikh et al., 2006*b*), whereas cortisol levels are more variable and often do not show a clear relationship to social status but may depend more on recent behaviors, experimental conditions, or other factors (Chen & Fernald, 2008; Clement et al., 2005; Fox, White, Kao, & Fernald, 1997; Greenwood, Wark, Fernald, & Hofmann, 2008; Maruska & Fernald, 2010*c*). Nevertheless, these fluctuations in circulating steroid levels that occur in response to the social environment likely play important roles in regulating gene expression and the display of appropriate behaviors. Although unexplored in *A. burtoni*, localized changes in steroid production in specific circuits or nuclei in the brain (i.e., neurosteroids) may also play an important role in regulating perception of social information and the output of adaptive behaviors (Do-Rego et al., 2006; Remage-Healey & Bass, 2006; Remage-Healey, London, & Schinger, 2009).

There are also many sex steroid receptor subtype-specific differences in mRNA levels between dominant and subordinate *A. burtoni* males that vary among different brain regions and the pituitary, suggesting a complex regulatory system even within a stable social phenotype (see Table 4.1) (Burmeister et al., 2007; Maruska, Zhang, et al., 2013; O'Connell & Hofmann, 2012). To test what role sex steroids might have in modulating male social behaviors, O'Connell and Hofmann (2012) used agonists and antagonists to different classes of sex steroid receptors and found that androgens and progestins modulate courtship behavior solely in dominant males, whereas estrogens influence aggressive behaviors independent of social status. There was also an apparent dissociation of the behavioral and physiological responses to sex steroid receptor antagonist treatment in dominant males because the robust changes in social behavior observed with antagonist treatment were not reflected in differences in gene expression of sex steroid receptors within the preoptic area (measured by quantitative PCR [qPCR]), whereas, in subordinate males, there were expression differences between control and experimental groups. Further, when the transcriptomes of the preoptic area from fish treated with either vehicle or an estrogen receptor antagonist were compared, there was a greater proportion of genes that changed expression levels in dominant (8.25 percent) compared to subordinate (0.56 percent) males. This study suggests, therefore, that social status acts as a permissive factor for sex steroid regulation of gene expression and complex behaviors, but that socially induced changes at one level of biological organization (e.g. behavioral, hormonal, gene expression) do not simply predict changes at other levels. Rather, there may be status-specific network modules that integrate various aspects of behavior, gene expression, and hormone profiles to regulate male sociality (O'Connell & Hofmann, 2012).

VI. Rapid Social Regulation of Gene Expression in Social Decision-Making Centers in Males

The recognition of a social opportunity likely influences several brain regions that process and ultimately relay the sensory information to the GnRH1 reproductive axis. In addition, however, there is certainly activation of other brain regions involved in the behavioral neural circuitry that prepares the male for his new role as a dominant territory owner, which includes the adoption of over a dozen new behaviors. For example, ascending males that take over a vacant territory must now defend it from other males by engaging in agonistic threat displays, chasing behavior, and fights, as well as begin to perform courtship displays to attract females for spawning. Social behaviors such as these are thought to be coordinated by conserved networks of neural circuits that continuously evaluate the salience of different inputs and contexts to produce adaptive behaviors. One of these defined networks is the *social behavior network* (SBN), a collection of six brain nuclei, or nodes (lateral septum, medial extended amygdala/bed nucleus of the stria terminalis, preoptic area, anterior hypothalamus, ventromedial hypothalamus, midbrain periaqueductal gray/

tegmentum), that are all reciprocally connected and implicated in the regulation of many social behaviors including aggression, parental care, mating and sexual behaviors, and communication (Goodson, 2005; Newman, 1999). The SBN was originally described in mammals (Newman, 1999), but homologous regions have been identified in reptiles (Crews, 2003), fishes (Goodson, 2005; Goodson & Bass, 2002), and birds (Goodson, 2005), and it therefore provides an important evolutionary framework for studying the neural basis of social behaviors (Desjardins, Klausner, & Fernald, 2010; O'Connell & Hofmann, 2011*a*, 2011*b*). Importantly, the nuclei of the SBN all express sex steroid receptors (Newman, 1999), suggesting that these nodes are also crucial neural substrates for integration of social signals with an animal's own internal hormonal state. Furthermore, the SBN is just one example of an identified brain network involved in social behavior, but other core neural networks, such as the *mesolimbic reward system* that are also conserved among vertebrates, may interact with the SBN and others to form a larger "*social decision-making network*" that regulates adaptive behavior (O'Connell & Hofmann, 2011*b*).

How can we identify which brain regions are participating in social behavior? Immediate early genes such as *egr-1, cfos, jun, arc*, and others have been useful tools for identifying activated neurons within the brain across vertebrate taxa (Clayton, 2000; Pfaus & Heeb, 1997). Importantly, however, the IEG response triggered by a specific behavioral context occurs rapidly (within minutes) and is an indicator of *changes* in relative neuronal activity that lead to downstream transcriptional changes in the cell (Clayton, 2000; Kovacs, 2008; Luckman, Dyball, & Leng, 1994). Thus, measuring IEG expression in response to specific novel social stimuli is a valuable first step that can provide important information on which brain regions are involved in the neural circuitry that ultimately leads to behavioral decisions and adaptive phenotypic change. How this IEG response is related or linked to changes in downstream transcriptional activity, however, remains enigmatic in most systems.

To test how the SBN in *A. burtoni* males transitioning from subordinate to dominant states might respond to this social opportunity, we measured mRNA levels of IEGs in microdissected brain regions as a proxy for neuronal activation (Maruska, Zhang, et al., 2013). We discovered that IEG mRNA levels (*egr-1, cfos*) were higher in all SBN nuclei in males that were given an opportunity to rise in social rank compared to stable subordinate and dominant individuals (Figure 4.7). Previous studies using IEGs to examine brain activation in response to social information show different patterns of expression depending on the social context, such that, for example, male aggression and male sexual behavior have distinctly different expression patterns across the SBN nodes (Goodson, 2005; Newman, 1999). In the case of *A. burtoni* ascent, however, the social opportunity is associated with simultaneous reproductive and aggressive/territorial contexts, possibly resulting in a combined activation pattern of all of the SBN nuclei. It is also possible that social opportunity simply causes a general and widespread response to initiate changes in neural and cognitive processing that may be required to maintain the new social rank. We also discovered several rapid (30 minute) region-specific changes in sex steroid receptor mRNA levels induced by social opportunity, most notably in estrogen receptor subtypes in brain areas that regulate social aggression and reproduction, suggesting that estrogenic signaling plays an important role in male social transitions. Several sex steroid receptor mRNA level changes also occurred in regions homologous to the mammalian septal formation and extended amygdala, two areas shared by SBN and reward circuits, suggesting an important functional role in the integration of social salience, hormonal state, and adaptive behaviors. This rapid transcriptional response suggests that the SBN is involved in the integration of social inputs with internal hormonal state to facilitate the transition to dominance, which ultimately leads to improved fitness for the previously reproductively suppressed individual.

VII. Transcriptional Changes Associated with Male–Male Interactions

What type(s) of sensory information from the dominant males might function to suppress other males' behavior and physiology? Chen and Fernald (2011) tested what role visual cues from a larger dominant male have on the social behavior, reproductive physiology, and stress response of a smaller male. Whereas just seeing a larger conspecific male through a clear barrier suppressed dominance behaviors and coloration patterns in the smaller male for a week, expression levels of reproductive and stress-related genes in the brain (GnRH1, corticotropin releasing factor system, arginine vasotocin), circulating androgens, and testes size did not differ from controls and, in fact, were more similar to dominant males. This experiment suggests that visual cues are important for regulation of behaviors but that additional sensory cues (e.g., chemosensory, acoustic, hydrodynamic, tactile) are needed for complete suppression of physiology and reproduction. Thus, males may use an opportunistic strategy and behave and look like a subordinate male to minimize aggressive attacks and potential injury from other males while simultaneously maintaining the physiology of a dominant male in anticipation of an opportunity to regain a territory and reproductive competence in the near future. This idea is further supported by the fact that even subordinate males that have been socially suppressed for 4–5 weeks still maintain activity of their entire HPG axis, including sperm production (Kustan et al., 2012; Maruska & Fernald, 2011*a*; Maruska et al., 2011).

Dominant males frequently engage in fights and border disputes with neighboring males to assert their dominance and position in the hierarchy, but how might these agonistic confrontations be received and processed in the brain? Desjardins and Fernald (2010) used IEGs (*egr-1, cfos*) as a proxy for brain activation to compare the response of dominant males that were allowed to fight with a conspecific male across a clear barrier, those that fought their own mirror image, and those that saw no opponent (control). Although there was no difference in aggressive behavior nor circulating androgens between males that fought a real opponent versus a mirror image, males that fought the mirror had higher levels of IEGs in the medial part of the dorsal telencephalon (Dm; homologous in part to the mammalian pallial amygdala) and the lateral part of the dorsal telencephalon (Dl; homologous in part to the mammalian hippocampus, medial pallium). This differential brain response suggests that the fish considers the mirror image as somehow different from the real opponent and may reflect cognitive distinction.

In contrast to the wealth of information on the physiological consequences of social ascent, relatively little is known about changes in gene expression that occur during social descent when a dominant male loses his territory and is forced to become subordinate (Parikh, Clement, & Fernald, 2006*a*; White et al., 2002; Maruska, Becker, Neboori, & Fernald, 2013). During social decline, coloration, behavioral, and endocrine changes occur relatively quickly; within minutes of losing a territory, previously dominant males show faded body coloration, turn off their eye-bars, perform more submissive behaviors such as fleeing, and have elevated plasma cortisol levels (Parikh et al., 2006*a*; White et al., 2002; Maruska et al., 2013). Several brain nuclei within the SBN are also rapidly activated when males fall in social rank, but the pattern of IEG expression differs from that observed when males rise in rank (Maruska et al., 2013). This suggests that the SBN quickly coordinates the perception of social cues about status that are of opposite valence, and translates them into adaptive phenotypic changes. In contrast to these rapid changes, physiological changes in GnRH1 soma size, *GnRH1* mRNA levels, and testes size take several weeks to decline to stable subordinate male levels (White et al., 2002). In another study, *GnRH1* mRNA levels in the brain were increased at 24 hours after social descent, possibly functioning as a short-term defense mechanism against status loss to help maintain the HPG axis in anticipation of a quick return to dominance (Parikh et al., 2006*a*). Thus, whereas social ascent is associated with rapid changes from the organismal to the molecular level, the molecular mechanisms responsible for social descent and suppression of the HPG axis seem to occur on a slower time scale, possibly as an adaptation to extend reproductive opportunities.

VIII. Transcriptional Changes Associated with Social Learning

A. burtoni exists in an unstable physical and social environment in which foraging and reproductive opportunities change frequently and are linked to an individual's dominance position. Thus, social learning and the act of acquiring necessary information from watching the behaviors of others are extremely important. For example, the importance of social information in the life of *A. burtoni* is highlighted by the discovery that males gain information by watching interactions between other males, and that they use transitive inference (the ability to infer relationships among items or individuals that have not been seen together) to determine their relative position in a dominance hierarchy (Grosenick, Clement, & Fernald, 2007). Furthermore, subordinate males will also change their behavior depending on audience composition, such that when a dominant male is out of view, they will display courtship and dominance behaviors and then cease these behaviors when he returns (Desjardins, Hofmann, & Fernald, 2012). These experiments suggest that males are aware of their own position in the hierarchy, as well as the relative status of other individuals, and can strategically modulate their behavior for reproductive and social advantage. This ability to gain information as a bystander is advantageous because it allows males to decide whether to engage in costly fights with novel competitors. It is not yet known, however, how this complex social information is processed in the brain and how it might influence an individual's molecular phenotype.

Gathering spatial information from the physical environment is also an important task for group-living territorial species such as *A. burtoni*, in which substrate for territories is often a limited resource but crucial for reproductive success. Thus, the ability to recognize suitable and vacant territories, as well as hiding places to avoid predators, is important for survival. To test how the brain might respond to learning a biologically relevant spatial task, male *A. burtoni* were trained to swim through a target hole in a clear barrier that separated them from access to a shelter and proximity to females. mRNA levels of IEGs (*egr-1, cfos, bdnf*) were then measured in socially relevant brain regions, including the SBN, after their tenth learning trial (Wood, Desjardins, & Fernald, 2011). Fish presented with this spatial task fell into three different categories: fish that could be trained (learners), fish that could not be trained (nonlearners), and fish that never attempted the task (nonattempters). Learners had higher IEG mRNA levels (*egr-1, bdnf*) in the brain region typically associated with spatial learning and memory (lateral part of the dorsal telencephalon, Dl; homologous in part to the mammalian hippocampus), lower plasma cortisol levels, and were more motivated to complete the task, whereas nonattempters had the lowest IEG levels in Dl and highest plasma cortisol levels. These results suggest that there could be a continuum in learning types in these fish that is reflected in differential IEG expression patterns in the brain and that stress may play an important role on performance in a spatial task.

IX. Social Regulation of the Brain: Differences in Gene Expression Between Dominant and Subordinate Male States

In addition to the distinct differences in gene expression along the reproductive axis between subordinate and dominant males mentioned earlier, there are also other gene expression differences between stable social states in which the timing of the changes is unknown because the changes have not yet been examined during the social transition. These studies further highlight the molecular phenotypic plasticity in this species and include gene expression differences related to the stress response, reproduction, growth, aggression, homeostatic mechanisms, and sensory perception; these are summarized here and in Table 4.1.

The hypothalamic-pituitary-interrenal axis and corticotropin-releasing factor (CRF) system play a crucial role in the adaptive response to stress across vertebrates (Denver, 2009). In many animals, chronic social stress, such as subordination, is associated with activation of the CRF stress axis that ultimately results in high circulating levels of the stress hormone cortisol (Denver, 2009; Sapolsky, 2005). In *A. burtoni*, subordinate males often have higher circulating cortisol levels compared to dominant males, but this is not always the case, as levels vary considerably among experimental paradigms and social composition (Fox et al., 1997; Greenwood et al., 2008; Maruska & Fernald, 2010*c*). In the brain, dominant male *A. burtoni* have higher mRNA levels of *CRF* compared to subordinate males, but levels of CRF-binding protein (*CRF-BP*) and the two CRF receptor subtypes (*CRF-R1, CRF-R2*) do not differ (Chen & Fernald, 2008). In the pituitary, where CRF protein stimulates the release of adrenocorticotropic hormone (ACTH) and β-endorphin, *CRF-R1* mRNA levels are two-fold higher in dominant males, whereas *CRF-BP* mRNA levels that may block or limit CRF action are two-fold higher in subordinate males (Chen & Fernald, 2008). These results suggest that during a prolonged period (3–4 weeks) of social stress, the CRF system in both the brain and pituitary is down-regulated in subordinate males, possibly as a homeostatic mechanism to maintain stable circulating cortisol levels.

The kisspeptin signaling system plays a crucial role in controlling both GnRH1 neuron activity and the reproductive axis, including puberty, across taxa (d'Anglemont de Tassigny & Colledge, 2010; Hameed, Jayasena, & Dhillo, 2011; Kauffman et al., 2007; Oakley, Clifton, & Steiner, 2009). Kisspeptins, a group of RFamide peptides encoded by the *Kiss1*, and in some species, the *Kiss2* gene, act via their cognate G-protein coupled receptor GPR54 (or kiss1r). In *A. burtoni*, dominant males have higher levels of *kiss1r* (renamed *gpr54-2b*; Tena-Sempere, Felip, Gomez, Zanuy, & Carrillo, 2012) in whole-brain samples compared to subordinate males, but there was no difference in *gpr54-2b* expression in preoptic area GnRH1 neurons between social states quantified via in situ hybridization (Grone et al., 2010). In a more recent study, however, *gpr54-2b* mRNA levels in microdissected preoptic areas were higher at 30 minutes after social opportunity compared to both dominant and subordinate males (Hu, Maruska, & Fernald, 2011), suggesting that the kisspeptin signaling system may be important during the social transition when suppressed males need to quickly up-regulate their reproductive behavior and physiology. The cellular identity of the neurons within the preoptic area that up-regulate *gpr54-2b* expression, however, remains unknown.

Somatostatin is a neuropeptide produced in the preoptic area of the brain that inhibits growth hormone secretion, regulates energy balance and metabolism, and functions as a neuromodulator to influence behaviors (Klein & Sheridan, 2008; Patel, 1999; Trainor & Hofmann, 2006). Dominance status in male *A. burtoni* is associated with larger somatostatin neurons and reduced growth rates (Hofmann et al., 1999; Hofmann & Fernald, 2000), and somatostatin can regulate aggressive behaviors (Trainor & Hofmann, 2006). Dominant males also have higher mRNA levels of somatostatin pre-propeptide and somatostatin receptor type 3 (*sstR3*) in the hypothalamus compared to subordinate males (Trainor & Hofmann, 2007). Furthermore, somatostatin receptor type 2 (*sstR2*) expression is positively correlated with body size in subordinate males but negatively correlated with body size in dominant males (Trainor & Hofmann, 2007), which is consistent with the inhibitory effects of somatostatin on somatic growth. Thus, both social status and body size may regulate gene expression of the somatostatin signaling system in this species in which it likely has multiple roles in regulating dominance behaviors, reproduction, and socially controlled growth.

In teleost fishes, the neuropeptide arginine vasotocin (AVT) (homolog of mammalian arginine vasopressin) is produced by parvocellular, magnocellular, and gigantocellular neurons in the preoptic area, and, in addition to its roles in the stress response, osmoregulation, sensory processing, and social affiliation (Balment, Lu, Weybourne, & Warne, 2006; Dewan, Maruska, & Tricas, 2008; Dewan, Ramey, & Tricas, 2011; Goodson & Bass, 2001; Maruska, 2009; Warne, 2002), is implicated in the control of territorial and reproductive behaviors (Dewan & Tricas, 2011; Oldfield & Hofmann, 2011; Santangelo & Bass, 2006, 2010; Thompson & Walton, 2004); for a detailed discussion of vasopressin and affiliation in voles see the chapter by Barrett and Young, elsewhere in this handbook. In *A. burtoni*, dominant males have higher levels of *AVT* mRNA expression (quantified by in situ hybridization) in the gigantocellular nucleus, whereas subordinate males have higher *AVT* levels in the parvocellular nucleus (Greenwood et al., 2008), suggesting that these AVT neuronal subpopulations may have different functions and regulatory mechanisms depending on social status (Figure 4.8). Furthermore, *AVT* expression in the gigantocellular nucleus was positively correlated with aggressive and reproductive behaviors, whereas expression in the parvocellular nucleus was negatively correlated with these same behaviors. *AVT* expression in the parvocellular nucleus was also positively correlated with the tendency of subordinate males to flee from dominant males. Collectively, these data reveal a complex relationship between *AVT* expression and social status within a single species that likely involves differential functions among the three primary AVT neuronal phenotypes.

In addition to the candidate gene studies just described, microarray experiments in *A. burtoni* have identified specific co-regulated gene modules and functional gene ontology categories associated with male social status (Renn, Aubin-Horth, & Hofmann, 2008). Many genes were differentially regulated in whole-brain samples between dominant and subordinate males, including those involved in cellular metabolism and those encoding structural proteins, synaptic vesicle elements, neuropeptides, cell cycle regulators, transcription factors, and neurotransmitter receptors. Some genes were expressed at higher levels in dominant males, whereas others were up-regulated in subordinate males (Figure 4.9). Interestingly, there was also considerable variation in transcript levels among individuals within a phenotype, as well as sex differences between males and females, a topic that deserves further investigation. This transcriptome-scale analysis in the brain has revealed complex patterns of gene expression associated with dominant and subordinate social phenotypes that warrant further exploration, including finer scale neuroanatomical resolution of status-dependent differences in transcriptional activity.

Social status differences in gene expression also exist in central and peripheral sensory processing regions in male *A. burtoni*, which has important implications for social communication. For example, in the olfactory bulbs, mRNA levels of

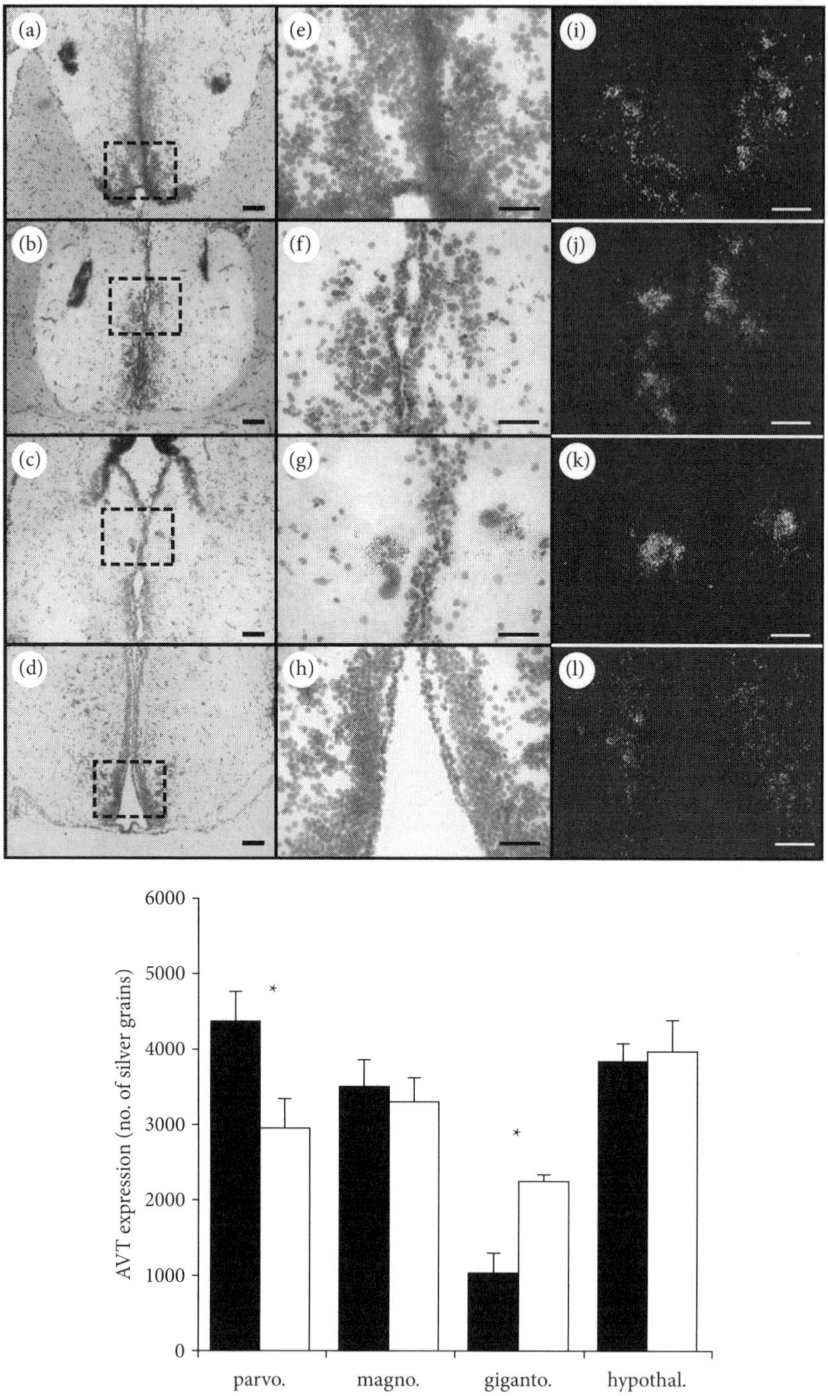

Fig. 4.8 Distribution of arginine vasotocin (*AVT*) mRNA expressing neurons in the preoptic area and hypothalamus of male *Astatotilapia burtoni* revealed by in situ hybridization. (A, E, I) *AVT* expression in the parvocellular population. (B, F, J) *AVT* expression in the magnocellular population. (C, G, K) *AVT* expression in the gigantocellular population. (D, H, L) *AVT* expression in the hypothalamic population. (A–D) Low-power bright field images of the four populations of AVT neurons. (E–H) Higher power images of the regions boxed in A–D, respectively. (I–L) Darkfield images corresponding to the same area as (E–H), respectively. Scale bars in (A–D) are 100 μm and in (E–L) are 50 μm. Bottom graph shows that *AVT* expression levels vary with social status such that levels in the parvocellular population are higher in subordinate males (*filled bars*), whereas levels in the gigantocellular population are higher in dominant males (*open bars*). Data are plotted as mean ± s.e.m. quantified from in situ hybridization, and statistical significance at $p < 0.05$ is indicated with an asterisk.

Modified from Greenwood et al. (2008).

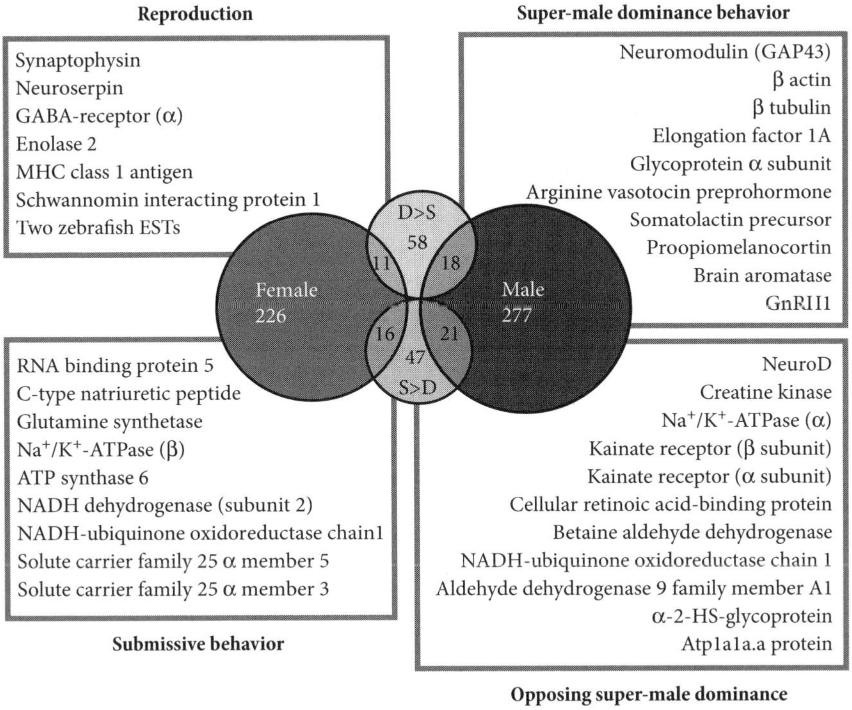

Fig. 4.9 Venn diagram depicting the relationship of sexually regulated and socially regulated genes in the brain of *Astatotilapia burtoni*. These relationships subdivide the gene classes to indicate modules of gene expression that potentially underlie reproduction (*upper left*), submissive behavior (*lower left*), super-male dominance (*upper right*), and opposing super-male dominance (*lower right*). Numbers indicate total unique sequences and unsequenced array features. Gene names given represent best-hit blast annotation for available sequences. The Venn diagram indicates regulation at a Bayesian posterior probability (BPP) of ≥0.99 (the specific BPP for regulation, down to 0.80, is indicated in supplementary material tables S1 and S2 of Renn et al. [2008]). D, dominant male; S, subordinate male. Modified from Renn et al. (2008).

GnRH-R1 are higher in subordinate males, whereas levels of the sex steroid receptor subtypes (*ARα, ARβ, ERα, ERβa, ERβb*) are higher in dominant compared to subordinate males, suggesting that olfactory sensitivity may change with social status (Maruska & Fernald, 2010*b*). mRNA levels of several steroid receptor subtypes (estrogen and corticosteroid receptors) in the main hearing organ of the inner ear (saccule) are also higher in subordinate compared to dominant males, indicating that social status may also be linked to auditory processing capabilities (Maruska & Fernald, 2010*c*; Maruska, Ung, et al., 2012). Since chemosensory and acoustic signaling are important components of the multimodal communicative repertoire of this species, this molecular plasticity has important functional implications. With the recently sequenced genome, it would be instructive as a future step for many of the studies mentioned here to use genetic tools to manipulate the expression levels of individual genes and quantify the resulting effects on behavior and physiology.

X. Social Regulation of Gene Expression in Females

When female *A. burtoni* are ready to reproduce, they must choose a dominant male and enter his territory to spawn. Females likely gather information on their prospective mate and the quality of his territory from multiple sensory channels (visual, chemosensory, acoustic, mechanosensory) prior to any mate choice decisions, but little is known about how this social information influences gene expression in the brain. In a study in which gravid females were first allowed to choose between two socially equivalent dominant males and were then shown a fight between these same two males in which the female's initial preferred male either won or lost, there were different patterns of IEG expression in distinct brain nuclei depending on the outcome of the fight (Desjardins et al., 2010). When females saw their preferred males win a fight, some nuclei within the SBN involved in reproduction (preoptic area and putative partial homolog of the ventromedial hypothalamus) showed higher mRNA levels of

IEGs (*egr-1, cfos*) (Figure 4.10). In contrast, when females saw their preferred male lose a fight, the putative teleost homolog of part of the septal formation (or lateral septum) (Vv, ventral nucleus of the ventral telencephalon), a region associated in part with anxiety-like behaviors, showed higher IEG levels. Although the exact neural circuits and cell types involved are not known, this study demonstrates that females use information from watching males interact that is reflected in specific behaviorally relevant nuclei in the brain.

In contrast to the wealth of information on how social interactions influence gene expression in male *A. burtoni*, relatively little is known in females (Renn, Carleton, Magee, Nguyen, & Tanner, 2009). Females do not typically hold territories but can perform many of the same aggressive and reproductive behaviors seen in males, including lateral displays, chases, frontal threats, digging, and courtship quivers, albeit at a much lower rate. Female *A. burtoni* also appear to form dominance hierarchies, with clear dominant and subordinate individuals that are distinguishable based on behavior and coloration patterns (Renn et al., 2009; Renn, Fraser, Aubin-Horth, Trainor, & Hofmann, 2012). For example, in all female groups, some individuals display male-typical aggressive and courtship behaviors, have higher circulating androgen levels, and have distinctive color patterns, similar to that of dominant males (Renn et al., 2012). Although hierarchical differences certainly exist in females, studies that specifically address how female social status might influence gene expression have not yet been conducted. It is difficult, however, to clearly separate social status from reproductive state in these females because they naturally cycle between a mouthbrooding parental care phase, which also includes a 2-week period of reduced food intake, and a gravid receptive phase every approximately 25–30 days. Whereas the size of GnRH1 neurons in the brain of males is regulated by social status, in females, cell size is correlated with reproductive state, such that spawning females have GnRH1 cells that are two-fold larger than those in females carrying a brood (White & Fernald, 1993). *GnRH1* mRNA levels in the brain of spawning females are also twice as high as the levels in brooding females (White et al., 2002). These studies suggest that although the neuroendocrine circuitry shows a plasticity parallel to that seen in males, the HPG axis in females is primarily regulated by internal reproductive state rather than by exogenous cues from the social environment, as it is in males. A recent study also measured mRNA levels of several neuropeptides and receptors involved in reproduction and feeding in whole brains of gravid, mouthbrooding, and food-deprived female *A. burtoni* (Grone, Carpenter, Lee, Maruska, & Fernald, 2012). This study found that the changes in plasma sex steroid levels and ovary size that occur during mouthbrooding are likely consequences of the food deprivation, whereas mRNA levels in the brain are probably regulated by different mechanisms.

Female reproductive-state differences in gene expression also occur among mouthbrooding, recovering (midcycle at approximately 2 weeks after brood release), and gravid individuals in central and peripheral sensory processing structures. For example, in the olfactory bulbs, mRNA levels of *GnRH-R1* are lower in recovering females compared to both mouthbrooding and gravid individuals, whereas levels of sex steroid receptor subtypes are generally lower in gravid females compared to mouthbrooding and recovering females (Maruska & Fernald, 2010*b*). These results suggest that the first-order olfactory processing center is a substrate for modulation by sex steroids and GnRH, which may function to regulate olfactory perception across the reproductive cycle. We also know that dominant males increase their urine release when exposed to gravid receptive females (Maruska & Fernald, 2012), but how this chemosensory information might influence gene expression in the female's brain or her behavior is not yet known. The main hearing organ of the inner ear (saccule) also appears to be a substrate for modulation by steroid hormones because mRNA levels of sex steroid and corticosteroid receptors vary with the female reproductive cycle (Maruska & Fernald, 2010*c*), and recent experiments show that females have greater hearing sensitivity in the frequency range of the males' courtship sounds when they are ready to spawn compared to when they are brooding (Maruska, Ung, et al., 2012). Thus, female physiology and their perception of the social environment appear to be tightly linked to their reproductive cycle.

XI. Conclusion

In this chapter, we have discussed how social information and position in a dominance hierarchy can modulate an animal's gene expression profiles and change its molecular phenotype, using examples from the model African cichlid fish *A. burtoni*. We provide evidence for the suitability of this model in future studies aimed at understanding the

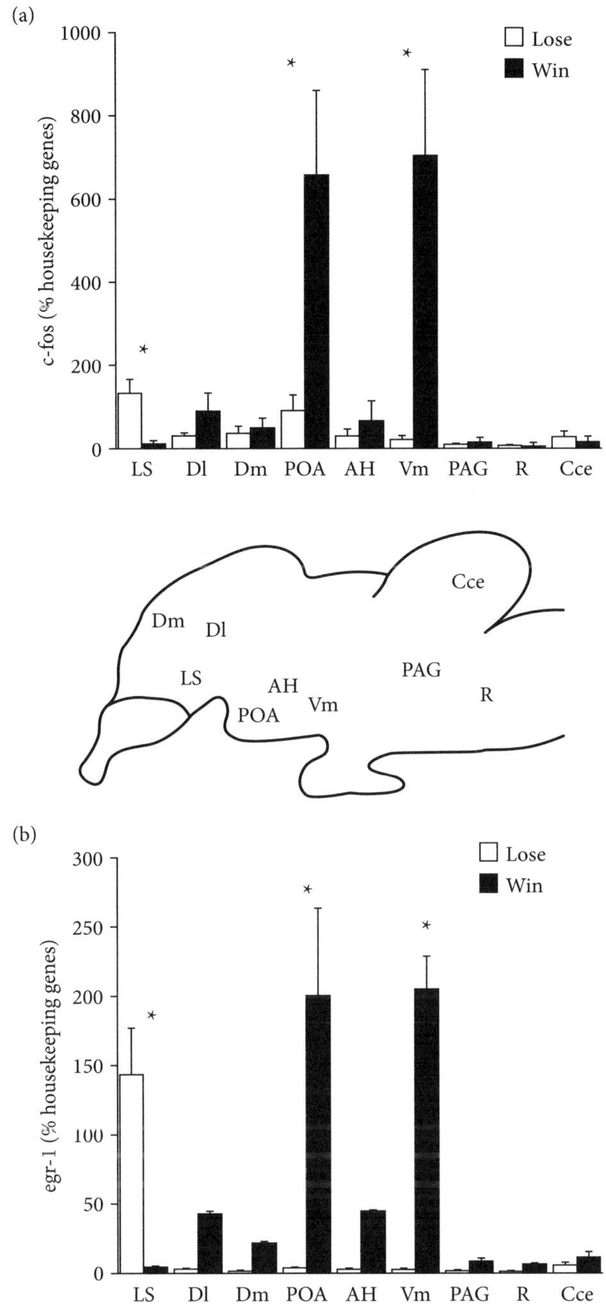

Fig. 4.10 Relative mRNA levels of immediate early genes *cfos* (a) and *egr-1* (b) for each of the six nodes of the social behavior network in *Astatotilapia burtoni*, plotted as a function of whether females saw their preferred males win (*filled bars*) or lose (*open bars*) a fight. Asterisks above pairs of values (mean + SE) indicate significant differences (t-tests, corrected for multiple comparisons). Between panels A and B is a schematic sagittal section of the *A. burtoni* brain showing the approximate locations of the microdissected brain regions (rostral is to the left). AH, anterior hypothalamus; Cce, cerebellum; Dm, medial part of the dorsal telencephalon; Dl, lateral part of the dorsal telencephalon; LS, lateral septum; PAG, periaqueductal gray; R, raphe nucleus; Vm, ventromedial hypothalamus.

Modified from Desjardins et al. (2010).

molecular basis and evolution of social behaviors, and we demonstrate that socially induced changes in gene expression can occur on more rapid time scales than previously appreciated. The phenotypic plasticity in *A. burtoni* makes evolutionary sense because individuals reversibly adapt their behavior and physiology to a changing physical and social environment, thus appropriately allocating resources between reproduction and growth to promote survival and fitness. Individuals that quickly recognize and seize status-gaining social opportunities, in which their physiological responses maximize adaptation on multiple levels (e.g., behavior, hormonal, cellular, molecular), will have a selective advantage. Although we and others have accumulated relevant data on how social information can rapidly alter patterns of behavior and activity of the reproductive axis from the brain to the testes, the details remain enigmatic: what are the regulatory mechanisms that mediate this plasticity? For example, what are the neural pathways from reception of an external social cue that lead to changes in gene expression? And what role do mechanisms such as epigenetics, chromatin remodeling, microRNA regulation, post-transcriptional and post-translational modification, and others play in social regulation of the brain and behavior? We have only begun to scratch the surface of taking advantage of this unique model system, and, with the recently sequenced genome of *A. burtoni*, along with that of other African cichlids, we propose that this species will be integral for addressing future questions on the proximate and ultimate mechanisms underlying complex social behaviors.

Acknowledgments

We thank members of the Fernald lab for discussions and Scott Juntti, Rosa Alcazar, Russ Carpenter, and Caroline Hu for valuable input on an earlier draft of this chapter. Supported in part by National Institutes of Health NIH F32NS061431 (KPM) and NIH NS 034950 (RDF), and National Science Foundation NSF IOS-0923588 (RDF).

References

Almeida, O. G., Miranda, A., Frade, P., Hubbard, P. C., Barata, E. N., & Canario, A. V. (2005). Urine as a social signal in the mozambique tilapia (*Oreochromis mossambicus*). *Chemical Senses, 30*(Suppl 1), i309-310. doi: 30/suppl_1/i309 [pii] 10.1093/chemse/bjh238

Arakawa, H., Blanchard, D. C., Arakawa, K., Dunlap, C., & Blanchard, R. J. (2008). Scent marking behavior as an odorant communication in mice. *Neuroscience and Biobehavioral Reviews, 32*(7), 1236–1248. doi: S0149-7634(08)00076-6 [pii] 10.1016/j.neubiorev.2008.05.012

Au, T. M., Greenwood, A. K., & Fernald, R. D. (2006). Differential social regulation of two pituitary gonadotropin-releasing hormone receptors. *Behavioral Brain Research, 170*(2), 342–346. doi: S0166-4328(06)00139-2 [pii] 10.1016/j.bbr.2006.02.027

Balment, R. J., Lu, W., Weybourne, E., & Warne, J. M. (2006). Arginine vasotocin a key hormone in fish physiology and behaviour: A review with insights from mammalian models. *General and Comparative Endocrinology, 147*(1), 9–16.

Blais, J., Plenderleith, M., Rico, C., Taylor, M. I., Seehausen, O., van Oosterhout, C., & Turner, G. F. (2009). Assortative mating among Lake Malawi cichlid fish populations is not simply predictable from male nuptial colour. *BMC Evolutionary Biology, 9*, 53. doi: 1471-2148-9-53 [pii] 10.1186/1471-2148-9-53

Bliss, S. P., Navratil, A. M., Xie, J., & Roberson, M. S. (2010). GnRH signaling, the gonadotrope and endocrine control of fertility. *Frontiers in Neuroendocrinology, 31*(3), 322–340. doi: S0091-3022(10)00029-4 [pii] 10.1016/j.yfrne.2010.04.002

Burmeister, S. S., Jarvis, E. D., & Fernald, R. D. (2005). Rapid behavioral and genomic responses to social opportunity. *PLoS Biology, 3*(11), e363.

Burmeister, S. S., Kailasanath, V., & Fernald, R. D. (2007). Social dominance regulates androgen and estrogen receptor gene expression. *Hormones and Behavior, 51*(1), 164–170. doi: S0018-506X(06)00269-8 [pii] 10.1016/j.yhbeh.2006.09.008

Chen, C. C., & Fernald, R. D. (2006). Distributions of two gonadotropin-releasing hormone receptor types in a cichlid fish suggest functional specialization. *Journal of Comparative Neurology, 495*(3), 314–323.

Chen, C. C., & Fernald, R. D. (2008). Sequences, expression patterns and regulation of the corticotropin-releasing factor system in a teleost. *General and Comparative Endocrinology*. doi: S0016-6480(08)00141-X [pii] 10.1016/j.ygcen.2008.04.003

Chen, C. C., & Fernald, R. D. (2011). Visual information alone changes behavior and physiology during social interactions in a cichlid fish (*Astatotilapia burtoni*). *PLoS One, 6*(5), e20313. doi: 10.1371/journal.pone.0020313 PONE-D-11-01580 [pii]

Clayton, D. F. (2000). The genomic action potential. *Neurobiology of Learning and Memory, 74*(3), 185–216. doi: 10.1006/nlme.2000.3967 S1074-7427(00)93967-4 [pii]

Clement, T. S., Parikh, V., Schrumpf, M., & Fernald, R. D. (2005). Behavioral coping strategies in a cichlid fish: The role of social status and acute stress response in direct and displaced aggression. *Hormones and Behavior, 47*(3), 336–342.

Crapon de Caprona, M. D. (1974). The effect of chemical stimuli from conspecifics on the behavior of *Haplochromis burtoni* (Cichlidae, Pisces). *Experientia, 30*, 1394–1395.

Crapon de Caprona, M. D. (1980). Olfactory communication in a cichlid fish, *Haplochromis burtoni. Zietschrift fur Tierpsychologie, 52*, 113–134.

Crews, D. (2003). The development of phenotypic plasticity: Where biology and psychology meet. *Developmental Psychobiology, 43*(1), 1–10. doi: 10.1002/dev.10115

Crowley, M. A., Rao, A., Wright, P. J., Illing, N., Millar, R. P., & Clarke, I. J. (1998). Evidence for differential regulation of multiple transcripts of the gonadotropin releasing hormone receptor in the ovine pituitary gland; effect of estrogen. *Molecular and Cellular Endocrinology, 146*, 141–149.

d'Anglemont de Tassigny, X., & Colledge, W. H. (2010). The role of kisspeptin signaling in reproduction. *Physiology (Bethesda), 25*(4), 207–217. doi: 25/4/207 [pii] 10.1152/physiol.00009.2010

Davis, M. R., & Fernald, R. D. (1990). Social control of neuronal soma size. *Journal of Neurobiology, 21*(8), 1180–1188.

Dellovade, T., Schwanzel-Fukuda, M., Gordan, J., & Pfaff, D. (1998). Aspects of GnRH neurobiology conserved across vertebrate forms. *General and Comparative Endocrinology,*

112(3), 276–282. doi: S0016-6480(98)97157-X [pii] 10.1006/gcen.1998.7157
Denver, R. J. (2009). Structural and functional evolution of vertebrate neuroendocrine stress systems. *Annals of the New York Academy of Sciences, 1163*, 1–16.
Desjardins, J. K., & Fernald, R. D. (2010). What do fish make of mirror images? *Biology Letters, 6*, 744–747.
Desjardins, J. K., Hofmann, H. A., & Fernald, R. D. (2012). Social context influences aggressive and courtship behavior in a cichlid fish. *PLoS ONE, 7*(7), e32781.
Desjardins, J. K., Klausner, J. Q., & Fernald, R. D. (2010). Female genomic response to mate information. *Proceedings of the National Academy of Sciences USA, 107*(49), 21176–21180. doi: 1010442107 [pii] 10.1073/pnas.1010442107
Dewan, A. K., Maruska, K. P., & Tricas, T. C. (2008). Arginine vasotocin neuronal phenotypes among congeneric territorial and shoaling reef butterflyfishes: Species, sex and reproductive season comparisons. Journal of Neuroendocrinology, *20*(12), 1382–1394. doi: JNE1798 [pii] 10.1111/j.1365-2826.2008.01798.x
Dewan, A. K., Ramey, M. L., & Tricas, T. C. (2011). Arginine vasotocin neuronal phenotypes, telencephalic fiber varicosities, and social behavior in butterflyfishes (Chaetodontidae): Potential similarities to birds and mammals. *Hormones and Behavior, 59*, 56–66.
Dewan, A. K., & Tricas, T. C. (2011). Arginine vasotocin neuronal phenotypes and their relationship to aggressive behavior in the territorial monogamous multiband butterflyfish, *Chaetodon multicinctus. Brain Res, 1401*, 74–84.
Dijkstra, P. D., Schaafsma, S. M., Hofmann, H. A., & Groothuis, T. G. G. (2012). 'Winner effect' without winning: Unresolved social conflicts increase the probability of winning a subsequent contest in a cichlid fish. *Physiology and Behavior, 105*, 489–492.
Do-Rego, J. L., Acharjee, S., Seong, J. Y., Galas, L., Alexandre, D., Bizet, P.,... Vaudry, H. (2006). Vasotocin and mesotocin stimulate the biosynthesis of neurosteroids in the frog brain. *Journal of Neuroscience, 26*(25), 6749–6760.
Fernald, R. D. (1977). Quantitative behavioral observations of *Haplochromis burtoni* under semi-natural conditions. *Animal Behavior, 25*, 643–653.
Fernald, R. D. (1984). Vision and behavior in an African cichlid fish. *American Scientist, 72*, 58–65.
Fernald, R. D. (2009). Social regulation of reproduction: What changes and why? *Hormones, Brain and Behavior, 1*, 683–691.
Fernald, R. D., & Hirata, N. R. (1977). Field study of *Haplochromis burtoni*: Quantitative behavioral observations. *Animal Behavior, 25*(4), 964–975.
Fernald, R. D., & Maruska, K. P. (2012). Social information changes the brain. *Proceedings of the National Academy of Sciences USA, 109*(Suppl 2), 17194–17199. doi: 10.1073/pnas.1202552109
Flanagan, C. A., Chen, C. C., Coetsee, M., Mamputha, S., Whitlock, K. E., Bredenkamp, N.,... Illing, N. (2007). Expression, structure, function, and evolution of gonadotropin-releasing hormone (GnRH) receptors GnRH-R1SHS and GnRH-R2PEY in the teleost, *Astatotilapia burtoni. Endocrinology, 148*(10), 5060–5071. doi: en.2006-1400 [pii] 10.1210/en.2006-1400
Fox, H. E., White, S. A., Kao, M. H., & Fernald, R. D. (1997). Stress and dominance in a social fish. *Journal of Neuroscience, 17*(16), 6463–6469.
Fraley, N. B., & Fernald, R. D. (1982). Social control of developmental rate in the African cichlid fish, *Haplochromis burtoni. Zietschrift fur Tierpsychologie, 60*, 66–82.
Francis, R. C., Jacobson, B., Wingfield, J. C., & Fernald, R. D. (1992). Hypertrophy of gonadotropin releasing hormone-containing neurons after castration in the teleost, *Haplochromis burtoni. Journal of Neurobiology, 23*, 1084–1093.
Francis, R. C., Soma, K., & Fernald, R. D. (1993). Social regulation of the brain-pituitary-gonadal axis. *Proceedings of the National Academy of Sciences USA, 90*, 7794–7798.
Gelez, H., & Fabre-Nys, C. (2006). Neural pathways involved in the endocrine response of anestrous ewes to the male or its odor. *Neuroscience, 140*(3), 791–800. doi: S0306-4522(06)00326-5 [pii] 10.1016/j.neuroscience.2006.02.066
Goodson, J. L. (2005). The vertebrate social behavior network: Evolutionary themes and variations. *Hormones and Behavior, 48*(1), 11–22.
Goodson, J. L., & Bass, A. H. (2001). Social behavior functions and related anatomical characteristics of vasotocin/vasopressin systems in vertebrates. [Research]. *Brain Research Reviews, 35*(3), 246–265.
Goodson, J. L., & Bass, A. H. (2002). Vocal-acoustic circuitry and descending vocal pathways in teleost fish: Convergence with terrestrial vertebrates reveals conserved traits. *Journal of Comparative Neurology, 448*(3), 298–322.
Greenwood, A. K., Butler, P. C., White, R. B., DeMarco, U., Pearce, D., & Fernald, R. D. (2003). Multiple corticosteroid receptors in a teleost fish: Distinct sequences, expression patterns, and transcriptional activities. *Endocrinology, 144*(10), 4226–4236.
Greenwood, A. K., & Fernald, R. D. (2004). Social regulation of the electrical properties of gonadotropin-releasing hormone neurons in a cichlid fish (*Astatotilapia burtoni*). *Biology of Reproduction, 71*(3), 909–918.
Greenwood, A. K., Wark, A. R., Fernald, R. D., & Hofmann, H. A. (2008). Expression of arginine vasotocin in distinct preoptic regions is associated with dominant and subordinate behaviour in an African cichlid fish. *Proceedings of the Royal Society B: Biological Sciences, 275*(1649), 2393–2402. doi: E752H37117235917 [pii] 10.1098/rspb.2008.0622
Grone, B. P., Carpenter, R. E., Lee, M., Maruska, K. P., & Fernald, R. D. (2012). Food deprivation explains effects of mouthbrooding on ovaries and steroid hormones, but not brain neuropeptide and receptor mRNAs, in an African cichlid fish. *Hormones and Behavior, 62*(1), 18–26. doi: S0018-506X(12)00130-4 [pii] 10.1016/j.yhbeh.2012.04.012
Grone, B. P., Maruska, K. P., Korzan, W. J., & Fernald, R. D. (2010). Social status regulates kisspeptin receptor mRNA in the brain of *Astatotilapia burtoni. General and Comparative Endocrinology, 169*(1), 98–107. doi: S0016-6480(10)00266-2 [pii] 10.1016/j.ygcen.2010.07.018
Grosenick, L., Clement, T. S., & Fernald, R. D. (2007). Fish can infer social rank by observation alone. *Nature, 445*(7126), 429–432. doi: nature05511 [pii] 10.1038/nature05511
Hameed, S., Jayasena, C. N., & Dhillo, W. S. (2011). Kisspeptin and fertility. *Journal of Endocrinology, 208*(2), 97–105. doi: JOE-10–0265 [pii]10.1677/JOE-10–0265
Harbott, L. K., Burmeister, S. S., White, R. B., Vagell, M., & Fernald, R. D. (2007). Androgen receptors in a cichlid fish,

Astatotilapia burtoni: Structure, localization, and expression levels. *Journal of Comparative Neurology, 504*(1), 57–73. doi: 10.1002/cne.21435

Heiligenberg, W., & Kramer, U. (1972). Aggressiveness as a function of external stimulation. *Journal of Comparative and Physiological Psychology, 77*, 332–340.

Heiligenberg, W., Kramer, U., & Schulz, V. (1972). The angular orientation of the black eye-bar in *Haplochromis burtoni* (Cichlidae, Pisces) and its relevance for aggressivity. *Zeitschrift für vergleichende Physiologie, 76*, 168–176.

Hirschenhauser, K., & Oliveira, R. F. (2006). Social modulation of androgens in male vertebrates: Meta-analyses of the challenge hypothesis. *Animal Behaviour, 71*(2), 265–277.

Hofmann, H. A., Benson, M. E., & Fernald, R. D. (1999). Social status regulates growth rate: Consequences for life-history strategies. *Proceedings of the National Academy of Sciences USA, 96*(24), 14171–14176.

Hofmann, H. A., & Fernald, R. D. (2000). Social status controls somatostatin neuron size and growth. *Journal of Neuroscience, 20*(12), 4740–4744.

Hu, C., Maruska, K. P., & Fernald, R. D. (2011). *Social regulation of kisspeptin signaling in Astaotilapia burtoni.* Paper presented at the North American Society for Comparative Endocrinology, Ann Arbor, Michigan.

Huffman, L. S., Mitchell, M. M., O'Connell, L. A., & Hofmann, H. A. (2012). Rising StARs: Behavioral, hormonal, and molecular responses to social challenge and opportunity. *Hormones and Behavior, 61*(4), 631–641. doi: S0018-506X(12)00047-5 [pii] 10.1016/j.yhbeh.2012.02.016

Kauffman, A. S., Park, J. H., McPhie-Lalmansingh, A. A., Gottsch, M. L., Bodo, C., Hohmann, J. G., . . . Rissman, E. F. (2007). The kisspeptin receptor GPR54 is required for sexual differentiation of the brain and behavior. *Journal of Neuroscience, 27*(33), 8826–8835. doi: 27/33/8826 [pii] 10.1523/JNEUROSCI.2099-07.2007

Klein, S. E., & Sheridan, M. A. (2008). Somatostatin signaling and the regulation of growth and metabolism in fish. *Molecular and Cellular Endocrinology, 286*, 148–154.

Kovacs, K. J. (2008). Measurement of immediate-early gene activation- c-fos and beyond. *Journal of Neuroendocrinology, 20*(6), 665–672. doi: JNE1734 [pii] 10.1111/j.1365-2826.2008.01734.x

Kustan, J. M., Maruska, K. P., & Fernald, R. D. (2012). Subordinate male cichlids retain reproductive competence during social suppression. *Proceedings of the Royal Society B: Biological Sciences, 279*, 434–443.

Leong, C.-Y. (1969). The quantitative effect of releasers on the attack readiness of the fish *Haplochromis burtoni. Zeitschrift Fur Tierpsychologie, 65*, 29–50.

Lethimonier, C., Madigou, T., Munoz-Cueto, J. A., Lareyre, J. J., & Kah, O. (2004). Evolutionary aspects of GnRHs, GnRH neuronal systems and GnRH receptors in teleost fish. *General and Comparative Endocrinology, 135*(1), 1–16. doi: S0016648003003538 [pii]

Levavi-Sivan, B., Safarian, H., Rosenfeld, H., Elizur, A., & Avitan, A. (2004). Regulation of gonadotropin-releasing hormone (GnRH)-receptor gene expression in tilapia: Effect of GnRH and dopamine. *Biology of Reproduction, 70*(6), 1545–1551. doi: 10.1095/biolreprod.103.021998 biolreprod.103.021998 [pii]

Lin, C. J., Wu, G. C., Lee, M. F., Lau, E. L., Dufour, S., & Chang, C. F. (2010). Regulation of two forms of gonadotropin-releasing hormone receptor gene expression in the protandrous black porgy fish, *Acanthopagrus schlegeli. Molecular and Cellular Endocrinology, 323*(2), 137–146. doi: S0303-7207(10)00208-X [pii] 10.1016/j.mce.2010.04.003

Luckman, S. M., Dyball, R. E., & Leng, G. (1994). Induction of c-fos expression in hypothalamic magnocellular neurons requires synaptic activation and not simply increased spike activity. *Journal of Neuroscience, 14*(8), 4825–4830.

Maruska, K. P. (2009). Sex and temporal variations of the vasotocin neuronal system in the damselfish brain. *General and Comparative Endocrinology, 160*(2), 194–204. doi: S0016-6480(08)00420-6 [pii] 10.1016/j.ygcen.2008.11.018

Maruska, K. P., Becker, L., Neboori, A.,, & Fernald, R. D. (2013). Social descent with territory loss causes rapid behavioral, endocrine, and transcriptional changes in the brain. *Journal of Experimental Biology, 216*, 3656–3666. doi:10.1242/jeb.088617

Maruska, K. P., Carpenter, R. E., & Fernald, R. D. (2012). Characterization of cell proliferation throughout the brain of the African cichlid fish *Astatotilapia burtoni* and its regulation by social status. *Journal of Comparative Neurology, 520*(15), 3471–3491. doi: 10.1002/cne.23100

Maruska, K. P., & Fernald, R. D. (2013). Social regulation of male reproductive plasticity in an African cichlid fish. *Integrative and Comparative Biology*, doi: 10.1093/icb/ict017

Maruska, K. P., & Fernald, R. D. (2010*a*). Behavioral and physiological plasticity: Rapid changes during social ascent in an African cichlid fish. *Hormones and Behavior, 58*, 230–240. doi: S0018-506X(10)00066-8 [pii] 10.1016/j.yhbeh.2010.03.011

Maruska, K. P., & Fernald, R. D. (2010*b*). Reproductive status regulates expression of sex steroid and GnRH receptors in the olfactory bulb. *Behavioral Brain Research, 213*, 208–217. doi: S0166-4328(10)00341-4 [pii] 10.1016/j.bbr.2010.04.058

Maruska, K. P., & Fernald, R. D. (2010*c*). Steroid receptor expression in the fish inner ear varies with sex, social status, and reproductive state. *BMC Neuroscience, 11*, 58.

Maruska, K. P., & Fernald, R. D. (2011*a*). Plasticity of the reproductive axis caused by social status change in an African cichlid fish: II. Testicular gene expression and spermatogenesis. *Endocrinology, 152*, 291–302.

Maruska, K. P., & Fernald, R. D. (2011*b*). Social regulation of gene expression in the hypothalamic-pituitary-gonadal axis. *Physiology, 26*, 412–423.

Maruska, K. P., & Fernald, R. D. (2012). Contextual chemosensory urine signaling in an African cichlid fish. *Journal of Experimental Biology, 215*, 68–74.

Maruska, K. P., Levavi-Sivan, B., Biran, J., & Fernald, R. D. (2011). Plasticity of the reproductive axis caused by social status change in an African cichlid fish: I. Pituitary gonadotropins. *Endocrinology, 152*, 281–290.

Maruska, K. P., Ung, U. S., & Fernald, R. D. (2012). The African cichlid fish *Astatotilapia burtoni* uses acoustic communication for reproduction: Sound production, hearing, and behavioral significance. *PLoS One, 7*(5), e37612. doi: 10.1371/journal.pone.0037612 PONE-D-12-03270 [pii]

Maruska, K. P., Zhang, A., Neboori, A., & Fernald, R. D. (2013). Social opportunity causes rapid transcriptional changes in the social behavior network of the brain in an African cichlid fish. *Journal of Neuroendocrinology, 25, 145-157.* doi: 10.1111/j.1365-2826.2012.02382.x

Meredith, M., & Fewell, G. (2001). Vomeronasal organ: Electrical stimulation activates Fos in mating pathways and in GnRH neurons. *Brain Research, 922*(1), 87–94. doi: S0006-8993(01)03153-5 [pii]

Millar, R. P. (2005). GnRHs and GnRH receptors. *Animal Reproduction Science, 88*(1–2), 5–28. doi: S0378-4320(05)00129-6 [pii] 10.1016/j.anireprosci.2005. 05.032

Moncaut, N., Somoza, G., Power, D. M., & Canario, A. V. (2005). Five gonadotrophin-releasing hormone receptors in a teleost fish: Isolation, tissue distribution and phylogenetic relationships. *Journal of Molecular Endocrinology, 34*(3), 767–779. doi: 34/3/767 [pii] 10.1677/jme.1.01757

Munchrath, L. A., & Hofmann, H. A. (2010). Distribution of sex steroid hormone receptors in the brain of an African cichlid fish, *Astatotilapia burtoni*. *Journal of Comparative Neurology, 518*(16), 3302–3326. doi: 10.1002/cne.22401

Nelissen, M. H. J. (1977). Sound production by *Haplochromis burtoni* (Gunther) and *Tropheus moorii* Boulenger (Pisces, Cichlidae). *Annales de la Société royale zoologique de Belgique, 106*, 155–166.

Newman, S. W. (1999). The medial extended amygdala in male reproductive behavior. A node in the mammalian social behavior network. *Annals of the New York Academy of Sciences, 877*, 242–257.

Nosil, P., & Harmon, L. J. (2009). Niche dimensionality and ecological speciation. In R. Butlin, J. Bridle & D. Schluter (Eds.), *Speciation and patterns of diversity* (pp. 127–154). Cambridge: Cambridge University Press.

Nosil, P., Harmon, L. J., & Seehausen, O. (2009). Ecological explanations for (incomplete) speciation. *Trends in Ecology and Evolution, 24*(3), 145–156. doi: S0169-5347(09)00024-X [pii] 10.1016/j.tree.2008.10.011

O'Connell, L. A., & Hofmann, H. A. (2011*a*). Genes, hormones, and circuits: An integrative approach to study the evolution of social behavior. *Frontiers in Neuroendocrinology*. doi: S0091-3022(10)00079-8 [pii] 10.1016/j.yfrne.2010.12.004

O'Connell, L. A., & Hofmann, H. A. (2011*b*). The vertebrate mesolimbic reward system and social behavior network: A comparative synthesis. *Journal of Comparative Neurology, 519*, 3599–3639. doi: 10.1002/cne.22735

O'Connell, L. A., & Hofmann, H. A. (2012). Social status predicts how sex steroid receptors regulate complex behavior across levels of biological organization. *Endocrinology, 153*, 1341–1351. doi: 10.1210/en.2011-1663

Oakley, A. E., Clifton, D. K., & Steiner, R. A. (2009). Kisspeptin signaling in the brain. *Endocrine Reviews, 30*(6), 713–743. doi: er.2009-0005 [pii] 10.1210/er.2009-0005

Oldfield, R. G., & Hofmann, H. A. (2011). Neuropeptide regulation of social behavior in a monogamous cichlid fish. *Physiology and Behavior, 102*, 296–303.

Oliveira, R. F., Hirschenhauser, K., Carneiro, L. A., & Canario, A. V. (2002). Social modulation of androgen levels in male teleost fish. *Comparative Biochemistry and Physiology B Biochemistry and Molecular Biology, 132*(1), 203–215.

Parikh, V. N., Clement, T., & Fernald, R. D. (2006*a*). Physiological consequences of social descent: Studies in *Astatotilapia burtoni*. *Journal of Endocrinology, 190*(1), 183–190.

Parikh, V. N., Clement, T. S., & Fernald, R. D. (2006*b*). Androgen level and male social status in the African cichlid, *Astatotilapia burtoni*. *Behavioral Brain Research, 166*(2), 291–295.

Patel, Y. C. (1999). Somatostatin and its receptor family. *Frontiers in Neuroendocrinology, 20*, 157–198.

Pfaus, J. G., & Heeb, M. M. (1997). Implications of immediate-early gene induction in the brain following sexual stimulation of female and male rodents. *Brain Research Bull, 44*(4), 397–407. doi: S0361-9230(97)00219-0 [pii]

Pfaus, J. G., Jakob, A., Kleopoulos, S. P., Gibbs, R. B., & Pfaff, D. W. (1994). Sexual stimulation induces Fos immunoreactivity within GnRH neurons of the female rat preoptic area: Interaction with steroid hormones. *Neuroendocrinology, 60*(3), 283–290.

Remage-Healey, L., & Bass, A. H. (2006). A rapid neuromodulatory role for steroid hormones in the control of reproductive behavior. *Brain Research, 1126*(1), 27–35. doi: S0006-8993(06)01871-3 [pii] 10.1016/j.brainres.2006.06.049

Remage-Healey, L., London, S. E., & Schinger, B. A. (2009). Birdsong and the neural production of steroids. *Journal of Chemical Neuroanatomy*, doi: S0891-0618(09)00116-1 [pii] 10.1016/j.jchemneu.2009.06.009

Renn, S. C., Aubin-Horth, N., & Hofmann, H. A. (2008). Fish and chips: Functional genomics of social plasticity in an African cichlid fish. *Journal of Experimental Biology, 211*(Pt 18), 3041–3056. doi: 211/18/3041 [pii] 10.1242/jeb.018242

Renn, S. C., Carleton, J. B., Magee, H., Nguyen, M. L., & Tanner, A. C. (2009). Maternal care and altered social phenotype in a recently collected stock of *Astatotilapia burtoni* cichlid fish. *Integrative and Comparative Biology, 49*(6), 660–673. doi: icp085 [pii] 10.1093/icb/icp085

Renn, S. C., Fraser, E. J., Aubin-Horth, N., Trainor, B. C., & Hofmann, H. A. (2012). Females of an African cichlid fish display male-typical social dominance behavior and elevated androgens in the absence of males. *Hormones and Behavior, 61*, 496–503.

Robinson, G. E., & Ben-Shahar, Y. (2002). Social behavior and comparative genomics: New genes or new gene regulation? *Genes, Brain and Behavior, 1*(4), 197–203.

Robinson, G. E., Fernald, R. D., & Clayton, D. F. (2008). Genes and social behavior. *Science, 322*(5903), 896–900. doi: 322/5903/896 [pii] 10.1126/science.1159277

Robison, R. R., Fernald, R. D., & Stacey, N. E. (1998). The olfactory system of a cichlid fish responds to steroidal compounds. *Journal of Fish Biology, 53*, 226–229.

Robison, R. R., White, R. B., Illing, N., Troskie, B. E., Morley, M., Millar, R. P., & Fernald, R. D. (2001). Gonadotropin-releasing hormone receptor in the teleost *Haplochromis burtoni*: Structure, location, and function. *Endocrinology, 142*, 1737–1743.

Sakamoto, H., Takahashi, H., Matsuda, K., Nishi, M., Takanami, K., Ogoshi, M., . . . Kawata, M. (2012). Rapid signaling of steroid hormones in the vertebrate nervous system. *Frontiers in Bioscience, 17*, 996–1019. doi: 3970 [pii]

Salzburger, W., Braasch, I., & Meyer, A. (2007). Adaptive sequence evolution in a color gene involved in the formation of the characteristic egg-dummies of male haplochromine cichlid fishes. *BMC Biology, 5*, 51.

Santangelo, N., & Bass, A. H. (2006). New insights into neuropeptide modulation of aggression: Field studies of arginine vasotocin in a territorial tropical damselfish. *Proceedings of the Royal Society B: Biological Sciences, 273*(1605), 3085–3092. doi: 10.1098/rspb.2006.3683

defense, and a lack of the option for polygyny may lead to enhanced reproductive success when a male and female take on a monogamous social structure (Kleiman, 1977; Wittenberger and Tilson, 1980; Mock and Fujioka, 1990; Dunbar, 1995). The characteristics of monogamy vary across the animal kingdom, but three dimensions of a monogamous mating system are widely shared: (1) an essentially exclusive mating relationship, (2) mutual parental care, and (3) preferential association with a specific opposite-sex partner (Wittenberger and Tilson, 1980; Dewsbury, 1987). Not all forms of monogamy necessarily meet each of these three features. Occasionally, males and females may take part in covert matings outside of the pair bond, but they remain "socially monogamous" to their partners. Similarly, monogamous systems do not always include biparental care or frequent social or sexual interactions. As Dewsbury (1987) illustrated in his review of monogamy, the diverse presentation of monogamous-pair bonds can be observed with examples from the rodent lineage (Dewsbury, 1987). A Mongolian gerbil female will seek out a male, develop a preference for him and his home range, and defend the shared territory from curious females (Agren, 1984). However, upon estrus, she leaves the territory in search of other mating partners, and the relationship is not restored until she comes out of estrus. Conversely, elephant shrews appear to mate exclusively and defend a common range (Kleiman, 1977; Rathbun, 1979), but "partners" rarely affiliate, do not show biparental care, and nest separately. Given the varied forms of monogamy across the animal kingdom, a model for the human condition must be wisely chosen and considered in terms of their naturalistic behaviors. Here, we argue that the prairie vole (Figure 5.1a, *Microtus* ochrogaster) provides a unique and valuable opportunity to investigate common underlying neurobiological systems regulating complex social behaviors.

B. Prairie voles as an animal model of complex sociality

Prairie voles display a rich social behavioral repertoire and meet all three of the criteria characteristic of monogamy; they maintain these behaviors in a laboratory setting, making this species well suited as an animal model to understand the biology of complex social behavior. In contrast to the vast majority of mammalian species, including the more commonly researched laboratory mice and rats, prairie voles are among the 3–5% of mammals that form socially monogamous relationships (Kleiman, 1977; Getz Carter, & Gavish, 1981). These pair bonds typically last for an entire lifetime and are characterized by biparental care, with the father contributing almost equally to caring for young (Carter, deVries, & Getz, 1995). Voles display a preference for familiar partners over unfamiliar females and for novel unmated over novel mated animals of the opposite sex, behaviors that encourage both the formation and maintenance of a monogamous mating strategy (Dewsbury, 1987; Ferguson, Fuentes, Sawrey, & Dewsbury, 1986; Shapiro, Austin, Ward, & Dewsbury, 1986). Upon formation of the bond, partners defend a mutual territory, and males display mating-induced aggression toward both male and female intruders (Winslow, Hastings, Carter, Harbaugh, & Insel, 1993; Gobrogge, Liu, Young, & Wang, 2009). Because pair bonding requires the integration of complex social abilities—including identification of socially salient cues, the long-term recognition of a familiar conspecific, and the motivation to preferentially associate with a partner (Lim, Bielsky, & Young, 2005)—the ability to form a pair bond may serve as a general readout of social cognitive performance.

Prairie voles are small hamster-sized rodents with a wide-ranging distribution in the central United States. In the summer breeding months, pairs nest and burrow together, as was discovered from early field studies that repeatedly jointly captured male-female pairs (Getz et al., 1981) and from radio-collared subjects (Figure 5.1b, (Getz and Hofmann, 1986; Hofmann, Getz, & Gavish, 1984; Ophir, Phelps, Sorin, & Wolf, 2008). Communal groups consisting of a male-female pair and their juvenile offspring form in the late autumn-winter months (Carter et al., 1995; Getz and Carter, 1996). Alloparental behavior is commonly performed by these older offspring (Carter et al., 1995), a common monogamous social strategy in family groups with reproductively inhibited juvenile and subadult offspring (Kleiman, 1977), although cases of reproductively active prairie vole offspring are also observed (Mcguire, Getz, Hofmann, Pizzuto, & Frase, 1993). Juveniles leave the nest and show high mortality rates in the summer, accounting for the absence of communal family groups during this season (Getz, Mcguire, Pizzuto, Hofmann, & Frase, 1993).

Prairie voles are well suited for laboratory studies in that they breed well, adapt to laboratory environments, and display pair bonding behavior in the lab, comparable to that observed in the field (Getz et al., 1981; Shapiro et al., 1986; Dewsbury, 1987). The prairie vole pair bond is assessed in

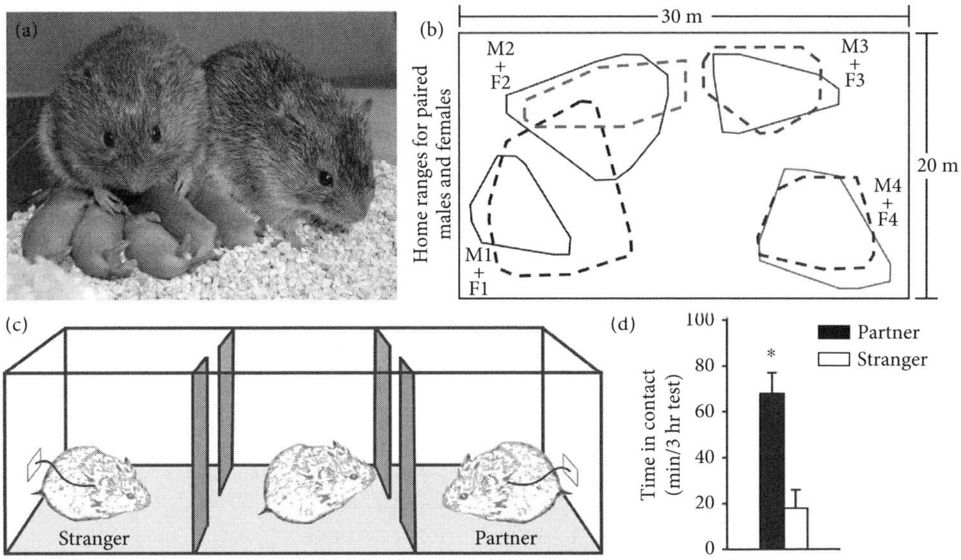

Fig. 5.1 Prairie vole model. (a) Typical prairie vole family units consist of a male-female biparental pair and their offspring. Prairie voles exhibit the three main characteristics of monogamy (essentially exclusive mating, biparental care, preferential association with partner). Photograph by Todd Ahern. (b) In the field, pair bonding can be measured with radiotelemetry and measuring the degree of home range overlap between partners (image from Ophir, Phelps, et al., 2008). (c) Partner preference arena. Subject animal freely roams three-chambered arena and time huddling with a opposite-sex "partner" or novel stimulus "stranger" is recorded over the three-hour test. (d) In the partner preference test, prairie voles spend significantly more time with their familiar "partner" over a novel "stranger."

the laboratory using the "partner preference test." This reliable behavior assay of social bonding permits study of the neural mechanisms that control attachment and monogamy. In the partner preference test, the experimental subject is housed with an opposite-sex "partner" for a defined length of time and subsequently tested for social preference for that "partner" over a novel "stranger" (Figure 5.1c; Williams, Catania, & Carter, 1992). Following the cohabitation, the "partner" and "stranger" stimulus animals are restricted via tethering to either end of a three-chambered arena and the subject has free access to all chambers. The amount of time the subject spends in close proximity or huddling with the partner is quantified over the three-hour test (Figure 5.1d). Using a computer coding system that has been validated against human scoring methods, high-throughput screening of up to 36 tests in a 24-hour period can be performed (Ahern, Modi, Burkett, & Young, 2009). In both sexes, mating facilitates the formation of a partner preference (operationally defined as spending twice as much time with the partner than the stranger), although lengthened cohabitation without mating can lead to a partner preference (Williams et al., 1992; DeVries and Carter, 1999). Pharmacological or other manipulations can be performed in the experimental subject during the cohabitation period in order to interrogate the neurobiological substrates of social-bond formation.

Another valuable feature of the prairie vole model is the ability to perform comparative studies. Differences in sociality between closely related species have revealed valuable insight to the factors underlying and characterizing monogamy. Research into the neural control of complex behavior has taken advantage of the divergent social structures within the *Microtus* genus by comparing the highly social prairie voles to the minimally social meadow (*M. pennsylvanicus*) and montane (*M. montanus*) voles. These closely related species are promiscuous, uniparental, and relatively asocial, providing a unique evolutionary comparison to the highly social organization of the prairie vole. Males and females defend separate areas and nest alone during the breeding season (Jannett, 1982). Promiscuous male home ranges encompass the territories of multiple females, and many males converge to ranges of females in estrus (Madison, 1980; Jannett, 1982). These species do not form pair bonds in the laboratory, as montane voles show no preference for a familiar opposite sex mate (Shapiro et al., 1986). As will be discussed, comparative studies of these behaviorally divergent species have helped to reveal the neural characteristics that give prairie voles the ability to form pair bonds.

II. Neuropeptidergic control of the pair bond and proximate behaviors

Early studies of the neural control of attachment looked toward the neuropeptides oxytocin and vasopressin because of their implications in a number of the social and sexual behaviors related to the initiation and maintenance of the pair bond. Vasopressin (AVP) and oxytocin (OT) are nonapeptides, differing by only two amino acids, produced in the paraventricular (PVN) and supraoptic nuclei (SON) of the hypothalamus, where they are released into the periphery to coordinate a variety of effects. Peripheral OT regulates the initiation of labor and stimulation of milk ejections, and peripheral AVP induces vasoconstriction and water reabsorbtion. Oxytocin is also released centrally from projections arising in the PVN and SON (Ross, Cole, et al., 2009), whereas central AVP arises from the hypothalamus in addition to sexually dimorphic neurons in the medial amygdala (MeA) and the bed nucleus of the stria terminalis (BNST) (Devries and Buijs, 1983). Before being implicated in monogamy, AVP was long known to play a role in facilitating social cognition (Dantzer, Bluthe, Koob, & Le Moal, 1987; de Wied and van Ree, 1989), and territorial flank-marking displays in golden hamsters (Ferris, Albers, Wesolowski, Goldman, & Luman, 1984).

Similarly, oxytocin was known to regulate social memory, promote affiliation (Van Wimersma Greidanus, Kroodsma, Pot, Stevens, & Maigret, 1990), and show pregnancy-induced increases in specific brain regions (Insel, 1990). Most importantly, it was linked to another important social relationship: the mother-infant bond. Central OT release coordinates peripheral events associated with birth with the behavioral changes required for the initiation of maternal care (Ross, Cole, et al., 2009). Virgin female rats will avoid or attack pups, but toward the end of pregnancy, rats will instinctively become maternal. This pregnancy-induced change in the dam's motivation to nurture are due in part to OT release since infusion of OT in virgin female rats induces maternal responses to novel pups (Pedersen, Ascher, Monroe, & Prange, 1982). In sheep, OT not only promotes maternal nurturing but also promotes a mother's bond specific to her own offspring (Kendrick, Keverne, & Baldwin, 1987). Using a variety of molecular techniques including pharmacological manipulations, immunochemical and histological mapping, viral vector technology, transgenic development, and genetic screens, the role of OT and AVP have been strongly related to pair bond formation and the behaviors related to monogamy (Young and Wang, 2004).

A. Species differences in neuropeptide circuitry

Comparative studies between *Microtus* species have uncovered drastic differences in the regulation and function of neuropeptidergic systems. Though both monogamous and promiscuous vole species display a conserved distribution of oxytocin peptide within the brain (Wang, Zhou, Hulihan, & Insel, 1996), the distribution of oxytocin receptors is strikingly different between prairie voles and their promiscuous relatives (Figure 5.2a,5.b). Prairie voles have high densities of OT receptor in the nucleus accumbens (NAcc), a brain region involved in reward, reinforcement, and addiction, as well as the BNST and medial prefrontal cortex (mPFC), whereas receptors are largely absent in NAcc in nonmonogamous species (Insel and Shapiro, 1992). In contrast, meadow and montane voles display elevated levels of OTR in the ventromedial hypothalamus and lateral septum (LS) compared to prairie voles (Insel and Shapiro, 1992).

In parallel to the OT circuitry among *Microtus* species, AVP projections are nearly identical between promiscuous and monogamous voles (Wang et al., 1996). Instead, divergent vasopressin V1a receptor (V1aR) distributions are thought to give rise to different behavioral phenotypes (Figure 5.2d,e). In prairie voles, V1aR binding is comparatively high in certain regions involved in socioemotional and reward processing, such as the anterior olfactory nuclei, cingulate cortex, amygdala, and ventral pallidum (Insel, Wang, & Ferris, 1994; Young, Winslow, Nilsen, & Insel, 1997; Lim, Murphy, & Young, 2004). Receptor binding in prairie and montane voles overlaps with *Avpr1a* mRNA, indicating that expression is regulated at a transcriptional level (Young et al., 1997). Some of this variation may be attributed to an upstream microsatellite region made of short repeat sequences more than 300 base pairs longer in monogamous vole species (Young, Nilsen, Waymire, MacGregor, & Insel, 1999). Although variation in the microsatellite may contribute to species differences in V1aR distribution, it does not explain species differences in social behavior among all vole species (Fink, Excoffier, & Heckel, 2006). Transgenic mice expressing the prairie vole vasopressin receptor gene (*Avpr1a*) neuroanatomical expression display a pattern similar to that of the prairie vole, and enhanced affiliation in response to AVP injections, whereas wild-type mice do not (Young et al., 1999).

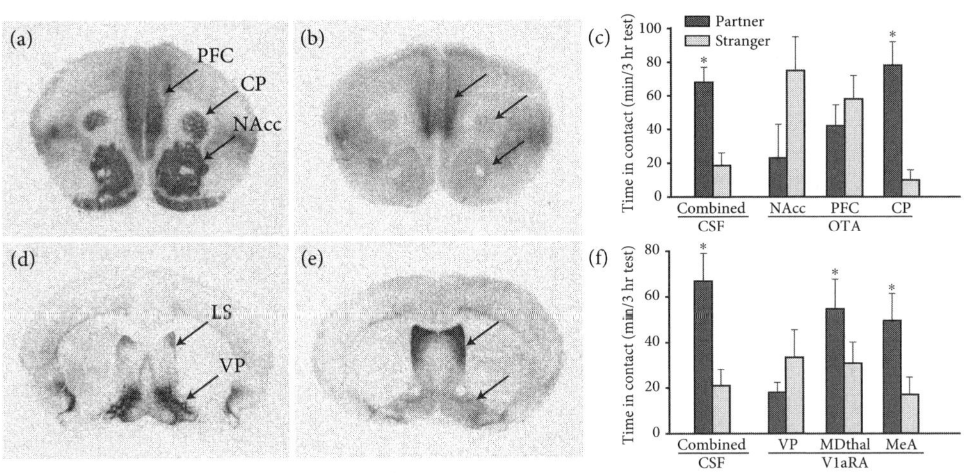

Fig. 5.2 Comparative differences in neuropeptide receptor expression and functional effects on behavior. (a,b) Monogamous prairie voles (a) have higher densities of OTR in the nucleus accumbens (NAcc) and caudate putamen (CP) than do nonmonogamous montane voles (b). Both species have OTR in the prefrontal cortex (PFC). (c) Infusion of a selective V1aR antagonist (V1aRA) into the VP, but not into the mediodorsal thalamus (MDthal) or medial amygdala (MeA), prevents mating-induced partner-preference formation in male prairie voles (Lim and Young, 2004). (d,e) Male prairie voles (d) have higher densities of V1aR in the ventral pallidum (VP) than do montane voles (e). (f) A selective OTR antagonist (OTA) infused bilaterally into the NAcc or PFC, but not the CP, blocks partner-preference formation in female prairie voles (Young et al., 2001). (modified from Young and Wang, 2004).

Similar differences in V1aR expression are found in comparisons of other groups of closely related, but behaviorally divergent species. Monogamous marmosets and California mice (*Peromyscus californicus*) display V1aR levels in the ventral pallidum that are absent in nonmonogamous rhesus macaques and white-footed mice (*P. leucopus*) (Wang, Toloczko, et al., 1997; Bester-Meredith, Young, & Marler, 1999). Thus, selection of species with specific neural V1aR patterns may have led to the convergent evolution of a monogamous social strategy across multiple taxa (Lim and Young, 2004). However, V1aR distribution may not explain the evolution of monogamy in all other species (Turner et al., 2010).

After the formation of a pair bond, central AVP changes are observed in monogamous, but not promiscuous, male voles. In male prairie voles, but not in females or meadow voles, vasopressin immunoreactive staining decreases in the lateral septum and lateral habenular nucleus after a mated cohabitation with a female and again after parturition (Bamshad, Novack, & deVries, 1993; Bamshad, Novack, & deVries, 1994). A decrease in immunoreactive fiber density may suggest a release of AVP peptide content from these terminals. Importantly, a concurrent increase in AVP mRNA in the BNST, a source of AVP to the LS and LH (Wang, Ferris, & deVries, 1994), suggests that AVP is released during pair bond formation, which may act to coordinate associated behaviors including partner preference, aggression, and paternal care. Activation of AVP circuitry through an injection of AVP into the brain has differing behavioral effects in monogamous and promiscuous species, because it enhances female-directed affiliation and resident-intruder aggression in prairie male voles but not meadow or montane vole males (Young et al., 1997; Young et al., 1999).

B. Partner preference

Originally inspired by the effects of oxytocin in maternal bonding, Sue Carter and colleagues explored the possibility that the same molecule might promote the formation of the pair bond between the female prairie vole and her male partner. In the first study to find functional effects of OT in voles, Witt (1990) reported enhanced grooming and huddling with a male partner after central OT administration to female prairie voles. They went on to investigate whether this enhancement of affiliation during cohabitation promotes bond formation. Pairing a female prairie vole with a male for six hours without mating is not sufficient for bond formation; however, infusion of OT into the brain during the shortened nonsexual cohabitation period pairing induces the female to bond with the male (Williams, Insel, Harbaugh, & Carter, 1994), and this effect is blocked by co-administration of an oxytocin antagonist (Cho, deVries, Williams, & Carter, 1999). During the partner preference test,

OT-treated females spend significantly more time with the familiar male partner than with a novel male. Furthermore, pretreatment with an OT antagonist alone will prevent OT- or mating-induced partner preference formation (Williams et al., 1994; Insel and Hulihan, 1995), demonstrating that OT is necessary for the formation of the bond. In fact, OT levels in microdialysates are elevated in females during cohabitation and mating with a male (Ross, Cole, et al., 2009). Activation of the oxytocinergic system thus accelerates the formation of the pair bond and overcomes the absence of mating during a shortened cohabitation period. Thus, the same neurobiological systems that promote maternal nurturing and mother-infant bonding play a critical role in regulating pair bonding in female prairie voles.

Although oxytocin may also contribute to pair bonding in male prairie voles (Cho et al., 1999), a more convincing role of AVP has been demonstrated in the biology of male social bonding. Both OT and AVP injections accelerate pair bonding, even with only a 1-hour cohabitation, in male prairie voles, and this enhancement is blocked by pretreatment with OT and AVP antagonists (Cho et al., 1999). However, only V1aR antagonists administered before mating block the formation of a pair bond in males (Winslow et al., 1993), indicating that AVP, but not OT, is required for male attachment formation.

Using site-specific pharmacological and viral manipulations, regions that show differences in receptor expression between *Mictrotus* species have also been identified as the critical sites of action for OT and AVP neurotransmission in pair bonding. OT antagonists administered directly to the nucleus accumbens (NAcc) or PFC, but not to the caudate-putamen, prevent the formation of pair bonds in female prairie voles (Figure 5.2c; Young, Gobrogge, Liu, & Wang, 2001). Pharmacological blockade of V1aR in the ventral pallidum or lateral septum, but not the medial amygdala or mediodorsal thalamus, inhibits the development of a partner preference in males (Figure 5.2f; Liu et al., 2001; Lim and Young, 2004). Increasing V1aR density in the ventral pallidum of male or OTR expression in the nucleus accumbens of female prairie voles, using viral vector gene transfer, promotes affiliation and accelerates pair bond formation (Pitkow et al., 2001; Ross, Freeman, et al., 2009). Overexpression of V1aR in the ventral pallidum of male prairie voles enhances mating-induced Fos neurotransmission, and, thus, likely enhances social processing during cohabitation (Lim & Young, 2004). Astonishingly, performing this viral manipulation on V1aR expression in the ventral pallidum of male meadow voles transforms social behavior and causes this normally nonmonogamous species to form partner preferences (Lim, Wang, et al., 2004), essentially recreating an evolutionary event in the lab.

C. Neuropeptide regulation of behaviors associated with pair bonding

In addition to the control of the pair bond itself, OT and AVP have been implicated in the behaviors associated with a monogamous mating strategy, including paternal care, alloparental care, spontaneous maternal behavior, and mating-induced aggression. During pregnancy and parturition, female rats exhibit plasticity in OT receptor (OTR) expression and are primed to respond to pups possibly through these changes in OT signaling (Insel, 1990). However, males do not experience pregnancy-related changes, and AVP signaling may, instead, have evolved to promote male parental care (De Vries & Villalba, 1997). A decrease in AVP fiber density in the lateral septum of prairie vole fathers is suggestive of vasopressin release, and is associated with a concurrent increase in the expression of paternal behaviors (Bamshad, Novak, & de Vries, 1993). However, as females enter postpartum estrus during this time, it is unknown whether this release is due to pup exposure or to the act of mating. In order to obtain a read-out of general neural activity controlling paternal care, Fos expression was mapped in response to pup exposure in male voles (Kirkpatrick, Kim, & Insel, 1994). The AOB, MeA, BNST, MPOA, and LS showed elevations in Fos activity, and all of these regions display V1aR expression. Additional work supports the notion that vasopressinergic activity in these areas controls paternal behavior. Site-specifically injecting AVP into the lateral septum of sexually naïve male prairie voles promotes paternal responsiveness to novel pups, whereas a V1aR antagonist reduces it (Wang et al., 1994). Sexually naïve male prairie voles contacted and crouched over pups more than did saline-injected controls, and antagonist-injected males groomed pups less. Though more evidence has pointed to a role of the lateral septum in mediating paternal care, males with bilateral lesions of the medial amygdala or olfactory bulbs display impairments in paternal behavior (Kirkpatrick, Carter, Newman, & Insel, 1994; Kirkpatrick, Kim, & Insel, 1994). AVP may also play a role in maternal behavior, maternal memory, and maternal aggression (Bosch and Neumann, 2008; Nephew and Bridges, 2008; Bosch, Pfortsch, Beiderbeck, Landgraf, & Neumann, 2010). Brattleboro rats are deficient in AVP and display inferior parental care (Engelmann and Landgraf, 1994). Similarly, injections of AVP promote maternal care in females (Pedersen et al., 1982).

Another social behavior in the monogamous vole repertoire, alloparenting by reproductively naïve juveniles in the communal nest, has also been linked to the OTR system. Within prairie vole juvenile females, levels of OTR expression in the NAcc correlate positively with alloparental behavior (Olazabal and Young, 2006). Although virgin adult females also display spontaneous maternal care of pups, the display of this behavior is highly variable within the species, with 40% attacking or ignoring pups (Roberts, Miller, Taymans, & Carter, 1998). However, some of this variability may be accounted for by OT receptor distribution, since females with high OTR expression in the NAcc display spontaneous maternal care, and OTR blockade in this region prevents spontaneous maternal behavior (Olazabal and Young, 2006). Interestingly, in adult female prairie voles, upregulation of OTR expression with viral vector gene transfer into the NAcc does not promote alloparental care (Ross, Freeman, et al., 2009). However, overexpression in prepubertal juvenile females enhances adult female alloparenting, suggesting a role for long-term developmental OTR signaling (Keebaugh and Young, 2011).

When prairie vole males form a pair bond, they display not only a social preference for their partner, but also selective aggression toward novel male and female conspecifics (Winslow et al., 1993), another behavior under control of neuropeptides. Central AVP injections stimulate male-directed aggression in paired males even in the absence of mating, whereas blocking V1aR prevents mating-induced aggression (Winslow et al., 1993). V1aR antagonists do not block aggression in established breeders, and, thus, they play a role in the initiation but not the expression of this behavior. More recently, anterior hypothalamic (AH) AVP has been implicated as necessary and sufficient in initiating selective aggression toward females (Gobrogge et al., 2009). AVP is released in the AH in pair-bonded males displaying selective aggression toward intruders, and the amount of aggression correlated with AVP levels in the microdialysates from the AH during the social encounters. Furthermore, AVP injections or overexpression of V1aR using viral gene transfer into the AH promote selective aggression in sexually naïve males toward a novel female. Given the role of AVP in promoting both territoriality, including mating-induced aggression in voles and flank- marking behavior in golden hamsters (Ferris et al., 1984), and bond formation primarily in males, it is possible that male pair bonding evolved through a system that supports territorial protection of his female partner and their offspring (Young and Flanagan-Cato, 2012).

III. Dopaminergic regulation of monogamy

OT and AVP appear to be tightly connected with dopaminergic signaling in prairie voles. Mesocorticolimbic dopamine pathways project from the ventral tegmental area to the NAcc, caudate putamen, and PFC (Wise, 2002). Oxytocin receptors are highly expressed in the NAcc of prairie voles, and this region projects to the ventral pallidum, the site of action of AVP. Mesolimbic dopamine is critical in processing the reward of behaviors critical to life such as feeding and mating, and it is also responsible for the addictive power of drugs of reward (Di Chiara, 2002; Koob and Nestler, 1997; Wise, 2002). In prairie voles, behavioral studies have shown that that dopaminergic activity in mesocorticolimbic regions modulates pair bond formation and maintenance, selective aggression, and paternal behavior (Aragona, Liu, Curtis, Stephan, & Wang, 2003; Aragona et al., 2006; Lonstein, 2002; Wang et al., 1999; Young et al., 2011). The dopamine system varies between species of *Microtus*, and it may contribute to species differences in behavior. In addition to the ventral tegmentum, dopamine cell bodies are also found in the BNST and MeA of prairie voles, but nonmongamous meadow voles exhibit very little labeling in these regions (Northcutt, Wang, & Lonstein, 2007). Furthermore, D2 dopamine receptor (D2R) levels are more densely expressed in the PFC of prairie vole brains (Smeltzer, Curtis, Aragona, & Wang, 2006), whereas D1 receptor (D1R) levels are higher in the PFC and NAcc of promiscuous meadow voles (Aragona et al., 2006). High expression of D1R may inhibit the display of affiliative behaviors in meadow voles, as administration of D1R antagonists promotes social affiliation in these animals (Aragona et al., 2006).

Mating induces a release of dopamine in the NAcc of both sexes in prairie voles (Gingrich, Liu, Cascio, Wang, & Insel, 2000; Aragona et al., 2003), and this release is critical for pair bond formation. Central dopamine blockade before cohabitation using haloperidol inhibits partner preference in both sexes (Wang et al., 1999; Aragona et al., 2003).

An enhancement of dopamine signaling with a low dose of the nonselective dopamine agonist apomorphine promotes pair bond formation after a shortened unmated cohabitation (Wang et al., 1999; Aragona et al., 2003). The facilitating effect of dopamine on pair bonding can be attributed to D2 receptor activation, as pharmacological activation of D2R in the NAcc shell accelerates pair bond formation, whereas blockade prevents it in both prairie vole sexes (Gingrich et al., 2000; Aragona et al., 2006). Conversely, activation of D1R in the NAcc shell inhibits bond formation, further suggesting an inhibitory role of D1R in affiliative behaviors (Aragona et al., 2006).

Neuroplasticity and experience-induced changes in the dopamine system may account for the transformation in behavior after bond formation. After two weeks of cohabitation, D1R levels increase in the NAcc, but not caudate putamen, of male prairie voles (Aragona et al., 2006). This elevation of D1R expression may play a role in maintaining the pair bond by inhibiting a male's motivation to affiliate with a stranger female by decreasing the rewarding aspects of novel sexual encounters. Moreover, the temporal specificity in D1R changes may induce the behavioral transition to selective aggression toward stranger females. Twenty-four hours of cohabitation is not sufficient to alter D1R expression or induce selective aggression, indicating D1R plays a role in the maintenance of the pair bond and proximate behaviors, but not the early formation (Aragona et al., 2006). In support of a role in aggression, blockade of NAcc D1R signaling abolishes stranger female-directed aggression in paired male prairie voles. Similarly, inhibiting kappa-opioid receptor signaling in the NAcc shell prevents mating-induced aggression in both sexes, suggesting the opioid system is also an important regulating mechanism in the maintenance of the pair bond and associated behaviors (Resendez, Kuenmuench, Krzyosinski, & Aragona, 2012).

IV. Social cognition, reward, and the neural circuitry of social bonding

OT and AVP are essential for social information processing and recognition of conspecifics in mice and rats (Bielsky Hu, Szegda, Westphal, & Young, 2004; Dantzer et al., 1987; Ferguson et al., 2000; Le Moal, Dantzer, Michaud, & Koob, 1987; Popik and van Ree, 1991; Winslow and Insel, 2004). An important integrator of incoming social olfactory information is the medial amygdala (MeA), which receives direct input from the olfactory bulbs, and is among the first nodes in the forebrain to process conspecific odors. In prairie voles, the MeA is required for the social drive to affiliate with a partner and display paternal care (Kirkpatrick, Carter, et al., 1994). This region processes incoming information during sociosexual encounters, as cohabitation, mating, and pup exposure elicit MeA c-fos activity (Cushing, Mogekwu, Le, Hoffman, & Carter, 2003; Kirkpatrick, Kim, et al., 1994; Lim and Young, 2004). OT activity in the MeA during an initial social encounter is necessary for the formation of a social memory (Ferguson, Aldag, Insel, & Young, 2001).

The MeA is also a critical extrahypothalamic site of vasopressin cell bodies, which project out to forebrain regions including the ventral pallidum and lateral septum (LS) (Devries and Buijs, 1983). The LS is reciprocally connected to the hippocampus and receives information about social cues, thus priming it to stamp in the memory of a social encounter. This region also projects to hypothalamic areas, including the MPOA, and may thus regulate male sexual behavior (Lonstein, 2002). Blockade of V1aR in the LS, via application of antisense oligonucleotides or antagonists, blocks social recognition in rats and mice (Bielsky, Hu, Ren, Terwilliger, & Young, 2005; Landgraf, Gerstberger, Montkowski, Probst, & Wotjak, 1995). Overexpression of V1aR in the LS of wildtype mice and rats enhances social recognition and the interexposure interval at which subjects can still recognize conspecifics (Bielsky et al., 2005; Landgraf et al., 1995). Further in support of a role in social memory, overexpression of V1aR in the LS restores social recognition in V1aR knockout mice (Bielsky et al., 2005).

In addition to acting in these social-information-processing regions, OT and AVP converge on areas involved in the mesolimbic dopaminergic reward pathway, and concurrent activation of the systems is necessary for bond formation. Pharmacological activation of D2-type dopamine receptors in the NAcc accelerates partner preference formation, an effect blocked by OT antagonist administration (Gingrich et al., 2000; Liu and Wang, 2003). Similarly, D2R blockade in the NAcc prevents oxytocinergic acceleration of bond formation (Liu & Wang, 2003). Direct interactions between the two systems may occur as both dopamine and OT induce release of the other (Melis, Argiolas, & Gessa, 1989; Pfister and Muir, 1989). Co-activation of AVP and dopamine systems may also be necessary for pair bonding in males, because meadow voles with an overexpression of V1aR in the ventral pallidum are induced to pair bond, but D2R blockade prevents this enhanced bond formation (Lim, Wang et al., 2004). In addition to dopaminergic reinforcement, new evidence has pointed to the necessity of opioid reward signaling in bond formation (Burkett, Spiegel, Inoue, Murphy, & Young, 2011). Blockade of mu-opioid receptors before cohabitation with a male blocks the formation of a partner preference in female prairie voles. Thus, it appears the concurrent activation of OT, AVP, dopamine, and opioids are all required in the formation of a pair bond.

OT and AVP are thought to promote bond formation by (a) increasing the saliency of incoming social information, (b) enhancing social motivation and approach, and (c) converging the olfactory signature of a mate on dopaminergic reward pathways (Figure 5.3). High concentrations of OTR in the NAcc and V1aR in the ventral pallidum in prairie voles may serve to link up the olfactory signature of a mate with dopaminergic and opioid-mediated reward and reinforcement, in essence yielding a conditioned partner preference (Burkett et al., 2011; Young and Wang, 2004; Young et al., 2005;). Mating-induced mesolimbic dopamine release resulting from somatosensory information from the genitalia during sexual activity (unconditioned stimulus) may assign a rewarding value to the neuropeptide-enhanced social cues of the partner's specific olfactory signature (conditioned stimulus) (Young and Wang, 2004). Although nonmonogamous mammals experience dopaminergic and opioid reward, the absence of the link with neuropeptidergic systems does not yield an enduring specific association with a partner. These neuroanatomical findings suggest that pair bonding is very similar to an addiction (Burkett et al., 2011). Indeed, when separated from the partner, voles exhibit behaviors characteristic of withdrawal and depression (Bosch, Nair, Ahern, Neumann, & Young, 2009; Pizzuto and Getz, 1998).

V. The link between stress, anxiety, social loss, and pair bonding

The stress axis is another system that appears to regulate prairie vole sociality. The expression of monogamy may be regulated by variability in environmental pressures, and one of the main pathways by which the environment shapes behaviors is through stress. In fact, a demanding dry environment with limited food and water has also been implicated in the evolution of a mating strategy in which prairie vole pairs can combine efforts and share resources (Carter et al., 1995 Getz, 1978; Mcguire et al., 1993; Wang and Novak, 1992). Conversely, *M. montanus* and *M. pennsylvanicus* live in thick grassy areas, and their diet consists of abundant and readily renewable grasses and sedges (Ostfeld, 1990), a resource-rich environment perhaps more amenable to promiscuity.

In response to a perceived or actual stressor, the hypothalamic-pituitary-adrenal (HPA) axis is activated beginning with hypothalamic corticotropin-releasing factor (CRF) stimulating the release of adrenocorticotropin releasing hormone (ACTH) from the anterior pituitary and subsequently inducing the production of corticosterone from the adrenal cortex. Prairie voles display basal corticosterone levels that are 5–10 times greater than rats and promiscuous voles, and are considered glucocorticoid-resistant with 10-fold lower glucocorticoid receptor affinities (Carter et al., 1995; Hastings, Orchinik, Aubourg, McEwen, 1999; Taymans et al., 1997), possibly suggestive of impaired negative feedback of the stress axis.

A. Stress axis in pair bond formation

Experimental activation of the HPA axis alters social behavior in male and female voles, although in divergent manners. In male prairie voles, stress activation, through either the psychological stress of forced swimming or exogenous corticosterone administration, accelerates the development of a partner preference (DeVries, De Vries, Tayman, & Carter, 1996). Similarly, males recently exposed to a forced swim test display an elevation of alloparental behavior (Bales, Kramer, Lewis-Reese, & Carter, 2006). Corticosterone not only facilitates bond formation but also appears necessary for pair bonding to occur. Adrenalectomy prevents partner preference formation in male prairie voles, and replacement of corticosterone rescues the expression of this behavior (DeVries et al., 1996). However, prairie voles display sexual dimorphism in the effects of stress on sociality, because stress inhibits, rather than promotes, bond formation in females. Corticosterone injections or the stress of forced swimming, inhibit the development of partner preference in female prairie voles (Devries, De Vries, Tayman, & Carter, 1995). Conversely, adrenalectomized females are able to form a bond with only 1 hour of cohabitation (Devries et al., 1995).

CRF innervates the NAcc, a region already extensively implicated in bond formation, and this accumbal CRF has been implicated in bond formation (Lim et al., 2007). CRF administered centrally

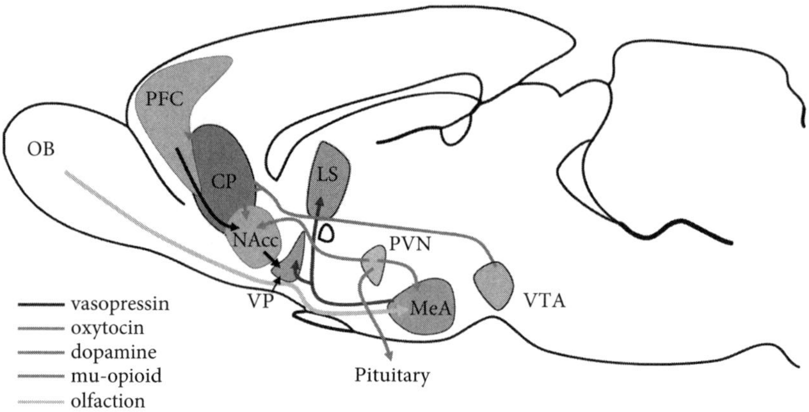

Fig. 5.3 Neurocircuitry of pair bond formation. A schematic of the proposed neural circuitry of pair bond formation in prairie voles. The neuropeptides OT and AVP carry important information on the olfactory signatures of a mate, enhance learning of socially relevant cues, and link information about the social cues of the partner with dopaminergic and opioid reward and reinforcement. Sociosexual interactions in female prairie voles stimulate the release of both OT from the paraventricular nucleus of the hypothalamus (PVN) and dopamine from the ventral tegmental area (VTA) into the NAcc. In males, AVP neurons in the extended amygdala release AVP into the ventral pallidum (VP) and the lateral septum (LS). Concurrently, olfactory signatures of the sexual partner are processed through the amygdala to ultimately lead to a condition partner preference (modified from Young, Murphy-Young, & Hammock, 2005).

at doses insufficient to induce anxiety-like behaviors accelerates bond formation in male prairie voles, an effect blocked by CRF1 and 2 receptor antagonists (DeVries, Guptaa, Cardillo, Cho, & Carter, 2002). CRF administration directly into the NAcc before cohabitation with a female accelerates partner preference formation in male prairie voles, but not in nonmonogamous meadow voles (Lim et al., 2007). CRF facilitation of bond formation was blocked by co-treatment with either CRF1 or CRF2 receptor antagonists.

The neural circuitry of the CRF system differs between monogamous and promiscuous vole species and displays sexual dimorphism, further suggesting a role of this peptide hormone in regulating social behavior (Lim, Nair, & Young, 2005). Monogamous prairie and pine voles display higher densities of CRFR2 and lower densities of CRFR1 in the NAcc than promiscuous meadow and montane voles. Within prairie voles, CRFR2 binding is greater in the BNST of males than that of females.

B. Stress axis in pair bond maintenance

In addition to influencing the willingness of an animal to form a pair bond, the systems controlling stress and anxiety may regulate the stability of a bond and are functionally altered after bond formation. Sexually naïve female prairie voles respond to novel males with a decrease in corticosterone, which presumably facilitates bond formation, but pair-bonded females display an increase in serum corticosterone, potentially functionally inhibiting bond formation (Devries et al., 1995). In male prairie voles, pairing with a female induces an increase in CRF mRNA in the BNST, and separation from the partner elevates circulating corticosterone and adrenal weight (Bosch et al., 2009). The increase in CRF mRNA may essentially prime the male to respond aversively to loss of the female partner. The CRF system, particularly that in the extended amygdala, displays parallel activation both in response to separation distress after partner removal and during withdrawal from drugs of reward (Burkett and Young, 2012). After loss of the bonded partner, male voles display passive stress-coping in the forced swim and tail suspension tests, and this social-loss-induced depression is prevented by blockade of CRF1 or 2 receptors using osmotic minipumps (Bosch et al., 2009). Loss of a bonded partner leads to a suite of traits associated with heightened stress and a depressive-like state, characteristics arguably similar to those of withdrawal (Burkett and Young, 2012). Social isolation induces a depression-like state in prairie voles, and isolated animals display decreased sucrose intake, elevated plasma corticosterone, and increased heart rate (Grippo, Lamv, Carter, & Porges, 2007). Meadow voles do not display an isolation-induced elevation in the stress response, suggesting this aversion to isolation is unique to animals preferring social contact (Stowe, Liu, Curtis, Freeman, & Wang, 2005). A separation from their female partners may induce proximity-seeking behavior in male voles toward their female partners, thus serving to stabilize the pair bond (Bosch et al., 2009).

C. Circuitry of OT and AVP within the stress axis

The circuitry by which stress affects pair bonding is largely unknown, but there may be overlap with neuropeptide systems. Along with CRF, vasopressin has long been known to stimulate ACTH release through the V1b receptor subtype (Whitnall, 1993). Stressors, including social defeat, induce AVP release in rats as shown by an increase in microdialysates (Wotjaket al., 1996), and heightened stress levels have been associated with increased AVP mRNA in the MeA and BNST in rats (Linfoot et al., 2009; Ueta, Dayanithi, & Fujihari, 2011). Brattleboro rats deficient in AVP display reduced anxiety-like behaviors (Landgraf et al., 1995). The anxiety-provoking aspects of AVP may be mediated through the V1a receptor subtype, in addition to V1bR in the pituitary, as V1aR antisense oligonucleotides or antagonists into the LS reduce anxiety-like behavior in rats, and V1aR overexpression in the LS increases expression of these behaviors (Bielsky et al., 2005; Landgraf et al., 1995; Liebsch, Wotjak, Landgraf, & Engelmann, 1996;). The systems may be reciprocally connected as corticosterone has been shown to increase V1aR expression in smooth muscle lines and centrally in the BNST, and there is a glucocorticoid response element in the 5' region of the *avpr1a* gene (Colson et al., 1992; Watters, Wilkinson, & Dorsa, 1996).

In contrast to AVP, OT appears to be primarily anxiolytic and is associated with decreased levels of anxiety-like behaviors in rodents and lower corticosterone levels (Amico, Cai, & Vollmer, 2008; Champagne and Meaney, 2007; Knobloch et al., 2012; Windle, Shanks, Lightman, & Ingram, 1997; Windle et al., 2004). It has been proposed that OT may act to disinhibit the approach to novelty and reduce the anxiety-provoking aspects of social stimuli, thus promoting the onset of social behaviors (Olazabol & Young, 2006). Recently, evidence for reciprocal regulation of OT and CRFR2 has been demonstrated in rats, with hypothalamic OT neurons expressing CRFR2 and striatal CRF neurons expressing OTR (Dabrowska et al., 2011). Further in support of a functional relationship between CRFR2 activation and OT release, female prairie voles raised by a single mother display a positive relationship between OT mRNA in the anterior PVN and CRFR2 expression in the dorsal raphe (Ahern and Young, 2009).

The molecules controlling stress and anxiety may have co-evolved a role in social behavior, possibly through modulating an animal's motivation to form a social bond (Lim et al., 2007). It has been hypothesized that stressful conditions may provide an evolutionary advantage for pair bonding in male but not in female prairie voles (DeVries et al., 1996). Females may benefit by staying in their natal nest and mating with non-family members (Getz et al., 1981; McGuire and Getz, 1995; Getz and Carter, 1996), whereas sexually mature males may be too aggressive toward conspecifics other than the partner to stay in the nest (DeVries et al., 1996). Males would then obtain the most reproductive success by forming new partnerships, defending a new home range, and providing care to offspring. The divergent roles of OT and AVP in mediating anxiety parallel the opposing influence of stress on pair bond formation in males and females, and may, in part, account for sex differences in the influence of neuropeptide systems on bonding in prairie voles.

VI. Within-species variability in social behavior

In addition to dramatic interspecies differences in vole social organization, significant natural variations in mating and parenting strategies also exist *within* the prairie vole species. In nature, breeding units vary between three types: male-female pairs, single-females, or communal groups composed of extended families (Getz and Carter, 1996). Within the communal nest, the size of the unit can vary between three and a dozen or more individual adults and their offspring (Getz and Carter, 1996). Groups are larger in late autumn-winter with an average of eight individuals as compared an average of three in spring-early autumn (Getz and Carter, 1996). Though most voles become "residents" and faithfully defend their partner and care for young, approximately 24% of females and 45% of males take on a "wandering" strategy, never forming a pair bond, (Getz and Carter, 1993). The percentage of wandering males in a population remains constant relative to population density (Getz and Carter, 1996). In the laboratory, voles vary significantly in their willingness to display spontaneous alloparental care, with up to 40% of females and 20% of male subjects attacking or ignoring pups (Roberts, Miller, et al., 1998).

Additionally, significant intraspecies variation in neuropeptide distribution has been reported in both wild and laboratory prairie vole populations. In a wild population of voles, V1aR densities can vary

studies, intranasal OT enhanced social cognitive function in male autistic individuals. Upon presentation of images of eyes in various affective states, the ability to read emotions from subtle facial cues was enhanced after OT infusion (Guastella et al., 2010). Similarly, intranasal OT enhanced gaze to the eye region in an autistic population (Andari et al., 2010). In the same study, socially reinforced learning improved with oxytocin infusion in a simulated socially interactive ball game. Participants were better able to discriminate between players that included them in the game over those that ignored them. Thus OT, OTR agonists, or drugs that stimulate OT release may enhance social functioning. Recently, intranasal OT was administered to prairie vole juveniles, and acutely increased social contact with conspecifics (Bales, Perkeybile, Conley, Lee, Guoynes, et al., 2012). However, consideration must be made for dosage and sex-specific effects as chronic low, but not high, dose of OT impaired male, but not female, prairie vole partner preference formation. Ultimately, social cognitive enhancers could act synergistically with applied behavioral therapies that use socially reinforced learning techniques to teach autistic individuals social rules, language, and social cues (Young, 2011; Modi and Young, 2012). Pair bonding in monogamous voles may be a useful behavioral paradigm to test for drugs that, if given to an socially compromised individual just prior to a social learning session, may enhance the effectiveness of social behavioral therapies. Thus, in addition to being useful for understanding the neural and genetic mechanisms underlying social behavior, pair bonding in voles may help with drug discovery to identify drugs that enhance social function.

VIII. Conclusion and Future Directions

Humans have a unique drive to engage in social interactions and display intricate social bonding, behaviors that, until recently, have been considered too complex to link to any neurobiological bases. The past few decades of study into the highly social prairie vole, once only a model for behavioral ecology and mammalogy, has helped to uncover the neural circuits driving sociality, and has generated hypotheses into our own behavior and physiology. In particular, the nonapeptides OT and AVP have been strongly linked to pair bonding and other behaviors associated with a monogamous social structure. These neuropeptides carry important information on the olfactory signatures of a mate, enhance learning of socially relevant cues, and link up information about a conspecific with dopaminergic reward and reinforcement. The stress axis and related neurotransmitter systems are important for regulating monogamous strategies in variable environments and for maintaining the expression of a pair bond. The systems controlling social attachment are also susceptible to experience early in life, pinpointing the neonatal period as a critical time in the development of pair bonding. OT and AVP signaling may both translate the experience of early social interactions and regulate the development of neural circuits controlling sociality throughout development.

The continued progress is making the prairie voles a premier model organism for social behavior may help reveal yet unanswered questions in the field such as: What are the underlying mechanisms behind individual variation in social attachment and functioning of the social brain? How do neural circuits controlling social information processing, social motivation, learning, reward, and stress and anxiety interact at cellular, developmental, and context-specific levels? What is the impact of neuropeptide systems on a temporal and region specific scale? The field of neuroendocrinology is also lacking in studies that bridge the wide gap between well-controlled laboratory-based neuroscience and evolutionarily relevant field studies The integration of modern molecular and pharmacological approaches with behavioral ecology would help to uncover mechanistic underpinnings of social behavior to explore the origin of diversity in an ecologically relevant context.

The processes underlying social bond formation in prairie voles and social cognition in humans display remarkable similarities. A burst in studies using intranasal neuropeptide administration, genetic association, and neuroimaging approaches in recent years has revealed striking human parallels to vole research and promising therapeutic strategies. Thus, insights from this highly social species have the potential to generate hypotheses and drive drug discovery for treating deficits in social ability in humans. Because there are presently no approved treatments that target the deficits in social reciprocity and social cognition seen in all cases of ASDs, there is a dire need to understand the pathophysiology of this disease and to identify treatable neurobiological targets. We are beginning to expand the genetic toolbox available for the prairie vole and thus its usefulness as an animal model (Donaldson and Young, 2008; McGraw et al., 2010; McGraw and Young, 2010; Keebaugh, Modi, Barrett, Jin, & Young, 2012). The prairie vole genome was recently sequenced, thus paving the way for more sophisticated temporal and target specific transgenic development using prairie vole specific gene sequences (McGraw and Young, 2010). For example, lentiviral-mediated gene transfer has been successfully used to create transgenic prairie voles expressing green-fluorescent protein (Donaldson, Yang, Chan, & Young, 2009). Future transgenic approaches to generate prairie voles with reduced or abolished neuropeptide receptors and impaired sociality may be viable systems with which to test drugs that target social learning in a socially compromised individual (Donaldson et al., 2009). The prairie vole provides a unique system for social neurobiology, and advancements in our ability to manipulate the genome of this behaviorally rich, nontraditional species can pave the way for a myriad of gene-behavior-environment studies.

References

Adkins-Regan, E. (2009). Neuroendocrinology of social behavior. *ILAR Journal*, 50(1), 5–14.

Agren, G. (1984). Pair formation in the mongolian gerbil. *Animal Behaviour*, 32, 528–535.

Ahern, T. H., Hammock, E. A., & Young, L. J. (2011). Parental division of labor, coordination, and the effects of family structure on parenting in monogamous prairie voles (Microtus ochrogaster). *Developmental Psychobiology*, 53(2): 118–131.

Ahern, T. H., Modi, M. E., Burkett, J. P., & Young, L. J. (2009). Evaluation of two automated metrics for analyzing partner preference tests. *Journal of Neuroscience Methods*, 182(2), 180–188.

Ahern, T. H. & Young, L. J. (2009). The impact of early life family structure on adult social attachment, alloparental behavior, and the neuropeptide systems regulating affiliative behaviors in the monogamous prairie vole (microtus ochrogaster). Front *Behavioral Neuroscience*, 3, 17.

Altemus, M., Cizza, G., & Gold, P. W. (1992). Chronic fluoxetine treatment reduces hypothalamic vasopressin secretion in vitro. *Brain Research 593*(2): 311–313.

Amico, J. A., Cai, H. M. and Vollmer, R. R. (2008). Corticosterone release in oxytocin gene deletion mice following exposure to psychogenic versus non-psychogenic stress. *Neuroscience Letters, 442*(3): 262–266.

Andari, E., Duhamel, J. R., Zalla, T., Herbrecht, E., Leboyer, M., & Sirigu, A. (2010). Promoting social behavior with oxytocin in high-functioning autism spectrum disorders. Proceedings of the National Academy of Science, *107*(9), 4389–4394.

Aragona, B. J., Liu, Y., Curtis, J. T., Stephan, F. K., & Wang Z. (2003). A critical role for nucleus accumbens dopamine in partner-preference formation in male prairie voles. *Journal of Neuroscience, 23*(8), 3483–3490.

Aragona, B. J., Liu, Y., Yu, Y. J., Curtis, J. T., Detwiler, J. M., Insel, T. R., & Wang, Z. (2006). Nucleus accumbens dopamine differentially mediates the formation and maintenance of monogamous pair bonds. *National Neuroscience, 9*(1), 133–139.

Bacchelli, E. & Maestrini, E. (2006). Autism spectrum disorders: molecular genetic advances. *American Journal of Medical Genetetics, Part C Seminars in Medical Genetetics, 142C*(1), 13–23.

Bales, K. L. & Carter, C. S. (2003). Sex differences and developmental effects of oxytocin on aggression and social behavior in prairie voles (Microtus ochrogaster). *Hormones and Behavior*, 44(3), 178–184.

Bales, K. L., Kramer, K. M., Lewis-Reese, A. D., & Carter, C. S. (2006). Effects of stress on parental care are sexually dimorphic in prairie voles. *Physiology and Behavior, 87*(2), 424–429.

Bales, K. L., Lewis-Reese, A. D., Pfeifer, L. A., Kramer, K. M., & Carter, C. S. (2007). Early experience affects the traits of monogamy in a sexually dimorphic manner. *Developmental Psychobiology, 49*(4), 335–342.

Bales, K. L. & Perkeybile, A. M. (2012). Developmental experiences and the oxytocin receptor system. *Hormones and Behavior, 61*(3), 313–319.

Bales, K. L., A. M. Perkeybile, O. G. Conley, M. H. Lee, C. D. Guoynes, G. M. Downing, C. R. Yun, M. Solomon, S. Jacob and S. P. Mendoza (2012). "Chronic Intranasal Oxytocin Causes Long-Term Impairments in Partner Preference Formation in Male Prairie Voles." *Biological Psychiatry*. (in press)

Bales, K. L., Pfeifer, L. A. & Carter, C. S. (2004). Sex differences and developmental effects of manipulations of oxytocin on alloparenting and anxiety in prairie voles. *Developmental Psychobiology, 44*(2), 123–131.

Bales, K. L., Plotsky, P. M., Young, L. J., Lim, M. M., Grotte, N., Ferrer, E., & Carter C.S. (2007). Neonatal oxytocin manipulations have long-lasting, sexually dimorphic effects on vasopressin receptors. *Neuroscience, 144*(1), 38–45.

Bales, K. L., van Westerhuyzen, J. A., Lewis-Reese, A. D., Grotte, N. D., Lanter, J. A., & Carter, C. S. (2007). Oxytocin has dose-dependent developmental effects on pair-bonding and alloparental care in female prairie voles. *Hormones and Behavior, 52*(2), 274–279.

Bamshad, M., Novak, M. A. & de Vries, G. J. (1993). Sex and species differences in the vasopressin innervation of sexually naive and parental prairie voles, Microtus ochrogaster and meadow voles, Microtus pennsylvanicus. *Journal of Neuroendocrinology, 5*(3), 247–255.

Bamshad, M., Novak, M. A., & de Vries, G. J. (1994). Cohabitation alters vasopressin innervation and paternal behavior in prairie voles (Microtus ochrogaster). *Physiology of Behavior, 56*(4), 751–758.

Barrett, C. E., Keebaugh, A. C., Ahern, T. H., Bass, C. E., Terwilliger, E. F., & Young, L. J. (in press). Variation in vasopressin receptor (Avpr1a) expression creates diversity in behaviors related to monogamy in prairie voles. *Hormones and Behavior.*

Bester-Meredith, J. K., Young, L. J., & Marler, C. A. (1999). Species differences in paternal behavior and aggression in peromyscus and their associations with vasopressin immunoreactivity and receptors. *Hormones and Behavior, 36*(1), 25–38.

Bielsky, I. F., Hu, S. B., Ren, X., Terwilliger, E. F., & Young, L. J. (2005). The V1a vasopressin receptor is necessary and sufficient for normal social recognition: a gene replacement study. *Neuron, 47*(4), 503–513.

Bielsky, I. F., Hu, S. B., Szegda, K. L., Westphal, H. & Young, L. J. (2004). Profound impairment in social recognition and reduction in anxiety-like behavior in vasopressin V1a receptor knockout mice. *Neuropsychopharmacology 29*(3), 483–493.
Blumstein, D. T., Ebensperger, L. A., Hayes, L. D., Vasquez, R. A., Ahern, T. H., Burger, J. R....Young, L.J (2010). Toward an integrative understanding of social behavior: new models and new opportunities. *Frontiers of Behavioral Neuroscience, 4*, 34.
Bosch, O. J., Nair, H. P., Ahern, T. H., Neumann, I. D., & Young, L. J. (2009). The CRF system mediates increased passive stress-coping behavior following the loss of a bonded partner in a monogamous rodent. *Neuropsychopharmacology 34*(6), 1406–1415.
Bosch, O. J. & Neumann, I. D. (2008). Brain vasopressin is an important regulator of maternal behavior independent of dams' trait anxiety. *Proceedings of the National Academy of Science, 105*(44), 17139–17144.
Bosch, O. J., Pfortsch, J., Beiderbeck, D. I., Landgraf, R.& Neumann, I. D. (2010). Maternal behaviour is associated with vasopressin release in the medial preoptic area and bed nucleus of the stria terminalis in the rat. *Journal of Neuroendocrinology, 22*(5), 420–429.
Buijs, R. M., Velis, D. N.,& Swaab, D. F. (1980). Ontogeny of vasopressin and oxytocin in the fetal rat: early vasopressinergic innervation of the fetal brain. *Peptides, 1*(4), 315–324.
Burkett, J. P., Spiegel, L. L., Inoue, K., Murphy, A. Z., & Young, L. J. (2011). Activation of mu-opioid receptors in the dorsal striatum is necessary for adult social attachment in monogamous prairie voles. *Neuropsychopharmacology, 36*(11), 2200–2210.
Burkett, J. P. & Young, L. J. (2012). The behavioral, anatomical and pharmacological parallels between social attachment, love and addiction. *Psychopharmacology (Berl), 224*(1): 1–26.
Caba, M., Rovirosa, M. J., & Silver, R. (2003). Suckling and genital stroking induces Fos expression in hypothalamic oxytocinergic neurons of rabbit pups. *Developmental Brain Research, 143*(2), 119–128.
Carter, C. S., Boone, E. M., Pournajafi-Nazarloo, H., & Bales, K. L. (2009). Consequences of early experiences and exposure to oxytocin and vasopressin are sexually dimorphic. *Developmental Neuroscience, 31*(4), 332–341.
Carter, C. S., DeVries, A. C., & Getz, L. L. (1995). Physiological substrates of mammalian monogamy: the prairie vole model. *Neuroscience Biobehavioral Review, 19*(2), 303–314.
CDC(2012). Prevalence of autism spectrum disorders—Autism and Developmental Disabilities Monitoring Network, 14 sites, United States, 2008. *MMWR Surveillance Summary, 61*(3), 1–19.
Champagne, F. A. & Meaney, M. J. (2007). Transgenerational effects of social environment on variations in maternal care and behavioral response to novelty. *Behavioral Neuroscience, 121*(6), 1353–1363.
Cho, M. M., DeVries, A. C., Williams, J. R., & Carter, C. S. (1999). The effects of oxytocin and vasopressin on partner preferences in male and female prairie voles (Microtus ochrogaster). *Behavioral Neuroscience, 113*(5), 1071–1079.
Coccaro, E. F., Kavoussi, R. J., Hauger, R. L., Cooper, T. B., & Ferris, C. F. (1998). Cerebrospinal fluid vasopressin levels: correlates with aggression and serotonin function in personality-disordered subjects. *Archives of General Psychiatry, 55*(8), 708–714.
Cole, F. R., & Batzli, G. O. (1979). "Nutrition and population dynamics of the prairie vole (Microtus ochrogaster) in central Illinois." *Journal of Animal Ecology, 48*, 455–470.
Colson, P., Ibarondo, J., Devilliers, G., Balestre, M. N., Duvoid, A., & Guillon, G. (1992). Upregulation of V1a vasopressin receptors by glucocorticoids. *American Journal of Physiology, 263*(6 Pt 1), E1054–1062.
Crews, D., & Moore, M. C. (1986). Evolution of mechanisms controlling mating behavior. *Science, 231*(4734), 121–125.
Cushing, B. S. & Kramer, K. M. (2005). Microtines: A model system for studying the evolution and regulation of social monogamy. *Acta Theriologica Sinica, 25*, 182–199.
Cushing, B. S., Levine, K., &. Cushing, N. L (2005). Neonatal manipulation of oxytocin influences female reproductive behavior and success *Hormones and Behavior, 47*(1), 22–28.
Cushing, B. S., Martin, J. O., Young, L. J. & Carter, C. S. (2001). The effects of peptides on partner preference formation are predicted by habitat in prairie voles. *Hormones and Behavior, 39*(1), 48–58.
Cushing, B. S., Mogekwu, N., Le, W. W., Hoffman, G. E., & Carter, C. S. (2003). Cohabitation induced Fos immunoreactivity in the monogamous prairie vole. *Brain Research, 965*(1–2), 203–211.
Dabrowska, J., Hazra, R. Ahern, T. H., Guo, J. D., McDonald, A. J., Mascagni, F....Rainnie, D.G (2011). Neuroanatomical evidence for reciprocal regulation of the corticotrophin-releasing factor and oxytocin systems in the hypothalamus and the bed nucleus of the stria terminalis of the rat: Implications for balancing stress and affect. *Psychoneuroendocrinology, 36*(9), 1312–1326.
Danielson, B. J. & Gaines, M. S. (1987). Spatial patterns in two synoptic species of microtines: Microtus ochrogaster and Synaptomys cooperi. *Journal of Mammalogy, 68*, 313–322.
Dantzer, R., Bluthe, R. M., Koob, G. F., & Le Moal, M. (1987). Modulation of social memory in male rats by neurohypophyseal peptides. *Psychopharmacology (Berl) 91*(3), 363–368.
de Kloet, C. S., Vermetten, E., Geuze, E., Wiegant, V. M., & Westenberg, H. G. (2008). Elevated plasma arginine vasopressin levels in veterans with posttraumatic stress disorder *Journal of Psychiatric Research, 42*(3), 192–198.
de Wied, D., & van Ree, J. M. (1989). Neuropeptides: Animal behaviour and human psychopathology. *European Archive of Psychiatry and Neurological Science 238*(5–6), 323–331.
DeVries, A. C., & Carter, C. S. (1999). Sex differences in temporal parameters of partner preference in prairie voles. *Canadian Journal of Zoology 77*(6), 885–889.
Devries, A. C., Devries, M. B., Taymans, S. & Carter, C. S. (1995). Modulation of pair bonding in female prairie voles (microtus-ochrogaster) by corticosterone. *Proceedings of the National Academy of Sciences of the United States of America, 92*(17), 7744–7748.
DeVries, A. C., DeVries, M. B., Taymans, S. E., & Carter, C. S. (1996). The effects of stress on social preferences are sexually dimorphic in prairie voles. *Proceedings of the National Academy of Science of the United States of America, 93*(21), 11980–11984.
DeVries, A. C., Guptaa, T., Cardillo, S., Cho, M., & Carter, C. S. (2002). Corticotropin-releasing factor induces social preferences in male prairie voles. *Psychoneuroendocrinology, 27*(6): 705–714.
De Vries, G. J., & Buijs, R. M. (1983). The origin of the vasopressinergic and oxytocinergic innervation of the rat-brain

with special reference to the lateral septum. *Brain Research, 273*(2), 307–317.

De Vries, G. J. and C. Villalba (1997). "Brain sexual dimorphism and sex differences in parental and other social behaviors." *Annals of the New York Acadamy of Sciences* 807: 273–286.

Dewsbury, D. A. (1987). *The comparative psychology of monogamy. American Zoology: Nebraska Symposium on Motivation* Lincoln, NE: University of Nebraska Press.

Di Chiara, G. (2002). Nucleus accumbens shell and core dopamine: Differential role in behavior and addiction. *Behavioral Brain Research, 137*(1–2), 75–114.

Donaldson, Z. R., Yang, S. H., Chan, A. W., & Young, L. J. (2009). Production of germline transgenic prairie voles (Microtus ochrogaster) using lentiviral vectors. *Biological Reproduction, 81*(6), 1189–1895.

Donaldson, Z. R., & Young, L. J. (2008). Oxytocin, vasopressin, and the neurogenetics of sociality. *Science, 322*(5903), 900–904.

Dunbar, R. (1995). The mating system of callitrichid primates: I. Conditions for the coevolution of pair bonding and twinning. *Animal Behaviour, 50*(4): 1057–1070.

Engelmann, M., & Landgraf, R. (1994). Microdialysis administration of vasopressin into the septum improves social recognition in Brattleboro rats. *Physiology & Behavior, 55*(1), 145–149.

Ferguson, B., Fuentes, S. M., Sawrey, D. K. & Dewsbury, D. A. (1986). Male-preferences for unmated versus mated females in 2 species of voles (Microtus-ochrogaster and microtus-montanus). *Journal of Comparative Psychology, 100*(3), 243–247.

Ferguson, J. N., Aldag, J. M., Insel, T. R., & Young, L. J. (2001). Oxytocin in the medial amygdala is essential for social recognition in the mouse. *Journal of Neuroscience, 21*(20), 8278–8285.

Ferguson, J. N., Young, L. J., Hearn, E. F., Matzuk, M. M., Insel, T. R., & Winslow, J. T. (2000). Social amnesia in mice lacking the oxytocin gene. *Nature Genetics, 25*(3), 284–288.

Ferris, C. F., Albers, H. E., Wesolowski, S. M., Goldman, B. D., & Luman, S. E. (1984). Vasopressin injected into the hypothalamus triggers a stereotypic behavior in golden hamsters. *Science, 224*(4648), 521–523.

Fink, S., Excoffier, L., & Heckel, G. (2006). Mammalian monogamy is not controlled by a single gene. *Proceedings of the National Academy of Science of the United States of America, 103*(29), 10956–10960.

Fitch, H. S. (1957). *Aspects of reproduction and development in the prairie voles (Microtus ochrogaster).* University of Kansas Publications, Museum of Natural History *10*, 129–161.

Francis, D., Diorio, J., Liu, D. & Meaney, M. J. (1999). Nongenomic transmission across generations of maternal behavior and stress responses in the rat. *Science, 286*(5442), 1155–1158.

Francis, D. D., Young, L. J., Meaney, M. J., & Insel, T. R. (2002). Naturally occurring differences in maternal care are associated with the expression of oxytocin and vasopressin (V1a) receptors: Gender differences. *Journal of Neuroendocrinology, 14*(5), 349–353.

Gaines, M. S., Fugate, C. L., Johnson, M. L., Johnson, D. C., Hisey, J. R., & Quadagno, D. M. (1985). Manipulation of aggressive behavior in male prairie voles (Microtus ochrogaster) implanted with testosterone in silastic tubing. *Canadian Journal of Zoology, 63*, 2525–2528.

Getz, L., Carter, C., & Gavish, L. (1981). The mating system of the prairie vole, microtus ochrogaster: Field and laboratory evidence for pair-bonding. *Behavioral Ecology, 8*(3), 189–194.

Getz, L. L. (1978). Speculation on social structure and population cycles of microtine rodents. *The Biologist, 60*(4), 134–147.

Getz, L. L., & Carter, C. S. (1993). Social organization of the prairie vole (*Microtus ochrogaster*). Journal of Mammalogy, 74, 44–58.

Getz, L. L., & Carter, C. S. (1996). Prairie vole partnerships. *American Science, 84*, 56–62.

Getz, L. L., & Hofmann, J. E. (1986). Social organization in free-living prairie voles, Microtus ochrogaster. *Behavioral Ecology and Sociobiology, 18*, 275–282.

Getz, L. L., Mcguire, B., Pizzuto, T., Hofmann, J. E., & Frase, B., (1993). Social-organization of the prairie vole (microtus-ochrogaster). *Journal of Mammalogy, 74*(1), 44–58.

Gingrich, B., Liu, Y., Cascio, C., Wang, Z., & Insel, T. R. (2000). Dopamine D2 receptors in the nucleus accumbens are important for social attachment in female prairie voles (Microtus ochrogaster). *Behavioral Neuroscience, 114*(1), 173–183.

Gobrogge, K. L., Liu, Y., Young, L. J., & Wang, Z. (2009). Anterior hypothalamic vasopressin regulates pair-bonding and drug-induced aggression in a monogamous rodent. *Proceedings of the National Academy of Science of the United states of America, 106*(45), 19144–19149.

Goldman, M., Marlow-O'Connor, M., Torres, I., & Carter, C. S. (2008). Diminished plasma oxytocin in schizophrenic patients with neuroendocrine dysfunction and emotional deficits. *Schizophrenia Research, 98*(1–3, 247–255.

Goldman, M. B., Robertson, G. L., & Hedeker, D. (1996). Oropharyngeal regulation of water balance in polydipsic schizophrenics. *Clinical Endocrinology (Oxf), 44*(1), 31–37.

Green, L., Fein, D., Modahl, C., Feinstein, C., Waterhouse, L., & Morris, M. (2001). Oxytocin and autistic disorder: alterations in peptide forms. *Biological Psychiatry, 50*(8), 609–613.

Gregory, S. G., Connelly, J. J., Towers, A. J., Johnson, J., Biscocho, D., Markunas, C. A.,... (2009). Genomic and epigenetic evidence for oxytocin receptor deficiency in autism. *BioMed Central Medicine, 7*, 62.

Grippo, A. J., Lamb, D. G., Carter, C. S., & Porges, S. W. (2007). Social isolation disrupts autonomic regulation of the heart and influences negative affective behaviors. *Biological Psychiatry, 62*(10), 1162–1170.

Guastella, A. J., Einfeld, S. L., Gray, K. M., Rinehart, N. J., Tonge, B. J., Lambert, T. J., & Hickie, I. B.... (2010). Intranasal oxytocin improves emotion recognition for youth with autism spectrum disorders. *Biological Psychiatry, 67*(7), 692–694.

Hammock, E. A., Lim, M. M., Nair, H. P., & Young, L. J. (2005). Association of vasopressin 1a receptor levels with a regulatory microsatellite and behavior. *Genes, Brain and Behavior, 4*(5), 289–301.

Hammock, E. A. & Young, L. J. (2002). Variation in the vasopressin V1a receptor promoter and expression: implications for inter- and intraspecific variation in social behaviour. *European Journal of Neuroscience 16*(3), 399–402.

Hastings, N. B., Orchinik, M., Aubourg, M. V., & McEwen, B. S. (1999). Pharmacological characterization of central and peripheral type I and type II adrenal steroid receptors in the prairie vole, a glucocorticoid-resistant rodent. *Endocrinology 140*(10), 4459–4469.

Hofmann, J. E., Getz, L. L., & Gavish, L. (1984). Home range overlap and nest cohabitation of male and female prairie voles. *American Midland Naturalist, 112*, 314–319.

Hollander, E., Bartz, J., Chaplin, W., Phillips, A., Sumner, J., Soorya, L.,...Wasserman, S. (2007). Oxytocin increases retention of social cognition in autism. *Biological Psychiatry, 61*(4), 498–503.

Hollander, E., Novotny, S., Hanratty, M., Yaffe, R., DeCaria, C. M., Aronowitz, B. R.,...& Mosovich, S. (2003). Oxytocin infusion reduces repetitive behaviors in adults with autistic and Asperger's disorders. *Neuropsychopharmacology 28*(1), 193–198.

Insel, T. R. (1990). Regional changes in brain oxytocin receptors post-partum: time-course and relationship to maternal behaviour. *Journal of Neuroendocrinology, 2*(4), 539–545.

Insel, T. R. (1991). Long-term neural consequences of stress during development: Is early experience a form of chemical imprinting? In B. J. Carroll (Ed.), *Psychopathology and the brain* (pp. 133–152). New York: Raven Press.

Insel, T. R., & Hulihan, T. J. (1995). A gender-specific mechanism for pair bonding: oxytocin and partner preference formation in monogamous voles. *Behavioral Neuroscience, 109*(4), 782–789.

Insel, T. R. & Shapiro, L. E. (1992). Oxytocin receptor distribution reflects social organization in monogamous and polygamous voles. *Proceedings of the National Academy of Science of the United States of America, 89*(13), 5981–5985.

Insel, T. R., Wang, Z. X., & Ferris, C. F. (1994). Patterns of brain vasopressin receptor distribution associated with social organization in microtine rodents. *Journal of Neuroscience, 14*(9), 5381–5392.

Insel, T. R., & Winslow, J. T. (1991). Central administration of oxytocin modulates the infant rat's response to social isolation. *European Journal of Pharmacology, 203*(1), 149–152.

Jannett, F. J. (1982). Nesting patterns of adult voles, Microtus-montanus, in field populations. *Journal of Mammalogy 63*(3), 495–498.

Jia, R., Tai, F., An, S., Zhang, X,.& Broders, H. (2009). Effects of neonatal paternal deprivation or early deprivation on anxiety and social behaviors of the adults in mandarin voles. *Behavioral Processes, 82*(3), 271–278.

Keebaugh, A. C., Modi, M. E., Barrett, C. E., Jin, C., & Young, L. J. (2012). Identification of variables contributing to superovulation efficiency for production of transgenic prairie voles (Microtus ochrogaster). *Reproductive Biology and Endocrinology, 10*(1), 54.

Keebaugh, A. C., & Young, L. J. (2011). Increasing oxytocin receptor expression in the nucleus accumbens of pre-pubertal female prairie voles enhances alloparental responsiveness and partner preference formation as adults. *Hormones & Behavior, 60*(5), 498–504.

Kendrick, K. M., Keverne, E. B., & Baldwin, B. A. (1987). Intracerebroventricular oxytocin stimulates maternal behaviour in the sheep. *Neuroendocrinology, 46*(1), 56–61.

Keri, S., Kiss, I., & Kelemen, O. (2009). Sharing secrets: oxytocin and trust in schizophrenia. *Social Neuroscience, 4*(4), 287–293.

Kim, S. J., Young, L. J., Gonen, D., Veenstra-VanderWeele, J., Courchesne, R., Courchesne, E.,...Insel, T.R. (2002). Transmission disequilibrium testing of arginine vasopressin receptor 1A (AVPR1A) polymorphisms in autism. *Molecular Psychiatry, 7*(5), 503–507.

Kirkpatrick, B., Carter, C. S., Newman S. W., & Insel, T. R. (1994). Axon-sparing lesions of the medial nucleus of the amygdala decrease affiliative behaviors in the prairie vole (Microtus ochrogaster): behavioral and anatomical specificity. *Behavioral Neuroscience, 108*(3), 501–513.

Kirkpatrick, B., Kim, J. W., & Insel, T. R. (1994). Limbic system fos expression associated with paternal behavior. *Brain Research, 658*(1–2), 112–118.

Kleiman, D. G. (1977). Monogamy in mammals. *Quarterly Review of Biology, 52*(1), 39–69.

Knobloch, H. S., Charlet, A., Hoffmann, L. C., Eliava, M., Khrulev, S., Cetin, A. H.,...Grinevich, V. (2012). Evoked axonal oxytocin release in the central amygdala attenuates fear response. *Neuron, 73*(3), 553–566.

Kojima, S., & Alberts, J. R. (2011). Oxytocin mediates the acquisition of filial, odor-guided huddling for maternally-associated odor in preweanling rats. *Hormones & Behavior, 60*(5), 549–558.

Koob, G. F., & Nestler, E. J. (1997). The neurobiology of drug addiction. *Journal of Neuropsychiatry and Clinical Neuroscience, 9*(3), 482–497.

Kramer, K. M., Cushing, B. S., & Carter, C. S. (2003). Developmental effects of oxytocin on stress response: single versus repeated exposure. *Physiology and Behavior, 79*(4–5), 775–782.

Landgraf, R., Gerstberger, R., Montkowski, A., Probst, J. C., Wotjak, C. T., Holsboer, F., & Engelmann, M (1995). V1 vasopressin receptor antisense oligodeoxynucleotide into septum reduces vasopressin binding, social discrimination abilities, and anxiety-related behavior in rats. *Journal of Neuroscience, 15*(6), 4250–4258.

Le Moal, M., Dantzer, R., Michaud, B., & Koob, G. F. (1987). Centrally injected arginine vasopressin (AVP) facilitates social memory in rats. *Neuroscience Letters, 77*(3), 353–359.

Lerer, E., Levi, S., Salomon, S., Darvasi, A., Yirmiya, N., & Ebstein, R. P. (2008). Association between the oxytocin receptor (OXTR) gene and autism: Relationship to Vineland Adaptive Behavior Scales and cognition. *Molecular Psychiatry, 13*(10), 980–988.

Liebsch, G., Wotjak, C. T., Landgraf, R., & Engelmann, M. (1996). Septal vasopressin modulates anxiety-related behaviour in rats. *Neuroscience Letters, 217*(2–3), 101–104.

Lim, M. M., Bielsky, I. F., & Young, L. J. (2005). Neuropeptides and the social brain: potential rodent models of autism. *International Journal of Developmental Neuroscience, 23*(2–3): 235–243.

Lim, M. M., Liu, Y., Ryabinin, A. E., Bai, Y., Wang, Z., & Young, L. J. (2007). CRF receptors in the nucleus accumbens modulate partner preference in prairie voles. *Hormones & Behavior, 51*(4), 508–515.

Lim, M. M., Murphy, A. Z. & Young, L. J. (2004). Ventral striatopallidal oxytocin and vasopressin V1a receptors in the monogamous prairie vole (Microtus ochrogaster). *Journal of Comparative Neurology, 468*(4), 555–570.

Lim, M. M., Nair, H. P., & Young, L. J. (2005). Species and sex differences in brain distribution of corticotropin-releasing factor receptor subtypes 1 and 2 in monogamous and promiscuous vole species. *Journal of Comparative Neurology, 487*(1), 75–92.

Lim, M. M., Wang, Z., Olazabal, D. E., Ren, X., Terwilliger, E. F., and Young, L. J., (2004b). Enhanced partner preference

in a promiscuous species by manipulating the expression of a single gene. *Nature, 429*(6993): 754–757.
Lim, M. M., & Young, L. J. (2004). Vasopressin-dependent neural circuits underlying pair bond formation in the monogamous prairie vole. *Neuroscience, 125*(1), 35–45.
Linfoot, I., Gray, M., Bingham, B., Williamson, M., Pinel, J. P., & Viau, V. (2009). Naturally occurring variations in defensive burying behavior are associated with differences in vasopressin, oxytocin, and androgen receptors in the male rat. *Progress in Neuro-Psychopharmacology & Biological Psychiatry, 33*(7), 1129–1140.
Liu, Y., J. T. Curtis and Z. Wang (2001). "Vasopressin in the lateral septum regulates pair bond formation in male prairie voles (Microtus ochrogaster)." *Behavioral Neuroscience* 115(4): 910–919.
Liu, Y., & Wang, Z. X. (2003). Nucleus accumbens oxytocin and dopamine interact to regulate pair bond formation in female prairie voles. *Neuroscience, 121*(3), 537–544.
Lonstein, J. S. (2002). Effects of dopamine receptor antagonism with haloperidol on nurturing behavior in the biparental prairie vole. *Pharmacology Biochemistry and Behavior, 74*(1), 11–19.
Lonstein, J. S., & De Vries, G. J. (2001). Social influences on parental and nonparental responses toward pups in virgin female prairie voles (Microtus ochrogaster). *Journal of Comparative Psychology, 115*(1), 53–61.
Lott, D. F. (1991). *Intraspecific variation in the social systems of wild vertebrates.* New York: Cambridge University Press.
Madison, D. L. (1980). Space use and social structure in Meadow Voles, Microtus pennsylvanicus. *Behavioral Ecology, 7*(1), 65–71.
McGraw, L. A., Davis, J. K., Lowman, J. J., ten Hallers, B. F., Koriabine, M., Young, L. J.,. Thomas, J.W. (2010). Development of genomic resources for the prairie vole (Microtus ochrogaster): construction of a BAC library and vole-mouse comparative cytogenetic map. *BioMed Central Genomics, 11*, 70.
McGraw, L. A., & Young, L. J. (2010). The prairie vole: an emerging model organism for understanding the social brain. *Trends in Neuroscience, 33*(2), 103–109.
McGuire, B. (1988). Effects of cross-fostering on parental behavior of meadow voles (Microtus pennsylvanicus). *Journal of Mammalogy, 69*(2), 332–341.
McGuire, B., & Getz, L. L. (1995). Communal nesting in prairie voles (Microtus ochrogaster): An evaluation of costs and benefits based on patterns of dispersal and settlement." *Canadian Journal of Zoology, 73*, 383–391.
Mcguire, B., Getz, L. L., Hofmann, J. E., Pizzuto, T., & Frase, B. (1993). Natal dispersal and philopatry in prairie voles (Microtus ochrogaster) in relation to population density, season, and natal social environment. *Behavavioral Ecology and Sociobiology, 32*, 293–302.
Meaney, M. J. (2001). Maternal care, gene expression, and the transmission of individual differences in stress reactivity across generations. *Annual Review of Neuroscience, 24*, 1161–1192.
Melis, M. R., Argiolas, A., & Gessa, G. L. (1989). Apomorphine increases plasma oxytocin concentration in male rats. *Neuroscience Letters, 98*(3), 351–355.
Meyer-Lindenberg, A., Domes, G., Kirsch, P., & Heinrichs, M. (2011). Oxytocin and vasopressin in the human brain: Social neuropeptides for translational medicine. *Nature Reviews Neuroscience, 12*(9), 524–538.
Mock, D. W., & Fujioka, M. (1990). Monogamy and long-term pair bonding in vertebrates. *Trends in Ecology and Evolution, 5*(2), 39–43.
Modi, M. E., & Young, L. J. (2011). D-cycloserine facilitates socially reinforced learning in an animal model relevant to autism spectrum disorders. *Biological Psychiatry 70*(3), 298–304.
Modi, M. E., & Young, L. J. (2012). The oxytocin system in drug discovery for autism: animal models and novel therapeutic strategies. *Hormones & Behavior, 61*(3), 340–350.
Murphy, M. R., Seckl, J. R., Burton, S., Checkley, S. A., & Lightman, S. L. (1987). Changes in oxytocin and vasopressin secretion during sexual activity in men." *The Journal of Clinical Endocrinology and Metabolism, 65*(4), 738–741.
Nair, H. P., & Young, L. J. (2006). Vasopressin and pair-bond formation: genes to brain to behavior. *Physiology 21*, 146–152.
Nelson, E., & Panksepp,J. (1996). Oxytocin mediates acquisition of maternally associated odor preferences in preweanling rat pups. *Behavioral Neuroscience 110*(3), 583–592.
Nephew, B. C., & Bridges, R. S. (2008). Arginine vasopressin V1a receptor antagonist impairs maternal memory in rats. *Physiology and Behavior, 95*(1–2), 182–186.
Northcutt, K. V., Wang, Z., & Lonstein, J. S. (2007). Sex and species differences in tyrosine hydroxylase-synthesizing cells of the rodent olfactory extended amygdala. *Journal of Comparative Neurology, 500*(1), 103–115.
Olazabal, D. E., & Young, L. J. (2006). Species and individual differences in juvenile female alloparental care are associated with oxytocin receptor density in the striatum and the lateral septum. *Hormones & Behavior, 49*(5), 681–687.
Oliveira, G., Ataide, A., Marques, C., Miguel, T. S., Coutinho, A. M., Mota-Vieira, L.,...et al. (2007). Epidemiology of autism spectrum disorder in Portugal: prevalence, clinical characterization, and medical conditions. *Developmental Medicine and Child Neurology, 49*(10), 726–733.
Ophir, A. G., Campbell, P., Hanna, K., & Phelps, S. M. (2008). Field tests of cis-regulatory variation at the prairie vole avpr1a locus: association with V1aR abundance but not sexual or social fidelity. *Hormones & Behavior, 54*(5), 694–702.
Ophir, A. G., Gessel, A., Zheng, D. J., & Phelps, S. M. (2012). Oxytocin receptor density is associated with male mating tactics and social monogamy. *Hormones & Behavior, 61*(3), 445–453.
Ophir, A. G., Phelps, S. M., Sorin, A. B., & Wolff, J. O. (2008). Social but not genetic monogamy is associated with greater breeding success in prairie voles. *Animal Behaviour, 75*, 1143–1154.
Ophir, A. G., Wolff, J. O., & Phelps, S. M. (2008). Variation in neural V1aR predicts sexual fidelity and space use among male prairie voles in semi-natural settings. *Proceedings of the National Academy of Science of the United States of America, 105*(4), 1249–1254.
Ostfeld, R. S. (1990). The ecology of territoriality in small mammals. *Trends in Ecology and Evolution, 5*(12), 411–415.
Pedersen, C. A., Ascher, J. A., Monroe, Y. L., & Prange, A. J., Jr. (1982). Oxytocin induces maternal behavior in virgin female rats. *Science, 216*(4546), 648–650.
Pfister, H. P. & Muir, J. L. (1989). Influence of exogenously administered oxytocin on central noradrenaline, dopamine and serotonin levels following psychological stress in nulliparous female rats (Rattus norvegicus). *International Journal of Neuroscience, 45*(3–4): 221–229.

Phelps, S. M. & Young, L. J. (2003). Extraordinary diversity in vasopressin (V1a) receptor distributions among wild prairie voles (Microtus ochrogaster): patterns of variation and covariation. *Journal of Comparative Neurology, 466*(4), 564–576.

Pitkow, L. J., Sharer, C. A., Ren, X., Insel, T. R., Terwilliger, E. F., & Young, L. J. (2001). Facilitation of affiliation and pair-bond formation by vasopressin receptor gene transfer into the ventral forebrain of a monogamous vole. *Journal of Neuroscience, 21*(18), 7392–7396.

Pizzuto, T. & Getz, L. L. (1998). Female prairie voles (Microtus ochrogaster) fail to form a new pair after loss of mate. *Behavioral Processes, 43*, 79–86.

Popik, P. & van Ree, J. M. (1991). Oxytocin but not vasopressin facilitates social recognition following injection into the medial preoptic area of the rat brain. *European Neuropsychopharmacology, 1*(4), 555–560.

Rathbun, G. B. (1979). The social structure and ecology of elephant-shrews. *Zeitschrift fur Tierpsychologie, 20*,1–76.

Resendez, S. L., Kuhnmuench, M., Krzywosinski, T., & Aragona, B. J. (2012). kappa-Opioid receptors within the nucleus accumbens shell mediate pair bond maintenance. *Journal of Neuroscience, 32*(20), 6771–6784.

Roberts, R. L., Miller, A. K., Taymans, S. E., & Carter, C. S. (1998). Role of social and endocrine factors in alloparental behavior of prairie voles (Microtus ochrogaster). *Canadian Journal of Zoology, 76*,1862–1868.

Roberts, R. L., Williams, J. R., Wang, A. K., & Carter, C. S. (1998). Cooperative breeding and monogamy in prairie voles: influence of the sire and geographical variation. *Animal Behavior, 55*(5), 1131–1140.

Ross, H. E., Cole, C. D., Smith, Y., Neumann, I. D., Landgraf, R., Murphy, A. Z., . . . et al. (2009). Characterization of the oxytocin system regulating affiliative behavior in female prairie voles. *Neuroscience 162*(4), 892–903.

Ross, H. E., Freeman, S. M., Spiegel, L. L., Ren, X., Terwilliger, E. F., & Young, L. J. (2009). Variation in oxytocin receptor density in the nucleus accumbens has differential effects on affiliative behaviors in monogamous and polygamous voles. *Journal of Neuroscience, 29*(5), 1312–1318.

Ross, H. E. & Young, L. J. (2009). Oxytocin and the neural mechanisms regulating social cognition and affiliative behavior. *Frontiers in Neuroendocrinology, 30*(4), 534–547.

Scantamburlo, G., Hansenne, M., Fuchs, S., Pitchot, W., Marechal, P., Pequeux, C., . . . Legros, J.J. (2007). Plasma oxytocin levels and anxiety in patients with major depression. *Psychoneuroendocrinology, 32*(4), 407–410.

Shapiro, L. E., Austin, D., Ward, S. E., & Dewsbury, D. A. (1986). Familiarity and female mate choice in 2 species of voles (microtus-ochrogaster and microtus-montanus). *Animal Behaviour, 34*, 90–97.

Shapiro, L. E., & Insel, T. R. (1990). Infant's response to social separation reflects adult differences in affiliative behavior: A comparative developmental study in prairie and montane voles. *Developmental Psychobiology, 23*(5), 375–393.

Smeltzer, M. D., Curtis, J. T., Aragona, B. J., & Wang, Z. (2006). Dopamine, oxytocin, and vasopressin receptor binding in the medial prefrontal cortex of monogamous and promiscuous voles. *Neuroscience Letters, 394*(2): 146–151.

Stone, A. I., & Bales, K. L. (2010). Intergenerational transmission of the behavioral consequences of early experience in prairie voles. *Behavioural Processes, 84*(3), 732–738.

Stowe, J. R., Liu, Y., Curtis, J. T., Freeman, M. E., & Wang, Z. (2005). Species differences in anxiety-related responses in male prairie and meadow voles: the effects of social isolation. *Physiology & Behavior, 86*(3), 369–378.

Stribley, J. M. & Carter, C. S. (1999). Developmental exposure to vasopressin increases aggression in adult prairie voles. *Proceedings of the National Academy of Science of the United States of America, 96*(22), 12601–12604.

Takayanagi, Y., Yoshida, M., Bielsky, I. F., Ross, H. E., Kawamata, M., Onaka, T., . . . Nishimon, K.(2005). Pervasive social deficits, but normal parturition, in oxytocin receptor-deficient mice. *Proceedings of the National Academy of Science of the United States of America 102*(44), 16096–16101.

Taymans, S. E., DeVries, A. C., DeVries, M. B., Nelson, R. J., Friedman, T. C., Castro, M., . . . Chrousos, G.P.(1997). The hypothalamic-pituitary-adrenal axis of prairie voles (Microtus ochrogaster): evidence for target tissue glucocorticoid resistance. *General and Comparative Endocrinology, 106*(1), 48–61.

Tribollet, E., S. Charpak, A. Schmidt, M. Dubois-Dauphin and J. J. Dreifuss (1989). Appearance and transient expression of oxytocin receptors in fetal, infant, and peripubertal rat brain studied by autoradiography and electrophysiology. *Journal of Neuroscience, 9*(5), 1764–1773.

Turner, L. M., Young, A. R., Rompler, H., Schoneberg, T., Phelps, S. M., & Hoekstra, H. E. (2010). Monogamy Evolves through Multiple Mechanisms: Evidence from V1aR in Deer Mice. *Molecular Biology and Evolution, 27*(6), 1269–1278.

Ueta, Y., Dayanithi, G., & Fujihara, H. (2011). Hypothalamic vasopressin response to stress and various physiological stimuli: visualization in transgenic animal models. *Hormones & Behavior, 59*(2), 221–226.

Uvnas-Moberg, K. (1997). Physiological and endocrine effects of social contact. *Annals of the New York Academy of Science, 807*, 146–163.

Uvnas-Moberg, K. (1998). Oxytocin may mediate the benefits of positive social interaction and emotions. *Psychoneuroendocrinology 23*(8), 819–835.

Uvnas-Moberg, K., Bruzelius, G. P. Alster, P., & Lundeberg, T. (1993). The antinociceptive effect of non-noxious sensory stimulation is mediated partly through oxytocinergic mechanisms." *Acta Physiology Scandinavia, 149*(2), 199–204.

van Londen, L., Goekoop, J. G., van Kempen, G. M., Frankhuijzen-Sierevogel, A. C., Wiegant, V. M., van der Velde, E. A., & De Wied, D. (1997). "Plasma levels of arginine vasopressin elevated in patients with major depression." *Neuropsychopharmacology* 17(4): 284–292.

Van Wimersma Greidanus, T. B., Kroodsma, J. M., Pot, M. L., Stevens, M., & Maigret, C. (1990). "Neurohypophyseal hormones and excessive grooming behaviour." *Eurupean Journal of Pharmacology* 187(1): 1–8.

Veenema, A. H. (2012). Toward understanding how early-life social experiences alter oxytocin-and vasopressin-regulated social behaviors. *Hormones & Behavior, 61*(3), 304–312.

Walum, H., Lichtenstein, P., Neiderhiser, J. M., Reiss, D., Ganiban, J. M., Spotts, E. L., . . . Westberg, L. (2012). Variation in the oxytocin receptor gene is associated with pair-bonding and social behavior. *Biological Psychiatry, 71*(5), 419–426.

Walum, H., Westberg, L., Henningsson, S., Neiderhiser, J. M., Reiss, D., Igl, W., . . . Lichetenstein, P. (2008). Genetic variation in the vasopressin receptor 1a gene (AVPR1A) associates with pair-bonding behavior in humans. *Proceedings of the National Academy of Sciences of the United States of America, 105*(37), 14153–14156.

Wang, Z., Ferris, C. F., & De Vries, G. J. (1994). Role of septal vasopressin innervation in paternal behavior in prairie voles (Microtus ochrogaster). *Proceedings of the National Academy of Science of the United States of America, 91*(1), 400–404.

Wang, Z., & Novak, M. A. (1992). Influence of the social environment on parental behavior and pup development of meadow voles (Microtus pennsylvanicus) and prairie voles (M. ochrogaster). *Journal of Comparative Psychology, 106*(2), 163–171.

Wang, Z., & Novak, M. A. (1994). Alloparental care and the influences of father presence on juvenile prarie voles, Microtus ochrogaster. *Animal Behaviour, 47*(2), 281–288.

Wang, Z., Toloczko, D., Young, L. J., Moody, K., Newman, J. D., & Insel, T. R. (1997). "Vasopressin in the forebrain of common marmosets (Callithrix jacchus): studies with in situ hybridization, immunocytochemistry and receptor autoradiography." *Brain Research* 768(1–2): 147–156.

Wang, Z., Young, L. J., Liu, Y., & Insel, T. R. (1997b). Species differences in vasopressin receptor binding are evident early in development: comparative anatomic studies in prairie and montane voles. *Journal of Comparative Neurology, 378*(4), 535–546.

Wang, Z., Yu, G., Cascio, C., Liu, Y., Gingrich, B., & Insel, T. R. (1999). Dopamine D2 receptor-mediated regulation of partner preferences in female prairie voles (Microtus ochrogaster): A mechanism for pair bonding? *Behavioral Neuroscience, 113*(3), 602–611.

Wang, Z., Zhou, L., Hulihan, T. J., & Insel, T. R. (1996). Immunoreactivity of central vasopressin and oxytocin pathways in microtine rodents: a quantitative comparative study. *Journal of Comparitive Neurology, 366*(4), 726–737.

Wassink, T. H., Piven, J., Vieland, V. J., Pietila, J., Goedken, R. J., Folstein, S. E., & Sheffield, V.C.. (2004). "Examination of AVPR1a as an autism susceptibility gene." *Molecular Psychiatry* 9(10): 968–972.

Watters, J. J., Wilkinson, C. W., & Dorsa, D. M. (1996). Glucocorticoid regulation of vasopressin V1a receptors in rat forebrain. *Molecular Brain Research, 38*(2): 276–284.

Wermter, A. K., Kamp-Becker, I., Hesse, P., Schulte-Korne, G., Strauch, K.,& Remschmidt, H. (2010). Evidence for the involvement of genetic variation in the oxytocin receptor gene (OXTR) in the etiology of autistic disorders on high-functioning level. *American Journal of Medical Genetics, Part B: Neuropsychiatric Genetics, 153B*(2), 629–639.

Whitnall, M. H. (1993). Regulation of the hypothalamic corticotropin-releasing hormone neurosecretory system. *Progress in Neurobiology, 40*(5), 573–629.

Whitnall, M. H., Key, S., Ben-Barak, Y., Ozato, K., & Gainer, H. (1985). Neurophysin in the hypothalamo-neurohypophysial system. II. Immunocytochemical studies of the ontogeny of oxytocinergic and vasopressinergic neurons. *Journal of Neuroscience, 5*(1), 98–109.

Williams, J. R., Catania, K. C., & Carter, C. S. (1992). Development of partner preferences in female prairie voles (Microtus ochrogaster): The role of social and sexual experience. *Hormones & Behavior, 26*(3): 339–349.

Williams, J. R., Insel, T. R., Harbaugh, C. R., & Carter, C. S. (1994). Oxytocin administered centrally facilitates formation of a partner preference in female prairie voles (Microtus ochrogaster). *Journal of Neuroendocrinology, 6*(3), 247–250.

Windle, R. J., Kershaw, Y. M., Shanks, N., Wood, S. A., Lightman, S. L. & Ingram, C. D. (2004). Oxytocin attenuates stress-induced c-fos mRNA expression in specific forebrain regions associated with modulation of hypothalamo-pituitary-adrenal activity. *Journal of Neuroscience, 24*(12), 2974–2982.

Windle, R. J., Shanks, N., Lightman, S. L., & Ingram, C. D. (1997). Central oxytocin administration reduces stress-induced corticosterone release and anxiety behavior in rats. *Endocrinology 138*(7), 2829–2834.

Winslow, J. T., Hastings, N., Carter, C. S., Harbaugh, C. R., & Insel, T. R. (1993). A role for central vasopressin in pair bonding in monogamous prairie voles." *Nature 365*(6446), 545–548.

Winslow, J. T., Hearn, E. F., Ferguson, J., Young, L. J., Matzuk, M. M., & Insel, T. R. (2000). Infant vocalization, adult aggression, and fear behavior of an oxytocin null mutant mouse. *Hormones & Behavior, 37*(2), 145–155.

Winslow, J. T., & Insel, T. R. (1993). Effects of central vasopressin administration to infant rats. *European Journal of Pharmacology, 233*(1), 101–107.

Winslow, J. T., & Insel, T. R. (2004). Neuroendocrine basis of social recognition. *Current Opinions in Neurobiology, 14*(2), 248–253.

Wise, R. A. (2002). Brain reward circuitry: insights from unsensed incentives. *Neuron, 36*(2): 229–240.

Witt, D. M., Carter, C. S., & Walton, D. M. (1990). Central and peripheral effects of oxytocin administration in prairie voles (Microtus ochrogaster). *Pharmacology Biochemistry & Behavior, 37*(1), 63–69.

Wittenberger, J. F., & Tilson, R. L. (1980). The evolution of monogamy: hypotheses and evidence. *Annual Review of Ecology, Evolution, and Systematics, 11*, 197–232.

Wotjak, C. T., Kubota, M., Liebsch, G., Montkowski, A., Holsboer, F., Neumann, I., & Landgraf, R.(1996). Release of vasopressin within the rat paraventricular nucleus in response to emotional stress: a novel mechanism of regulating adrenocorticotropic hormone secretion? *Journal of Neuroscience 16*(23), 7725–7732.

Wu, S., Jia, M., Ruan, Y., Liu, J., Guo, Y., Shuang, M.,. Zhang, D. (2005). Positive association of the oxytocin receptor gene (OXTR) with autism in the Chinese Han population. *Bioogicall Psychiatry, 58*(1), 74–77.

Yamamoto, Y., Cushing, B. S., Kramer, K. M., Epperson, P. D., Hoffman, G. E., & Carter, C. S. (2004). Neonatal manipulations of oxytocin alter expression of oxytocin and vasopressin immunoreactive cells in the paraventricular nucleus of the hypothalamus in a gender-specific manner. *Neuroscience, 125*(4), 947–955.

Yirmiya, N., Rosenberg, C., Levi, S., Salomon, S., Shulman, C., Nemanov, L.,. . .Ebstein, R.P(2006). "Association between the arginine vasopressin 1a receptor (AVPR1a) gene and autism in a family-based study: mediation by socialization skills." *Molecular Psychiatry* 11(5): 488–494.

Young, K. A., Gobrogge, K. L., Liu, Y., & Wang, Z. (2001). The neurobiology of pair bonding: insights from a socially monogamous rodent. *Front Neuroendocrinology, 32*(1), 53–69.

Young, L. J. (2011). Can understanding social preferences in rodents lead to novel pharmacotherapies for social anxiety and avoidance in psychiatric disorders? *Neuropsychopharmacology 36*(11), 2151–2152.

Young, L. J., & Flanagan-Cato, L. M. (2012). Editorial comment: oxytocin, vasopressin and social behavior. *Hormones and Behavior, 61*(3), 227–229.

Young, L. J. & Hammock, E. A. D. (2007). On switches and knobs, microsatellites and monogamy. *Trends in Genetics, 23*(5), 209–212.
Young, L. J., Murphy Young, A. Z., & Hammock, E. A. (2005). Anatomy and neurochemistry of the pair bond. *Journal of Comparitive Neurology, 493*(1), 51–57.
Young, L. J., Nilsen, R., Waymire,K. G., MacGregor, G. R., & Insel T. R. (1999). Increased affiliative response to vasopressin in mice expressing the V1a receptor from a monogamous vole. *Nature 400*(6746), 766–768.
Young, L. J. and Z. Wang (2004). "The neurobiology of pair bonding." *Nature Neurosci ence* 7(10): 1048–1054.
Young, L. J., J. T. Winslow, R. Nilsen and T. R. Insel (1997). "Species differences in V1a receptor gene expression in monogamous and nonmonogamous voles: behavioral consequences." *Behavioral Neuroscience* **111**(3): 599–605.

CHAPTER

6 Gene-by-Environment Interactions in Primates

Christina S. Barr

Abstract

Because of their complex social structures, behaviors, and genetic similarities to humans, nonhuman primates are useful for studying how genetic factors influence alcohol consumption. The neurobiological systems that influence addiction vulnerability may do so by acting on alcohol response, reward pathways, behavioral dyscontrol, and vulnerability to stress and anxiety. Rhesus macaques show individual differences in alcohol response and temperament, and such differences are influenced by genetic variants that are similar functionally to those present in humans. Genes in which variation moderates these phenotypes provide opportunities for modeling how genetic and environmental factors (i.e., stress exposure, individual's sex, or alcohol response) interact to influence alcohol consumption. Studies in primates may also reveal selective factors that have driven maintenance or fixation of alleles that increase risk for alcohol use disorders in modern humans.

Key Words: gene–environment interaction, G×E, stress, alcohol, primates, monoamine oxidase A (MAOA-LPR), serotonin transporter (5-HTTLPR), corticotropin releasing hormone (CRH-248C/T and -2232 C/G), neuropeptide Y (NPY-1002 T/G), μ-opioid receptor (OPRM1 C77G)

Many forms of psychopathology and/or psychiatric illness can occur through the pathways of altered reward sensitivity, impulsivity, and anxious responding. All of these are heritable traits, with neurogenetic underpinnings. In many cases, although the genetic factors that contribute to these traits may be adaptive, they can, under the environmental conditions that we face as modern humans, contribute to psychopathological outcomes. This review discusses the roles of temperament, reactivity, and behavioral differences as they relate to survival and reproductive success and presents data generated in nonhuman primates that show us potential mechanisms by which genes that predict adaptive traits can lead to psychiatric problems or psychopathology.

In evolutionary biology, a behavior is adaptive if it makes the organism more fit to survive and reproduce in comparison to other members of the same species. Several decades ago, Maynard Smith applied "game theory" to animal behavior and found that aggressivity and fearfulness are traits that tend to balance each other in any social population. He described the "hawk-dove game" in which "hawks" (aggressive individuals, who were proactive and adopted fight-or-flight responses to stress) and "doves" (who were fearful, cooperative, and adopted a freeze-and-hide strategy with stress) were likely to co-occur in the same species. He determined that both "hawk" and "dove" strategies could potentially be adaptive, perhaps especially so in certain environmental contexts. Whereas "hawks" would be predicted to do better when food is abundant and population density is high (aka, they are better off fighting for access to mates than foraging), "doves" would likely outcompete "hawks" when the opposite were true ("doves" would be better at getting food during periods of resource scarcity and avoiding danger during times of increased conflict). The two types of temperaments must be balanced in any given population because if there is a population comprised completely of "hawks" there will be an excessive level of aggressive encounters among members of the population, reducing the likelihood that any one individual will survive to successfully reproduce; if the population is one comprised by only "doves," then they will be killed off by some other "hawk" species. In other words, the presence of both types of "temperaments" may be critical to survival of the species while, at the same time, selective pressure at the individual level will depend on life history variables, sex, or environmental context. It then stands to reason that the genetic factors that underlie these two alternative strategies would likely be subject to balancing selection and that, as a result, both types of traits or strategies will be observed among individuals of any social species (Figure 6.1).

As stated earlier, certain traits are known to predict vulnerability to developing psychopathology. Although individual and breed-based differences in temperament have been studied in a variety of species, of particular relevance to the human condition may be those studies performed in nonhuman primates. Early studies of temperament differences in nonhuman primates indicated that rhesus macaques (*Macaca mulatta*) could be classified as being either "uptight" or "laid back" (Higley & Suomi, 1989; Suomi, 1982). Vervet monkeys (*Cercopithecus aethiops*) can be similarly characterized on the basis of behavior in the social group and responses to challenge (Bradwejn et al., 1992; Palmour, Mulligan, Howbert, & Ervin, 1997). In the social group, calm vervets are more active, are groomed more often, and compete more effectively for resources. Uptight monkeys, on the other hand, are more isolated and exhibit extremely submissive behaviors (cowering or crouching). When placed in a single cage, the uptight vervets exhibit stereotypical behaviors, whereas those that are laid back sit quietly, exploring their environment.

There is ample evidence that both the intermediate phenotypes that predict temperament and temperament itself are linked to neurobiological substrates in a variety of animal species. The Old World monkeys (e.g., macaques and vervets), who are only 25 mya diverged from humans, give us evidence for these relationships. It has been shown that anxious (excitable) macaques exhibit increased behavioral and endocrine responses to stress in the laboratory. Field studies show that, in addition to being inhibited, excitable animals have higher

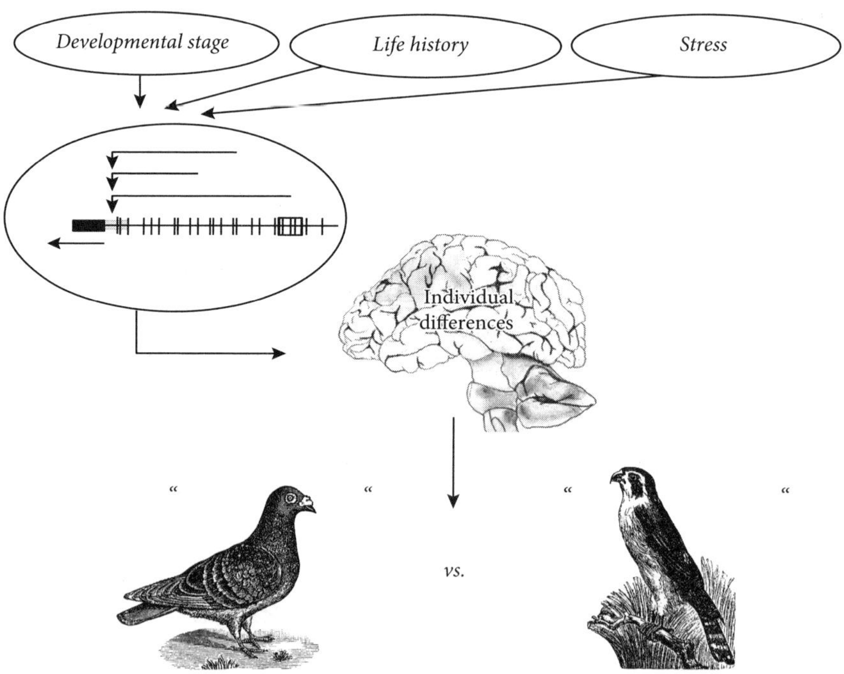

Fig. 6.1 Genetic Complexity Combined with Environmental Plasticity at Genes Important in Regulating Behavior May Favor "Alternative Strategies." This figure illustrates how factors such as developmental stage, life history variables, and stress exposure can differentially regulate genes expressed in brain in a genotype-dependent manner to promote diversity in behavioral strategies within a given population.

cerebrospinal fluid (CSF) levels of corticotropin releasing hormone (CRH) and a high degree of electroencephalographic (EEG) laterality (Kalin & Shelton, 2003; Kalin, Shelton, & Davidson, 2000), both of which have been documented in anxious or depressed human subjects. Other field studies show that impulsive (confident) macaques engage in risky behaviors and aggressive encounters and migrate from their natal troops at a younger age (Mehlman et al., 1994; 1995). They also have lower CSF levels of the serotonin metabolite, 5-hydroxyindoleacetic acid (5-HIAA) (Westergaard et al., 2003). In both instances, these traits appear to be fairly consistent across time and situation (Higley et al., 1996*a*; Higley, Suomi, & Linnoila, 1996*b*; Shannon et al., 2005; Kalin & Shelton, 2003). Using a tool that assesses individual responses to an unfamiliar conspecific, studies have shown that both anxious and impulsive behaviors appear to be heritable in vervet monkeys (Fairbanks et al., 2004). Other studies show anxiety to be heritable in rhesus monkeys (Williamson et al., 2003).

I. Nonhuman Primates to Model Stress Effects and G×E Interactions

Stress is a universal condition of life; if it is chronic, severe, or occurring during critical developmental windows, it contributes to a variety of disease vulnerabilities, particularly disorders of the brain (McEwen, 2006). In humans, there are known links between stress and a variety of psychiatric disorders, including depression, posttraumatic stress disorder (PTSD), anxiety disorders, and substance use disorders (Neigh, Gillespie & Nemeroff, 2009; Sinha & Li, 2007). Genetic variants that influence stress reactivity could determine whether an individual will be vulnerable to such disorders (Caspi et al., 2002; 2003). This may occur by determining environmental sensitivity, magnitude, and/or duration of stress response, or the *types* of responses exhibited by an individual following stress exposure. Genetic variation that determines whether individuals habituate or sensitize to repeated or chronic stress would also be expected to play a major role (Spinelli et al., 2012).

Rhesus macaques provide an opportunity to examine gene-by-environment (G×E) interactions in a controlled, prospective manner (Barr et al., 2003*b*; 2004*d*; Barr & Goldman, 2006). Not only can variables like diet, light–dark cycle, and other factors that are contributors to "noise" in human genetic studies be controlled, but environmental factors that might contribute to G×E interactions can also be applied in a controlled manner. Such factors commonly studied in behavioral research include alcohol exposure (prenatal or in adolescence/adulthood), acute challenge (e.g., the presentation of an intruder or separation from attachment sources), response to treatment (pharmacogenetics, including alcohol response), or early life stress (variable foraging demand or peer/nursery rearing) (reviewed in Barr & Goldman, 2006).

Peer rearing is probably the most extensively reported environmental factor for rhesus macaque G×E studies. As with other primate species, rhesus macaque mothers invest much of their energy into defending, comforting, and caring for their infants, and this maternal buffering appears to be critical to normal infant development (Suomi, 1982). In the so-called peer rearing condition, subjects are removed from their parents at birth and reared with other age-matched infants so that they develop in the absence of adult influence (Chamove, Rosenblum, & Harlow, 1973; Harlow & Suomi, 1974). Peer-reared monkeys develop strong bonds with their age mates and use them as a base from which to explore. When compared to their mother-reared counterparts, however, peer-reared subjects exhibit evidence of insecure attachment, higher levels of anxiety, and lower levels of exploration in novel settings (Suomi, 1982). Because their peers do not necessarily punish inappropriate behavior, they also can have impaired development of the behavioral inhibition system, and, as in humans, macaques that have been exposed to early adversity (in the form of peer-rearing) show long-lasting differences in brain function and behavior (Higley, Hasert, Suomi, & Linnoila, 1991; Spinelli et al., 2009; 2010).

What follows is a discussion of some of the key neurogenetic studies that have been performed in the rhesus macaque, showing how genetic factors that could be adaptive in certain contexts can also render individuals more or less sensitive to particular environmental variables, ultimately driving potentially maladaptive responses. The behavioral systems on which this chapter focuses will include reward sensitivity, stress reactivity, and impulsivity. For some of these, both putative molecular mechanisms and generalizability of environmental factors (aka, stress vs. alcohol exposure, which is a stressor) will be briefly discussed. The chapter will close with a review of the presented findings in terms of how they could simultaneously promote evolutionary success while making modern humans more vulnerable to psychopathology and the addictions, with the additional treatment of the issue of sexually dichotomous G×E interactions.

II. Candidate Gene Studies Performed in Rhesus Macaques

A. Reward Sensitivity and OPRM1 Genotype Responses to Both Natural and Artificial Rewards

The reward systems are critical to survival and reproduction because they are involved in driving ingestion of food, social interactions, and sexual activity, to name a few. During times when food availability is scarce, these are particularly important because they relate to resource acquisition and control. Of interest, some of the behaviors that can increase the likelihood of gaining access to limited resources (i.e., aggression) also involve the reward pathways.

The endogenous opioid system is one system that is activated in response to both natural and artificial rewards. In multiple primate species, there are nonsynonymous single-nucleotide polymorphisms (SNPs) in the first exon of the OPRM1 gene that encodes the opioid mu receptor, which produce amino acid changes in the ligand-binding domain of the receptor (Bond et al., 1998; Miller et al., 2004). Among these are the A118G SNP in humans and the C77G SNP in rhesus macaques, both of which, based on early in vitro work, appeared to increase the affinity of the receptor for its endogenous ligand (ß-endorphin) by approximately three-fold. Studies examining intermediate phenotypes likely to be under the control of this receptor (e.g., hypothalamus-pituitary-adrenal [HPA] axis activity) also suggest gain-of-function roles for these polymorphisms (Barr et al., 2007; Chong et al., 2006; Ray & Hutchison, 2004; Schwandt et al., 2011).

In both humans and rhesus, the minor alleles for these SNPs also predict increased alcohol-induced stimulation. This is highly relevant for both risk and treatment approaches for alcohol use and addiction because, despite the fact that numerous studies have demonstrated that the opioid receptor antagonist naltrexone (NTX) is effective in clinical treatment of alcohol dependence, one large negative trial called into question the efficacy of this compound for the treatment of alcohol problems. More recent studies indicated that the OPRM1 genotype might be deterministic for NTX response (Anton et al., 2008; Oslin et al., 2003), and several labs replicated these findings in rhesus macaques, verifying that the OPRM1 genotype is one factor that predicts naltrexone response (Barr et al., 2010; Vallender, Ruidi-Bettschen, Miller, & Platt, 2010).

The OPRM1 SNPs present in human and in rhesus macaques not only predict responses to artificial rewards such as alcohol, but also to natural rewards as well. In rhesus, the 77G allele predicts increased expression of attachment of an infant to its mother, and this is particularly true following repeated exposures to periods of maternal separation (Barr et al., 2008*a*). This finding has also been replicated in a study performed using a sample of human children aged 9–13, in which various measures of attachment to the caregiver were increased as a function of parental inconsistency or unavailability (Copeland et al., 2011). These studies provide examples of how the macaque can be used to both confirm findings derived from human neurogenetic or pharmacogenetic studies and, more importantly, provide a powerful model for informing analyses that should be performed in humans.

B. Impulsivity in a High-Risk Environment: CRH and MAOA

Impulsivity is a trait that is observed in many psychopathological conditions, from the personality disorders to attention deficit hyperactivity disorder (ADHD) to the addictions. It is a coping mechanism that, at certain life history stages and in certain environmental contexts, might be an adaptive one. The CRH system is one that is critical to behavioral responses to stress, and studies utilizing experimental manipulations of CRH system activity suggest that naturally occurring CRH gene variation may mediate individual variability in behavioral traits that are key to determining an individual's coping style. One of the most consistent behavioral correlates of CRH system activity is the way an organism approaches novelty and unfamiliar conspecifics (Kalin et al., 2000; Korte, Koolhaas, Wingfield, & McEwen, 2005). Individuals that readily seek out and investigate novel stimuli are considered "exploratory" or "bold"; those more likely to show fear or withdrawal when confronted with new objects or individuals are described as more "inhibited" or "shy" (Kagan, Resnick, & Snidman, 1988).

The CRH haplotype has been shown to predict behavioral inhibition in children (Smoller et al., 2003), and studies in rhesus macaques suggest that human CRH variation may moderate risk for alcohol use disorders, perhaps through the pathway of behavioral inhibition. A rhesus polymorphism (-2232 C/G) that has similar in vitro functional effects to some CRH haplotypes reported in humans

(Wagner et al., 2006) predicts decreased CSF levels of CRH, an intermediate phenotype demonstrated in individuals or strains characterized as being particularly extroverted, aggressive, or bold. Infant macaques carrying the G allele are characterized as being more exploratory and bold (Barr et al., 2008*b*), and, following adolescence, males that are G allele carriers exhibit a bolder and more active response to an unfamiliar conspecific. In rhesus macaques, this allele and early environmental stress in the form of peer-rearing produce additive effects on high-risk alcohol consumption and suggests that CRH variation may influence early or uncontrolled alcohol consumption in human populations.

Another gene at which variation is linked to impulsivity and impulsive aggression in both animal models and in humans is the monoamine oxidase A (MAOA) gene. Monoamine oxidase A degrades the monoamine transmitters (dopamine, norepinephrine, and serotonin) and therefore can influence synaptic concentrations of these neurotransmitters. A variable number tandem repeat (VNTR) polymorphism in the transcriptional control region for the human MAOA gene (MAOA-LPR) has been shown to produce differential transcriptional activity in vitro (Sabol, Hu, & Hamer, 1998). In humans, the MAOA-LPR low-activity allele predicts decreased prefrontocortical and increased amygdalar responses to emotional stimuli, suggesting impaired ability to control emotional responses during arousal (Meyer-Lindenberg et al., 2006). The low-activity MAOA-LPR allele has also been associated with trait-like, alcohol-independent antisocial behavior in alcohol-dependent populations (Samochowiec et al., 1999; Tikkanen et al., 2009), and various studies have shown that early environmental factors interact with genotype to predict antisocial behavior, aggressiveness, and violence (Caspi et al., 2002; Sjoberg et al., 2008), traits that are present in early-onset, type 2 alcoholics. Other studies show that MAOA genotype is associated with alcohol dependence, age of onset of dependence, and comorbid drug abuse (Contini, Marques, Farcia, Hutz, & Bau, 2006). The role of MAOA genotype in human aggression and violence is also discussed in detail by Buckholtz and Meyer-Lindenberg elsewhere in this volume.

The low-activity allele of rhMAOA-LPR contributes additively with peer-rearing to increase impulsivity, aggression, and alcohol consumption (Newman et al., 2005). These studies in rhesus macaques support the notion that MAOA gene promoter variation may specifically increase the risk for conduct or personality disorders in which impaired impulse control and aggressivity are features, particularly among those exposed to early stress or lack of parental influence.

C. Stress Reactivity in a Stressful Environment: 5-HTTLPR, CRH, and NPY

Of particular interest for the study of G×E interactions is variation within regulatory or coding regions of genes encoding stress-responsive signaling molecules, which may contribute to stress vulnerability or resiliency. Perhaps the most extensively studied gene in psychiatric genetics and in nonhuman primate neurobehavioral studies is that encoding the serotonin transporter (Bennett et al., 2002). The serotonin transporter is a protein critical to regulating serotonin function in the brain since serotonin's action in the synapse is terminated by reuptake. In mice, targeted disruption of the serotonin transporter gene results in increased adrenocorticotrophic hormone (ACTH) and corticosterone responses to immobilization stress, as well as increased anxiety during the elevated plus maze and light/dark exploration tasks (Holmes, Murphy, Gold, & Crawley, 2003; Lanfumey, Mannoury La Cour, Froger, & Hamon et al., 2000; Li et al., 1999). Gene expression studies demonstrate that monkeys with high levels of stress reactivity have lower gene expression levels for the serotonin transporter (Bethea et al., 2004; 2006).

In humans, there is a common, functional repeat length variant in the regulatory region for the serotonin transporter gene (5-HTT). Variation of this serotonin transporter linked polymorphic region (5-HTTLPR) predicts certain personality traits related to anxiety, depression, and aggression, such as neuroticism, harm avoidance, and disagreeableness (Lesch et al., 1997; Mazzanti et al., 1998; see also the chapter by Buckholtz and Meyer-Lindenberg in this volume). There is variation in the serotonin transporter gene regulatory region in many nonhuman primate species (Lesch et al., 1997; Wendland et al., 2005). In rhesus, a 21-bp insertion/deletion polymorphism, rh5-HTTLPR, has been shown to alter transcriptional efficiency (Bennett et al., 2002), resulting in decreased serotonin transporter mRNA levels in brains of l/s macaques (Lopez & Higley, 2002), which may be further regulated by epigenetic mechanisms (Kinnally et al., 2010). The 5-HTTLPR polymorphism has been studied extensively in both rhesus and human G×E studies.

Studies in multiple, independent laboratories demonstrate that the rh5-HTTLPR s allele predicts anxiety and responses to stress in rhesus macaques (Bethea et al., 2004; Champoux et al., 2002). It has also been shown that rh5HTT-LPR genotype interacts with duration and number of stress exposures to predict the patterns and types of behavioral responses that macaque infants exhibit during early development (Spinelli et al., 2012). Other studies in rhesus show that 5-HTTLPR genotype can interact with controlled exposures to prenatal stress/alcohol or early life adversity to result in long-lasting differences in stress reactivity, sensation seeking, aggression, and alcohol consumption (Barr et al., 2004b; Kraemer, Moore, Newman, Barr, & Schneider, 2008; Schneider et al., 2010; Schneider, Moore, Barr, Larson, & Kraemer, 2011; Schwandt et al., 2010; Spinelli et al., 2007). Alcohol consumption appears to decrease anxiety in various primate species. Consistent with this, studies have demonstrated that peer-reared female rhesus macaques carrying the s allele exhibit higher levels of alcohol preference than do l/l animals (Barr et al., 2004*b*), suggesting that they may be seeking alcohol for its anxiolytic effects. It may be that independent effects of genotype on anxiety and alcohol response converge to increase alcohol consumption among these subjects (Barr, 2013). These studies in rhesus have provided support for the notion that 5-HTTLPR variation interacts with stressful life experiences to moderate risk for stress-related alcohol consumption and other related disorders in humans (Caspi et al., 2003; Caspi, Hariri, Holmes, Uher, & Mofitt, 2010; Caspi & Moffitt, 2006).

The CRH system is one that influences not only how individuals approach novel stimuli, as just discussed, but is also one that is critical for physiological and behavioral adaptation to stress. However, chronic overactivity of this system can lead to stress-related pathology (Goldman & Barr, 2002; Korte et al., 2005; McEwen, 2006; Sapolsky, 2001). In fact, dysregulation of the CRH system has been linked to a variety of stress-related psychiatric disorders, including depression, PTSD, and alcohol dependence (Gold & Chrousos, 2002; Hundt, Zimmerman, Pottig, Spring, & Holsboer, 2001; Southwick, Vythilingam, & Charney, 2005). Studies performed in animal models have shown that an up-regulated CRH system can produce anxiety- and/or depression-like phenotypes (Jaferi & Bhatnagar, 2007; Kalin et al., 2000; Servatius et al., 2005; Strome et al., 2002). Those performed in rodents show that CRH system up-regulation (driven either by genetic variation or environmental factors) leads to escalated alcohol drinking (Hansson et al., 2006; Nie et al., 2004; Sommer et al., 2008). From this body of work, it is inferred that up-regulated activity of the CRH system is critical for the transition from impulsive to compulsive alcohol use and, therefore, addiction. Based on these findings, CRHR1 antagonists have been proposed for the treatment of alcohol dependence (Egli, 2005; Heilig & Koob, 2007).

Most studies demonstrating a role for CRH system up-regulation in driving anxiety, stress reactivity, and alcohol seeking have been performed in rodents. However, the relative levels of expression and distributions of key mediators of stress responses differ between rodents and catarrhine primates (Sanchez, Young, Plotsky, & Insel, 1999). As such, demonstrating a link between increased CRH system activity and individual differences in behavior in a primate species provides critical support for the notion that rodent findings may translate to humans. There is a CRH promoter SNP in rhesus macaques (-248 C/T) (Barr et al., 2009), which is located in a region that is under purifying selection and has been demonstrated empirically to be critical for regulation of CRH expression (King & Nicholson, 2007). In vitro studies show that the -248 T allele increases cAMP-stimulated CRH promoter activity, but also disrupts glucocorticoid-mediated repression, indicating that this particular SNP would result in augmented CRH expression in response to stress. When effects of -248 C/T genotype on stress responding are examined, the T allele predicts increased behavioral and endocrine responses to stress among monkeys with a history of early stress exposure. A similar pattern is observed for alcohol consumption. The fact that there is a SNP in the corresponding human region (-201 C>T) (http://genome.ucsc.edu/) suggests that CRH promoter variation could interact with environmental stressors to increase stress responding and alcohol consumption in humans.

The neuropeptide Y (NPY) system is one whose regulation mediates stress adaptation and is, therefore, a candidate system in which functional genetic variation may impact stress resilience (Sommer et al., 2010; Zhou et al., 2008). In response to protracted or repeated periods of stress exposure, NPY is released in key regions of the brain, a mechanism proposed to be important for countering the effects of stress and, in particular, that is mediated through the CRH system (Heilig & Koob, 2007). There is considerable evidence suggesting that NPY regulates stress-induced alcohol consumption as well (Badia-Elder et al., 2001; Badia-Elder, Stewart,

Powrozek, Murphy, & Li, 2003). Mice that are deficient for the Npy gene consume more ethanol, whereas consumption is reduced in transgenic mice overexpressing Npy (Thorsell et al., 2007). In humans, linkage to the chromosomal region containing the NPY gene has been demonstrated (Reich et al., 1998), and there have been associations of NPY variation with both alcohol consumption and dependence (Lappalainen et al., 2002; Mottagui-Tabar et al., 2005). Other studies, however, have failed to replicate this association (Zhu et al., 2003; Zill, Preuss, Koller, Bondy, & Soyka, 2008). This could potentially be indicative of the fact that NPY variation could modify alcohol consumption via interactions with other variables.

Studies performed in rhesus macaques support this and point to an interaction between genotype and environmental stress. Using the peer-rearing primate model of early adversity, a loss-of-function variant in rhesus macaques (NPY -1002 T>G) influences CSF levels of NPY, behavioral arousal during periods of stress, and alcohol consumption. In contrast, individuals with no prior stress exposure are unaffected by genotype. The NPY variant also predicts increased alcohol consumption as a function of repeated cycles of alcohol exposure, but only among peer-reared subjects. This suggests a high degree of prior stress exposure to be required for the G allele to produce an effect, raising the possibility that human NPY variation might increase risk for alcohol dependence specifically among individuals with traumatic life experiences or high cumulative levels of stress exposure. In support of this argument, the only studies to have reported a link between functional NPY variation and alcohol dependence involved the use of late-onset alcoholics (Mottagui-Tabar et al., 2005) or samples highly represented by war veterans (Lappalainen et al., 2002). Overall, the data from rhesus macaques suggest a role for NPY gene variation in the susceptibility to alcohol-related disorders and may further implicate the NPY system as a treatment target in selected individuals.

III. Role of Genetic Selection: Understanding the Origins of Human Psychopathology

A number of research groups have been investigating those genetic variations in the rhesus macaque that contribute to the expression of traits that have been linked with human alcohol problems and other psychiatric disorders (i.e., stress reactivity, behavioral dyscontrol, aggression, and reward seeking/sensitivity). What has emerged from this body of work is the fact that, in many cases, the variants identified and studied in the macaque are functionally similar to those present in human populations, and some findings suggest a convergent evolution and that these variants may have been maintained by selection in both species (Barr et al., 2008*a*; 2008*b*; Vallender et al., 2008). These data reinforce how the macaque model has proven itself useful for learning about how relatively common genetic variants, which are associated with traits that may be adaptive in certain environmental contexts, can also increase vulnerability to stress-related or alcohol problems.

Although the field of behavioral genetics is growing rapidly, most of its research is concerned with the identification of "disease alleles" or gene variation underlying what is considered pathological behavior. Its methods and findings, however, can be applied to a long-standing goal of evolutionary anthropology: to understand how changes in allele frequency can affect divergences in primate behavior. Several studies have identified associations between specific alleles and natural features of behavior and life history strategies. For example, the loss-of-function short (s) allele of the serotonin transporter gene promoter length polymorphism (5-HTTLPR), which increases risk for developing depression in the face of adversity, has a functional equivalent in the rhesus macaque (see Bennett et al., 2002). In macaques, this allele is associated with increased endocrine and behavioral stress reactivity as a function of stress exposure, often in a sexually dichotomous manner (Barr et al., 2004b; Schwandt et al., 2010; Spinelli et al., 2007). Therefore, this variant appears to increase risk for developing psychopathology, particularly in the context of stress. Despite this, these variants have been maintained in both humans and in rhesus (in addition to some other nonhuman primate species). Moreover, in human populations in which the s allele is rare, another loss-of-function variant on the L allele background (LA>LG) is present at a higher frequency (Hu et al., 2006). In humans, there is also a VNTR in the second intron, which appears to be functional (Fiskerstrand et al., 1999). This VNTR is present in a number of primate and nonprimate species and is polymorphic in a number of hominoid species (Soeby, Larsen, Olsen, Rasmussen, & Werge, 2005).

Although SNPs are not necessarily conserved across species, there are instances in which functionally similar SNPs occur in the human and rhesus macaque (Barr et al., 2008*a*; 2008*b*; Miller et al., 2004; Vallender et al., 2008). The serotonin transporter gene regulatory region is tremendously variable within species, and it has been demonstrated that gain-of-function *5-HTT* SNPs have arisen and been maintained in both rhesus and humans, suggesting that both gain- and loss-of-function variants may be under selection in primates (Vallender et al., 2008). It is of interest that *5-HTT* variation not only predicts individual differences in impulse control and stress reactivity (Barr et al., 2004b and 2004c; Bennett et al., 2002; Champoux et al., 2002; Schwandt et al., 2010), but is also associated with adaptive traits in free-ranging macaques, such as earlier male dispersal (Trefilov, Berard, Krawczak, & Schmidtke, 2000) and male reproductive timing (Krawczak et al., 2005), and may thus also be adaptive in humans (Homberg & Lesch, 2011).

An impulsive individual that readily approaches novel objects or conspecifics may do well in certain social situations but may face higher risk of predation or attack than a more cautious, harm-avoidant individual. Such behaviors might, therefore, be predicted to confer selective advantage at particular developmental or life history stages and in certain environmental contexts. Moreover, because of differences in their behavioral and physiologic responses to stress, the types of stress-related pathology to which bold, proactive individuals and harm-avoidant, reactive individuals are vulnerable are distinct. Whereas the latter are at risk for internalizing disorders, such as depression and anxiety, the former are more likely to develop externalizing conditions, primarily characterized by impaired impulse control (Korte et al., 2005). Such traits are known to impart risk for alcohol use disorders. In humans, anxiety is a risk factor for developing alcohol problems, and stress exposures can lead to craving and relapse (Barr & Goldman, 2006; Sinha & Li, 2007). It is also known that impulsivity or behavioral dyscontrol can predispose individuals to early and uncontrolled alcohol intake (Barr & Goldman, 2006; Goldman, Oroszi, & Ducci, 2005).

The CRH locus is one at which variation would be predicted to increase stress adaptation or modify behavior in a manner that is adaptive but that may also moderate risk for stress-related disorders in modern humans. In macaques, the two most common haplotypes are yin-yang or alternative haplotypes (Barr et al., 2008*b*; see also Reuter and Montag elsewhere in this volume for a discussion of yin-yang in human positive emotionality). The maintenance of these divergent haplotypes over time is suggestive of the fact that they have been subject to selection, such that at least one of the alleles on each background is being selected—possibly in a particular environmental context—while the rest are hitchhiking. Several studies in humans (Baerwald et al., 1999; Shimmin et al., 2007) have shown evidence for selection at the CRH locus in which, similar to the rhesus macaque, alternative, yin-yang haplotype clades are observed. As in the rhesus macaque, the major human CRH haplotypes have been shown to vary in terms of their in vitro promoter activity, and among the observed differences are those pertaining to glucocorticoid-sensitivity (Wagner et al., 2006). In rhesus macaques, carriers of a CRH -2232 G allele consume higher levels of alcohol when tested with age-matched peers in a social group, a proposed model for high-risk, impulsivity-related alcohol consumption (Barr et al., 2008*b*). They also exhibit lower levels of the serotonin metabolite, 5-HIAA, a neurochemical endophenotype observed both in macaques exposed to early life stress and among individuals with early-onset, type 2 alcoholism (Higley et al., 1991; 1996*a*). It may be that, in humans, genetic variation that alters CRH system function could influence multiple behavioral dimensions (i.e., both neuroticism and extraversion) and that variants that place an individual at the extremes of these spectra (i.e., inhibited and anxious/stress reactive vs. bold/impulsive and novelty seeking) could increase the risk for developing alcohol use disorders. Of note, studies that examine effects of CRHR1 haplotype demonstrate both evidence for selection (Nelson et al., 2010) and a moderating effect of haplotype as it relates to stress-induced alcohol drinking (Blomeyer et al., 2008; Nelson et al., 2010).

As another example, in both rhesus and in humans, there are nonsynonymous SNPs in that portion of the OPRM1 gene that encodes the N-terminal domain of the receptor (C77G in rhesus macaque and A118G in human, as discussed earlier), and these SNPs have been observed to confer similar functions in vivo (Barr et al., 2007; Chong et al., 2006; Ray & Hutchison, 2004). In humans, the 118G allele is suspected to increase the likelihood that an individual will abuse alcohol because it increases alcohol-induced dopamine release and subjective euphoria (Ramchandani et al., 2011; Ray & Hutchison, 2004). We have shown that rhesus carrying the 77G allele exhibit increased alcohol-induced stimulation (a marker for the euphorogenic effects of alcohol) and that

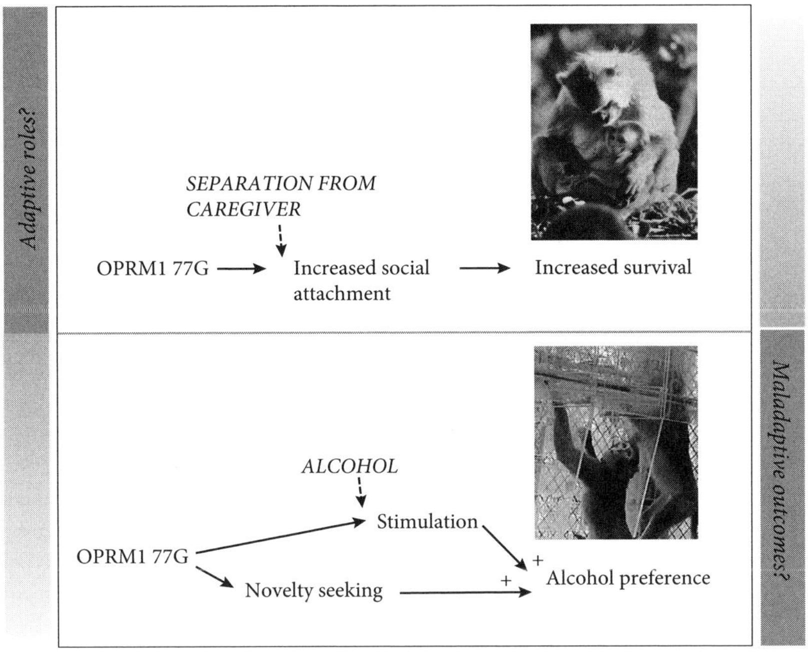

Fig. 6.2 Genetic Variation at the OPRM1 Gene Could Have Conferred Selective Advantage in Ancestral or Ancient Populations (Adaptive Response?), but Also Increases Risk for Alcohol Problems in Modern Humans (Maladaptive Outcome). Macaque infants carrying a functional allele at the OPRM1 gene, which influences reward sensitivity, exhibit increased maternal attachment, especially as a function of repeated maternal separation. They also exhibit increased alcohol preference and consumption, likely through the pathways of novelty seeking and alcohol response.

G allele carriers also consume more alcohol in the laboratory (Barr et al., 2007). It would stand to reason that OPRM1 variation might predict sensitivity to natural rewards as well. Based on the fact that these two variants confer similar functional effects, that both are observed at relatively high frequencies, and, furthermore, that there is an extended region of LD with the A118G allele in humans (Luo et al., 2008; Pang, Wang, Wang, Goh, & Lee, 2009; Zhang et al., 2006), it might be hypothesized that they have evolved as the result of similar selective pressures in the two species. Data to directly address this hypothesis are presently not available. However, studies performed in the macaque demonstrate this variant to predict behaviors that could theoretically be under selection. The 77G allele predicts aggressive behavior (Miller et al., 2004), and G allele carriers form stronger attachment bonds with their mothers during infancy (Barr et al., 2008; Higham et al., 2011), especially as a function of repeated maternal separation. It is of interest that the effects reported to occur during repeated exposures to maternal separation and reunion are similar to those that might be observed during periods of alcohol intake and withdrawal. Similar effects of OPRM1 genotype on the expression of social attachment have recently been demonstrated in human children, showing increased quality of parent–child relations as a function of parental unavailability or inconsistency (Copeland et al., 2011). These types of studies highlight how traits that could have conferred selective advantage at some point in the evolutionary history of humans can increase risk for addictive disorders in modern society (Figure 6.2).

IV. Sexually Dichotomous G×E Interactions

From an evolutionary perspective, the roots of psychopathology may lie in the different strategies that have evolved for coping with environmental challenge (Korte et al., 2005). Because selective pressures differ between the two sexes, males and females exhibit differences in their responses to environmental challenge (Eme, 2007; Wood & Eagly, 2002). In some ways, the strategies adopted by males and females parallel those described in the hawk–dove model: an aggressive/bold strategy (hawk) opposed by a nonaggressive, cautious strategy (dove) (Smith & Price, 1973). The hawk–dove model maps well onto the proposed genetic structure of externalizing and internalizing human psychiatric disorders (Kendler, Prescott, Myers, &

Neale, 2003), including type 1 and type 2 alcohol dependence. This, combined with the observation that females are generally more prone to internalizing disorders, such as anxiety and affective disorders, whereas males are more likely to develop externalizing disorders, such as antisocial behavior and substance use disorders (Cale & Lilenfeld, 2002; Williams et al., 1995), underscores the notion of psychopathology as an outcome of the response to stress and suggests that, in addition to genetic and environmental variables, an individual's sex is likely to play an important role.

The serotonin transporter-linked polymorphism is one for which variation results in sexually dichotomous qualitative and quantitative G×E interactions, and rhesus studies were the first to demonstrate these. Even prior to pubertal development, the 5-HTTLPRs allele predicted increased stress response, but only among females with histories of early adversity (Barr et al., 2004*b*). Responding to social threat is one domain in which males and females are likely to adopt different adaptive solutions, such as those ascribed to the hawk–dove model. Recently, it has been shown that male adolescents with the *s* allele who were exposed to early life stress are more likely to respond aggressively toward an unfamiliar conspecific, a risky response that, under certain circumstances, would also be adaptive (Schwandt et al., 2010). Of relevance to the psychiatric disorders, it may be that the *s* allele confers risk to both internalizing and externalizing disorders in a sex-dependent manner.

V. Conclusion

Because of their complex social structures, behaviors, and genetic similarities to humans, nonhuman primates are useful for studying how genetic factors influence alcohol consumption. The neurobiological systems that influence vulnerability to psychopathology may do so by acting on reward pathways, behavioral dyscontrol, and vulnerability to stress and anxiety. Rhesus macaques show individual differences in reactivity and temperament, and such differences are influenced by genetic variants that are similar functionally to those present in humans. These polymorphisms also predict alcohol consumption in certain environmental contexts. Candidate gene-based studies performed in nonhuman primates appear to have translational value for investigating effects of genetic variation on traits that increase risk for psychopathology and psychiatric disorders, including the addictions (Barr et al., 2004*a*; 2004*b*; 2004*c*; 2009; 2010; Lindell et al., 2010). This body of work has not only been critically important in arguments for the validity of human G×E studies (Caspi et al., 2010), but it has also provided a solid foundation for supporting studies aimed at identifying novel genetic variants in rhesus macaques that may be good candidates for modeling how genetic and environmental variables interact to influence alcohol consumption in modern humans.

VI. Future Directions

In 2009, a high-profile meta-analysis was performed that called into the question the validity of all of the 5-HTTLPR × stress findings that had been reported in the human literature (Risch et al., 2009). The authors of this chapter and those who came out against candidate gene-based and G×E studies in response to its publication claimed that the field needed to look toward whole-genome linkage as a tool. Those who came out in defense of the verity of G×E interactions argued not only that with whole-genome studies comes a substantial loss of power, but also cited many of the animal studies, including those performed in nonhuman primates (Caspi et al., 2010). This debate is ongoing and calls out for studies that might verify the validity of G×E interaction results by elucidating mechanisms by which they might occur. The emergence of genomics and cell biology technologies and approaches are among those that may be critically informative for determining the mechanisms that underlie G×E interactions. These studies may also inform us of genetic variation that promotes stress sensitivity or resilience, permitting the identification of candidate loci for the performance of future G×E interaction studies.

One mechanism by which stress could interact with genotype is via epigenetic mechanisms. The emergence of next-generation sequencing technologies has broadened our potential for the discovery of epigenetic effects. Several companies offer platforms that permit high-throughput sequencing and, by combining immunoprecipitation with these approaches, the effects of stress on levels of histone binding or sites of DNA modification (two forms of epigenetic regulation) can be elucidated (Figure 6.3). It is also possible that genetic factors might be identified that directly promote or inhibit these processes.

Another mechanism by which stress might interact with genetic variation is via corticosteroid receptor-induced regulation of gene expression. Cortisol exerts pleiotropic effects by altering gene expression in brain regions containing high concentrations of glucocorticoid receptors, which can alone

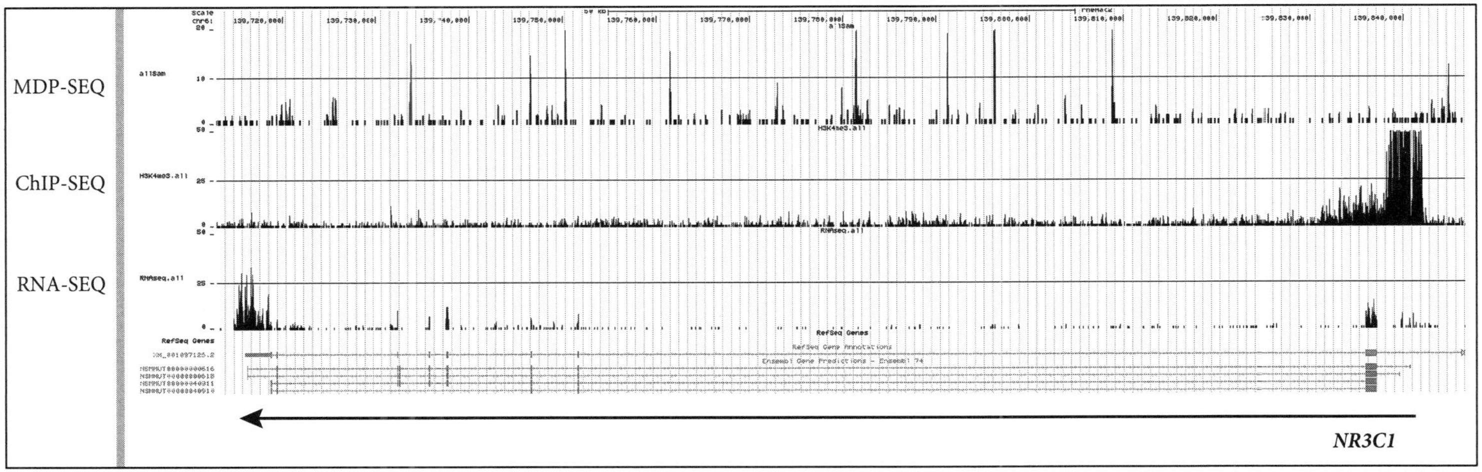

Fig. 6.3 Outputs from Data Generated Using Next-Generation Sequencing Strategies to Identify Stress and Epigenetic Effects in Postmortem Brain Tissue. Data generated using MDP-SEQ (CpG methylation), ChIP-SEQ (chromatin immunoprecipitation using an antibody directed against H3K4me3, which marks active promoters), and RNA-SEQ (RNA sequencing) are shown for a known stress-sensitive gene, NR3C1, which encodes the glucocorticoid receptor. Such strategies may be critically important for identification of underlying mechanisms and/or novel loci for studying gene-by-environment (G×E) interactions.

or in combination with other factors work to influence expression of genes containing glucocorticoid response elements, or GREs. Of interest, all of the gene variants presented in this chapter (5-HTTLPR, CRH -2232 C/G and -248 C/T, MAOA-LPR, NPY -1002 T/G, and OPRM1 C77G) have GREs in their regulatory regions, making them stress responsive, and, in many of the cases in which we've identified gene-by-stress or gene-by-alcohol interactions, glucocorticoid response is either enhanced or disrupted via polymorphism within GREs. Unlike histone proteins, transcription factors are not tightly associated with DNA. However, one might still be able to use next-generation sequencing technologies to identify corticosteroid-sensitive genes. By performing X-ChIP-SEQ (with antiglucocorticoid receptor antibodies) combined with mRNA-SEQ, there may be potential to identify genes that are sensitive to glucocorticoid receptor-mediated activation or repression and to determine if this regulation occurs in a genotype-dependent manner.

Finally, various other nongenomic approaches can be used to assist in elucidating mechanisms through which G×E interactions might occur. The development of induced pluripotent stem cells (IPSCs) could be critically important in determining how G×E interactions occur in brain. These cells, if induced to develop into neurons, could be tremendously informative for examining hormone effects (both sex and stress steroids) and effects of drug or alcohol exposure. The fact that cells can be derived from peripheral tissues affords the opportunity to examine effects of any or all of these variables that are specific to the individual's genetic makeup and, therefore, can be used to examine both within- and between-species differences.

References

Anton, R. F., Oroszi, G., O'Malley, S., Couper, D., Swift, R., Pettinati, H., & Goldman, D. (2008). An evaluation of mu-opioid receptor (OPRM1) as a predictor of naltrexone response in the treatment of alcohol dependence: Results from the Combined Pharmacotherapies and Behavioral Interventions for Alcohol Dependence (COMBINE) study. *Archives of General Psychiatry, 65*, 135–144.

Badia-Elder, N. E., Stewart, R. B., Powrozek, T. A., Murphy, J. M., & Li, T. K. (2003). Effects of neuropeptide Y on sucrose and ethanol intake and on anxiety-like behavior in high alcohol drinking (HAD) and low alcohol drinking (LAD) rats. *Alcoholism: Clinical and Experimental Research, 27*(6), 894–899.

Badia-Elder, N. E., Stewart, R. B., Powrozek, T. A., Roy, K. F., Murphy, J. M., & Li, T. K. (2001). Effect of neuropeptide Y (NPY) on oral ethanol intake in Wistar, alcohol- preferring (P), and -nonpreferring (NP) rats. *Alcoholism: Clinical and Experimental Research, 25*(3), 386–390.

Baerwald, C. G., Mok, C. C., Fife, M. S., Tickly, M., Lau, C. S., Wordsworth, B. P., Ollier, B., Panayi, G. S., & Lanchbury, J. S. (1999). Distribution of corticotropin-releasing hormone promoter polymorphism in different ethnic groups: Evidence for natural selection in human populations. *Immunogenetics, 49*, 894–899.

Barr, C. S. (2013). Non-human primate models of alcohol-related phenotypes: The influences of genetic and environmental factors. *Current Topics in Behavioral Neurosciences, 13*, 223–249.

Barr, C. S., Chen, S. A., Schwandt, M. L., Lindell, S. G., Sun, H., Suomi, S. J., & Heilig, M. (2010). Suppression of alcohol preference by naltrexone in the rhesus macaque: A critical role of genetic variation at the micro-opioid receptor gene locus. *Biological Psychiatry, 67*, 78–80.

Barr, C. S., Dvoskin, R. L., Gupte, M., Sommer, W., Sun, H., Schwandt, M. L., Lindell, S. G., Kasckow, J. W., Suomi, S. J., Goldman, D., Higley, J. D., & Heilig, M. (2009). Functional CRH promoter variation drives stress-induced alcohol consumption in primates *Proceedings of the National Academy of Sciences, 1806*, 14593–14598.

Barr, C. S., Dvoskin, R. L., Yuan, Q., Lipsky, R. H., Gupte, M., Hu, X., Zhou, Z., Schwandt, M. L., Lindell, S. G., McKee, M., Becker, M. L., Kling, M. A., Gold, P. W., Higley, J. D., Heilig, M., Suomi, S. J., & Goldman, D. (2008b). CRH haplotype as a factor influencing cerebrospinal fluid levels of corticotropin-releasing hormone, hypothalamic-pituitary-adrenal axis activity, temperament, and alcohol consumption in rhesus macaques. *Archives of General Psychiatry, 65*(8), 934–944.

Barr, C. S., & Goldman, D. (2006). Nonhuman primate models of inheritance of vulnerability to alcohol abuse and addiction. *Addiction Biology, 11*, 374–385.

Barr, C. S., Newman, T. K., Becker, M. L., Parker, C. C., Champoux, M., Lesch, K. P., Goldman, D., Suomi, S. J., & Higley, J. D. (2003b). The utility of the non-human primate; model for studying gene by environment interactions in behavioral research. *Genes, Brain and Behavior, 2*, 336–340.

Barr, C. S., Newman, T. K., Lindell, S., Shannon, C., Champoux, M., Lesch, K. P., Suomi, S. J., Goldman, D., & Higley, J. D. (2004a). Interaction between serotonin transporter gene variation and rearing condition in alcohol preference and consumption in female primates. *Archives of General Psychiatry, 61*, 1146–1152.

Barr, C. S., Newman, T. K., Schwandt, M., Shannon, C., Dvoskin, R. L., Lindell, S. G., Taubman, H., Thompson, B., Champoux, M., Lesch, K. P., Goldman, D., Suomi, S. J., & Higley, J. D. (2004b). Sexual dichotomy of an interaction between early adversity and the serotonin transporter gene promoter variant in rhesus macaques. *Proceedings of the National Academy of Sciences USA, 101*, 12358–12363.

Barr, C. S., Newman, T. K., Shannon, C., Parker, C., Dvoskin, R. L., Becker, M. L., Schwandt, M., Champoux, M., Lesch, K. P., Goldman, D., Suomi, S. J., & Higley, J. D. (2004c). Rearing condition and rh5-HTTLPR interact to influence limbic-hypothalamic-pituitary-adrenal axis response to stress in infant macaques. *Biological Psychiatry, 55*, 733–738.

Barr, C. S., Schwandt, M. L., Lindell, S. G., Chen, S. A., Suomi, S. J., Goldman, D., Heilig, M., & Higley, J. D. (2007). Mu opioid receptor gene variation is associated with alcohol response and consumption in rhesus monkeys. *Archives of General Psychiatry, 64*, 369–376.

Barr, C. S., Schwandt, M. L., Lindell, S. G., Higley, J. D., Maestripieri, D., Goldman, D., Suomi, S. J., & Heilig, M.

(2008a). A functional OPRM1 variant is associated with attachment behavior in infant rhesus macaques. *Proceedings of the National Academy of Sciences, 105*, 5277–5281.

Barr, C. S., Schwandt, M. L., Newman, T. K., & Higley, J. D. (2004d). The use of adolescent nonhuman primates to model human alcohol intake: Neurobiological, genetic and environmental variables. *Annals of the New York Academy of Sciences, 1021*, 221–233.

Bennett, A. J., Lesch, K.-P., Heils, A., Long, J. C., Lorenz, J. G., Shoaf, S. E., Champoux, M., Suomi, S. J., Linnoila, M. V., & Higley, J. D. (2002). Early experience and serotonin transporter gene variation interact to influence primate CNS function. *Molecular Psychiatry, 7*, 118–122.

Bethea, C. L., Streicher, J. M., Coleman, K., Pau, F. K. Y., Moessner, R., & Cameron, J. L. (2004). Anxious behavior and fenfluramine-induced prolactin secretion in young rhesus macaques with different alleles of the serotonin reuptake transporter polymorphism (5HTTLPR). *Behavior Genetics, 34*, 295–307.

Bethea, C. L., Streicher, J. M., Mirkes, S. J., Sanchez, R. L., Reddy, A. P., & Cameron, J. L. (2006). Serotonin-related gene expression in female monkeys with individual sensitivity to stress. *Neuroscience, 132*, 151–166.

Blomeyer, D., Treutlein, J., Esser, G., Schmidt, M. H., Schumann, G., & Laucht, M. (2008). Interaction between CRHR1 gene and stressful life events predicts adolescent heavy alcohol use. *Biological Psychiatry, 63*, 146–151.

Bond, C., LaForge, K. S., Tian, M. T., Melia, D., Zhang, S. W., Borg, L., Gong, J. H., Schluger, J., Strong, J. A., Leal, S. M., Tischfield, J. A., Kreek, M. J., & Yu, L. (1998). Single-nucleotide polymorphism in the human mu opioid receptor gene alters beta-endorphin binding activity: Possible implications for opiate addiction. *Proceedings of the National Academy of Sciences USA, 95*, 9608–9613.

Bradwejn, J., Kosycki, D., Deutetre-Couteax, A., Bourin, M., Palmour, R. M., & Ervin, F. R. (1992). The cholecystokinin hypothesis of panic and anxiety disorders: A review. *Journal of Psychopharmacology, 6*, 345–351.

Cale, E. M., & Lilienfeld, S. O. (2002). Sex differences in psychopathy and antisocial personality disorder: A review and integration. *Clinical Psychology Review, 22*, 1179–1207.

Caspi, A., McClay, J., Moffitt, T. E., Mill, J., Martin, J., Craig, I. W., Taylor, A., & Poulton, R. (2002). Role of genotype in the cycle of violence in maltreated children. *Science, 297*, 851–854.

Caspi, A., Sugden, K., Moffitt, T. E., Taylor, A., Craig, I. W., Harrington, H., McClay, J., Mill, J., Martin, J., Braithwaite, A., & Poulton, R. (2003). Influence of life stress on depression: Moderation by a polymorphism in the 5-HTT gene. *Science, 301*, 386–389.

Caspi, A., Hariri, A. R., Holmes, A., Uher, R., & Mofitt, T. E. (2010). Genetic sensitivity to the environment; the case of the serotonin transporter gene and its implications for studying complex diseases and traits. *American Journal of Psychiatry, 167*(5), 509–527.

Caspi, A., & Moffitt, T. E. (2006). Gene-environment interactions in psychiatry: Joining forces with neuroscience. *Nature Reviews Neuroscience, 7*(7), 583–590.

Chamove, A. S., Rosenblum, L. A., & Harlow, H. F. (1973). Monkeys (Macaca mulatta) raised only with peers. *Animal Behavior, 21*, 316–325.

Champoux, M., Bennett, A., Shannon, C., Higley, J. D., Lesch, K. P., & Suomi, S. J. (2002). Serotonin transporter gene polymorphism, differential early rearing, and behavior in rhesus monkey neonates. *Molecular Psychiatry, 7*, 1058–1063.

Chong, R. Y., Oswald, L., Yang, X., Uhart, M., Lin, P. I., & Wand, G. S. (2006). The mu-opioid receptor polymorphism A118G predicts cortisol responses to naloxone and stress. *Neuropsychopharmacology, 31*, 204–211.

Contini, V., Marques, F. Z., Farcia, C. E., Hutz, M. H., & Bau, C. H. (2006). MAOA-uVNTR polymorphism in a Brazilian sample; further support for the association with impulsive behaviors and alcohol dependence. *American Journal of Medical Genetics Part B: Neuropsychiatric Genetics, 141*, 305–308.

Copeland, W. E., Sun, H., Costello, E. J., Angold, A., Heilig, M. A., & Barr, C. S. (2011). Child –opioid receptor gene variant influences parent-child relations. *Neuropsychopharmacology, 36*(6), 1165–1170.

Egli, M. (2005). Can experimental paradigms and animal models be used to discover clinically effective medications for alcoholism? *Addiction Biology, 10*, 309–319.

Eme, R. F. (2007). Sex differences in child-onset, life-course-persistent conduct disorder: A review of biological influences. *Clinical Psychology Review, 27*, 607–627.

Fairbanks, L. A., Newman, T. K., Bailey, J. N., Jorgensen, M. J., Breidenthal, S. E., Ophoff, R. A., Comuzzie, A. G., Marin, L. J., & Rogers, J. (2004). Genetic contributions to social impulsivity and aggressiveness in vervet monkeys. *Biological Psychiatry, 55*, 642–647.

Fiskerstrand, C. E., Lovejoy, E. A., & Quinn, J. P. (1999). An intronic polymorphic domain often associated with susceptibility to affective disorders has allele dependent differential enhancer activity in embryonic stem cells. *FEBS Letters, 458*, 171–174.

Gold, P. W., & Chrousos, G. P. (2002). Organization of the stress system and its dysregulation in melancholic and atypical depression: High vs low CRH/NE states. *Molecular Psychiatry, 7*, 254–275.

Goldman, D., & Barr, C. S. (2002). Restoring the addicted brain. *New England Journal of Medicine, 347*, 843–845.

Goldman, D., Oroszi, G., & Ducci, F. (2005). The genetics of addictions: Uncovering the genes. *Nature Reviews Genetics, 6*, 521–532.

Hansson, A. C., Cippitelli, A., Sommer, W. H., Fedeli, A., Bjork, K., Soverchia, L., Terasmaa, A., Massi, M., Heilig, M., & Ciccocioppo, R. (2006). Variation at the rat Crhr1 locus and sensitivity to relapse into alcohol seeking induced by environmental stress. *Proceedings of the National Academy of Sciences USA, 103*, 15236–15241.

Harlow, H. F., & Suomi, S. J. (1974). Induced depression in monkeys. *Behavioral Biology, 12*, 273–296.

Heilig, M., & Koob, G. F. (2007). A key role for corticotropin-releasing factor in alcohol dependence. *Trends in Neuroscience, 30*(8), 399–406.

Holmes, A., Li, Q., Murphy, D. L., Gold, E., & Crawley, J. N. (2003). Abnormal anxiety-related behavior in serotonin transporter null mutant mice: The influence of genetic background. *Genes, Brain and Behavior, 26*, 365–380.

Higham, J. P., Barr, C. S., Hoffman, C. L., Mandalaywala, T. M., Parker, K. J., & Maestripieri, D. (2011). Interaction between OPRM1 genotype and reproductive status on CSF levels of oxytocin in free-ranging rhesus macaques. *Behavioral Neuroscience, 125*(2), 131–136, 2011.

Higley, J. D., Hasert, M. F., Suomi, S. J., & Linnoila, M. (1991). Nonhuman primate model of alcohol abuse: Effects of early experience, personality, and stress on alcohol consumption. *Proceedings of the National Academy of Sciences USA, 88*, 7261–7265.

Higley, J. D., King, S. T., Hasert, M. F., Campoux, M., Suomi, S. J., & Linnoila, M. (1996a). Stability of interindividual differences in serotonin function and its relationship to severe aggression and competent social behavior in rhesus macaque females. *Neuropsychopharmacology, 14*, 67–76.

Higley, J. D., & Suomi, S. J. (1989). Temperamental reactivity in non-human primates. In D. Kohnstamm, H. E. Bates, & M. K. Rothbart (Eds.), *Temperament in childhood.* Chichester, UK: John Wiley, pp. 153–167.

Higley, J. D., Suomi, S. J., & Linnoila, M. (1996b). A nonhuman primate model of type II excessive alcohol consumption? Part 1. Low cerebrospinal fluid 5-hydroxyindoleacetic acid concentrations and diminished social competence correlate with excessive alcohol consumption. *Alcoholism: Clinical and Experimental Research, 20*, 629–642.

Homberg, J. R., & Lesch, K. P. (2011). Looking on the bright side of serotonin transporter gene variation. *Biological Psyhciatry, 69*, 513–519.

Hu, X. Z., Lipsky, R. H., Zhu, G., Akhtar, L., Taubman, J., Greenberg, B. D., Xu, K., Arnold, P. D., Richter, M. A., Kennedy, J. L., Murphy, D. L., & Goldman, D. (2006). Additional functional variation at the SLC6A4 gene. *American Journal of Human Genetics, 78*, 815–826.

Hundt, W., Zimmermann, U., Pottig, M., Spring, K., & Holsboer, F. (2001). The combined dexamethasone-suppression/CRH stimulation test in alcoholics during and after acute withdrawal. *Alcoholism: Clinical and Experimental Research, 25*, 687–691.

Jaferi, A., & Bhatnagar, S. (2007). Corticotropin-releasing hormone receptors in the medial prefrontal cortex regulate HPA activity and anxiety-related behavior regardless of prior stress experience. *Brain Research, 1186*, 212–223.

Kagan, J., Resnick, J. S., & Snidman, N. (1988). Biological basis of childhood shyness. *Science, 240*, 167–171.

Kalin, N. H., & Shelton, S. E. (2003). Nonhuman primate models to study anxiety, emotion regulation and psychopathology. *Annals of the New York Academy of Sciences, 1008*, 189–200.

Kalin, N. H., Shelton, S. E., & Davidson, R. J. (2000). Cerebrospinal fluid corticotropin releasing hormone levels are stable in monkeys with patterns of brain activity associated with fearful temperament. *Biological Psychiatry, 47*, 579–585.

Kendler, K. S., Prescott, C. A., Myers, J., & Neale, M. C. (2003). The structure of genetic and environmental risk factors for common psychiatric and substance use disorders in men and women. *Archives of General Psychiatry, 60*, 929–937.

Kinnally, E. L., Capitanio, J. P., Leibel, R., Deng, L., Leduc, C., Haghighi, F., & Mann, J. J. (2010). Epigenetic regulation of serotonin transporter expression and behavior in infant rhesus macaques. *Genes, Brain and Behavior, 9*, 575–582.

King, B. R., & Nicholson, R. C. (2007). Advances in understanding corticotrophin-releasing hormone gene expression. *Frontiers in Bioscience, 12*, 581–590.

Korte, S. M., Koolhaas, J. M., Wingfield, J. C., & McEwen, B. S. (2005). The Darwinian concept of stress: Benefits of allostasis and costs of allostatic load and the trade offs in health and disease. *Neuroscience and Behavioral Reviews, 29*, 3–38.

Kraemer, G. W., Moore, C. F., Newman, T. K., Barr, C. S., & Schneider, M. L. (2008). Moderate levels of fetal alcohol exposure and serotonin transporter gene promoter polymorphism affect neonatal temperament and LHPA axis regulation in monkeys. *Biological Psyhciatry, 63*(3), 317–324.

Krawczak, M., Trefilov, A., Berard, J., Bercovitch, F., Kessler, M., Sauermann, U., Croucher, P., Nürnberg, P., Widdig, A., & Schmidtke, J. (2005). Male reproductive timing in rhesus macaques is influenced by the HTTLPR promoter polymorphism of the serotonin transporter gene. *Biology Reproduction, 72*, 1109–1113.

Lanfumey, L., Mannoury La Cour, C., Froger, N., & Hamon M. (2000). 5-HT-HPA interactions in two models of transgenic mice relevant to major depression. *Neurochemical Research, 25*, 1199–1206.

Lappalainen, J., Kranzler, H. R., Malison, R., Price, L. H., Van Dyck, C., Rosenheck, R. A., Cramer, J., Southwick, S., Charney, D., Krystal, J., & Gelernter, J. (2002). A functional neuropeptide Y Leu7Pro polymorphism associated with alcohol dependence in a large population sample from the United States. *Archives of General Psychiatry, 59*(9), 825–831.

Lesch, K. P., Meyer, J., Glatz, K., Flügge, G., Hinney, A., Hebebrand, J., Klauck, S., Poustka, A., Poustka, F., Bengel, D., Mössner, R., Riederer, P., & Heils, A. (1997). The 5-HT transporter gene-linked polymorphic region (5-HTTLPR) in evolutionary perspective: Alternative biallelic variation in rhesus monkeys. *Journal of Neural Transmission, 104*, 1259–1266.

Li, Q., Wichems, C., Heils, A., Van de Dar, L. D., Lesch, K.-P., & Murphy, D. L. (1999). Reduction of 5-HT1A binding sites in 5-HT transporter knockout mice *Journal of Pharmacology and Experimental Therapeutics, 291*, 999–1007.

Lindell, S. G., Schwandt, M. L., Sun, H., Sparenborg, J. D., Bjoerk, K., Kasckow, J. W., Sommer, W. H., Goldman, D., Higley, J. D., Suomi, S. J., Heilig, M., & Barr, C. S. (2010). Functional NPY variation as a factor in stress resilience in rhesus macaques. *Archives of General Psychiatry, 67*(4), 423–431.

Lopez, J. F., & Higley, J. D. (2002). The effect of early experience on brain corticosteroid and serotonin receptors in rhesus monkeys. *Biological Psyhciatry, 51*, 294.

Luo, X., Zuo, L., Kranzler, H., Zhang, H., Wang, S., & Gelernter, J. (2008). Multiple OPR genes influence personality traits in substance dependent and healthy subjects in two American populations. *American Journal of Medical Genetics Part B: Neuropsychiatric Genetics, 147B*, 1028–1039.

Mazzanti, C. M., Lappalainen, J., Long, J. C., Bengel, D., Naukkarinen, H., Eggert, M., Virkkunen, M., Linnoila, M., & Goldman, D. (1998). Role of the serotonin transporter promoter polymorphism in anxiety-related traits. *Archives of General Psychiatry, 55*, 936–940.

McEwen, B. S. (2006). Protective and damaging effects of stress mediators: Central role of the brain. *Dialogues in Clinical Neuroscience, 8*, 367–381.

Mehlman, P. T., Higley, J. D., Faucher, I., Lilly, A. A., Taub, D. M., Vickers, J., Suomi, S. J., & Linnoila, M. (1994). Low CSF 5-HIAA concentrations and severe aggression and impaired impulse control in nonhuman primates. *American Journal of Psychiatry, 151*, 1485–1491.

Mehlman, P. T., Higley, J. D., Faucher, I., Lilly, A. A., Taub, D. M., Vickers, J., Suomi, S. J., & Linnoila, M. (1995). Correlation of CSF 5-HIAA concentration with sociality and the timing of emigration in free-ranging primates. *American Journal of Psychiatry, 152*, 907–913.

Meyer-Lindenberg, A., Buckholtz, J. W., Koachana, B., Hariri, A., Pezawas, L., Blasi, G., Wabnitz, A., Honea, R., Verchinski, B., Callicott, J. H., Egan, M., Mattay, V., & Weinberger, D. R. (2006). Neural mechanisms of genetic risk for impulsivity and violence in humans. *Proceedings of the National Academy of Sciences USA, 103*(16), 6269–6274.

Miller, G. M., Bendor, J., Tiefenbacher, S., Yang, H., Novak, M. A., & Madras, B. K. (2004). A mu-opioid receptor single nucleotide polymorphism in rhesus monkey: Association with stress response and aggression. *Molecular Psychiatry, 9*(1), 99–108.

Mottagui Tabar, S., Prince, J. A., Wahlestedt, C., Zhu, G., Goldman, D., & Heilig, M. (2005). A novel single nucleotide polymorphism of the neuropeptide Y (NPY) gene associated with alcohol dependence. *Alcoholism: Clinical and Experimental Research, 29*(5), 702–707.

Neigh, G. N., Gillespie, C. F., & Nemeroff, C. B. (2009). The neurobiological toll of child abuse and neglect. *Trauma Violence Abuse, 10*, 389–410.

Nelson, E. C., Agrawal, A., Pergadia, M. L., Wang, J. C., Whitfield, H. B., Saccone, F. S., Kern, J., Grant, J. D., Schrage, A. J., Rice, J. P., Montgomery, G. W., Heath, A. C., Goate, A. M., Martin, N. G., & Madden, P. A. (2010). H2 haplotype at chromosome 17q21.31 protects against childhood sexual abuse-associated risk for alcohol consumption and dependence. *Addiction Biology, 15*, 1–11.

Newman, T. K., Syagailo, Y. V., Barr, C. S., Wendland, J. R., Champoux, M., Graessle, M., Suomi, S. J., Higley, J. D., & Lesch, K. P. (2005). Monoamine oxidase A gene promoter variation and rearing experience influences aggressive behavior in rhesus monkeys. *Biological Psychiatry, 57*, 167–172.

Nie, Z., Schweitzer, P., Roberts, A. J., Madamba, S. G., Moore, S. D., & Siggins, G. R. (2004). Ethanol augments GABAergic transmission in the central amygdala via CRF1 receptors. *Science, 303*, 1512–1514.

Oslin, D. W., Berrettinini, W., Kranzler, H. R., Pettinati, H., Gelernter, J., Volpicelli, J. R., & O'Brien, C. P. (2003). A functional polymorphism of the mu-opioid receptor gene is associated with naltrexone response in alcohol-dependent patients. *Neuropsychopharmacology, 28*, 146–155.

Palmour, R. M., Mulligan, J., Howbert, J. J., & Ervin, F. (1997). Of monkeys and men: Vervets and the genetics of human-like behaviors. *American Journal of Human Genetics, 61*, 481–488.

Pang, G., Wang, J., Wang, Z., Goh, C., & Lee, C. (2009). The G allele of SNP E1/A118G at the μ-opioid receptor gene locus shows genomic evidence of recent positive selection. *Pharmacogenomics, 10*, 1101–1109.

Ramchandani, V. A., Umhau, J., Pavon, F. J., Ruiz-Velasco, V., Margas, W., Sun, H., Damadzic, R., Eskay, R., Schoor, M., Thorsell, A., Schwandt, M. L., Sommer, W. H., George, D. T., Parsons, L. H., Herscovitch, P., Hommer, D., Heilig, M. (2011). A genetic determinant of striatal dopamine response to alcohol in men. *Molecular Psychiatry, 16*, 809–817.

Ray, L. A., & Hutchison, K. E. (2004). A polymorphism of the mu-opioid receptor gene (OPRM1) and sensitivity to the effects of alcohol in humans. *Alcoholism: Clinical and Experimental Research, 28*, 1789–1795.

Reich, T., Edenberg, H. J., Goate, A., Williams, J. T., Rice, J. P., Van Eerdewegh, P., Foroud, T., Hesselbrock, V., Schuckit, M. A., Bucholz, K. K., Porjesz, B., Li, T. K., Conneally, P. M., Nurnberger, J., Tischfield, J. A., Crowe, R. R., Cloninger, C. R., Wu, W., Shears, S., Carr, K., Crose, C., Willig, C., & Begleiter, H. (1998). Genome-wide search for genes affecting the risk for alcohol dependence. *American Journal of Medical Genetics, 81*(3), 207–215.

Risch, N., Herrell, R., Lehner, T., Liang, K. Y., Eaves, L., Hoh, J., Griem, A., Kovacs, M., Ott, J., & Merikangas, K. R. (2009). Interaction between the serotonin transporter gene, stressful life events and risk of depression: A meta-analysis. *Journal of the American Medical Association, 301*, 2462–2471.

Sabol, S. Z., Hu, S., & Hamer, D. (1998). A functional polymorphism in the monoamine oxidase A gene promoter. *Human Genetics, 103*, 273–279.

Samochowiec, J., Lesch, K. P., Rottmann, M., Smolka, M., Syagailo, Y. V., Okladnova, O., Rommlspacher, H., Winterer, G., Schmidt, L. F., & Sander, T. (1999). Association of a regulatory polymorphism in the promoter region of the monoamine oxidase A gene with antisocial alcoholism. *Psychiatry, 86*, 67–72.

Sanchez, M. M., Young, L. J., Plotsky, P. M., & Insel, T. R. (1999). Autoradiographic and in situ hybridization localization of corticotropin-releasing factor 1 and 2 receptors in nonhuman primate brain. *Journal of Comparative Neurology, 408*, 365–377.

Sapolsky, R. M. (2001). Physiological and pathophysiological implications of social stress in animals. In B. S. McEwen & H. M. Goodman (Eds.), *Coping with the environment: Neural and endocrine mechanisms*. New York: Oxford University Press, pp. 517–532.

Schneider, M. L., Moore, C. F., Barr, C. S., Larson, J. A., & Kraemer, G. W. (2011). Moderate prenatal alcohol exposure and serotonin genotype interact to alter CNS serotonin function in rhesus monkey offspring. *Alcoholism: Clinical and Experimental Research, 35*(5), 912–920.

Schneider, M. L., Moore, C. F., Larson, J. A., Barr, C. S., Dejesus, O. T., & Roberts, A. D. (2010). Timing of moderate level prenatal alcohol exposure influences gene expression of sensory processing behavior in rhesus monkeys. *Frontiers in Integrative Neuroscience, 3*, 30–38.

Schwandt, M. L., Lindell, S. G., Higley, J. D., Suomi, S. J., Heilig, M., & Barr C. S. (2011). OPRM1 gene variation influences HPA axis function in response to a variety of stressors in rhesus macaques. *Psychoneuroendocrinology, 36*, 1303–1311.

Schwandt, M. L., Lindell, S. G., Sjoberg, R. L., Chisholm, K. L., Higley, J. D., Suomi, S. J., Heilig, M., & Barr, C. S. (2010). Gene-environment interactions and response to social intrusion in male and female rhesus macaques. *Biological Psychiatry, 67*(4), 323–330.

Servatius, R. J., Beck, K. D., Moldow, R. L., Salameh, G., Tumminello, T. P., & Short, K. R. (2005). A stress-induced anxious state in male rats: CRH induces persistent changes in associative learning and startle reactivity. *Biological Psychiatry, 57*, 865–872.

Shannon, C., Schwandt, M. L., Champoux, M., Shoaf, S. E., Suomi, S. J., Linnoila, M., & Higley, J. D. (2005). Maternal absence and stability of individual differences in CSF 5-HIAA concentrations in rhesus monkey infants. *American Journal of Psychiatry, 162*, 1658–1664.

Shimmin, L. C., Natarajan, S., Ibarguen, H., Montasser, M., Kim, D. K., Hanis, C. L., Boerwinkle, E., Wadhwa, P. D., & Hixson, J. E. (2007). Corticotropin releasing hormone (CRH) gene variation: Comprehensive resequencing for variant and molecular haplotype discovery in monosomic hybrid cell lines. *DNA Sequence, 18*(6), 432–442.

Sinha, R., & Li, C. S. (2007). Imaging stress- and cue-induced drug and alcohol craving: Association with relapse and clinical implications. *Drug & Alcohol Review, 26*(1), 25–31.

Sjoberg, R. L., Ducci, F., Barr, C. S., Newman, T. K., Dell'Osso, L., Virkkunen, M., & Goldman, D. (2008). A non-additive interaction of a functional MAOA VNTR and testosterone predicts antisocial behavior. *Neuropsychopharmacology, 33*(2), 425–430.
Smith, M. J., & Price, G. R. (1973). The logic of animal conflict. *Nature, 246*, 15–18.
Smoller, J. W., Rosenbaum, J. F., Biederman, J., Kennedy, J., Dai, D., Racette, S. R., Laird, N. M., Kagan, J., Snidman, N., Hirshfeld-Becker, D., Tsuang, M. T., Sklar, P. B., & Slaugenhaupt, S. A. (2003). Association of a genetic marker at the corticotropin-releasing hormone locus with behavioral inhibition. *Biological Psychiatry, 54*(12), 1376–1381.
Sommer, W. H., Lidström, J., Sun, H., Passer, D., Eskay, R., Parker, S. C., Witt, S. H., Zimmermann, U. S., Nieratschker, V., Rietschel, M., Margulies, E. H., Palkovits, M., Laucht, M., & Heilig, M. (2010). Human NPY promoter variation rs16147:T>C as a moderator of prefrontal NPY gene expression and negative affect. *Human Mutation, 31*(8), 1594–1608.
Sommer, W. H., Rimondini, R., Hansson, A. C., Hipskind, P. A., Gehlert, D. R., Barr, C. S., & Heilig, M. (2008). Up-regulation of voluntary alcohol intake, behavioral sensitivity to stress and amygdala Crhr1 expression following a history of dependence. *Biological Psychiatry, 63*(2), 139–145.
Southwick, S. M., Vythilingam, M., & Charney, D. S. (2005). The psychobiology of depression and resilience to stress: Implications for prevention and treatment. *Annual Review of Clinical Psychology, 1*, 255–291.
Spinelli, S., Chefer, S., Carson, R. E., Jagoda, E., Lang, L., Heilig, M., Barr, C. S., Suomi, S. J., Higley, J. D., & Stein, E. (2010). Effects of early-life stress on serotonin 1a receptors in juvenile rhesus monkeys measured by PET. *Biological Psychiatry, 67*, 1146–1153.
Spinelli, S., Chefer, S., Suomi, S. J., Higley, J. D., Barr, C. S., & Stein, E. (2009). Early life stress induces long-term morphologic changes in primate brain. *Archives of General Psychiatry, 66*, 658–665.
Spinelli, S., Schwandt, M. L., Heilig, M., Suomi, S. J., Higley, J. D., & Barr, C. S. (2012). HTTLPR genotype is associated with behavioral response to repeated stress in infant rhesus macaques. *Development and Psychopathology, 24*, 157–165.
Spinelli, S., Schwandt, M. L., Lindell, S. G., Newman, T. K., Heilig, M., Higley, J. D., Suomi, S. J., Goldman, D., & Barr, C. S. (2007). Association between the rh-5HTTLPR polymorphism and behavior in rhesus macaques during social separation stress. *Developmental Psychopathology, 19*(4), 977–987.
Soeby, K., Larsen, S. A., Olsen, L., Rasmussen, H. B., & Werge, T. (2005). Serotonin transporter: Evolution and impact of polymorphic transcriptional regulation. *American Journal of Human Genetics, 136*, 53–57.
Strome, E. M., Wheler, G. H., Higley, J. D., Loriaux, D. L., Suomi, S. J., & Doudet, D. J. (2002). Intracerebroventricular corticotropin-releasing factor increases limbic glucose metabolism and has social context-dependent behavioral effects in nonhuman primates. *Proceedings of the National Academy of Sciences USA, 99*, 15749–15754.
Suomi, S. J. (1982). Abnormal behavior in and primate models of psychopathology. In J. D. Fobes & J. E. King (Eds.), *Primate behavior*. New York: Academic Press, pp. 171–215.
Thorsell, A., Repunte-Canonigo, V., O'Dell, L. E., Chen, S. A., King, A. R., Lekic, D., Koob, G. F., & Sanna, P. P. (2007). Viral vector-induced amygdala NPY overexpression reverses increased alcohol intake caused by repeated deprivations in Wistar rats. *Brain, 130*, 1330–1337.
Tikkanen, R., Sjoberg, R. L., Ducci, F., Goldman, D., Holi, M., Tiihonen, J., & Virkkunen, M. (2009). Effects of MAOA-genotype, alcohol consumption and aging on violent behavior. *Alcoholism: Clinical and Experimental Research, 33*, 428–443.
Trefilov, A., Berard, J., Krawczak, M., & Schmidtke, J. (2000). Natal dispersal in rhesus macaques is related to serotonin transporter gene promoter variation. *Behavioral Genetics, 30*, 295–301.
Vallender, E. J., Priddy, C. M., Hakim, S., Yang, H., Chen, G. L., & Miller, G. M. (2008). Functional variation in the 3′ UTR of the serotonin transporter in human and rhesus macaque. *Genes, Brain and Behavior, 7*, 690–697.
Vallender, E. J., Ruidi-Bettschen, D., Miller, G. M., & Platt, D. M. (2010). A pharmacogenetic model of naltrexone-induced attenuation of alcohol consumption in rhesus monkeys. *Drug Alcohol Depend, 109*, 252–256.
Wagner, U., Wahle, M., Moritz, F., Wagner, U., Hantzschel, H., & Baerwald, C. G. (2006). Promoter polymorphisms regulating corticotrophin-releasing hormone transcription in vitro. *Hormone and Metabolic Research, 38*(2), 69–75.
Wendland, J. R., Lesch, K. P., Newman, T. K., Timme, A., Gachot-Neveu, H., Thierry, B., & Suomi, S. J. (2005). Differential functional variability of serotonin transporter and monoamine oxidase A genes in macaque species displaying contrasting levels of aggression-related behavior. *Behavioral Genetics, 30*, 1–10.
Westergaard, G. C., Suomi, S. J., Chavanne, T. J., Houser, L., Hurley, A., Cleveland, A., Snoy, P. J., & Higley, H. D. (2003). Physiological correlates of aggression and impulsivity in free-ranging female primates. *Neuropsychopharmacology, 28*, 1045–1055.
Williams, J. B., Spitzer, R. L., Linzer, M., Kroenke, K., Hahn, S. R., deGruy, F. V., & Lazev, A. (1995). Gender differences in depression in primary care. *American Journal of Obstetrics and Gynecology, 173*, 654–659.
Williamson, D. E., Coleman, K., Bacanu, S., Devlin, B. J., Rogers, J., Ryan, N. D., & Cameron, J. L. (2003). Heritability of fearful-anxious endophenotypes in infant rhesus macaques: A preliminary report. *Biological Psychiatry, 53*, 284–291.
Wood, W., & Eagly, A. H. (2002). A cross-cultural analysis of the behavior of women and men: Implications for the origins of sex differences. *Psychol Bulletin, 128*, 699–727.
Zhang, H., Luo, X., Kranzler, H. R., Lappalainen, J., Yang, B. Z., Krupitsky, E., Zvartau, E., & Gelernter, J. (2006). Association between two mu-opioid receptor gene (OPRM1) haplotype blocks and drug or alcohol dependence. *Human Molecular Genetics, 15*, 807–819.
Zhou, Z., Zhu, G., Hariri, A. R., Enoch, M. A., Scott, D., Sinha, R., Virkkunen, M., Mash, D. C., Lipsky, R. H., Hu, X. Z., Hodgkinson, C. A., Xu, K., Buzas, B., Yuan, Q., Shen, P. H., Ferrell, R. E., Manuck, S. B., Brown, S. M., Hauger, R. L., Stohler, C. S., Zubieta, J. K., & Goldman, D. (2008). Genetic variation in human NPY expression affects stress response and emotion. *Nature, 452*(7190), 997–1001.
Zhu, G., Pollak, L., Mottagui-Tabar, S., Wahlestedt, C., Taubman, J., Virkkunen, M., Goldman, D., & Heilig, M. (2003). NPY leu7pro and alcohol dependence in Finnish and Swedish Populations. *Alcoholism: Clinical and Experimental Research, 27*(1), 19–24.
Zill, P., Preuss, U. W., Koller, G., Bondy, B., & Soyka, M. (2008). Analysis of single nucleotide polymorphisms and haplotypes in the neuropeptide Y gene: No evidence for association with alcoholism in a German population sample. *Alcoholism: Clinical and Experimental Research, 32*(3), 430–434.

CHAPTER

7 Genetic Perspectives on the Neurochemistry of Human Aggression and Violence

Joshua W. Buckholtz *and* Andreas Meyer-Lindenberg

Abstract

Violence is a devastating social phenomenon that is costly both to affected individuals and to society at large. Pathological aggression, especially reactive/impulsive aggression, is a cardinal symptom common to several psychiatric disorders—including antisocial personality disorder, borderline personality disorder, and psychopathy—that are associated with risk for violence. Thus, understanding the factors that predispose people to impulsive violence represents a crucial goal for psychology, neuroscience, and psychiatry. Although we are far from a full understanding of the etiopathophysiology of violence, impulsive aggression is heritable, suggesting that genetic mechanisms may be important for determining individual variation in susceptibility. This chapter synthesizes available preclinical and human data to propose a compelling neurogenetic mechanism for violence, specifically arguing that a genetically determined excess in serotonin signaling during a critical developmental period leads to dysregulation within a key corticolimbic circuit for emotional arousal and regulation, inhibitory control, and social cognition.

Key Words: aggression, violence, serotonin, corticolimbic, amygdala, genetic, MAOA, 5-HTTLPR, tryptophan hydroxylase, 5-HT_{1b}

I. The Cost of Violence

In the next year, it is estimated that nearly 17,000 individuals will die by the hand of another in the United States alone (Corso, Mercy, Simon, Finkelstein, & Miller, 2007). An additional 2.2 million Americans will require medical treatment for a (non self-inflicted) violence-related injury (Corso et al., 2007). Apart from the emotional and psychological toll that these acts of interpersonal violence will place on the victims and their families, society as a whole will shoulder an astounding economic burden as well. Each violent fatality costs approximately $1.3 million in combined medical costs and lost productivity, and each nonfatal violent assault leads to more than $80,000 in aggregate costs (Corso et al., 2007). All told, estimates for the total cost of violence place this figure at nearly 3.3 percent of the Unites States' gross domestic product (GDP), and economic assessments for other countries reveal similarly astonishing figures (Waters et al., 2004).[1]

One remarkable aspect of these costs is that they appear to be driven by a relatively small number of violent individuals. For example, an examination of crime statistics in the United Kingdom reveals that approximately 10 percent of criminal offenders commit nearly 50 percent of the total crimes committed, including violent crime (*Understanding the prolific... offender programme*, 2007). Remarkably, this closely mirrors extant data on the familial aggregation of crime within communities. Within any given community, it is thought that approximately 50 percent of the crime (including violent crime) is committed by 10 percent of the families in that community. This familial component to criminal behavior echoes the known familiality of antisocial traits and conduct and suggests an inherited predisposition to criminality generally and violence in particular (Farrington & Barnes, 1996; Farrington, Jolliffe, Loeber, Stouthamer-Loeber, & Kalb, 2001). Clearly, violence is complex, and susceptibility to violent behavior is multiply determined, with biological, sociological, and cultural factors all playing contributing (and interacting) roles. However, there is clear evidence that individual differences in violence can be shaped by genetic factors, and these genetic factors, in turn, likely exert their effects on susceptibility to violence by influencing the brain (Moffitt, 2005). Thus, understanding the myriad factors that contribute to risk for violence—especially, how these factors act and interact to affect brain chemistry, structure, function and connectivity—represents a crucial goal for psychiatry, psychology, and neuroscience. In this review, we detail evidence that genetic variation in brain monoamine signaling pathways leads to synapse-level changes in monoaminergic tone that alter the development of distributed neural circuits for emotional arousal, inhibitory control, reward drive, and social cognition. These systems-wide, genetically mediated neurodevelopmental changes provide a reactive substrate for the malign influence of other causal factors for violence, enhancing the likelihood of their reaching full expression in an individual's behavior.

II. Defining Our Terms: Violence as Pathological Aggression

Given that human violence is a complex and heterogeneous phenomenon, it is potentially useful to situate it within well-characterized models of animal behavior. In particular, ethological models of aggression make a useful distinction between aggressive behaviors that are "functional" and those that are "pathological" (Natarajan, Vires, Saaltink, Boer, & Koolhaas, 2008; Takahashi, Quadros, de Almeida, & Miczek, 2011). Functional aggression is contextually appropriate and constrained by species-specific rules and rituals; we can perhaps view a human ortholog of this functional aggression in the ritualized violence of professional football or hockey. By contrast, pathological aggressive behaviors are out of context, unstructured, and do not adhere to norms that would otherwise serve to mitigate their damaging impact to both victim and community. Thus, in human behavioral terms, the aggression of a hockey fight and the aggression of a street mugging can be considered to be categorically distinct. Although they may well be matched with respect to the intensity of the aggression on display, it is readily apparent that the potential for serious injury and attendant social costs (both economic and nonfiduciary) is significantly greater for the street mugging compared to the hockey fight. Considered in this light, only the latter expression of aggression would be construed as violence. Additionally, human pathological aggression is often divided further into reactive-impulsive and proactive-instrumental subtypes that mirror ethological distinctions between defensive and predatory aggression (Blair, 2001; Polman, Orobio de Castro, Koops, van Boxtel, & Merk, 2007; Vitiello & Stoff, 1997). These two forms of aggression are associated with distinct sociocognitive alterations and biological mechanisms and may have at least partially distinct genetic diatheses (Baker, Raine, Liu, & Jacobson, 2008; Bezdjian, Tuvblad, Raine, & Baker, 2011; Blair, 2001; Brendgen, Vitaro, Boivin, Dionne, & Pérusse, 2006; Hubbard, Dodge, Cillessen, Coie, & Schwartz, 2001; Tuvblad, Raine, Zheng, & Baker, 2009).

It is worth noting that pathological aggression—particularly, the reactive-impulsive subtype—is a symptom common to a range of psychiatric diagnoses, including antisocial personality disorder (APD), borderline personality disorder (BPD), psychopathy, and bipolar disorder, as well as to certain subtypes of major depression (e.g. angry/irritable). Thus, pathological aggression is fundamentally a transdiagnostic construct and could potentially prove useful in efforts to devise a more empirically based diagnostic classification scheme for psychiatric illness. Studying the neurobiology of aggression will improve what we know about the common and unique biological alterations that give rise to these disorders and, in so doing, provide insight into the latent pathobiological structure of externalizing psychopathology. Further, common genetic risk factors may underpin common aggression-linked neurobiological alterations in these disorders and thus may account for some of their shared genetic variance (Krueger & South, 2009). Therefore, studying heritable risk mechanisms for aggression may reveal much about the overall genetic architecture of externalizing psychopathology.

III. Pathological Aggression and Serotonin: Rodent Pharmacology

In considering potential neurobiological pathways that could account for individual variation in pathological aggression, the strongest empirical support, by far, implicates the neurotransmitter serotonin (5-hydroxytryptamine; 5-HT). Early studies suggested an inhibitory role for serotonin on aggressive behavior. For instance, pharmacological inhibition of serotonin synthesis (e.g., by parachlorophenylalanine; PCPA) facilitates some forms of aggression (Valzelli & Bernasconi, 1979; Valzelli, Bernasconi, & Dalessandro, 1981*a*; Valzelli, Garattini, Bernasconi, & Sala, 1981*b*; Vergnes, Depaulis, & Boehrer, 1986; Vergnes & Kempf, 1981), and directionally consistent effects are seen following dietary manipulations that either reduce or enhance central 5-HT levels (Gibbons, Barr, Bridger, & Leibowitz, 1981; 1979; Vergnes & Kempf, 1981). Moreover, treatment with drugs that enhance serotonergic transmission—such as selective serotonin reuptake inhibitors (SSRIs)—reduces aggressive behaviors that occur spontaneously in certain rat strains (Gibbons et al., 1979; Stark, Fuller, & Wong, 1985) and rescues aggressive behaviors induced by pharmacological inhibition of serotonin synthesis (Kostowski, Valzelli, Kozak, & Bernasconi, 1984; Molina, Ciesielski, Gobaille, Isel, & Mandel, 1987), by stress-induction procedures such as social isolation or footshock (Datla, Mitra, & Bhattacharya, 1991), or by lesioning serotonergic neurons of the dorsal raphe. Although there are substantial inconsistencies in this literature—due in part to species differences, drug dosing and route of administration issues, and interstudy variability in the operationalization of aggressive behavior—a recent meta-analysis of more than 200 reported effects (from 84 studies) revealed that pharmacological enhancement of serotonergic transmission (e.g., by administering SSRIs or 5-HT precursors) does significantly reduce aggressive behavior across a wide range of species (Carrillo, Ricci, Coppersmith, & Melloni, 2009). Finally, in vitro measurements of 5-HT levels and turnover also correlate (typically, inversely) with aggression (Daruna & Kent, 1976; Valzelli & Bernasconi, 1979) in rodents, consistent with the behavioral consequences of pharmacologically manipulating serotonergic function outlined earlier. Together, this early rodent data stoked considerable interest in serotonin as an important neurochemical modulator of aggression and laid the foundation for a "low 5-HT" hypothesis of human violence.

IV. Pathological Aggression and Serotonin: Human Neurochemistry

Spurred in part by the rodent work just highlighted, subsequent human and nonhuman primate research focused on identifying peripheral and central serotonergic biomarkers for violence. Data from these studies were largely consonant with the notion that lowered serotonergic function is an enduring trait-like central correlate of impulsive aggression. Starting in the 1970s, chromatographic analysis of cerebrospinal fluid (CSF) provided investigators with a means of assessing central neurochemistry in living human subjects for the first time. Many of these studies assayed levels of 5-hydroxyindoleacetic acid (5-HIAA), serotonin's primary metabolite via oxidative catabolism by monoamine oxidase (MAO). As a putative measure of serotonin turnover, 5-HIAA levels are thought to be an indirect gauge of overall central serotonergic tone. Although CSF 5-HIAA is necessarily an integral measure that cannot be parsed for regional specificity, such levels do appear to significantly predict human frontal 5-HIAA concentrations in vitro (Stanley, Traskman-Bendz, & Dorovini-Zis, 1985; Wester et al., 1990) and orbitofrontal cortex metabolism in vivo (Doudet et al. 1995; Williams et al., 2004), although the variance explained is moderate. These data suggest that CSF 5-HIAA is a relevant measure of in vivo serotonergic function that may be related to variation in prefrontal activity.

Asberg and colleagues were the first to link impulsive violence to decreased CSF 5-HIAA in a study of depressed patients with a history of violent suicide attempts (Asberg, Träskman, & Thorén, 1976). Subsequently, Brown et al. (1982) examined a cohort of patients diagnosed with personality disorders who presented with a history of aggressive behavior and violent suicide attempts but without marked affective psychopathology. Extending the Asberg finding, aggressive traits were inversely correlated with CSF 5-HIAA even in the absence of affective illness, suggesting a specific relationship to violent behavior (Brown et al., 1982). Importantly, Linnoila and coworkers (1983) found that the association between 5-HIAA levels and aggression is selective for reactive aggression: lowered CSF 5-HIAA was seen in impulsively violent individuals, but not in those who had committed acts of instrumental or premeditated violence. The 5-HIAA–impulsive violence linkage is largely supported by a number of studies that have since replicated this association using a range of aggression measures in adolescents presenting with impulsive violence (Kruesi, 1989; Kruesi et al., 1990; 1992) and in adults with aggressive traits (Brown et al., 1982), with impulsive-antisocial psychopathic traits in forensic samples (Soderstrom, Blenno, Manhem, & Forsman, 2001; Soderstrom,

Blennow, Sjodin, & Forsman, 2003), in arsonists (Virkkunen, Nuutila, Goodwin, & Linnoila, 1987), in convicted murderers (Lidberg, Tuck, Asberg, Scalia-Tomba, & Bertillson, 1985), and in type II (antisocial) alcoholics (Roy & Linnoila, 1989; Roy, Virkkunen, & Linnoila, 1987; Virkkunen, Goldman, Nielsen, & Linnoila, 1995; Virkkunen & Linnoila, 1990; 1993).

Pharmacological challenge approaches have also revealed diminished responsiveness of the serotonergic system in individuals with impulsive aggression. Administration of the serotonin-releasing agent fenfluramine causes a reliable increase in peripheral prolactin concentrations, an effect that is likely mediated by enhanced stimulation of hypothalamic 5-$HT_{2a/2c}$ receptors (Coccaro, Berman, Kavoussi, & Hauger, 1996*a*). Variability in prolactin response is thus thought to index central serotonergic reactivity. In analog to the findings with CSF 5-HIAA, there is considerable evidence that impulsive aggression is associated with lowered serotonin responsivity. For example, reduced prolactin response to fenfluramine (PRL-fen) is negatively correlated with both self-reported aggression history and behavioral (task-based) measures of aggressive responding in personality-disordered patients (Cleare & Bond, 1997; Coccaro et al., 1996*a*; Coccaro, Lee, & Kavoussi, 2010*a*). This same relationship holds in both psychiatric inpatients (Herpertz, Sass, & Favazza, 1997), in community volunteers with impulsive and aggressive personality traits (Evans, Platts, Lightman, & Nutt, 2000; Manuck, Flory, Muldoon, & Ferrell, 2002; Manuck et al., 1998), and in individuals diagnosed with antisocial personality disorder (O'Keane et al., 1992), and it has also been found using a measure of impulsive-antisocial psychopathic traits in forensic inpatients (Dolan & Anderson, 2003). In addition, these changes in serotonergic function appear to be associated with the development of aggression throughout the life course. In a longitudinal study of boys diagnosed with disruptive behavioral disorders, higher PRL-fen at ages 7–11 was found to protect against the subsequent expression of antisocial conduct and aggression in adolescence (Halperin et al., 2006) and was associated with lower rates of antisocial personality disorder diagnosis in early adulthood (Flory, Newcorn, Miller, Harty, & Halperin, 2007). Intriguingly, fenfluramine administration increases the rate of cingulate and ventromedial prefrontal cortical glucose metabolism in healthy individuals; this increase is markedly blunted in impulsive-aggressive inpatients, further suggestive of serotonergically mediated prefrontal cortical dysfunction in impulsive aggression (Siever et al., 1999).

Finally, a range of peripheral biomarkers for serotonergic function has been assessed for relationships to aggression. Platelet 3H-paroxetine binding, thought to be indicative of serotonin transporter protein [5-HTT] concentration, is inversely correlated with life history of aggression in personality disordered patients (Coccaro et al., 1996*a*; Coccaro, Lee, & Kavoussi, 2010*b*) and with measures of sensation-seeking, impulsivity, and aggression in cocaine abusers (Patkar et al., 2003). In addition, platelet 3H-imipramine (another 5-HTT ligand, although less selective) binding is lower—and inversely correlated with aggressive symptoms—in children diagnosed with attention-deficit hyperactivity disorder and comorbid conduct disorder, compared to those without comorbid conduct disorder (Stadler, Schmeck, Nowraty, Müller, & Poustka, 2004; Stoff, Pollock, Vitiello, Behar, & Bridger, 1987). Consistent with these relationships, blood platelet uptake of 3H-5HT is lower in individuals with episodic aggression and negatively correlated with impulsivity (Brown et al., 1989). Notably, measures of whole-blood 5-HT levels are positively correlated with impulsivity and aggression. In a large epidemiological sample, Moffitt and colleagues (1998) reported higher whole-blood serotonin levels in men (but not in women) with a history of committing violent crimes. This relationship held even after controlling for an impressive number of potentially confounding variables, including illicit drug history, socioeconomic status, plasma tryptophan levels, and overall criminal offending (Moffit et al., 1998). Moreover, it was stable across multiple measures of violence. Higher whole-blood serotonin levels are also seen in incarcerated adolescents with a diagnosis of conduct disorder and are positively correlated with violent offending and social skill impairment in these individuals (Askenazy, Caci, Myquel, Darcourt, & Lecrubier, 2000; Unis et al., 1997). Although these findings might seem to run counter to a "low 5-HT" model of violence, higher whole-blood levels would be consistent with both reduced uptake and reduced catabolism (one parameter that would impact measures of turnover) and suggest that the relationship between 5-HT and aggression may be more nuanced than early "low 5-HT" hypotheses would have implied.

V. Pathological Aggression and Serotonin: Nonhuman Primate Models

An informative series of nonhuman primate studies provides a substantial complement to this line of clinical work. Manifest aggressive behavior is difficult to assess directly in human studies, which typically rely on participant report either during life history interviews or via questionnaire measures. By contrast, the repertoire of aggressive behaviors in nonhuman primate species, such as the rhesus macaque, is well characterized, straightforward to measure, and can be observed directly and in real time (for a detailed discussion of this work, with a focus on gene–environment interactions, see the chapter by Barr, elsewhere in this handbook). Consistent with human data, low CSF 5-HIAA levels are selectively associated with pathological (but not with functional) aggression in primates, are positively correlated with deficits in impulse control, and are a robust predictor of premature mortality due to violence (Higley & Linnoila, 1997; Higley et al., 1992; 1996*a*; 1996*b*; 1996*c*; Mehlman et al., 1994). Of note, monkeys with low CSF 5-HIAA also show profound social maladaptation and appear less able to maintain stable, long-term social relationships (Higley et al., 1996*b*; Mehlman et al., 1995; 1997), a pattern of interpersonal dysfunction that is characteristic of human externalizing psychopathology (Higley, Suomi, & Linnoila, 1996). Also in line with human findings, impulsive aggression is associated with blunted serotonergic responsivity in nonhuman primates: PRL-fen is inversely correlated with impulsivity (Tiefenbacher, Davenport, Novak, Pouliot, & Meyer, 2003) and with aggression (Manuck, Kaplan, Rymeski, Fairbanks, & Wilson, 2003) in monkeys. These data confirm that altered serotonergic function is associated with pathological aggression in well-operationalized and ecologically valid animal models of violence.

VI. Role of Serotonergic Function in the Heritability of Aggression

As we alluded to previously, genetic factors appear to play a significant role in predisposing aggressive traits and behaviors. Antisocial behavior is a familial trait, suggesting that heritable factors are involved in its intergenerational transmission (Cervantes & Delville, 2011; Smith & Farrington, 2004), and twin studies confirm that genetic factors account for a large proportion of the population variance in antisocial traits and behaviors (Moffitt, 2005). However, antisociality and aggression, although related, are not synonymous (Eley, Lichtenstein, & Moffitt, 2003), and it is therefore important to study the genetic structure of aggression per se (i.e., separate from antisocial behavior or criminality more generally). Here, too, twin and family studies are useful in parsing the etiology of risk. Aggression phenotypes in mice (Miczek, Maxson, Fish, & Faccidomo, 2001) and nonhuman primates (Fairbanks et al., 2004; Suomi, 2003) show evidence of heritability, and variation in higher order personality traits with conceptual links to aggression, such as Anger and Angry Hostility (Cates, Houston, Vavak, Crawford, & Uttley, 1993), is under significant genetic control. Moderate to strong heritability has been demonstrated for aggressive symptoms in adolescents diagnosed with conduct disorder (Gelhorn et al., 2005; 2006) and in adults with a history of aggression (Coccaro, Kavoussi, Cooper, & Hauger, 1997). Of relevance to the distinction between antisocial behavior and aggression made earlier, aggressive antisocial behavior shows significantly higher heritability than does nonaggressive antisocial behavior (Button, Scourfield, Martin, & McGuffin, 2004; Eley et al., 2003). Genetic contributions to aggression appear early (van Beijsterveldt, Verhulst, Molenaar, & Boomsma, 2004) and are seen stably at ages 3, 7, and 10 (Baker, Jacobson, Raine, Lozano, & Bezdjian, 2007; Hudziak et al., 2003). Moreover, the developmental transition of antisocial behavior from childhood to adolescence is heritable, with the strongest evidence for a genetic influence on the developmental stability of aggressive, but not nonaggressive, antisocial behavior. Intriguingly, the genetic correlation between reactive (impulsive) and proactive (instrumental) aggression in preadolescent boys appears to be incomplete, suggesting the existence of partially non-overlapping genetic risk factors (Baker et al., 2008). Also in line with the notion of genetic heterogeneity within a broad aggression construct, genetic factors appear to be differentially correlated with distinct aggression subtypes in adults, such as physical versus nonphysical (Yeh, Coccaro, & Jacobson, 2010). On the whole, these findings provide compelling support for an etiological model of aggression that includes a prominent role for genetic factors in determining variability in risk.

If the alterations in serotonergic function described here constitute a biological trait, or endophenotype, that conveys meaningful information about the causal architecture of risk for aggression (i.e., rather than highlighting a

important moderating role in this interaction, as some studies suggest that MAOA-H women are at greater risk following early life stress (Aslund et al., 2011; Nilsson et al., 2011; Sjöberg et al., 2007). On the whole, however, genetic association studies with the MAOA u-VNTR do support a model wherein excessive serotonin signaling—caused by a genetically mediated reduction in MAOA activity—predisposes the development of impulsive aggression, especially in men and particularly in concert with environmental risk factors.

This conclusion is bolstered by very consistent evidence that impulsive aggression is predicted by lowered MAOA activity measured both in platelets (af Klinteberg, Schalling, Edman, Oreland, & Asberg, 1987; Belfrage, Lidberg, & Oreland, 1992; Davis, Yu, Boulton, Wormith, & Addington, 1983; Eklund, Alm, & af Klinteberg, 2005; Mukasa, Nakamura, Yamada, Inoue, & Nakazawa, 1990; Perris, Eisemann, von Knorring, Oreland, & Perris, 1984; Prochazka, Anderberg, Oreland, Knorring, & Agren, 2003; Skondras, Markianos, Botsis, Bistolaki, & Christodoulou, 2004; Stålenheim, 2001) and in brain (Alia-Klein et al., 2008). In addition, a number of putatively functional variants reside in the MAOA gene and in the upstream promoter region; both positive association signals and nonreplications could reflect an interacting or confounding influence of these variants (Balciuniene et al., 2002; Philibert et al., 2011, Shumay & Fowler, 2010). Thus, although one may question the specific functional role of the 30 bp upstream VNTR that has been the strongest focus of research to date (Balciuniene et al., 2002; Cirulli & Goldstein, 2007; Fowler et al., 2007), it seems clear that heritable variation in MAOA activity leading to altered serotonin levels comprises a prominent component of the risk architecture for impulsive violence.

VIII. Other Serotonin Genes: SERT, TPH, 5-HT$_{1b}$

Further support for a "high 5-HT" model of risk for aggression is derived from studies of genes encoding multiple nodes within the serotonergic pathway. In this section, we detail additional evidence that excessive serotonin signaling contributes to risk for violence by examining preclinical and human associations between impulsive aggression and polymorphic variation in genes involved in 5-HT clearance (SERT), 5-HT synthesis (TPH2), and 5-HT release (5-HT$_{1b}$).

A. 5-HT Clearance: SERT

The human serotonin transporter gene (SLC6A4; SERT) also contains a functional promoter VNTR (often termed the 5-HTTLPR) that impacts transporter expression and has been linked to adverse psychiatric outcomes. Possessing 14 copies of a 22 bp repeat cassette (the 5-HTTLPR "short" allele) is associated with lower transporter availability and increased risk for negative emotionality phenotypes compared to the "long" allele (16 repeats), particularly following exposure to childhood adversity (Canli & Lesch, 2007; Caspi et al., 2003; Holmes & Hariri, 2003; Karg, Burmeister, Shedden, & Sen, 2011). Although the study of this variant has most commonly occurred in the context of anxiety and depression, a parallel literature has arisen showing associations to aggressive behavior and related traits. Consistent with prior data linking impulsive aggression to reduced SERT availability in platelets (Coccaro et al., 2010*b*; Coccaro, Kavoussi, Sheline, Lish, & Csernansky, 1996*b*; Marazziti et al., 2010) and in brain (Frankle et al., 2005; Lindström et al., 2004; Sekine et al., 2006), individuals carrying the low expressing short allele are at significantly increased risk for violence. The short allele has been tied to impulsive violence in alcoholics and suicide attempters (Baca-García et al., 2005; Hallikainen et al., 1999); to recurrent physical violence in forensic inpatients (Retz, Retz-Juninger, Supprian, Thome, & Rösler, 2004) and incarcerated criminals (Liao, Hong, Shih, & Tsai, 2004); to antisocial personality disorder diagnosis (Lyons-Ruth et al., 2007); to increased impulsivity on performance-based measures (Paaver et al., 2007; Walderhaug, Herman, Magnusson, Morgan, & Landrø, 2010); and to enhanced antisocial, novelty seeking, and hostility traits (Gerra et al, 2005; Gonda et al., 2009) in community-based samples of adults. In addition, the short allele genotype predicts higher levels of aggressive behavior in children (Beitchman et al., 2006; Haberstick, Smolen, & Hewitt, 2006) and increased aggressive behavior and impulsive-antisocial traits in adolescents (Fowler et al., 2009; Sadeh et al., 2010; Sakai et al., 2006). Finally, interactions between short allele status and childhood maltreatment in determining adult antisocial behavior have been found in both humans (Li & Lee, 2010) and nonhuman primates (Barr et al., 2003; Schwandt et al., 2010). Curiously, in contrast to the MAOA knockout mice, SERT knockout mice show decreased aggression (Holmes, Murphy, & Crawley, 2003), and it is also worth noting that reducing transporter function in adults has been shown to reduce aggression (Silva et al., 2010). Thus, while on the whole the associations highlighted here support a "high 5-HT" model of aggression, as we highlight below, it is important to consider temporal and context-dependent aspects of the 5-HT–aggression linkage.

B. 5-HT Synthesis: TPH2

5-HT is synthesized from the amino acid L tryptophan in a two-step process that commences with the hydroxylation of L-tryptophan to 5-hydroxytryptophan (the rate-limiting step in this process) by the enzyme tryptophan hydroxylase (TPH). Given the importance of TPH in regulating 5-HT availability, a number of studies have examined the impact of genetic variation in TPH activity on aggression-related phenotypes. Work in mice suggested a role for enhanced 5-HT synthesis in mouse strains showing high levels of aggression, as TPH activity was found to be higher in these animals (Popova, 2006). Early human genetic studies also supported genetic variation in TPH as a risk factor for impulsive violence: an intron 7 single-nucleotide polymorphism (SNP) (A218C) in the TPH gene was linked, in a directionally consistent fashion, to lower 5-HIAA (Jönsson et al., 1997), a blunted PRL-fen response (Manuck et al., 1999), increased impulsivity and criminality (New et al., 1998), and heightened levels of trait anger and aggression (Hennig, Reuter, Netter, Burk, & Landt, 2005; Manuck et al., 1999; Rujescu et al., 2002). However, the interpretation of those findings is somewhat complicated by the subsequent discovery of a brain-specific TPH isoform (TPH2) (Walther & Bader, 2003), encoded by a gene (TPH2) that is independent of the one on which the A218C SNP resides. Notably, Zhang and colleagues (2004) identified a SNP (C1473G, coding for a proline → arginine substitution) in the murine tph2 gene that selectively affects 5-HT synthesis in brain. The arginine variant, which leads to lower 5-HT synthesis and reduced midbrain TPH2 activity (Osipova, Kulikov, & Popova, 2009), is associated with reduced aggression in mice (Kulikov, Osipova, Naumenko, & Popova, 2005; Osipova et al., 2009). Furthermore, a common haplotype in the human TPH2 gene linked to lower CSF 5-HIAA levels and autoaggression (suicide) is also enriched in patients with borderline personality disorder and predicts impulsive-aggressive symptoms and traits in these individuals (Perez-Rodriguez et al., 2010). These findings are suggestive of a role for TPH2 variation in human impulsive violence. However, unlike the MAOA and SERT VNTRs described earlier, little is known regarding the functional molecular effects of this putative risk haplotype. Moreover, few studies to date have directly examined associations between TPH2 variation and aggression-related phenotypes in humans, and inconsistencies abound in the extant TPH2 psychiatric association literature more generally.

C. 5-HT Release: HTR_{1b}

HTR_{1b}, encoding the 5-HT terminal autoreceptor 5-HT_{1b}, was initially proposed as an aggression susceptibility gene by an informative series of rodent knockout and behavioral pharmacology studies. Genetic deletion of the 5-HT_{1b} receptor in mice produces a dramatically aggressive phenotype (Saudou et al., 1994), and these mice also show profound deficits in impulse control (Bouwknecht et al., 2001). Consistent with the apparent proaggression effects of reducing 5-HT_{1b} function, 5-HT_{1b} agonists decrease offensive (Bell, Donaldson, & Gracey, 1995) and frustration-induced (de Almeida & Miczek, 2002) aggression in rats and attenuate anabolic steroid-induced aggression in hamsters (Grimes & Melloni, 2005). Furthermore, there appears to be some degree of regional specificity with respect to this action: De Almeida and colleagues (2006) found that microinjections of a 5-HT_{1b} agonist in ventral orbitofrontal cortex—but not more lateral aspects of PFC—dose-dependently decreased aggressive responding in mice. Similar antiaggressive effects were found following microinjections of a 5-HT_{1b} agonist into 5-HTergic neurons of the dorsal raphe (Bannai, Fish, Faccidomo, & Miczek, 2007), and—intriguingly, given the linkage between 5-HT and type II alcoholism—these injections reduced escalated aggression seen after alcohol administration (Faccidomo, Bannai, & Miczek, 2008). Of note, this receptor appears to play a role in impulsive choice behavior because as the 5-$HT_{1a/b}$ agonist eltoprazine decreases impulsive responding in a delay aversion paradigm in rats (van den Bergh, Bloemarts, Groenink, Olivier, & Oosting, 2006). The behavioral effects of these genetic and pharmacological manipulations likely result from 5-HT_{1b}-mediated reductions in phasic 5-HT transmission, due to its role in negatively regulating 5-HT release from presynaptic terminals (de Boer & Koolhaas, 2005); however, the modulation of other transmitter systems by postsynaptic 5-HT_{1b} heteroreceptors may also be involved (Olivier & van Oorschot, 2005).

Human genetic association data also support a role for HTR_{1b} variation in predisposing impulsive violence. An early study reported that a synonymous coding variant in HTR_{1b} (G861C) was associated with increased risk for antisocial behavior in alcoholics (Lappalainen et al., 1998), and the same allele linked to antisocial alcoholism in that study (861C) was subsequently found to be enriched in individuals with substance abuse disorders and major depression (Huang et al., 2003). Of note, the other allele at that locus (861G) is part of a haplotype that is associated with increased transcription in vitro (Duan et al., 2003) and with higher PFC 5-HT_{1b} receptor availability in postmortem human brains (Huang, Grailhe, Arango, Hen, & Mann, 1999), possibly because it is in Linkage Disequilibrium (LD) with a true functional marker (Duan et al., 2003). Zouk (2007) found that the -161T allele predicts increased aggressive behavior in those who complete suicide, which is all the more interesting given that this allele was linked to decreased transcriptional activity using a reporter gene assay in human cortical neuron cells (Duan et al., 2003). More recently, work by Jensen and colleagues (2009) identified a common polymorphic variant (rs13212041) in the 3′ UTR of HTR_{1b} that impacts microRNA-mediated repression of HTR_{1b} expression. In that study, the A allele at that SNP was associated with relatively greater repression of HTR_{1b} by miR-96. Critically, the authors found that this same allele predicted higher levels of antisocial behavior in young adults, and this association was selective for aggressive antisocial behaviors compared to nonaggressive antisocial behaviors (e.g., rule-breaking, cheating) (Jensen et al., 2009). Similar associations were subsequently reported to anger and hostility traits in rs13212041-A carrying community volunteers (Conner et al., 2010). Together, the extant human data suggest that genetic risk for aggression attributable to HTR_{1b} variation is linked to genetically mediated decreases in HTR_{1b} expression. Because lower HTR_{1b} expression would lead to diminished autoreceptor control over 5-HT release, the human genetic findings described here accord with the notion that dysregulated (enhanced) 5-HT transmission is involved in propensity to violence.

IX. Summary

So far, we have outlined multiple lines of evidence—transgenic, pharmacological, and human genetic—which converge to nominate variation in 5-HT signaling pathway genes as a heritable susceptibility factor for impulsive violence. However, despite the profound importance of such studies, these approaches do not permit discovery of the systems-level pathomechanisms that might be responsible for conveying risk in humans. In the next section, we describe a strategy that marries human functional imaging and classic genetic association (imaging genetics) to yield evidence about the specific neurobiological pathways that translate genetic risk to manifest illness.

X. Systems-Level Mechanisms for Genetic Risk: Imaging Genetics

The development of noninvasive neuroimaging techniques has permitted a novel and complementary approach to studying the pathobiology of aggression: the intermediate phenotype strategy. Risk allele effect sizes are typically small for taxonomic clinical phenotypes (such as *Diagnostic and Statistical Manual of Mental Disorders* (DSM-IV)-based diagnoses), personality traits, and behavioral variables, perhaps in part owing to the subtle changes in protein function or transcript expression produced by such variants and the complexity of cognition and behavior. These small effect sizes may, in turn, account for the inconsistencies and nonreplications that plague psychiatric genetics generally. However, the penetrance of genetic variants—the likelihood that any given individual with a certain genotype will express a putatively associated phenotype—varies markedly depending on the level of the phenotype being examined. By studying quantitative biological phenotypes that are closer to the direct impact of the genetic variant under study (intermediate phenotypes), we are able to increase our ability to detect subtle genetic effects (Meyer-Lindenberg et al., 2006). Critically, the use of neuroimaging allows investigators to probe the brain at the systems level to give a functional readout of the impact of a genetic variant on brain function, structure, and connectivity. Thus, this strategy both enhances our ability to uncover genetic effects and affords us the means of discovering neurobiological genetic risk mechanisms that are meaningfully relatable to specific domains of human cognition and behavior.

The past 10 years has seen an explosion in the use of multimodal genetic imaging to characterize risk mechanisms in psychiatry, and this approach has proven highly fruitful for the study of impulsive aggression. Across studies, a compelling picture has begun to emerge that implicates dysfunction within corticolimbic circuitry for affective arousal, emotion regulation, inhibitory control, and social cognition in genetic risk for violence. In this section, we review the available data, which converge to suggest that genetically mediated enhancements in serotonergic neurotransmission lead to dysregulation within this circuitry.

XI. Brain Structure

In an early study of genetic effects on brain structure, Good and colleagues found morphometric changes in the orbitofrontal cortex and amygdala of women with Turner syndrome (45,X) and those with partial deletions of the X-chromosome encompassing the MAOA gene (Good et al., 2003). These women were also relatively impaired at facial expression recognition, an amygdala-mediated function (Adolphs, Tranel, Damasio, & Damasio, 1994). Since then, a number of studies have found that individuals carrying MAOA-L or SERT short alleles show profound alterations in the structure of key limbic and paralimbic regions involved in socioaffective processing. For example, we have shown previously that MAOA-L allele carriers exhibit dramatically reduced gray matter volume in the amygdala and insula and throughout the cingulate gyrus, and a sex-specific increase (men only) in lateral orbitofrontal volume (Meyer-Lindenberg, 2006). This pattern of structural change has since been independently replicated (in the cingulate and orbitofrontal cortex) (Alia-Klein, 2011; Cerasa, 2008) and extended using measurements of cortical surface thickness (Cerasa et al., 2010). Similar effects on structure have been found to result from allelic variation at the 5-HTTLPR locus. Short allele carriers show reduced gray matter volume in amygdala and cingulate (Frodl et al., 2004; 2008; Pezawas et al., 2005; 2008) and additionally in hippocampus (Frodl et al., 2004; 2008). Of note, Jedema and colleagues recently reported analogous structural findings in rhesus monkeys, with carriers of an orthologous rh5-HTTLPR short allele demonstrating morphological deficits in amygdala, cingulate, and hippocampus (Jedema et al., 2010). One study to date has examined structural effects of TPH2 variation and found smaller amygdala and hippocampal volumes in individuals carrying a T allele at a 5′ promoter SNP (G-703T; rs4570625) (Inoue et al., 2010), which exists in strong linkage disequilibrium with a variant that affects TPH2 promoter activity in vitro (Scheuch et al., 2007). Currently, no data exist currently on the impact of HTR_{1b} variation on brain morphology. Together, the findings described here strongly implicate enhanced serotonin signaling due to reduced clearance (via transport or catabolism) in contributing to changes in the structure of limbic and paralimbic cortical regions.

XII. Brain Function

Using experimental paradigms that index multiple aspects of cognitive and affective processing, functional imaging studies make clear that the genetically determined structural alterations in corticolimbic circuitry just detailed are accompanied by profound changes in the function of this circuitry. In two early studies using a conflict resolution task (Fan, Fossella, Sommer, Wu, & Posner, 2003) and an inhibitory control task (Passamonti et al., 2005; 2008), MAOA-L carriers showed decreased engagement of anterior cingulate and ventral prefrontal cortex (PFC), the degree of which predicted scores on a measure of trait impulsivity. Subsequently, we found highly significant genotype-related differences in multiple domains of function that are linked to impulsive aggression. During aversive facial emotion processing (angry and fearful faces), MAOA-L subjects showed exaggerated amygdala and insula activation, with diminished recruitment of regulatory regions of PFC (orbitofrontal and anterior cingulate cortices). In the domain of emotional memory, we found a specific effect on aversive memory retrieval, such that the MAOA-L allele was associated with greater activation in amygdala and hippocampus during the recall of negatively valenced visual scenes. Last, MAOA-L males, but not females, showed reduced activation in dorsal cingulate (a region of maximal structural change) during a go/no-go task, replicating earlier studies in a significantly larger sample (Meyer-Lindenberg et al., 2006). Lee and Ham (2008) replicated the finding of enhanced amygdala activation in MAOA-L carriers using a face emotion processing paradigm, and Alia-Klein and colleagues (2009) have shown that increased amygdala activation in MAOA-L individuals is associated with higher levels of self-reported anger reactivity. Notably, two studies suggest that MAOA-L carriers show enhanced responsivity in corticolimbic circuitry to social rejection (Eisenberg, Way, Taylor, Welch, & Lieberman, 2007; Sebastian et al., 2010), and this exaggerated reactivity is associated with trait aggression and interpersonal hypersensitivity (Eisenberg et al., 2007).

With respect to 5-HTTLPR, there is a wealth of data showing that short allele carriers have enhanced amygdala function. This effect has been demonstrated during aversive face emotion processing (Canli et al., 2006; Hariri et al., 2002; 2005, Pezawas et al., 2005; Munafo, Brown, & Hariri, 2008; von dem Hagen, Passamonti, Nutland, Sambrook, & Calder, 2011), in response to negatively valenced words (Canli et al., 2005b), at rest (Canli et al., 2006; Rao et al., 2007), during negative mood induction (Fortier et al., 2010), in response to threat in nonhuman primates (Kalin et al., 2008), and during economic decision making (Roiser et al., 2009). In addition, short allele carriers show alterations in cingulate function during performance monitoring (Holmes, Bogdan, & Pizzagalli, 2010) and response inhibition (Passamonti et al., 2008), mirroring the effect of the MAOA-L allele on cingulate activity during cognitive control. Several studies have investigated the effect of TPH2 variation on limbic reactivity, each of which showed enhanced amygdala activation during aversive face emotion processing in rs4570625-T allele carriers, which accords with the structural finding reported by Inoue and colleagues (2010) (see also Brown et al., 2005; Canli, Congdon, Todd Constable, & Lesch, 2008; Canli et al., 2005*a*). Notably, the rs4570625-T allele has been shown to predict relatively increased TPH2 promoter activity in vitro (when considered as part of a 2-SNP haplotype with rs11178997). To date, no studies have examined potential consequences of HTR_{1b} variation on brain function.

On the whole, these data show that individuals who carry polymorphic variants that are putatively associated with increased serotonergic tone demonstrate a consistent pattern of altered functional responses within corticolimbic circuitry for affective arousal, social cognition, and inhibitory control. Specifically, this pattern is characterized by exaggerated amygdala activation in the context of reduced regulatory control by aspects of medial PFC (mPFC)—especially the anterior cingulate. Given the plethora of evidence that mPFC and amygdala are wired as a negative regulatory circuit (Price, 2005; Quirk, Likhtik, Pelletier, & Pare, 2003; Sotres-Bayon, Cain, & LeDoux, 2006) and recent data supporting the notion that this circuit is disrupted in impulsive aggression (Coccaro, Sripada, Yanowitch, & Phan, 2011), these data point to the possibility that altered connectivity within this circuit might be an important consequence of genetic risk.

XIII. Brain Connectivity

We sought to investigate whether the MAOA-L allele was, in fact, associated with perturbed connectivity between nodes within a medial prefrontal–amygdala network for affective arousal and control. Using functional connectivity analyses, we found that MAOA-L men showed stronger coupling between amygdala and ventromedial PFC (vmPFC). The degree of functional connectivity was correlated with the magnitude of amygdala hyperactivity in MAOA-L men ($R > .3$), suggesting regulation, and positively correlated with a trait measure of emotional reactivity (Buckholtz et al., 2008). As direct anatomical connections between vmPFC and amygdala are sparse (Ongur & Price, 2000; Ghashghaei & Barbas, 2002; Ghashghaei, Hilgetag, & Barbas, 2007), we hypothesized that another region with direct connections to both regions might be involved. In a subsequent analysis, we found that activity in perigenual anterior cingulate—robustly interconnected with both amygdala (Carmichael & Price, 1996) and vmPFC (Ongur & Price, 2000)—strongly tracked the magnitude of amygdala–vmPFC coupling. These findings suggested that vmPFC is brought online as a second-level emotion regulation node to provide compensatory support to perigenual cingulate, the structure and function of which is compromised in individual MAOA-L males. To test this model, we used path analysis to parse the directional connections within this network. This analysis suggested that vmPFC regulates amygdala activation indirectly, via an input to perigenual cingulate, which sends an direct inhibitory projection to amygdala (Buckholtz et al., 2008). To date, one other study has found similar MAOA u-VNTR genotype-dependent alterations in corticolimbic connectivity (Dannlowski et al., 2009). Strikingly, 5-HTTLPR appears to have a similar impact on this circuitry. Short allele carriers demonstrate enhanced amygdala–VMPFC coupling (Friedel et al., 2009; Heinz et al., 2005) in the context of reduced amygdala–perigenual cingulate connectivity (Lemogne et al., 2011; Pezawas et al., 2005), which is in turn negatively correlated with a trait measure of affective reactivity. Alterations in cortical regulation of subcortical reactivity in short allele carriers may be due to changes in structural connectivity: short-allele carriers show relatively degraded white matter architecture in the uncinate fasciculus, a fiber tract that connects mPFC and amygdala (Pacheco et al., 2009). This genetically mediated disruption in corticolimbic connectivity is also linked to deleterious affective biases, suggesting a role for genetic risk in maladaptive decision-making behavior (Roiser et al., 2009).

In summary, these data confirm that genetic variation in serotonin signaling predicts a consistent pattern of change in brain structure, function, and connectivity. Individuals who carry alleles that are associated with enhanced serotonin signaling show reduced gray matter volume in amygdala and mPFC, exaggerated reactivity to aversive stimuli in amygdala in concert with compromised engagement of regulatory control by mPFC, and degraded functional and structural connectivity within this same amygdala–mPFC circuit for affective arousal and regulation. Such findings are consistent with prior associations of these alleles to impulsive aggression and accord with available data suggesting dysregulation of this circuit in disorders that are characterized by high levels of impulsive aggression. A wealth of data highlight mPFC as critical in providing inhibitory cortical feedback to amygdala (Quirk et al., 2003; Rosenkranz & Grace, 1999), underlying its essential role in fear extinction (Quirk et al., 2003; Sotres-Bayon et al., 2006), emotion regulation (Ochsner, Bunge, Gross, & Gabrieli, 2002), and temperamental variation (Pezawas et al., 2005). Importantly, this region is highly sensitive to modulation by serotonin because it has the greatest density of serotonin receptors in human cortex (Varnäs, Halldin, & Hall, 2004). Thus, the available data suggest a compelling and neurobiologically plausible system-level mechanism for this association: derangement of corticolimbic circuit development by excess serotonin signaling during ontogeny. In the next section, we marshal several lines of preclinical evidence to provide support for this hypothesis.

XIV. Genetic Derangement of Corticolimbic Circuitry: Developmental Mechanisms

Although the findings just described converge to implicate a genetic predisposition toward excess serotonin signaling in impulsive aggression, this model is at odds with what is known about the impact of acutely manipulating serotonin on impulsivity and aggression. For example, pharmacological enhancement of serotonergic transmission (e.g., by administering SSRIs or 5-HT precursors) reduces aggressive behavior (Carrillo et al., 2009) across a wide range of species, and lowering 5-HT levels via dietary tryptophan depletion increases impulsive and aggressive behavior in humans (Cools et al., 2005; Crockett, Clark, & Robbins, 2009; Robinson, Cools, & Sahakian, 2011; Walderhaug et al., 2002). So how is it possible that, on the one hand, genetic variants that predict increased synaptic serotonin are associated with increased impulsive aggression, whereas on the other, acute reductions in serotonin appear to induce it as well? Preclinical data provide insight into this apparent contradiction and suggest that developmental timing is a crucial factor in determining the effect that 5-HT has on brain development and behavior.

In their original study of MAOA knockout mice, Cases and colleagues (1996) reported a nine-fold increase in 5-HT levels in pups, which normalized after postnatal week 1 as MAO-B expression first becomes detectable. In addition to the characteristic behavioral phenotype, MAOA knockout mice showed a striking morphological phenotype as well: barrel field formation in somatosensory cortex was dramatically altered in MAOA knockouts, and these changes persisted into adulthood. Critically, these morphological deficits were restored by blocking 5-HT synthesis in neonates, suggesting that this dysregulated pattern of development in MAOA knockouts was induced by excessive 5-HT signaling in early life (Cases et al., 1995). Indeed, there does appear to be a critical period for the adverse effects of excess 5-HT on cortical development in these animals. Normal development is restored in MAOA knockouts by pharmacological attenuation of 5-HT, but not catecholamine synthesis, in the first week of life. Furthermore, transient inhibition of MAOA activity during the first postnatal week reproduces these effects in wild-type rats (Cases et al., 1996). In a subsequent study, these investigators administered monoamine oxidase inhibitors at different times throughout development to precisely identify the critical developmental window for adverse effects of excess serotonin on cortical development: sensitivity to excess serotonin was found to be maximal in the period between birth and 4 days (Vitalis et al., 1998). These findings accord well with other data showing that transient pharmacological inhibition of the serotonin transporter during a critical developmental window produces heightened emotional reactivity in adults (Ansorge, Morelli, & Gingrich, 2008; Ansorge, Zhou, Lira, Hen, & Gingrich, 2004) and highlight the temporally specific nature of excess 5-HT signaling on brain development and adult behavior.

Notably, a separate line of research suggests the involvement of 5-HT_{1b} in the developmental impact of excess 5-HT. Both retinogeniculate and thalamocortical axon development is retarded in MAOA knockouts and in 5-HTT knockouts; however, a normal morphological phenotype is observed in both MAOA/5-HT_{1b} and 5-HTT/5-HT_{1b} double-knockouts, suggesting that 5-HT_{1b} hyperstimulation may play a role in aberrant axon patterning in MAOA knockouts (Murphy et al., 2003; Rebsam, Seif, & Gaspar, 2002; Salichon et al., 2001; Upton et al., 2002). Recent work suggests that 5-HT-induced alterations in axon pathfinding may result from 5-HT_{1b}-mediated changes in the response of axons to chemical guidance cues. Although thalamocortical axons are typically attracted to netrin-1, excessive 5-HT signaling through 5-HT_{1b} receptors causes axons to instead be repelled from netrin-1 sources (Bonnin, Torii, Wang, Rakic, & Levitt, 2007). Together, these data provide a compelling demonstration that excess 5-HT signaling during a precise temporal window in early life has a profound and lasting impact on brain development and adult behavior. Furthermore, they offer a plausible mechanism that could explain why genetically determined variation in 5-HT tone contributes to risk for psychiatric phenotypes that are characterized by impulsivity and emotional reactivity, even as adult administration of 5-HT-enhancing drugs appears to ameliorate these same symptoms.

XV. Conclusion

In this review, we highlight a key role for genetically determined individual differences in serotonergic tone in susceptibility to impulsive violence. We propose that genetic predisposition to violence is mediated at the neural level, at least in part, by the adverse effects of excessive 5-HT signaling during a critical sensitive period on the development of brain circuitry for emotional arousal, affect regulation, and inhibitory control. However, we do not wish to ignore the essential role of environment in determining how, or whether, this predisposition is expressed. Given that the heritability of antisocial behavior and aggression is far from complete, it is crucial to consider the role of nonheritable family, community, and social factors that act independently from and interactively with genetic risk to determine behavioral outcomes. In particular, forms of early life stress such as childhood maltreatment and neglect are strongly associated with adolescent and adult antisocial behavior and aggression (Dodge, Bates, & Pettit, 1990; Lansford et al., 2006; Lau & Weisz, 2003). Critically, there is ample evidence that genetic factors for antisocial aggression are significantly more penetrant on a background of early life stress (Hicks, South, Dirago, Iacono, & McGue, 2009), and there is compelling support for a specific interaction between several of the genes identified earlier and childhood maltreatment in determining adult aggression (Caspi et al., 2002; Kim-Cohen et al., 2006). Although the precise cellular and systems-level mechanisms underlying this interaction are not yet known, it is tempting to speculate given the available data. For instance, quantitative DNA methylation profiling of the MAOA promoter shows that individual differences in methylation appear early, are observed even in monozygotic twins, and change dynamically over time (Wong et al., 2010), suggesting environmental regulation of MAOA. Consistent with this, bioinformatic analysis of the MAOA promoter has revealed the presence of two CpG islands, which are known targets of epigenetic methylation. The MAOA u-VNTR is contained within one of these, and CpG sites have been identified within the polymorphic region itself (Shumay & Fowler, 2010). Furthermore, this region contains a glucocorticoid response element (GRE), located approximately 150 bp from the u-VNTR, which regulates MAOA expression in response to stress hormone signaling (Ou, Chen, & Shih, 2006). Thus, variation at the u-VNTR locus could conceivably modulate the response of stress on MAOA expression by altering the structure of cis-regulatory sites in the MAOA promoter region. In addition, stress up-regulates forebrain expression of the S-100 calcium binding protein p11, a molecule that appears to play an important role in affective psychopathology (Svenningsson & Greengard, 2007; Zhang et al., 2008). This is intriguing, given evidence that p11, in turn, up-regulates cell-surface expression of the 5-HT_{1b} receptor (Svenningsson et al., 2006). Thus, early life stress could potentiate the impact of susceptibility alleles that enhance synaptic 5-HT levels by increasing the density of available 5-HT_{1b} receptors, leading to alterations in axon patterning as described earlier.

Systems-level neuroscience approaches can also help to further specify interacting effects of genes and environment. For example, acute alcohol use is implicated in approximately one-half of violent crimes and sexual assaults, and individuals with alcohol abuse or dependence are significantly more likely to be involved in violence compared to individuals without a psychiatric disorder. Furthermore, since several twin and adoption studies have failed to observe a significant disposition toward violent behavior independent of increased risk to develop alcohol dependence, a common hereditary factor might predispose individuals to both alcohol dependence and violent behavior. We have recently reviewed evidence (Heinz, Beck, Meyer-Lindenberg, Sterzer, & Heinz, 2011) that not only shows an impact of acute and chronic alcohol intake on serotonin metabolism and turnover across species, but also suggests that alcohol effects may be partially mediated by the same circuit of prefrontal control over limbic structures such as amygdala, outlined earlier. Similarly, urban living is a risk factor for violent acts, which are much more frequent in this environment than in rural areas. Even though it is unclear which specific aspects of the urban experience might be causal for this, neuroimaging during social evaluative stress (Vitiello & Stoff, 1997) shows a specific impact of urban living on the activation of the same amygdala–cingulate circuit that is found impacted on by MAOA and 5-HTTLPR, suggesting that early developmental effects (perhaps themselves stress-mediated) on serotonergic neurotransmission might play a role here also. In both cases, systems neuroscience cannot prove the relevance of serotonergic neurotransmission, but it suggests a clear research agenda to clarify that contribution by showing that a risk factor impacts on a specific circuit in which serotonin plays a key role.

Finally, although we have highlighted the role of genetically determined individual difference in 5-HT tone in susceptibility to impulsive aggression, this is likely only one of many potential genetic pathways to violence. Dopamine (DA) is known from both preclinical and human imaging work to play a role in impulsivity and aggression (Buckholtz et al., 2010*a*; 2010*b*; Couppis & Kennedy, 2008), and genetic variation in DA signaling has also been associated with aggression and related psychopathology (Grigorenko et al., 2010). The social neuropeptides vasopressin and oxytocin make attractive targets as well, given their importance for regulating aggression and social behavior in animal models (Ferris, 2005; Heinrich & Domes, 2008). Of note, variability in human neuropeptide signaling appears to affect the same key corticolimbic circuitry for affective arousal and control that we and others have shown to be dysregulated as a consequence of genetically determined alterations in 5-HT (Meyer-Lindenberg et al., 2009; Tost et al., 2010). To conclude, there is much work yet to be done if we are to fully understand the what, where, when, and how of susceptibility to impulsive violence. Future studies that incorporate environmental assessments, measurements of genome and epigenome-wide variation, and validated systems-level phenotypes into epidemiologically valid longitudinal designs will be required to complete our understanding of the complex neurogenetic architecture of this devastating and costly phenomenon.

Note

1 Data circa 2000. More recent data from the Centers for Disease Control suggest that current numbers are even higher. http://www.cdc.gov/violenceprevention/violentdeaths

References

Abney, M., McPeek, M. S., & Ober, C. (2001). Broad and narrow heritabilities of quantitative traits in a founder population. *American Journal of Human Genetics, 68*, 1302–1307.

Adolphs, R., Tranel, D., Damasio, H., & Damasio, A. (1994). Impaired recognition of emotion in facial expressions following bilateral damage to the human amygdala. *Nature, 372*, 669–672.

af Klinteberg, B., Schalling, D., Edman, G., Oreland, L., & Asberg, M. (1987). Personality correlates of platelet monoamine oxidase (MAO) activity in female and male subjects. *Neuropsychobiology, 18*, 89–96.

Alia-Klein, N., et al. (2008). Brain, monoamine oxidase A activity predicts trait aggression. *Journal of Neuroscience, 28*, 5099–5104.

Alia-Klein, N., et al. (2009). Neural mechanisms of anger regulation as a function of genetic risk for violence. *Emotion, 9*, 385–396.

Alia-Klein, N., et al. (2011). Gene x disease interaction on orbitofrontal gray matter in cocaine addiction. *Archives of General Psychiatry, 68*, 283–294.

Ansorge, M. S., Morelli, E., & Gingrich, J. A. (2008). Inhibition of serotonin but not norepinephrine transport during development produces delayed, persistent perturbations of emotional behaviors in mice. *Journal of Neuroscience, 28*, 199–207.

Ansorge, M. S., Zhou, M., Lira, A., Hen, R., & Gingrich, J. A. (2004). Early-life blockade of the 5-HT transporter alters emotional behavior in adult mice. *Science, 306*, 879–881.

Asberg, M., Träskman, L., & Thorén, P. (1976). 5-HIAA in the cerebrospinal fluid. A biochemical suicide predictor? *Archives of General Psychiatry, 33*, 1193–1197.

Askenazy, F., Caci, H., Myquel, M., Darcourt, G., & Lecrubier, Y. (2000). Relationship between impulsivity and platelet serotonin content in adolescents. *Psychiatry Research, 94*, 19–28.

Aslund, C., et al. (2011). Maltreatment, MAOA, and delinquency: Sex differences in gene-environment interaction in a large population-based cohort of adolescents. *Behavioral Genetics, 41*, 262–272.

Baca-García, E., et al. (2005). A pilot genetic study of the continuum between compulsivity and impulsivity in females: The serotonin transporter promoter polymorphism. *Progress in Neuropsychopharmacology, and Biological Psychiatry, 29*, 713–717.

Baker, L. A., Jacobson, K. C., Raine, A., Lozano, D. I., & Bezdjian, S. (2007). Genetic and environmental bases of childhood antisocial behavior: A multi-informant twin study. *Journal of Abnormal Psychology, 116*, 219–235.

Baker, L. A., Raine, A., Liu, J., & Jacobson, K. C. (2008). Differential genetic and environmental influences on reactive and proactive aggression in children. *Journal of Abnormal Child Psychology, 36*, 1265–1278.

Balciuniene, J., Emilsson, L., Oreland, L., Pettersson, U., & Jazin, E. (2002). Investigation of the functional effect of monoamine oxidase polymorphisms in human brain. *Human Genetics, 110*, 1–7.

Bannai, M., Fish, E. W., Faccidomo, S., & Miczek, K. A. (2007). Anti-aggressive effects of agonists at 5-HT1B receptors in the dorsal raphe nucleus of mice. *Psychopharmacology, 193*, 295–304.

Barnett, J. H., Xu, K., Heron, J., Goldman, D., & Jones, P. B. (2011). Cognitive effects of genetic variation in monoamine neurotransmitter systems: A population-based study of COMT, MAOA, and 5HTTLPR. *American Journal of Medical Genetics Part B: Neuropsychiatric Genetics, 156*, 158–167.

Barr, C. S., et al. (2003). The utility of the non-human primate; model for studying gene by environment interactions in behavioral research. *Genes, Brain, and Behavior, 2*, 336–340.

Beaver, K. M., DeLisi, M., Vaughn, M. G., & Barnes, J. C. (2010). Monoamine oxidase A genotype is associated with gang membership and weapon use. *Comprehensive Psychiatry, 51*, 130–134.

Beitchman, J. H., et al. (2006). Serotonin transporter polymorphisms and persistent, pervasive childhood aggression. *The American Journal of Psychiatry, 163*, 1103–1105.

Beitchman, J. H., Mik, H. M., Ehtesham, S., Douglas, L., & Kennedy, J. L. (2004). MAOA and persistent, pervasive childhood aggression. *Molecular Psychiatry, 9*, 546–547.

Belfrage, H., Lidberg, L., & Oreland, L. (1992). Platelet monoamine oxidase activity in mentally disordered violent offenders. *Acta Psychiatrica Scandinavica, 85*, 218–221.

Bell, R., Donaldson, C., & Gracey, D. (1995). Differential effects of CGS 12066B and CP-94,253 on murine social and agonistic behaviour. *Pharmacology, Biochemistry and Behavior, 52*, 7–16.

Bezdjian, S., Tuvblad, C., Raine, A., & Baker, L. A. (2011). The genetic and environmental covariation among psychopathic personality traits, and reactive and proactive aggression in childhood. *Child Development, 82*, 1267–1281.

Blair, R. J. (2001). Neurocognitive models of aggression, the antisocial personality disorders, and psychopathy. *Journal of Neurology, Neurosurgery and Psychiatry, 71*, 727–731.

Bonnin, A., Torii, M., Wang, L., Rakic, P., & Levitt, P. (2007). Serotonin modulates the response of embryonic thalamocortical axons to netrin-1. *Nature Neuroscience, 10*, 588–597.

Bouwknecht, J. A., et al. (2001). Absence of 5-HT(1B) receptors is associated with impaired impulse control in male 5-HT(1B) knockout mice. *British Psychological Society, 49*, 557–568.

Brendgen, M., Vitaro, F., Boivin, M., Dionne, G., & Pérusse, D. (2006). Examining genetic and environmental effects on reactive versus proactive aggression. *Developmental Psychology, 42*, 1299–1312.

Brown, C. S., et al. (1989). Blood platelet uptake of serotonin in episodic aggression. *Psychiatry Research, 27*, 5–12.

Brown, G. L., et al. (1982). Aggression, suicide, and serotonin: Relationships to CSF amine metabolites. *The American Journal of Psychiatry, 139*, 741–746.

Brown, S. M., et al. (2005). A regulatory variant of the human tryptophan hydroxylase-2 gene biases amygdala reactivity. *Molecular Psychiatry, 10*, 884–888, 805.

Brunner, H. G., Nelen, M., Breakefield, X. O., Ropers, H. H., & van Oost, B. A. (1993*a*). Abnormal behavior associated with a point mutation in the structural gene for monoamine oxidase A. *Science, 262*, 578–580.

Brunner, H. G., et al. (1993*b*). X-linked borderline mental retardation with prominent behavioral disturbance: Phenotype, genetic localization, and evidence for disturbed monoamine metabolism. *American Journal of Human Genetics, 52*, 1032–1039.

Buckholtz, J. W., et al. (2008). Genetic variation in MAOA modulates ventromedial prefrontal circuitry mediating individual differences in human personality. *Molecular Psychiatry, 13*, 313–324.

Buckholtz, J. W., et al. (2010*a*). Dopaminergic network differences in human impulsivity. *Science, 329*, 532.

Buckholtz, J. W., et al. (2010*b*). Mesolimbic dopamine reward system hypersensitivity in individuals with psychopathic traits. *Nature Neuroscience, 13*, 419–421.

Button, T. M. M., Scourfield, J., Martin, N., & McGuffin, P. (2004). Do aggressive and non-aggressive antisocial behaviors in adolescents result from the same genetic and environmental effects? *American Journal of Medical Genetics Part B: Neuropsychiatric Genetics, 129B*, 59–63.

Canli, T., Congdon, E., Gutknecht, L., Constable, R. T., & Lesch, K. P. (2005*a*). Amygdala responsiveness is modulated by tryptophan hydroxylase-2 gene variation. *Journal of Neural Transmission, 112*, 1479–1485.

Canli, T., Congdon, E., Todd Constable, R., & Lesch, K. P. (2008). Additive effects of serotonin transporter and tryptophan hydroxylase-2 gene variation on neural correlates of affective processing. *Biological Psychology, 79*, 118–125.

Canli, T., & Lesch, K. P. (2007). Long story short: The serotonin transporter in emotion regulation and social cognition. *Nature Neuroscience, 10*, 1103–1109.

Canli, T., et al. (2005*b*). Beyond affect: A role for genetic variation of the serotonin transporter in neural activation during a cognitive attention task. *Proceedings of the National Academy of Sciences USA, 102*, 12224–12229.

Canli T., et al. (2006). *Neural correlates of epigenesis. Proceedings of the National Academy of Sciences USA, 103*, 16033–16038.

Carmichael, S. T., & Price, J. L. (1996). Connectional networks within the orbital and medial prefrontal cortex of macaque monkeys. *Journal of Comparative Neurology, 371*, 179–207.

Carrillo, M., Ricci, L. A., Coppersmith, G. A., & Melloni, R. H. (2009). The effect of increased serotonergic neurotransmission on aggression: A critical meta-analytical review of preclinical studies. *Psychopharmacology, 205*, 349–368.

Cases, O., et al. (1995). Aggressive behavior and altered amounts of brain serotonin and norepinephrine in mice lacking MAOA. *Science, 268*, 1763–1766.

Cases, O., et al. (1996). Lack of barrels in the somatosensory cortex of monoamine oxidase A-deficient mice: Role of a serotonin excess during the critical period. *Neuron, 16*, 297–307.

Caspi, A., et al. (2002). Role of genotype in the cycle of violence in maltreated children. *Science, 297*, 851–854.

Caspi, A., et al. (2003). Influence of life stress on depression: Moderation by a polymorphism in the 5-HTT gene. *Science, 301*, 386–389.

Cates, D. S., Houston, B. K., Vavak, C. R., Crawford, M. H., & Uttley, M. (1993). Heritability of hostility-related emotions, attitudes, and behaviors. Journal of *Behavioral Medicine, 16*, 237–256.

Cerasa, A., et al. (2008). MAO A VNTR polymorphism and variation in human morphology: A VBM study. *Neuroreport, 19*, 1107–1110.

Cerasa, A., et al. (2010). Morphological correlates of MAO A VNTR polymorphism: New evidence from cortical thickness measurement. *Behavioral Brain Research, 211*, 118–124.

Cervantes, M. C., & Delville, Y. (2011). Developmental predictors of an impulsive-aggressive phenotype. *Developmental Psychobiology, 53*, 343–358.

Cirulli, E. T., & Goldstein, D. B. (2007). In vitro assays fail to predict in vivo effects of regulatory polymorphisms. *Human Molecular Genetics, 16*, 1931–1939.

Clarke, A. S., et al. (1995). Evidence for heritability of biogenic amine levels in the cerebrospinal fluid of rhesus monkeys. *British Psychological Society, 38*, 572–577.

Cleare, A. J., & Bond, A. J. (1997). Does central serotonergic function correlate inversely with aggression? A study using D-fenfluramine in healthy subjects. *Psychiatry Research, 69*, 89–95.

Coccaro, E. F., Berman, M. E., Kavoussi, R. J., & Hauger, R. L. (1996*a*). Relationship of prolactin response to d-fenfluramine to behavioral and questionnaire assessments of aggression in personality-disordered men. *British Psychological Society, 40*, 157–164.

Coccaro, E. F., Kavoussi, R. J., Cooper, T. B., & Hauger, R. L. (1997). Central serotonin activity and aggression: Inverse relationship with prolactin response to d-fenfluramine, but not CSF 5-HIAA concentration, in human subjects. *The American Journal of Psychiatry, 154*, 1430–1435.

Coccaro, E. F., Kavoussi, R. J., Sheline, Y. I., Lish, J. D., & Csernansky, J. G. (1996*b*). Impulsive aggression in personality disorder correlates with tritiated paroxetine binding in the platelet. *Archives of General Psychiatry, 53*, 531–536.

Coccaro, E. F., Lee, R., & Kavoussi, R. J. (2010*a*). Aggression, suicidality, and intermittent explosive disorder: Serotonergic correlates in personality disorder and healthy control subjects. *Neuropsychopharmacology, 35*, 435–444.

Coccaro, E. F., Lee, R., & Kavoussi, R. J. (2010*b*). Inverse relationship between numbers of 5-HT transporter binding sites and life history of aggression and intermittent explosive disorder. *Journal of Psychiatric Research, 44*, 137–142.

Coccaro, E. F., Sripada, C. S., Yanowitch, R. N., & Phan, K. L. (2011). Corticolimbic function in impulsive aggressive behavior. *British Psychological Society, 69*, 1153–1159.

Conner, T. S., et al. (2010). Functional polymorphisms in the serotonin 1B receptor gene (HTR1B) predict self-reported anger and hostility among young men. *American Journal of Medical Genetics Part B: Neuropsychiatric Genetics, 153B*, 67–78.

Contini, V., Marques, F. Z. C., Garcia, C. E. D., Hutz, M. H., & Bau, C. H. D. (2006). MAOA-uVNTR polymorphism in a Brazilian sample: Further support for the association with impulsive behaviors and alcohol dependence. *American Journal of Medical Genetics Part B: Neuropsychiatric Genetics, 141B*, 305–308.

Cools, R., et al. (2005). Tryptophan depletion disrupts the motivational guidance of goal-directed behavior as a function of trait impulsivity. *Neuropsychopharmacology, 30*, 1362–1373.

Corso, P., Mercy, J., Simon, T., Finkelstein, E., & Miller, T. (2007). Medical costs and productivity losses due to interpersonal and self-directed violence in the United States. *American Journal of Preventive Medicine, 32*, 474–482.

Couppis, M. H., & Kennedy, C. H. (2008). The rewarding effect of aggression is reduced by nucleus accumbens dopamine receptor antagonism in mice. *Psychopharmacology, 197*, 449–456.

Crockett, M. J., Clark, L., & Robbins, T. W. (2009). Reconciling the role of serotonin in behavioral inhibition and aversion: Acute tryptophan depletion abolishes punishment-induced inhibition in humans. *Journal of Neuroscience, 29*, 11993–11999.

Dannlowski, U., et al. (2009). Reduced amygdala-prefrontal coupling in major depression: Association with MAOA genotype and illness severity. *International Journal of Neuropsychopharmacology, 12*, 11–22.

Daruna, J. H., & Kent, E. W. (1976). Comparison of regional serotonin levels and turnover in the brain of naturally high and low aggressive rats. *Brain Research, 101*, 489–501.

Datla, K. P., Mitra, S. K., & Bhattacharya, S. K. (1991). Serotonergic modulation of footshock induced aggression in paired rats. *Indian Journal of Experimental Biology, 29*, 631–635.

Davis, B. A., Yu, P. H., Boulton, A. A., Wormith, J. S., & Addington, D. (1983). Correlative relationship between biochemical activity and aggressive behaviour. *Progress in Neuropsychopharmacology, and Biological Psychiatry, 7*, 529–535.

De Almeida, R. M. M., & Miczek, K. A. (2002). Aggression escalated by social instigation or by discontinuation of reinforcement ("frustration") in mice: Inhibition by anpirtoline: A 5-HT1B receptor agonist. *Neuropsychopharmacology, 27*, 171–181.

De Almeida, R. M. M., et al. (2006). 5-HT(1B) receptors, ventral orbitofrontal cortex, and aggressive behavior in mice. *Psychopharmacology, 185*, 441–450.

de Boer, S. F., & Koolhaas, J. M. (2005). 5-HT1A and 5-HT1B receptor agonists and aggression: A pharmacological challenge of the serotonin deficiency hypothesis. *European Journal of Pharmacology, 526*, 125–139.

Deckert, J., et al. (1999). Excess of high activity monoamine oxidase A gene promoter alleles in female patients with panic disorder. *Human Molecular Genetics, 8*, 621–624.

Denney, R. M., Koch, H., & Craig, I. W. (1999). Association between monoamine oxidase A activity in human male skin fibroblasts and genotype of the MAOA promoter-associated variable number tandem repeat. *Human Genetics, 105*, 542–551.

Derringer, J., Krueger, R. F., Irons, D. E., & Iacono, W. G. (2010). Harsh discipline, childhood sexual assault, and MAOA genotype: An investigation of main and interactive effects on diverse clinical externalizing outcomes. *Behavioral Genetics, 40*, 639–648.

Dodge, K. A., Bates, J. E., & Pettit, G. S. (1990). Mechanisms in the cycle of violence. *Science, 250*, 1678–1683.

Dolan, M. C., & Anderson, I. M. (2003). The relationship between serotonergic function and the psychopathy checklist: Screening version. *Journal of Psychopharmacology, 17*, 216–222.
Doudet, D., et al. (1995). Cerebral glucose metabolism, CSF 5-HIAA levels, and aggressive behavior in rhesus monkeys. *The American Journal of Psychiatry, 152*, 1782–1787.
Duan, J., et al. (2003). Polymorphisms in the 5'-untranslated region of the human serotonin receptor 1B (HTR1B) gene affect gene expression. *Molecular Psychiatry, 8*, 901–910.
Ducci, F., et al. (2006). A functional polymorphism in the MAOA gene promoter (MAOA-LPR) predicts central dopamine function and body mass index. *Molecular Psychiatry, 11*, 858–866.
Edwards, A. C., et al. (2010). MAOA-uVNTR and early physical discipline interact to influence delinquent behavior. *Journal of Child Psychology and Psychiatry, 51*, 679–687.
Eisenberger, N. I., Way, B. M., Taylor, S. E., Welch, W. T., & Lieberman, M. D. (2007). Understanding genetic risk for aggression: Clues from the brain's response to social exclusion. *British Psychological Society, 61*, 1100–1108.
Eklund, J., Alm, P. O., & af Klinteberg, B. (2005). Monoamine oxidase activity and tri-iodothyronine level in violent offenders with early behavioural problems. *Neuropsychobiology, 52*, 122–129.
Eley, T. C., Lichtenstein, P., & Moffitt, T. E. (2003). A longitudinal behavioral genetic analysis of the etiology of aggressive and nonaggressive antisocial behavior. *Development and Psychopathology, 15*, 383–402.
Enoch, M.-A., Steer, C. D., Newman, T. K., Gibson, N., & Goldman, D. (2010). Early life stress, MAOA, and gene-environment interactions predict behavioral disinhibition in children. *Genes, Brain, and Behavior, 9*, 65–74.
Evans, J., Platts, H., Lightman, S., & Nutt, D. (2000). Impulsiveness and the prolactin response to d-fenfluramine. *Psychopharmacology, 149*, 147–152.
Faccidomo, S., Bannai, M., & Miczek, K. A. (2008). Escalated aggression after alcohol drinking in male mice: Dorsal raphé and prefrontal cortex serotonin and 5-HT(1B) receptors. *Neuropsychopharmacology, 33*, 2888–2899.
Fairbanks, L. A., et al. (2004). Genetic contributions to social impulsivity and aggressiveness in vervet monkeys. *British Psychological Society, 55*, 642–647.
Fan, J., Fossella, J., Sommer, T., Wu, Y., & Posner, M. I. (2003). Mapping the genetic variation of executive attention onto brain activity. *Proceedings of the National Academy of Sciences USA, 100*, 7406–7411.
Farrington, D., & Barnes, G. (1996). The concentration of offending in families. *Legal and Criminological Psychology, 1*, 47–63.
Farrington, D. P., Jolliffe, D., Loeber, R., Stouthamer-Loeber, M., & Kalb, L. M. (2001). The concentration of offenders in families, and family criminality in the prediction of boys' delinquency. *Journal of Adolescence, 24*, 579–596.
Fergusson, D. M., Boden, J. M., Horwood, L. J., Miller, A. L., & Kennedy, M. A. (2011). MAOA, abuse exposure and antisocial behaviour: 30-year longitudinal study. *The British Journal of Psychiatry: The Journal of Mental Science, 198*, 457–463.
Ferris, C. F. (2005). Vasopressin/oxytocin and aggression. *Novartis Foundation Symposium, 268*, 190–198; discussion 198–200, 242–253.
Flory, J. D., Newcorn, J. H., Miller, C., Harty, S., & Halperin, J. M. (2007). Serotonergic function in children with attention-deficit hyperactivity disorder: Relationship to later antisocial personality disorder. *The British Journal of Psychiatry: The Journal of Mental Science, 190*, 410–414.
Foley, D. L., et al. (2004). Childhood adversity, monoamine oxidase a genotype, and risk for conduct disorder. *Archives of General Psychiatry, 61*, 738–744.
Fortier, E., et al. (2010). Early impact of 5-HTTLPR polymorphism on the neural correlates of sadness. *Neuroscience, Letters, 485*, 261–265.
Fowler, J. S., et al. (2007). Evidence that brain MAO A activity does not correspond to MAO A genotype in healthy male subjects. *British Psychological Society, 62*, 355–358.
Fowler, T., et al. (2009). Psychopathy trait scores in adolescents with childhood ADHD: The contribution of genotypes affecting MAOA, 5HTT and COMT activity. *Psychiatric Genetics, 19*, 312–319.
Frankle, W. G., et al. (2005). Brain, serotonin transporter distribution in subjects with impulsive aggressivity: A positron emission study with [11C]McN 5652. *The American Journal of Psychiatry, 162*, 915–923.
Frazzetto, G., et al. (2007). Early trauma and increased risk for physical aggression during adulthood: The moderating role of MAOA genotype. *PLoS ONE, 2*, e486.
Friedel, E., et al. (2009). 5-HTT genotype effect on prefrontal-amygdala coupling differs between major depression and controls. *Psychopharmacology, 205*, 261–271.
Frodl, T., et al. (2004). Reduced hippocampal volumes associated with the long variant of the serotonin transporter polymorphism in major depression. *Archives of General Psychiatry, 61*, 177–183.
Frodl, T., et al. (2008). Reduced gray matter brain volumes are associated with variants of the serotonin transporter gene in major depression. *Molecular Psychiatry, 13*, 1093–1101.
Garpenstrand, H., Annas, P., Ekblom, J., Oreland, L., & Fredrikson, M. (2001). Human fear conditioning is related to dopaminergic and serotonergic biological markers. *Behavioral Neuroscience, 115*, 358–364.
Garpenstrand, H., et al. (2002). A regulatory monoamine oxidase a promoter polymorphism and personality traits. *Neuropsychobiology, 46*, 190–193.
Gelhorn, H. L., et al. (2005). Genetic and environmental influences on conduct disorder: Symptom, domain and full-scale analyses. *Journal of Child Psychology and Psychiatry, 46*, 580–591.
Gelhorn, H., et al. (2006). Common and specific genetic influences on aggressive and nonaggressive conduct disorder domains. Journal of the American Academy of Child and Adolescent Psychiatry, 45, 570–577.
Gerra, G., et al. (2005). Serotonin transporter promoter polymorphism genotype is associated with temperament, personality traits and illegal drugs use among adolescents. *Journal of Neural Transmission, 112*, 1397–1410.
Ghashghaei, H. T., & Barbas, H. (2002). Pathways for emotion: Interactions of prefrontal and anterior temporal pathways in the amygdala of the rhesus monkey. *Neuroscience, 115*, 1261–1279.
Ghashghaei, H. T., Hilgetag, C. C., & Barbas, H. (2007). Sequence of information processing for emotions based on the anatomic dialogue between prefrontal cortex and amygdala. *NeuroImage, 34*, 905–923.

Gibbons, J. L., Barr, G. A., Bridger, W. H., & Leibowitz, S. F. (1979). Manipulations of dietary tryptophan: Effects on mouse killing and brain serotonin in the rat. *Brain Research*, *169*, 139–153.

Gibbons, J. L., Barr, G. A., Bridger, W. H., & Leibowitz, S. F. (1981). L-Tryptophan's effects on mouse killing, feeding, drinking, locomotion, and brain serotonin. *Pharmacology, Biochemistry and Behavior*, *15*, 201–206.

Gonda, X., et al. (2009). Association of the s allele of the 5-HTTLPR with neuroticism-related traits and temperaments in a psychiatrically healthy population. *European Archives of Psychiatry and Clinical Neuroscience*, *259*, 106–113.

Good, C. D., et al. (2003). Dosage-sensitive X-linked locus influences the development of amygdala and orbitofrontal cortex, and fear recognition in humans. *Brain*, *126*, 2431–2446.

Grigorenko, E. L., et al. (2010). Aggressive behavior, related conduct problems, and variation in genes affecting dopamine turnover. *Aggressive Behavior*, *36*, 158–176.

Grimes, J. M., & Melloni, R. H. (2005). Serotonin-1B receptor activity and expression modulate the aggression-stimulating effects of adolescent anabolic steroid exposure in hamsters. *Behavioral Neuroscience*, *119*, 1184–1194.

Guindalini, C., et al. (2005). Association of MAO A polymorphism and alcoholism in Brazilian females. *Psychiatric Genetics*, *15*, 141–144.

Haberstick, B. C., Smolen, A., & Hewitt, J. K. (2006). Family-based association test of the 5HTTLPR and aggressive behavior in a general population sample of children. *British Psychological Society*, *59*, 836–843.

Hakamata, Y., et al. (2005). No association between monoamine oxidase A promoter polymorphism and personality traits in Japanese females. *Neuroscience, Letters*, *389*, 121–123.

Hallikainen, T., et al. (1999). Association between low activity serotonin transporter promoter genotype and early onset alcoholism with habitual impulsive violent behavior. *Molecular Psychiatry*, *4*, 385–388.

Halperin, J. M., et al. (2006). Elevated childhood serotonergic function protects against adolescent aggression in disruptive boys. *Journal of the American Academy of Child and Adolescent Psychiatry*, *45*, 833–840.

Hariri, A. R., et al. (2002). Serotonin transporter genetic variation and the response of the human amygdala. *Science*, *297*, 400–403.

Hariri, A. R., et al. (2005). A susceptibility gene for affective disorders and the response of the human amygdala. *Archives of General Psychiatry*, *62*, 146–152.

Heinrichs, M., & Domes, G. (2008). Neuropeptides and social behaviour: Effects of oxytocin and vasopressin in humans. *Progress in Brain Research*, *170*, 337–350.

Heinz, A., et al. (2005). Amygdala-prefrontal coupling depends on a genetic variation of the serotonin transporter. *Nature Neuroscience*, *8*, 20–21.

Heinz, A. J., Beck, A., Meyer-Lindenberg, A., Sterzer, P., & Heinz, A. (2011). Cognitive and neurobiological mechanisms of alcohol-related aggression. *Nature Reviews Neuroscience*, *12*, 400–413.

Hennig, J., Reuter, M., Netter, P., Burk, C., & Landt, O. (2005). Two types of aggression are differentially related to serotonergic activity and the A779C TPH polymorphism. *Behavioral Neuroscience*, *119*, 16–25.

Herpertz, S., Sass, H., & Favazza, A. (1997). Impulsivity in self-mutilative behavior: Psychometric and biological findings. *Journal of Psychiatric Research*, *31*, 451–465.

Hicks, B. M., South, S. C., Dirago, A. C., Iacono, W. G., & McGue, M. (2009). Environmental adversity and increasing genetic risk for externalizing disorders. *Archives of General Psychiatry*, *66*, 640–648.

Higley, J. D., & Linnoila, M. (1997). Low central nervous system serotonergic activity is traitlike and correlates with impulsive behavior. A nonhuman primate model investigating genetic and environmental influences on neurotransmission. *Annals of the New York Academy of Sciences*, *836*, 39–56.

Higley, J. D., et al. (1992). Cerebrospinal fluid monoamine and adrenal correlates of aggression in free-ranging rhesus monkeys. *Archives of General Psychiatry*, *49*, 436–441.

Higley, J. D., et al. (1993). Paternal and maternal genetic and environmental contributions to cerebrospinal fluid monoamine metabolites in rhesus monkeys (*Macaca mulatta*). *Archives of General Psychiatry*, *50*, 615–623.

Higley, J. D., et al. (1996*a*). CSF testosterone and 5-HIAA correlate with different types of aggressive behaviors. *British Psychological Society*, *40*, 1067–1082.

Higley, J. D., et al. (1996*b*). Stability of interindividual differences in serotonin function and its relationship to severe aggression and competent social behavior in rhesus macaque females. *Neuropsychopharmacology*, *14*, 67–76.

Higley, J. D., et al. (1996*c*). Excessive mortality in young free-ranging male nonhuman primates with low cerebrospinal fluid 5-hydroxyindoleacetic acid concentrations. *Archives of General Psychiatry*, *53*, 537–543.

Higley, J. D., Suomi, S. J., & Linnoila, M. (1996). A nonhuman primate model of type II alcoholism? Part 2. Diminished social competence and excessive aggression correlates with low cerebrospinal fluid 5-hydroxyindoleacetic acid concentrations. *Alcoholism: Clinical and Experimental Research*, *20*, 643–650.

Holmes, A., & Hariri, A. R. (2003). The serotonin transporter gene-linked polymorphism and negative emotionality: Placing single gene effects in the context of genetic background and environment. *Genes, Brain, and Behavior*, *2*, 332–335.

Holmes, A., Murphy, D. L., & Crawley, J. N. (2003). Abnormal behavioral phenotypes of serotonin transporter knockout mice: Parallels with human anxiety and depression. *British Psychological Society*, *54*, 953–959.

Holmes, A. J., Bogdan, R., & Pizzagalli, D. A. (2010). Serotonin transporter genotype and action monitoring dysfunction: A possible substrate underlying increased vulnerability to depression. *Neuropsychopharmacology*, *35*, 1186–1197.

Huang, Y.-Y., et al. (2003). Substance abuse disorder and major depression are associated with the human 5-HT1B receptor gene (HTR1B) G861C polymorphism. *Neuropsychopharmacology*, *28*, 163–169.

Huang, Y.-Y., et al. (2004). An association between a functional polymorphism in the monoamine oxidase a gene promoter, impulsive traits and early abuse experiences. *Neuropsychopharmacology*, *29*, 1498–1505.

Huang, Y. Y., Grailhe, R., Arango, V., Hen, R., & Mann, J. J. (1999). Relationship of psychopathology to the human serotonin 1B genotype and receptor binding kinetics in postmortem brain tissue. *Neuropsychopharmacology*, *21*, 238–246.

Hubbard, J. A., Dodge, K. A., Cillessen, A. H., Coie, J. D., & Schwartz, D. (2001). The dyadic nature of social information processing in boys' reactive and proactive aggression. *Journal of Personality and Social Psychology, 80*, 268–280.

Hudziak, J. J. et al. (2003). Individual differences in aggression: Genetic analyses by age, gender, and informant in 3-, 7-, and 10-year-old Dutch twins. *Behavioral Genetics, 33*, 575–589.

Inoue, H., et al. (2010). Effect of tryptophan hydroxylase-2 gene variants on amygdalar and hippocampal volumes. Brain Research, 1331, 51–57.

Jacob, C. P., et al. (2005). Cluster B personality disorders are associated with allelic variation of monoamine oxidase A activity. *Neuropsychopharmacology, 30*, 1711–1718.

Jedema, H. P., et al. (2010). Cognitive impact of genetic variation of the serotonin transporter in primates is associated with differences in brain morphology rather than serotonin neurotransmission. *Molecular Psychiatry, 15*, 512–22, 446.

Jensen, K. P., et al. (2009). A common polymorphism in serotonin receptor 1B mRNA moderates regulation by miR-96 and associates with aggressive human behaviors. *Molecular Psychiatry, 14*, 381–389.

Jernej, B., & Cicin-Sain, L. (1990). Platelet serotonin level in rats is under genetic control. *Psychiatry Research, 32*, 167–174.

Jönsson, E. G., et al. (1997). Tryptophan hydroxylase and catechol-O-methyltransferase gene polymorphisms: Relationships to monoamine metabolite concentrations in CSF of healthy volunteers. *European Archives of Psychiatry and Clinical Neuroscience, 247*, 297–302.

Jönsson, E. G., et al. (2000). A promoter polymorphism in the monoamine oxidase A gene and its relationships to monoamine metabolite concentrations in CSF of healthy volunteers. *Journal of Psychiatric Research, 34*, 239–244.

Kalin, N. H., et al. (2008). The serotonin transporter genotype is associated with intermediate brain phenotypes that depend on the context of eliciting stressor. *Molecular Psychiatry, 13*, 1021–1027.

Karg, K., Burmeister, M., Shedden, K., & Sen, S. (2011). The serotonin transporter promoter variant (5-HTTLPR), stress, and depression meta-analysis revisited: Evidence of genetic moderation. *Archives of General Psychiatry, 68*, 444–454.

Kim, J. J., et al. (1997). Selective enhancement of emotional, but not motor, learning in monoamine oxidase A-deficient mice. *Proceedings of the National Academy of Sciences USA, 94*, 5929–5933.

Kim-Cohen, J., et al. (2006). MAOA, maltreatment, and gene-environment interaction predicting children's mental health: New evidence and a meta-analysis. *Molecular Psychiatry, 11*, 903–913.

Kinnally, E. L., et al. (2009). Parental care moderates the influence of MAOA-uVNTR genotype and childhood stressors on trait impulsivity and aggression in adult women. *Psychiatric Genetics, 19*, 126–133.

Koller, G., Bondy, B., Preuss, U. W., Bottlender, M., & Soyka, M. (2003). No association between a polymorphism in the promoter region of the MAOA gene with antisocial personality traits in alcoholics. *Alcohol and Alcoholism, 38*, 31–34.

Kostowski, W., Valzelli, L., Kozak, W., & Bernasconi, S. (1984). Activity of desipramine, fluoxetine and nomifensine on spontaneous and p-CPA-induced muricidal aggression. *Pharmacological Research Communications, 16*, 265–271.

Krueger, R. F., & South, S. C. (2009). Externalizing disorders: Cluster 5 of the proposed meta-structure for DSM-V and ICD-11. *Psychological Medicine, 39*, 2061–2070.

Kruesi, M. J. (1989). Cruelty to animals and CSF 5HIAA. *Psychiatry Research, 28*, 115–116.

Kruesi, M. J., et al. (1990). Cerebrospinal fluid monoamine metabolites, aggression, and impulsivity in disruptive behavior disorders of children and adolescents. *Archives of General Psychiatry, 47*, 419–426.

Kruesi, M. J., et al. (1992). A 2-year prospective follow-up study of children and adolescents with disruptive behavior disorders. Prediction by cerebrospinal fluid 5-hydroxyindoleacetic acid, homovanillic acid, and autonomic measures? *Archives of General Psychiatry, 49*, 429–435.

Kulikov, A. V., Osipova, D. V., Naumenko, V. S., & Popova, N. K. (2005). Association between Tph2 gene polymorphism, brain tryptophan hydroxylase activity and aggressiveness in mouse strains. *Genes, Brain, and Behavior, 4*, 482–485.

Lansford, J. E., et al. (2006). Developmental trajectories of externalizing and internalizing behaviors: Factors underlying resilience in physically abused children. *Development and Psychopathology, 18*, 35–55.

Lappalainen, J., et al. (1998). Linkage of antisocial alcoholism to the serotonin 5-HT1B receptor gene in 2 populations. *Archives of General Psychiatry, 55*, 989–994.

Lau, A. S., & Weisz, J. R. (2003). Reported maltreatment among clinic-referred children: Implications for presenting problems, treatment attrition, and long-term outcomes. *Journal of the American Academy of Child and Adolescent Psychiatry, 42*, 1327–1334.

Lee, B.-C., et al. (2008). An interaction between the norepinephrine transporter and monoamine oxidase A polymorphisms, and novelty-seeking personality traits in Korean females. *Progress in Neuropsychopharmacology, and Biological Psychiatry, 32*, 238–242.

Lee, B.-T., & Ham, B.-J. (2008). Monoamine oxidase A-uVNTR genotype affects limbic brain activity in response to affective facial stimuli. *Neuroreport, 19*, 515–519.

Lee, S. S. (2011). Deviant peer affiliation and antisocial behavior: Interaction with Monoamine Oxidase A (MAOA) genotype. *Journal of Abnormal Child Psychology, 39*, 321–332.

Lemogne, C., et al. (2011). Cognitive appraisal and life stress moderate the effects of the 5-HTTLPR polymorphism on amygdala reactivity. *Human Brain Mapping*, doi:10.1002/hbm.21150

Li, J. J., & Lee, S. S. (2010). Latent class analysis of antisocial behavior: Interaction of serotonin transporter genotype and maltreatment. *Journal of Abnormal Child Psychology, 38*, 789–801.

Liao, D.-L., Hong, C.-J., Shih, H.-L., & Tsai, S.-J. (2004). Possible association between serotonin transporter promoter region polymorphism and extremely violent crime in Chinese males. *Neuropsychobiology, 50*, 284–287.

Lidberg, L., Tuck, J. R., Asberg, M., Scalia-Tomba, G. P., & Bertilsson, L. (1985). Homicide, suicide and CSF 5-HIAA. *Acta Psychiatrica Scandinavica, 71*, 230–236.

Lindström, M. B., et al. (2004). Impulsivity related to brain serotonin transporter binding capacity in suicide attempters. *European Neuropsychopharmacology, 14*, 295–300.

Linnoila, M., et al. (1983). Low cerebrospinal fluid 5-hydroxyindoleacetic acid concentration differentiates impulsive from nonimpulsive violent behavior. *Life Science, 33*, 2609–2614.

Lyons-Ruth, K., et al. (2007). Serotonin transporter polymorphism and borderline or antisocial traits among low-income young adults. *Psychiatric Genetics, 17*, 339–343.

Manuck, S. B., et al. (1998). Aggression, impulsivity, and central nervous system serotonergic responsivity in a nonpatient sample. *Neuropsychopharmacology, 19*, 287–299.

Manuck, S. B., et al. (1999). Aggression and anger-related traits associated with a polymorphism of the tryptophan hydroxylase gene. *British Psychological Society, 45*, 603–614.

Manuck, S. B., Flory, J. D., Ferrell, R. E., Mann, J. J., & Muldoon, M. F. (2000). A regulatory polymorphism of the monoamine oxidase-A gene may be associated with variability in aggression, impulsivity, and central nervous system serotonergic responsivity. *Psychiatry Research, 95*, 9–23.

Manuck, S. B., Flory, J. D., Muldoon, M. F., & Ferrell, R. E. (2002). Central nervous system serotonergic responsivity and aggressive disposition in men. *Physiology and Behavior, 77*, 705–709.

Manuck, S. B., Kaplan, J. R., Rymeski, B. A., Fairbanks, L. A., & Wilson, M. E. (2003). Approach to a social stranger is associated with low central nervous system serotonergic responsivity in female cynomolgus monkeys (Macaca fascicularis). *American Journal of Primatology, 61*, 187–194.

Marazziti, D., et al. (2010). Impulsivity, gender, and the platelet serotonin transporter in healthy subjects. *Neuropsychiatric Disease and Treatment, 6*, 9–15.

McDermott, R., Tingley, D., Cowden, J., Frazzetto, G., & Johnson, D. D. P. (2009). Monoamine oxidase A gene (MAOA) predicts behavioral aggression following provocation. *Proceedings of the National Academy of Sciences, 106*, 2118–2123.

Mehlman, P. T., et al. (1994). Low CSF 5-HIAA concentrations and severe aggression and impaired impulse control in nonhuman primates. *The American Journal of Psychiatry, 151*, 1485–1491.

Mehlman, P. T., et al. (1995). Correlation of CSF 5-HIAA concentration with sociality and the timing of emigration in free-ranging primates. *The American Journal of Psychiatry, 152*, 907–913.

Mehlman, P. T., et al. (1997). CSF 5-HIAA, testosterone, and sociosexual behaviors in free-ranging male rhesus macaques in the mating season. *Psychiatry Research, 72*, 89–102.

Meltzer, H. Y., & Arora, R. C. (1988). Genetic control of serotonin uptake in blood platelets: A twin study. *Psychiatry Research, 24*, 263–269.

Mertins, V., Schote, A. B., Hoffeld, W., Griessmair, M., & Meyer, J. (2011). Genetic susceptibility for individual cooperation preferences: The role of Monoamine Oxidase A Gene (MAOA) in the voluntary provision of public goods. *PLoS ONE, 6*, e20959.

Meyer-Lindenberg, A., et al. (2006). Neural mechanisms of genetic risk for impulsivity and violence in humans. *Proceedings of the National Academy of Sciences USA, 103*, 6269–6274.

Meyer-Lindenberg, A., et al. (2009). Genetic variants in AVPR1A linked to autism predict amygdala activation and personality traits in healthy humans. *Molecular Psychiatry, 14*, 968–975.

Miczek, K. A., Maxson, S. C., Fish, E. W., & Faccidomo, S. (2001). Aggressive behavioral phenotypes in mice. *Behavioral Brain Research, 125*, 167–181.

Moffitt, T. E. (2005). Genetic and environmental influences on antisocial behaviors: Evidence from behavioral-genetic research. *Advances in Genetics, 55*, 41–104.

Moffitt, T. E., et al. (1998). Whole blood serotonin relates to violence in an epidemiological study. *British Psychological Society, 43*, 446–457.

Molina, V., Ciesielski, L., Gobaille, S., Isel, F., & Mandel, P. (1987). Inhibition of mouse killing behavior by serotonin-mimetic drugs: Effects of partial alterations of serotonin neurotransmission. *Pharmacology, Biochemistry and Behavior, 27*, 123–131.

Mukasa, H., Nakamura, J., Yamada, S., Inoue, M., & Nakazawa, Y. (1990). Platelet monoamine oxidase activity and personality traits in alcoholics and methamphetamine dependents. *Drug and Alcohol Dependence, 26*, 251–254.

Munafo, M., Brown, S., & Hariri, A. (2008). Serotonin transporter (5-HTTLPR) genotype and amygdala activation: A meta-analysis. *Biological Psychiatry, 63*, 852–857.

Murphy, D. L., et al. (2003). Experimental gene interaction studies with SERT mutant mice as models for human polygenic and epistatic traits and disorders. *Genes, Brain, and Behavior, 2*, 350–364.

Natarajan, D., Vries, H., Saaltink, D.-J., Boer, S. F., & Koolhaas, J. M. (2008). Delineation of violence from functional aggression in mice: An ethological approach. *Behavioral Genetics, 39*, 73–90.

New, A. S., et al. (1998). Tryptophan hydroxylase genotype is associated with impulsive-aggression measures: A preliminary study. *American Journal of Medical Genetics, 81*, 13–17.

Newman, T. K., et al. (2005). Monoamine oxidase A gene promoter variation and rearing experience influences aggressive behavior in rhesus monkeys. *British Psychological Society, 57*, 167–172.

Nilsson, K. W., et al. (2006). Role of monoamine oxidase A genotype and psychosocial factors in male adolescent criminal activity. *British Psychological Society, 59*, 121–127.

Nilsson, K. W., et al. (2007). The monoamine oxidase A (MAO-A) gene, family function and maltreatment as predictors of destructive behaviour during male adolescent alcohol consumption. *Addiction, 102*, 389–398.

Nilsson, K. W., et al. (2011). MAOA genotype, family relations and sexual abuse in relation to adolescent alcohol consumption. *Addiction Biology, 16*, 347–355.

Ober, C., Abney, M., & McPeek, M. S. (2001). The genetic dissection of complex traits in a founder population. *American Journal of Human Genetics, 69*, 1068–1079.

Ochsner, K. N., Bunge, S. A., Gross, J. J., & Gabrieli, J. D. (2002). Rethinking feelings: An FMRI study of the cognitive regulation of emotion. *Journal of Cognitive Neuroscience, 14*, 1215–1229.

O'Keane, V., et al. (1992). Blunted prolactin responses to d-fenfluramine in sociopathy. Evidence for subsensitivity of central serotonergic function. *The British Journal of Psychiatry: The Journal of Mental Science, 160*, 643–646.

Olivier, B., & van Oorschot, R. (2005). 5-HT1B receptors and aggression: A review. *European Journal of Pharmacology, 526*, 207–217.

Ongur, D., & Price, J. L. (2000). The organization of networks within the orbital and medial prefrontal cortex of rats, monkeys and humans. *Cerebral Cortex, 10*, 206–219.

Osipova, D. V., Kulikov, A. V., & Popova, N. K. (2009). C1473G polymorphism in mouse tph2 gene is linked to tryptophan hydroxylase-2 activity in the brain, intermale

aggression, and depressive-like behavior in the forced swim test. *Journal of Neuroscience, Research, 87*, 1168–1174.
Ou, X.-M., Chen, K., & Shih, J. C. (2006). Glucocorticoid and androgen activation of monoamine oxidase A is regulated differently by R1 and Sp1. *Journal of Biological Chemistry, 281*, 21512–21525.
Paaver, M., et al. (2007). Platelet MAO activity and the 5-HTT gene promoter polymorphism are associated with impulsivity and cognitive style in visual information processing. *Psychopharmacology, 194*, 545–554.
Pacheco, J., et al. (2009). Frontal-limbic white matter pathway associations with the serotonin transporter gene promoter region (5-HTTLPR) polymorphism. *Journal of Neuroscience, 29*, 6229–6233.
Parsian, A., Cloninger, C. R., Sinha, R., & Zhang, Z. H. (2003). Functional variation in promoter region of monoamine oxidase A and subtypes of alcoholism: Haplotype analysis. *American Journal of Medical Genetics Part B: Neuropsychiatric Genetics, 117B*, 46–50.
Passamonti, L., et al. (2005). Monoamine oxidase-A genetic variations influence brain, activity associated with inhibitory control: New insight into the neural correlates of impulsivity. *Biological Psychiatry*.
Passamonti, L., et al. (2008). Genetically dependent modulation of serotonergic inactivation in the human prefrontal cortex. *NeuroImage, 40*, 1264–1273.
Patkar, A. A., et al. (2003). Relationship between platelet serotonin uptake sites and measures of impulsivity, aggression, and craving among African-American cocaine abusers. *American Journal on Addictions, 12*, 432–447.
Perez-Rodriguez, M. M., et al. (2010). Tryptophan-hydroxylase 2 haplotype association with borderline personality disorder and aggression in a sample of patients with personality disorders and healthy controls. *Journal of Psychiatric Research, 44*, 1075–1081.
Perris, C., Eisemann, M., Knorring, von, L., Oreland, L., & Perris, H. (1984). Personality traits and monoamine oxidase activity in platelets in depressed patients. *Neuropsychobiology, 12*, 201–205.
Pezawas, L., et al. (2005). 5-HTTLPR polymorphism impacts human cingulate-amygdala interactions: A genetic susceptibility mechanism for depression. *Nature Neuroscience, 8*, 828–834.
Pezawas, L., et al. (2008). Evidence of biologic epistasis between BDNF and SLC6A4 and implications for depression. *Molecular Psychiatry, 13*, 709–716.
Philibert, R. A., et al. (2011). Gene environment interactions with a novel variable Monoamine Oxidase A transcriptional enhancer are associated with antisocial personality disorder. *Biological Psychology, 87*, 366–371.
Polman, H., Orobio de Castro, B., Koops, W., van Boxtel, H. W., & Merk, W. W. (2007). A meta-analysis of the distinction between reactive and proactive aggression in children and adolescents. *Journal of Abnormal Child Psychology, 35*, 522–535.
Popova, N. K. (2006). From genes to aggressive behavior: The role of serotonergic system. *Bioessays, 28*, 495–503.
Popova, N. K., et al. (2001). Behavioral characteristics of mice with genetic knockout of monoamine oxidase type A. *Neuroscience, and Behavioral Physiology, 31*, 597–602.
Price, J. L. (2005). Free will versus survival: Brain systems that underlie intrinsic constraints on behavior. *Journal of Comparative Neurology, 493*, 132–139.
Prochazka, H., Anderberg, U. M., Oreland, L., Knorring, L. V., & Agren, H. (2003). Self-rated aggression related to serum testosterone and platelet MAO activity in female patients with the fibromyalgia syndrome. *World Journal of Biological Psychiatry, 4*, 35–41.
Quirk, G. J., Likhtik, E., Pelletier, J. G., & Pare, D. (2003). Stimulation of medial prefrontal cortex decreases the responsiveness of central amygdala output neurons. *Journal of Neuroscience, 23*, 8800–8807.
Rao, H., et al. (2007). Genetic variation in serotonin transporter alters resting brain function in healthy individuals. *British Psychological Society, 62*, 600–606.
Rebsam, A., Seif, I., & Gaspar, P. (2002). Refinement of thalamocortical arbors and emergence of barrel domains in the primary somatosensory cortex: A study of normal and monoamine oxidase a knock-out mice. *Journal of Neuroscience, 22*, 8541–8552.
Reif, A., et al. (2007). Nature, and nurture predispose to violent behavior: Serotonergic genes and adverse childhood environment. *Neuropsychopharmacology, 32*, 2375–2383.
Reti, I. M., et al. (2011). Monoamine oxidase A regulates antisocial personality in whites with no history of physical abuse. *Comprehensive Psychiatry, 52*, 188–194.
Retz, W., Retz-Junginger, P., Supprian, T., Thome, J., & Rösler, M. (2004). Association of serotonin transporter promoter gene polymorphism with violence: Relation with personality disorders, impulsivity, and childhood ADHD psychopathology. *Behavioral Sciences and the Law, 22*, 415–425.
Robinson, O. J., Cools, R., & Sahakian, B. J. (2011). Tryptophan depletion disinhibits punishment but not reward prediction: Implications for resilience. *Psychopharmacology*, doi: 10.1007/s00213-011-2410-5
Roiser, J. P., et al. (2009). A genetically mediated bias in decision making driven by failure of amygdala control. *Journal of Neuroscience, 29*, 5985–5991.
Rosenkranz, J. A., & Grace, A. A. (1999). Modulation of basolateral amygdala neuronal firing and afferent drive by dopamine receptor activation in vivo. *Journal of Neuroscience, 19*, 11027–11039.
Roy, A., & Linnoila, M. (1989). CSF studies on alcoholism and related behaviours. *Progress in Neuropsychopharmacology, and Biological Psychiatry, 13*, 505–511.
Roy, A., Virkkunen, M., & Linnoila, M. (1987). Reduced central serotonin turnover in a subgroup of alcoholics? *Progress in Neuropsychopharmacology, and Biological Psychiatry, 11*, 173–177.
Rujescu, D., et al. (2002). Association of anger-related traits with SNPs in the TPH gene. *Molecular Psychiatry, 7*, 1023–1029.
Sabol, S. Z., Hu, S., & Hamer, D. (1998). A functional polymorphism in the monoamine oxidase A gene promoter. *Human Genetics, 103*, 273–279.
Sadeh, N., et al. (2010). Serotonin transporter gene associations with psychopathic traits in youth vary as a function of socioeconomic resources. *Journal of Abnormal Psychology, 119*, 604–609.
Saito, T., et al. (2002). Analysis of monoamine oxidase A (MAOA) promoter polymorphism in Finnish male alcoholics. *Psychiatry Research, 109*, 113–119.
Sakai, J. T., et al. (2006). Case-control and within-family tests for an association between conduct disorder and 5HTTLPR. *American Journal of Medical Genetics Part B: Neuropsychiatric Genetics, 141B*, 825–832.

Salichon, N., et al. (2001). Excessive activation of serotonin (5-HT) 1B receptors disrupts the formation of sensory maps in monoamine oxidase a and 5-ht transporter knock-out mice. *Journal of Neuroscience, 21*, 884–896.

Saudou, F., et al. (1994). Enhanced aggressive behavior in mice lacking 5-HT1B receptor. *Science, 265*, 1875–1878.

Scheuch, K., et al. (2007). Characterization of a functional promoter polymorphism of the human tryptophan hydroxylase 2 gene in serotonergic raphe neurons. *British Psychological Society, 62*, 1288–1294.

Schwandt, M. L., et al. (2010). Gene-environment interactions and response to social intrusion in male and female rhesus macaques. *Biological Psychiatry, 67*, 323–330.

Scott, A. L., Bortolato, M., Chen, K., & Shih, J. C. (2008). Novel monoamine oxidase A knock out mice with human-like spontaneous mutation. *Neuroreport, 19*, 739–743.

Sebastian, C. L., et al. (2010). Effects of age and MAOA genotype on the neural processing of social rejection. *Genes, Brain, and Behavior, 9*, 628–637.

Sekine, Y., et al. (2006). Brain, serotonin transporter density and aggression in abstinent methamphetamine abusers. *Archives of General Psychiatry, 63*, 90–100.

Shih, J. C., et al. (1999). Ketanserin and tetrabenazine abolish aggression in mice lacking monoamine oxidase A. *Brain Research, 835*, 104–112.

Shumay, E., & Fowler, J. S. (2010). Identification and characterization of putative methylation targets in the MAOA locus using bioinformatic approaches. *Epigenetics: Official Journal of the DNA Methylation Society, 5*, 325–342.

Siever, L. J., et al. (1999). d,l-Fenfluramine response in impulsive personality disorder assessed with [18F]fluorodeoxyglucose positron emission tomography. *Neuropsychopharmacology, 20*, 413–423.

Silva, H., et al. (2010). Fluoxetine response in impulsive-aggressive behavior and serotonin transporter polymorphism in personality disorder. *Psychiatric Genetics, 20*, 25–30.

Sjöberg, R. L., et al. (2007). Adolescent girls and criminal activity: Role of MAOA-LPR genotype and psychosocial factors. *American Journal of Medical Genetics Part B: Neuropsychiatric Genetics, 144B*, 159–164.

Skondras, M., Markianos, M., Botsis, A., Bistolaki, E., & Christodoulou, G. (2004). Platelet monoamine oxidase activity and psychometric correlates in male violent offenders imprisoned for homicide or other violent acts. *European Archives of Psychiatry and Clinical Neuroscience, 254*, 380–386.

Smith, C. A., & Farrington, D. P. (2004). Continuities in antisocial behavior and parenting across three generations. *Journal of Child Psychology and Psychiatry, 45*, 230–247.

Soderstrom, H., Blennow, K., Manhem, A., & Forsman, A. (2001). CSF studies in violent offenders. I. 5-HIAA as a negative and HVA as a positive predictor of psychopathy. *Journal of Neural Transmission, 108*, 869–878.

Soderstrom, H., Blennow, K., Sjodin, A.-K., & Forsman, A. (2003). New evidence for an association between the CSF HVA:5-HIAA ratio and psychopathic traits. *Journal of Neurology, Neurosurgery and Psychiatry, 74*, 918–921.

Sotres-Bayon, F., Cain, C. K., & LeDoux, J. E. (2006). Brain, mechanisms of fear extinction: Historical perspectives on the contribution of prefrontal cortex. *Biological Psychiatry,*

Stadler, C., Schmeck, K., Nowraty, I., Müller, W. E., & Poustka, F. (2004). Platelet 5-HT uptake in boys with conduct disorder. *Neuropsychobiology, 50*, 244–251.

Stålenheim, E. G. (2001). Relationships between attempted suicide, temperamental vulnerability, and violent criminality in a Swedish forensic psychiatric population. *European Psychiatry, 16*, 386–394.

Stanley, M., Traskman-Bendz, L., & Dorovini-Zis, K. (1985). Correlations between aminergic metabolites simultaneously obtained from human CSF and brain. *Life Science, 37*, 1279–1286.

Stark, P., Fuller, R. W., & Wong, D. T. (1985). The pharmacologic profile of fluoxetine. *Journal of Clinical Psychiatry, 46*, 7–13.

Stoff, D. M., Pollock, L., Vitiello, B., Behar, D., & Bridger, W. H. (1987). Reduction of (3H)-imipramine binding sites on platelets of conduct-disordered children. *Neuropsychopharmacology, 1*, 55–62.

Suomi, S. J. (2003). Gene-environment interactions and the neurobiology of social conflict. *Annals of the New York Academy of Science, 1008*, 132–139.

Svenningsson, P., & Greengard, P. (2007). p11 (S100A10)—an inducible adaptor protein that modulates neuronal functions. *Current Opinion in Pharmacology, 7*, 27–32.

Svenningsson, P., et al. (2006). Alterations in 5-HT1B receptor function by p11 in depression-like states. *Science, 311*, 77–80.

Syagailo, Y. V., et al. (2001). Association analysis of the functional monoamine oxidase A gene promoter polymorphism in psychiatric disorders. *American Journal of Medical Genetics, 105*, 168–171.

Takahashi, A., Quadros, I. M., de Almeida, R. M. M., & Miczek, K. A. (2011). Brain, serotonin receptors and transporters: Initiation vs. termination of escalated aggression. *Psychopharmacology, 213*, 183–212.

Tiefenbacher, S., Davenport, M. D., Novak, M. A., Pouliot, A. L., & Meyer, J. S. (2003). Fenfluramine challenge, self-injurious behavior, and aggression in rhesus monkeys. *Physiology and Behavior, 80*, 327–331.

Tochigi, M., et al. (2006). Combined analysis of association between personality traits and three functional polymorphisms in the tyrosine hydroxylase, monoamine oxidase A, and catechol-O-methyltransferase genes. *Neuroscience, Research, 54*, 180–185.

Tost, H., et al. (2010). A common allele in the oxytocin receptor gene (OXTR) impacts prosocial temperament and human hypothalamic-limbic structure and function. *Proceedings of the National Academy of Sciences, 107*, 13936–13941.

Tuvblad, C., Raine, A., Zheng, M., & Baker, L. A. (2009). Genetic and environmental stability differs in reactive and proactive aggression. *Aggressive Behavior, 35*, 437–452.

Understanding the prolific and other priority offender programme. (2007). London: Home Office.

Unis, A. S., et al. (1997). Platelet serotonin measures in adolescents with conduct disorder. *British Psychological Society, 42*, 553–559.

Upton, A. L., et al. (2002). Lack of 5-HT(1B) receptor and of serotonin transporter have different effects on the segregation of retinal axons in the lateral geniculate nucleus compared to the superior colliculus. *National Safety Council, 111*, 597–610.

Valzelli, L., & Bernasconi, S. (1979). Aggressiveness by isolation and brain serotonin turnover changes in different strains of mice. *Neuropsychobiology, 5*, 129–135.

Valzelli, L., Bernasconi, S., & Dalessandro, M. (1981*a*). Effect of tryptophan administration on spontaneous and

P-CPA-induced muricidal aggression in laboratory rats. *Pharmacological Research Communications, 13*, 891–897.
Valzelli, L., Garattini, S., Bernasconi, S., & Sala, A. (1981*b*). Neurochemical correlates of muricidal behavior in rats. *Neuropsychobiology, 7*, 172–178.
Van Beijsterveldt, C. E. M., Verhulst, F. C., Molenaar, P. C. M., & Boomsma, D. I. (2004). The genetic basis of problem behavior in 5-year-old Dutch twin pairs. *Behavioral Genetics, 34*, 229–242.
Van den Bergh, F. S., Bloemarts, E., Groenink, L., Olivier, B., & Oosting, R. S. (2006). Delay aversion: Effects of 7-OH-DPAT, 5-HT1A/1B-receptor stimulation and D-cycloserine. *Pharmacology, Biochemistry and Behavior, 85*, 736–743.
Vanyukov, M. M., et al. (2004). Haplotypes of the monoamine oxidase genes and the risk for substance use disorders. *American Journal of Medical Genetics Part B: Neuropsychiatric Genetics, 125B*, 120–125.
Varnäs, K., Halldin, C., & Hall, H. (2004). Autoradiographic distribution of serotonin transporters and receptor subtypes in human brain. *Human Brain Mapping, 22*, 246–260.
Vergnes, M., Depaulis, A., & Boehrer, A. (1986). Parachlorophenylalanine-induced serotonin depletion increases offensive but not defensive aggression in male rats. *Physiology and Behavior, 36*, 653–658.
Vergnes, M., & Kempf, E. (1981). Tryptophan deprivation: Effects on mouse-killing and reactivity in the rat. *Pharmacology, Biochemistry and Behavior, 14* (Suppl 1), 19–23.
Virkkunen, M., & Linnoila, M. (1990). Serotonin in early onset, male alcoholics with violent behaviour. *Annals of Medicine, 22*, 327–331.
Virkkunen, M., & Linnoila, M. (1993). Brain, serotonin, type II alcoholism and impulsive violence. *Journal of Studies on Alcohol and Drugs*, Suppl 11, 163–169.
Virkkunen, M., Goldman, D., Nielsen, D. A., & Linnoila, M. (1995). Low brain serotonin turnover rate (low CSF 5-HIAA) and impulsive violence. *Journal of Psychiatry and Neuroscience, 20*, 271–275.
Virkkunen, M., Nuutila, A., Goodwin, F. K., & Linnoila, M. (1987). Cerebrospinal fluid monoamine metabolite levels in male arsonists. *Archives of General Psychiatry, 44*, 241–247.
Vitalis, T., et al. (1998). Effects of monoamine oxidase A inhibition on barrel formation in the mouse somatosensory cortex: Determination of a sensitive developmental period. *Journal of Comparative Neurology, 393*, 169–184.
Vitiello, B., & Stoff, D. M. (1997). Subtypes of aggression and their relevance to child psychiatry. *Journal of the American Academy of Child and Adolescent Psychiatry, 36*, 307–315.
Von dem Hagen, E. A. H., Passamonti, L., Nutland, S., Sambrook, J., & Calder, A. J. (2011). The serotonin transporter gene polymorphism and the effect of baseline on amygdala response to emotional faces. *Neuropsychologia, 49*, 674–680.
Walderhaug, E., et al. (2002). Lowering of serotonin by rapid tryptophan depletion increases impulsiveness in normal individuals. *Psychopharmacology, 164*, 385–391.
Walderhaug, E., Herman, A. I., Magnusson, A., Morgan, M. J., & Landrø, N. I. (2010). The short (S) allele of the serotonin transporter polymorphism and acute tryptophan depletion both increase impulsivity in men. *Neuroscience Letters, 473*, 208–211.
Walther, D. J., & Bader, M. (2003). A unique central tryptophan hydroxylase isoform. *Biochemical Pharmacology, 66*, 1673–1680.
Waters, H., et al. (2004). *The economic dimensions of interpersonal violence* (pp. 1–70). Geneva: Department of Injuries and Violence Prevention, World Health Organization.
Weiss, L. A., Abney, M., Cook, E. H., & Ober, C. (2005). Sex-specific genetic architecture of whole blood serotonin levels. *American Journal of Human Genetics, 76*, 33–41.
Wester, P., et al. (1990). Ventricular cerebrospinal fluid monoamine transmitter and metabolite concentrations reflect human brain neurochemistry in autopsy cases. *Journal of Neurochemistry, 54*, 1148–1156.
Widom, C. S., & Brzustowicz, L. M. (2006). MAOA and the "cycle of violence": Childhood abuse and neglect, MAOA genotype, and risk for violent and antisocial behavior. *British Psychological Society, 60*, 684–689.
Williams, L. M., et al. (2009). A polymorphism of the MAOA gene is associated with emotional brain markers and personality traits on an antisocial index. *Neuropsychopharmacology, 34*, 1797–1809.
Williams, W., et al. (2004). Glucose utilization in the medial prefrontal cortex correlates with serotonin turnover rate and clinical depression in alcoholics. *Psychiatry Research, 132*, 219–224.
Wong, C. C. Y., et al. (2010). A longitudinal study of epigenetic variation in twins. *Epigenetics: Official Journal of the DNA Methylation Society, 5*, 516–526.
Yeh, M. T., Coccaro, E. F., & Jacobson, K. C. (2010). Multivariate behavior genetic analyses of aggressive behavior, subtypes. *Behavioral Genetics, 40*, 603–617.
Yu, Y. W.-Y., et al. (2005). Association study of a functional MAOA-uVNTR gene polymorphism and personality traits in Chinese young females. *Neuropsychobiology, 52*, 118–121.
Zalsman, G., et al. (2005). Relationship of MAO-A promoter (u-VNTR) and COMT (V158M) gene polymorphisms to CSF monoamine metabolites levels in a psychiatric sample of Caucasians: A preliminary report. *American Journal of Medical Genetics Part B: Neuropsychiatric Genetics, 132B*, 100–103.
Zhang, J., et al. (2010). A cis-phase interaction study of genetic variants within the MAOA gene in major depressive disorder. *Biological Psychiatry, 68*, 795–800.
Zhang, L., et al. (2008). p11 is up-regulated in the forebrain of stressed rats by glucocorticoid acting via two specific glucocorticoid response elements in the p11 promoter. *National Safety Council, 153*, 1126–1134.
Zhang, X., Beaulieu, J.-M., Sotnikova, T. D., Gainetdinov, R. R., & Caron, M. G. (2004). Tryptophan hydroxylase-2 controls brain serotonin synthesis. *Science, 305*, 217.
Zouk, H., et al. (2007). The effect of genetic variation of the serotonin 1B receptor gene on impulsive aggressive behavior and suicide. *American Journal of Medical Genetics Part B: Neuropsychiatric Genetics, 144B*, 996–1002.

CHAPTER

8 Oxytocin and Vasopressin Gene Variation and the Neural Basis of Social Behaviors

Lars Westberg *and* Hasse Walum

Abstract

Experimental studies in rodents and humans show that the neuropeptides oxytocin and vasopressin are important regulators of behaviors related to social interactions. Evidence for positive effects of oxytocin treatment on symptoms of psychiatric disorders characterized by impaired social functioning has emerged. Numerous studies report associations between various social behaviors, the risk of autism, and polymorphisms in *OXTR* and *AVPR1A*. This chapter provides an overview of these genetic association studies. Although many of the published findings are inconclusive and need replication in independent samples, the chapter concludes that variants of *OXTR* and *AVPR1A* seem to moderate individual variation in different aspects of social behavior. The challenges for future studies include replication of current findings, identification of the functional variants, and characterization of the neural mechanisms mediating the gene–behavior associations, as well as exploration of the pharmacogenetic potential of *OXTR* and *AVPR1A* in future clinical trials.

Key Words: hormones, autism, schizophrenia, bonding behavior, polymorphism, genetic

In social species, the ability to interact with and bond to other individuals is closely related to fitness. Humans display large amounts of variation in social behavior, and some of the most severe psychiatric disorders are characterized by social deficits. Our knowledge of the biology of human social behaviors and, therefore, disorders disrupting these behaviors is sparse. However, by implementing information gained from animal studies that have generated insights into the neural regulation of social motivation and social information processing, neural circuits possibly also relevant to humans can be revealed. Experimental research in rodents and other mammals has unveiled the neuropeptides oxytocin and vasopressin as crucial for maternal behavior, pair-bonding behavior, social recognition, aggression, and sexual behaviors, as well as for the regulation of stress and anxiety (for a detailed discussion of this work, see the chapter by Barrett and Young, in this volume). After the discovery that neuropeptides can be delivered to the human brain by the use of nasal spray (Born et al., 2002), a considerable number of experimen tal pharmacological studies using intranasal treatment with oxytocin and vasopressin have been conducted, and many intriguing findings of positive effects on a wide spectrum of social behaviors in humans have been revealed. It may be argued, however, that treatment effects of exogenous neuropeptides do not prove that endogenous oxytocin and vasopressin are important for human social interactions in daily life. Human genetics studies could, however, provide ultimate evidence for such importance if they reveal true effects of variants in genes involved in the actions of oxytocin and vasopressin on human social behaviors. To date, a rather extensive number of studies have investigated genetic variants in the genes encoding the receptors for oxytocin (*OXTR*) and vasopressin (*AVPR1A*). The aim of this chapter is to give an overview of the and stimulated agonistic facial motor patterns. In women, vasopressin increased the perception of friendliness in the faces of unfamiliar women and stimulated affiliative facial motor patterns. To date, one study has investigated the *AVPR1A* in relation to aggressive behaviors, assessed with Buss–Durkee Hostility Inventory, in 161 patients with bipolar disorder (Vogel et al., 2012). Patients carrying shorter RS3 repeats displayed more impulsive aggression than did those carrying longer repeats.

IV. AVPR1A and Drug Dependence

Considering the fact that the effects of vasopressin on social bonding behaviors are mediated by effects on the reward system (Burkett & Young, 2012; Young & Wang, 2004), the reports of associations between *AVPR1A* SNPs and drug dependence are intriguing. Levran and co-workers investigated 1,350 SNPs in 130 genes in a case-control study of 202 former heroin addicts and 167 healthy controls (Levran et al., 2009). One of their strongest associations observed was between heroin dependence and rs3759292 located upstream of the *AVPR1A*, close to the RS1 and RS3 repeats. Furthermore, in a study of drug use disorders, 1,536 SNPs in 106 genes were investigated in a discovery sample of 398 substance abuse cases and 359 control subjects (Maher et al., 2011). Interestingly, the strongest associations with substance use were found with three SNPs in the 3′-region of *AVPR1A* (rs1587097, rs10784339, and rs11174811). As the risk for substance use disorder is associated with deviations in social behavior, including marital instability, the authors investigated whether the association between *AVPR1A* and drug use is mediated by spousal satisfaction. The rs11174811 SNP was associated with spousal satisfaction in males but not in females. Spousal satisfaction was also an important mediator of the association between rs11174811 and drug use disorder diagnosis in males. Moreover, the association between rs11174811 and drug use was, to some extent, replicated in a clinical (n = 1,399) and an epidemiological (n = 2,231) sample. The direction of the association was consistent across the clinical sample but reversed in the epidemiological sample. This discrepancy between clinical and population-based samples may be explained by the *flip-flop phenomena* (Lin et al., 2007) (discussed later) and/or differences in the sample ascertainment, such as disorder severity.

V. AVPR1A and Brain Function

Although several studies indicate that the *AVPR1A* repeat polymorphisms are associated with social behavioral phenotypes, their functional implications have, so far, only been sparsely investigated. Knafo and co-workers reported an association between relatively shorter RS3 repeats and lower levels of *AVPR1A* mRNA in postmortem hippocampus (Knafo et al., 2008). In line with these findings, in vitro studies indicate that shorter RS3, but also RS1, repeats decreased relative promoter activity in a human neuroblastoma cell line (Tansey et al., 2011). Moreover, the rs11174811 in the 3′-untranslated region (UTR) was found to be associated with expression levels of *AVPR1A* in brain tissue (Maher et al., 2011), which is in line with an in vitro study demonstrating this SNP to affect gene expression by disrupting a microRNA binding site (Nossent et al., 2011).

To investigate to what extent the *AVPR1A* repeat polymorphisms influence human brain function, Meyer-Lindenberg and co-workers (see also the chapter by Buckholtz and Meyer-Lindenberg on human aggression and violence elsewhere in this handbook) investigated RS1 and RS3 in 121 healthy individuals studied with functional magnetic resonance imaging (fMRI) in an emotional face-matching paradigm (Meyer-Lindenberg et al., 2009). The 334 allele of RS3 was associated with overactivation of the left and right amygdala. For RS1, the 320 allele, relative to all other alleles, showed significantly less activity in left amygdala. When comparing long and short variants of RS1 and RS3, significant effects on amygdala activation with opposite directionality were seen. For RS3, longer variants were associated with significantly stronger activation of amygdala, whereas, for RS1, shorter variants showed stronger activity. None of the associations between *AVPR1A* variants and amygdala activation seemed to be dependent on sex.

VI. Concluding Remarks About *AVPR1A* and Human Social Behaviors

Although there are several promising findings of associations between *AVPR1A* polymorphisms and different aspects of social behaviors, the inconsistencies between studies make it, as yet, not possible to draw any final conclusions. However, some intriguing but tentative suggestions appears when comparing the associations found so far regarding the RS3 repeat. Confusion often arises when trying to figure out if this is the same allele that associates with phenotypes between studies. This is because many studies have used various sets of primers, resulting in different lengths of their PCR products. However, attempts to compare studies (Levin et al., 2009; Tansey et al., 2011) despite these complications reveal that relatively shorter RS3 alleles, which have been associated with lower *AVPR1A* mRNA expression in the brain (Knafo et al., 2008) and lower transcriptional activity in vitro (Tansey et al., 2011) are associated with increased risk of autism (Kim et al., 2002; Wassink et al., 2004), less altruistic behavior (Knafo et al., 2008), less prepulse inhibition (Levin et al., 2009), less amygdala activation to social recognition (Meyer-Lindenberg et al., 2009), and increased aggressive behavior (Vogel et al., 2012), as well as with a decreased ability to bond to a partner in a relationship and increased risk of having marital problems (Walum et al., 2008). This is in line with the vole studies showing associations between repeat length and decreased number of V1aR and less ability to bond with a partner and to conduct related social behavior, discussed in greater detail by Barrett and Young elsewhere in this handbook. Noteworthy in this context is also that, in several of the above-mentioned studies, one of the most common of the shorter alleles (which has been referred to as either allele 334 or allele 327 in different studies depending on which primers were used) seems to explain most of the associations seen with the short-long dichotomization (Avinun et al., 2011; Avinun et al., 2012; Kim et al., 2002; Meyer-Lindenberg et al., 2009; Ukkola et al., 2009; Walum et al., 2008). As reported (Thibonnier et al., 2000), the RS3 repeat is a complex polymorphism featuring two adjacent dinucleotide repeats that are both polymorphic. This means that genotype designation based on PCR product lengths do not capture the full extent of the genetic variation of this repeat polymorphism. It is possible that more accurate genotyping using an alternative method would increase the strength of the association or refine the genotypes implicated. Further molecular genetic studies are needed before we understand more precisely how the RS3 repeat polymorphism affects expression of the *AVPR1A*.

Even though only a few studies so far have investigated *AVPR1A* SNPs in relation to behavioral phenotypes, intriguing findings already have been revealed. Most striking is the association between an SNP associated with expression levels of *AVPR1A* in brain tissue and drug dependence in several samples (Maher et al., 2011). Furthermore, the fact that the same SNP was also associated with spousal satisfaction in males and that another SNP in the promoter of *AVPR1A* was associated with the ability to develop peer relationships (Yang et al., 2010b) is intriguing and provides further evidence for the importance of *AVPR1A* for bonding behaviors. To what extent these associations are dependent or independent of the RS3 repeat deserves further investigation. Future studies of *AVPR1A* in relation to social behaviors should include both repeat polymorphisms and SNPs.

VII. Oxytocin and *OXTR*

Oxytocin is best known for its role in peripheral circulation, particularly in contraction of the uterus during labor and ejection of milk during lactation, but it is also implicated in a wide range of social behaviors when centrally released. In rodents, the nonapeptide oxytocin, by acting on the oxytocin receptor, promotes social memory formation, maternal behavior, and partner preference; it also, in some contexts, modulates aggressive and sexual behaviors, as well as the response to stress and anxiogenic stimuli (Donaldson & Young, 2008; Gabor et al., 2012; Lee, H.J., et al., 2009; Meyer-Lindenberg et al., 2011; Young & Wang, 2004; see also chapter by Barrett and Young elsewhere in this volume). Recent human studies indicate that many aspects of social behaviors, including reduction of social problems in patients with psychiatric disorders, are affected by oxytocin treatment. This work has been reviewed comprehensively elsewhere (Bartz et al., 2011; Bos et al., 2012; Green & Hollander, 2010; Guastella & MacLeod, 2012; Macdonald & Macdonald, 2010; Meyer-Lindenberg et al., 2011; Zink & Meyer-Lindenberg, 2012).

The human gene encoding the oxytocin receptor (*OXTR*) is located on chromosome 3p25 (Kimura et al., 1992), and a large number of genetic variants situated in or close to this gene have been reported (see Table 8.1 and Figure 8.1). As reviewed and discussed here, most positive associations to social behaviors have so far been reported to SNPs located in the large intron 3, but intriguing findings have also been reported with SNPs located in potential regulatory regions upstream and downstream of the gene.

VIII. OXTR and Autism Spectrum Disorders

Although most genome-wide linkage analyses of ASD did not find any peak near *OXTR* on chromosome 3p25, one of the largest linkage studies performed to date, which included 314 families,

Table 8.1 (Continued)

SNP	Position	Phenotype	N of SNPs	N of subjects	Age group	Ethnicity	Paper
rs2254298	intron 3	Brain structure size; Brain activation	1	343	Adults	Caucasian	(Tost et al., 2011)
rs2254298	intron 3	Depression; Anxiety	1	92	Children; Adolescents	Caucasian	(Thompson et al., 2011)
rs2254298	intron 3	Depression; Separation anxiety	2	185 depressive patients	Adults	Caucasian	(Costa et al., 2009)
rs2254298	intron 3	Plasma oxytocin	2	352: 272 parents and 80 nonparents	Adults	Israeli/Jewish	(Feldman et al., 2012)
rs53576	intron 3	Autism	4	195 patients plus family members	Children	Han Chinese	(Wu et al., 2005)
rs53576	intron 3	Brain structure size; Brain activation	1	228	Adult	Caucasian	(Tost et al., 2010)
rs53576	intron 3	Depression; Separation anxiety	2	185 depressive patients	Adults	Caucasian	(Costa et al., 2009)
rs53576	intron 3	Emotional regulation	1	99 Koreans;152 Americans	Adults	Americans; Koreans	(Kim et al., 2011)
rs53576	intron 3	Emotional support seeking	1	134 Koreans; 140 Americans	Adults	Americans; Koreans	(Kim et al., 2010)
rs53576	intron 3	Empathy; Emotion recognition; Stress reactivity	1	192	Adults	Mixed	(Rodrigues et al., 2009)
rs53576	intron 3	Expression of prosociality	1	116 observers; 23 targets	Adults	Mixed	(Kogan et al., 2011)
rs53576	intron 3	Maternal sensitivity	1	159	Adults	Caucasian	(Bakermans-Kranenburg & van Ijzendoorn 2008)
rs53576	intron 3	Maternal sensitivity	1	201	Adults	Mixed	(Sturge-Apple et al., 2012)

rs53576	intron 3	Moderation of oxytocin treatment effect on response to infant faces	1	56	Adults	Mixed	(Marsh et al., 2012)
rs53576	intron 3	Optimism; Personal Control: Self-Esteem; Depressive symptoms	1	344	Adults	Mixed	(Saphire-Bernstein et al., 2011)
rs53576	intron 3	Physiological reactivity to infant crying	1	40	Adults	Caucasian	(Riem et al., 2011)
rs53576	intron 3	Positive affect; Intelligence	3	285; 117	Adults; Adolescents	Caucasian	(Lucht et al., 2009)
rs53576	intron 3	Schizophrenia	4	406 patients; 406 controls	Adults	Caucasian	(Montag et al., 2012)
rs53576	intron 3	Social auditory processing	1	45	Adults	Caucasian	(Tops et al., 2011)
rs53576	intron 3	Social cognition in ADHD	5	112	Children	Caucasian	(Park et al., 2010)
rs53576	intron 3	Stress response	1	92	Elderly	Mixed	(Norman et al., 2012)
rs53576	intron 3	Stress response after social support	1	194	Adults	Caucasian	(Chen et al., 2011b)
rs237899	intron 3	Alcohol dependence	GWAS	1,192 patients; 692 controls	Adult	Mixed	(Edenberg et al., 2010)
rs237899	intron 3	Alcohol dependence and depression	GWAS	467patients; 407 controls	Adult	Mixed	(Edenberg et al., 2010)
rs237900	intron 3	Harm Avoidance	2	99	Adult	Caucasian	(Stankova et al., 2012)
rs237902	exon 3	Harm Avoidance	2	99	Adult	Caucasian	(Stankova et al., 2012)

(Continued)

Table 8.1 (Continued)

SNP	Position	Phenotype	N of SNPs	N of subjects	Age group	Ethnicity	Paper
rs237902	exon 3	Novelty Seeking; Cooperativeness	1	204	Adult	Caucasian	(Comings et al., 2000)
rs2228485	exon 3	Emotion Recognition	3	76	Adolescents	Caucasian	(Lucht et al., 2013)
rs4564970	intron 1	Alcohol induced aggression—Experimental	12	116	Adults	Caucasian	(Johansson et al., 2012a)
rs4564970	intron 1	Alcohol induced aggression and anger—Self-assessment	3	3,577	Adults	Caucasian	(Johansson et al., 2012b)
rs4564970	intron 1	Autism	20	436 patients plus family members	Children	Caucasian	(Tansey et al., 2010)
rs2301261	intron 1	Autism	GWAS	1,558 families	Children	Caucasian	(Anney et al., 2010)
rs2268498	5′	Moral judgements	1	154	Adults	Caucasian	(Walter et al., 2012)
rs2268498	5′	Negative emotionality	1	750	Adults	Caucasian	(Montag et al., 2011)
rs1488467	5′	Alcohol induced aggression—Experimental	12	116	Adults	Caucasian	(Johansson et al., 2012a)
rs1488467	5′	Alcohol induced aggression and anger—Self-assessment	3	3,577	Adults	Caucasian	(Johansson et al., 2012b)
rs2270465	5′	Autism	22	100 patients plus family members	Children	Caucasian	(Wermter et al., 2010)
rs75775	5′	Altruism	9	684	Adults	Caucasian	(Apicella et al., 2010)
rs75775	5′	Autism	GWAS	3,115 patients; 7,031 controls plus family members	Children	Caucasian	(Wang et al., 2009)

CATSS, Child and Adolescent Twin Study in Sweden; TCHAD, Twin Study of Child and Adolescent Development; TOSS, Twin Offspring Study

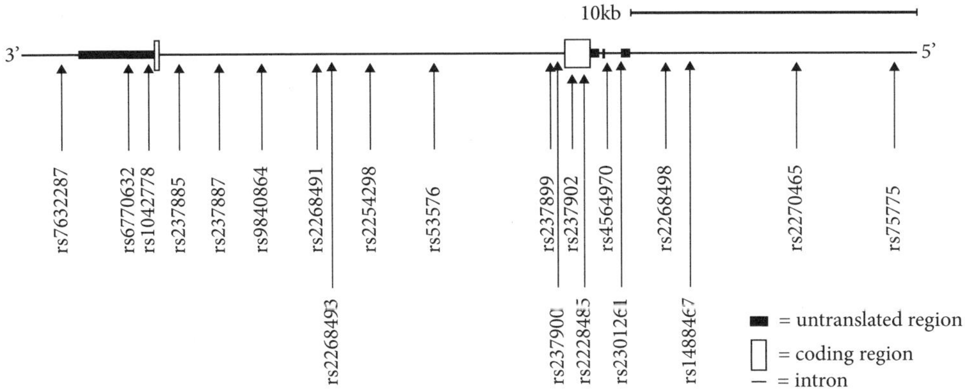

Fig. 8.1 Genetic variants of the oxytocin receptor gene (*OXTR*) positively associated with social behaviors. The *OXTR* contains four exons and three introns. Exons 3 and 4 encode the amino acids of the oxytocin receptor. Exons 1 and 2 comprise the 5′ untranslated region and exon 4 contains the 3′ untranslated region, including the polyadenylation signal.

highlighted a linkage peak directly over the *OXTR*, establishing *OXTR* as a positional candidate gene for this disorder (Ylisaukko-oja et al., 2006).

Wu and co-workers conducted the first association study of *OXTR* in relation to the risk of autism. They investigated four SNPs in 195 Chinese families and found associations of two markers in the third intron of *OXTR*, the A alleles of rs2254298 and rs53576, respectively (Wu et al., 2005). Further evidence for an association between the rs2254298 A allele and ASD, as well as with autistic traits, was found in a Japanese case–control sample (Liu et al., 2010) and in males of mixed ethnicity from the normal population (Chen & Johnson, 2011), respectively. In Caucasian families, the opposite allele, rs2254298G, was associated with ASD risk in one study (Jacob et al., 2007) and contributed to a risk haplotype in another study (Lerer et al., 2008). However, two family-based studies of Caucasian subjects failed to find any association between rs2254298 and risk of autism (Campbell et al., 2011; Wermter et al., 2010). The association between the A allele of rs53576 and autism by Wu and co-workers was weakly supported in a study of Japanese patients (Liu et al., 2010). The same allele was also part of the risk haplotype for autism in a Caucasian family sample (Wermter et al., 2010), whereas the A allele of rs53576 was associated with better social ability in attention deficit hyperactivity disorder (ADHD) patients (Park et al., 2010). Other studies did not find any evidence for an association between rs53576 and risk of autism (Campbell et al., 2011; Chen & Johnson, 2011; Jacob et al., 2007). Another SNP located in intron 3, rs2268493, which is in tight linkage disequilibrium (LD) with rs2254298, was associated with risk of autism in two studies of Caucasian families (Campbell et al., 2011; Yrigollen et al., 2008). Conversely, Liu et al. did not report any associations between autism risk and rs2268493 in Japanese patients, but instead found associations with other SNPs in intron 3 (e.g., rs2268491 and rs2254298) (Liu et al., 2010). These inconsistencies between studies may be due to different LD patterns between populations and ethnic groups and/or due to heterogeneity of ASD phenotypes. In an attempt to investigate if 14 *OXTR* variants influence 15 different clinical phenotypes of ASD, Egawa and co-workers revealed an association, surviving multiple testing correction, between rs237887 and the imitation subscale in ASD patients (Egawa et al., 2013). Taken together, the numerous nominally positive findings could indicate that one or more functional variants are located in the third intron of *OXTR*. Intriguingly, in this context, this large intron comprises a genomic element that may be involved in transcriptional suppression of the gene (Mizumoto et al., 1997), and SNPs located in or adjacent to this element may therefore affect gene expression.

In addition to the suggestive evidence for associations of SNPs located in intron 3 of *OXTR*, polymorphisms in and near the 3′ UTR have also been implicated in ASD risk (Campbell et al., 2011; Lerer et al., 2008; Tansey et al., 2010; Walum et al., 2012). Lerer et al. (2008) and Campbell et al. (2011) described associations between the G allele of the 3′-UTR SNP rs1042778 and ASD risk (Campbell et al., 2011; Lerer et al., 2008). Furthermore, the three largest candidate gene studies of *OXTR* conducted so far all found positive associations with rs7632287 located 3′ of *OXTR* (Campbell et al.,

2011; Tansey et al., 2010; Walum et al., 2012). It is puzzling, however, that the studies by Tansey et al. and Campbell et al. show associations between autism and the major (G) allele, whereas our data (Walum et al., 2012) suggest associations between increased number of autism symptoms, as well as related social problems (in addition to decreased ability to bond to a partner in adulthood; see discussion below) and the minor (A) allele. This may be explained by differences between studies regarding ethnicity and measures of autism symptoms. Furthermore, whereas the associations in our study were restricted to female subjects, sex-specific analysis was not conducted in the two previous studies of the rs7632287. Moreover, these kinds of "flip-flop" associations can occur when the investigated variant is correlated, through interactive effects or LD, with a causal variant at another locus (Lin et al., 2007). The rs7632287 SNP is located in the 3′ region of *OXTR,* and in silico analysis suggests that rs7632287 afflicts transcription factor binding to the 3′ end of the gene that may be important for gene expression levels (Tansey et al., 2010). However, further investigations are needed before we know if this SNP has any importance in the protein function per se or if it is in LD with a functional variant.

Although none of the genomewide association studies (GWAS) of autism conducted so far has highlighted the *OXTR* at chromosome 3p25, two of them described suggestive evidence for association of *OXTR* markers. Wang et al. (2009) reported an association between autism and rs75775 (P = .003) (Wang et al., 2009), whereas Anney et al. (2010) described an association with rs2301261 (P = .016 in the whole sample; P = .007 in Caucasians only) (Anney et al., 2010). Interestingly, both these genetic variants are located upstream of the encoding sequence of *OXTR* and hence may have implications for the gene's expression regulation. Although the first attempt to replicate the findings from the GWAS failed (Campbell et al., 2011), other candidate gene studies provide evidence for associations between ASD and SNPs located in the upstream part of *OXTR*. In the study by Tansey and coworkers (2010), rs4564970, which is located in the vicinity of rs2301261, was associated with risk of autism in their Irish subsample, but not in the total sample (see also Lerer et al., 2008). Moreover, the only single SNP association with autism in the study by Wermter et al. of 100 families was found with rs2270465, which is also located upstream of *OXTR* (Wermter et al., 2010).

Noteworthy in this context are two reports of patients with autism carrying a deletion of the region including *OXTR*, *CAV3*, and three neighboring genes (Gregory et al., 2009; Hedges et al., 2012). Gregory and co-workers also described altered patterns of methylation in the *OXTR* promoter in individuals with ASD and a decreased expression of *OXTR* in postmortem brains of individuals with ASD. Therefore, there may be multiple modes of disrupting *OXTR* that result in decreased expression of the oxytocin receptor and an increased risk for ASD.

IX. *OXTR* and Schizophrenia

There are converging lines of evidence that oxytocin may be involved in the pathophysiology of schizophrenia, as well as having antipsychotic treatment effects (Macdonald & Feifel, 2012; Meyer-Lindenberg et al., 2011). In line with this notion, Montag and co-workers recently reported associations between the rs53576 and rs237885 and risk of schizophrenia (Montag et al., 2012). These two SNPs were not included in a study of a Japanese case–control sample which, however, reported nominal association between the intron 3 SNP rs9840864 and schizophrenia risk (Watanabe et al., 2012). Furthermore, rs237885 and other SNPs in the intron 3 were associated with severity of symptoms, as well as with response to antipsychotic treatment in a sample of patients with schizophrenia (Souza et al., 2010).

X. *OXTR* and Social Behavior in Healthy Subjects

A. OXTR and Bonding Behavior

In analogy with our findings of an association between *AVPR1A* polymorphisms and pair-bonding behavior in men, we recently reported findings proposing a role of *OXTR* for social bond formation in women (Walum et al., 2012) (for a discussion of the oxytocin receptor gene and social behavioral in voles, see the chapter by Barrett and Young elsewhere in this volume). We investigated twelve SNPs of the *OXTR* in the Twin Offspring Study (TOSS) sample, also used in the study of *AVPR1A* (Walum et al., 2008), as well as in a separate sample (the Twin Study of Child and Adolescent Development [TCHAD]) comprising a measure of pair-bonding behavior, the Relationship Quality Survey (RQS), in young men and women. Intriguingly, rs7632287 (but none of the other SNPs) was associated with self-reported pair-bonding behavior in both samples, and the association was specific for women.

Further analysis of the TOSS sample revealed associations between rs7632287 and risk of marital crisis during the last year, as well as with measures of marital quality as perceived by the husbands. Taken together, these findings indicate that oxytocin influences pair-bonding behavior in women, which is in line with oxytocin's role in female voles.

To gain a better understanding of underlying factors relevant for the pair-bonding phenotype in adults, we took advantage of the longitudinal design of the TCHAD study (Walum et al., 2012). To this end, we used the Child Behavioral Checklist inventory (CBCL) reported by parents in the first wave of the TCHAD study (at age 8–9). The parent-reported CBCL social problems subscale (a scale assessing a variety of difficulties that children may display in their relationships with peers and adults), but no other CBCL subscale, was significantly associated with behavior in romantic relationships during adulthood. Interestingly, further analysis revealed that girls carrying the A allele of rs7632287, which was associated with lower pair-bonding scores in adult women, experienced more social problems than did girls not carrying this allele. Because the CBCL social problems scale previously has been associated with measures of the autism spectrum (Hallerod et al., 2010), it is intriguing that (as mentioned earlier) symptoms of autism measured in children collected from the normal population (Anckarsater et al., 2011) were associated with the A allele of rs7632287, specifically in girls (Walum et al., 2012).

In the most comprehensive investigation of oxytocin and related genetic variants in relation to parental behaviors, Feldman and co-workers showed that parents with high plasma oxytocin levels touched their infants more frequently and engaged in longer episodes of gaze synchrony with their infants (episodes when parent gazes at infant and infant gazes at parent) than did parents with low levels (Feldman et al., 2012). Furthermore, parents carrying rs1042778 TT or rs2254298 GG genotypes had lower plasma oxytocin levels. In line with these findings, parents homozygous for the TT risk genotype of the rs1042778 provided less touch than did parents carrying the G allele (Feldman et al., 2012). Other researchers found that mothers who are rs53576 A-allele carriers demonstrated lower maternal sensitivity relative to GG genotype carriers (Bakermans-Kranenburg & van Ijzendoorn, 2008; Sturge-Apple et al., 2012). Another study of reproductive-aged women without children revealed greater heart-rate response to baby cries in those who were homozygous G-allele carriers, relative to A-allele carriers of rs53576 (Riem et al., 2011). This result was interpreted to indicate greater sensitivity to the baby's needs and emotional state among GG carriers. Furthermore, Marsh and co-workers showed that infants' faces were more strongly preferred following oxytocin inhalation relative to placebo. When the participants were separated according to genotype, this effect was only observed for participants homozygous for the rs53576 G allele. Interestingly, parallel effects were not seen for adults' faces (Marsh et al., 2012). In a study of *OXTR* variation and infant-attachment behavior, 176 healthy infants between the ages of 12 and 16 months were genotyped for rs53576 and rs2254298 and assessed with the Strange Situation procedure. The A allele of rs2254298 was associated with attachment security in non-Caucasian, but not in Caucasian, infants (Chen et al., 2011a). Overall, the results from studies of attachment behaviors indicate a more efficient oxytocinergic function in rs53576G homozygotes compared to A-allele carriers. This is, as described later, a view supported by many (but not all) studies of this SNP in relation to sensitivity to emotions and stress, as well as to various prosocial and social cognitive abilities.

B. *OXTR*, Prosocial Behaviors, and Social Cognition

In two separate studies, Kim and co-workers (2010, 2011) investigated to what extent the effect of rs53576 genotype is moderated by the social environment. For this purpose, they focused on emotional support-seeking since, in times of distress, this behavior differs between cultures. In a group of Koreans and Americans, the rs53576 genotype was not associated with differential support-seeking behavior in Korean subjects, but was associated with less support-seeking behavior in American homozygous A-allele carriers, compared to American G-allele carriers (Kim et al., 2010). Furthermore, in a similar study of emotion regulation, Caucasian Americans carrying the GG genotype reported using emotional suppression less than those carrying the AA genotype, whereas Koreans showed the opposite pattern. These findings suggest that rs53576 influences the sensitivity to input from cultural norms regarding emotion regulation (Kim et al., 2011). In addition, Kogan and co-workers reported that individuals homozygous for the G allele were rated by naïve observers to be more prosocial than were carriers of the A allele (Kogan et al., 2011). Moreover, the GG carriers displayed greater nonverbal affiliative cues associated with prosociality (e.g., head nods, smiles) than did A allele carriers, a behavior which in turn ultimately accounted for how prosocial the observers judged the subjects to be. Thus, individual differences in rs53576 are associated with behavioral manifestations of prosociality, which ultimately guide the judgments others make about the individual (Kogan et al., 2011). There is also evidence that *OXTR* SNPs may moderate processing of social information from facial expressions of emotions, as well as from the human voice. In two studies using the Reading the Mind in the Eyes Test, measuring the ability to detect emotion from pictures of human faces showing only the eyes, T-allele carriers of the synonymous SNP rs2228485 (Lucht et al., 2013) and GG homozygotes of rs53576 (Rodrigues et al., 2009) had fewer numbers of incorrect answers than did carriers of other genotypes. In line with Rodrigues et al., another study showed that rs53576 GG homozygotic women reported less difficulty hearing and understanding people in noisy environments than did women who were A-allele carriers (Tops et al., 2011). Recently, rs2268498, located in the promoter region of *OXTR*, was associated with moral judgment, specifically blameworthiness of accidentally committed harm (Walter et al., 2012).

In 2005, Kosfeld and co-workers published a seminal paper showing that intranasal oxytocin treatment increased trust, as measured in an economic game (Kosfeld et al., 2005). This finding has been replicated and extended in several studies (see Baumgartner et al., 2008; Bartz et al., 2011). Based on these findings, Israel and co-workers investigated to what extent *OXTR* variants are associated with economic behavior (Israel et al., 2009). They genotyped 15 SNPs of the *OXTR* in approximately 200 individuals who had performed the dictator game and the Social Values Orientation task. Several of the SNPs located in the intron 3 associated nominally with different behavioral measures, but three SNPs (rs1042778, rs2268490, and rs237887) showed significant associations with performance on both games. The most significant association was observed with rs1042778, and this SNP was also associated with altruistic behavior in an independent sample of 98 female subjects (Israel et al., 2009). Furthermore, in a study using the trust game, individuals carrying the GG genotype of rs53576 showed higher trust behavior than did individuals carrying the A allele (AA/AG) (Krueger et al., 2012). The same SNP also interacted with perceived threat to predict engagement in charitable activities in a study of real-world prosocial behavior (Poulin et al., 2012). However, when nine *OXTR* polymorphisms (including rs1042778, rs237887, and rs53576) were investigated in 684 individuals from the Swedish population who had performed the dictator and trust games with real monetary consequences (Apicella et al., 2010), neither the associations reported by Israel et al. nor any association with rs53576 was seen. Instead, the most significant association was found between rs75775, located upstream of *OXTR*, and altruistic behavior in men. However, this association did not remain significant after correction for multiple hypothesis testing.

XI. OXTR, Stress, Anxiety, and Depression

In addition to its direct effects on social behaviors, oxytocin also regulates the response to stress and anxiogenic stimuli (Neumann & Landgraf, 2012; Onaka et al., 2012). In a stressful situation, the hypothalamus-pituitary-adrenal axis is activated to secrete corticotropin-releasing hormone, adrenocorticotropic hormone (ACTH), and cortisol, in order to adapt to the situation. Studies in rodents argue that oxytocin is a potent regulator of cortisol release, which is supported by intranasal treatment studies showing that oxytocin dampens the cortisol release triggered during public speaking (Heinrichs et al., 2003; Quirin et al., 2011), as well as during a couple conflict (Ditzen et al., 2009).

In line with these findings, a number of studies suggest that *OXTR* variants may regulate stress reactivity, anxiety, and depression and depressed mood. Rodrigues and co-workers found that rs53576 A-allele carriers exhibited higher physiological and dispositional stress reactivity than did GG individuals, based on heart rate response during a startle anticipation task and an affective reactivity scale (Rodrigues et al., 2009). In line with these findings, the A allele has been shown to associate with lower self-reported positive affect (Lucht et al., 2009) and lower levels of optimism, mastery, and self-esteem, as well as with higher levels of depressive symptomatology (Saphire-Bernstein et al., 2011) (see, however (Cornelis et al., 2012). Two studies further investigated potential associations between rs53576 and physiological measures of the stress response. Chen and co-workers explored the interaction between this polymorphism and social support on cortisol responses to stress and found that men carrying the G allele and who received social support from a female partner or friend showed lower cortisol responses to stress compared to other men (Chen et al., 2011), indicating that genetic variation of the oxytocin system modulates the effectiveness of positive social interaction as a protective buffer against a stressful experience. In older adults (50–68 years of age), GG carriers displayed lower awakening cortisol levels and less variation in salivary cortisol across the day as compared to A-allele carriers (Norman et al., 2012). Moreover, men carrying the GG genotype showed lower levels of resting sympathetic cardiac control, but, in contrast to Rodrigues and co-workers' findings, they showed higher levels of sympathetic reactivity to psychological stress as compared to the A-allele carriers (Norman et al., 2012). These discrepancies may be explained by the use of different stress paradigms and measures of stress response between studies. As the participants of the two studies also were of different ages, variations in autonomic heart function due to aging (Kaye & Esler, 2008) may also be a contributing factor.

In contrast to most of the hitherto described findings regarding rs53576 and stress reactivity, Costa and co-workers (2009) showed a positive association between the GG genotype of rs53576 and unipolar depression (Costa et al., 2009), as well as Attachment Style Questionnaire factors that have been previously associated with depression and adult separation anxiety. In a sample collected from the normal population, girls having high early adversity and the A allele of rs2254298 reported the highest levels of symptoms of depression, physical anxiety, and social anxiety (Thompson et al., 2011). More recently, the intronic rs237900 and the synonymous rs237902 SNPs were associated with Harm Avoidance, measured with Temperament and Character Inventory (TCI), specifically in women (Stankova et al., 2012). In a previous study of personality traits measured with TCI, the rs237902 was associated with Novelty Seeking and Cooperativeness, but not with Harm Avoidance (Comings et al., 2000). In a study of both rs2268498, located in the promoter of *OXTR,* and 5-HTTLPR, located in the promoter of the serotonin transporter gene, participants carrying both the homozygous LL variant of the 5-HTTLPR and the TT genotype of rs2268498 showed lowest scores on the personality dimensions Fear and Sadness, as well as on the underlying factor Negative Emotionality (Montag et al., 2011). The report by Unternaehrer et al. is further complicating the picture of the role of *OXTR* in stress regulation by showing that acute psychosocial stress changes DNA methylation of the *OXTR* (Unternaehrer et al., 2012). Taken together, there is fairly consistent support for the view that the GG genotype of rs53576 is associated with a more efficient oxytocinergic function. However, as discussed further later, more investigations of other SNPs in the gene and studies of their functional effects, along with investigations of epigenetic mechanisms, are warranted.

XII. OXTR and Aggression

Oxytocin has been associated with aggressive behaviors both in rodent and human studies. In rodents, the effects of oxytocin on aggressive behaviors are highly context- and species-dependent (Albers, 2012; Bosch & Neumann, 2012; Caldwell et al., 2008; Ferris et al., 2008; Lee H.J., et al., 2009). In line with these findings, intranasally administered oxytocin has, in addition to its prosocial effects, been shown to increase defensive aggression during specific conditions (De Dreu et al., 2010; De Dreu, 2012). Furthermore, cerebrospinal fluid (CSF) levels of oxytocin were inversely correlated with life history of aggression (Lee R., et al., 2009) and suicide intent (Jokinen et al., 2012).

To explore whether *OXTR* variants modulate the effect of alcohol on experimentally induced aggression, 116 Finnish men were assessed for aggressive behavior using the Response Choice Aggression Paradigm (Zeichner et al., 1999) after drinking either alcohol or placebo and were thereafter genotyped for twelve *OXTR* polymorphisms (Johansson et al., 2012a). Two SNPs, rs1488467 and rs4564970, showed nominally significant interactions with alcohol manipulation on aggressive behaviors. Alcohol increased aggression in individuals with a CG genotype on the SNPs in question, whereas it did not have such an effect for those homozygous for the G allele. Only the interaction between rs4564970 and alcohol remained significant after correcting for multiple comparisons (Johansson et al., 2012a). In a follow-up study using a population-based sample of 3,557 individuals, alcohol consumption was positively correlated with aggressive behavior, and an interactive effect between rs4564970 and alcohol consumption on aggressive behavior, as well as with trait anger, was seen (Johansson et al., 2012b). In line with the experimental study, the correlation between alcohol consumption and aggressive behavior was more pronounced for carriers of at least one C allele than for those homozygous for the G allele. Importantly, the association did not seem to be due to an effect of genotype on the subjects' drinking patterns (Johansson et al., 2012b). Furthermore, in the experimental study (Johansson et al., 2012a), carriers of the rs1042778 G allele exhibited nominally higher levels of aggression than did those who were homozygous for the T allele. Whereas no associations were seen between rs1042778 and aggressive behavior, trait anger, or anger control in the population-based sample (Johansson et al., 2012b), the G allele was more prevalent among boys displaying extreme aggressive behavior compared to controls (Malik et al., 2012). Furthermore, the closely located SNPs rs6770632 and rs237885 were associated with childhood-onset aggression (Malik et al., 2012) and with callous-unemotional traits (Beitchman et al., 2012), respectively.

XIII. *OXTR* and Drug Dependence

Because oxytocin exerts its actions on bonding behavior by affecting the reward system through receptors in the nucleus accumbens (NAcc; Young & Wang, 2004), which has been associated with drug-taking behavior (Burkett & Young, 2012), the *OXTR* is a reasonable candidate for drug dependence disorder. In a GWAS of alcohol dependence including more than 1 million SNPs in 1,884 individuals, one SNP located in intron 3 of the *OXTR*, rs237899, was significantly associated with alcohol dependence (2.0E–04, this also when restricting the analysis to cases with an early onset: 1.1E–03). This association was, however, not replicated in an independent family sample (Edenberg et al., 2010). In a subsample of the GWAS sample, rs237899 was associated with comorbid depressive syndrome and alcohol dependence (Edwards et al., 2012), indicating that *OXTR* may be of importance for a subset of patients.

XIV. *OXTR* and Brain Function

To date, a limited number of studies have explored how intranasal oxytocin spray modulates human brain activity using fMRI (Zink & Meyer-Lindenberg, 2012). In line with the rodent studies, amygdala seems to play a key role for the regulation of fear and emotion also in humans (see also the chapters by Barrett and Young and by Buckholtz and Meyer-Lindenberg elsewhere in this volume).

Some studies have, to date, explored associations between *OXTR* SNPs and brain function. Inspired by the behavioral genetics findings seen regarding the rs53576 SNP, Tost et al. investigated the effects of rs53576 on brain structure and function (Tost et al., 2010). The A allele was associated with decreased gray matter volume in hypothalamus, with the strongest effect seen in men. A significant gene-by-sex interaction effect for rs53576 was seen in the right amygdala, in which men carrying the A allele displayed a relative increase in gray matter volume. Furthermore, carriers of the rs53576 A allele showed a significant structural coupling between gray matter volumes of the hypothalamus and of the dorsal anterior cingulate cortex (ACG), as well as of the amygdala. Based on the structural imaging results, the effect of rs53576 on amygdala activation and interregional coupling during the processing of emotionally salient social cues was investigated. Subjects homozygous for the rs53576 A allele showed the lowest amygdala activations during the perceptual processing of facial emotion, and homozygotes for the G allele showed the highest. Furthermore, a significant increase in the functional coupling between the hypothalamus and amygdala was observed in rs53576 A-allele carriers. No significant sex-by-genotype interaction effects were observed for either amygdala activation or hypothalamus–amygdala coupling (Tost et al., 2010).

Inoue and co-workers investigated seven *OXTR* SNPs in 208 healthy adults from the Japanese population and reported an association between the rs2254298 A allele and larger bilateral amygdala volume (Inoue et al., 2010). A similar association between the rs2254298 A allele and larger amygdala volume was also seen in 55 Caucasian girls (Furman et al., 2011). Importantly, this association between the rs2254298 A allele and amygdala size was seen when using a regionally focused morphometric approach using manual tracing (Furman et al., 2011; Inoue et al., 2010) but not when using voxel-based morphometry analysis (VBM) (Furman et al., 2011; Tost et al., 2011; Yamasue et al., 2011). In a whole-brain VBM analysis, Furman and co-workers revealed that the A allele was associated with greater volume of the posterior brainstem, as well as with smaller volume of dorsal ACG (dACG), a region of prefrontal cortex important for regulation of limbic function (Furman et al., 2011). Also, Yamasue and co-workers found an association between the rs2254298 A allele and a smaller dACG volume (Yamasue et al., 2011). Tost and co-workers showed that men being rs2254298 A-allele carriers had smaller hypothalamus gray matter volume and a significant increase in structural coupling between the hypothalamus and dACG (Tost et al., 2011). Furthermore, in their fMRI study, Tost et al. also found deficient deactivation of the dACG during emotion processing in rs2254298 A allele carriers.

Overall, rather convincing evidence is provided for an association between the A allele of rs2254298 and increased amygdala volume (Furman et al., 2011; Inoue et al., 2010), as well as with decreased dACG volume or function (Furman et al., 2011; Tost et al., 2011; Yamasue et al., 2011), independent of ethnicity. In addition, some first intriguing associations between rs53576 and brain activation patterns during emotion processing have been reported (Tost et al., 2010). These findings may hence pinpoint some the neural mechanisms mediating the effects of *OXTR* variants on social behavior.

XV. Concluding Remarks about *OXTR* and Human Social Behaviors

In this chapter many promising and interesting associations between *OXTR* polymorphisms and social behaviors, as well as risk of psychiatric disorders such as autism, schizophrenia, and alcoholism, have been reported (see Table 8.1). Most of these findings are, however, inconclusive and warrant replication in independent samples. Further complicating the interpretations of the findings is the fact that few, if any, functional *OXTR* variants have so far been identified. The polymorphisms of *OXTR* associated with social behaviors are distributed throughout the gene, and many of these SNPs are in LD with each other. Thus, in order to estimate whether an association between a given SNP and a social trait may be dependent on another SNP, the LD patterns between *OXTR* SNPs need to be more fully characterized. However, it is already known that some of the SNPs showing associations with behavioral traits are not in LD with each other and are located in separate parts of the gene containing regulatory sequences implicated in gene regulation (i.e., in upstream regions of the gene, in intron 3, and in downstream regions of the gene) (Figure 8.1). Therefore, it is tempting to suggest that *OXTR* may contain multiple functional variants that may affect *OXTR* function in different ways and independently of each other. If this notion is true, it may at least partly explain inconsistencies between studies. Furthermore, before the functional variants are pinpointed, researchers should avoid conducting studies of single SNPs and instead investigate comprehensive SNP sets. Finally, although some preliminary functional studies have been conducted of *OXTR* SNPs (Dixon et al., 2007; Edenberg et al., 2010; Park et al., 2010; Montag et al., 2011; Tansey et al., 2010), it should be a high priority for future research to identify the functional variants in *OXTR*.

XVI. Conclusion and Future Directions

Altogether, the chances seem good that future studies may prove that *OXTR* and *AVPR1A* variants—and hence oxytocin and vasopressin actions—actually are of importance for how we interact and bond to each other in everyday life. As the studies conducted so far did not, however, provide such definite proofs, we discuss here some of their weaknesses but also raise issues that researchers should consider in their future investigations of oxytocin- and vasopressin-related genes in relation to social behaviors and psychiatric disorders.

The majority of associations presented in this chapter are rather weak, and corresponding p-values are most often quite high. Thus, many of the associations neither survived strict correction for multiple testing nor could be replicated in independent samples. The major reasons for this are probably that many studies suffer from limited sample sizes as well as ethnic stratification (see Table 8.1), which increases the risk for spurious associations (Hamer & Sirota, 2000; Sullivan, 2007). Many of the investigated behavioral phenotypes, such as the diagnosis of autism, are very complex and heterogeneous, making it hard to reveal effects of common genetic variants with weak effects in moderately sized samples. Studies of such complex traits may therefore benefit from assessment of endophenotypes, specific symptoms, more narrowly defined subgroups of patients with no homogenenous clinical profiles, or phenotypes that are conceptualized as continuous rather than dichotomous variables, in order to attain a better correlation between gene function and phenotype. Some of the studies of *OXTR* in autism have already successfully applied such approaches (Campbell et al., 2011; Egawa et al., 2013; Lerer et al., 2008; Walum et al., 2012). Furthermore, it is well known from the rodent literature that many of oxytocin's and vasopressin's behavioral actions are sexually dimorphic. As this probably is also true for humans, studies of genes related to the function to oxytocin and/or vasopressin should, if possible, include sex-specific analysis. Several studies have shown this to be a successful approach (Stankova et al., 2012; Tost et al., 2010; Walum et al., 2008; Walum et al., 2012).

Future studies should also investigate *OXTR* and *AVPR1A* in relation to additional aspects of social behaviors, including social recognition, emotion recognition, and psychopathic personality traits, as well as to the risk of psychiatric disorders such as schizophrenia and social phobia. Furthermore, pharmacogenetic studies are warranted to reveal if the treatment efficacy of the neuropeptides is modulated by genetic variation of *OXTR* and *AVPR1A*. The neuronal circuits affected by the *OXTR* and *AVPR1A* variants should be further explored using various imaging genetics approaches. Furthermore, the initial reports of epigenetic mechanisms, as well as of rare mutations affecting the function of oxytocin- and vasopressin-related genes, need more attention in upcoming studies. Other genes essential for the function of oxytocin and vasopressin, such as the genes encoding the neuropeptides themselves and *CD38*, an important regulator of oxytocin release, have already been shown to associate with social behaviors in some recent studies (Ebstein et al., 2012; Feldman et al., 2012; Higashida et al., 2012). Thus, polymorphisms of these and other genes regulating the actions of oxytocin and vasopressin, such as *SIM1* and *ARNT2*—transcription factors necessary for the development of the paraventricular nucleus, as well as for the expression of the neuropeptides—deserve to be included in future investigations. As a final note, we hope that the striking creativity regarding the designs of many of the studies presented in this chapter will inspire researchers in their future investigations of genetic variants in relation to social behaviors.

References

Albers, H. E. (2012). The regulation of social recognition, social communication and aggression: vasopressin in the social behavior neural network. *Horm Behav, 61*(3), 283–292.

Anckarsater, H., Lundstrom, S., Kollberg, L., Kerekes, N., Palm, C., Carlstrom, E., et al. (2011). The Child and Adolescent Twin Study in Sweden (CATSS). *Twin Res Hum Genet, 14*(6), 495–508.

Anney, R., Klei, L., Pinto, D., Regan, R., Conroy, J., Magalhaes, T. R., et al. (2010). A genome-wide scan for common alleles affecting risk for autism. *Hum Mol Genet, 19*(20), 4072–4082.

Apicella, C. L., Cesarini, D., Johannesson, M., Dawes, C. T., Lichtenstein, P., Wallace, B., et al. (2010). No association between oxytocin receptor (OXTR) gene polymorphisms and experimentally elicited social preferences. *PLoS One, 5*(6), e11153.

Avinun, R., Ebstein, R. P., & Knafo, A. (2012). Human maternal behaviour is associated with arginine vasopressin receptor 1A gene. *Biol Lett, 8*(5), 894–896.

Avinun, R., Israel, S., Shalev, I., Gritsenko, I., Bornstein, G., Ebstein, R. P., et al. (2011). AVPR1A variant associated with preschoolers' lower altruistic behavior. *PLoS One, 6*(9), e25274.

Bachner-Melman, R., Dina, C., Zohar, A. H., Constantini, N., Lerer, E., Hoch, S., et al. (2005). AVPR1a and SLC6A4 gene polymorphisms are associated with creative dance performance. *PLoS Genet, 1*(3), e42.

Bailey, A., Le Couteur, A., Gottesman, I., Bolton, P., Simonoff, E., Yuzda, E., et al. (1995). Autism as a strongly genetic disorder: evidence from a British twin study. *Psychol Med, 25*(1), 63–77.

Bakermans-Kranenburg, M. J., & van Ijzendoorn, M. H. (2008). Oxytocin receptor (OXTR) and serotonin transporter (5-HTT) genes associated with observed parenting. *Soc Cogn Affect Neurosci, 3*(2), 128–134.

Bartz, J. A., Zaki, J., Bolger, N., & Ochsner, K. N. (2011). Social effects of oxytocin in humans: context and person matter. *Trends Cogn Sci, 15*(7), 301–309.

Baumgartner, T., Heinrichs, M., Vonlanthen, A., Fischbacher, U., & Fehr, E. (2008). Oxytocin shapes the neural circuitry of trust and trust adaptation in humans. *Neuron, 58*(4), 639–650.

Beitchman, J. H., Zai, C. C., Muir, K., Berall, L., Nowrouzi, B., Choi, E., et al. (2012). Childhood aggression, callous-unemotional traits and oxytocin genes. *Eur Child Adolesc Psychiatry, 21*(3), 125–132.

Bisceglia, R., Jenkins, J. M., Wigg, K. G., O'Connor, T. G., Moran, G., & Barr, C. L. (2012). Arginine vasopressin 1a receptor gene and maternal behavior: evidence of association and moderation. *Genes Brain Behav, 11*(3), 262–268.

Born, J., Lange, T., Kern, W., McGregor, G. P., Bickel, U., & Fehm, H. L. (2002). Sniffing neuropeptides: a transnasal approach to the human brain. *Nat Neurosci, 5*(6), 514–516.

Bos, P. A., Panksepp, J., Bluthe, R. M., & van Honk, J. (2012). Acute effects of steroid hormones and neuropeptides on human social-emotional behavior: a review of single administration studies. *Front Neuroendocrinol, 33*(1), 17–35.

Bosch, O. J., & Neumann, I. D. (2012). Both oxytocin and vasopressin are mediators of maternal care and aggression in rodents: from central release to sites of action. *Horm Behav, 61*(3), 293–303.

Burkett, J. P., & Young, L. J. (2012). The behavioral, anatomical and pharmacological parallels between social attachment, love and addiction. *Psychopharmacology (Berl), 224*(1), 1–26.

Caldwell, H. K., Lee, H. J., Macbeth, A. H., & Young, W. S., 3rd. (2008). Vasopressin: behavioral roles of an "original" neuropeptide. *Prog Neurobiol, 84*(1), 1–24.

Campbell, D. B., Datta, D., Jones, S. T., Batey Lee, E., Sutcliffe, J. S., Hammock, E. A., et al. (2011). Association of oxytocin receptor (OXTR) gene variants with multiple phenotype domains of autism spectrum disorder. *J Neurodev Disord, 3*(2), 101–112.

Chen, F. S., Barth, M. E., Johnson, S. L., Gotlib, I. H., & Johnson, S. C. (2011a). Oxytocin Receptor (OXTR) Polymorphisms and Attachment in Human Infants. *Front Psychol, 2*, 200.

Chen, F. S., & Johnson, S. C. (2011). An Oxytocin Receptor Gene Variant Predicts Attachment Anxiety in Females and Autism-Spectrum Traits in Males. *Social Psychological and Personality Science, 3*(1), 93–99.

Chen, F. S., Kumsta, R., von Dawans, B., Monakhov, M., Ebstein, R. P., & Heinrichs, M. (2011b). Common oxytocin receptor gene (OXTR) polymorphism and social support interact to reduce stress in humans. *Proc Natl Acad Sci U S A, 108*(50), 19937–19942.

Cherkas, L. F., Oelsner, E. C., Mak, Y. T., Valdes, A., & Spector, T. D. (2004). Genetic influences on female infidelity and number of sexual partners in humans: a linkage and association study of the role of the vasopressin receptor gene (AVPR1A). *Twin Res, 7*(6), 649–658.

Comings, D. E., Gade-Andavolu, R., Gonzalez, N., Wu, S., Muhleman, D., Blake, H., et al. (2000). A multivariate analysis of 59 candidate genes in personality traits: the temperament and character inventory. *Clin Genet, 58*(5), 375–385.

Cornelis, M. C., Glymour, M. M., Chang, S. C., Tchetgen, E. J., Liang, L., Koenen, K. C., et al. (2012). Oxytocin receptor (OXTR) is not associated with optimism in the Nurses' Health Study. *Mol Psychiatry, 17*(12), 1157–1159.

Costa, B., Pini, S., Gabelloni, P., Abelli, M., Lari, L., Cardini, A., et al. (2009). Oxytocin receptor polymorphisms and adult attachment style in patients with depression. *Psychoneuroendocrinology, 34*(10), 1506–1514.

De Dreu, C. K. (2012). Oxytocin modulates cooperation within and competition between groups: an integrative review and research agenda. *Horm Behav, 61*(3), 419–428.

De Dreu, C. K., Greer, L. L., Handgraaf, M. J., Shalvi, S., Van Kleef, G. A., Baas, M., et al. (2010). The neuropeptide oxytocin regulates parochial altruism in intergroup conflict among humans. *Science, 328*(5984), 1408–1411.

Ditzen, B., Schaer, M., Gabriel, B., Bodenmann, G., Ehlert, U., & Heinrichs, M. (2009). Intranasal oxytocin increases positive communication and reduces cortisol levels during couple conflict. *Biol Psychiatry, 65*(9), 728–731.

Dixon, A. L., Liang, L., Moffatt, M. F., Chen, W., Heath, S., Wong, K. C., et al. (2007). A genome-wide association study of global gene expression. *Nat Genet, 39*(10), 1202–1207.

Donaldson, Z. R., & Young, L. J. (2008). Oxytocin, vasopressin, and the neurogenetics of sociality. *Science, 322*(5903), 900–904.

Edenberg, H. J., Koller, D. L., Xuei, X., Wetherill, L., McClintick, J. N., Almasy, L., et al. (2010). Genome-wide association study of alcohol dependence implicates a region on chromosome 11. *Alcohol Clin Exp Res, 34*(5), 840–852.

Edwards, A. C., Aliev, F., Bierut, L. J., Bucholz, K. K., Edenberg, H., Hesselbrock, V., et al. (2012). Genome-wide association study of comorbid depressive syndrome and alcohol dependence. *Psychiatr Genet, 22*(1), 31–41.

Egawa, J., Watanabe, Y., Endo, T., Tamura, R., Masuzawa, N., & Someya, T. (2013). Association between OXTR and clinical phenotypes of autism spectrum disorders. *Psychiatry Res, 208*(1), 99–100.

Feldman, R., Zagoory-Sharon, O., Weisman, O., Schneiderman, I., Gordon, I., Maoz, R., et al. (2012a). Sensitive parenting is associated with plasma oxytocin and polymorphisms in the OXTR and CD38 genes. *Biol Psychiatry, 72*(3), 175–181.

Ferris, C. F., Stolberg, T., Kulkarni, P., Murugavel, M., Blanchard, R., Blanchard, D. C., et al. (2008). Imaging the neural circuitry and chemical control of aggressive motivation. *BMC Neurosci,9*, 111.

Folstein, S., & Rutter, M. (1977). Genetic influences and infantile autism. *Nature, 265*(5596), 726–728.

Furman, D. J., Chen, M. C., & Gotlib, I. H. (2011). Variant in oxytocin receptor gene is associated with amygdala volume. *Psychoneuroendocrinology, 36*(6), 891–897.

Gabor, C. S., Phan, A., Clipperton-Allen, A. E., Kavaliers, M., & Choleris, E. (2012). Interplay of oxytocin, vasopressin, and sex hormones in the regulation of social recognition. *Behav Neurosci, 126*(1), 97–109.

Geller, B., Tillman, R., Badner, J. A., & Cook, E. H., Jr. (2005). Are the arginine vasopressin V1a receptor microsatellites related to hypersexuality in children with a prepubertal and early adolescent bipolar disorder phenotype? *Bipolar Disord, 7*(6), 610–616.

Green, J. J., & Hollander, E. (2010). Autism and oxytocin: new developments in translational approaches to therapeutics. *Neurotherapeutics, 7*(3), 250–257.

Gregory, S. G., Connelly, J. J., Towers, A. J., Johnson, J., Biscocho, D., Markunas, C. A., et al. (2009). Genomic and epigenetic evidence for oxytocin receptor deficiency in autism. *BMC Med, 7*, 62.

Guastella, A. J., Kenyon, A. R., Alvares, G. A., Carson, D. S., & Hickie, I. B. (2010). Intranasal arginine vasopressin enhances the encoding of happy and angry faces in humans. *Biol Psychiatry, 67*(12), 1220–1222.

Guastella, A. J., Kenyon, A. R., Unkelbach, C., Alvares, G. A., & Hickie, I. B. (2011). Arginine Vasopressin selectively enhances recognition of sexual cues in male humans. *Psychoneuroendocrinology, 36*(2), 294–297.

Guastella, A. J., & MacLeod, C. (2012). A critical review of the influence of oxytocin nasal spray on social cognition in humans: evidence and future directions. *Horm Behav, 61*(3), 410–418.

Hallerod, S. L., Larson, T., Stahlberg, O., Carlstrom, E., Gillberg, C., Anckarsater, H., et al. (2010). The Autism--Tics, AD/HD and other Comorbidities (A-TAC) telephone interview: convergence with the Child Behavior Checklist (CBCL). *Nord J Psychiatry, 64*(3), 218–224.

Hamer, D., & Sirota, L. (2000). Beware the chopsticks gene. *Mol Psychiatry, 5*(1), 11–13.

Higashida, H., S.Yokoyama, S., et al. (2012). CD38 and its role in oxytocin secretion and social behavior. *Hormones and Behavior, 61*(3), 351–358.

Hedges, D. J., Hamilton-Nelson, K. L., Sacharow, S. J., Nations, L., Beecham, G. W., Kozhekbaeva, Z. M., et al. (2012). Evidence of novel fine-scale structural variation at autism spectrum disorder candidate loci. *Mol Autism, 3*, 2.

Heinrichs, M., Baumgartner, T., Kirschbaum, C., & Ehlert, U. (2003). Social support and oxytocin interact to suppress cortisol and subjective responses to psychosocial stress. *Biol Psychiatry, 54*(12), 1389–1398.

Inoue, H., Yamasue, H., Tochigi, M., Abe, O., Liu, X., Kawamura, Y., et al. (2010). Association between the oxytocin receptor gene and amygdalar volume in healthy adults. *Biol Psychiatry, 68*(11), 1066–1072.

Israel, S., Lerer, E., Shalev, I., Uzefovsky, F., Riebold, M., Laiba, E., et al. (2009). The oxytocin receptor (OXTR) contributes to prosocial fund allocations in the dictator game and the social value orientations task. *PLoS One, 4*(5), e5535.

Jacob, S., Brune, C. W., Carter, C. S., Leventhal, B. L., Lord, C., & Cook, E. H., Jr. (2007). Association of the oxytocin receptor gene (OXTR) in Caucasian children and adolescents with autism. *Neurosci Lett, 417*(1), 6–9.

Johansson, A., Bergman, H., Corander, J., Waldman, I. D., Karrani, N., Salo, B., et al. (2012a). Alcohol and aggressive behavior in men--moderating effects of oxytocin receptor gene (OXTR) polymorphisms. *Genes Brain Behav, 11*(2), 214–221.

Johansson, A., Westberg, L., Sandnabba, K., Jern, P., Salo, B., & Santtila, P. (2012b). Associations between oxytocin receptor gene (OXTR) polymorphisms and self-reported aggressive behavior and anger: Interactions with alcohol consumption. *Psychoneuroendocrinology, 37*(9), 1546–1556.

Jokinen, J., Chatzittofis, A., Hellstrom, C., Nordstrom, P., Uvnas-Moberg, K., & Asberg, M. (2012). Low CSF oxytocin reflects high intent in suicide attempters. *Psychoneuroendocrinology, 37*(4), 482–490.

Kaye, D. M., & Esler, M. D. (2008). Autonomic control of the aging heart. *Neuromolecular Med, 10*(3), 179–186.

Kim, H. S., Sherman, D. K., Mojaverian, T., Sasaki, J. Y., Park, J., Suh, E. M., et al. (2011). Gene-Culture Interaction: Oxytocin Receptor Polymorphism (OXTR) and Emotion Regulation. *Social Psychological and Personality Science, 2*(6), 665–672.

Kim, H. S., Sherman, D. K., Sasaki, J. Y., Xu, J., Chu, T. Q., Ryu, C., et al. (2010). Culture, distress, and oxytocin receptor polymorphism (OXTR) interact to influence emotional support seeking. *Proc Natl Acad Sci U S A, 107*(36), 15717–15721.

Kim, S. J., Young, L. J., Gonen, D., Veenstra-VanderWeele, J., Courchesne, R., Courchesne, E., et al. (2002). Transmission disequilibrium testing of arginine vasopressin receptor 1A (AVPR1A) polymorphisms in autism. *Mol Psychiatry, 7*(5), 503–507.

Kimura, T., Tanizawa, O., Mori, K., Brownstein, M. J., & Okayama, H. (1992). Structure and expression of a human oxytocin receptor. *Nature, 356*(6369), 526–529.

Knafo, A., Israel, S., Darvasi, A., Bachner-Melman, R., Uzefovsky, F., Cohen, L., et al. (2008). Individual differences in allocation of funds in the dictator game associated with length of the arginine vasopressin 1a receptor RS3 promoter region and correlation between RS3 length and hippocampal mRNA. *Genes Brain and Behavior, 7*(3), 266–275.

Kogan, A., Saslow, L. R., Impett, E. A., Oveis, C., Keltner, D., & Rodrigues Saturn, S. (2011). Thin-slicing study of the oxytocin receptor (OXTR) gene and the evaluation and expression of the prosocial disposition. *Proc Natl Acad Sci U S A, 108*(48), 19189–19192.

Kosfeld, M., Heinrichs, M., Zak, P. J., Fischbacher, U., & Fehr, E. (2005). Oxytocin increases trust in humans. *Nature, 435*(7042), 673–676.

Krueger, F., Parasuraman, R., Iyengar, V., Thornburg, M., Weel, J., Lin, M., et al. (2012). Oxytocin receptor genetic variation promotes human trust behavior. *Front Hum Neurosci, 6*, 4.

Lee, H. J., Macbeth, A. H., Pagani, J. H., & Young, W. S., 3rd. (2009). Oxytocin: the great facilitator of life. *Prog Neurobiol, 88*(2), 127–151.

Lee, R., Ferris, C., Van de Kar, L. D., & Coccaro, E. F. (2009). Cerebrospinal fluid oxytocin, life history of aggression, and personality disorder. *Psychoneuroendocrinology, 34*(10), 1567–1573.

Lerer, E., Levi, S., Salomon, S., Darvasi, A., Yirmiya, N., & Ebstein, R. P. (2008). Association between the oxytocin receptor (OXTR) gene and autism: relationship to Vineland Adaptive Behavior Scales and cognition. *Mol Psychiatry, 13*(10), 980–988.

Levin, R., Heresco-Levy, U., Bachner-Melman, R., Israel, S., Shalev, I., & Ebstein, R. P. (2009). Association between arginine vasopressin 1a receptor (AVPR1a) promoter region polymorphisms and prepulse inhibition. *Psychoneuroendocrinology, 34*(6), 901–908.

Levran, O., Londono, D., O'Hara, K., Randesi, M., Rotrosen, J., Casadonte, P., et al. (2009). Heroin addiction in African Americans: a hypothesis-driven association study. *Genes Brain Behav, 8*(5), 531–540.

Lichtenstein, P., Carlstrom, E., Rastam, M., Gillberg, C., & Anckarsater, H. (2010). The genetics of autism spectrum disorders and related neuropsychiatric disorders in childhood. *Am J Psychiatry, 167*(11), 1357–1363.

Lin, P. I., Vance, J. M., Pericak-Vance, M. A., & Martin, E. R. (2007). No gene is an island: the flip-flop phenomenon. *Am J Hum Genet, 80*(3), 531–538.

Liu, X., Kawamura, Y., Shimada, T., Otowa, T., Koishi, S., Sugiyama, T., et al. (2010). Association of the oxytocin receptor (OXTR) gene polymorphisms with autism spectrum disorder (ASD) in the Japanese population. *J Hum Genet, 55*(3), 137–141.

Lucht, M. J., Barnow, S., Sonnenfeld, C., Rosenberger, A., Grabe, H. J., Schroeder, W., et al. (2009). Associations between the oxytocin receptor gene (OXTR) and affect, loneliness and intelligence in normal subjects. *Prog Neuropsychopharmacol Biol Psychiatry, 33*(5), 860–866.

Lucht, M. J., Barnow, S., Sonnenfeld, C., Ulrich, I., Grabe, H. J., Schroeder, W., et al. (2013). Associations between the oxytocin receptor gene (OXTR) and "mind-reading" in humans--an exploratory study. *Nord J Psychiatry, 67*(1), 15–21.

Ma, D. Q., Cuccaro, M. L., Jaworski, J. M., Haynes, C. S., Stephan, D. A., Parod, J., et al. (2007). Dissecting the locus heterogeneity of autism: significant linkage to chromosome 12q14. *Mol Psychiatry, 12*(4), 376–384.

Macdonald, K., & Feifel, D. (2012). Oxytocin in schizophrenia: a review of evidence for its therapeutic effects. *Acta Neuropsychiatr, 24*(3), 130–146.

Macdonald, K., & Macdonald, T. M. (2010). The peptide that binds: a systematic review of oxytocin and its prosocial effects in humans. *Harv Rev Psychiatry, 18*(1), 1–21.

Maher, B. S., Vladimirov, V. I., Latendresse, S. J., Thiselton, D. L., McNamee, R., Kang, M., et al. (2011). The AVPR1A gene and substance use disorders: association, replication, and functional evidence. *Biol Psychiatry, 70*(6), 519–527.

Malik, A. I., Zai, C. C., Abu, Z., Nowrouzi, B., & Beitchman, J. H. (2012). The role of oxytocin and oxytocin receptor gene variants in childhood-onset aggression. *Genes Brain Behav, 11*(5), 545–551.

Marsh, A. A., Yu, H. H., Pine, D. S., Gorodetsky, E. K., Goldman, D., & Blair, R. J. (2012). The influence of oxytocin administration on responses to infant faces and potential moderation by OXTR genotype. *Psychopharmacology (Berl), 224*(4), 469–476.

Meyer-Lindenberg, A., Domes, G., Kirsch, P., & Heinrichs, M. (2011). Oxytocin and vasopressin in the human brain: social neuropeptides for translational medicine. *Nat Rev Neurosci, 12*(9), 524–538.

Meyer-Lindenberg, A., Kolachana, B., Gold, B., Olsh, A., Nicodemus, K. K., Mattay, V., et al. (2009). Genetic variants in AVPR1A linked to autism predict amygdala activation and personality traits in healthy humans. *Molecular Psychiatry, 14*(10), 968–975.

Mizumoto, Y., Kimura, T., & Ivell, R. (1997). A genomic element within the third intron of the human oxytocin receptor gene may be involved in transcriptional suppression. *Mol Cell Endocrinol, 135*(2), 129–138.

Montag, C., Brockmann, E. M., Bayerl, M., Rujescu, D., Muller, D. J., & Gallinat, J. (2012). Oxytocin and oxytocin receptor gene polymorphisms and risk for schizophrenia: A case-control study. *World J Biol Psychiatry.*

Montag, C., Fiebach, C. J., Kirsch, P., & Reuter, M. (2011). Interaction of 5-HTTLPR and a variation on the oxytocin receptor gene influences negative emotionality. *Biol Psychiatry, 69*(6), 601–603.

Morley, A. P., Narayanan, M., Mines, R., Molokhia, A., Baxter, S., Craig, G., et al. (2012). AVPR1A and SLC6A4 polymorphisms in choral singers and non-musicians: a gene association study. *PLoS One, 7*(2), e31763.

Neumann, I. D., & Landgraf, R. (2012). Balance of brain oxytocin and vasopressin: implications for anxiety, depression, and social behaviors. *Trends Neurosci, 35*(11), 649–659.

Norman, G. J., Hawkley, L., Luhmann, M., Ball, A. B., Cole, S. W., Berntson, G. G., et al. (2012). Variation in the oxytocin receptor gene influences neurocardiac reactivity to social stress and HPA function: a population based study. *Horm Behav, 61*(1), 134–139.

Nossent, A. Y., Hansen, J. L., Doggen, C., Quax, P. H., Sheikh, S. P., & Rosendaal, F. R. (2011). SNPs in microRNA binding sites in 3'-UTRs of RAAS genes influence arterial blood pressure and risk of myocardial infarction. *Am J Hypertens, 24*(9), 999–1006.

Onaka, T., Takayanagi, Y., & Yoshida, M. (2012). Roles of oxytocin neurones in the control of stress, energy metabolism, and social behaviour. *J Neuroendocrinol, 24*(4), 587–598.

Park, J., Willmott, M., Vetuz, G., Toye, C., Kirley, A., Hawi, Z., et al. (2010). Evidence that genetic variation in the oxytocin receptor (OXTR) gene influences social cognition in ADHD. *Prog Neuropsychopharmacol Biol Psychiatry, 34*(4), 697–702.

Poulin, M. J., Holman, E. A., & Buffone, A. (2012). The neurogenetics of nice: receptor genes for oxytocin and vasopressin interact with threat to predict prosocial behavior. *Psychol Sci, 23*(5), 446–452.

Prichard, Z. M., Mackinnon, A. J., Jorm, A. F., & Easteal, S. (2007). AVPR1A and OXTR polymorphisms are associated with sexual and reproductive behavioral phenotypes in humans. Mutation in brief no. 981. Online. *Hum Mutat, 28*(11), 1150.

Quirin, M., Kuhl, J., & Dusing, R. (2011). Oxytocin buffers cortisol responses to stress in individuals with impaired emotion regulation abilities. *Psychoneuroendocrinology, 36*(6), 898–904.

Riem, M. M., Pieper, S., Out, D., Bakermans-Kranenburg, M. J., & van Ijzendoorn, M. H. (2011). Oxytocin receptor gene and depressive symptoms associated with physiological reactivity to infant crying. *Soc Cogn Affect Neurosci, 6*(3), 294–300.

Rodrigues, S. M., Saslow, L. R., Garcia, N., John, O. P., & Keltner, D. (2009). Oxytocin receptor genetic variation relates to empathy and stress reactivity in humans. *Proc Natl Acad Sci U S A, 106*(50), 21437–21441.

Ronald, A., Larsson, H., Anckarsater, H., & Lichtenstein, P. (2011). A twin study of autism symptoms in Sweden. *Mol Psychiatry, 16*(10), 1039–1047.

Saphire-Bernstein, S., Way, B. M., Kim, H. S., Sherman, D. K., & Taylor, S. E. (2011). Oxytocin receptor gene (OXTR) is related to psychological resources. *Proc Natl Acad Sci U S A, 108*(37), 15118–15122.

Souza, R. P., de Luca, V., Meltzer, H. Y., Lieberman, J. A., & Kennedy, J. L. (2010). Schizophrenia severity and clozapine treatment outcome association with oxytocinergic genes. *Int J Neuropsychopharmacol, 13*(6), 793–798.

Stankova, T., Eichhammer, P., Langguth, B., & Sand, P. G. (2012). Sexually dimorphic effects of oxytocin receptor gene (OXTR) variants on Harm Avoidance. *Biol Sex Differ, 3*(1), 17.

Stevenson, E. L., & Caldwell, H. K. (2012). The vasopressin 1b receptor and the neural regulation of social behavior. *Horm Behav, 61*(3), 277–282.

Sturge-Apple, M. L., Cicchetti, D., Davies, P. T., & Suor, J. H. (2012). Differential susceptibility in spillover between interparental conflict and maternal parenting practices: evidence for OXTR and 5-HTT genes. *J Fam Psychol, 26*(3), 431–442.

Sullivan, P. F. (2007). Spurious genetic associations. *Biol Psychiatry, 61*(10), 1121–1126.

Tansey, K. E., Brookes, K. J., Hill, M. J., Cochrane, L. E., Gill, M., Skuse, D., et al. (2010). Oxytocin receptor (OXTR) does not play a major role in the aetiology of autism: genetic and molecular studies. *Neurosci Lett, 474*(3), 163–167.

Tansey, K. E., Hill, M. J., Cochrane, L. E., Gill, M., Anney, R. J., & Gallagher, L. (2011). Functionality of promoter microsatellites of arginine vasopressin receptor 1A (AVPR1A): implications for autism. *Mol Autism, 2*(1), 3.

Thibonnier, M., Graves, M. K., Wagner, M. S., Auzan, C., Clauser, E., & Willard, H. F. (1996). Structure, sequence, expression, and chromosomal localization of the human V1a vasopressin receptor gene. *Genomics, 31*(3), 327–334.

Thibonnier, M., Graves, M. K., Wagner, M. S., Chatelain, N., Soubrier, F., Corvol, P., et al. (2000). Study of V(1)-vascular vasopressin receptor gene microsatellite polymorphisms in human essential hypertension. *J Mol Cell Cardiol, 32*(4), 557–564.

Thompson, R. J., Parker, K. J., Hallmayer, J. F., Waugh, C. E., & Gotlib, I. H. (2011). Oxytocin receptor gene polymorphism (rs2254298) interacts with familial risk for psychopathology to predict symptoms of depression and anxiety in adolescent girls. *Psychoneuroendocrinology, 36*(1), 144–147.

Thompson, R. R., George, K., Walton, J. C., Orr, S. P., & Benson, J. (2006). Sex-specific influences of vasopressin on human social communication. *Proc Natl Acad Sci U S A, 103*(20), 7889–7894.

Tops, M., van Ijzendoorn, M. H., Riem, M. M., Boksem, M. A., & Bakermans-Kranenburg, M. J. (2011). Oxytocin receptor gene associated with the efficiency of social auditory processing. *Front Psychiatry, 2*, 60.

Tost, H., Kolachana, B., Hakimi, S., Lemaitre, H., Verchinski, B. A., Mattay, V. S., et al. (2010). A common allele in the oxytocin receptor gene (OXTR) impacts prosocial temperament and human hypothalamic-limbic structure and function. *Proc Natl Acad Sci U S A, 107*(31), 13936–13941.

Tost, H., Kolachana, B., Verchinski, B. A., Bilek, E., Goldman, A. L., Mattay, V. S., et al. (2011). Neurogenetic effects of OXTR rs2254298 in the extended limbic system of healthy Caucasian adults. *Biol Psychiatry, 70*(9), e37–39.

Ukkola, L. T., Onkamo, P., Raijas, P., Karma, K., & Jarvela, I. (2009). Musical aptitude is associated with AVPR1A-haplotypes. *PLoS One, 4*(5), e5534.

Unternaehrer, E., Luers, P., Mill, J., Dempster, E., Meyer, A. H., Staehli, S., et al. (2012). Dynamic changes in DNA methylation of stress-associated genes (OXTR, BDNF) after acute psychosocial stress. *Transl Psychiatry, 2*, e150.

Uzefovsky, F., Shalev, I., Israel, S., Knafo, A., & Ebstein, R. P. (2012). Vasopressin selectively impairs emotion recognition in men. *Psychoneuroendocrinology, 37*(4), 576–580.

Walter, N. T., Montag, C., Markett, S., Felten, A., Voigt, G., & Reuter, M. (2012). Ignorance is no excuse: moral judgments are influenced by a genetic variation on the oxytocin receptor gene. *Brain Cogn, 78*(3), 268–273.

Walum, H., Lichtenstein, P., Neiderhiser, J. M., Reiss, D., Ganiban, J. M., Spotts, E. L., et al. (2012). Variation in the oxytocin receptor gene is associated with pair-bonding and social behavior. *Biol Psychiatry, 71*(5), 419–426.

Walum, H., Westberg, L., Henningsson, S., Neiderhiser, J. M., Reiss, D., Igl, W., et al. (2008). Genetic variation in the vasopressin receptor 1a gene (AVPR1A) associates with pair-bonding behavior in humans. *Proc Natl Acad Sci U S A, 105*(37), 14153–14156.

Wang, K., Zhang, H., Ma, D., Bucan, M., Glessner, J. T., Abrahams, B. S., et al. (2009). Common genetic variants on 5p14.1 associate with autism spectrum disorders. *Nature, 459*(7246), 528–533.

Wassink, T. H., Piven, J., Vieland, V. J., Pietila, J., Goedken, R. J., Folstein, S. E., et al. (2004). Examination of AVPR1a as an autism susceptibility gene. *Mol Psychiatry, 9*(10), 968–972.

Watanabe, Y., Kaneko, N., Nunokawa, A., Shibuya, M., Egawa, J., & Someya, T. (2012). Oxytocin receptor (OXTR) gene and risk of schizophrenia: Case-control and family-based analyses and meta-analysis in a Japanese population. *Psychiatry Clin Neurosci, 66*(7), 622.

Wermter, A. K., Kamp-Becker, I., Hesse, P., Schulte-Korne, G., Strauch, K., & Remschmidt, H. (2010). Evidence for the involvement of genetic variation in the oxytocin receptor gene (OXTR) in the etiology of autistic disorders on high-functioning level. *Am J Med Genet B Neuropsychiatr Genet, 153B*(2), 629–639.

Vogel, F., Wagner, S., Baskaya, O., Leuenberger, B., Mobascher, A., Dahmen, N., et al. (2012). Variable number of tandem repeat polymorphisms of the arginine vasopressin receptor 1A gene and impulsive aggression in patients with borderline personality disorder. *Psychiatr Genet, 22*(2), 105–106.

Wu, S., Jia, M., Ruan, Y., Liu, J., Guo, Y., Shuang, M., et al. (2005). Positive association of the oxytocin receptor gene (OXTR) with autism in the Chinese Han population. *Biol Psychiatry, 58*(1), 74–77.

Yamasue, H., Suga, M., Yahata, N., Inoue, H., Tochigi, M., Abe, O., et al. (2011). Reply to Neurogenetic Effects of OXTR rs2254298 in the Extended Limbic System of Healthy Caucasian Adults. *Biol Psychiatry, 70*(9), e41–42.

Yang, S. Y., Cho, S. C., Yoo, H. J., Cho, I. H., Park, M., Yoe, J., et al. (2010a). Family-based association study of microsatellites in the 5' flanking region of AVPR1A with autism spectrum disorder in the Korean population. *Psychiatry Res, 178*(1), 199–201.

Yang, S. Y., Cho, S. C., Yoo, H. J., Cho, I. H., Park, M., Kim, B. N., et al. (2010b). Association study between single nucleotide polymorphisms in promoter region of AVPR1A and Korean autism spectrum disorders. *Neurosci Lett, 479*(3), 197–200.

Yirmiya, N., Rosenberg, C., Levi, S., Salomon, S., Shulman, C., Nemanov, L., et al. (2006). Association between the arginine vasopressin 1a receptor (AVPR1a) gene and autism in a family-based study: mediation by socialization skills. *Mol Psychiatry, 11*(5), 488–494.

Ylisaukko-oja, T., Alarcon, M., Cantor, R. M., Auranen, M., Vanhala, R., Kempas, E., et al. (2006). Search for autism loci by combined analysis of Autism Genetic Resource Exchange and Finnish families. *Ann Neurol, 59*(1), 145–155.

Young, L. J., & Wang, Z. (2004). The neurobiology of pair bonding. *Nat Neurosci, 7*(10), 1048–1054.

Yrigollen, C. M., Han, S. S., Kochetkova, A., Babitz, T., Chang, J. T., Volkmar, F. R., et al. (2008). Genes controlling affiliative behavior as candidate genes for autism. *Biol Psychiatry, 63*(10), 911–916.

Zeichner, A., Frey, F. C., Parrott, D. J., & Butryn, M. F. (1999). Measurement of laboratory aggression: a new response-choice paradigm. *Psychol Rep, 85*(3 Pt 2), 1229–1237.

Zink, C. F., & Meyer-Lindenberg, A. (2012). Human neuroimaging of oxytocin and vasopressin in social cognition. *Horm Behav, 61*(3), 400-409.

CHAPTER

9 Modeling the Genetics of Social Cognition in the Laboratory

Richard P. Ebstein, Songfa Zhong, Robin Chark, Poh San Lai, *and* Soo Hong Chew

Abstract

This chapter examines recent advances in the genetics of social cognition, discussing evidence from twin studies that confirm the relevance of genetic hard wiring in understanding many social phenotypes, with important implications for the social sciences and for genome-wide association studies (GWAS) that may identify specific genes contributing to a wide range of social phenotypes, genoeconomics, and individual and social decision making. Stressing the importance of phenotype definition and precise measurement as key to success in GWAS, the authors argue that laboratory-based behavioral economic paradigms using ethnically homogenous student populations generate the best prospects for successful GWAS. Also discussed is the neurochemical/neurogenetic architecture of behavioral economic games that measure individual and social decision making and the considerable progress made in unraveling the neurogenetics of human parenting and the beginning of a political attitudes neuroscience. The authors' own GWAS is used to present a set of guidelines for future research directions.

Key Words: decision making, social cognition, behavioral economics, genetics, genoeconomics, neuroeconomics

Human beings are not only predictive machines of conspecific behaviors but also chart their own course with marvelous accuracy through the maze of social relationships that constitute the social world. Our social brains are in tune and often synchronous with the social brains of other humans, allowing us to infer and communicate our and others' intentions to a remarkable degree. We not only infer the actions of other individuals, but we also infer the actions of groups of individuals, allowing us a latitude of behavioral actions surpassing all other species on the planet. Notably, we need to respond appropriately to the intentions and actions of others so that we respond to trust by displaying trustworthiness and are willing to help others in times of need by displaying altruism toward other people. In exchange, we expect others to reciprocate our good acts. Moreover, when we observe acts of unfairness and perceived selfishness, we are willing to punish others at a cost to ourselves with no apparent accruing of personal benefit. We are not, however, perfect, and we also often display considerable envy, spite, and jealousy when observing the good fortune of our compatriots. How do we manage to negotiate the sea of social relationships that surround us, to make the right choices, and to emerge from all these complex encounters in a manner that maximizes our own resources but also allows others to benefit?

The molecular brain mechanisms underpinning our ability to navigate the complex relationships that embrace our environment remain largely unknown. Indeed, a major challenge to neuroscience is to understand the biology of individual and social decision making from genomics to mental constructs. In this postgenomic era when the human genome has been sequenced, we ask how the DNA code translates into the ability to maintain and foster complex decision-making processes that often include the social relationships that so uniquely characterize *Homo sapiens*. This question is the focus of this chapter: an attempt to gain an inkling of understanding of how the sequence of four bases in DNA partially account not only for dyadic human relationships but for the very structure of human societies across the globe. Human beings, despite their ability to learn from parents, teachers, and peers and to adjust their behaviors during a long developmental period from birth throughout adolescence, are not, however, a tabula rasa. To a not inconsiderable extent, we are hardwired toward achieving a level of individual and social decision-making skills that allows us to interact successfully with conspecifics, to find a mate, to parent our offspring, and to achieve our goals even in a complex, globalized economy. When these skills are inadequate due to both environmental and genetic deficits, human beings fail to function effectively, and such deficits in decision-making processes lead to considerable cost to individuals and society as a whole.

This review first discusses the overall evidence from twin studies that heredity significantly contributes to human variation in how individual and social decisions are made. We then move on to discuss current research strategies (candidate genes and genome-wide associations) employed to identify specific genes associated with individual and social phenotypes. Most importantly, we discuss novel approaches to defining decision-making phenotypes and how such phenotypes can be measured by the paradigms from behavioral and experimental economics. Of particular interest are the phenotypes relevant to other-regarding or social behaviors such as altruism, trust, sense of fairness, and competitiveness. Social psychology focuses on parenting, mate selection, and pair bonding, and we give due mention to some recent advances in understanding their genetics.

Genetics has recently made some notable incursions into the world of business and management. Heredity partially explains individual differences in financial risk-taking behavior, leadership qualities, and even consumer behavior. Moreover, specific genes have been provisionally identified for some of those phenotypes. In the meantime, political science is also coming under the influence of genomics, and political attitudes (liberal vs. conservative) as well as the nature of society itself (individualistic vs. collectivist) are being scrutinized using the tools of molecular genetics. Indeed, we suggest that the social sciences (economics, political science, and sociology) could be informed by the genomics revolution in the life sciences, and much can be learned about human behavior, both singularly and aggregately, by incorporating genes. That being said, we are not suggesting a *reductio ad absurdum* scenario whereby human beings are slaves to their genes and genetic heritage. But, similar to all other species on the planet, our behaviors have evolved, and genetic mechanisms have been selected for that enable us to efficiently function as a species exquisitely fine-tuned to render decisions that maximize our fitness as both individuals and social creatures designed to interact with conspecifics. These genetic mechanisms can be studied, and we believe this study will inform a deeper understanding of the nature of man and his relationship to society.

I. Twin Studies

A. *Estimating Heritability*

Heritability is the proportion of the phenotypic variance in the population that is attributed to genetic variation. Heritability is usually studied by comparing monozygotic (MZ, identical) and dizygotic (DZ, fraternal) twins. Extensive twin studies leave little doubt that genes make a significant contribution to the molding of the human social brain and that heredity explains a good deal of the differences between people. Quantitative behavioral studies can estimate the proportion of variance in a trait due to genetic (heritability) and shared and nonshared environmental factors. Shared environmental factors are those influences that, for example, make children in the same family similar to each other, presumably due to parental and home influences. Nonshared environmental factors are what make children in the same family different from each other. The main way to study this is to examine MZ and DZ twins. Monozygotic twins share their entire DNA, whereas DZ twins share on the average 50 percent of their DNA. If MZ twins are more similar than DZ, there is evidence for a role for genes in determining the trait. If MZ twins are not exactly the same, there is evidence for nonshared environment.

For prosociality (Knafo & Plomin, 2006) and empathy (Knafo, Zahn-Waxler, Van Hulle, Robinson, & Rhee, 2008; Rushton, 2004), church attendance and religiosity (Kendler & Myers, 2009), propensity to marriage (Johnson, McGue,

Krueger, & Bouchard, 2004), values (Knafo & Spinath, 2011; Schermer, Vernon, Maio, & Jang, 2011), and parenting style (Harlaar et al., 2008; Kendler, 1996; Neiderhiser et al., 2004), it is shown that genes clearly matter (Ebstein, Israel, Chew, Zhong, & Knafo, 2010). Genes also clearly matter for individual decision making, such as in attitudes to financial risk taking (Cesarini, Dawes, Johannesson, Lichtenstein, & Wallace, 2009; Le, Miller, Slutske, & Martin, 2010; Zhong, Chew et al., 2009).

B. Social Attitudes to Abortion and Gay Rights

Just how informative the twin method can be in uncovering the role of genes in shaping our social brains is especially revealed when this approach is used to estimate the contribution of heredity to attitudes at the heart of many sociological and political science studies. Religiosity and attitudes to politically "hot" topics such as abortion and gay rights are potent topics in the news, and quantitative behavioral analysis is a powerful tool toward understanding individual differences. Longitudinal twin studies show that the variance in religiosity during the teen years, as measured by church attendance as well as by more detailed pencil-and-paper questionnaires, is largely due to shared environment. However, as we age, the role of genetic factors increases, and the role played by the shared environment shrinks to just about zero (Kendler & Myers, 2009; Koenig, McGue, & Iacono, 2008). Altogether, as individuals mature and become adults, they moderate their social milieu mostly due to their genes. As children and teens, parents have a greater influence on religiosity, and genes take a back seat; but their influence emerges strongly in early adulthood to explain a substantial portion of the variance in variables such as church attendance, including elements of prayer and faith.

Twin studies also identify genetic and environmental interactions (G×E interactions), and a good example is the moderating effect of religiosity on smoking and alcohol use (Kendler, Gardner, & Prescott, 1997; Kendler & Myers, 2009; Koopmans, Slutske, van Baal, & Boomsma, 1999; Timberlake et al., 2006). High levels of religiousness were found to attenuate the additive genetic component for smoking initiation and were associated with lower prevalence of smoking initiation. Genetic influences accounted for 40 percent of the variance in alcohol use initiation in nonreligious females, compared to 0 percent in religiously raised females; shared environmental influences accounted for 54 percent of the variance for nonreligious females and 88 percent of the variance in religious females (Koopmans et al., 1999).

In the United States, attitudes toward abortion and gay rights continue to play a salient role in local and national politics, and twin studies provide important insights into how such opinions are formed and how difficult they may be to change. The traditional social science model attributes similar family attitudes to these loaded issues to be a result of familial socialization (Legge, 1983) or psychological/societal factors (Herek & Capitanio, 1996). In an intriguing study, Eaves and Hatemi (2008) modeled not only the social transmission of attitudes toward abortion and gay rights, but also, notably, the biological basis for these beliefs in a large sample of adult twins, siblings, and their parents (nuclear family). Their results showed a surprisingly low nongenetic effect of parental attitudes ("vertical cultural inheritance") on the development of adult values in their children, whereas there was a substantial role of genes (50–70 percent) overall in the transmission of attitudes toward both abortion and gay rights. Overall, social learning from the parental attitude accounted for less than 5 percent of the total variance.

Notably, it is the choice of mate (apparently assortative mating and not spousal interaction) that has an important impact on polarization of social values, consistent with research in the social sciences, and that attests to primary phenotypic assortment (spousal correlations) for attitudes on many salient social issues. Indeed, it is the social structures involving mate selection that may represent the mechanism by which parents impact their children's long-term attitudes toward social issues such as gay rights or abortion. The choice of mate is an example of gene–environment correlation (rGE). If genes are important in determining the traits (i.e., attitudes toward abortion and gay rights) on which assortment occurs, then the choice of mate is a form of rGE in which parental genes impact the social behavior and preferences of their offspring. Remarkably, by this scenario, a social act (choosing a mate) increases the genetic impact of parents on their children.

C. Hard and Soft Social Issues

Interestingly, in the social sciences, certain political issues are "hard" (e.g., educational reform) and most often require evaluation and reflection, whereas other attitudes, including abortion and gay rights, are "soft" or "easy" and evoke an instant "gut feeling" in eliciting particularly strong opinions. Tesser (1993) provides evidence that attitudes with stronger genetic influences are manifested more quickly, are more stable, and are more likely to lead to assortative mating. Attitudes toward abortion and gay rights appear to fall into this category, suggesting that they are inherently more difficult to change and likely resonate more strongly with holders of more polarized opinions. This being said, it is worth noting that attitudes toward these issues are not written in stone, and there have been marked changes in public opinion toward gay rights and abortion in the past half-century. These considerations underscore an important feature of heritability of complex traits: the influence of time, culture, and specific contexts. Heritability is a dynamic concept, shifting slowly and reflecting often-unknown tides of change. Examples illustrating this somewhat counterintuitive aspect of heritability are provided by studies examining the significant gene contributions to height (Yang et al., 2010) and IQ (Flynn, 2000). Although measurements of these two phenotypes indicate significant recent changes, they clearly do not evidence evolutionary selection. The realization that people's attitudes on social and political issues are formed and maintained to some degree by "hardwiring" and are encoded in our DNA has important implications for understanding current political trends, especially in countries such as the United States where these issues often influence elections at a national level. The subject of same-sex marriage is one of the most volatile political issues in the United States (Sherkat, Powell-Williams, Maddox, & de Vries, 2011), and it is even more remarkable that attitudes toward this issue are, to a not inconsiderable extent, heritable. As stated in a recent article on this subject "there is substantial, entrenched, opposition to same-sex marriage, and opposition to same-sex-marriage in the younger cohorts is rooted strongly in religious and political identifications. Values are changing, but conflicts over same-sex-marriage are not going away any time soon." (p. 177). We suggest the notion that one key concept needed toward understanding future changes in attitudes toward gay rights is the realization that some of the variance in such opinions is genetically determined and hence perhaps resistant to easy modification.

D. *What Twins Tell Us About the Epigenome*

The current wave of intense study of the epigenome and its implications for understanding complex behaviors has presented a unique opportunity to use twin studies, particularly of MZ twins (see Bell & Spector, 2011, for a current review). Epigenetic studies in twins can answer some important questions. First, is there epigenetic heritability across the genome? Second, to what extent do epigenetic patterns truly contribute to complex traits? Simply put, any phenotypic differences within MZ twin pairs presumably are a result of different environmental exposure between twins. Epigenetic study of traits that are discordant in MZ twins is a powerful research strategy for understanding disease- or trait-specific epigenetic changes. Moreover, it has been suggested that epigenetics is perhaps the key to understanding the heritability of complex traits (Petronis, 2010).

An epigenetic MZ/DZ twin study (Kaminsky et al., 2009) suggests that MZ methylation patterns are more similar in DZ twins across tissues. Interestingly, the most heritable CpG sites were correlated with functional regions and promoters, suggesting that functionally relevant methylation patterns are under strong genetic control. In an intriguing study by Fraga et al. (2005), it was observed that twins are epigenetically indistinguishable during the early years of life but, as they age and their lifestyles diverge, MZ twins exhibit differences in their epigenetic signatures that affect their gene expression portrait. A recent *Science* article (Freund et al., 2013) underscores the role of environment in shaping behavior. By studying inbred mice in an enriched environment, the authors showed that factors unfolding or emerging during development contribute to individual differences in structural brain plasticity and behavior, including exploratory behavior and hippocampal neurogenesis.

II. Association Studies

A. *Candidate Gene Approach*

Human behavioral genetics studies have, in the past two or more decades, employed several strategies in an effort to identify genes associated with human traits. The most obvious strategy is the so-called *candidate gene approach* that tests polymorphic genes hypothesized to contribute vulnerability to particular diseases or traits. A good example are genes that encode dopamine receptors that, based on the dopamine hypothesis of schizophrenia (Snyder, 1976), were hypothesized to be likely candidates for conferring vulnerability to schizophrenia. Treatment of schizophrenia depends on blocking the DRD_2 receptor, making this gene a prime target for association studies. Yet, genetic studies only weakly support such a contention. Indeed, a recent functional genomics study did not flag DRD_2 as a likely candidate contributing to schizophrenia, albeit other top candidates did emerge using this innovative design (Ayalew et al., 2012).

The failure to replicate many candidate gene studies has been extensively commented on (Ioannidis, Ntzani, Trikalinos, & Contopoulos-Ioannidis, 2001; Ioannidis, Trikalinos, Ntzani, & Contopoulos-Ioannidis, 2003). Many reasons explain the difficulty in replicating those first studies. Complex traits reflect a large number of genetic and environmental factors and their interactions. Common alleles contributing to such traits are probabilistic and not deterministic in comparison to rare Mendelian disease alleles. Hence, such alleles only slightly contribute to the traits they are associated with. Moreover, to identify specific genes associated with complex traits, a phenotype with at least moderate heritability is a prerequisite.

Perhaps the most important reason for poor replication of candidate gene studies is limited sample size, which increases the chance of false-positive results. Small studies overestimate the true effect of gene variants. As Ioannidis et al. have observed (2003), the magnitude of the genetic effect differed significantly in large versus smaller studies, and the largest studies generally yielded more conservative results. Only in nine (16 percent) meta-analyses was the genetic association significant and replicated without hints of heterogeneity or bias. Lest we think that these problems of replication are somehow unique to candidate gene studies, the reader is referred to an intriguing discussion suggesting that most findings across biomedicine are likely false (Ioannidis, 2005), and similar issues have been recently raised in social psychology (Carpenter, 2012; Simmons, Nelson, & Simonsohn, 2011).

Other reasons for failure to replicate (see Li & Meyre, 2012, among many reviews) include lack of correction for multiple testing; population stratification; genotype misclassification; inappropriate statistical analyses; gene × gene (G×G) and G×E interactions; genetic, phenotypic, and ascertainment heterogeneity; inheritance models; and subjective interpretation of data (the Proteus phenomenon; Ioannidis & Trikalinos, 2005; Pfeiffer, Bertram, & Ioannidis, 2011).

Despite the bad reputation now attached to candidate gene studies, we suggest that there is a danger of throwing out the baby with the bath water because there have been some important success stories using the candidate gene approach. A good example is the role of the dopamine D_4 receptor (DRD4) variable number tandem repeat (VNTR) in contributing vulnerability to attention deficit hyperactivity disorder (ADHD; Smalley et al., 1998). In the first study of *DRD4* and ADHD, the number of subjects examined was less than 100. Nevertheless, despite the odds of this finding being a false positive, it has been subsequently replicated, and meta-analyses attest to its robust association with ADHD (Faraone, Doyle, Mick, & Biederman, 2001; Gizer, Ficks, & Waldman, 2009; Ptacek, Kuzelova, & Stefano, 2011).

Another good example of a successful candidate gene strategy is the association between *SORL1* and Alzheimer's disease (Lee et al., 2007; Reitz et al., 2011; Rogaeva et al., 2007). The generation of the Aβ toxic neuropeptide derived from the amyloid precursor protein (APP) via the endocytic pathway is a main cause of late-onset Alzheimer's disease (AD). This concept is supported by prior reports that the expression of several candidate proteins within these pathways (such as SORL1 and VPS35) is reduced in brain tissue from individuals with AD, and reductions in the expression of some of these proteins are associated with increased Aβ production. However, it was unclear whether these changes were causal or simply reactive to AD. Based on these considerations, *SORL1* association with AD was tested (Rogaeva et al., 2007). The samples from six datasets were divided into a "discovery cohort" composed of families with late-onset familial AD (FAD) and a "replication cohort" composed of discordant sibships and collections of individuals with AD and normal controls matched for age, gender, and ethnic origin. Association was demonstrated between *SORL1* single-nucleotide polymorphisms (SNPs) and haplotypes with AD; importantly, this finding was subsequently confirmed in a later meta-analysis (Reitz et al., 2011). The *SORL1* story underscores the usefulness and value of a well-designed candidate gene approach in two ways. First, the *SORL1* success story is based on robust evidence regarding the molecular basis of the disease itself and, second, on expression studies that considerably strengthened the hypothesized role for the candidate gene in the pathology of AD.

B. *Imaging Genomics*

So-called *imaging genomics* (Hariri & Weinberger, 2003; see also the chapter by Hyde, Bogdan and Hariri in this handbook) is making an important contribution to understanding the action of common polymorphisms at the neural level and the molecular mechanisms underlying variance in complex behaviors. Many of these functional polymorphisms will impact information processing in the brain. Functional neuroimaging, including positron emission tomography (PET), functional magnetic resonance imaging (fMRI), and electroencephalography/magnetoencephalography (EEG/MEG), because of its ability to individually evaluate information processing in distinct brain regions, is uniquely placed as a strategy for explaining aspects of functional genomics in the brain. One of the first demonstrations of the power of this approach was a seminal study by Egan et al. (2003) who showed that a common *BDNF* polymorphism, which predicts secretion of this hormone, mediates aspects of human memory and hippocampal function.

There are many other examples of how a neurogenetics approach coupled with brain imaging can explain the neural basis of complex traits and the underlying molecular architecture. One focus of the current review is social decision making, and two peptide hormones—oxytocin (OT) and arginine vasopressin (AVP)—play key roles in social information processing, not only in voles but also in humans (Insel, 2010; Young, Lim, Gingrich, & Insel, 2001). Indeed, a number of candidate gene studies suggest provisional association between the oxytocin receptor (*OXTR*) gene and facets of the social brain (Ebstein, Knafo, Mankuta, Chew, & Lai, 2012). Among the *OXTR* SNPs tentatively identified with social traits, one of these, rs2254298 (G>A), has also been associated with differences in amygdala volume (Furman, Chen, & Gotlib, 2011). Similarly, the arginine vasopressin 1A receptor (AVPR1a) promoter-region polymorphism, also associated with normal and abnormal human social behavior, predicts amygdala activation (Meyer-Lindenberg et al., 2009; for a detailed discussion of OXTR and AVPR1a in voles, see the chapter by Barrett and Young, and in humans, by Westberg and Walum, elsewhere in this handbook).

Overall, a number of imaging genetics studies have demonstrated how amygdala activation is modulated by different genes that agree with their neurochemical relationship with the amygdala (Blasi et al., 2009; Canli & Lesch, 2007; Drabant et al., 2006; Fakra et al., 2009; Hariri & Whalen, 2011; Manuck et al., 2010; Rasch et al., 2009; Roiser et al., 2009; Smolka et al., 2005).

The imaging genomics strategy can be especially important in confirming gene associations and is a great aid in strengthening the robustness of provisional findings. A good example is the story of *ANK3*, which has been implicated in genome-wide association studies (GWAS) of bipolar disorder (Chen et al., 2011) and schizophrenia (Ripke et al., 2011). *ANK3* belongs to a family of proteins (Ankyrins) that are involved in cellular functions such as motility, activation, proliferation, and the maintenance of specialized membrane domains. In an innovative study, Roussos et al. (2012) found that mRNA expression of multiple nodes of Ranvier-expressed genes was decreased in schizophrenia. Specifically, the *ANK3* rs9804190 C allele was associated with lower *ANK3* mRNA expression levels and with higher risk for schizophrenia in the case–control cohort. It was also associated with cognitive impairments with respect to working memory and executive function, as well as with increased prefrontal activation during a working memory task in healthy individuals. Altogether, such a combined strategy goes beyond "mere" GWAS and offers a path to confirming promising gene associations with complex disorders.

III. Genome-Wide Association Studies

A. *Discovery Through Sample Size*

The advent of GWAS in the past 5 or more years has led to an enormous acceleration in the discovery of genes contributing to complex traits, and more than 1,000 loci have been convincingly associated. Importantly, the establishment of international consortiums generating large combined cohorts enables the identification of variants with modest effect sizes (odds ratio [OR] <2). For example, it is remarkable that it has also been possible to carry out a GWAS of hippocampal volume on approximately 9,000 dementia-free subjects, and at least two SNPs clocked in at p values of 10^{-11} (Stein et al., 2012). This study was leveraged by combining the resources and findings from ten independent groups across three continents.

In schizophrenia research, the recent meta-analysis carried out by the Schizophrenia Psychiatric GWAS Consortium (Ripke et al., 2011) is a mega-study looking at approximately 22,000 European schizophrenic subjects in the discovery sample (17 separate studies) and a stage-two replication sample of close to 30,000 patients—a remarkable achievement. The combined analysis showed GWAS significance levels for seven loci. It is important to emphasize that the individual cohorts collected in many countries were relatively modest in scope, and it was only the combined sample that reached such mega proportions.

B. Missing Heritability

One of the mysteries of the GWAS revolution is that, for most traits that have so far been studied, the associated SNPs from GWAS only appeared to explain a very small fraction of the heritability—the enigma of the so-called *missing heritability* (Maher, 2008; Manolio et al., 2009). A number of possibilities have been suggested to explain the missing heritability. Among these suggestions, the most likely is that complex traits like height are characterized by a large number of common polymorphisms, each only explaining a very small percent of the variance. Other explanations include a few rare variants with large effects and copy number variations. In a recent study, Yang and his colleagues (Yang et al., 2010; Yang, Lee, Goddard, & Visscher, 2011) implemented a method for estimating the total amount of phenotypic variance due to variance of all SNPs on the chip array and found that, for height, almost 45 percent of the variance could be explained by all these common variants, in comparison to 90 percent heritability in twin studies and only a few percent heritability when only individual SNPs producing significant GWAS levels were considered (the difference between the heritability estimates from twin studies and the heritability explained by these "hit" SNPs is the so-called missing heritability). As the authors noted, most of the missing heritability is, in fact, there but "hiding." A similar observation has been reported for IQ (Davies et al., 2011). It should be noted that, as discussed earlier, the classical estimation of heredity by twin studies is about the influence of all the SNPs, whereas the method used by Yang et al. focuses on the influence of all the SNPs covered in the GWAS chip array, which offers an alternative and complementary approach to measuring the contribution of genes to complex traits. Emerging technologies such as next-generation sequencing (NGS) are also offering unprecedented opportunities to detect an entire spectrum of genetic variants in individuals at the whole-genome level (Altshuler et al., 2010; Shendure & Ji, 2008; The 1000 Genomes Project Consortium 2010, 2012). Most rare SNP associations with medium to large effects may be missed by current GWAS methods, and multiple rare SNPs (frequency of less than 1 percent), together with thousands of common small-effect variants, may play a bigger role than previously thought in accounting for this "missing heritability" in complex traits (Dai, Jiang, & Dong, 2012; Siu, Zhu, Jin, & Xiong, 2011).

In a recent review, Marian (2012) noted further important sources that could account for missing heritability, including G×G and G×E interactions. Moreover, since SNPs identified in GWAS are often not the actual functional alleles but merely in linkage disequilibrium (LD) with the risk alleles, such "hits" tend to underestimate the true heritability (Spencer, Hechter, Vukcevic, & Donnelly, 2011). This is because LD is very rarely 100 percent. Some of the missing heritability may be due to such epistatic effects. Lander and his colleagues have termed this "phantom heritability" (Zuk, Hechter, Sunyaev, & Lander, 2012).

Unfortunately, G×G interactions (epistasis) are difficult to estimate from GWAS. Modeling gene–gene interaction is very complex. Even for the simplest case, where gene–gene interaction is restricted to SNP–SNP interaction (or two-locus interaction), the number of interaction models is huge. Second, whole-genome screening is very time-consuming. For example, exhaustive investigation of SNP–SNP interaction leads to 1.25 × 1,011 statistical tests for datasets from 500 k chips. Nevertheless, new statistical treatments are being developed to overcome these difficulties (see, e.g., Emily, 2012).

One strategy in examining the role of epistasis in human social behavior can be implemented in studies of lower mammals. A good example of this approach is demonstrated in a quantitative trait locus (QTL) genetic analysis of nest building in mice (Sauce, de Brito, & Peripato, 2012). Nest building is a complex social trait in mice important for fitness and survival of pups. Altogether, 15 direct-effect QTLs were found, explaining from 4 to 13 percent of the phenotypic variation, mostly with nonadditive effect. Notably, 71 significant epistatic interactions were discovered, accounting for 28.4 to 75.5 percent of the variation. Thus, the genetic contributions to nest-building behavior appear to reflect primarily epistatic processes. Indeed, the authors suggest that fitness-related phenotypes like nest-building reflect a genetic architecture with small direct effects and a larger number of epistatic interactions.

research to help investigators assemble, visualize, and ultimately test multidimensional phenomics hypotheses. It is important to realize that phenotyping has become the key rate- and cost-limiting factor in human genetics. Importantly, phenotypes need to be assessed in a well-operationalized fashion. Moreover, wherever possible, phenotypes should be defined from the perspective of a relevant biological context. It is also notable that many phenotypes used in the social as well as the medical sciences do not align readily with genetic determinants because such phenotypes were not conceptualized with genetics in mind. This would certainly include many survey questions in the social sciences that are very distant from any underlying biology and, hence, we suggest, are particularly difficult to match to genes. On the other hand, the phenomics strategy explicitly calls for efforts to redefine phenotypes as multilevel combinations of measures that may offer more realistic constraints on the mechanistic paths leading from genome to syndrome. For phenotypes of individual and social decision making, one suggestion that is worth examining is to extract, for example, a combined "prosociality" or "risk attitude" index based on multiple game assessments.

In considering the desiderata of a good phenotype, measurement properties are of paramount importance (Bilder et al., 2009). Phenotypes need to show not only high internal consistency (e.g., Cronbach's α) but also ensure that a meaningful and coherent construct is being measured to begin with. Phenotypes should be stable over time and organismic states. Moreover, the sensitivity of the phenotype measurements should be strong across widely varying levels of phenotype expression. Another consideration is how well the phenotype is suitable to high-throughput phenotyping because highly informative, intensive phenotyping now exceeds the cost and time of genotyping. For example, a "comprehensive" examination of cognitive abilities may require 10–20 hours, and briefer assessments routinely fail to provide broad phenotypic coverage, fail to specify adequately the phenotypes of interest, or both. Whether more cost-effective phenotyping procedures (e.g., internet-based) can preserve the needed information for genetic analysis remains to be seen.

VII. Schizophrenia Consortium

The work of the International Schizophrenia Consortium illustrates what we believe is the road to take in identifying genes contributing to the molecular genetic architecture of complex traits characterizing individual and other-regarding decision making. First, it is crucial that a meticulous definition of the phenotype be employed, with no shortcuts. Otherwise, there is a danger of a garbage-in garbage-out GWAS based on poorly constructed and biologically distant and implausible phenotypes that are badly matched at the molecular level. Such a phenotype-weak study is likely to lead to false-negative findings and an overly pessimistic view that impedes future progress. Second, it is crucial to assemble individual groups of investigators willing to contribute to the larger consortium, yet with each one/group toiling in their own "backyard" with well-documented phenotypic definitions and ascertainment procedures in place. Moreover, such individual groups provide a cultural and ethnic diversity to the ultimate GWAS meta-analysis so that results can be generalized across populations. Finally, these individual studies are pooled to attain the power needed for finding genes of moderate to small effect sizes characteristic of complex traits. Notably, in the psychiatric genetics field, potential candidate genes flagged in GWAS are often further authenticated by drawing on results from imaging genomics, expression studies, and pathway analyses, as was carried out so elegantly by Ayalew et al. (2012) and is discussed in the next paragraph.

A beautiful example of a novel approach in the field of complex behavioral phenotypes is the recent article by Ayalew and his colleagues (2012). They used a so-called *translational convergent functional genomics* (CFG) design to identify and prioritize genes involved in schizophrenia by gene-level integration of GWAS data with other genetic and gene expression studies in humans and animal models. Notably, they showed how the top candidate genes identified by their approach can be used to generate a genetic risk prediction score (GRPS) to assist in schizophrenia diagnosis. The first step in their analysis was to comprehensively identify candidate genes, pathways, and mechanisms for schizophrenia, integrating all available evidence. They used data from GWAS investigations and integrated those data with gene expression studies from relevant animal model brain and blood gene expression data. The second step was to integrate, as part of this comprehensive approach, other genetic data—human genetic data (linkage, copy number variant [CNV], or association), as well as relevant mouse model genetic evidence.

VIII. Decision Making

A. *Individual Decision Making: Risk Attitude*

Recently, genoeconomics (Navarro, 2009) has gained traction with the publication of a number of candidate gene studies using behavioral economic paradigms as QTLs (Apicella et al., 2010; Avinun et al., 2011; Chew, Ebstein, & Zhong, 2012; Dreber & Apicella, 2009; Eisenegger et al., 2010; Israel et al., 2009; Kuhnen & Chiao, 2009; McDermott, Tingley, Cowden, Frazzetto, & Johnson, 2009; Roe et al., 2009; Zhong, Chark, Ebstein, & Chew, 2012; Zhong, Israel, Xue, Ebstein, & Chew, 2009; Zhong, Israel, Xue, Sham, et al., 2009). The first such molecular genetic investigation employing a classical behavioral economic phenotype, the dictator game, was carried out by us and is discussed in a later section (Knafo, Israel, et al., 2008).

The studies to date of economic risk employed only a small number of subjects and must be considered exploratory and provisional. Nevertheless, they do add, albeit incrementally, to our knowledge of the neurochemical substrate underlying economic risk attitude. One of the first studies was by Dreber et al. (Dreber & Apicella, 2009; Dreber et al., 2010). The study was quite small and included only 94 Harvard undergraduates who were administered an investment game with real monetary payoffs (Gneezy & Potters, 1997). However, the choice of a candidate gene was hypothesis-driven, based on more than a decade of evidence showing in some (but not all) studies a relationship between the dopamine D_4 receptor exon 3 VNTR and novelty seeking, risky behavior, and impulsivity (Ebstein, 2006; Ebstein et al., 1996; Munafo, Yalcin, Willis-Owen, & Flint, 2008). Inter alia, it is worth noting that although the evidence for a role of this gene in the personality trait of novelty seeking is mixed, some imaging studies support the involvement of DRD4 in novelty-related brain events (Marco-Pallares, Nager, et al., 2010; Strobel et al., 2004). Additionally, dopamine pathways are central to brain reward mechanisms, and the DRD4 gene has been implicated in magnitude-related reward effects in the anterior insula and cingulate cortex (Camara et al., 2010). As predicted, Dreber et al. found that carriers of the 7 repeat invest more money in a financial risk game relative to noncarriers; in contrast, there was no association as a function of allelic genotype in the DRD2 gene. Kuhnen and Chiao (2009) reported similar results for 65 subjects when they reported that DRD4 7-repeat carriers took 25 percent more risk than did noncarriers. They also observed that subjects who carried the serotonin transporter promoter-region (5-HTTLPR) short allele were less risk-taking, which is consistent with a large literature suggesting a role of this polymorphism in anxiety-related personality traits (Canli & Lesch, 2007). However, two studies failed to replicate the association between DRD4 and risk taking (Carpenter, Garcia, & Lum, 2011; Frydman, Camerer, Bossaerts, & Rangel, 2011). Although the Carpenter et al. study seems contrary to the conclusions about risk reported in Kuhnen and Chiao (2009), Dreber et al. (Dreber & Apicella, 2009), and Frydman et al. (2011), the risk tasks used in those studies were all somewhat different, perhaps explaining the inconsistent results obtained. In the Kuhnen and Chiao (2009) and Dreber et al. (Dreber & Apicella, 2009) studies, the tasks were actually more similar to the loss task in the Carpenter et al. (2011) study, and, for loss, a marginally significant effect was observed for DRD4 7R. Whether there is a true association between DRD4 and risk taking will become clearer as more evidence accumulates for meta-analysis. The next study implemented a pharmacogenetic approach toward evaluating the role of the DRD4 gene in risky behavior. Eisenegger et al. (2010) used a gambling task and demonstrated an interaction between L-DOPA administration (that increases dopamine [DA] release in the brain) and the DRD4 7-repeat. They observed increased gambling in subjects who carried the 4/7 genotype and who received L-DOPA, but not in those who received L-DOPA and who carry the 4/4 genotype. In this study, introducing a pharmacological arm helped sharpen the distinction between 4/4 and 4/7 carriers, thus underscoring the power of pharmacogenomics as a strategy in association studies.

Another well-characterized, widely studied, and functional polymorphism that has been associated with financial risk is the monoamine oxidase-A (MAOA) promoter region repeat polymorphism (Frydman et al., 2011; Zhong, Israel, Xue, Ebstein, et al., 2009). Our group (Zhong, Israel, Xue, Ebstein, et al., 2009) observed subjects' incentivized choices in a controlled laboratory setting and investigated their task performance in association with MAOA. We found that subjects with the high MAOA activity (4-repeat) allele were characterized by a preference for riskier choices (i.e., in favor of the long-shot lottery and also less insurance purchasing) compared to subjects with the low MAOA activity (3-repeat) allele. Our study was the first to link attitude toward long-shot risks to a specific gene. It complements recent findings on the neurobiological basis of economic risk taking discussed earlier. In a second study, the authors (Frydman et al., 2011) followed the strategy first developed by our group (Avinun et al., 2011; Israel et al., 2009; Knafo, Israel, et al., 2008; Zhong, Israel, Xue, Ebstein, et al., 2009) and combined methods from neuroeconomics and behavioral genetics to investigate the impact that the genes encoding for MAOA, 5-HTTLPR, and DRD4 have on these two computations. Consistent with previous literature, and specifically our own study, they found that carriers of the MAOA 4-repeat polymorphism were more likely to take financial risks. The computational choice model, rooted in established decision theory, showed that 4-repeat carriers exhibited such behavior because they are able to make better financial decisions under risk and not because they are more impulsive. In contrast, and as mentioned earlier, they found no behavioral or computational differences among the 5-HTTLPR and DRD4 polymorphisms.

The study by Roe et al. (2009) found an association between two SNPs in the *VMAT2* gene and risk attitude using a behavioral economic paradigm but failed to find association with the *MAOA* VNTR. However, the study included only 67 subjects of mixed ethnicity and some nonstudents.

It is of interest to note that some of the VNTRs (MAOA, DRD4, DAT1, and 5-HTTLPR) just discussed are not genotyped in the usual GWAS that rely on mass SNP arrays. Hence, it behooves the careful investigator, especially in the nascent field of genoeconomics, to examine these polymorphisms using more "custom" genotyping.

B. *A Neurochemical Model of Risk*

Prospect theory proposes that people have diminishing sensitivity in valuing increases in the size of monetary outcomes for both gains and losses. For decision making under risk, this implies a tendency to be risk tolerant over losses while being generally risk averse over gains. In a first study of its kind, our group suggested a neurochemistry-based framework for understanding an economic model for PT—specifically, the diminishing valuation sensitivity hypothesis. We propose that DA tone modulates the sensitivity toward valuation of gains, whereas serotonin (5-HT) tone modulates the sensitivity toward valuation of losses. Consequently, higher DA tone would yield a more concave valuation function over gains, whereas higher 5-HT tone would yield a more convex valuation function over losses. Using a neurogenetics strategy to test our neurochemical model (Zhong, Israel, Xue, Sham, et al., 2009), we found that subjects with the 9-repeat allele of DAT1 (lower DA tone; van Dyck et al., 2005) dopamine transporter polymorphism are more risk-tolerant over gains than are subjects with the 10-repeat allele, and that subjects with the 10-repeat allele of STin2 (higher 5-HT tone; Hranilovic et al., 2004) are more risk tolerant over losses than are subjects with the 12-repeat allele. Overall, our results support the implications of our model and provide the first neurogenetics evidence that risk attitudes are partially hardwired in differentiating between gain- and loss-oriented risks (see Figure 9.1).

C. *Imaging Genomics of the Neurochemical Model of Risk*

Our group was one of the first to use a neurogenetic strategy/behavioral economic paradigm combined with functional imaging to reveal the underlying neurochemical pathways of risk attitude (Zhong, Chark, et al., 2012) to test the neurochemical model just described. We found enhanced striatal activation responding to increases in the magnitude of utility for risks over gains and to increases in the magnitude of disutility for risks over losses, whereas increased amygdala activation correlates only with the disutility for risks over losses. Stratifying brain activation by genotype, we found that DAT1 contributes to individual differences in striatal response for gain-oriented risks, whereas a polymorphism in the 5-HT transporter (STin2) partially accounts for individual differences in amygdala responses for loss-oriented risks. No association was observed for COMT and 5-HTTLPR in either region.

Together, our results suggest the role of the amygdala and corresponding serotonergic pathway in evaluating losses. This further corroborates the hypothesis of 5-HT being linked to DA in an "opponent partnership" (Daw, Kakade, & Dayan, 2002) (see Figure 9.2).

IX. Other-Regarding or Social Decision Making

Other-regarding preferences can be measured in the laboratory using "classical" behavioral economic paradigms. The reader is referred to the excellent article by Fehr and Camerer (2007) for a detailed discussion of behavioral economic paradigms

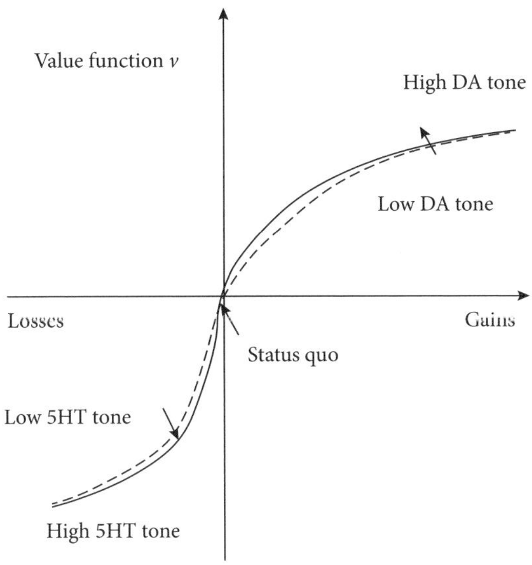

Fig. 9.1 Figure from Zhong, S., Israel, S., Xue, H., Ebstein, R. P., & Chew, S. H. (2009). Loss–gain differentiation. Prospect theory assumes a loss-averse value function that is concave over gains, convex over losses, and vanishes at the status quo, represented by 0. Our hypothesis is that DA (5-HT) tone modulates sensitivity toward incremental gain (loss) and the higher the DA (5-HT) tone, the lower (higher) the sensitivity toward incremental gain (loss).

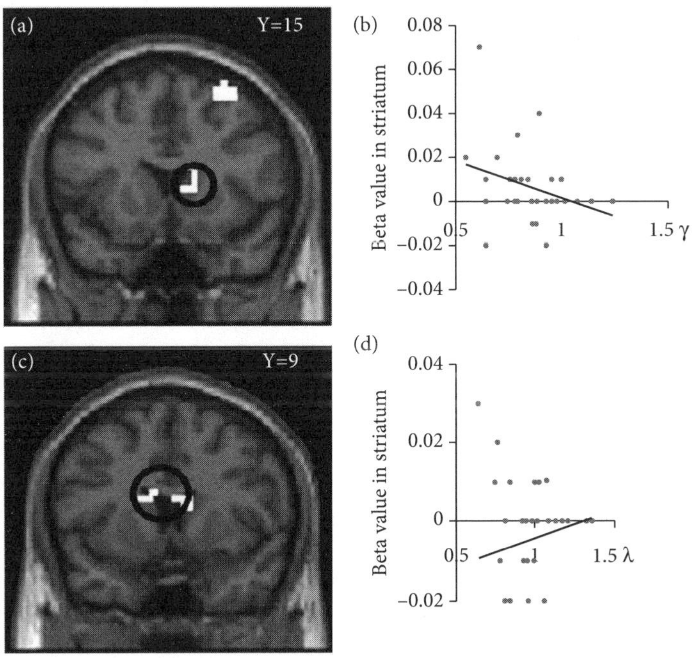

Fig. 9.2 Striatal activation correlates with utility for prospects. Figure from Zhong et al. (2012). (a) Coronal section showing the striatum whose activation correlates significantly with increases in the magnitude of the utility for prospects. (b) The β value of striatum activation is significantly negatively correlated with the index γ of the utility function over gains, in which lower values of γ represent greater degrees of risk aversion. (c) Coronal section of striatum whose activation correlates significantly with increased magnitude of utility for hazards. (d) The β value of striatum activation is not significantly positively correlated with the index λ of the utility function over losses, in which higher values of λ represent greater degrees of risk aversion.

widely used in measuring social norms. These "classic" games include the prisoner's dilemma, public goods, ultimatum, dictator, third-party punishment, trust, and gift exchange games. Of course, such preferences can also be examined using survey information, and the relationship between survey results and laboratory-based studies are of considerable interest (Dohmen et al., 2006; Fehr, 2009).

Some social behaviors, such as parenting, can also be studied in the laboratory. Behaviors such as pair bonding and mate preferences can feasibly be studied only in field experiments and surveys. In the next section, we discuss in detail the trust game and some of the important applications of this game in neuroeconomics and as an important laboratory-based paradigm in genetic analyses.

A. *Trust Game*

The trust game is a singular success of neuroeconomics strategy in understanding the molecular architecture of human decision making. It is based on a seminal article by Kosfeld et al. (2005) in which intranasally administered OT was demonstrated to enhance trust but not trustworthiness. Importantly, this experiment wove together two important research strategies in revealing human preferences—behavioral economics and pharmacology. Moreover, the choice of OT as a neuropeptide likely to enhance trusting behavior is based on a large body of animal (aka "translational") evidence showing the importance of this nonapeptide as a paramount social hormone in mammals (Insel, 2010).

In the trust game, the first player is endowed with a certain amount of money, say \$20, while the second player is endowed with nothing. In the first stage, the first player decides how much to send (S) to an anonymous and randomly matched second player (\$20 – S). For every dollar the first player sends, the second player receives three times (3S) as much. In the second stage, the second player decides how much (B) out of 3S to send back to the first player. At the end, the first player receives (\$20 – S + B), being the amount he or she keeps plus the amount the second player sends back, whereas the second player receives three times the amount sent deducting the amount sent back (3S – B). This is the design employed by Kosfeld et al. in their seminal article (Kosfeld et al., 2005). A modification of the trust game is the strategic method (Camerer, 2003; Camerer & Fehr, 2004), in which the second player states his or her response to each of several multiple possible choices from the first player. Every participant plays both roles of first and second players without any feedback. At the payment stage with real money, there is a random determination of the specific role—first or second mover—for each pair of subjects. The amount sent by the first player is used as a measure of trust, whereas the average return amount from the second player is a measure of trustworthiness. The strategic method allows more observations to be generated, whereas the standard method is more likely to induce positive or negative emotion when responding to actual choice. Experimental evidence shows no significant difference in behavior between the two approaches (Camerer, 2003; Camerer & Fehr, 2004).

An important application of the trust game has been published by the Montague group, who examined play in an iterated trust game in a group of subjects with borderline personality disorder (BPD) (King-Casas et al., 2008). Nonclinical subjects were characterized by a linear relation between anterior insula response and both magnitude of monetary offer received from their partner and the amount of money repaid to their partner. Intriguingly, activity in the anterior insula of BPD participants was related only to the magnitude of repayment sent back to their partner. Importantly, these results suggest that subjects with BPD are not sensitive to social norms. Importantly, this game-theoretic strategy toward understanding psychopathology may provide a new window into characterizing and studying other psychiatric disorders.

Trust game play has been examined with interesting results via MRI in a number of other studies of nonclinical subjects. For example, Koscik and Tranel (2011) found that the amygdala is required in the development of normal interpersonal trust. The study of van den Bos and colleagues (2009) examined the neural basis of reciprocity using a trust game and imaging. Krueger et al. (2007) examined the underlying brain mechanisms of conditional and unconditional trust in social reciprocal exchange. They used a hyperscanning approach in which two strangers interacted online with one another in a sequential reciprocal trust game while their brains were simultaneously scanned. Delgado and colleagues (2005) investigated whether prior social and moral information about potential trading partners affects the neural circuitry of trust. The trust game has also been used to reveal prejudices, as in the study carried out by Stanley and his collaborators (2012). Their results suggest that the amygdala may represent emotionally relevant social group information as a subset of the general detection function it serves, whereas the striatum is involved in representing race-based reputations that shape trust decisions.

Taken together, these studies ingeniously use a behavioral economic paradigm to measure trust, which, coupled with cutting-edge tools from the neurosciences, provides a powerful insight into human decision making.

B. Dictator Game

The dictator game is a one-shot game with two players. Player 1 is given a fixed sum of money and has to decide how to split it between himself and an anonymous Player 2 endowed with no money. Player 2 simply receives the money from Player 1 and does not have an active role in the game. Because the recipient is completely powerless, "dictators" are unconstrained by fear of reprisal or other strategic considerations, and their allotment can be seen as a measure of altruism (Forsythe, Horowitz, Savin, & Sefton, 1994; Kahneman, Knetsch, & Thaler, 1986). In short, the dictator game offers a simple, straightforward measure of altruism and generosity, a laboratory-based distillation of the essence of prosocial behavior.

Similar to other behavioral economic paradigms, the dictator game lends itself exquisitely to imaging and psychophysiological studies (Gunther Moor et al., 2011; Meshulam, Winter, Ben-Shakhar, & Aharon, 2012; Moor et al., 2012; Weiland, Hewig, Hecht, Mussel, & Miltner, 2012). Combining imaging and the dictator game helps identify neural pathways underlying altruism and sense of fairness. Fewer studies have combined neurogenetics × imaging × games, but such approaches have the potential to shine a spotlight with greater certainty on the neurotransmitter pathways and elements of synaptic transmission in decision making (see, e.g., Zhong, Chark, et al., 2012; Zhong, Israel, Xue, Sham, et al., 2009).

Weiland et al. (2012) examined subjects via MRI while playing the ultimatum and dictator games. Their study supports the idea that altruistic motives primarily drive fair offers in the dictator game, denoted in that report as altruistic fairness. In an interesting study, the dictator game was used to examine how individuals can regulate their emotions, indexed by skin conductance and post-game self-report questionnaires, in response to fair and unfair offers (Meshulam et al., 2012). The authors were able to demonstrate that people could choose their own mental state when incentivized to do so. Angry but not happy emotions could be manipulated. In a study combining an imaging protocol and pharmacological manipulation, albeit not the standard dictator game paradigm, van Ijzendoorn and his colleagues examined charitable giving to UNICEF in a laboratory setting and found that asymmetric frontal brain activity significantly predicted donating (van Ijzendoorn, Huffmeijer, Alink, Bakermans-Kranenburg, & Tops, 2011). Additionally, frontal α asymmetry moderated the interactive effect of OT and parental love withdrawal on charitable giving.

Games lend themselves to precise, reproducible, and objective cross-cultural studies in the laboratory, allowing the focused evaluation of human social behavior among various ethnic and population groups (Henrich et al., 2005; Herrmann, Thoni, & Gachter, 2008; Norenzayan & Shariff, 2008). One of the most interesting of such studies (Henrich et al., 2005) is a cross-cultural study of behavior in three representative games measuring other-regarding behavior: the ultimatum, public goods, and dictator games. The study was carried out in no less than 15 small-scale societies characterized by a diverse set of economic and cultural variables to find that, first, the neoclassical economic model of purely selfish behavior toward maximizing profit cannot explain the behavior in the societies studied. Second, a larger than expected amount of behavioral variability is observed. Third, the greater the market integration and the higher the monetary reward to cooperation in everyday life, the greater the level of prosociality displayed in experimental games. Fourth, and surprisingly, individual-level economic and demographic variables do not explain the variance in game behavior, either within or across groups. Fifth, in many cases, experimental play appears to resonate with the social exchanges occurring in day-to-day living.

One of the more intriguing applications of economic games is arriving at an understanding of the evolution of human cooperation (Gintis, Henrich, Bowles, Boyd, & Fehr, 2008; Hagen & Hammerstein, 2006; Price, 2008). A recent article on this debate by Marlowe and his collaborators (2011) shows that strong reciprocity, including generous offers and third-party punishment (3PP; see Fehr & Fischbacher, 2004), takes place mainly in complex societies characterized by dilemmas involving collective action. Hence, they argue that "spiteful" second-party punishment (2PP; as modeled by the ultimatum game), motivated by the basic emotion of anger, is more universal than

3PP and that it is sufficient to explain the origins of human cooperation.

Economic games can also be used to trace the ontogeny of human fairness from childhood through adolescence (Almas, Cappelen, Sorensen, & Tungodden, 2010). Using modified versions of the dictator game, these authors demonstrated that children's level of self-interest was stable across adolescence, whereas their concept of fairness markedly shifted in the same period. In particular, they observed an enhanced focus on the value of meritocratic fairness in evaluating the allocation of funds, a view that requires the player to make a distinction between various origins of inequality. Such studies of the development of fairness would lend themselves to both a neurogenetic and imaging strategies but, to our knowledge, have yet to be reported.

A final word on the use of the dictator game involves the exploration of psychopathology and the potential value of economic games in understanding decision making in mental illness. For example, one study examined decision making in the ultimatum and dictator games as a function of subjects' level of schizotypy across the full schizotypal continuum (van't Wout & Sanfey, 2011). The results showed that higher levels of schizotypal symptoms, particularly positive and disorganized schizotypy, were related to proposing higher offers to both human and nonhuman computer partners. Additionally, the number of interpersonal schizotypal symptoms was associated with an increased acceptance rate of very unfair offers from human partners, possibly reflecting a blunted emotional response to such offers.

Another study using the ultimatum and dictator games compared performance of schizophrenic patients to nonpatient controls (Wischniewski & Brune, 2011). The study found that even though both groups increasingly rejected unfair offers as the level of unfairness increased, patients accepted significantly more unfair offers than did controls.

Ebstein and colleagues carried out the first molecular genetic study examining individual differences in dictator game allocations (Knafo, Israel, et al., 2008). This study showed a significant association between the length of the RS3 AVPR1a promoter region repeat and allocations in the dictator game. The long repeats were associated with greater allocations. They also showed a relationship between RS length and transcription of AVPR1a in postmortem hippocampal specimens, and the long repeat predicted greater AVPR1a mRNA levels. The decreased expression of the RS3 short repeat has been replicated in an in vitro expression system (Tansey et al., 2011). A partial replication of this finding was then demonstrated in children using a somewhat different classification scheme for the RS3 repeat (Avinun et al., 2011). In the "dictator kids" publication, Avinun et al. classified the RS3 repeat by comparing the second most common allele versus all others (327 or 334 base pair [bp] vs. all others). The second most common allele has been associated with autism (Kim et al., 2002) and poor pair bonding in humans (Walum et al., 2008); for a detailed discussion of the AVPR1a gene and social behavior, see Westberg and Walum, elsewhere in this handbook. Interestingly, Meyer-Lindenberg and colleagues have shown that the RS3 repeat predicts amygdala activation using both classification schemes (Meyer-Lindenberg et al., 2009). Further evidence for functionality of the AVPR1a RS3 repeat was obtained from a study of prepulse inhibition (Levin et al., 2009).

Identification of an association between AVPR1a and dictator game allocations immediately suggested the possible involvement of OT as well, since both nonapeptide hormones contribute to social and affiliative behaviors in animals and people. Hence, we examined the role of the oxytocin receptor gene, OXTR, in dictator game giving (Israel et al., 2009). Three tagging SNPs across the gene region showed significant association with both the dictator and the social values orientation games (Van Lange, 1999). However, in a study of Swedish twins, no association was observed between OXTR and either the dictator or the trust games (Apicella et al., 2010). More recently, it was shown in healthy male students participating in a trust game that a commonly occurring genetic variation (rs53576) in the OXTR gene is reliably associated with trust behavior rather than a general increase in trustworthy or risk behaviors (Krueger et al., 2012). Individuals homozygous for the G allele (GG) showed higher trust behavior than did individuals who were A allele carriers (AA/AG).

Evidence for involvement of both the OXTR and AVPR1a genes in real-world prosocial activity was reported by Poulin and his colleagues (Poulin, Holman, & Buffone, 2012). They examined OXTR rs53576, as well as AVPR1a polymorphisms rs1 and rs3 in a national sample of US residents (n = 348). These polymorphisms interacted with perceived threat to predict engagement in volunteer work or charitable activities and commitment to civic duty. Perceived threat was assessed at all waves of the life-events study with a reverse-coded, six-item version of Janoff-Bulman's World Assumptions Scale (Janoff-Bulman, 1989). Items on the scale included "There is more good in the world than bad" and "Human nature is basically good"; participants responded to each item using a scale from 1, *strongly disagree*, to 5, *strongly agree*. Specifically, greater perceived threat predicted engagement in fewer charitable activities for individuals with A/A and A/G genotypes of OXTR rs53576, but not for G/G individuals. Similarly, greater perceived threat predicted lower commitment to civic duty for individuals with one or two short alleles for AVPR1a rs1, but not for individuals with only long alleles.

The rs53576 SNP is of interest because it has been associated with a variety of social behaviors and cognition such as parenting (Bakermans-Kranenburg & van Ijzendoorn, 2008), empathy (Rodrigues, Saslow, Garcia, John, & Keltner, 2009; Wu, Li, & Su, 2012), loneliness (Lucht et al., 2009), emotional support seeking (Kim et al., 2010), social auditory processing (Tops, van Ijzendoorn, Riem, Boksem, & Bakermans-Kranenburg, 2011), psychological resources (Saphire-Bernstein, Way, Kim, Sherman, & Taylor, 2011), social support (Chen et al., 2011), and neurocardiac reactivity to social stress (Norman et al., 2012). On the clinical side, this SNP has been associated with facets of disorders of social cognition including autism (Wu et al., 2005; but also see Jacob et al., 2007), depression (Costa et al., 2009), and ADHD (Park et al., 2010).

C. *Ultimatum Game*

In the ultimatum game, two players are offered a chance to divide a certain sum of money. The proposer proposes how to split the sum. The responder can accept or reject the deal. If the proposal is accepted, there is a deal. If the proposal is rejected, neither player gets anything. The rational solution, suggested by game theory, is for the proposer to offer the smallest possible share and for the responder to accept it. However, when humans play the game, the most frequent outcome is a fair share. Overall, considerable evidence shows that, in the ultimatum game, substantial offers are often made and sometimes rejected (Camerer & Thaler, 1995). Offers average 30–40 percent of the total and even splits are generally the mode. Offers of less than 20 percent are generally rejected.

A study by Zhong and colleagues examined the role of the dopamine D_4 receptor VNTR in contributing to ultimatum game behavior (Zhong et al., 2010). The DRD4 exon3 VNTR is a well-characterized functional polymorphism known to be associated with ADHD and personality traits including novelty seeking and self-report altruism. Applying a neurogenetic approach, these investigators found that DRD4 is significantly associated with fairness preference. Additionally, the interaction among this gene, season of birth, and gender is highly significant. This is the first result to link preference for reciprocal fairness to a specific gene and suggests that G×E interactions contribute to economic decision making.

Other biological markers have also been associated with ultimatum game behavior. Low serum levels of omega-3 polyunsaturated fatty acids have been linked to impulse control and hostility. Emanuele et al. (2009) examined the serum omega-3 and omega-6 fractions in relation to ultimatum bargaining behavior. Participants were 60 economy students, and the results of this study suggest that a depletion of the serum omega-3 fatty acids is associated with rejections of unfair ultimatum offers in an experimental neuroeconomic setting.

The neurotransmitter 5-HT has also been shown to partially mediate ultimatum game behavior (Crockett, 2009; Crockett, Clark, Tabibnia, Lieberman, & Robbins, 2008). Participants with depleted 5-HT levels rejected a greater proportion of unfair offers, but not fair offers, without showing changes in mood, fairness judgment, basic reward processing, or response inhibition. These results suggest that 5-HT may play a critical role in regulating emotion during social decision making. Interestingly, low platelet 5-HT may serve as a reliable biomarker to identify people who are more likely to reject unfair ultimatum offers in the ultimatum game (Emanuele, Brondino, Bertona, Re, & Geroldi, 2008). These data point to a role for the serotonergic system in human sense of fairness modeled in the ultimatum game.

In an innovative study by Takahashi and colleagues (2012), it was shown that individuals who were characterized as "peaceful" (characterized by high levels of straightforwardness and trust) rejected unfair offers more often than did individuals who were characterized as "aggressive" (characterized by high levels of impulsivity or hostility) (Takahashi et al., 2012). Moreover, PET scanning associated reduced 5-HT transporter levels in the dorsal raphe nucleus with honesty and trustfulness, the traits associated with retaliatory behavior in the ultimatum game. As the authors note "higher central serotonin transmission might allow us to behave adroitly and opportunistically, being good at playing games while pursuing self-interest" (p. 4281).

Weiland and colleagues examined neural correlates of preferences modeled in the dictator and ultimatum games (Weiland et al., 2012). Functional MRI analysis revealed that almost a third of fair offers in the ultimatum game were correlated with greater activity in prefrontal areas, especially in regions presumed to be involved in reward and theory of mind. A limitation of the study is that the number of fair and unfair offers differed among participants and conditions. Because participants could choose freely among the six alternative offers, there was no way to prevent such a bias without risking the free-choice condition. As a result, some participants had to be excluded from analysis, and therefore only 28 percent of the fair offers were included. This observation suggests that egoistic motives are mainly responsible for fair offers, which the authors call *strategic fairness*. Fair offers in the ultimatum game hence are suggested to be related to strategic processing and are mainly motivated by egoistic motives to emphasize reward. This result was corroborated by the significant activity in the striatum in comparing the ultimatum game with the dictator game. Fair offers in the dictator game, however, were related to greater activity in the dorsal anterior cingulate cortex and the posterior cingulate cortex. This lends support to the idea that altruistic motives are driving fair offers in the dictator game, which the authors designate *altruistic fairness*. In another investigation of the neural substrate underlying fairness, responder behavior was examined to unfair offers in the ultimatum game with conditions that differed in their intentionality constraints (Guroglu, van den Bos, Rombouts, & Crone, 2010). Intentionality is the perception of fairness that is influenced by the intentions of the interaction partner. It is context-dependent information that highly influences fairness judgments. Brain activity underlying rejection versus acceptance of unfair offers appeared highly dependent on intentionality. These results highlight the significance of intentionality considerations in fairness-related social decision-making processes. In another study of patients with brain lesions, it was shown that the ventromedial prefrontal cortex (vmPFC) encodes the expected value of abstract, future goals in a common neural currency that takes into account both reward and social signals in order to optimize economic decision making in the ultimatum game (Moretti, Dragone, & di Pellegrino, 2009).

D. Prisoner's Dilemma

The prisoner's dilemma is an exemplary game that has been intensively studied in game theory and has considerable importance in economics, psychology, and evolutionary studies. In the game, two players choose to either cooperate or defect and receive a payoff that depends on the interaction of their respective choices. It creates a situation in which cooperation is in the best interest of the group but defection is in the best interest of the individual. The iterated version of the prisoner's dilemma game is a model for relationships based on reciprocal altruism.

A series of interesting studies examining the neural basis of social cooperation using the prisoner's dilemma paradigm have been carried out by Rilling and his colleagues (Rilling et al., 2002; Rilling et al., 2011; Rilling et al., 2007; Rilling et al., 2008; Rilling, Sanfey, Aronson, Nystrom, & Cohen, 2004*a*; Rilling et al., 2004*b*; Wood, Rilling, Sanfey, Bhagwagar, & Rogers, 2006). A sequential-choice prisoner's dilemma game was employed in which player 1 chooses and player 2 is then able to view player 1's choice before making his own choice. Each of the four outcomes is associated with a different payoff. In one of the first studies of its kind, Rilling and colleagues showed that mutual cooperation was associated with consistent activation in brain areas that have been linked with reward processing: nucleus accumbens, the caudate nucleus, ventromedial frontal/orbitofrontal cortex, and rostral anterior cingulate cortex. Based on these observations, the authors propose that activation of this neural network positively reinforces reciprocal altruism, thereby motivating subjects to resist the temptation to selfishly accept but not reciprocate favors. In an attempt to identify the neural substrate of cooperation, Rilling and colleagues used a tryptophan-depletion paradigm. Tryptophan depletion produced significant reductions in the level of cooperation shown by participants when playing prisoner's dilemma (Wood et al., 2006). This effect was accompanied by a significantly diminished probability of cooperative responding given previous mutually cooperative behavior. These data further strengthen the notion that 5-HT plays a significant role in the acquisition of socially cooperative behavior.

In a more recent study, researchers manipulated brain neurochemistry in subjects by administering intranasally either AVP or OT (Rilling et al., 2011). Imaging results showed that, compared to both AVP and placebo, OT enhanced the caudate nucleus reaction to reciprocated cooperation. Oxytocin also increased left amygdala activation following reciprocated cooperation (the reader is referred to the article by Baas et al., 2004, regarding the significance of amygdala lateralization). At the level of behavior, intranasal OT led to increased rates of cooperation following unreciprocated cooperation in the previous round compared with AVP. AVP strongly enhanced cooperation following a cooperative gesture by the partner compared with both placebo and OT. In response to reciprocated cooperation, AVP enhanced activation in an area including vasopressin circuitry implicated in affiliative behaviors in nonhuman species. Finally, both OT and AVP increased amygdala functional connectivity with the anterior insula. It was suggested that increased connectivity between these two regions may increase the amygdala's ability to evoke visceral somatic responses that modulate decision making. As the authors noted, these findings extend our knowledge of the neural and behavioral effects of OT and AVP to the context of genuine social interactions.

E. Ingroup Versus Outgroup Paradigm

A recent meta-analysis investigating the effects of OT on intergroup trust concluded that OT administration enhances ingroup trust but does not decrease outgroup trust (Van IJzendoorn & Bakermans-Kranenburg, 2012). The studies reviewed in the meta-analysis, most notably those by De Dreu et al. (2010; 2011), were set in contexts of intergroup competition. Although consistent with this finding, results from the study of Israel et al. (Israel, Weisel, Ebstein, & Bornstein, 2012) demonstrate that when set in a context of intergroup cooperation, OT not only improves contributions to the ingroup but also increases contribution rates to the collective interest of all groups. These results suggest that group-oriented prosocial effects of OT may be sensitive to contextual cues, which may differ depending on the inherent structure of the social dilemma. Importantly, these prosocial effects do not extend to AVP, suggesting an OT-specific neuropeptide influence on cooperative intergroup behavior.

Additionally, within the framework of the nested social dilemma, participants receiving OT in the Israel et al. study were not only more willing to give to others, but they also expected that others would reciprocate their cooperative behavior. This effect is supported by the positive effect of OT on the expectation that others would contribute to both local and global interests and by the closer pairing of expectations to behavior in the OT condition only.

In today's increasingly interconnected world, deciding with whom and at what level to cooperate is a matter of increasing importance because large-scale cooperation often comes at the expense of cooperating at the local level and vice versa (Dietz, Ostrom, & Stern, 2003). Studies employing the nested social dilemma paradigm help to delineate the behavioral effects of OT and AVP in these types of social dilemmas and provide additional evidence for the positive effects of OT in facilitating cooperative behavior, both within and across groups.

We are aware of three additional investigations that bear on the role of OT in ingroup–outgroup altruism (Declerck, Boone, & Kiyonari, 2013; Declerck, Boone, & Kiyonari, 2010; Radke & de Bruijn, 2012). However, none of these three studies actually implements a true ingroup–outgroup game paradigm, although they draw inferences from single-player games including the dictator, ultimatum, and prisoner's dilemma games. We therefore believe their conclusions need to be interpreted cautiously when their results are compared to the nested social dilemma game. In the Radke and de Bruijn study (2012) subjects played a one-shot dictator game and a modified ultimatum game. The results suggest that OT decreases the adherence to fairness norms in social settings where others are likely to be *perceived* (herein lies the rub) as not belonging to one's ingroup. The authors interpret their results as not supporting the prosocial conception of OT, and they corroborate recent ideas that the effects of OT are more nuanced than assumed in the past. In the first Declerck et al. study (2010), participants received OT or a placebo and played two economic games: a coordination game (with strong incentives to cooperate) and a prisoner's dilemma (with weak cooperative incentives). Oxytocin enhanced cooperation only when social information was present, and this effect was significantly more pronounced in the coordination game. When social information was lacking, OT surprisingly decreased cooperation. In the second Declerck et al. study (2013), a one-shot prisoner's dilemma game design was used. The outcome of these mixed-motive games are known to be highly dependent on values and on social information that might reveal the partner's intent. Oxytocin and social information interact significantly to affect the behavior of individuals with a pro-self value orientation: following prior contact with the game partner, OT enhances cooperative behavior, whereas in anonymous conditions it exacerbates their intrinsic self-interested behavior. These effects of OT do not hold for individuals with a prosocial value orientation, whose cooperation levels appear to be more influenced by prior contact with the game partner.

F. Competitiveness

The gender gap in wages is an interesting observation in the labor market (Goldin, 1990). One behavioral interpretation has to do with gender difference in competitiveness. In an experimental setting (Gneezy, Leonard, & List, 2009; Niederle & Vesterlund, 2007), subjects were asked to perform simple arithmetic calculations with real monetary incentive. When asked to choose either a tournaments payment scheme or a piece-rate payment scheme, the majority of male subjects choose the tournaments scheme, whereas the majority of female subjects choose the piece-rate scheme.

The role of socialization in competitiveness has been recently examined by Gneezy and colleagues (Gneezy et al., 2009). They investigated gender-specific competitiveness in a patriarchal society, the Maasai in Tanzania, and a matrilineal society, the Khasi in India. Notably, among the Khasi, women chose the competitive environment more often than did men. As with findings in contemporary Western societies, Maasai men opt to compete at roughly twice the rate as do Maasai women. These findings underpin socialization as an important determinant of the observed gender difference in behavior, suggest that culture can trump genetic and hormonal influences, and highlight how culture may contribute to observed gender differences.

However, not all differences in ultimatum game behavior or other examples of revealed preferences are due to socialization. As shown in the papers by Chen et al. (Chen, Katuscak, & Ozdenoren, 2013) and by Pearson and Schipper (2013), hormonal factors indexed by menstrual cycle phase play a role in women's preferences. Regarding men, those who reject low offers ($5 out of $40) have significantly higher testosterone levels than those who accept low offers (Burnham, 2007). A within-subject repeated measures study found that men were 27 percent less generous toward strangers with money they controlled when their testosterone levels were artificially raised, compared to when they were on placebo (Zak et al., 2009).

Despite the growing literature documenting a significant role for sex hormones in decision making related to our sense of fairness, little is known about the underlying genetic mechanisms mediating these effects. Using a genetic association approach, we examined in a first study (Chew, Ebstein, & Zhong, 2011) the neurogenetic basis for gender difference in ultimatum game behavior. Using a Chinese and an Israeli sample, we genotyped participants for three well-characterized repeat polymorphisms of sex-hormone receptor genes: androgen receptor gene (AR), estrogen receptor alpha gene (ERα), and estrogen receptor beta gene (ERβ). Our findings were twofold. For the Chinese sample, males who carried the AR short CAG repeat allele (which encodes a polyglutamine stretch, the length of which correlates inversely with androgen activity) stated higher minimal acceptable offers than did males who only carried the long repeat allele. On the other hand, females who carried the CA long repeat allele of ERβ (which is associated with less estrogen activity) stated higher minimal acceptable offers than did females who only carried the short repeat allele. Our findings for the Israeli sample replicated the role of ERβ among female subjects but not the role of AR for male subjects.

X. Parenting

In the past several years, significant insights have been gained regarding the neuroendocrinological and neurogenetic mechanisms underlying human parenting. In particular, the role of the OT–vasopressin and the dopaminergic neural pathways in these processes has been revealed. Studies mostly used a neurogenetic approach and genotyped subjects for the DRD4 receptor and OXTR and AVPR1a genes. Relevant to the current review are studies of parenting in a controlled laboratory setting or experimental sessions during home visits (Avinun, Ebstein, & Knafo, 2012; Avinun et al., 2011; Bakermans-Kranenburg, Pijlman, Mesman, & Juffer, 2008; Bakermans-Kranenburg & van Ijzendoorn, 2006; 2008; 2011; Bakermans-Kranenburg, Van, Caspers, & Philibert, 2011; Feldman, Gordon, & Zagoory-Sharon, 2011; Gervai, 2009; Kaitz et al., 2010; Knafo, Israel, & Ebstein, 2011; Lee et al., 2010; Luijk et al., 2011; Naber, van Ijzendoorn, Deschamps, van Engeland, & Bakermans-Kranenburg, 2010; Propper, Willoughby, Halpern, Carbone, & Cox, 2007; Sheese, Voelker, Rothbart, & Posner, 2007; Van Ijzendoorn & Bakermans-Kranenburg, 2006; van Ijzendoorn, Bakermans-Kranenburg, & Mesman, 2008).

Additional evidence for a role of OT in parenting comes from the studies by Feldman and her colleagues (Feldman, Gordon, Schneiderman, Weisman, & Zagoory-Sharon, 2010; Gordon, Zagoory-Sharon, Leckman, & Feldman, 2010) who measured plasma OT levels in new fathers and mothers across the transition to parenthood in relation to maternal and paternal typical parenting behaviors. More recently, this group has added a neurogenetic perspective to their studies of parenting (Feldman et al., 2012).

Another notable study examined the role of OXTR and the influence of depressive symptoms in explaining differences in physiological reactivity to infant crying (Riem, Pieper, Out, Bakermans-Kranenburg, & van Ijzendoorn, 2011). Heart rate responses of healthy females without children were measured during the presentation of three episodes of infant cry sounds. Participants with the presumably more efficient variant of the oxytonergic system gene (OXTR GG) had more pronounced physiological reactivity to repeated cry sounds, except when they showed more symptoms of depression. This is the first study to suggest effects of OXTR genotype on physiological reactivity to infant crying. Depressive symptoms may, however, suppress the effect of the OXTR GG genotype. (For a general review of the role of OT in mother–infant relations, the reader is referred to a recent review by Galbally, Lewis, Ijzendoorn, and Permezel, 2011.)

To summarize, laboratory-based studies of parenting, similar to laboratory-based studies of decision making, have gained considerable traction and are paving the way for a deeper understanding of the biological and especially genetic underpinnings of human parenting. Most of these studies are relatively small-scale, reflecting the constraining logistics of laboratory measures of parent–child interactions. Such studies are labor intensive and not easily scaled up to the numbers currently needed for a GWAS approach. Nevertheless, it might be considered that grants covering many participating investigators could generate the numbers required for high-throughput genotyping approaches.

XI. Political Arena

Liberalism is trust of the people tempered by prudence; Conservatism is distrust of the people tempered by fear.

– *William E. Gladstone*

There is considerable family resemblance in political attitudes (Alwin, Cohen, & Newcomb, 1991), which has generally been attributed to social transmission within families. However, considerable evidence also points to a genetic contribution to political attitudes (Eaves, Eysenck, & Martin, 1989), More recently, longitudinal and extended pedigree studies further underscore the important role of genes in shaping political attitudes (Eaves & Hatemi, 2008; Hatemi et al., 2010). Family, twin, and adoption studies set the stage for searching for specific genes that contribute to political attitudes (Fowler, Baker, & Dawes, 2008*b*). To our knowledge, the first report identifying a specific gene with a political phenotype is by Fowler and colleagues (Dawes & Fowler, 2009; Fowler, Baker, & Dawes, 2008*a*; Fowler et al., 2008*b*; Fowler & Dawes, 2008; Settle, Dawes, Christakis, & Fowler, 2010). Using the National Longitudinal Study of Adolescent Health, Fowler et al. (Dawes & Fowler, 2009) showed that individuals with the A2 allele of the D_2 dopamine receptor gene were significantly more likely to identify themselves as a partisan (one holding party identification) than were those with the A1 allele. The baseline model shows that the A2 allele is significant, and the odds of an individual with one A2 allele being a partisan are 1.2 times greater than for someone with no A2 alleles, and the odds of an individual with two A2 alleles being partisan are 1.4 times greater than for someone with no A2 alleles. The A2 allele remains significant in the model even when SES measures and ORs are nearly identical. Furthermore, they find that this gene's association with partisanship also mediates an indirect association between the A2 allele and voter turnout. These results were the first to identify a specific gene that may be partly responsible for the tendency to join political groups, and they may help to explain correlations in parent and child partisanship and the persistence of partisan behavior over time.

Toward understanding how DRD2 may influence partisanship, the authors hypothesized that improved ability to form social attachments and/or improved cognitive function, both of which have been shown to be associated with the A2 allele, may increase the likelihood that an individual with the A2 allele will form and/or maintain an attachment with a political party.

In a second study, these authors examined a second DA receptor, DRD4 (Settle et al., 2010).

They showed that among those who carry the 7R allele of the DRD4 gene, the number of friends a person had in adolescence was positively associated with liberal self- identification in early adulthood. Among those who did not carry the 7R allele, there was no relationship between number of friends and ideology. Moreover, the 7R allele was not directly associated with the reported number of friends, nor was it directly associated with ideology. Instead, it was the combination of this specific gene variant with a specific social environment that may have contributed to the development of a liberal political ideology.

Hatemi and colleagues carried out the first genome-wide study of political attitudes (Hatemi et al., 2011). This "linkage" study was carried out using microsatellite markers, which, it should be noted, provide a less dense coverage of the map than the usual GWAS SNP-based studies. Hence, their power to detect regions of small effect sizes was limited despite the large number of subjects (approximately 13,000 from about 2,700 families) examined. Subjects were of Caucasian ancestry and inventoried for political attitudes using a 50-item scale. Their results revealed three peaks with LOD scores of greater than 3.0, indicating genome-wide level linkage and one peak at "suggestive" linkage (greater than 2.5; Lander & Kruglyak, 1995). The four most significant QTLs accounted for between 9 and 12.9 percent of the total phenotypic variation on the Conservatism–Liberalism attitude factor scale.

These provisional linkage sites from the Hatemi et al. study explain a surprisingly relatively high percentage of the variance for this trait. Moreover, three regions reached genome-level significance, and a fourth region clocked in as "suggestive." At first glance, this phenotype might appear to be distant from the underlying biology. However, a closer look at political attitudes reveals something quite interesting. It is intriguing that ideology is impacted by fear, threat, and anxiety (Lupia & Menning, 2009). Oxley et al. (2008) found physiological differences in threat reaction in those individuals who held more conservative attitudes toward outgroup members, and Hatemi and colleagues (Hatemi, McDermott, Eaves, Kendler & Neale, 2013) found that individual differences in fear disposition were a significant predictor of political attitudes. Interestingly, the relationship between social fear and outgroup attitudes was largely a function of shared genes (Jost et al., 2007). Moreover, certain serotonergic markers are associated with political participation (Fowler & Dawes, 2008) and power seeking (Madsen, 1986), further strengthening the link between political attitudes and anxiety/neuroticism based on the role of 5-HT as a paramount neurotransmitter associated with withdrawal as opposed to approach behavior.

Taken together, these considerations suggest the notion that some political attitudes are driven by "motivated reasoning" (Westen, Blagov, Harenski, Kilts, & Hamann, 2006), with "hard" social issues perhaps reflecting "gut feelings" and largely conforming to Kahneman's system 1 reasoning. These kinds of attitudes are apparently closer to genes than are soft social issues and system 2 reasoning. Interestingly, some political science investigations are using imaging studies to advance a deeper understanding of the underlying cognitive and emotional processes involved in political attitudes (Tingley, 2006).

We note that although the Hatemi study (Hatemi et al., 2011) observed several chromosomal regions that reached genome-wide significance levels (LOD score of more than 3.0) for political attitude, no such success was reported by the Benjamin study (Benjamin et al., 2012), which also examined political attitudes using a high-throughput SNP approach. The reasons for this discrepancy are at present unclear, but the relative success of the Hatemi study detracts somewhat from the argument put forward by Benjamin et al. that social phenotypes such as political or economic attitudes need tens of thousands of subjects to find genes contributing to these traits. It would be exciting if Hatemi could genotype his sample using a SNP platform.

XII. Conclusion

The take-home message from this review is that it is not only possible, but indeed eminently desirable, to measure individual and social decision making in the laboratory, especially when the goal of the research has a primarily genetic perspective. Laboratory-based decision-making paradigms are, we believe, the best phenotypes for molecular genetic association studies of social behavior. The choice of phenotype is critical to the success of GWAS, and identifying phenotypes that are closest to the underlying neurobiological structure is crucial. We suggest that employing phenotypes that are distant from any supporting biology are less likely to generate significant findings regardless of the number of subjects inventoried. Indeed, the recent failure of a GWAS of "educational attainment" and economic/political preferences is a case in point (Benjamin et al., 2012).

Often, the tradeoff in GWAS design is between pencil-and-paper questionnaire phenotypes that are relatively cheap to ascertain and laboratory-based paradigms from behavioral economics that are more costly both in time and money but that, we suggest, minimize measurement error and are reproducible across studies. Few, if any, GWAS have been reported for behavioral phenotypes, especially for nonclinical subjects, to judge which approach is more likely to capture significant associations between genes and decision making.

Indeed, we venture to suggest a litmus test of what constitutes a good nonclinical behavioral phenotype that is likely to generate significant GWAS results. We would simply ask whether the phenotype lends itself to fMRI studies. If so, then the phenotype is likely matching up with more basic neurobiology, and hence we suggest it is likely to successfully capture GWAS at a 5×10^{-8} level of significance. For example, risk attitude and social decision making have been modeled by behavioral economic paradigms in many imaging studies, suggesting that these games are indeed closely tapping into the underlying neural pathways. Moreover, imaging studies using behavioral economic paradigms are well suited to the addition of a genetic component, further tightening the link between biology and phenotype. Altogether, what we call "gutsy" phenotypes and "hard" social issues are more likely to have a narrower genetic base than more reflective system 2 reasoning and soft social issues.

A final word regarding the study base of GWAS of individual and social decision making is that we understand the convenience of leveraging existing survey studies that may have been ongoing for decades and have been genotyped for public health issues to extract information for GWAS of political and economic attitudes. Yet such studies are heterogeneous across a wide variety of demographic, socioeconomic, and cognitive variables. We believe that such heterogeneity, often unobserved, generates so much noise that the genetic signals are difficult if not impossible to detect. The alternative is to focus on study groups of homogenous student populations with a narrow age and ethnic background, which we suggest are much more promising to identify genes for social and individual decision making. Student populations of similar ethnic background should be characterized by a minimum signal-to-noise ratio compared to population samples and hence are the desirable target for first GWAS of decision making.

XIII. Future Directions

For future directions, we thought it would be informative to present some of our work-in-progress. We are located in the National University of Singapore, giving us potential access to a Singapore population of Chinese, Malay, Indian, and Caucasian. The overarching aim of our research is the molecular genetic architecture and biological basis of individual and social decision making. In our research, we chose behavioral economic paradigms implemented using a laboratory-based, incentivized design. We recruited only Han Chinese university students to minimize population stratification in genetic studies. We have carried out initially three waves of subject recruitment. The first wave of subjects was done in Singapore, and we recruited approximately 1,150 Han Chinese students. We further recruited a replication sample of 660 Han Chinese students from Beijing University who have a similar genetic background to our core sample of Singaporean Han but who are characterized by somewhat different cultural/environmental exposures. We have recruited additional groups from Singapore and Beijing and now have approximately 3,400 subjects for GWAS.

Blood samples have been obtained from the first two waves of students and saliva samples for the second Singapore replication sample. For the first Singapore wave, the blood samples were taken so that DNA, RNA, and hormone levels could be ascertained. All subjects are genotyped using Illumina OmniExpress Bead Chips at the Genome Institute of Singapore in collaboration with Dr. C. C. Khor. Genome-wide genotyping data from 731,442 SNPs genome-wide were available. In addition to SNPs, we also undertook genotyping of a number of microsatellite and VNTR markers that are of considerable interest in behavioral genetics such as the DRD4 exon3 VNTR, DAT1, 5-HTTLPR, AVPR1a RS3, and MAOA.

Furthermore, we positioned ourselves to examine hormone levels (plasma OT, cortisol, and others) and RNA expression in 1,150 of these subjects. Additionally, we are undertaking epigenetic studies of peripheral DNA by examining methylation patterns for candidate genes, as well as overall methylation levels.

The first results from our ongoing study are encouraging. From the core decision-making tasks (risk and social), we observed several promising

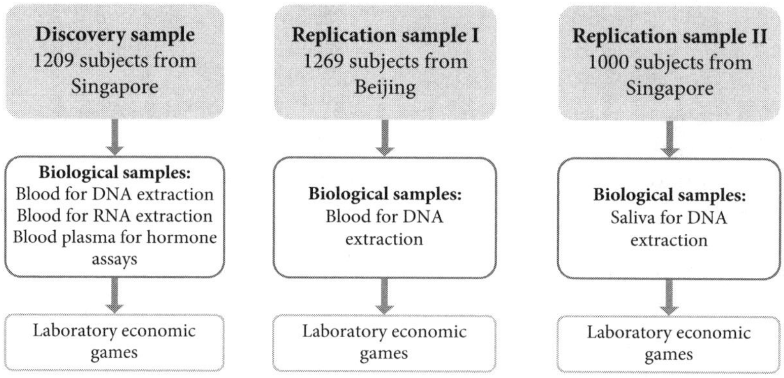

Fig. 9.3 Work flow of B²ESS (http://b2ess.nus.edu.sg/) GWAS study of human financial decision making.

"hits" at p values from 10^{-7} to 10^{-10} including one finding, which we are now preparing for submission (on altruistic giving measured by the modified dictator game of Andreoni and Miller [2002], with our top SNP reaching significance levels of 6.87×10^{-10} and located in a biologically plausible gene). We are also testing our top hit candidate SNP in an imaging paradigm designed to further validate the role of this SNP in altruistic giving under MRI.

We also have two papers (one accepted, one under review) related to plasma OT levels. The accepted publication describes a significant U-shaped relationship between plasma OT levels and behavior in the trust game, such that the extremes of plasma OT levels are associated with trust behavior (Zhong, Monakhov, et al., 2012). In the second paper, we observe a relationship between plasma OT levels and an impulse-purchasing scale (Zhang et al., in preparation). Our results indicate a significant interaction between plasma OT levels, OXTR genotype, and gender in predicting scores on the impulse-purchasing scale. Significant findings between OXTR SNPs and plasma OT levels with a social phenotype (mother–infant interactions) have been recently reported (Feldman et al., 2012). Finally, we have a work in progress in which we observe a relationship between the DRD4 gene and altruism measured in the dictator game, but the association is contingent on religious affiliation. This study is the first to examine the interaction effect of gene and culture using incentivized economic games, suggesting that both situational factors and cultural factors could be interacting with a gene to shape human behavior.

This description of our group's activities needs to be taken with caution since our results have yet to be vetted by peer review; they should be considered provisional at best. Nevertheless, we bring them to the attention of the reader to underscore how we have built our study to maximize the information we can gain from our study population. We have both positioned ourselves to carry out one of the first GWAS on approximately 3,400 student subjects from a homogenous ethnic group but also to use a candidate gene and biomarker approach. We have yet to begin analyzing our epigenetic and expression studies. We are optimistic that a study in depth of a few thousand subjects is a more productive strategy than studying tens of thousands of subjects superficially (see Figure 9.3).

Acknowledgments

Financial support is acknowledged from the Templeton Foundation, the AXA Foundation (project title "The Biology of Decision Making Under Risk"), the Singapore Ministry of Education (project title "Biological Economics and Decision Making"), and the National University of Singapore.

References

Almas, I., Cappelen, A. W., Sorensen, E. O., & Tungodden, B. (2010). Fairness and the development of inequality acceptance. *Science, 328*(5982), 1176–1178. doi: 328/5982/1176 [pii] 10.1126/science.1187300

Altshuler, D. M., Lander, E. S., Ambrogio, L., Bloom, T., Cibulskis, K., Fennell, T. J.,...Sougnez, C. L. (2010). A map of human genome variation from population scale sequencing. *Nature, 467*(7319), 1061–1073.

Alwin, D. F., Cohen, R. L., & Newcomb, T. M. (1991). *Political attitudes over the life span: The Bennington women after fifty years.* Madison: University of Wisconsin Press.

Andreoni, J., & Miller, J. (2002). Giving according to GARP: An experimental test of the consistency of preferences for altruism. *Econometrica, 70*(2), 737–753.

Apicella, C. L., Cesarini, D., Johannesson, M., Dawes, C. T., Lichtenstein, P., Wallace, B.,...Westberg, L. (2010). No association between oxytocin receptor (OXTR) gene polymorphisms and experimentally elicited social preferences. *PLoS One, 5*(6), e11153. doi: 10.1371/journal.pone.0011153

Avinun, R., Ebstein, R. P., & Knafo, A. (2012). Human maternal behaviour is associated with arginine vasopressin receptor 1A gene. *Biology Letter,* doi: Rsbl.2012.0492 [pii] 10.1098/rsbl.2012.0492

Avinun, R., Israel, S., Shalev, I., Gritsenko, I., Bornstein, G., Ebstein, R. P., & Knafo, A. (2011). AVPR1A variant associated with preschoolers' lower altruistic behavior. *PloS One, 6*(9), e25274. doi: 10.1371/journal.pone.0025274

Axelrod, R., & Hamilton, W. D. (1981). The evolution of cooperation. *Science, 211*(4489), 1390–1396.

Ayalew, M., Le-Niculescu, H., Levey, D. F., Jain, N., Changala, B., Patel, S. D.,... Niculescu, A. B. (2012). Convergent functional genomics of schizophrenia: From comprehensive understanding to genetic risk prediction. *Molecular Psychiatry,* doi: 10.1038/mp.2012.37

Baas, D., Aleman, A., & Kahn, R. S. (2004). Lateralization of amygdala activation: A systematic review of functional neuroimaging studies. *Brain Research Reviews, 45*(2), 96–103.

Bakermans-Kranenburg, M. J., Van, I. M. H., Pijlman, F. T., Mesman, J., & Juffer, F. (2008). Experimental evidence for differential susceptibility: Dopamine D4 receptor polymorphism (DRD4 VNTR) moderates intervention effects on toddlers' externalizing behavior in a randomized controlled trial. *Developmental Psychology, 44*(1), 293–300.

Bakermans-Kranenburg, M. J., & van Ijzendoorn, M. H. (2006). Gene-environment interaction of the dopamine D4 receptor (DRD4) and observed maternal insensitivity predicting externalizing behavior in preschoolers. *Developmental Psychobiology, 48*(5), 406–409.

Bakermans-Kranenburg, M. J., & van Ijzendoorn, M. H. (2008). Oxytocin receptor (OXTR) and serotonin transporter (5-HTT) genes associated with observed parenting. *Social Cognitive and Affective Neuroscience, 3*(2), 128–134.

Bakermans-Kranenburg, M. J., & van Ijzendoorn, M. H. (2011). Differential susceptibility to rearing environment depending on dopamine-related genes: New evidence and a meta-analysis. *Developmental Psychopathology, 23*(1), 39–52. doi: S0954579410000635 [pii] 10.1017/S0954579410000635

Bakermans-Kranenburg, M. J., van, I. M. H., Caspers, K., & Philibert, R. (2011). DRD4 genotype moderates the impact of parental problems on unresolved loss or trauma. *Attachment & Human Development, 13*(3), 253–269. doi: 936575953 [pii] 10.1080/14616734.2011.562415

Bakken, T. E., Roddey, J. C., Djurovic, S., Akshoomoff, N., Amaral, D. G., Bloss, C. S.,... Carlson, H. (2012). Association of common genetic variants in GPCPD1 with scaling of visual cortical surface area in humans. *Proceedings of the National Academy of Sciences of the United States of America, 109*(10), 3985–3990. doi: 10.1073/pnas.1105829109

Basson, J., Simino, J., & Rao, D. C. (2012). Between candidate genes and whole genomes: Time for alternative approaches in blood pressure genetics. *Current Hypertension Reports, 14*(1), 46–61. doi: 10.1007/s11906-011-0241-8

Bearden, C. E., & Freimer, N. B. (2006). Endophenotypes for psychiatric disorders: Ready for primetime? *Trends in Genetics, 22*(6), 306–313. doi: S0168-9525(06)00121-1 [pii] 10.1016/j.tig.2006.04.004

Beauchamp, J., Cesarini, D., Rosenquist, J., Fowler, J., & Christakis, N. (2009). A genome wide association study of educational attainment. Paper presented at the Institute for the Study of Labor (IZA) Workshop: Genes, Brains, and the Labor Market in Bonn, Switzerland. www.neuroeconomics.nyu.edu/papers/Cesarini1.pdf.

Bekkers, R., & Wiepking, P. (2011). A literature review of empirical studies of philanthropy. *Nonprofit and Voluntary Sector Quarterly, 40*(5), 924–973.

Bel, E. H. (2004). Clinical phenotypes of asthma. *Current Opinion in Pulmonary Medicine, 10*(1), 44–50.

Bell, J. T., & Spector, T. D. (2011). A twin approach to unraveling epigenetics. *Trends in Genetics,* doi: S0168-9525(10)00242-8 [pii] 10.1016/j.tig.2010.12.005

Benjamin, D., Chabris, C., Glaeser, E., Gudnason, V., Harris, T., Laibson, D.,... Purcell, S. (2007). Genoeconomics. In M. Weinstein, J. W. Vaupel, & K. W. Wachter, eds., *Biosocial surveys* (pp. 304–335). Washington, DC: National Academies Press.

Benjamin, D. J., Cesarini, D., van der Loos, M. J., Dawes, C. T., Koellinger, P. D., Magnusson, P. K.,... Visscher, P. M. (2012). The genetic architecture of economic and political preferences. *Proceedings of the National Academy of Sciences of the United States of America,* doi: 10.1073/pnas.1120666109

Berg, J., Dickhaut, J., & McCabe, K. (1995). Trust, reciprocity, and social history. *Games and Economic Behavior, 10*(1), 122.

Bilder, R. M., Sabb, F. W., Cannon, T. D., London, E. D., Jentsch, J. D., Parker, D. S.,... Freimer, N. B. (2009). Phenomics: The systematic study of phenotypes on a genome-wide scale. *Neuroscience, 164*(1), 30–42. doi: 10.1016/j.neuroscience.2009.01.027

Bis, J. C., Decarli, C., Smith, A. V., van der Lijn, F., Crivello, F., Fornage, M.,... Seshadri, S. (2012). Common variants at 12q14 and 12q24 are associated with hippocampal volume. *Nature Genetics, 44*(5), 545–551. doi: 10.1038/ng.2237

Blasi, G., Lo Bianco, L., Taurisano, P., Gelao, B., Romano, R., Fazio, L.,... Bertolino, A. (2009). Functional variation of the dopamine D2 receptor gene is associated with emotional control as well as brain activity and connectivity during emotion processing in humans. *Journal of Neuroscience, 29*(47), 14812–14819. doi: 29/47/14812 [pii] 10.1523/JNEUROSCI.3609-09.2009

Burnham, T. C. (2007). High-testosterone men reject low ultimatum game offers. *Proceedings of Biological Science, 274*(1623), 2327–2330.

Camara, E., Kramer, U. M., Cunillera, T., Marco-Pallares, J., Cucurell, D., Nager, W.,... Munte, T. F. (2010). The effects of COMT (Val108/158Met) and DRD4 (SNP -521) dopamine genotypes on brain activations related to valence and magnitude of rewards. *Cerebral Cortex, 20*(8), 1985–1996. doi: Bhp263 [pii] 10.1093/cercor/bhp263

Camerer, C. (2003). *Behavioral game theory: Experiments in strategic interaction.* Princeton, NJ: Princeton University Press.

Camerer, C., & Thaler, R. H. (1995). Anomalies: Ultimatums, dictators and manners. *The Journal of Economic Perspectives, 9*(2), 209–219.

Camerer, C. F., & Fehr, E. (2004). Measuring social norms and preferences using experimental games: A guide for social scientists. In J. Henrich et al., eds., *Foundations of human sociality: Economic experiments and ethnographic evidence from fifteen small-scale societies* (pp. 55–95). Oxford/New York: Oxford University Press.

Canli, T., & Lesch, K. P. (2007). Long story short: The serotonin transporter in emotion regulation and social cognition. *Nature Neuroscience, 10*(9), 1103–1109.
Carpenter, J. P., Garcia, J. R., & Lum, J. K. (2011). Dopamine receptor genes predict risk preferences, time preferences, and related economic choices. *Journal of Risk and Uncertainty, 42*(3), 233–261.
Carpenter, S. (2012). Psychology research. Psychology's bold initiative. [News]. *Science, 335*(6076), 1558–1561. doi: 10.1126/science.335.6076.1558
Carter, C. S. (1998). Neuroendocrine perspectives on social attachment and love. *Psychoneuroendocrinology, 23*(8), 779–818.
Cesarini, D., Dawes, C. T., Johannesson, M., Lichtenstein, P., & Wallace, B. (2009). Genetic variation in preferences for giving and risk taking. *Quarterly Journal of Economics, 124*(2), 809–842.
Chammah, A. M., & Rapoport, A. (1965). *Prisoner's dilemma: A study in conflict and cooperation.* Ann Arbor: University of Michigan Press.
Chen, D. T., Jiang, X., Akula, N., Shugart, Y. Y., Wendland, J. R., Steele, C. J., . . . McMahon, F. J. (2011). Genome-wide association study meta-analysis of European and Asian-ancestry samples identifies three novel loci associated with bipolar disorder. *Molecular Psychiatry,* doi: 10.1038/mp.2011.157
Chen, F. S., Kumsta, R., von Dawans, B., Monakhov, M., Ebstein, R. P., & Heinrichs, M. (2011). Common oxytocin receptor gene (OXTR) polymorphism and social support interact to reduce stress in humans. *Proceedings of the National Academy of Sciences of the United States of America,* doi: 10.1073/pnas.1113079108
Chen, Y., Katuscak, P., & Ozdenoren, E. (2013). Why can't a woman bid more like a man? *Games and Economic Behavior, 77,* 181–213.
Chew, S., Ebstein, R., & Zhong, S. (2011). Sex-hormone genes and gender difference in ultimatum game: Experimental evidence from China and Israel. Retrieved from http://papers.ssrn.com/sol3/papers.cfm?abstract_id=1975735.
Chew, S. H., Ebstein, R. P., & Zhong, S. (2012). Ambiguity aversion and familiarity bias: Evidence from behavioral and gene association studies. *Journal of Risk and Uncertainty, 44,* 1–18.
Churchland, P. S., & Winkielman, P. (2012). Modulating social behavior with oxytocin: How does it work? What does it mean? *Hormones and Behavior, 61*(3), 392–399. doi: 10.1016/j.yhbeh.2011.12.003
Costa, B., Pini, S., Gabelloni, P., Abelli, M., Lari, L., Cardini, A., . . . Martini, C. (2009). Oxytocin receptor polymorphisms and adult attachment style in patients with depression. *Psychoneuroendocrinology, 34,* 1506–1514.
Crockett, M. J. (2009). The neurochemistry of fairness: Clarifying the link between serotonin and prosocial behavior. *Annals of the New York Academy of Sciences, 1167,* 76–86. doi: NYAS04506 [pii] 10.1111/j.1749-6632.2009.04506.x
Crockett, M. J., Clark, L., Tabibnia, G., Lieberman, M. D., & Robbins, T. W. (2008). Serotonin modulates behavioral reactions to unfairness. *Science, 320*(5884), 1739.
Dai, Y., Jiang, R., & Dong, J. (2012). Weighted selective collapsing strategy for detecting rare and common variants in genetic association study. *BMC Genetics, 13*(1), 7.
Davies, G., Tenesa, A., Payton, A., Yang, J., Harris, S. E., Liewald, D., . . . Luciano, M. (2011). Genome-wide association studies establish that human intelligence is highly heritable and polygenic. *Molecular Psychiatry, 16*(10), 996–1005.
Daw, N. D., Kakade, S., & Dayan, P. (2002). Opponent interactions between serotonin and dopamine. *Neural Networks, 15*(4–6), 603–616.
Dawes, C. T., & Fowler, J. H. (2009). Partisanship, voting, and the dopamine D2 receptor gene. *Journal of Politics, 71,* 1157–1171.
de Bono, M., & Bargmann, C. I. (1998). Natural variation in a neuropeptide Y receptor homolog modifies social behavior and food response in C. *elegans. Cell, 94*(5), 679–689.
De Dreu, C. K. W., Greer, L. L., Handgraaf, M. J. J., Shalvi, S., Van Kleef, G. A., Baas, M., . . . Feith, S. W. W. (2010). The neuropeptide oxytocin regulates parochial altruism in intergroup conflict among humans. *Science, 328*(5984), 1408–1411.
De Dreu, C. K. W., Greer, L. L., Van Kleef, G. A., Shalvi, S., & Handgraaf, M. J. J. (2011). Oxytocin promotes human ethnocentrism. *Proceedings of the National Academy of Sciences, 108*(4), 1262.
De Martino, B., Kumaran, D., Seymour, B., & Dolan, R. J. (2006). Frames, biases, and rational decision-making in the human brain. *Science, 313*(5787), 684–687.
de Quervain, D. J., Fischbacher, U., Treyer, V., Schellhammer, M., Schnyder, U., Buck, A., & Fehr, E. (2004). The neural basis of altruistic punishment. *Science, 305*(5688), 1254–1258.
de Waal, F. B. (2008). Putting the altruism back into altruism: The evolution of empathy. *Annual Review of Psychology, 59,* 279–300.
Declerck, C., Boone, C., & Kiyonari, T. (2013). The effect of oxytocin on cooperation in a prisoner's dilemma depends on the social context and a person's social value orientation. *Social Cognitive and Affective Neuroscience,* doi: 10.1093/scan/nst040
Declerck, C. H., Boone, C., & Kiyonari, T. (2010). Oxytocin and cooperation under conditions of uncertainty: The modulating role of incentives and social information. *Hormones and Behavior, 57*(3), 368–374. doi: S0018-506X(10)00018-8 [pii] 10.1016/j.yhbeh.2010.01.006
Delgado, M. R., Frank, R. H., & Phelps, E. A. (2005). Perceptions of moral character modulate the neural systems of reward during the trust game. *Nature Neuroscience, 8*(11), 1611–1618.
Dietz, T., Ostrom, E., & Stern, P. C. (2003). The struggle to govern the commons. *Science, 302*(5652), 1907.
Dohmen, T., Falk, A., Huffman, D., Schupp, J., Sunde, U., & Wagner, G. (2006). Individual risk attitudes: New evidence from a large, representative, experimentally-validated survey. CEPR Discussion paper No. 5517. London: Centre for Economic Policy Research London.
Donaldson, Z. R., & Young, L. J. (2008). Oxytocin, vasopressin, and the neurogenetics of sociality. *Science, 322*(5903), 900–904.
Drabant, E. M., Hariri, A. R., Meyer-Lindenberg, A., Munoz, K. E., Mattay, V. S., Kolachana, B. S., . . . Weinberger, D. R. (2006). Catechol O-methyltransferase val158met genotype and neural mechanisms related to affective arousal and regulation. *Archives of General Psychiatry, 63*(12), 1396–1406. doi: 63/12/1396 [pii] 10.1001/archpsyc.63.12.1396
Dreber, A., & Apicella, C. L. (2009). The 7R polymorphism in the dopamine D4 receptor gene (DRD4) is associated

with financial risk-taking in men. *Evolution and Human Behavior, 30*, 85–92.
Dreber, A., Rand, D. G., Wernerfelt, N., Garcia, J. R., Vilar, M. G., Lum, J. K., & Zeckhauser, R. (2010). Dopamine and risk choices in different domains: Findings among serious tournament bridge players. *Journal of Risk and Uncertainty*, 1–20.
Dreher, J. C. (2007). Sensitivity of the brain to loss aversion during risky gambles. *Trends in Cognitive Science, 11*(7), 270–272.
Eaves, L. J., Eysenck, H. J., & Martin, N. G. (1989). *Genes, culture and personality: An empirical approach*. New York: Academic Press.
Eaves, L. J., & Hatemi, P. K. (2008). Transmission of attitudes toward abortion and gay rights: Effects of genes, social learning and mate selection. *Behavior Genetics, 38*(3), 247–256.
Ebstein, R. P. (2006). The molecular genetic architecture of human personality: Beyond self-report questionnaires. *Molecular Psychiatry, 11*(5), 427–445.
Ebstein, R. P., Israel, S., Chew, S. H., Zhong, S., & Knafo, A. (2010). Genetics of human social behavior. *Neuron, 65*(6), 831–844. doi: S0896-6273(10)00137-6 [pii] 10.1016/j.neuron.2010.02.020
Ebstein, R. P., Knafo, A., Mankuta, D., Chew, S. H., & Lai, P. S. (2012). The contributions of oxytocin and vasopressin pathway genes to human behavior. *Hormones and Behavior, 61*(3), 359–379. doi: 10.1016/j.yhbeh.2011.12.014
Ebstein, R. P., Novick, O., Umansky, R., Priel, B., Osher, Y., Blaine, D.,...Belmaker, R. H. (1996). Dopamine D4 receptor (D4DR) exon III polymorphism associated with the human personality trait of Novelty Seeking. *Nature Genetics, 12*(1), 78–80.
Ecker, C. (2011). Autism biomarkers for more efficacious diagnosis. *Biomarkers in Medicine, 5*(2), 193–195. doi: 10.2217/bmm.11.13
Egan, M. F., Kojima, M., Callicott, J. H., Goldberg, T. E., Kolachana, B. S., Bertolino, A.,...Weinberger, D. R. (2003). The BDNF val66met polymorphism affects activity-dependent secretion of BDNF and human memory and hippocampal function. *Cell, 112*(2), 257–269.
Eisenegger, C., Knoch, D., Ebstein, R. P., Gianotti, L. R., Sandor, P. S., & Fehr, E. (2010). Dopamine receptor D4 polymorphism predicts the effect of L-DOPA on gambling behavior. *Biological Psychiatry, 67*(8), 702–706. doi: S0006-3223(09)01123-8 [pii] 10.1016/j.biopsych.2009.09.021
Emanuele, E., Brondino, N., Bertona, M., Re, S., & Geroldi, D. (2008). Relationship between platelet serotonin content and rejections of unfair offers in the ultimatum game. *Neuroscience Letter, 437*(2), 158–161.
Emanuele, E., Brondino, N., Re, S., Bertona, M., & Geroldi, D. (2009). Serum omega-3 fatty acids are associated with ultimatum bargaining behavior. *Physiology and Behavior, 96*, 180–183.
Emily, M. (2012). IndOR: A new statistical procedure to test for SNP-SNP epistasis in genome-wide association studies. *Statistics in Medicine*, doi: 10.1002/sim.5364
Fakra, E., Hyde, L. W., Gorka, A., Fisher, P. M., Munoz, K. E., Kimak, M.,...Hariri, A. R. (2009). Effects of HTR1A C(-1019)G on amygdala reactivity and trait anxiety. *Archives in General Psychiatry, 66*(1), 33–40. doi: 66/1/33 [pii] 10.1001/archpsyc.66.1.33
Faraone, S. V., Doyle, A. E., Mick, E., & Biederman, J. (2001). Meta-analysis of the association between the 7-repeat allele of the dopamine D(4) receptor gene and attention deficit hyperactivity disorder. *American Journal of Psychiatry, 158*(7), 1052–1057.
Fehr, E. (2009). On the economics and biology of trust. *Journal of the European Economic Association, 7*(2–3), 235–266.
Fehr, E., & Camerer, C. F. (2007). Social neuroeconomics: the neural circuitry of social preferences. *Trends Cogn Sci, 11*(10), 419–427.
Fehr, E., & Fischbacher, U. (2004). Third party punishment and social norms. *Evolution and Human Behavior, 25*(2), 63–87.
Fehr, E., & Gachter, S. (2002). Altruistic punishment in humans. *Nature, 415*(6868), 137–140.
Feldman, R., Gordon, I., Schneiderman, I., Weisman, O., & Zagoory-Sharon, O. (2010). Natural variations in maternal and paternal care are associated with systematic changes in oxytocin following parent-infant contact. *Psychoneuroendocrinology*, doi: S0306-4530(10)00029-6 [pii] 10.1016/j.psyneuen.2010.01.013
Feldman, R., Gordon, I., & Zagoory-Sharon, O. (2011). Maternal and paternal plasma, salivary, and urinary oxytocin and parent-infant synchrony: Considering stress and affiliation components of human bonding. *Developmental Science, 14*(4), 752–761. doi: 10.1111/j.1467-7687.2010.01021.x
Feldman, R., Zagoory-Sharon, O., Weisman, O., Schneiderman, I., Gordon, I., Maoz, R.,...Ebstein, R. P. (2012). Sensitive parenting is associated with plasma oxytocin and polymorphisms in the oxytocin receptor (OXTR) and CD38 genes. *Biological Psychiatry, 72*, 175–181.
Flynn, J. R. (2000). IQ gains, WISC subtests and fluid g: G theory and the relevance of Spearman's hypothesis to race. *Novartis Found Symposium, 233*, 202–216.
Forsythe, R., Horowitz, J., Savin, N. E., & Sefton, M. (1994). Fairness in simple bargaining experiments *Games and Economic Behavior, 6*, 347–369.
Fowler, J. H., Baker, L. A., & Dawes, C. T. (2008*a*). The genetic basis of political participation. *American Political Science Review, 102*(2), 233–248.
Fowler, J. H., Baker, L. A., & Dawes, C. T. (2008*b*). Genetic variation in political participation. *American Political Science Review, 102*(02), 233–248.
Fowler, J. H., & Dawes, C. T. (2008). Two genes predict voter turnout. *Journal of Politics, 70*(03), 579–594.
Fraga, M. F., Ballestar, E., Paz, M. F., Ropero, S., Setien, F., Ballestar, M. L.,...Esteller, M. (2005). Epigenetic differences arise during the lifetime of monozygotic twins. *Proceedings of the National Academy of Science of the United States of America, 102*(30), 10604–10609.
Freund, J., Brandmaier, A. M., Lewejohann, L., Kirste, I., Kritzler, M., Kruger, A.,...Kempermann, G. (2013). Emergence of individuality in genetically identical mice. *Science, 340*(6133), 756–759. doi: 10.1126/science.1235294
Frydman, C., Camerer, C., Bossaerts, P., & Rangel, A. (2011). MAOA-L carriers are better at making optimal financial decisions under risk. *Proceedings of the Royal Society B: Biological Sciences, 278*(1714), 2053–2059. doi: Rspb.2010.2304 [pii] 10.1098/rspb.2010.2304
Furman, D. J., Chen, M. C., & Gotlib, I. H. (2011). Variant in oxytocin receptor gene is associated with amygdala volume. *Psychoneuroendocrinology, 36*(6), 891–897. doi: 10.1016/j.psyneuen.2010.12.004

Royal Society B:. Biological Sciences, 271(1535), 177–182. doi: 10.1098/rspb.2003.2586

Oxley, D. R., Smith, K. B., Alford, J. R., Hibbing, M. V., Miller, J. L., Scalora, M., . . . Hibbing, J. R. (2008). Political attitudes vary with physiological traits. *Science, 321*(5896), 1667–1670.

Park, J., Willmott, M., Vetuz, G., Toye, C., Kirley, A., Hawi, Z., . . . Kent, L. (2010). Evidence that genetic variation in the oxytocin receptor (OXTR) gene influences social cognition in ADHD. *Progress in Neuropsychopharmacology and Biological Psychiatry,* doi: S0278-5846(10)00122-3 [pii] 10.1016/j.pnpbp.2010.03.029

Pearson, M., & Schipper, B. C. (2013). Menstrual cycle and competitive bidding. *Games and Economic Behavior, 78*, 1–20.

Peripato, A. C., De Brito, R. A., Matioli, S. R., Pletscher, L. S., Vaughn, T. T., & Cheverud, J. M. (2004). Epistasis affecting litter size in mice. *Journal of Evolutionary Biology, 17*(3), 593–602. doi: 10.1111/j.1420-9101.2004.00702.x JEB702 [pii]

Peters, J., & Buchel, C. (2010). Episodic future thinking reduces reward delay discounting through an enhancement of prefrontal-mediotemporal interactions. *Neuron, 66*(1), 138–148. doi: 10.1016/j.neuron.2010.03.026

Petronis, A. (2010). Epigenetics as a unifying principle in the aetiology of complex traits and diseases. *Nature, 465*(7299), 721–727. doi: Nature09230 [pii] 10.1038/nature09230

Pfeiffer, T., Bertram, L., & Ioannidis, J. P. (2011). Quantifying selective reporting and the Proteus phenomenon for multiple datasets with similar bias. *PloS One, 6*(3), e18362. doi: 10.1371/journal.pone.0018362

Poulin, M. J., Holman, E. A., & Buffone, A. (2012). The neurogenetics of nice: Receptor genes for oxytocin and vasopressin interact with threat to predict prosocial behavior. *Psychology Science,* doi: 0956797611428471 [pii] 10.1177/0956797611428471

Prevost, C., Pessiglione, M., Metereau, E., Clery-Melin, M. L., & Dreher, J. C. (2010). Separate valuation subsystems for delay and effort decision costs. *Journal of Neuroscience: The Official Journal of the Society for Neuroscience, 30*(42), 14080–14090. doi: 10.1523/JNEUROSCI.2752-10.2010

Price, M. E. (2008). The resurrection of group selection as a theory of human cooperation. *Social Justice Research, 21*(2), 228–240.

Propper, C., Willoughby, M., Halpern, C. T., Carbone, M. A., & Cox, M. (2007). Parenting quality, DRD4, and the prediction of externalizing and internalizing behaviors in early childhood. *Developmental Psychobiology, 49*(6), 619–632.

Ptacek, R., Kuzelova, H., & Stefano, G. B. (2011). Dopamine D4 receptor gene DRD4 and its association with psychiatric disorders. *Medical Science Monitor, 17*(9), RA215-220. doi: 881925 [pii]

Radke, S., & de Bruijn, E. R. (2012). The other side of the coin: Oxytocin decreases the adherence to fairness norms. *Frontiers in Human Neuroscience, 6*, 193. doi: 10.3389/fnhum.2012.00193

Rangel, A., Camerer, C., & Montague, P. R. (2008). A framework for studying the neurobiology of value-based decision making. *Nature Reviews Neuroscience, 9*(7), 545–556.

Rasch, B., Spalek, K., Buholzer, S., Luechinger, R., Boesiger, P., Papassotiropoulos, A., & de Quervain, D. J. (2009). A genetic variation of the noradrenergic system is related to differential amygdala activation during encoding of emotional memories. *Proceedings of the National Academy of Science of the United States of America, 106*(45), 19191–19196. doi: 0907425106 [pii] 10.1073/pnas.0907425106

Reitz, C., Cheng, R., Rogaeva, E., Lee, J. H., Tokuhiro, S., Zou, F., . . . Mayeux, R. (2011). Meta-analysis of the association between variants in SORL1 and Alzheimer disease. *Archives of Neurology, 68*(1), 99–106. doi: 10.1001/archneurol.2010.346

Riem, M. M., Pieper, S., Out, D., Bakermans-Kranenburg, M. J., & van Ijzendoorn, M. H. (2011). Oxytocin receptor gene and depressive symptoms associated with physiological reactivity to infant crying. *Social Cognitive and Affective Neuroscience, 6*(3), 294–300. doi: 10.1093/scan/nsq035

Rietveld, C. A., Medland, S. E., Derringer, J., Yang, J., Esko, T., Martin, N. W., et al. (2013). GWAS of 126,559 individuals identifies genetic variants associated with educational attainment. *Science, 340*(6139), 1467–1471.

Rilling, J., Gutman, D., Zeh, T., Pagnoni, G., Berns, G., & Kilts, C. (2002). A neural basis for social cooperation. *Neuron, 35*(2), 395–405.

Rilling, J. K., Demarco, A. C., Hackett, P. D., Thompson, R., Ditzen, B., Patel, R., & Pagnoni, G. (2011). Effects of intranasal oxytocin and vasopressin on cooperative behavior and associated brain activity in men. *Psychoneuroendocrinology,* doi: 10.1016/j.psyneuen.2011.07.013

Rilling, J. K., Glenn, A. L., Jairam, M. R., Pagnoni, G., Goldsmith, D. R., Elfenbein, H. A., & Lilienfeld, S. O. (2007). Neural correlates of social cooperation and non-cooperation as a function of psychopathy. *Biological Psychiatry, 61*(11), 1260–1271.

Rilling, J. K., Goldsmith, D. R., Glenn, A. L., Jairam, M. R., Elfenbein, H. A., Dagenais, J. E., . . . Pagnoni, G. (2008). The neural correlates of the affective response to unreciprocated cooperation. *Neuropsychologia, 46*(5), 1256–1266.

Rilling, J. K., Sanfey, A. G., Aronson, J. A., Nystrom, L. E., & Cohen, J. D. (2004*a*). The neural correlates of theory of mind within interpersonal interactions. *Neuroimage, 22*(4), 1694–1703.

Rilling, J. K., Sanfey, A. G., Aronson, J. A., Nystrom, L. E., & Cohen, J. D. (2004*b*). Opposing BOLD responses to reciprocated and unreciprocated altruism in putative reward pathways. *Neuroreport, 15*(16), 2539–2543.

Ripke, S., Sanders, A. R., Kendler, K. S., Levinson, D. F., Sklar, P., Holmans, P. A., . . . Gejman, P. V. (2011). Genome-wide association study identifies five new schizophrenia loci. *Nature Genetics, 43*(10), 969–976. doi: 10.1038/ng.940

Rodrigues, S. M., Saslow, L. R., Garcia, N., John, O. P., & Keltner, D. (2009). Oxytocin receptor genetic variation relates to empathy and stress reactivity in humans. *Proceedings of the National Academy of Science of the United States of America, 106*(50), 21437–21441. doi: 0909579106 [pii] 10.1073/pnas.0909579106

Roe, B. E., Tilley, M. R., Gu, H. H., Beversdorf, D. Q., Sadee, W., Haab, T. C., & Papp, A. C. (2009). Financial and psychological risk attitudes associated with two single nucleotide polymorphisms in the nicotine receptor (CHRNA4) gene. *PLoS One, 4*(8), e6704. doi: 10.1371/journal.pone.0006704

Rogaeva, E., Meng, Y., Lee, J. H., Gu, Y., Kawarai, T., Zou, F., . . . St George-Hyslop, P. (2007). The neuronal sortilin-related receptor SORL1 is genetically associated with Alzheimer disease. *Nature Genetics, 39*(2), 168–177. doi: 10.1038/ng1943

Roiser, J. P., de Martino, B., Tan, G. C., Kumaran, D., Seymour, B., Wood, N. W., & Dolan, R. J. (2009). A genetically mediated bias in decision making driven by failure of amygdala control. *Journal of Neuroscience, 29*(18), 5985–5991.

Roussos, P., Katsel, P., Davis, K. L., Bitsios, P., Giakoumaki, S. G., Jogia, J., . . . Haroutunian, V. (2012). Molecular and genetic evidence for abnormalities in the nodes of Ranvier in schizophrenia. *Archives of General Psychiatry, 69*(1), 7–15. doi: 10.1001/archgenpsychiatry.2011.110

Rushton, J. P. (2004). Genetic and environmental contributions to pro-social attitudes: A twin study of social responsibility. *Proceedings of the Royal Society B: Biological Sciences, 271*(1557), 2583–2585.

Saphire-Bernstein, S., Way, B. M., Kim, H. S., Sherman, D. K., & Taylor, S. E. (2011). Oxytocin receptor gene (OXTR) is related to psychological resources. *Proceedings of the National Academy of Sciences of the United States of America, 108*(37), 15118–15122. doi: 10.1073/pnas.1113137108

Sauce, B., de Brito, R. A., & Peripato, A. C. (2012). Genetic architecture of nest building in mice LG/J x SM/J. *Frontiers in Genetics, 3*, 90. doi: 10.3389/fgene.2012.00090

Sauer, C., Montag, C., Worner, C., Kirsch, P., & Reuter, M. (2012). Effects of a common variant in the CD38 gene on social processing in an oxytocin challenge study: Possible links to autism. *Neuropsychopharma cology: Official Publication of the American College of Neuropsychopharmacology*, doi: 10.1038/npp.2011.333

Schermer, J. A., Vernon, P. A., Maio, G. R., & Jang, K. L. (2011). A behavior genetic study of the connection between social values and personality. *Twin Research in Human Genetics, 14*(3), 233–239. doi: 10.1375/twin.14.3.233 10.1375/twin.14.3.233 [pii]

Schulze, T. G., & McMahon, F. J. (2004). Defining the phenotype in human genetic studies: Forward genetics and reverse phenotyping. *Human Heredity, 58*(3–4), 131–138. doi: 83539 [pii] 10.1159/000083539

Settle, J. E., Dawes, C. T., Christakis, N. A., & Fowler, J. H. (2010). Friendships moderate an association between a dopamine gene variant and political ideology. *Journal of Politics, 72*(4), 1189–1198.

Shamay-Tsoory, S. G. (2011). The neural bases for empathy. *Neuroscientist, 17*(1), 18–24. doi: 1073858410379268 [pii] 10.1177/1073858410379268

Sheese, B. E., Voelker, P. M., Rothbart, M. K., & Posner, M. I. (2007). Parenting quality interacts with genetic variation in dopamine receptor D4 to influence temperament in early childhood. *Developmental Psychopathology, 19*(4), 1039–1046.

Shendure, J., & Ji, H. (2008). Next-generation DNA sequencing. *Nature Biotechnology, 26*(10), 1135–1145. doi: Nbt1486 [pii] 10.1038/nbt1486

Sherkat, D. E., Powell-Williams, M., Maddox, G., & de Vries, K. M. (2011). Religion, politics, and support for same-sex marriage in the United States, 1988–2008. *Social Science Research, 40*(1), 167–180. doi: 10.1016/j.ssresearch.2010.08.009

Shyn, S. I., Shi, J., Kraft, J. B., Potash, J. B., Knowles, J. A., Weissman, M. M., . . . Hamilton, S. P. (2011). Novel loci for major depression identified by genome-wide association study of sequenced treatment alternatives to relieve depression and meta-analysis of three studies. *Molecular Psychiatry, 16*(2), 202–215. doi: Mp2009125 [pii] 10.1038/mp.2009.125

Sigmund, K., & Hauert, C. (2002). Primer: Altruism. *Current Biology, 12*(8), R270–272.

Silk, J. B., Brosnan, S. F., Vonk, J., Henrich, J., Povinelli, D. J., Richardson, A. S., . . . Schapiro, S. J. (2005). Chimpanzees are indifferent to the welfare of unrelated group members. *Nature, 437*(7063), 1357–1359.

Simmons, J. P., Nelson, L. D., & Simonsohn, U. (2011). False-positive psychology. *Psychological Science, 22*(11), 1359–1366.

Siu, H., Zhu, Y., Jin, L., & Xiong, M. (2011). Implication of next-generation sequencing on association studies. *BMC Genomics, 12*(1), 322.

Smalley, S. L., Bailey, J. N., Palmer, C. G., Cantwell, D. P., McGough, J. J., Del'Homme, M. A., . . . Nelson, S. F. (1998). Evidence that the dopamine D4 receptor is a susceptibility gene in attention deficit hyperactivity disorder [see comments]. *Molecular Psychiatry, 3*(5), 427–430.

Smith, V. L. (1962). An experimental study of competitive market behavior. *Journal of Political Economy*, 111–137.

Smolka, M. N., Schumann, G., Wrase, J., Grusser, S. M., Flor, H., Mann, K., . . . Heinz, A. (2005). Catechol-O-methyltransferase val158met genotype affects processing of emotional stimuli in the amygdala and prefrontal cortex. *Journal of Neuroscience, 25*(4), 836–842. doi: 25/4/836 [pii] 10.1523/JNEUROSCI.1792-04.2005

Snyder, S. H. (1976). The dopamine hypothesis of schizophrenia: Focus on the dopamine receptor. *American Journal of Psychiatry, 133*(2), 197–202.

Spencer, C., Hechter, E., Vukcevic, D., & Donnelly, P. (2011). Quantifying the underestimation of relative risks from genome-wide association studies. *PLoS Genetics, 7*(3), e1001337. doi: 10.1371/journal.pgen.1001337

Stanley, D. A., Sokol-Hessner, P., Fareri, D. S., Perino, M. T., Delgado, M. R., Banaji, M. R., & Phelps, E. A. (2012). Race and reputation: Perceived racial group trustworthiness influences the neural correlates of trust decisions. *Philosophical Transactions of the Royal Society of London B: Biological Sciences, 367*(1589), 744–753. doi: Rstb.2011.0300 [pii] 10.1098/rstb.2011.0300

Stanovich, K. E., & West, R. F. (2000). Individual differences in reasoning: Implications for the rationality debate? *Behavioral and Brain Sciences, 23*(5), 645–665.

Stein, J. L., Medland, S. E., Vasquez, A. A., Hibar, D. P., Senstad, R. E., Winkler, A. M., . . . Thompson, P. M. (2012). Identification of common variants associated with human hippocampal and intracranial volumes. *Nature Genetics, 44*(5), 552–561. doi: Ng.2250 [pii] 10.1038/ng.2250

Strobel, A., Debener, S., Anacker, K., Muller, J., Lesch, K. P., & Brocke, B. (2004). Dopamine D4 receptor exon III genotype influence on the auditory evoked novelty P3. *Neuroreport, 15*(15), 2411–2415.

Takahashi, H., Takano, H., Camerer, C. F., Ideno, T., Okubo, S., Matsui, H., . . . Suhara, T. (2012). Honesty mediates the relationship between serotonin and reaction to unfairness. *Proceedings of the National Academy of Science of the United States of America, 109*(11), 4281–4284. doi: 1118687109 [pii] 10.1073/pnas.1118687109

Tansey, K. E., Hill, M. J., Cochrane, L. E., Gill, M., Anney, R. J., & Gallagher, L. (2011). Functionality of promoter microsatellites of arginine vasopressin receptor 1A (AVPR1A): Implications for autism. *Molecular Autism, 2*(1), 3. doi: 2040-2392-2-3 [pii] 10.1186/2040-2392-2-3

Tesser, A. (1993). The importance of heritability in psychological research: The case of attitudes. *Psychological Review, 100*(1), 129–142.
The 1000 Genomes Project Consortium (2010, October 28). A map of human genome variation from population-scale sequencing. *Nature, 467*, 1061–1073.
The 1000 Genomes Project Consortium (2012, November 1). An integrated map of genetic variation from 1,092 human genomes. *Nature, 491*, 56–65.
Thompson, P. M., Martin, N. G., & Wright, M. J. (2010). Imaging genomics. *Current Opinions in Neurology, 23*(4), 368–373. doi: 10.1097/WCO.0b013e32833b764c
Timberlake, D. S., Rhee, S. H., Haberstick, B. C., Hopfer, C., Ehringer, M., Lessem, J. M.,... Hewitt, J. K. (2006). The moderating effects of religiosity on the genetic and environmental determinants of smoking initiation. *Nicotine and Tobacco Research, 8*(1), 123–133. doi: U61620554156Q647 [pii] 10.1080/14622200500432054
Tingley, D. (2006). Neurological imaging as evidence in political science: A review, critique, and guiding assessment. *Social Science Information, 45*(1), 5–33.
Tom, S. M., Fox, C. R., Trepel, C., & Poldrack, R. A. (2007). The neural basis of loss aversion in decision-making under risk. *Science, 315*(5811), 515–518.
Tops, M., van Ijzendoorn, M. H., Riem, M. M., Boksem, M. A., & Bakermans-Kranenburg, M. J. (2011). Oxytocin receptor gene associated with the efficiency of social auditory processing. *Frontiers in Psychiatry/Frontiers Research Foundation, 2*, 60. doi: 10.3389/fpsyt.2011.00060
van 't Wout, M., & Sanfey, A. G. (2011). Interactive decision-making in people with schizotypal traits: A game theory approach. *Psychiatry Research, 185*(1–2), 92–96. doi: S0165-1781(10)00297-0 [pii] 10.1016/j.psychres.2010.05.013
van den Bos, W., van Dijk, E., Westenberg, M., Rombouts, S. A., & Crone, E. A. (2009). What motivates repayment? Neural correlates of reciprocity in the trust game. *Social Cognitive and Affective Neuroscience, 4*, 294–304.
van Dyck, C. H., Malison, R. T., Jacobsen, L. K., Seibyl, J. P., Staley, J. K., Laruelle, M.,... Gelernter, J. (2005). Increased dopamine transporter availability associated with the 9-repeat allele of the SLC6A3 gene. *Journal of Nuclear Medicine, 46*(5), 745–751.
van Ijzendoorn, M. H., & Bakermans-Kranenburg, M. J. (2006). DRD4 7-repeat polymorphism moderates the association between maternal unresolved loss or trauma and infant disorganization. *Attachment and Human Development, 8*(4), 291–307.
van IJzendoorn, M. H., & Bakermans-Kranenburg, M. J. (2012). A sniff of trust: Meta-analysis of the effects of intranasal oxytocin administration on face recognition, trust to in-group, and trust to out-group. *Psychoneuroendocrinology, 37*, 438–443.
van IJzendoorn, M. H., Bakermans-Kranenburg, M. J., & Mesman, J. (2008). Dopamine system genes associated with parenting in the context of daily hassles. *Genes Brain and Behavior, 7*(4), 403–410.
van Ijzendoorn, M. H., Huffmeijer, R., Alink, L. R., Bakermans-Kranenburg, M. J., & Tops, M. (2011). The impact of oxytocin administration on charitable donating is moderated by experiences of parental love-withdrawal. *Frontiers in Psychology, 2*, 258. doi: 10.3389/fpsyg.2011.00258
van Lange, P. A. M. (1999). The pursuit of joint outcomes and equality in outcomes: An integrative model of social value orientation. *Journal of Personality and Social Psychology, 77*(2), 337–349.
Walum, H., Westberg, L., Henningsson, S., Neiderhiser, J. M., Reiss, D., Igl, W.,... Lichtenstein, P. (2008). Genetic variation in the vasopressin receptor 1a gene (AVPR1A) associates with pair-bonding behavior in humans. *Proceedings of the National Academy of Science of the United States of America, 105*(37), 14153–14156.
Wang, K., Li, M., & Hakonarson, H. (2010). Analysing biological pathways in genome-wide association studies. *Nature Reviews Genetics, 11*(12), 843–854.
Warneken, F., & Tomasello, M. (2009*a*). The roots of human altruism. *British Journal of Psychology, 100*(Pt 3), 455–471. doi: 311766 [pii] 10.1348/000712608X379061
Warneken, F., & Tomasello, M. (2009*b*). Varieties of altruism in children and chimpanzees. *Trends in Cognitive Science, 13*(9), 397–402. doi: S1364-6613(09)00149-1 [pii] 10.1016/j.tics.2009.06.008
Weiland, S., Hewig, J., Hecht, H., Mussel, P., & Miltner, W. H. (2012). Neural correlates of fair behavior in interpersonal bargaining. *Social Neuroscience*, doi: 10.1080/17470919.2012.674056
Westen, D., Blagov, P. S., Harenski, K., Kilts, C., & Hamann, S. (2006). Neural bases of motivated reasoning: An fMRI study of emotional constraints on partisan political judgment in the 2004 US presidential election. *Journal of Cognitive Neuroscience, 18*(11), 1947–1958.
Wilhelm, M. O. (2006). New data on charitable giving in the PSID. *Economics Letters, 92*(1), 26–31.
Winawer, M. R. (2006). Phenotype definition in epilepsy. *Epilepsy Behavior, 8*(3), 462–476. doi: S1525-5050(06)00022-9 [pii] 10.1016/j.yebeh.2006.01.012
Wischniewski, J., & Brune, M. (2011). Moral reasoning in schizophrenia: An explorative study into economic decision making. *Cognitive Neuropsychiatry*, 1–16. doi: 932770295 [pii] 10.1080/13546805.2010.539919
Wittmann, M., Lovero, K. L., Lane, S. D., & Paulus, M. P. (2010). Now or later? Striatum and insula activation to immediate versus delayed rewards. *Journal of Neuroscience, Psychology, and Economics, 3*(1), 15–26. doi: 10.1037/a0017252
Wojczynski, M. K., & Tiwari, H. K. (2008). Definition of phenotype. *Advances in Genetics, 60*, 75–105. doi: 10.1016/S0065-2660(07)00404-X
Wong, M. Y., Day, N. E., Luan, J. A., & Wareham, N. J. (2004). Estimation of magnitude in gene-environment interactions in the presence of measurement error. *Statistics in Medicine, 23*(6), 987–998. doi: 10.1002/sim.1662
Wood, A. C., Rijsdijk, F., Saudino, K. J., Asherson, P., & Kuntsi, J. (2008). High heritability for a composite index of children's activity level measures. *Behavior Genetics, 38*(3), 266–276. doi: 10.1007/s10519-008-9196-1
Wood, R. M., Rilling, J. K., Sanfey, A. G., Bhagwagar, Z., & Rogers, R. D. (2006). Effects of tryptophan depletion on the performance of an iterated Prisoner's Dilemma game in healthy adults. *Neuropsychopharmacology, 31*(5), 1075–1084.
Wu, N., Li, Z., & Su, Y. (2012). The association between oxytocin receptor gene polymorphism (OXTR) and trait empathy. *Journal of Affective Disorders, 138*(3), 468–472. doi: 10.1016/j.jad.2012.01.009

Wu, S., Jia, M., Ruan, Y., Liu, J., Guo, Y., Shuang, M.,... Zhang, D. (2005). Positive association of the oxytocin receptor gene (OXTR) with autism in the Chinese Han population. *Biological Psychiatry, 58*(1), 74–77.

Xu, L., Liang, Z. Y., Wang, K., Li, S., & Jiang, T. (2009). Neural mechanism of intertemporal choice: From discounting future gains to future losses. *Brain Research, 1261*, 65–74. doi: S0006-8993(08)03136-3 [pii] 10.1016/j.brainres.2008.12.061

Yang, J., Benyamin, B., McEvoy, B. P., Gordon, S., Henders, A. K., Nyholt, D. R.,... Visscher, P. M. (2010). Common SNPs explain a large proportion of the heritability for human height. *Nature Genetics, 42*(7), 565–569. doi: Ng.608 [pii] 10.1038/ng.608

Yang, J., Lee, S. H., Goddard, M. E., & Visscher, P. M. (2011). GCTA: A tool for genome-wide complex trait analysis. *American Journal of Human Genetics, 88*(1), 76–82. doi: 10.1016/j.ajhg.2010.11.011

Young, L. J., Lim, M. M., Gingrich, B., & Insel, T. R. (2001). Cellular mechanisms of social attachment. *Hormones and Behavior, 40*(2), 133–138.

Zak, P. J., Kurzban, R., Ahmadi, S., Swerdloff, R. S., Park, J., Efremidze, L.,... Matzner, W. (2009). Testosterone administration decreases generosity in the ultimatum game. *PLoS One, 4*(12), e8330. doi: 10.1371/journal.pone.0008330

Zhang, X., Monakhov, M., Lai, P. S., Liu, J., Tong, T., Chew, S. H., & Ebstein, R. P. (in preparation). The role of oxytocin in women's impulsive purchasing behavior.

Zhong, S., Chark, R., Ebstein, R. P., & Chew, S. H. (2012). Imaging genetics for utility of risks over gains and losses. *Neuroimage, 59*(1), 540–546. doi: S1053-8119(11)00792-0 [pii] 10.1016/j.neuroimage.2011.07.031

Zhong, S., Chew, S. H., Set, E., Zhang, J., Xue, H., Sham, P. C.,... Israel, S. (2009). The heritability of attitude toward economic risk. *Twin Research and Human Genetics, 12*(1), 103–107.

Zhong, S., Israel, S., Shalev, I., Xue, H., Ebstein, R. P., & Chew, S. H. (2010). Dopamine D4 receptor gene associated with fairness preference in ultimatum game. *PLoS One, 5*(11), e13765. doi: 10.1371/journal.pone.0013765

Zhong, S., Israel, S., Xue, H., Ebstein, R. P., & Chew, S. H. (2009). Monoamine oxidase A gene (MAOA) associated with attitude towards longshot risks. *PLoS One, 4*(12), e8516. doi: 10.1371/journal.pone.0008516

Zhong, S., Israel, S., Xue, H., Sham, P. C., Ebstein, R. P., & Chew, S. H. (2009). A neurochemical approach to valuation sensitivity over gains and losses. *Proceedings of the Royal Society B: Biological Sciences, 276*(1676), 4181–4188. doi: Rspb.2009.1312 [pii] 10.1098/rspb.2009.1312

Zhong, S., Monakhov, M., Mok, H. P., Tong, T., Lai, P. S., Chew, S. H., & Ebstein, R. P. (2012). U-shaped relation between plasma oxytocin levels and behavior in the trust game. *PLoS One, 7*(12), e51095. doi: 10.1371/journal.pone.0051095

Zuk, O., Hechter, E., Sunyaev, S. R., & Lander, E. S. (2012). The mystery of missing heritability: Genetic interactions create phantom heritability. *Proceedings of the National Academy of Science of the United States of America, 109*(4), 1193–1198. doi: 1119675109 [pii] 10.1073/pnas.1119675109

CHAPTER

10 Gene-by-Environment Mouse Models for Mood Disorders

Maria Razzoli, Alessandro Bartolomucci, *and* Valeria Carola

Abstract

Much of the impact of genes on mood disorders likely depends on interactions between genes and the environment. Recent studies demonstrating an interaction between specific genes (e.g., serotonin transporter and brain-derived neurotrophic factor genes) and life stressful events (early and/or adult) in the modulation of several mood disorders have compelled researchers to incorporate information about adverse environmental experiences into the study of genetic risk factors; these same gene-by-environment (G×E) interactions have been identified in mouse models. Notably, G×E interactions not yet described in humans (e.g., serotonin 1A receptor gene) have been uncovered, providing helpful indications to discover similar interactions in humans. Accurate knowledge of the modality of expression of gene-by-stress interaction may help design prevention protocols aimed at identifying susceptibility to mood disorders on the basis of genetic predisposition and exposure to environmental stressful conditions, thus providing patients with appropriate pharmacological and psychological support.

Key Words: gene-by-environment, maternal environment, anxiety, depression, serotonin transporter, brain-derived neurotrophic factor, early environment, adult environment, mouse models, serotonin 1A receptor

Psychiatric disorders are known to be influenced by both genetic and environmental factors. Twin studies suggest that genetic factors contribute to about 30–40 percent of the variance in incidence of mood disorders, whereas environmental factors explain the remaining 60–70 percent. However, it is becoming clear that few genetic and environmental factors act alone, although they regularly act in concert to determine predisposition to psychiatric disorders.

A typical example of this concept is the study of the effect of functional polymorphism in the serotonin transporter (5-HTT, SLC6A4) gene promoter (called serotonin transporter gene-linked polymorphic region, 5-HTTLPR) on human behavior. When this polymorphism was identified, the short variant (S allele) was associated with a very small but significant increase in anxiety-related personality traits such as neuroticism and disagreeableness (Greenberg et al., 2000; Lesch et al., 1996). However, a much bigger effect was identified when the polymorphism was studied together with alteration of early and/or adult environmental conditions (exposure to life stress). Specifically, it has been shown that human subjects carrying one or two of the short variants (S/S or S/L individuals) demonstrated increased risk for major depression when exposed to either childhood maltreatment or stressful life events (Caspi et al., 2003; Kaufman et al., 2004 Risch et al., 2009; for review, see Uher et al., 2011). Studies in healthy subjects have also shown evidence for a similar

5-HTTLPR-by-stress effect on physiological and behavioral traits, including heart rate reactivity and anxiety (Caspi, Hariri, Holmes, Uher, & Moffitt, 2010). These studies have precipitated far-reaching changes in the psychiatric genetics field and have compelled researchers to incorporate information about adverse environmental experiences into the study of genetic risk factors.

Conversely, because of its complexity, the interaction between genes and environment is not always an easy phenomenon to study, and, even more, to interpret. A good example of such complexity is given by the study of a common single-nucleotide polymorphism (SNP) in the brain-derived neurotrophic factor (BDNF) gene, consisting of a methionine (*Met*) substitution for valine (*Val*) at codon 66 (Val66Met) interfering with BDNF activity-dependent secretion (Egan et al., 2003). In an elegant study, Gatt et al. (2009) refute the view that one of the BDNF alleles (*Val/Met*) may induce either "risk" or "protection" in the presence of environmentally adverse conditions, suggesting instead that both alleles may contribute to distinct mechanisms of risk when an individual is also exposed to early life stressors. They observed an increased level of syndromal anxiety in individuals carrying the *Val/Val* allele and exposed to early-life traumatic events. On the contrary, higher syndromal depression was observed in individuals exposed to similar environmental conditions but carrying the *Val/Met* allele. They explain this phenomenon as dependent on a presumably different effect of this interaction on brain circuits that mediate either syndromal anxiety or syndromal depression.

Imaging studies have partially elucidated the neural circuits involved in these gene-by-environment (G×E) risk factors (Canli et al., 2006; Heinz et al., 2007; Gatt et al., 2009), but the identification of the molecular mechanisms implicated is slow and difficult to perform in humans. This kind of research can take advantage of animal models, in which the manipulation of genetic and environmental components is easier to perform. Thus, the mouse model can be considered a powerful tool to model and study G×E interactions. Some of the advantages of using mouse models are listed here:

1. The powerful tissue- and temporal-specific transgenic tools available in the mouse make this system ideal to model the disruption/alteration of a specific human gene and analyze the behavioral and pathological outcome. In this context, screening mutations that, in the presence of adverse environmental conditions, make individuals more susceptible to develop pathological adult phenotypes is possible in mouse on a large scale (high-throughput screening).
2. Human studies aimed at identifying molecular mechanisms involved in psychiatric disorders are complicated by several factors, including lack of experimental control over the subject's environment, genetic background, and inaccessibility of brain tissues required for the analysis. The use of the mouse model overcomes these problems.
3. Applying protocols of manipulation of the perinatal and adult environment is easily, routinely, and successfully done in mouse. In addition, in mouse, it is possible to identify the outcome of each single stressful experience. This is difficult to perform in humans due to the usual combination of multiple traumatic life events that individuals experience during their lives.

In the next sections, we briefly review the experimental procedures most commonly used to manipulate the genetic and environmental components in mouse. We hope that this chapter provides interested researchers a useful overview of suitable techniques or methods for their own G×E interaction studies in mouse. However, we recommend that readers refer to specific literature (Claessens et al., 2011, Millstein & Holmes, 2007) for more detailed descriptions of these procedures (both genetic and environmental) and their effects on behavioral, physiological, and biological phenotypes. We further describe several examples of established mouse models of gene-by-stress risk factors for mood disorders.

I. Identifying Interactions Between Genes and Environment in the Mouse: What We Need to Manipulate

A. Genetic Component

The possibility of easily manipulating the genome makes the mouse an ideal candidate to model human diseases in which the modification/alteration of gene function is involved. Indeed, in the mouse, sophisticated transgenic techniques allow researchers to study the effect of gene expression level alterations, as well as to control gene expression in a temporary and tissue-specific manner.

A.1 TRANSGENIC LINES OF MICE

Transgenic lines of mice are usually generated to investigate the consequences of gene overexpression. The production of these lines is based on the introduction of exogenous genes/DNA sequences (transgenes) that the host randomly integrates in several copies into the genome (Gordon & Ruddle, 1981; Harbers, Jahner, & Jaenisch, 1981). The use of classic transgenic techniques is limited by the impossibility of having spatial and temporal control over gene expression in transgenic mice (Hogan, Beddington, Costantini, & Lacy, 1994). This is an important issue since, in humans, gene deregulation is frequently limited to specific cell populations and to specific time windows. To address this need, several conditional strategies have been developed to control the expression of transgenes (tissue/time-specific conditional) by the simple, benign inclusion of a drug or an inert small molecule (Lewandoski, 2001).

A.2 KNOCKOUT LINES OF MICE

Knockout mice are generated to investigate the effect of gene loss of function or gain of (altered) function. To produce these lines, a specific genetic locus is targeted and made nonfunctional either by the deletion of DNA sequence information from the targeted locus or by the insertion of irrelevant DNA sequence information. Similar to the transgenic lines, several knockout lines of mice are available in which it is possible to produce temporal or cell- and/or tissue-specific ablation of gene expression (Evans & Kaufman, 1981; Smithies, Gregg, Boggs, Koraleswksi, & Kucherlapati, 1985; Thomas, Folger, & Capecchi, 1986).

A.3 KNOCKIN LINES OF MICE

In knockin mice, an exact gene locus has been modified, either by adding sequence information that is usually absent in the endogenous genetic locus or by conducting one-for-one substitution of DNA sequence information (Roebroek, Wu, & Bram, 2003). An interesting extension of this technique is based on the "humanization" procedure, in which the murine gene is replaced by its human counterpart. "Humanization" of animal models is considered a very promising tool in deciphering physiological processes and increasing the predictive validity of animal models.

Several issues should be considered when choosing the most suitable mouse "genetic" model to be used in a G×E study. It is desirable to choose the mouse genetic model that best mimics changes in gene expression observed in humans. For example, when studying the interaction between a specific functional SNP and the environment, since SNPs usually induce alterations of gene expression level, but not the total abolition of expression, it would be desirable to use a genetic mouse model in which the gene expression level is modified (e.g., heterozygous knockout mice) but not completely abolished. Additionally, it is important to have a clear knowledge of human tissue and temporal gene expression in order to choose the best genetic mouse model with a similar expression pattern.

B. Environmental Component: Perinatal Environment

Different experimental protocols have been applied to manipulate the early pre -and postnatal environment in mouse models (for a detailed analysis of postnatal influences related to nutrition and epigenetic mechanisms affecting cognition and emotion, see the chapter by Naninck, Lucassen, and Korosi elsewhere in this Handbook). These paradigms have been shown to induce different and variable short- and long-term effects on mouse emotionality and stress response, depending on the type and timing of experimental manipulation (Bartolomucci et al., 2004*b*; Levine, 1957; Maccari & Morley-Fletcher, 2007; Miczek, Yap, & Covington III, 2008; Millstein & Holmes, 2007; Veenema, 2009). The most common paradigms used to manipulate the early environment are (a) prenatal perturbation of the maternal environment, applying stressful conditions to the mother and indirectly to the offspring in uterus; (b) postnatal perturbation of the mother–pup relationship, with resulting alteration in maternal care received, nutritional factors provided, and environmental temperature the pups are subjected to; and (c) perturbation of the postnatal physical and/or social environmental conditions, with effects on both mother (perturbation of maternal behavior and stress response) and offspring.

Maternal prenatal stress. Different kinds of stress protocols against the pregnant mother can be used in paradigms aimed to induce stress in the offspring during the prenatal period. Among them, restraint stress, exposure to a cold environment (4 °C), overnight food deprivation, swimming stress in room-temperature water, exposure to constant light, and social stress induced by overcrowded housing conditions are frequently used (Darnaudéry & Maccari, 2008; Maccari & Morley-Fletcher, 2007).

Separation from the mother. During the first postnatal weeks, mouse pups have an exclusive relationship with their mothers, and they are completely dependent on their mothers' care. A substantial proportion of paradigms manipulating the early postnatal environment aim to perturb this relationship. Models of this class range from short separation from the mother with only slightly changes nutritional intake and deregulation of body temperature, to longer maternal separations associated with large or severe alteration/reduction of nutritional factors provided and substantial deregulation of body temperature.

Early handling. In this protocol, the pups are separated daily from the mother for a short period of time (maximally 15 minutes). During this separation, they are placed either in a novel or familiar environment with/without control/balance of temperature (D'Amato et al., 1992; D'Amato, Cabib, Ventura, & Orsini, 1998).

Maternal separation. This procedure is similar to early handling, but the period of separation of pups from their mother is longer, varying from 1 to 8 hours per day (Millstein, Ralph, Yang, & Holmes, 2006; Romeo et al., 2003; van Heerden, Russell, Korff, Stein, & Illing, 2010).

Early deprivation. During this manipulation, a pup is separated (isolated) from its litter as well as from its dam. This protocol is widely considered to induce more severe effects than simple maternal separation in which the separation is only from the mother and not from the rest of the litter (Millstein & Holmes, 2007).

Single episode of maternal separation. Pups exposed to this protocol/manipulation experience a single episode of separation for a very long period (usually 24 hours). This procedure can be applied at different time points during he postnatal period (Macri & Laviola, 2004).

Early weaning. Mice exposed to this protocol are weaned precociously at 14 days of age rather than at the conventional time of 21–28 days. This induces a dramatic change in nutritional levels and maternal care received (Kikusui & Mori, 2009) by the pups.

B.1 QUALITY OF MATERNAL ENVIRONMENT

Several paradigms manipulating the early postnatal environment are based on varying the environment provided by the mother to her pups, including the amount of maternal care, the composition of nutritional factors (through milk), and hormonal "stimulation."

Naturally occurring variations in maternal environment. In rodents, by using genetically distinct inbred strains, it is possible to observe stable differences in maternal care during the early postnatal period. For example C57BL/6J and DBA/2J female mice are considered "good mothers," providing high levels of maternal care and spending a lot of time in the nest, whereas BALB/cByJ and 129S2 female mice are usually considered "bad mothers," showing low levels of care and spending little time in the nest (Caldji, Diorio, & Meaney, 2003; Carola, Frazzetto, & Gross, 2006; Carola, Mirabeau, & Gross, 2011*b*; Shoji & Kato, 2006). These differences go along with variations in the nutritional (e.g., quantity and gross composition of milk; Ragueneau 1987) and hormonal factors/state (e.g., stress hormones, Savignac et al., 2011) provided to the pups from their mother. A singular case has been described for F1 hybrid females derived from reciprocal intercrosses between the two inbred strains C57BL/6J and BALB/cByJ. These females are genetically identical, but they exhibit different maternal behavior depending on the mother's strain (Calatayud, Coubard, & Belzung, 2004; Carola et al., 2006; 2008; 2011*b*; Carola & Gross, 2010; Carola, Pascucci, Puglisi-Allegra, Cabib, & Gross, 2011*a*; Chevalet, Le Pape, & Lassalle, 1987).

Cross-fostering. According to this technique, offspring are removed from their biological mother at birth and raised by a foster mother until weaning. A cross-fostering protocol is usually applied to study the impact of postnatal environment on the physiological and behavioral pattern of offspring (Bartolomucci et al., 2004*b*; Priebe et al., 2005).

Repeated cross-fostering. In this experimental rearing paradigm, the offspring changes caregiver every 24 hours from postnatal day 1 to 4. In this protocol, each dam is shifted to four different litters and each litter is shifted to four different dams. This protocol is applied to create an "unstable" maternal environment for the pups (D'Amato et al., 2011*a*).

Double-mothering. In this protocol, the amount of maternal care provided to the pups is experimentally intensified by placing a lactating and/or nonlactating female together with the dam and its litter from birth until weaning (D'Amato et al., 2011*b*).

B.2 CHANGES OF THE PHYSICAL AND/OR SOCIAL ENVIRONMENTAL CONDITIONS

Several protocols aim to alter postnatal environmental conditions using manipulation of the social environment. In one of these paradigms, *enrichment of social* environment is obtained by housing together three females with their offspring (Branchi, 2009). This arrangement increases the possibility of social interactions for the offspring. Conversely, other paradigms are based on the *impoverishment/worsening* of the environment. For example, in several studies, mouse mothers and their offspring have been exposed to soiled bedding from the cage of unfamiliar adult males (adverse stimuli) or to reduced amounts of nesting-bedding material (Brunson et al., 2005; Ivy, Brunson, Sandman, & Baram, 2008; Ivy et al., 2010; Rice, Sandan, Lenjavi, & Baram, 2008).

C. Environmental Component: Adult Environment

Several models of adult stress are used in the laboratory, including some based on variation of both physical and social environmental conditions (stressful environment) and others based on variation of solely the social environment (adult social stress).

The first category includes models such as *unpredictable chronic mild stress,* which is considered to mimic stressful socioenvironmental events that have a role in precipitating human psychiatric syndromes (Ducottet, Aubert, & Belzung, 2004; Willner, 2005). Mice exposed to this protocol are repeatedly subjected to various stressors according to a "random" schedule. Stressors that have been used include, for example: altered bedding conditions (sawdust change, substitution of sawdust with 21°C water, rat or cat feces directly placed in the home cage), cage tilting (45 degrees) or shaking, cage exchange (mice placed in the empty cage of another male), induced defensive posture (repeated slight grips on the back until the mouse showed a defensive posture), and altered length and time of light–dark cycle

Models of adult social stress can be categorized into one of two broadly defined groups: *dyadic interactions* and *social hierarchy* or *colony* models (Bartolomucci, 2007; Bartolomucci & Leopardi, 2009*a*; Bartolomucci et al., 2009*b*)

Resident/intruder paradigm. This paradigm consists of the placement of a male into a territory in which he will become resident and which he will defend from unfamiliar intruders of the same sex (Ginsburg & Alle, 1942; Miczek, 1979). The intruder is a conspecific taken out of his home cage and introduced into the resident cage. A fight will occur, leading to the definition of a hierarchy in which the two mice will acquire either dominant or subordinate status relative to each other. Physiological consequences of the fight, such as changes in corticosterone levels, body weight, and resting body temperature, will be remarkable in both winners and losers (Koolhaas et al., 2011; Sgoifo et al., 1999). Given these responses, the resident/intruder test is considered a very useful tool to investigate the acute stress response. Furthermore, on repeated or chronic application, this procedure allows researchers to establish stable hierarchical social relationship between dyads of male rodents and the development of valid animal models for human stress-related disease (Bartolomucci, 2007).

Sensory contact paradigm. This protocol is designed to induce a depression-like state in mice based on chronic social confrontations (Kudryavtseva, 1991). In this model, the subordinate animal is subjected to multiple defeats by a different dominant individual on a daily basis for up to 30 days. The subordinate animal is then brought to experience chronic defeat and continuous sensory contact with an unfamiliar conspecific. Most of the consequences of this procedure on the subordinate animal involve decreased activity and increased anxiety, up-regulation of hypothalamic-pituitary-adrenal (HPA) axis function, and alterations in brain neurochemistry, which together are comparable to depressive-like states (Berton et al., 2006).

Chronic psychosocial stress. This model is based on stable resident/intruder dyads, which live in chronic sensory contact for up to 6 weeks, interspaced with daily physical aggressive encounters (Bartolomucci et al., 2001; 2004*a*). At the beginning of the stress protocol, the social relations between the resident and the intruder mouse undergo dynamic changes, leading to either the resident or the intruder becoming dominant. The main difference compared to the sensory contact model is that, in the chronic psychosocial stress protocol, the same dyads are maintained during the length of the procedure, thus allowing a clearer investigation of the stress effect on social status. Despite the fact that mice exposed to both sensory contact and chronic psychosocial stress develop similar depression-like changes, the physiological effects of animals exposed to these two models are not fully overlapping (e.g., increased body weight and vulnerability to obesity in subordinate mice in the chronic psychosocial stress model; Bartolomucci et al., 2004*a*; 2009*b*; 2010; Dadomo et al., 2011) versus body weight loss in subordinate mice in the sensory contact model (Berton et al., 2006; Kudryavtseva, 1991).

Mouse model of social stress based on differential housing. In this stress paradigm, group-housed male mice (residents and experimental animals) experience repeated social defeat stress due to the daily introduction of an aggressive intruder mouse into their home cage. Experimental groups may undergo this procedure for several days to weeks depending on the protocol (Avitsur, Padgett, & Sheridan, 2006; Padgett et al., 1998). This procedure is particularly useful when a large number of defeated mice is required. However, the outcomes are open to a certain degree of variability, due to the individual aggressiveness of each intruder aggressor, as well as to intrinsic individual differences in the resident group.

II. Mouse Models of the Gene-by-Stress Risk Factor for Mood Disorders

A. Serotonin Transporter

One of the best studied and modeled G×E interaction in mouse is the 5-HTTLPR-by-stress interaction, originally described by Caspi et al. 2003 in humans. Mouse studies focused on this interaction have all taken advantage of homozygous and/or heterozygous 5-HTT knockout mice (Bengel et al., 1998; Lesch et al., 1996) as a model for 5-HTTLPR. These studies have provided useful information on both behavioral phenotypes and biological functions modulated by this interaction, dramatically increasing knowledge of this phenomenon.

A.1 EARLY STRESS

In 2007, Carroll et al. studied for the first time the effects of early environment on behavioral phenotypes in 5-HTT knockout mice. In this study, mouse pups were exposed daily to three mild footshocks over a period of 150 seconds from postnatal day 7 to 13. In adulthood, these mice were tested on a battery of anxiety- and depression-related behavioral tests (elevated plus-maze, light–dark, open field, and forced swimming test). Although these authors observed a genotype effect of 5-HTT mutation in anxiety and depression tests, no interaction between genotype and environmental stressors was detected.

A more recent study (Carola et al., 2008) showed that the amount of maternal care during the first 2 weeks of life modulates the effect of heterozygous 5-HTT knockout mutations on anxiety and depression-like behavior. Mice were raised by F1 hybrid mothers that were genetically identical but nevertheless provided high or low maternal care (C57BL/6 × BALB/c vs. BALB/c × C57BL/6, respectively). Heterozygous 5-HTT knockout mice showed comparable levels of avoidance and behavioral despair to wild-type littermates when exposed to high levels of maternal care, but when subjected to low maternal care these mice displayed significantly more avoidance and behavioral despair than controls. Moreover, together with such depression-like behavior, this last group showed enhanced fear in response to partially conditioned cues, suggesting enhanced cognitive processing of ambiguous threatening cues. This study suggests that environmental stressors interfering with critical social cues such as maternal care can better model childhood maltreatment with respect to physical stressors such as foot shocks.

In a similar manner, Heiming et al. (2009) have assessed anxiety-related behavior in heterozygous and homozygous 5-HTT knockout mice after exposure to adverse environmental stimuli across both pre- and postnatal periods. Pregnant and lactating mothers were exposed to an aversive stimulus by having the bedding from the cage of an unfamiliar adult male introduced into their cages. This represents an olfactory cue that is aversive to both mother and offspring, due to the habit of adult male mice to kill pups that are not their own. Homozygous 5-HTT knockout and wild-type control mice in adulthood were assayed in anxiety-related behavioral tests. A significant effect of genotype was observed in the open field, elevated plus maze, and dark–light tests, with homozygous 5-HTT knockouts showing increased anxiety-like behavior compared to wild-type mice. The exposure to an aversive stimulus (environmental effect) was effective only in homozygous knockout mice and not in wild-type mice, suggesting an increased susceptibility of the knockout to threatening olfactory cues compared to wild-type.

The sensibility of heterozygous 5-HTT knockout mice to perinatal factors/environment is also demonstrated by recent work from Van den Hove et al. (2011), in which a prenatal stress protocol was applied to both wild-type and heterozygous 5-HTT knockout mice. This manipulation increased depressive-like behavior only in heterozygous mutant but not in wild-type mice.

Taken together, these studies corroborate human data and demonstrate the susceptibility of mice carrying a heterozygous/homozygous mutation in the 5-HTT gene to adverse prenatal and/or early postnatal environment.

A.2 ADULT STRESS

The first study to incorporate adult stress in the analysis of behavioral phenotypes in 5-HTT knockout mice used a chronic mild stress procedure, which involved daily exposure to several stressors (forced swimming, wet bedding, light cycle reversal, unfamiliar cage mate, tilted cage, restraint stress, no bedding material, etc.) (Joeyen-Waldorf, Edgar, & Sibille, 2009). In this study, chronic stress induced an increased level of anxiety behavior in the novelty suppressed feeding test, and this effect was potentiated in both heterozygous and homozygous 5-HTT knockout female, but not male, mice.

A similar anxiogenic phenotype was elicited in homozygous 5-HTT knockout mice when they were exposed to a winner or loser experience in a 3-day resident-intruder paradigm (Jansen et al., 2010). The mere experience of adverse social exposure, irrespective of winning or losing the social confrontations, induced increased anxiety-related behavior in homozygous but not in heterozygous 5-HTT knockout mice when tested in the elevated plus maze, dark–light test, and open field.

Following the application of a more severe psychosocial stressor, it was also possible to highlight enhanced stress susceptibility in heterozygous 5-HTT knockout mice (Bartolomucci et al., 2010). Three weeks of continuous sensory contact and brief daily defeat by a dominant unfamiliar CD1 induced a social avoidance response that was greater in heterozygous 5-HTT knockout mice compared to wild-type littermates. Importantly, heterozygous 5-HTT knockout mice were not different from stressed wild-type littermates in several physiological parameters like daily social defeat–induced hyperthermia, body weight gain, and plasma corticosterone, suggesting a role for 5-HTT genotype in the stress coping strategies rather than in the modulation of the direct physiological consequences of stress. These data, together with the results from a similar but shorter social stress procedure (Jansen et al., 2010) also highlight the importance of the length/salience of the stress exposure when addressing the contribution of a particular genotype to stress resilience/susceptibility. More specifically, a 50 percent decrease in 5-HTT function was shown to be linked to increased susceptibility in the continuous chronic psychosocial procedure (Bartolomucci et al., 2010), but not in the intermittent 3-day resident/intruder procedure (Jansen et al., 2010).

A.3 PHYSIOLOGICAL AND MOLECULAR SUBSTRATES OF THE 5-HTT × STRESS RISK FACTOR

Some of the studies described earlier that aimed to model the behavioral outcome of the 5-HTTLPR-by-environment interaction observed in humans also provided useful information on the biomolecular substrates implicated in this interaction. In the study by Carola et al. (2008) showing the susceptibility of heterozygous 5-HTT knockout mice to develop high levels of anxiety and depression-related behavior as a consequence of low maternal care, molecular substrates that correlated with the effect of genotype, early environment, and G×E were also investigated. Specifically, these authors observed: (a) a significant effect of the environment (maternal care) on levels of γ-aminobutyric acid type A ($GABA_A$) receptors (as measured by ligand-binding to the benzodiazepine receptor) in the central nucleus of the amygdala. Mice exposed to low maternal care during early postnatal weeks showed a reduction in the expression level of these receptors compared to mice exposed to high levels of maternal care; (b) a significant genotype effect on serotonin turnover as measured by the ratio of the serotonin metabolite 5-hydroxy-indole-acetic acid (5-HIAA) to serotonin (5-HT) in the hippocampus; specifically, heterozygous 5-HTT knockout mice showed a decreased turnover compared to wild-type littermates regardless of environmental exposure; and (c) a significant interaction of genotype and environment on levels of BDNF mRNA in the CA1 region of the hippocampus, with heterozygous 5-HTT knockouts exposed to low maternal care showing the highest BDNF levels, compared to the experimental groups.

In a follow-up study, the same group (Carola et al., 2011*a*) then investigated the molecular basis of these changes in young mice during the early postnatal period, when high levels of maternal care are usually provided to mouse pups (postnatal day 10). In mouse pups, BDNF mRNA in the hippocampus was elevated in offspring receiving low maternal care regardless of genotype. This result suggests that the low-functioning 5-HTT variant may act to maintain elevated BDNF expression induced by low maternal care. Thus, 5-HTT genotype may not modulate the immediate impact of environment on neurophysiology but rather control its perseverance. At the same time, a significant interaction of genotype and environment was observed on levels of excitatory neurotransmitter receptors (of the α-amino-3-hydroxy-5-methyl-4-isoxazolepropionic acid [AMPA] class) in the hippocampus at postnatal day 10. Heterozygous 5-HTT knockout mice exposed to low maternal care showed increased levels of these receptors, compared to the other experimental groups. This difference was not detected in adulthood (V. Carola, unpublished observations) and thus enhanced excitatory neurotransmission in hippocampus may serve as an immediate substrate for early stress and 5-HTT variation that subsequently triggers long-term interacting genotype and environment effects in this or other brain circuits.

Recently, Van den Hove et al. (2011) have also tried to elucidate the molecular changes underlying the G×E effect they observed in depression-like behavior of heterozygous 5-HTT knockout female mice. They performed a genome-wide gene expression profiling study on hippocampal tissue collected from these mice. This analysis revealed that genotype, environment (prenatal stress paradigm), and G×E interaction differentially affected the expression of numerous genes and related pathways within the female hippocampus. In summary, they observed (a) distinct genotype and environmental effects on mitogen activated protein kinase (MAPK) and neurotrophin signaling; and (b) genotype-by-early environment interaction effect on cytokine and Wnt signalling.

Finally, the neural substrates of the 5-HTT × adult stress risk factor have been also recently investigated. Bartolomucci et al. (2010) observed a significant genotype effect on serotonin turnover in most brain regions, whereas a significant G×E interaction effect was revealed in frontal cortex serotonin homeostasis, with the lowest serotonin turnover in heterozygous 5-HTT knockout mice experiencing chronic social defeat. This result stands in contrast with serotonin turnover data from experiments with 5-HTT × early stress (Carola et al., 2008), where no such interaction was observed. Why frontal cortical serotonin turnover is more plastic in adulthood than in childhood following stress needs to be clarified but could reflect differences in the principal circuits engaged by stress at these ages or the increased passage of time since experience of the childhood stressor.

B. Brain Derived Neurotrophic Factor

Experiments aimed at modeling the interaction between the BDNF polymorphism (Val66Met; Egan et al., 2003) and environmental stress in humans (Gatt et al., 2009; Gerritsen et al., 2012; Miller et al., 2012) have been conducted in mice and used different transgenic and knockout mouse lines, including heterozygous BDNF knockout mice (Ernfors, Lee, & Jaenisch, 1994), forebrain conditional BDNF knockout mice (BDNF deletion selectively in forebrain; Monteggia et al., 2004), transgenic mice overexpressing BDNF in forebrain excitatory neurons (Huang et al., 1999), and transgenic mice overexpressing the dominant-negative truncated splice variant of BDNF receptor trkB in cortical and hippocampal neurons (Saarelainen et al., 2000).

B.I EARLY STRESS

To date, only one study has examined the ability of genetic variation in BDNF mouse gene to moderate the long-term effects of early environment on adult mouse behavior (Carola et al., 2010). In the first experiment of this study, mouse pups were allowed to develop in the presence of two different environmental maternal conditions, which were obtained using F1 hybrid mothers derived from reciprocal intercrosses between the two inbred strains C57BL/6J and BALB/cByJ (genetically identical mothers characterized by different maternal styles; Calatayud et al., 2004; Carola et al., 2006, 2008, 2011*a*; 2011*b*; Carola & Gross, 2010; Chevalet et al, 1987), both prenatally and postnatally. In the second experiment, the maternal environment was changed only postnatally. F1 hybrid pups obtained from crosses between the BALB/cByJ mothers and C57BL/6J fathers were cross-fostered to either C57BL/6J (high levels of maternal care) or BALB/cByJ (low levels of maternal care) mothers at birth. Similarly, both experiments observed a significant G×E interaction effect, with heterozygous BDNF knockout mice showing increased sensitivity to the maternal environment compared to wild-type mice. Specifically, heterozygous BDNF knockout mice receiving high and low levels of maternal care showed, respectively, low and high levels of avoidance behavior as measured in the open field and elevated plus maze tests. However, a stronger interaction between BDNF genotype and maternal environment was observed in the first experiment, in which both the pre- and postnatal environments were different between the groups, suggesting that BDNF also plays an important role in prenatal maternal programming.

B.2 ADULT STRESS

To evaluate the susceptibility of heterozygous BDNF knockout mice (Ernfors et al., 1994) to adult stress, Advani et al. (2009) exposed these mice to either acute or subchronic mild stress. During acute stress, mice received three injections of saline over 24 hours, whereas mice experiencing subchronic stress were daily injected with saline for 7 consecutive days. The effect of both stress paradigms was then evaluated on depression-like behavior and plasma stress-induced corticosterone. After both mild acute and subchronic stress, heterozygous BDNF knockout male mice (but none of the other groups) exhibited increased immobility in the forced swimming test and increased corticosterone levels. Surprisingly, this effect was absent in heterozygous BDNF knockout females, suggesting the presence of a protective effect in this gender.

However, Ibarguen-Vargas et al. (2009), using the same genetic model, failed to observe a similar G×E interaction in mice exposed to unpredictable chronic mild stress protocol for a total of 8 weeks. At the end of the stress protocol, the mice were behaviorally tested in a battery of tests, including a 24-hour home food consumption test, an actimeter test, a resident/intruder aggression test, a novelty suppressed feeding test, and a tail suspension test. No evidence for an association between a deficit in BDNF and enhanced vulnerability to unpredictable chronic mild stress was recorded.

Conditional knockout mice with inducible BDNF deletion in forebrain (Monteggia et al., 2004) were used by Autry et al. (2009). These authors evaluated the effect of an unpredictable chronic stress paradigm (Willner, 2005) on both wild-type and BDNF inducible knockout mice. Mice were exposed to either one or two stressors chosen from among food or water deprivation, periods of overnight illumination, 45-degree cage tilt, single housing, and bedding soiled with water or rat feces. Each stress session could last from 4 to 12 hours and took place every 24 hours. The entire stress procedure lasted for 52 days and, at the end of this protocol, the mice were tested for anxiety and depression-like behaviors. After chronic stress, an increased level of anxiety-like behavior in open field test, a decreased consumption of sucrose in the sucrose consumption test, and a heightened stress-induced corticosterone level was observed in BDNF-inducible knockout female, but not in male, mice.

Govindarajan et al. (2006) analyzed the susceptibility of transgenic mice overexpressing BDNF in forebrain excitatory neurons (Huang et al., 1999) to a chronic stress paradigm. The protocol consisted in exposing mice to an immobilization procedure for 2 hours per day for a total of 10 days. The experimental groups were tested for anxiety-like behavior in the open field test at the end of the stress protocol. Surprisingly, this stress protocol increased anxiety-like behavior only in wild-type but not in transgenic mice. This result suggests a protective role of the forebrain BDNF overexpression.

Recently, Razzoli et al. (2011) used transgenic mice overexpressing the dominant-negative truncated splice variant of BDNF receptor trkB in cortical and hippocampal neurons (Saarelainen et al., 2000) for their G×E study. In this report, both wild-type and transgenic mice were exposed to a chronic sensory contact model of stress (Bartolomucci et al., 2010), and the behavioral consequences of this procedure were assessed 4 weeks after the last social defeat experience (Berton et al., 2006; Tsankova et al., 2006). The exposure to chronic social stress model induced higher levels of social avoidance in transgenic defeated mice with respect to wild-type littermates.

In conclusion, unlike the animal studies we described for the 5-HTT gene, mouse studies investigating possible interactions between BDNF mutation and environmental stress produced mixed results.

C. Serotonin 1A Receptor

Whereas the studies just reviewed illustrate a translational approach from human to animal work, we now turn to the first example of studies exhibiting the reverse direction. This is a case of a mouse model suggesting the existence of gene-by-early environment (early stress) interaction that has not yet identified in humans, based on the interaction between a functional SNP of the serotonin 1A receptor (HTR1A; Armbruster et al., 2011; Chipman, Jorm, Tan, & Easteal, 2010; Lemonde et al., 2003; Rothe et al., 2004; Strobel et al., 2003) and early stress. Even though the presence/existence of this interaction has been extensively investigated, no evidence of it is available at the moment. In contrast, mouse studies have revealed a possible interaction between HTR1A gene and early environment on anxiety-related behavior, but only on a specific aspect of this behavior. Specifically, these studies have used homozygous and/or heterozygous 5-HT1A knockout mice (Lemonde et al., 2003; Rothe et al., 2004; Strobel et al., 2003) as a model of 5-HTR1A human polymorphism.

The first study to expose heterozygous 5-HT1A knockout mouse pups to early environmental manipulation was by Carola et al. (2006). Similarly to that previously described for other experiments from the same research group (Carola et al., 2008; Carola & Gross, 2010), in this experiment, heterozygous mutant mice were exposed to different maternal environments ("good" vs. "bad" mothers) using F1 hybrid females derived from reciprocal intercrosses between the two inbred strains C57BL/6J and BALB/cByJ. Wild-type and 5-HT1A heterozygous mutant mice exposed to the two different environmental conditions were then tested for anxiety-related behavior in adulthood. Although distinct significant effects of the environment and gene were detected, no evidence for a G×E interaction was observed.

However, a follow-up study did discover a significant G×E interaction, which was observed when homozygous 5-HT1A knockout mice were exposed to early handling (Zanettini et al., 2010). It is known that this procedure can induce a reduction of anxiety-like behavior and stress responses in wild-type mice in adulthood (D'Amato et al., 1992; 1998). In this work, the effect of the environmental manipulation was stronger in homozygous 5-HT1A knockout mice than in wild-type mice, where it was possible to observe a reduction of the levels of social anxiety-related behavior measured in a social interaction test (Brodkin, Hagemann, Nemetski, & Silver, 2004).

This study restricts the effect of the interaction between early environment and mutation in the 5-HT1A gene to a specific type of anxiety: social anxiety. Future studies investigating similar interactions in humans could take advantage of these results by directing their attention to the specific type of behavioral outcome/phenotype to be analyzed and heavily considering social anxiety as possible field of 5-HT1A G×E interaction.

III. Conclusion and Future Direction

Mouse models such as those described here were useful not only for expanding our understanding of the behavioral features of mood disorders, but also and most importantly for identifying the molecular and physiological mechanisms implicated in G×E interaction effects.

One important future step would be to expand the number of environmental stressors modeled in the mouse. Stressful events like exposure to neglect and physical abuse during early development in the mouse, for example, need still to be modeled to better mimic the human condition. Some paradigms (such as maternal deprivation) have better face validity for stressful early postnatal environments than other paradigms (such as footshock) do, and researchers should strive to continue to develop improved behavioral paradigms. Moreover, only few of the studies reviewed here have combined both behavioral and molecular analyses. The identification of such biological changes and how these changes are modulated by different type of stressors, as well as the developmental window when the stress is received, are all important considerations when developing, testing, and validating therapeutic tools targeted to each specific interaction/situation. Finally, it would be desirable to improve the exchange of scientific results between researchers investigating G×E interaction in humans with those investigating similar interactions in animal models in order to accelerate the discovery of new therapeutic and intervention protocols for mood disorders.

References

Advani, T., Koek, W., & Hensler, J. G. (2009). Gender differences in the enhanced vulnerability of BDNF+/- mice to mild stress. *International Journal of Neuropsychopharmacology, 12*, 583–588.

Armbruster, D., Mueller, A., Strobel, A., Lesch, K. P., Brocke, B., & Kirschbaum, C. (2011). Predicting cortisol stress responses in older individuals: Influence of serotonin receptor 1A gene (HTR1A) and stressful life events. *Hormones and Behavior, 60*, 105–111.

Autry, A. E., Adachi, M., Cheng, P., & Monteggia, L. M. (2009). Gender-specific impact of brain-derived neurotrophic factor signaling on stress-induced depression-like behavior. *Biological Psychiatry, 66*, 84–90.

Avitsur, R., Padgett, D. A., & Sheridan, J. F. (2006). Social interactions, stress, and immunity. *Neurologic Clinics, 24*, 483–491.

Bartolomucci, A. (2007). Social stress, immune functions and disease in rodents. *Frontiers in Neuroendocrinology, 28*, 28–49.

Bartolomucci, A., Cabassi, A., Govoni, P., Ceresini, G., Cero, C., Berra, D., et al. (2009*b*). Metabolic consequences and vulnerability to diet-induced obesity in male mice under chronic social stress. *PLoS ONE, 4*, e4331.

Bartolomucci, A., Carola, V., Pascucci, T., Puglisi-Allegra, S., Cabib, S., Lesch, K. P., et al. (2010). Increased vulnerability to psychosocial stress in heterozygous serotonin transporter knockout mice. *Disease Models and Mechanisms, 3*, 459–470.

Bartolomucci, A., Gioiosa, L., Chirieleison, A., Ceresini, G., Parmigiani, S., & Palanza, P. (2004*b*). Cross fostering in mice: Behavioral and physiological carry-over effects in adulthood. Genes, Brain and Behavior, 3, 115–122.

Bartolomucci, A., & Leopardi, R. (2009*a*). Stress and depression: Preclinical research and clinical implications. *PLoS One*, *4*, e4265.

Bartolomucci, A., Palanza, P., Gaspani, L., Limiroli, E., Panerai, A. E., Ceresini, G., et al. (2001). Social status in mice: Behavioral, endocrine and immune changes are context dependent. *Physiology and Behavior*, *73*, 401–410.

Bartolomucci, A., Pederzani, T., Sacerdote, P., Panerai, A. E., Parmigiani, S., & Palanza, P. (2004*a*). Behavioral and physiological characterization of male mice under chronic psychosocial stress. *Psychoneuroendocrinology*, *29*, 899–910.

Bengel, D., Murphy, D. L., Andrews, A. M., Wichems, C. H., Feltner, D., & Heils, A. (1998). Altered brain serotonin homeostasis and locomotor insensitivity to 3, 4-methylenedioxymethamphetamine ("Ecstasy") in serotonin transporter-deficient mice. *Molecular Pharmacology*, *53*, 649–655.

Berton, O., McClung, C. A., Dileone, R. J., Krishnan, V., Renthal, W., Russo, S. J., et al. (2006). Essential role of BDNF in the mesolimbic dopamine pathway in social defeat stress. *Science*, *311*, 864–868.

Branchi, I. (2009). The mouse communal nest: Investigating the epigenetic influences of the early social environment on brain and behavior development. *Neuroscience and Biobehavioral Reviews*, *33*, 551–559.

Brodkin, E. S., Hagemann, A., Nemetski, S. M., & Silver, L. M. (2004). Social approach-avoidance behavior of inbred mouse strains towards DBA/2 mice. *Brain Research*, *1002*, 151–157.

Brunson, K. L., Kramar, E., Lin, B., Chen, Y., Colgin, L. L., Yanagihara, T. K., et al. (2005). Mechanisms of late-onset cognitive decline after early-life stress. *Journal of Neuroscience*, *25*, 9328–9338.

Calatayud, F., Coubard, S., & Belzung, C. (2004). Emotional reactivity in mice may not be inherited but influenced by parents. *Physiology and Behavior*, *80*, 465–474.

Caldji, C., Diorio, J., & Meaney, M. J. (2003). Variations in maternal care alter GABA(A) receptor subunit expression in brain regions associated with fear. *Neuropsychopharmacology*, *28*, 1950–1959.

Canli, T., Qiu, M., Omura, K., Congdon, E., Haas, B. W., Amin, Z., et al. (2006). Neural correlates of epigenesis. *Proceedings of the National Academy of Sciences USA*, *103*, 16033–16038.

Carola, V., Frazzetto, G., & Gross, C. (2006). Identifying interactions between genes and early environment in the mouse. *Genes, Brain and Behavior*, *5*, 189–199.

Carola, V., Frazzetto, G., Pascucci, T., Audero, E., Puglisi-Allegra, S., Cabib, S., et al. (2008). Identifying molecular substrates in a mouse model of the serotonin transporter x environment risk factor for anxiety and depression. *Biological Psychiatry*, *63*, 840–846.

Carola, V., & Gross, C. (2010). BDNF moderates early environmental risk factors for anxiety in mouse. *Genes, Brain and Behavior*, *9*, 379–389.

Carola, V., Mirabeau, O., & Gross C. T. (2011*b*). Hidden Markov model analysis of maternal behavior patterns in inbred and reciprocal hybrid mice. *PLoS One*, *6*, e14753.

Carola, V., Pascucci, T., Puglisi-Allegra, S., Cabib, S., & Gross, C. (2011*a*). Effect of the interaction between the serotonin transporter gene and maternal environment on developing mouse brain. *Behavioral Brain Research*, *217*, 188–194.

Carroll, J. C., Boyce-Rustay, J. M., Millstein, R., Yang, R., Wiedholz, L. M., Murphy, D. L., et al. (2007). Effects of mild early life stress on abnormal emotion-related behaviors in 5-HTT knockout mice. *Behavioral Genetics*, *37*, 214–222.

Caspi, A., Hariri, A., Holmes, A., Uher, R., & Moffitt, T. (2010). Genetic sensitivity to the environment: The case of the serotonin transporter gene and its implications for studying complex diseases and traits. *American Journal of Psychiatry*, *167*, 509–527.

Caspi, A., Sugden, K., Moffitt, T. E., Taylor, A., Craig, I. W., Harrington, H., et al. (2003). Influence of life stress on depression: Moderation by a polymorphism in the 5-HTT gene. *Science*, *301*, 386–389.

Chevalet, P., Le Pape, G., & Lassalle, J. M. (1987). Mise en evidence d'effets grand-maternels sur les conduites maternelles de la souris. *Behavioral Processes*, *15*, 123–130.

Chipman, P., Jorm, A. F., Tan, X. Y., & Easteal, S. (2010). No association between the serotonin-1A receptor gene single nucleotide polymorphism rs6295C/G and symptoms of anxiety or depression, and no interaction between the polymorphism and environmental stressors of childhood anxiety or recent stressful life events on anxiety or depression. *Psychiatric Genetics*, *20*, 8–13.

Claessens, S. E., Daskalakis, N. P., van der Veen, R., Oitzl, M. S., de Kloet, E. R., & Champagne, D. L. (2011). Development of individual differences in stress responsiveness: An overview of factors mediating the outcome of early life experiences. *Psychopharmacology (Berlin)*, *214*, 141–154.

Dadomo, H., Sanghez, V., Di Cristo, L., Lori, A., Ceresini, G., Malinge, I., et al. (2011). Vulnerability to chronic subordination stress-induced depression-like disorders in adult 129SvEv male mice. *Progress in Neuropsychopharmacology and Biological Psychiatry*, *35*, 1461–1471.

D'Amato, F. R., Cabib, S., Puglisi-Allegra, S., Patacchioli, F. R., Cigliana, G., Maccari, S., et al. (1992). Effects of acute and repeated exposure to stress on the hypothalamo-pituitary-adrenocortical activity in mice during postnatal development. *Hormones and Behavior*, *26*, 474–485.

D'Amato, F. R., Cabib, S., Ventura, R., & Orsini, C. (1998). Long-term effects of postnatal manipulation on emotionality are prevented by maternal anxiolytic treatment in mice. *Developmental Psychobiology*, *32*, 225–234.

D'Amato, F. R., Mazzacane, E., Capone, F., & Pavone, F. (1999). Effects of postnatal manipulation on nociception and morphine sensitivity in adult mice. *Brain Research. Developmental Brain Research*, *117*, 15–20.

D'Amato, F. R., Zanettini, C., Lampis, V., Coccurello, R., Pascucci, T., Ventura, R., et al. (2011*a*). Unstable maternal environment, separation anxiety, and heightened CO2 sensitivity induced by gene-by-environment interplay. *PLoS One*, *6*, e18637.

D'Amato, F. R., Zanettini, C., Sgobio, C., Sarli, C., Carone, V., Moles, A., et al. (2011l). Intensification of maternal care by double-mothering boosts cognitive function and hippocampal morphology in the adult offspring. *Hippocampus*, *21*, 298–308.

Darnaudéry, M., & Maccari, S. (2008). Epigenetic programming of the stress response in male and female rats by prenatal restraint stress. *Brain Research Reviews*, *57*, 571–585.

Ducottet, C., Aubert, A., & Belzung, C. (2004). Susceptibility to subchronic unpredictable stress is related to individual reactivity to threat stimuli in mice. *Behavioral Brain Research, 155*, 291–299.
Egan, M. F., Kojima, M., Callicott, J. H., Goldberg, T. E., Kolachana, B. S., Bertolino, A., et al. (2003). The BDNF val66met polymorphism affects activity-dependent secretion of BDNF and human memory and hippocampal function. *Cell, 112*, 257–269.
Ernfors, P., Lee, K. F., & Jaenisch, R. (1994). Mice lacking brain-derived neurotrophic factor develop with sensory deficits. *Nature, 368*, 147–150.
Evans, M. J., & Kaufman, M. H. (1981). Establishment in culture of pluripotential cells from mouse embryos. *Nature, 292*, 154–156.
Gatt, J. M., Nemeroff, C. B., Dobson-Stone, C., Paul, R. H., Bryant, R. A., Schofield, P. R. et al. (2009). Interactions between BDNF Val66Met polymorphism and early life stress predict brain and arousal pathways to syndromal depression and anxiety. *Molecular Psychiatry, 14*, 681–695.
Gerritsen, L., Tendolkar, I., Franke, B., Vasquez, A. A., Kooijman, S., Buitelaar, J., et al. (2012). BDNF Val66Met genotype modulates the effect of childhood adversity on subgenual anterior cingulate cortex volume in healthy subjects. *Molecular Psychiatry, 17*, 597–603.
Ginsburg, B., & Allee, W. C. (1942). Some effects of conditioning on social dominance and subordination in inbred strains of mice. *Physiological Zoology, 15*, 485–506.
Gordon, J. W., & Ruddle, F. H. (1981). Integration and stable germ line transmission of genes injected into mouse pronuclei. *Science, 214*, 1244–1246.
Govindarajan, A., Rao, B. S., Nair, D., Trinh, M., Mawjee, N., Tonegawa, S., et al. (2006). Transgenic brain-derived neurotrophic factor expression causes both anxiogenic and antidepressant effects. *Proceedings of the National Academy of Sciences USA, 103*, 13208–13213.
Greenberg, B. D., Tolliver, T. J., Huang, S. J., Li, Q., Bengel, D., & Murphy, D. L. (1999). Genetic variation in the serotonin transporter promoter region affects serotonin uptake in human blood platelets. *American Journal of Medical Genetics, 88*, 83–87.
Harbers, K., Jahner, D., & Jaenisch, R. (1981). Microinjection of cloned retroviral genomes into mouse zygotes: Integration and expression in the animal. *Nature, 293*, 540–542.
Heiming, R. S., Jansen, F., Lewejohann, L., Kaiser, S., Schmitt, A., Lesch, K.P., & Sachser, N. (2009). Living in a dangerous world: The shaping of behavioral profile by early environment and 5-HTT genotype. *Frontiers in Behavioral Neuroscience, 3*, 26.
Heinz, A., Smolka, M. N., Braus, D. F., Wrase, J., Beck, A., Flor, H., et al. (2007). Serotonin transporter genotype (5-HTTLPR): Effects of neutral and undefined conditions on amygdala activation. *Biological Psychiatry, 61*, 1011–1014.
Hogan, B., Beddington, R., Costantini, F., & Lacy, E. (1994). *Manipulating the mouse embryo. A laboratory manual.* Plainview, NY: Cold Spring Harbor Laboratory Press.
Huang, Z. J., Kirkwood, A., Pizzorusso, T., Porciatti, V., Morales, B., Bear, M. F., et al. (1999). BDNF regulates the maturation of inhibition and the critical period of plasticity in mouse visual cortex. *Cell, 98*, 739–755.
Ibarguen-Vargas, Y., Surget, A., Vourc'h, P., Leman, S., Andres, C. R., Gardier, A. M., et al. (2009). Deficit in BDNF does not increase vulnerability to stress but dampens antidepressant-like effects in the unpredictable chronic mild stress. *Behavioral Brain Research, 202*, 245–251.
Ivy, A. S., Brunson, K. L., Sandman, C., & Baram, T. Z. (2008). Dysfunctional nurturing behavior in rat dams with limited access to nesting material: A clinically relevant model for early-life stress. *Neuroscience, 154*, 1132–1142.
Ivy, A. S., Rex, C. S., Chen, Y., Dube, C., Maras, P. M., Grigoriadis, D. E., et al. (2010). Hippocampal dysfunction and cognitive impairments provoked by chronic early-life stress involve excessive activation of CRH receptors. *Journal of Neuroscience, 30*, 13005–13015.
Jansen, F., Heiming, R. S., Lewejohann, L., Touma, C., Palme, R., Schmitt, A., et al. (2010). Modulation of behavioural profile and stress response by 5-HTT genotype and social experience in adulthood. *Behavioral Brain Research, 207*, 21–29.
Joeyen-Waldorf, J., Edgar, N., & Sibille, E. (2009). The roles of sex and serotonin transporter levels in age- and stress-related emotionality in mice. *Brain Research, 1286*, 84–93.
Kaufman, J., Yang, B. Z., Douglas-Palumberi, H., Houshyar, S., Lipschitz, D., Krystal, J. H., et al. (2004). Social supports and serotonin transporter gene moderate depression in maltreated children. *Proceedings of the National Academy of Science USA, 101*, 17316–17321.
Kikusui, T., & Mori, Y. (2009). Behavioural and neurochemical consequences of early weaning in rodents. *Journal Neuroendocrinology, 21*, 427–431.
Koolhaas, J. M., Bartolomucci, A., Buwalda, B., de Boer S. F., Flügge, G., Korte, S. M., et al. (2011). Stress revisited: A critical evaluation of the stress concept. *Neuroscience and Biobehavioral Reviews, 35*, 1291–1301.
Kudryavtseva, N. N. (1991). The sensory contact model for the study of aggressive and submissive behaviors in male mice. *Aggressive Behavior, 17*, 285–291.
Lemonde, S., Turecki, G., Du Bakish, D. L., Hrdina, P. D., Bown, C. D., Sequeira, A., et al. (2003). Impaired repression at a 5-Hydroxytryptamine 1A receptor gene polymorphism associated with major depression and suicide. *Journal of Neuroscience, 23*, 8788–8795.
Lesch, K. P., Bengel, D., Heils, A., Sabol, S. Z., Greenberg, B. D., Petri, S., et al. (1996). Association of anxiety-related traits with a polymorphism in the serotonin transporter gene regulatory region. *Science, 274*, 1527–1531.
Levine, S. (1957). Infantile experience and resistance to physiological stress. *Science, 126*, 405.
Lewandoski, M. (2001). Conditional control of gene expression in the mouse. *Nature Reviews Genetics, 2*, 743–755.
Maccari, S., & Morley-Fletcher, S. (2007). Effects of prenatal restraint stress on the hypothalamus-pituitary-adrenal axis and related behavioural and neurobiological alterations. *Psychoneuroendocrinology, 32*(Suppl 1), S10-S15.
Macri, S., & Laviola, G. (2004). Single episode of maternal deprivation and adult depressive profile in mice: Interaction with cannabinoid exposure during adolescence. *Behavioral Brain Research, 154*, 231–238.
Miczek, K. A. (1979). A new test for aggression in rats without aversive stimulation: Differential effects of d-amphetamine and cocaine. *Psychopharmacology (Berlin), 28*(60), 253–259.
Miczek, K. A., Yap, J. J., & Covington, H. E. III (2008). Social stress, therapeutics and drug abuse: Preclinical

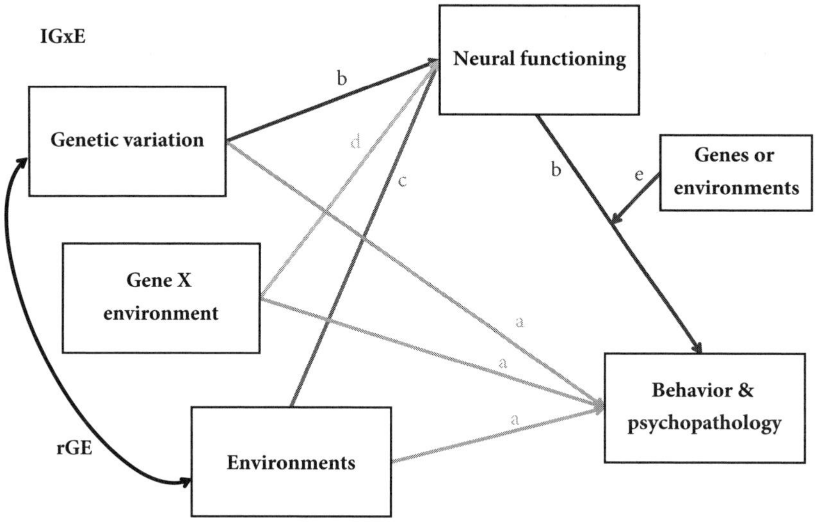

Fig. 11.1 A conceptual model of IG×E: To understand how IG×E might be modeled conceptually and statistically, we demonstrate the relationships of the variables through highlighting traditional G×E and imaging genetics paths, as well as new paths possible in IG×E studies. The A paths model typical G×E relationships. B paths model the paths from an ideal imaging genetics study. The C path models the direct effect of the environment on neural functioning demonstrated in epigenetic studies. The D path models a gene–environment interaction predicting neural functioning (IG×E effect). In this interaction, a gene would be more predictive or have a greater effect on a neural phenotype in some environments but not others (or the reverse: the environment would be predictive of neural functioning for those with one genetic variant but not another). The E path represents another interaction: the possibility that genetic variation or an environmental variable could interact with neural functioning to predict behavior. For example, those with a variant in a gene affecting endocannabinoid signaling show greater correlation between reward-related brain reactivity and a measure of impulsivity (Hariri et al., 2009). Additionally, those with low social support have a greater relationship between their threat-related neural reactivity and trait anxiety (Hyde, Gorka, Manuck & Hariri, 2011). Interactions involving the environment could be between gene and environment predicting neural function (D path) or between gene and neural functioning predicting behavior (E path), but, in typical G×E studies, both of these interactions would be equivalent even though they are likely to be due to very different mechanisms. Note that indirect and mediated pathways can be connected between many of the variables (e.g., G×E to behavior through neural functioning); thus, an ideal IG×E finding would be that the G×E interaction term predicts behavior through neural functioning. Finally, within a structural equation model (SEM) modeling a continuous interaction, the covariance between a genetic variant and an environment can be modeled to reflect the rGE between the specific genetic variant and specific environment.

Derijk, Vanleeuwen, Klok, & Zitman, 2008). Even after controlling for main effects documenting that childhood emotional neglect and MR rs5522 Val allele carrier status independently confer heightened amygdala reactivity, an interaction between prior childhood emotional neglect and MR genotype emerged (Figure 11.2). Specifically, there was a positive association between emotional neglect and threat-related amygdala reactivity in Iso allele homozygotes only. In contrast, Val allele carriers display heightened amygdala reactivity relative to Iso allele homozygotes only in the context of low prior childhood emotional neglect. Thus, Val allele carriers have an HPA axis profile mirroring that of maltreated individuals (Lupien et al., 2009) and also show similar patterns in heightened amygdala reactivity to threat (Maheu et al., 2010; Tottenham et al., 2011), even in the context of no prior maltreatment. This may reflect a physiological ceiling in Val allele carriers whereby maltreatment cannot further exacerbate amygdala reactivity and further suggests that Iso allele homozygotes may be more sensitive to environmental circumstances, for better or worse (Belsky et al., 2009). The dysregulated HPA axis profile and heightened amygdala reactivity of Val allele carriers may leave them more vulnerable to the development of stress-related illness, even in the absence of extreme stress experience.

This study highlights several advantages of an IG×E approach through carefully incorporating measures of environmental experience into neurogenetics research (Hyde et al., 2011). Even after accounting for main effects of genotype and prior emotional neglect, the G×E interaction explained additional variance in amygdala reactivity. With IG×E, inclusion of an environmental measure and its interaction with genotype adds explanatory variance not captured within a traditional neurogenetics approach consistent with the generally enhanced predictive validity of G×E research over

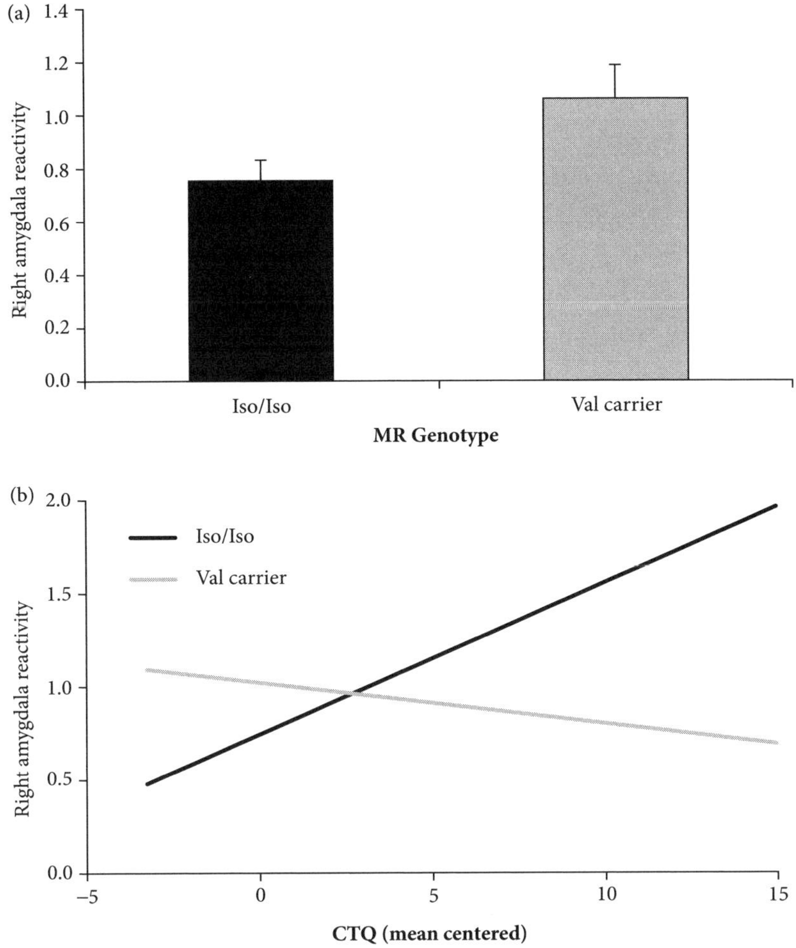

Fig. 11.2 Threat-related amygdala reactivity is predicted by MR (*NR3C2*) Iso/Val (rs5522) genotype, as well as by its interaction with prior childhood emotional neglect. (a) Main effect of MR genotype. Val carriers have elevated threat-related amygdala reactivity relative to Iso/Iso homozygotes. (b) MR genotype interacts with prior childhood emotional neglect to predict additional variability in threat-related amygdala reactivity. Childhood emotional neglect is only positively associated with amygdala reactivity in Iso homozygotes. Val carriers display heightened amygdala reactivity in the context of low neglect, but the genotype groups do not differ at high levels of neglect.

that investigating only main effects of genetic variants (Caspi et al., 2003; Ursini et al., 2011). More important, modeling G×E informs us about how genetically driven variation in brain shapes individual differences in risk for psychopathology across different environments. Ideally, more neurogenetics studies will begin to include manipulations of the environment to exert greater experimental control and minimize potential issues such as gene-by-environment correlation or self-report biases (Bogdan, Santesso, et al., 2011; Cousijn et al., 2010). Last, the incorporation of experiential measures into neurogenetics research also offers the potential to detect masked effects, such as when a genetic variant only has an association with a phenotype under specific circumstances (e.g., 5-HTTLPR and depression only in the context of stressful life events) (Caspi et al., 2003).

IV. The Importance of Epistasis

Epistasis refers to the interaction between two or more polymorphisms, such that the observed phenotype differs from what would be expected by either polymorphism alone. Much like the incorporation of the environment into neurogenetics research, capturing epistatic (i.e., gene-by-gene, G×G) interactions has the potential to clarify relationships between genetic variation and brain function. For example, if an individual possessed a monoamine transporter variant that resulted in reduced reuptake (i.e., enhanced synaptic neurotransmitter concentrations) and also a genetic variant of a presynaptic inhibitory autoreceptor that conferred a loss of binding (i.e., less presynaptic inhibition), the effects of enhanced synaptic neurotransmitter concentrations conferred by the transporter variant would be compounded by the reduced negative feedback conferred by the autoreceptor variant, resulting in greater postsynaptic signaling than either variant would confer alone. Conversely, if an individual had polymorphisms conferring reduced transporter availability but enhanced inhibitory autoreceptor function, there may be no net effect on synaptic signaling because the inhibitory autoreceptors would be more sensitive to what little neurotransmitter is available in the synapse. However, these types of interactions have not been addressed in most studies, although examples given later emphasize the importance of these interactions within neurogenetics research.

Several epistatic relationships have been documented with the *COMT* Val158Met polymorphism in relation to prefrontal function and working memory performance (Honea et al., 2009; Meyer-Lindenberg et al., 2006; Nicodemus et al., 2007; Tan, Chen, Sust, et al., 2007). In one such case, Buckholtz and colleagues (2007) reported an interaction between the rs951436 polymorphism of the gene encoding regulator of G protein signaling 4 (RGS4), which regulates striatal dopaminergic signaling, and the *COMT* polymorphism, such that the effects of one genotype was dependent on the other. Specifically, the A allele of rs951436 was associated with working memory–related prefrontal inefficiency in Val158 allele homozygotes, but with enhanced efficiency in 158Met allele carriers. Thus, by examining genetic variation within two genes coding for distinct proteins influencing DA function, greater clarity was gained regarding the conditional relationship between genetic variation and behaviorally relevant brain function. These results highlight how significant effects at one polymorphic locus can be moderated and potentially even masked by genetic variation in a disparate locus within a common signaling pathway. Critically, had this epistatic relationship not been modeled, this study may have found no relationship between genetic variation across either gene and neural function.

V. Biologically Informed Multilocus Profiles

The vast majority of neurogenetics research has examined single polymorphic loci to predict differences in brain, behavior, and psychopathology. Importantly, a single functional polymorphism confers differences in a single protein's function and/or expression within the backdrop of multiple genes and resulting proteins comprising a neural system. With increasing knowledge of the effects of polymorphisms on basic brain chemistry, it is possible to construct biologically informed multilocus profiles that represent the cumulative effect of multiple polymorphic loci of known functionality on a specific signaling mechanism.

The utility of such an approach was recently demonstrated in a study showing that five functional dopaminergic polymorphisms (i.e., DAT1 nine-repeat, DRD4 seven-repeat, DRD2 -141C Del, DRD2 Taq1A C (A2), and COMT 158Met alleles) predicted nearly 11% of the variance in reward-related ventral striatal reactivity when combined into a single variable, whereas none of the variants alone significantly contributed to reward-related reactivity (Nikolova et al., 2011). Thus, the use of profile scores representing multiple polymorphisms without independent significant effects can account for significant proportions of variability, presumably by better characterizing genetically driven variability in overall signaling. As our knowledge of the function of these polymorphisms increases, it may even be possible to move beyond simple additive approaches to have polymorphisms differentially weighted based on functional knowledge.

This multilocus profile approach holds tremendous potential for neurogenetics research, whereby known functional polymorphisms may be collectively harnessed to represent function across a specific neural system and could relatively easily be applied across neural systems. For example, this approach could be used to understand how individual differences in HPA axis function influence brain activation, behavior, and psychopathology (Figure 11.3). Complementing this hypothesis-driven, biologically informed strategy are statistical approaches that allow for the hypothesis-free evaluation of multiple sites of genetic variation. For example, novel approaches such as regression trees (Hizer, Wright, & Garcia, 2004), recursive partitioning (Gruenewald, Seeman, Ryff, Karlamangla, & Singer, 2006), and machine learning (Mitchell et al., 2008) have potential to identify genetic interactions empirically.

VI. Furthering Our Understanding of Molecular Mechanisms

Because functional genetic polymorphisms can represent individual differences in brain chemistry

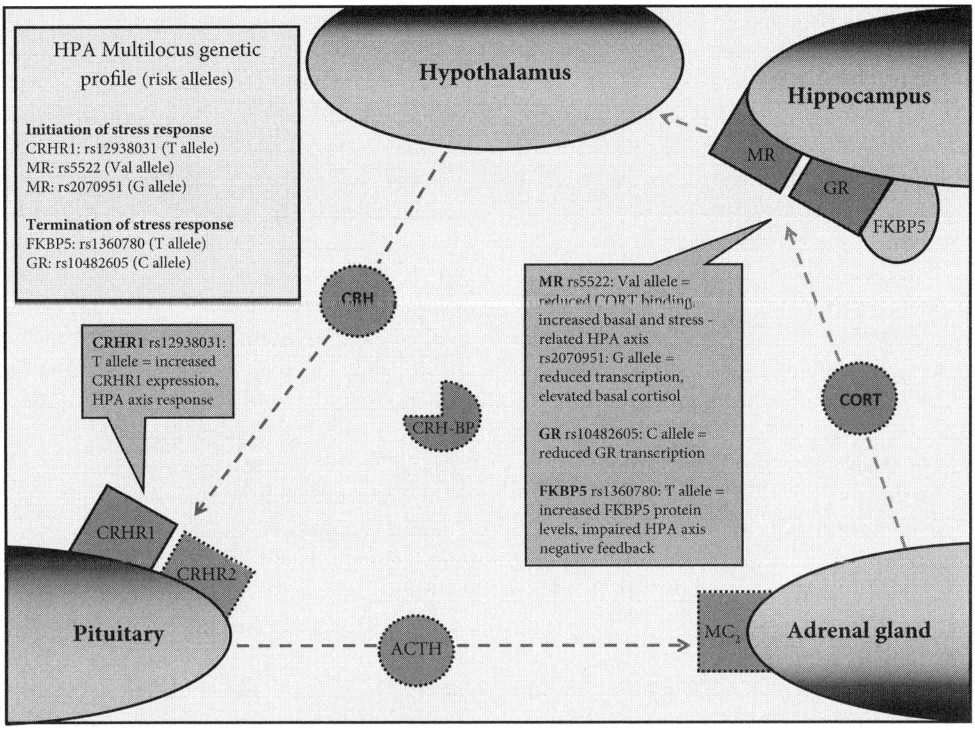

Fig. 11.3 Functional polymorphisms in genes coding for regulatory components of the hypothalamic-pituitary-adrenal (HPA) axis can be combined into a biologically informed multilocus genetic profile. Because the HPA axis is a central regulator of the body's biological response to stress (Ulrich-Lai & Herman, 2009), it is an ideal candidate for G×E research. Briefly, in response to environmental challenge, the hypothalamus secretes corticotropin-releasing hormone (CRH), which binds to type 1 (CRHR1) and type 2 (CRHR2) receptors located in the pituitary gland (among other locations). Corticotropin-releasing hormone binding stimulates the secretion of adrenocorticotropin hormone (ACTH), which travels to the adrenal gland and stimulates cortisol (CORT) release. Cortisol binds to mineralocorticoid receptors (MR) and glucocorticoid receptors (MR) in the hippocampus (among other locations) to shut down activation of the HPA axis and return the body to homeostasis. Genetic variants in *CRHR1*, MR (*NR3C2*), GR (*NR3C1*), and *FKBP5* may be integrated into a single multilocus genetic profile representing enhanced stress responsiveness (e.g., *CRHR1* rs12938031 T allele, MR (*NR3C2*), rs5522 Val allele, MR (*NR3C2*) rs2070951 G allele) and impaired negative feedback (i.e., return to homeostasis) of the HPA axis (e.g., *FKBP5* rs1360780 T allele; GR (*NR3C1*) rs10482605 C allele; see Binder, 2009; Derijk et al., 2008; Kumsta et al., 2009;). Additionally, it is important to note that the GR was recently shown to also have an excitatory role on HPA axis activity, suggesting that attention is warranted based on the neural phenotypes of interest (Kolber et al., 2008). Many other proteins impact this system but are not included (e.g., AVP, NE). Also not included are other variants within included genes (e.g., *FKBP5*: rs9296158, rs9470080, rs3800373, rs7748266, rs9394309, all in high LD with rs1360780; see Binder, 2009; Velders et al., 2011; and *CRHR1* rs110402, rs242924; Tyrka et al., 2009) that have been shown to influence HPA axis function. As continuing research clarifies the impact of these polymorphisms and potential haplotype associations, this profile can be expanded and/or refined.

Gray lines represents excitatory mechanisms. Black lines represents inhibitory mechanisms (note, however, that while MR binding inhibits HPA axis function, research suggests that differences in binding here are critical to the onset of the stress response). Dotted boundaries represent proteins for which there are limited data linking polymorphisms to HPA axis function.

CRH, corticotropin-releasing hormone; CRHR1, corticotropin-releasing hormone type 1 receptor; CRHR1, corticotropin-releasing hormone type 2 receptor; CRH-BP, corticotropin-releasing hormone-binding protein (can block CRH-mediated secretion of ACTH); ACTH, adrenocorticotropin hormone; MC_2, melanocortin receptor type 2 (ACTH receptor); MR, mineralocorticoid receptor; GR, glucocorticoid receptor; FKBP5, FK506 binding protein 5.

and associated neural signaling pathways, neurogenetics can inform our ultimate goal of identifying the detailed and complex biological mechanisms that give rise to the diversity of human behavior and related risk for psychopathology. However, to truly gain mechanistic knowledge, a more comprehensive neurogenetics approach needs to more widely incorporate assessments (e.g., ligand PET; Willeit & Praschak-Rieder, 2010) and/or manipulations (e.g., pharmacological challenge; King & Liberson, 2009) of brain chemistry in humans and work alongside nonhuman animal research, which allows for even more detailed and controlled assessment and manipulation of such brain chemistry. In the next section, we first illustrate how nonhuman animal research is uniquely positioned to target specific molecular mechanisms that would otherwise be inaccessible in humans. We then discuss recent developments in the field of epigenetics, which are already informing one of the major questions of modern science: how does experience shape biology? Last, we touch on the promise of genome-wide association studies (GWAS) to identify novel or understudied proteins critical to neural and behavioral phenotypes of interest.

VII. Integration with Nonhuman Animal Models

Traditionally, nonhuman animal research within psychiatry has sought to model psychopathology and related intermediate behaviors by documenting changes in response to a behavioral, environmental, pharmacologic, and/or genetic manipulation. Because of the ability to manipulate and directly measure brain chemistry and gene function, studies using animal models can target specific molecular mechanisms otherwise inaccessible in human studies. In addition, neurogenetics research in nonhuman animal models offers unique opportunities to control both genetic background and environmental experience, neither of which is practical or ethically feasible in human studies (Hariri & Holmes, 2006). The presence of orthologous genetic variants and the development of novel manipulations (e.g., transgenic lines, optogenetics) provide researchers using nonhuman animal models with opportunities to identify specific molecular mechanisms for behaviors related to human psychopathology, which also represent potential novel treatment targets (Barr et al., 2003; Fenno, Yizhar, & Deisseroth, 2011; Holmes, 2001).

Orthologous genetic variants, which are similar but not identical to human polymorphisms, allow for more direct comparison of nonhuman animals and humans based on shared genetic differences that confer similar effects on gene transcription and/or protein function. For instance, an orthologue of the 5-HTTLPR exists in rhesus macaque monkeys (i.e., rh5-HTTLPR) and, like the human variant, the short allele of rh5-HTTLPR is associated with relatively decreased transcription (Bennett et al., 2002). This functional genetic orthologue has allowed researchers to examine how various factors, such as maternal separation (Barr et al., 2004) and chronic selective serotonin reuptake inhibitor (SSRI) administration (Bogdan, Fitzgibbon, et al., 2011), influence behavior and brain chemistry in experimentally controlled settings (Barr et al., 2003). For instance, it has been shown that following maternal separation, monkeys carrying the rh5-HTTLPR short allele display greater anxiety, agitation, and exaggerated HPA axis response (Barr et al., 2004; Caspi et al., 2010; for a detailed discussion of this work, please see the essay "Gene-Environment Interactions in Monkeys" by Barr). Moreover, much like humans, rh5-HTTLPR short allele carriers have greater metabolic activity in the amygdala (Kalin et al., 2008), engage in less eye contact with high-status conspecifics (e.g., dominant male monkeys), and are more risk-averse in their presence (Caspi et al., 2010; Watson, Ghodasra, & Platt, 2009). Emerging neuroimaging work with the rh5-HTTLPR has begun to establish mechanisms by which these associations may emerge. For instance, results from human studies suggest that the behavioral consequences of the short 5-HTTLPR allele may arise via neurodevelopmental influences on brain morphometry (Canli et al., 2005; Pezawas et al., 2005), which has been supported in the rhesus model (Jedema et al., 2010). Moreover, the rhesus model also has confirmed human findings indicating that the effects of the 5-HTTLPR on adult brain function and behavior may be mediated by downstream modulation of other 5-HT signaling mechanisms (e.g., 5-HT_{1A} autoreceptor downregulation; David et al., 2005; Oler et al., 2009). Thus, the availability of orthologous genetic variants allows for testing and validation of observed neurogenetic pathways of individual differences in behavior, genetic disease risk in the context of adverse environments, and, perhaps most importantly, the discovery of specific molecular mechanisms mediating neurogenetic effects on behavior and psychopathology.

Similarly, transgenic mouse models have provided invaluable insight into the genetic and molecular components of complex behavior and related psychopathology. For example, 5-HTT knockout mice have informed our understanding of how the short allele in humans may contribute to depression. A series of studies using this knockout model have demonstrated that loss of 5-HTT function results in a multitude of downstream effects, including regionally specific up- and down-regulation of both pre- and postsynaptic 5-HT receptors, which ultimately determine the modulatory effects of 5-HT on neuronal signaling (Kim et al., 2005). Moreover, conditional transgenic models, which, unlike the constitutive (i.e., from birth) 5-HTT model allow experimenters to control the exact timing of gene knockout, have demonstrated that similar differences in 5-HT signaling (but conferred by autoreceptor negative feedback and not transporter reuptake) must occur early in development to alter adult brain function and behavior (Gross et al., 2002). These latter results suggest that 5-HTT function and resulting effects on brain and behavior vary substantially across development. Thus, consistent with the monkey research reviewed earlier, these rodent data suggest that the short allele of the 5-HTTLPR may exert its influence on behavior and risk for depression through neurodevelopmental mechanisms influencing downstream 5-HT signaling mechanisms, as well as corticolimbic structure that may be exacerbated by stressful experiences. These convergent patterns across mouse, monkey, and man, help detail the molecular mechanisms that shape the effects of the 5-HTTLPR on brain chemistry, circuitry, behavior, and risk for psychopathology (Caspi et al., 2010).

VIII. Epigenetics

Broadly, epigenetics refers to cell-specific changes in gene expression that are caused by factors (e.g., methylation affecting gene transcription accessibility) other than the underlying static DNA sequence (Meaney, 2010; Roth & Sweatt, 2011; Zhang & Meaney, 2010). Studies conducted primarily in nonhuman animals have shown that experiences, especially those occurring early in life, can elicit epigenetic modifications that trigger a cascade of cellular signaling changes that more broadly affect brain structure and function, as well as behavior (for a detailed discussion of such mechanisms in rodents, see the chapter "Consequences of Early-Life Experiences on Cognition and Emotion: A Role for Nutrition and Epigenetic Mechanisms" by Naninck, Lucassen, and Korosi). Paramount among epigenetics research is that conducted by Michael Meaney and colleagues who have shown that, in rats, maternal care of offspring affects later adult behavior through epigenetic regulation of HPA axis reactivity to stress (Meaney, 2010; Weaver et al., 2004). Specifically, rat pups who receive elevated maternal licking and grooming and arched-back nursing (LG-ABN) have increased serotonin levels, which leads to elevated nerve growth factor inducible protein A (NGFI-A) expression. NGFI-A expression leads to decreased methylation and increased acetylation of the promoter region of the glucocorticoid receptor (GR) gene in hippocampal neurons. This decreased methylation and increased acetylation increases GR gene expression, resulting in more GRs in the hippocampus. Because negative feedback regulation of the HPA axis (i.e., return to homeostasis) is achieved through GR-cortisol binding in the hippocampus, the increased GR expression characteristic of rats who received elevated LG-ABN care results is a stress-resilient phenotype that is better able to return to homeostasis following the extinction of a stressor. Interestingly, these epigenetic changes persist throughout the rat's lifespan and promote adult behavior that is characterized by relative stress resilience and increased subsequent maternal care, whereby high LG-ABN mothers beget relatively stress-resilient pups that become high LG-ABN mothers by experience-dependent mechanisms (Meaney, 2010; Weaver et al., 2004). Moreover, results from a recent human postmortem study (McGowan et al., 2009) are remarkably consistent with this rodent work, suggesting striking conservation of epigenetic mechanisms across species and highlighting the synergy between nonhuman animal and human neurogenetic research.

Evidence that epigenetic factors are dependent on genotype has also begun to emerge (Ursini et al., 2011). The Val158 allele of *COMT* rs4680, described above, results in a CpG island (a cytosine-guanine pair connected by a phosphate backbone) that is absent in the 158Met allele. Because methylation typically occurs at CpG islands, the Val158 allele confers a methylation site that is absent in 158Met alleles. Interestingly, Ursini and colleagues (2011) show that methylation at this site is negatively associated with lifetime stress and positively correlated with working memory performance. Moreover, an interaction between stress and methylation occurs wherein greater stress and lower methylation is associated with reduced COMT mRNA and protein expression, as well as less efficient prefrontal cortex (PFC) activation. These results suggest that the high COMT activity conferred by the Val158 allele can be reduced through stress-related methylation and show, for the first time in humans, that environment-related methylation within functional genetic variants is an important regulator of gene expression and behaviorally relevant brain function, thus demonstrating true G×E effects via epigenetic influence. More broadly, in concert with epistatic relationships documented with *COMT*, these findings also suggest that even more complex relationships (e.g.,

G×G×E) are likely to exist and be mediated by epigenetic mechanisms.

Collectively, studies of epigenetic regulation inform the mechanisms through which experiences can have direct, long-lasting, and even heritable effects on biology. In turn, such effects can translate into important differences in brain circuitry and behavior. These biological mechanisms indicate that the impact of genetic variation on relative risk and resilience for psychopathology will be experience- and context-dependent (Masten, 2001). One major question currently confronting human epigenetics research is how faithfully peripheral blood measures of methylation, which are readily accessible in humans, map onto regionally specific patterns of methylation in the brain. Although there is some evidence that these peripheral measures correlate with brain methylation in nonhuman animal models (Ursini et al., 2011), how well they represent methylation in the human brain, which can be regionally dependent (Xin et al., 2010), is uncertain, and this question is vital to extending epigenetic research to human studies.

IX. Identification of Novel Proteins and Pathways

Genome-wide association studies (GWAS), which test for associations between a phenotype and genetic variation across the entire genome, are hypothesis-free investigations unconstrained by prior evidence that have the potential to identify novel genes that could play important, potentially unexpected, mechanistic roles within distinct neural systems (Cichon et al., 2009). Because of the focus on targets that are more proximal to functional genetic variation (i.e., neural structure and function) than related behavioral or clinical phenotypes, neurogenetics GWAS has the potential to overcome some of the impedances to traditional psychiatric GWAS (e.g., low penetrance, diagnostic heterogeneity, self-report bias). Moreover, the potential mechanistic role of any novel genes identified would be more tractable because of their direct association with known neural systems (e.g., Hibar et al., 2012; Stein et al., 2012).

This potential also comes with tremendous challenge. Obtaining adequate sample sizes to detect the small effects of common polymorphisms at genome-wide significance levels (usually $p < 5 \times 10-8$) within neuroimaging studies is a herculean task that only the largest ongoing investigations approach. Moreover, such a stringent statistical approach will likely not identify most, if any, causal variants, even in the largest studies, because of these small effects. As such, some of the suggestions provided for confronting the challenge of small effects in neurogenetics research may be particularly useful for integration with GWAS. For instance, much like the biologically informed multilocus profile approach suggested previously, data-driven profiles (e.g., the summation of all variants reaching an a priori threshold of significance) could be developed from GWAS data and applied to independent datasets for replication. Indeed, such approaches have been shown to at least partially address the conundrum of hidden heritability, suggesting that genetic variance may simply be "hidden" below the threshold of genome-wide significance (Gibson, 2010; Park et al., 2010; Purcell et al., 2009; Yang et al., 2010). Moreover, incorporating other factors (e.g., G×E, epistasis) known to affect the phenotype of interest will aid the detection of polymorphisms that only may be associated with the phenotype under certain circumstances (Ege et al., 2011). Although neurogenetics GWAS holds great potential, until recently, only one underpowered study existed (e.g., $n = 30$) (Liu et al., 2010). Given the increasing economy of genome-wide genotyping, investigators of even underpowered studies may wish to collect DNA to pool with other investigators for GWAS analyses and/or to use for candidate gene investigations based on new findings introduced by future GWAS. Pooling DNA across site and study, while using the same imaging paradigms, could allow for more statistically powerful and feasible neurogenetics GWAS. Two recent and exciting neurogenetics GWAS studies (Hibar et al., 2012; Stein et al., 2012) exemplify this approach: through combining samples (Hibar et al., 2012) and using consortium approaches to data collection and analysis (Stein et al., 2012), they have been able to examine and map neural structure/volume to genetic variation via genome-wide methods.

Last, with regard to mechanisms, it is important to mention recent developments in our basic understanding of gene transcription and translation that will undoubtedly impact all research in genetics. Challenging canonical models of how DNA is ultimately translated to proteins, recent work has documented widespread sequence differences between RNA and DNA (Li et al., 2011). Thus, it is less clear how, if at all, DNA sequence variation, which is the foundation of neurogenetics research, affects protein function and downstream neural and behavioral phenotypes. However, this work has sparked much controversy among geneticists with concerns that these findings may have resulted from sequencing or other errors (Hayden, 2011). In parallel to the research on DNA–RNA sequence variation, a number of studies have documented that epigenetic fingerprints are themselves heritable (Meaney, 2010; Roth & Sweatt, 2011; Zhang & Meaney, 2010). Thus, sequence variation may not necessarily impact biology as would be predicted simply from the DNA to RNA to protein chain. Maintaining awareness of such paradigmatic shifts in basic genetics and other biological phenomena is paramount for conducting neurogenetics research that will usefully advance our understanding of genes, brain, behavior, and risk for psychopathology.

X. Clinical Relevance

The ultimate goal of neurogenetics research is to generate knowledge that can be used to improve mental health and promote resiliency. The clinical potential of neurogenetics research lies in its ability to both predict individual differences in brain, behavior, and risk for psychopathology and inform the development of novel strategies for treatment and prevention. As a first step toward such clinical utility, emerging neurogenetics research has begun to show that genetically conferred differences in brain function mediate relationships between polymorphisms and variability in behavior conferring risk for psychopathology (Fakra et al., 2009; Furmark et al., 2008). By establishing specific pathways mediating relationships between genes, behavior, and, possibly, clinical symptoms, neurogenetics research can then usefully inform and even direct the search for novel therapeutic targets. Because these targets are born of genetic polymorphisms, they can be further tailored to specific individuals in the broader context of personalized medicine.

XI. Predicting Behavior

Traditionally, neurogenetics research has shown that genetic differences are associated with differences in the brain that have previously been linked to behavioral differences and/or risk for psychopathology. For example, in early neurogenetics work, the short allele of the 5-HTTLPR was linked to relatively increased threat-related amygdala reactivity; in other studies not incorporating a genetic component, elevated amygdala reactivity has been consistently associated with behavioral responsiveness to stress and threat, as well as to the pathophysiology of mood and anxiety disorders. Thus, these independent results suggest that elevated amygdala reactivity associated with the 5-HTTLPR short allele may lead to increased stress/threat responsiveness and, especially in the context of provocation (e.g., trauma), contribute to the emergence of mood and anxiety disorders (Caspi et al., 2010).

The novel application of appropriate statistical techniques, such as mediation analyses (Preacher & Hayes, 2008), in neurogenetics research holds promise for establishing meaningful links between genes, brain, and behavior by modeling indirect pathways between genetic variation and behavior (or psychopathology) via the brain (Hyde et al., 2011). We are aware of only two human neurogenetics studies that incorporate mediation analyses (Fakra et al., 2009; Furmark et al., 2008). In one recent study by our research group (Fakra et al., 2009), we examined a common functional single nucleotide polymorphism (SNP; rs6295) in the promoter region of the 5-HT_{1A} gene (*HTR1A* C-1019G). The G allele of rs6295 is associated with increased gene expression and resulting 5-HT_{1A} autoreceptor density, and hence increased capacity for negative feedback inhibition and subsequently decreased serotonin signaling (Czesak, Lemonde, Peterson, Rogaeva, & Albert, 2006; Lemonde et al., 2003; Parsey et al., 2006). In this study, path analyses revealed that variation in the rs6295 SNP indirectly accounted for more than 9 percent of the variance in trait anxiety *through its effects* on threat-related amygdala reactivity. Consistent with the effects of other functional polymorphisms resulting in relatively increased serotonin signaling (e.g., 5-HTTLPR short allele; Hariri et al., 2002), individuals homozygous for the C allele of rs6295, which presumably results in increased serotonin via decreased negative feedback, exhibited greater amygdala reactivity in comparison with G allele carriers. Importantly, however, there was no statistically significant *direct* link between the polymorphism and anxiety (Fakra et al., 2009). Thus, this example provides evidence of how neurogenetics research can detect indirect associations between genes and behavior through the brain, even when no large direct gene–behavior link is evident. As such, the indirect pathway (i.e., gene-brain-behavior) inherently contained within neurogenetics research can uniquely advance our understanding of both etiologic and pathophysiologic mechanisms in psychiatry (Hyde et al., 2011).

XII. Treatment and Prevention

By deconstructing the molecular mechanisms underlying gene-brain-behavior pathways, especially in collaboration with animal models and in vitro research, neurogenetics has the potential to spur discovery of novel therapeutic targets. For example, combining the above evidence linking HTR1A rs6295 with anxiety through amygdala reactivity with prior work demonstrating effects of the polymorphism on the capacity for negative feedback inhibition suggests that targeting 5-HT_{1A} autoreceptors, perhaps as an adjuvant to SSRI treatment, may produce greater clinical effect. In fact, a recent study in a transgenic mouse model of 5-HT_{1A} autoreceptor function demonstrated that reducing autoreceptor levels prior to SSRI administration converted nonresponders into responders (Richardson-Jones et al., 2010). Thus, neurogenetics research with *HTR1A* rs6295 has identified a novel therapeutic target (i.e., antagonism of 5-HT_{1A} autoreceptors) and a marker that could be used to individually tailor treatment (i.e., C allele homozygotes).

Another example of how neurogenetics can inform treatment and prevention comes from research on TREK1, a background potassium channel. Inspired by a TREK1 knockout mouse study showing that deletion of TREK1 results in a depression-resistant phenotype (Heurteaux et al., 2006), human studies have linked variation in the human TREK1 gene (*KCNK2*) to depression (Liou et al., 2009) and blunted striatal response to reward (a neural profile associated with depression; Dillon et al., 2010), as well as to antidepressant treatment response (Perlis et al., 2008). More recently, these findings have led researchers to develop antidepressant medications that antagonize TREK1. One such compound designed to inhibit TREK1 has been associated with a positive antidepressant response, hippocampal neurogenesis, and increased serotonergic signaling in rodents (Mazella et al., 2010). Although the potential of this novel treatment mechanism has yet to be tested in humans, it does further document how neurogenetics research can spur therapeutic advancements by identifying novel targets. These examples highlight the great potential of neurogenetics to inform targets for novel interventions and ways to better personalize psychiatric treatment. However, they also underline the importance of additional, translational research in this area.

XIII. Conclusion

The field of neurogenetics has informed our understanding of the neurobiological pathways that lead to differences in brain, behavior, and risk for psychopathology. We reviewed three challenges currently confronting neurogenetics research: (1) conducting research on individual variables of small effects, (2) absence of detailed mechanisms, and (3) a need for clinical translation of research findings. To confront challenges of small effects, we discussed how incorporating the environment and epistatic interactions into neurogenetics models, as well as the construction of multilocus genetic profiles, can more accurately represent biological function. We examined how nonhuman animal and epigenetics research can shed light on detailed molecular mechanisms and biological pathways through which the environment interacts with genotype to shape brain, behavior, and risk for psychopathology. We also touched on the promise of GWAS to further our basic neuroscience understanding by identifying novel proteins involved in neural function. Last, we explored how neurogenetics research is beginning to provide clinically informative findings that promise to inform ongoing efforts to improve treatment. Collectively, we hope to have clearly illustrated that neurogenetics research is uniquely positioned to revolutionize our understanding of the emergence of individual differences in genes, brain, behavior, and psychopathology, as well as our ability to manipulate these systems for the benefit of individuals and society.

Acknowledgments

This manuscript is largely based on Bogdan, Hyde, & Hariri, *Molecular Psychiatry*, 2012 as well as Hyde, L. W., Bogdan, R., & Hariri, A. R. (2011). Understanding risk for psychopathology through imaging gene-environment interactions. *Trends in Cognitive Sciences*, *15*, 417–427.

References

Barr, C. S., Newman, T. K., Becker, M. L., Parker, C. C., Champoux, M., Lesch, K. P., et al. (2003). The utility of the non-human primate: Model for studying gene by environment interactions in behavioral research. *Genes, Brain, and Behavior*, *2*, 336–340.

Barr, C. S., Newman, T. K., Shannon, C., Parker, C., Dvoskin, R. L., Becker, M. L., et al. (2004). Rearing condition and rh5-HTTLPR interact to influence limbic-hypothalamic-pituitary-adrenal axis response to stress in infant macaques. *Biological Psychiatry*, *55*, 733–738.

Belsky, J., Jonassaint, C., Pluess, M., Stanton, M., Brummett, B., & Williams, R. (2009). Vulnerability genes or plasticity genes? *Molecular Psychiatry*, *14*, 746–754.

Belsky, J., & Pluess, M. (2009). Beyond diathesis stress: Differential susceptibility to environmental influences. *Psychological Bulletin*, *135*, 885–908.

Bennett, A. J., Lesch, K. P., Heils, A., Long, J. C., Lorenz, J. G., Shoaf, S. E., et al. (2002). Early experience and serotonin transporter gene variation interact to influence primate CNS function. *Molecular Psychiatry, 7*, 118–122.

Bertolino, A., Rubino, V., Sambataro, F., Blasi, G., Latorre, V., Fazio, L., et al. (2006). Prefrontal-hippocampal coupling during memory processing is modulated by COMT val-158met genotype. *Biological Psychiatry, 60*, 1250–1258.

Bigos, K. L., Pollock, B. G., Aizenstein, H. J., Fisher, P. M., Bies, R. R., & Hariri, A. R. (2008). Acute 5-HT reuptake blockade potentiates human amygdala reactivity. *Neuropsychopharmacology, 33*, 3221–3225.

Bigos, K. L., & Weinberger, D. R. (2010). Imaging genetics—days of future past. *NeuroImage, 53*, 804–809.

Binder, E. B. (2009). The role of FKBP5, a co-chaperone of the glucocorticoid receptor in the pathogenesis and therapy of affective and anxiety disorders. *Psychoneuroendocrinology, 34 Suppl 1*, S186–S195.

Blakely, R., & Veenstra-Vanderweele, J. (2011). Genetic indeterminism, the 5-HTTLPR, and the paths forward in neuropsychiatric genetics. *Archives of General Psychiatry, 68*, 457–458.

Bogdan, R., Fitzgibbon, H., Woolverton, W. L., Bethea, C. L., Iyo, A., Stockmeier, C. A., et al. (2011). 5-HTTLPR genotype and gender, but not chronic fluoxetine administration, are associated with cortical TREK1 protein expression in rhesus macaques. *Neuroscience Letters, 503*, 83–86.

Bogdan, R., Hyde, L.W., and Hariri, A.R. (2013). A neurogenetics approach to understanding individual differences in brain, behavior, and risk for psychopathology. *Molecular Psychiatry, 18*, 288–299.

Bogdan, R., Perlis, R. H., Fagerness, J., & Pizzagalli, D. A. (2010). The impact of mineralocorticoid receptor ISO/VAL genotype (rs5522) and stress on reward learning. *Genes, Brain and Behavior, 9*, 658–667.

Bogdan, R., Santesso, D. L., Fagerness, J., Perlis, R. H., & Pizzagalli, D. A. (2011). Corticotropin-releasing hormone receptor type 1 (CRHR1) genetic variation and stress interact to influence reward learning. *Journal of Neuroscience, 31*, 13246–13254.

Bogdan, R., Williamson, D. E., & Hariri, A. R. (2012). Mineralocorticoid receptor iso/val (rs5522) genotype moderates the association between prior childhood emotional neglect and amygdala reactivity. *American Journal of Psychiatry, 169*, 515–522.

Brody, G. H., Beach, S. R. H., Philibert, R. A., Chen, Y., & Murry, V. M. B. (2009). Prevention effects moderate the association of 5 HTTLPR and youth risk behavior initiation: Gene x environment hypotheses tested via a randomized prevention design. *Child Development, 80*, 645–661.

Buckholtz, J. W., Sust, S., Tan, H. Y., Mattay, V. S., Straub, R. E., Meyer-Lindenberg, A., et al. (2007). fMRI evidence for functional epistasis between COMT and RGS4. *Molecular Psychiatry, 12*, 893–895, 885.

Buckholtz, J. W., Treadway, M. T., Cowan, R. L., Woodward, N. D., Li, R., Ansari, M. S., et al. (2010). Dopaminergic network differences in human impulsivity. *Science, 329*, 532.

Canli, T., Omura, K., Haas, B. W., Fallgatter, A., Constable, R. T., & Lesch, K. P. (2005). Beyond affect: A role for genetic variation of the serotonin transporter in neural activation during a cognitive attention task. *Proceedings of the National Academy of Sciences, 102*, 12224–12229.

Canli, T., Qui, M., Omura, K., Congdon, E., Haas, B. W., Amin, Z., et al. (2006). Neural correlates of epigenesis. *Proceedings of the National Academy of Sciences, 103*, 16033–16038.

Cardon, L. R., & Palmer, L. J. (2003). Population stratification and spurious allelic association. *Lancet, 361*, 598–604.

Casey, B. J., Glatt, C. E., Tottenham, N., Soliman, F., Bath, K., Amso, D., et al. (2009). Brain-derived neurotrophic factor as a model system for examining gene by environment interactions across development. *Neuroscience, 164*, 108–120.

Caspi, A., Hariri, A. R., Holmes, A., Uher, R., & Moffitt, T. E. (2010). Genetic sensitivity to the environment: The case of the serotonin transporter gene and its implications for studying complex diseases and traits. *American Journal of Psychiatry, 167*, 509–527.

Caspi, A., & Moffitt, T. E. (2006). Gene-environment interactions in psychiatry: Joining forces with neuroscience. *Nature Reviews Neuroscience, 7*, 583–590.

Caspi, A., Sugden, K., Moffitt, T. E., Taylor, A., Craig, I. W., Harrington, H., et al. (2003). Influence of life stress on depression: Moderation by a polymorphism in the 5-HTT gene. *Science, 301*, 386–389.

Charney, D. S., Barlow, D. H., Botteron, K., Cohen, J. D., Goldman, D., Gur, R. E., et al. (2002). Neuroscience research agenda to guide development of a pathophysiologically based classification system. In D. J. Kupfer, M. B. First, & D. A. Regier (Eds.), *A research agenda for DSM-V* (pp. 31–83). Washington, DC: American Psychiatric Association.

Chen, J., Lipska, B. K., Halim, N., Ma, Q. D., Matsumoto, M., Melhem, S., et al. (2004). Functional analysis of genetic variation in catechol-O-methyltransferase (COMT): Effects on mRNA, protein, and enzyme activity in postmortem human brain. *American Journal of Human Genetics, 75*, 807–821.

Cichon, S., Craddock, N., Daly, M., Faraone, S. V., Gejman, P. V., Kelsoe, J., et al. (2009). Genome wide association studies: History, rationale, and prospects for psychiatric disorders. *American Journal of Psychiatry, 166*, 540–556.

Costello, E. J., Compton, S. N., Keeler, G., & Angold, A. (2003). Relationships between poverty and psychopathology. *JAMA, 290*, 2023–2029.

Cousijn, H., Rijpkema, M., Qin, S., van Marle, H. J. F., Franke, B., Hermans, E. J., et al. (2010). Acute stress modulates genotype effects on amygdala processing in humans. *Proceedings of the National Academy of Sciences, 107*, 9867–9872.

Czesak, M., Lemonde, S., Peterson, E. A., Rogaeva, A., & Albert, P. R. (2006). Cell-specific repressor or enhancer activities of Deaf-1 at a serotonin 1A receptor gene polymorphism. *Journal of Neuroscience, 26*, 1864–1871.

David, S. P., Murthy, N. V., Rabiner, E. A., Munafò, M. R., Johnstone, E. C., Jacob, R., et al. (2005). A functional genetic variation of the serotonin (5-HT) transporter affects 5-HT1A receptor binding in humans. *Journal of Neuroscience, 25*, 2586–2590.

De Raedt, R., Leyman, L., Baeken, C., Van Schuerbeek, P., Luypaert, R., Vanderhasselt, M. A., et al. (2010). Neurocognitive effects of HF-rTMS over the dorsolateral prefrontal cortex on the attentional processing of emotional information in healthy women: An event-related fMRI study. *Biological Psychology, 85*, 487–495.

Derijk, R., Vanleeuwen, N., Klok, M., & Zitman, F. (2008). Corticosteroid receptor-gene variants: Modulators of the stress-response and implications for mental health. *European Journal of Pharmacology, 585*, 492–501.

DeRijk, R. H., Wust, S., Meijer, O. C., Zennaro, M. C., Federenko, I. S., Hellhammer, D. H., et al. (2006). A common polymorphism in the mineralocorticoid receptor modulates stress responsiveness. *Journal of Clinical Endocrinology & Metabolism, 91*, 5083–5089.

Dillon, D. G., Bogdan, R., Fagerness, J., Holmes, A. J., Perlis, R. H., & Pizzagalli, D. A. (2010). Variation in TREK1 gene linked to depression-resistant phenotype is associated with potentiated neural responses to rewards in humans. *Human Brain Mapping, 31*, 210–221.

Drabant, E. M., Ramel, W., Edge, M. D., Hyde, L. W., Kuo, J. R., Goldin, P. R. Hariri, A. R., & Gross, J. J. (2012). Neural mechanisms underlying 5-HTTLPR related sensitivity to acute stress. *American Journal of Psychiatry, 169*, 397–405.

Eaves, L., Silberg, J., & Erkanli, A. (2003). Resolving multiple epigenetic pathways to adolescent depression. *Journal of Child Psychology and Psychiatry and Allied Disciplines, 44*, 1006–1014.

Egan, M. F., Goldberg, T. E., Kolachana, B. S., Callicott, J. H., Mazzanti, C. M., Straub, R. E., et al. (2001). Effect of COMT Val108/158 Met genotype on frontal lobe function and risk for schizophrenia. *Proceedings of the National Academy of Sciences, 98*, 6917–6922.

Ege, M. J., Strachan, D. P., Cookson, W. O., Moffatt, M. F., Gut, I., Lathrop, M., et al. (2011). Gene-environment interaction for childhood asthma and exposure to farming in Central Europe. *Journal of Allergy and Clinical Immunology, 127*, 138–144.

Ellis, B. J., & Boyce, W. T. (2011). Differential susceptibility to the environment: Toward an understanding of sensitivity to developmental experiences and context. *Development and Psychopathology, 23*, 1–5.

Fakra, E., Hyde, L. W., Gorka, A., Fisher, P. M., Munoz, K. E., Kimak, M., et al. (2009). Effects of HTR1A C(1019)G on amygdala reactivity and trait anxiety. *Archives of General Psychiatry, 66*, 33–40.

Fenno, L., Yizhar, O., & Deisseroth, K. (2011). The development and application of optogenetics. *Annual Review Neuroscience, 34*, 389–412.

Flint, J., & Munafò, M. R. (2007). The endophenotype concept in psychiatric genetics. *Psychological Medicine, 37*, 163–180.

Franklin, T. B., & Mansuy, I. M. (2010). Epigenetic inheritance in mammals: Evidence for the impact of adverse environmental effects. *Neurobiology of Disease, 39*, 61–65.

Furmark, T., Appel, L., Henningsson, S., Ahs, F., Faria, V., Linnman, C., et al. (2008). A link between serotonin-related gene polymorphisms, amygdala activity, and placebo-induced relief from social anxiety. *Journal of Neuroscience, 28*, 13066–13074.

Gerritsen, L., Tendolkar, I., Franke, B., Vasquez, A. A., Kooijman, S., Buitelaar, J., et al. (2012). BDNF Val66Met genotype modulates the effect of childhood adversity on subgenual anterior cingulate cortex volume in healthy subjects. *Molecular Psychiatry, 17*, 597–603.

Gibson, G. (2010). Hints of hidden heritability in GWAS. *Nature Genetics, 42*, 558–560.

Gillihan, S. J., Rao, H., Brennan, L., Wang, D. J., Detre, J. A., Sankoorikal, G. M., et al. (2011). Serotonin transporter genotype modulates the association between depressive symptoms and amygdala activity among psychiatrically healthy adults. *Psychiatry Research, 193*, 161–167.

Gross, C., Zhuang, X., Stark, K., Ramboz, S., Oosting, R., Kirby, L., et al. (2002). Serotonin1A receptor acts during development to establish normal anxiety-like behaviour in the adult. *Nature, 416*, 396–400.

Gruenewald, T. L., Seeman, T. E., Ryff, C. D., Karlamangla, A. S., & Singer, B. H. (2006). Combinations of biomarkers predictive of later life mortality. *Proceedings of the National Academy of Sciences, 103*, 14158–14163.

Hariri, A. R., & Holmes, A. (2006). Genetics of emotional regulation: The role of the serotonin transporter in neural function. *Trends in Cognitive Sciences, 10*, 182–191.

Hariri, A. R. (2009). The neurobiology of individual differences in complex behavioral traits. *Annual Review of Neuroscience, 32*, 225–247.

Hariri, A. R. (2011). The what, where and when of COMT. *Biological Psychiatry, 70*, 215–215.

Hariri, A. R., Drabant, E. M., Munoz, K. E., Kolachana, B. S., Mattay, V. S., Egan, M. F., et al. (2005). A susceptibility gene for affective disorders and the response of the human amygdala. *Archives of General Psychiatry, 62*, 146–152.

Hariri, A. R., Gorka, A., Hyde, L. W., Kimak, M., Halder, I., Ducci, F., Ferrell, R. E., Goldman, D., and Manuck, S. B. (2009). Divergent effects of genetic variation in endocannabinoid signaling on threat- and reward-related human brain function. *Biological Psychiatry, 66*, 9–16.

Hariri, A. R., Mattay, V. S., Tessitore, A., Kolachana, B., Fera, F., Goldman, D., et al. (2002). Serotonin transporter genetic variation and the response of the human amygdala. *Science, 297*, 400–403.

Hasler, G., & Northoff, G. (2011). Discovering imaging endophenotypes for major depression. *Molecular Psychiatry, 16*, 604–619.

Hayden, E. C. (2011). Evidence of altered RNA stirs debate. *Nature, 473*, 432.

Heils, A., Teufel, A., Petri, S., Stober, G., Riederer, P., Bengel, D., et al. (1996). Allelic variation of human serotonin transporter gene expression. *Journal of Neurochemistry, 66*, 2621–2624.

Heiming, R. S., & Sachser, N. (2010). Consequences of serotonin transporter genotype and early adversity on behavioral profile—pathology or adaptation? *Frontiers in Neuroscience, 4*, 187.

Heurteaux, C., Lucas, G., Guy, N., El Yacoubi, M., Thümmler, S., Peng, X. D., et al. (2006). Deletion of the background potassium channel TREK-1 results in a depression-resistant phenotype. *Nature Neuroscience, 9*, 1134–1141.

Hibar, D. P., Stein, J. L., Ryles, A. B., Kohannim, O., Jahanshad, N., Medland, S. E., et al. (2012). Genome-wide association identified genetic variants associated with lentiform nucleus volume in N=1345 young and elderly subjects. *Brain Imaging and Behavior.* doi 10.1007/s11682-012-9199-7

Hizer, S. E., Wright, T. M., & Garcia, D. K. (2004). Genetic markers applied in regression tree prediction models. *Animal Genetics, 35*, 50–52.

Holmes, A. (2001). Targeted gene mutation approaches to the study of anxiety-like behavior in mice. *Neuroscience and Biobehavioral Reviews, 25*, 261–273.

Holmes, A., le Guisquet, A. M., Vogel, E., Millstein, R. A., Leman, S., & Belzung, C. (2005). Early life genetic, epigenetic and environmental factors shaping emotionality in rodents. *Neuroscience & Biobehavioral Reviews, 29,* 1335–1346.

Holtzheimer, P. E., & Mayberg, H. S. (2011). Deep brain stimulation for psychiatric disorders. *Annual Review of Neuroscience, 34,* 289–307.

Honea, R., Verchinski, B. A., Pezawas, L., Kolachana, B. S., Callicott, J. H., Mattay, V. S., et al. (2009). Impact of interacting functional variants in COMT on regional gray matter volume in human brain. *NeuroImage, 45,* 44–51.

Hyde, L. W., Bogdan, R., & Hariri, A. R. (2011). Understanding risk for psychopathology through imaging gene-environment interactions. *Trends in Cognitive Sciences, 15,* 417–427.

Hyde, L. W., Gorka, A., Manuck, S. B., and Hariri, A. R. (2011). Perceived social support moderates the link between threat-related amygdala reactivity and trait anxiety. *Neuropsychologia, 49,* 651–656.

Jaffee, S. R. (2011). Genotype-environment correlations: Definitions, methods of measurement, and implications for research on adolescent psychopathology. In K. S. Kendler, S. R. Jaffee & D. Romer (Eds.), *The dynamic genome and mental health* (pp. 79–102). New York: Oxford.

Jaffee, S. R., Caspi, A., Moffitt, T. E., Dodge, K. A., Rutter, M., Taylor, A., et al. (2005). Nature X nurture: Genetic vulnerabilities interact with physical maltreatment to promote conduct problems. *Development and Psychopathology, 17,* 67–84.

Jang, J. H., Jung, W. H., Choi, J. S., Choi, C. H., Kang, D. H., Shin, N. Y., et al. (2011). Reduced prefrontal functional connectivity in the default mode network is related to greater psychopathology in subjects with high genetic loading for schizophrenia. *Schizophrenia Research, 127,* 58–65.

Jedema, H. P., Gianaros, P. J., Greer, P. J., Kerr, D. D., Liu, S., Higley, J. D., et al. (2010). Cognitive impact of genetic variation of the serotonin transporter in primates is associated with differences in brain morphology rather than serotonin neurotransmission. *Molecular Psychiatry, 15,* 512–522.

Kalin, N. H., Shelton, S. E., Fox, A. S., Rogers, J., Oakes, T. R., & Davidson, R. J. (2008). The serotonin transporter genotype is associated with intermediate brain phenotypes that depend on the context of eliciting stressor. *Molecular Psychiatry, 13,* 1021–1027.

Karg, K., Burmeister, M., Shedden, K., & Sen, S. (2011). The serotonin transporter promoter variant (5-HTTLPR), stress, and depression meta-analysis revisited: Evidence of genetic moderation. *Archives of General Psychiatry, 68,* 444–454.

Kaufman, J., Yang, B. Z., Douglas-Palumberi, H., Houshyar, S., Lipschitz, D., Krystal, J. H., et al. (2004). Social supports and serotonin transporter gene moderate depression in maltreated children. *Proceedings of the National Academy of Sciences, 101,* 17316–17321.

Kempton, M. J., Salvador, Z., Munafò, M. R., Geddes, J. R., Simmons, A., Frangou, S., et al. (2011). Structural neuroimaging studies in major depressive disorder: Meta-analysis and comparison with bipolar disorder. *Archives of General Psychiatry, 68,* 675–690.

Kendler, K. S. (2011). A conceptual overview of gene-environment interaction and correlation in a developmental context. In K. S. Kendler, S. R. Jaffee, & D. Romer (Eds.), *The dynamic genome and mental health* (pp. 5–28). New York: Oxford.

Kendler, K. S., Kuhn, J. W., Vittum, J., Prescott, C. A., & Riley, B. (2005). The interaction of stressful life events and a serotonin transporter polymorphism in the prediction of episodes of major depression: A replication. *Archives of General Psychiatry, 62,* 529–535.

Kilpatrick, D. G., Koenen, K. C., Ruggiero, K. J., Acierno, R., Galea, S., Resnick, H. S., et al. (2007). The serotonin transporter genotype and social support and moderation of posttraumatic stress disorder and depression in hurricane-exposed adults. *American Journal of Psychiatry, 164,* 1693–1699.

Kim, D. K., Tolliver, T. J., Huang, S. J., Martin, B. J., Andrews, A. M., Wichems, C., et al. (2005). Altered serotonin synthesis, turnover and dynamic regulation in multiple brain regions of mice lacking the serotonin transporter. *Neuropharmacology, 49,* 798–810.

King, A. P., & Liberzon, I. (2009). Assessing the neuroendocrine stress response in the functional neuroimaging context. *NeuroImage, 47,* 1116–1124.

Kolber, B. J., Roberts, M. S., Howell, M. P., Wozniak, D. F., Sands, M. S., & Muglia, L. J. (2008). Central amygdala glucocorticoid receptor action promotes fear-associated CRH activation and conditioning. *Proceedings of the National Academy of Sciences, 105,* 12004–12009.

Kumsta, R., Moser, D., Streit, F., Koper, J. W., Meyer, J., & Wust, S. (2009). Characterization of a glucocorticoid receptor gene (GR, NR3C1) promoter polymorphism reveals functionality and extends a haplotype with putative clinical relevance. *American Journal of Medical Genetics B, 150,* 476–482.

Kupfer, D. J., & Regier, D. A. (2011). Neuroscience, clinical evidence, and the future of psychiatric classification in DSM-5. *American Journal of Psychiatry, 168,* 672–674.

Lemonde, S., Turecki, G., Bakish, D., Du, L., Hrdina, P. D., Bown, C. D., et al. (2003). Impaired repression at a 5-hydroxytryptamine 1A receptor gene polymorphism associated with major depression and suicide. *Journal of Neuroscience, 23,* 8788–8799.

Lenroot, R. K., & Giedd, J. N. (2011). Annual research review: Developmental considerations of gene by environment interactions. *Journal of Child Psychology and Psychiatry, 52,* 429–441.

Lesch, K. P., Bengel, D., Heils, A., Sabol, S. Z., Greenberg, B. D., Petri, S., et al. (1996). Association of anxiety-related traits with a polymorphism in the serotonin transporter gene regulatory region. *Science, 274,* 1527–1531.

Leve, L. D., Kerr, D. C. R., Shaw, D., Ge, X., Neiderhiser, J. M., Scaramella, L. V., et al. (2010). Infant pathways to externalizing behavior: Evidence of genotype×environment interaction. *Child Development, 81,* 340–356.

Li, M., Wang, I. X., Li, Y., Bruzel, A., Richards, A. L., Toung, J. M., et al. (2011). Widespread RNA and DNA sequence differences in the human transcriptome. *Science, 333,* 53–58.

Li, T., Sham, P. C., Vallada, H., Xie, T., Tang, X., Murray, R. M., et al. (1996). Preferential transmission of the high activity allele of COMT in schizophrenia. *Psychiatric Genetics, 6,* 131–133.

Liou, Y. J., Chen, T. J., Tsai, S. J., Yu, Y. W., Cheng, C. Y., & Hong, C. J. (2009). Support for the involvement of the

KCNK2 gene in major depressive disorder and response to antidepressant treatment. *Pharmacogenetics and Genomics, 19*, 735–741.
Liu, X., Akula, N., Skup, M., Brotman, M. A., Leibenluft, E., & McMahon, F. J. (2010). A genome-wide association study of amygdala activation in youths with and without bipolar disorder. *Journal of the American Academy of Child and Adolescent Psychiatry, 49*, 33–41.
Lonsdorf, T. B., Golkar, A., Lindstom, K. M., Fransson, P., Schalling, M., Ohman, A., et al. (2011). 5-HTTLPR and COMTval158met genotype gate amygdala reactivity and habituation. *Biological Psychology, 87*, 106–112.
Lupien, S. J., McEwen, B. S., Gunnar, M. R., & Heim, C. (2009). Effects of stress throughout the lifespan on the brain, behaviour and cognition. *Nature Reviews Neuroscience, 10*, 434–445.
Maheu, F. S., Dozier, M., Guyer, A. E., Mandell, D., Peloso, E., Poeth, K., et al. (2010). A preliminary study of medial temporal lobe function in youths with a history of caregiver deprivation and emotional neglect. *Cognitive, Affective, & Behavioral Neuroscience, 10*, 34–49.
Manuck, S. B. (2009). The reaction norm in gene× environment interaction. *Molecular Psychiatry, 15*, 881–882.
Manuck, S. B., & McCaffery, J. M. (2010). Genetics of stress: Gene-stress correlation and interaction. In A. Steptoe (Ed.), *Handbook of behavioral medicine: Methods and applications* (pp. 454–478). New York: Springer.
Masten, A. S. (2001). Ordinary magic. Resilience processes in development. *American Psychologist, 56*, 227–238.
Mazella, J., Petrault, O., Lucas, G., Deval, E., Beraud-Dufour, S., Gandin, C., et al. (2010). Spadin, a sortilin-derived peptide, targeting rodent TREK-1 channels: A new concept in the antidepressant drug design. *PLoS Biology, 8*, e1000355.
McClelland, G. H., & Judd, C. M. (1993). Statistical difficulties of detecting interactions and moderator effects. *Psychological Bulletin, 114*, 376–390.
McGowan, P. O., Sasaki, A., D'Alessio, A. C., Dymov, S., Labonté, B., Szyf, M., et al. (2009). Epigenetic regulation of the glucocorticoid receptor in human brain associates with childhood abuse. *Nature Neuroscience, 12*, 342–348.
Meaney, M. J. (2010). Epigenetics and the biological definition of gene x environment interactions. *Child Development, 81*, 41–79.
Meyer-Lindenberg, A. (2011). Neurogenetic mechanisms of gene-environment interactions. In K. A. Dodge & M. Rutter (Eds.), *Gene-environment interactions in developmental psychopathology* (pp. 71–87). New York: Guilford.
Meyer-Lindenberg, A., Kohn, P. D., Kolachana, B., Kippenhan, S., McInerney-Leo, A., Nussbaum, R., et al. (2005). Midbrain dopamine and prefrontal function in humans: Interaction and modulation by COMT genotype. *Nature Neuroscience, 8*, 594–596.
Meyer-Lindenberg, A., Nichols, T., Callicott, J. H., Ding, J., Kolachana, B., Buckholtz, J., et al. (2006). Impact of complex genetic variation in COMT on human brain function. *Molecular Psychiatry, 11*, 867–877.
Meyer-Lindenberg, A., & Weinberger, D. R. (2006). Intermediate phenotypes and genetic mechanisms of psychiatric disorders. *Nature Reviews Neuroscience, 7*, 818–827.
Mier, D., Kirsch, P., & Meyer-Lindenberg, A. (2010). Neural substrates of pleiotropic action of genetic variation in COMT: A meta-analysis. *Molecular Psychiatry, 15*, 918–927.
Mitchell, T. M., Shinkareva, S. V., Carlson, A., Chang, K. M., Malave, V. L., Mason, R. A., et al. (2008). Predicting human brain activity associated with the meanings of nouns. *Science, 320*, 1191–1195.
Munafò, M. R., Brown, S. M., & Hariri, A. R. (2008). Serotonin transporter (5-HTTLPR) genotype and amygdala activation: A meta-analysis. *Biological Psychiatry, 63*, 852–857.
Munafò, M. R., Freimer, N. B., Ng, W., Ophoff, R., Veijola, J., Miettunen, J., et al. (2009). 5-HTTLPR genotype and anxiety-related personality traits: A meta-analysis and new data. *American Journal of Medical Genetics B, 150*, 271–281.
Nicodemus, K. K., Kolachana, B. S., Vakkalanka, R., Straub, R. E., Giegling, I., Egan, M. F., et al. (2007). Evidence for statistical epistasis between catechol-O-methyltransferase (COMT) and polymorphisms in RGS4, G72 (DAOA), GRM3, and DISC1: Influence on risk of schizophrenia. *Human Genetics, 120*, 889–906.
Nikolova, Y. S., Ferrell, R. E., Manuck, S. B., & Hariri, A. R. (2011). Multilocus genetic profile for dopamine signaling predicts ventral striatum reactivity. *Neuropsychopharmacology, 36*, 1940–1947.
Oler, J. A., Fox, A. S., Shelton, S. E., Christian, B. T., Murali, D., Oakes, T. R., et al. (2009). Serotonin transporter availability in the amygdala and bed nucleus of the stria terminalis predicts anxious temperament and brain glucose metabolic activity. *Journal of Neuroscience, 29*, 9961–9966.
Paloyelis, Y., Asherson, P., Mehta, M. A., Faraone, S. V., & Kuntsi, J. (2010). DAT1 and COMT effects on delay discounting and trait impulsivity in male adolescents with attention deficit/hyperactivity disorder and healthy controls. *Neuropsychopharmacology, 35*, 2414–2426.
Park, J. H., Wacholder, S., Gail, M. H., Peters, U., Jacobs, K. B., Chanock, S. J., et al. (2010). Estimation of effect size distribution from genome-wide association studies and implications for future discoveries. *Nature Genetics, 42*, 570–575.
Parsey, R., Oquendo, M., Ogden, R., Olvet, D., Simpson, N., Huang, Y., et al. (2006). Altered serotonin 1A binding in major depression: A [carbonyl-C-11]WAY100635 positron emission tomography study. *Biological Psychiatry, 59*, 106–113.
Perlis, R. H., Moorjani, P., Fagerness, J., Purcell, S., Trivedi, M. H., Fava, M., et al. (2008). Pharmacogenetic analysis of genes implicated in rodent models of antidepressant response: Association of TREK1 and treatment resistance in the STAR*D study. *Neuropsychopharmacology, 33*, 2810–2819.
Pezawas, L., Meyer-Lindenberg, A., Drabant, E. M., Verchinski, B. A., Munoz, K. E., Kolachana, B. S., et al. (2005). 5-HTTLPR polymorphism impacts human cingulate-amygdala interactions: A genetic susceptibility mechanism for depression. *Nature Neuroscience, 8*, 828–834.
Pizzagalli, D. A. (2011). Frontocingulate dysfunction in depression: Toward biomarkers of treatment response. *Neuropsychopharmacology, 36*, 183–206.
Pizzagalli, D. A., Sherwood, R. J., Henriques, J. B., & Davidson, R. J. (2005). Frontal brain asymmetry and reward responsiveness: A source-localization study. *Psychological Science, 16*, 805–813.
Preacher, K. J., & Hayes, A. F. (2008). Asymptotic and resampling strategies for assessing and comparing indirect effects

in multiple mediator models. *Behavioral Research Methods, 40*, 879–891.

Purcell, S. M., Wray, N. R., Stone, J. L., Visscher, P. M., O'Donovan, M. C., Sullivan, P. F., et al. (2009). Common polygenic variation contributes to risk of schizophrenia and bipolar disorder. *Nature, 460*, 748–752.

Reiss, D., & Leve, L. D. (2007). Genetic expression outside the skin: Clues to mechanisms of genotype x environment interaction. *Development and Psychopathology, 19*, 1005–1027.

Richardson-Jones, J. W., Craige, C. P., Guiard, B. P., Stephen, A., Metzger, K. L., Kung, H. F., et al. (2010). 5-HT1A autoreceptor levels determine vulnerability to stress and response to antidepressants. *Neuron, 65*, 40–52.

Risch, N., Herrell, R., Lehner, T., Liang, K. Y., Eaves, L., Hoh, J., et al. (2009). Interaction between the serotonin transporter gene (5-HTTLPR), stressful life events, and risk of depression. *JAMA, 301*, 2462–2471.

Roth, T. L., & Sweatt, J. D. (2011). Annual research review: Epigenetic mechanisms and environmental shaping of the brain during sensitive periods of development. *Journal of Child Psychology and Psychiatry, 52*, 398–408.

Rutter, M., & Dodge, K. A. (2011). Gene-environment interactions: The state of science. In K. A. Dodge & M. Rutter (Eds.), *Gene-environment interaction in developmental psychopathology* (pp. 87–101). New York: Guilford.

Rutter, M., Moffitt, T. E., & Caspi, A. (2006). Gene-environment interplay and psychopathology: Multiple varieties but real effects. *Journal of Child Psychology and Psychiatry, 47*, 226–261.

Santesso, D. L., Evins, A. E., Frank, M. J., Schetter, E. C., Bogdan, R., & Pizzagalli, D. A. (2009). Single dose of a dopamine agonist impairs reinforcement learning in humans: Evidence from event-related potentials and computational modeling of striatal-cortical function. *Human Brain Mapping, 30*, 1963–1976.

Serretti, A., Kato, M., De Ronchi, D., & Kinoshita, T. (2006). Meta-analysis of serotonin transporter gene promoter polymorphism (5-HTTLPR) association with selective serotonin reuptake inhibitor efficacy in depressed patients. *Molecular Psychiatry, 12*, 247–257.

Sjöberg, R. L., Nilsson, K. W., Wargelius, H. L., Leppert, J., Lindström, L., & Oreland, L. (2006). Adolescent girls and criminal activity: Role of MAOA-LPR genotype and psychosocial factors. *American Journal of Medical Genetics B, 144*, 159–164.

Stein, D. J., Newman, T. K., Savitz, J., & Ramesar, R. (2006). Warriors versus worriers: The role of COMT gene variants. *CNS Spectrums, 11*, 745–748.

Stein, J. L., Medland, M. L., Vasquez, A. A., Hibar, D. P., Senstad, R. E., Winkler, A. M., et al. (2012). Identification of common variants associated with human hippocampal and intracranial volumes. *Nature Genetics, 44*, 552–561.

Tan, H. Y., Chen, Q., Goldberg, T. E., Mattay, V. S., Meyer-Lindenberg, A., Weinberger, D. R., et al. (2007). Catechol-O-methyltransferase Val158Met modulation of prefrontal-parietal-striatal brain systems during arithmetic and temporal transformations in working memory. *Journal of Neuroscience, 27*, 13393–13401.

Tan, H. Y., Chen, Q., Sust, S., Buckholtz, J. W., Meyers, J. D., Egan, M. F., et al. (2007). Epistasis between catechol-O-methyltransferase and type II metabotropic glutamate receptor 3 genes on working memory brain function. *Proceedings of the National Academy of Sciences, 104*, 12536–12541.

Thomason, M. E., & Thompson, P. M. (2011). Diffusion imaging, white matter, and psychopathology. *Annual Review of Clinical Psychology, 7*, 63–85.

Tottenham, N., Hare, T. A., Millner, A., Gilhooly, T., Zevin, J., & Casey, B. J. (2011). Elevated amygdala response to faces following early deprivation. *Developmental Science, 14*, 190–204.

Tyrka, A. R., Price, L. H., Gelernter, J., Schepker, C., Anderson, G. M., & Carpenter, L. L. (2009). Interaction of childhood maltreatment with the corticotropin-releasing hormone receptor gene: Effects on hypothalamic-pituitary-adrenal axis reactivity. *Biological Psychiatry, 66*, 681–685.

Uher, R. (2011). Gene-environment interactions. In K. S. Kendler, S. R. Jaffee, & D. Romer (Eds.), *The dynamic genome and mental health* (pp. 29–58). New York: Oxford.

Uher, R., & McGuffin, P. (2010). The moderation by the serotonin transporter gene of environmental adversity in the etiology of depression: 2009 update. *Molecular Psychiatry, 15*, 18–22.

Ulrich-Lai, Y. M., & Herman, J. P. (2009). Neural regulation of endocrine and autonomic stress responses. *Nature Reviews Neuroscience, 10*, 397–409.

Ursini, G., Bollati, V., Fazio, L., Porcelli, A., Iacovelli, L., Catalani, A., et al. (2011). Stress-related methylation of the catechol-o-methyltransferase val158 allele predicts human prefrontal cognition and activity. *Journal of Neuroscience, 31*, 6692–6698.

van Leeuwen, N., Kumsta, R., Entringer, S., de Kloet, E. R., Zitman, F. G., DeRijk, R. H., et al. (2010). Functional mineralocorticoid receptor (MR) gene variation influences the cortisol awakening response after dexamethasone. *Psychoneuroendocrinology, 35*, 339–349.

Velders, F. P., Kuningas, M., Kumari, M., Dekker, M. J., Uitterlinden, A. G., Kirschbaum, C., et al. (2011). Genetics of cortisol secretion and depressive symptoms: A candidate gene and genome wide association approach. *Psychoneuroendocrinology, 36*, 1053 1061.

von dem Hagen, E. A., Passamonti, L., Nutland, S., Sambrook, J., & Calder, A. J. (2011). The serotonin transporter gene polymorphism and the effect of baseline on amygdala response to emotional faces. *Neuropsychologia, 49*, 674–680.

Watson, K. K., Ghodasra, J. H., & Platt, M. L. (2009). Serotonin transporter genotype modulates social reward and punishment in rhesus macaques. *PLoS One, 4*, e4156.

Weaver, I. C. G., Cervoni, N., Champagne, F. A., D'Alessio, A. C., Sharma, S., Seckl, J. R., et al. (2004). Epigenetic programming by maternal behavior. *Nature Neuroscience, 7*, 847–854.

Weder, N., Yang, B. Z., Douglas-Palumberi, H., Massey, J., Krystal, J. H., Gelernter, J., et al. (2009). MAOA genotype, maltreatment, and aggressive behavior: The changing impact of genotype at varying levels of trauma. *Biological Psychiatry, 65*, 417–424.

Wenten, M., Gauderman, W. J., Berhane, K., Lin, P. C., Peters, J., & Gilliland, F. D. (2009). functional variants in the catalase and myeloperoxidase genes, ambient air pollution, and respiratory-related school absences: An example of epistasis in gene-environment interactions. *American Journal of Epidemiology, 170*, 1494–1501.

Whitfield-Gabrieli, S., Thermenos, H. W., Milanovic, S., Tsuang, M. T., Faraone, S. V., McCarley, R. W., et al.

(2009). Hyperactivity and hyperconnectivity of the default network in schizophrenia and in first-degree relatives of persons with schizophrenia. *Proceedings of the National Academy of Sciences*, *106*, 1279–1284.

Widom, C. S., & Brzustowicz, L. M. (2006). MAOA and the "cycle of violence:" childhood abuse and neglect, MAOA genotype, and risk for violent and antisocial behavior. *Biological Psychiatry*, *60*, 684–689.

Willeit, M., & Praschak-Rieder, N. (2010). Imaging the effects of genetic polymorphisms on radioligand binding in the living human brain: A review on genetic neuroreceptor imaging of monoaminergic systems in psychiatry. *NeuroImage*, *53*, 878–892.

Xin, Y., Chanrion, B., Liu, M. M., Galfalvy, H., Costa, R., Ilievski, B., et al. (2010). Genome-wide divergence of DNA methylation marks in cerebral and cerebellar cortices. *PLoS One*, *5*, e11357.

Yang, J., Benyamin, B., McEvoy, B. P., Gordon, S., Henders, A. K., Nyholt, D. R., et al. (2010). Common SNPs explain a large proportion of the heritability for human height. *Nature Genetics*, *42*, 565–569.

Yehuda, R. (2002). Post-traumatic stress disorder. *New England Journal of Medicine*, *346*, 108–114.

Zhang, T. Y., & Meaney, M. J. (2010). Epigenetics and the environmental regulation of the genome and its function. *Annual Review of Psychology*, *61*, 439–466, C431–C433.

CHAPTER

12 Copy Number Variation in Neuropsychiatric Disorders

James J. H. Rucker *and* Peter McGuffin

Abstract

It has long been known that the human genome is subject to deletion and duplication of genetic material by various molecular mechanisms. Until recently, such events were assumed to be relatively rare phenomena. It is now known that submicroscopic deletions or duplications called *copy number variants* (CNVs) are a major source of genomic variation. Rare CNVs (defined as occurring in less than 1 percent of the population) have been implicated in schizophrenia and autism. Measured in terms of odds ratios, individual CNVs have been shown to have large effects, some increasing the risk of disorder several-fold. But they are incompletely penetrant, no one CNV is either necessary or sufficient to cause the disorder. The findings are less clear-cut with bipolar disorder but, here, too, rare CNVs probably play a role. In unipolar depression, initial evidence suggests an overall increase in rare CNVs that disrupt exons, the coding regions of genes.

Key Words: copy number variants, CNVs, schizophrenia, autism, bipolar disorder, depression

The study of human genetic variation and its contribution to neuropsychiatric disorders has rapidly gained momentum since the end of the past century. The first (nearly) complete annotated sequence of the human genome was published in 2001 (Lander et al., 2001, Venter, Adams, Myers, Li, & Mural, 2001), and, at about the same time, microarray chips were developed that allowed thousands of genetic variants to be detected at once on a single glass slide. These two advances made what previously had been a time-consuming and expensive process both fast and economically viable to perform on large numbers of samples. As a result, genome-wide association studies (GWAS) soon became feasible, and a landmark paper published in 2007 presented results on seven common disorders (Wellcome Trust Case Control Consortium, 2007), one of which was bipolar disorder. Subsequently, many GWAS on other neuropsychiatric and physical disorders have followed and with this has come vast quantities of new genetic data.

It has long been known that the human genome is subject to deletion and duplication of genetic material by various molecular mechanisms; however, such events were assumed to be relatively rare phenomena, confined mostly to infrequently occurring, sporadic diseases (Perry et al., 2008). The new wave of genetic data from GWAS indicated that these events were much more common than was assumed, coming as a surprise to many geneticists. Of particular note was the frequent presence of submicroscopic deletions and duplications of genetic material, now termed *copy number variation.* This chapter discusses copy number variants (CNVs), the current state of evidence for their role in the pathogenesis of different neuropsychiatric disorders, and the application of such knowledge in clinical settings.

I. Definitions

A CNV is currently usually defined as a deletion or duplication of genomic material between

1,000 base pairs and 3,000,000 base pairs in length (Feuk, Carson, & Scherer, 2006). In practice, this definition is arbitrary, and the size limit of what constitutes a CNV often varies according to different research groups. The upper limit of 3,000,000 base pairs is the approximate size of the smallest abnormality that can be seen on a karyotype (i.e., a stained preparation of the 23 pairs of chromosomes) under the light microscope. Since chromosomes are paired, deletion CNVs may consist of one copy (hemizygous deletion) or no copies (homozygous deletion) of a stretch of genetic material, with a homozygous deletion being likely to be more deleterious than a hemizygous deletion. Duplication CNVs may exist as three copies, four copies, or more than four copies. Table 12.1 illustrates current CNV nomenclature. Note that the X chromosome follows the same convention, except that it is based on whether the sample is male (in which case there is only one copy of the X chromosome) or female (in which there are two). Thus, for example, a copy number of 2 in a male X chromosome sample would denote a duplication, 1 normal copy number, and 0 a homozygous deletion. For females, the convention is the same as for autosomes. Total DNA copy numbers of above four are difficult to detect with any certainty.

As with other types of genomic variants, CNVs occur with varying degrees of frequency throughout the population. Copy number variants are conventionally termed as either common (occurring at a frequency of more than 1 in 100 individuals in a population) or rare (less than 1 in 100 individuals). The rarest CNVs are of particular interest, first because they are likely to be evolutionarily under negative selection pressure and thus pertinent to the etiology of disease states, and second, because they are presumably enriched in so-called de novo CNVs. A de novo CNV is defined as a CNV present in a child but not in the parents of that child. Thus, the CNV has occurred during the creation of sperm or egg or after conception. Many neuropsychiatric diseases occur sporadically (i.e., they do not occur in a family with other members affected by the same disease). Sporadically occurring cases of disease could, in part, be explained by de novo CNVs. Copy number variants may also be referred to as "singleton" CNVs, which refers to a CNV that is only seen once in a dataset. Such CNVs are also, presumably, rare. For the purposes of this chapter, we focus on rare CNVs and their etiology in neuropsychiatric disorders. Although common CNVs may also play a role, current evidence suggests that their effects are modest (Conrad et al., 2010).

II. Copy Number Variants in Neuropsychiatric Disorders

Neuropsychiatric disorders such as schizophrenia are notable within medicine for the lack of any objective biological markers to guide diagnosis. Although psychiatrists have sought to cluster these disorders by observable behavior and reported symptomatology, there is little reason to expect this to translate into neat delineations at the genetic level. Thus, it is not likely that any particular CNV will segregate solely with any particular disorder.

Table 12.1. Copy number variant nomenclature for autosomal (non-sex) chromosomes.

Copy number state	Total DNA copy number	Description	Associated genotypes (each genotype is arbitrarily denoted A and B)
1	0	Deletion of two copies (homozygous deletion)	Null
2	1	Deletion of one copy (hemizygous deletion)	A, B
3	2	Normal state	AA, AB, BB
4	2	Normal state with loss of heterozygosity (AB)	AA, BB
5	3	Single copy duplication (hemizygous duplication)	AAA, AAB, ABB, BBB
6	4	Double copy duplication (homozygous duplication)	AAAA, AAAB, AABB, ABBB, BBBB

The current state of the evidence favors the idea that individual CNVs are associated with more than one disorder (an example of genetic pleiotropy). This may indicate that the role of CNVs and other genetic variants is in the manifestation of a general liability to neuropsychiatric disorder, but that the processes finally defining which disorder develops are complex gene–environment interactions over the lifespan of the individual. Put another way, the conceptual gap between gene and observed behavior is vast, and the etiological links between them likely to be abstruse. Although there are many intriguing hypotheses about the links between different neuropsychiatric disorders and the genes they associate with, and some biological evidence to support the role of different gene products in brain development and function, the biological etiology of neuropsychiatric disorders remains largely a mystery. Copy number variants probably represent a small, yet significant, piece of a larger puzzle.

In the investigation of the neurobiology of neuropsychiatric disorders, where the etiology is generally unknown, GWAS have distinct advantages and disadvantages. It is valuable to be able to test all regions of the genome for association simultaneously (a so-called hypothesis-free approach). However, in performing many simultaneous statistical tests for association, many false-positive results will be obtained unless correction for multiple testing is applied. This will, however, tend to run the risk of type 2 errors missing some true associations with small effect sizes. In contrast, studies using prior hypotheses about which genes may be involved are more likely to detect small effect sizes, if they exist. These approaches can reveal valuable mechanistic data, but will, of course, miss associations in areas of the genome not tested.

III. Specific Disorders

A. Schizophrenia

Schizophrenia is a common (~0.8 percent lifetime incidence) neuropsychiatric disorder usually striking in early adulthood and characterized by problems with perception, inferential thinking, goal-directed behavior, and emotional expression, as well as, cardinally, a lack of insight into the disorder. It is a clinically heterogeneous disorder with no one symptom or sign necessary for the diagnosis. Although some individuals recover fully, many follow a chronic course of illness with substantial psychosocial deficits and increased morbidity and mortality, especially from suicide (Harris & Barraclough, 1998).

Twin studies in schizophrenia estimate a substantial heritability of approximately 80–85 percent, with first-degree relatives of sufferers having a 10-fold increased risk of developing the disorder (for a review see [Gottesman, 1991]). Although past genetic studies involving linkage, single nucleotide polymorphisms, and candidate genes have discovered some interesting regions in the genome, often these have not been replicated (for a review see, for example [Schwab & Wildenauer, 2009] and [Sullivan, 2005]). A substantial proportion of heritability remains to be explained.

Four notable studies of rare CNVs in schizophrenia were published in 2008. The first report, by Walsh et al. in April 2008, was a comparatively small study of 150 patients with schizophrenia and 268 controls of the same ethnicity as the cases. They also repeated their experiment on 92 patients with childhood-onset schizophrenia, a particularly severe form of the disease (Walsh et al., 2008). Among CNVs over gene-coding regions a threefold enrichment was found in the schizophrenia cohort, with those with an onset of illness below 18 years more than four times as likely to harbor a CNV as controls. Within the childhood-onset schizophrenia cohort, 28 percent harbored a rare CNV over a gene-coding region, significantly greater than in the control sample. The authors commented on the range of genes covered by the variants detected, which were also significantly more likely to be associated with biological pathways known to be involved with neuronal development and function than chance, specifically neuregulin signaling, ERK/MAPK signaling, synaptic long-term potentiation, axonal guidance signaling, integrin signaling, and glutamate receptor signaling. In contrast, within the control group, no such association was observed, with genes involving any biological pathway equally as likely to be observed. The authors looked at a number of gene-coding CNVs in detail, for example, those covering *ERBB4*, which is a large, complex gene known to bind and activate neuregulin proteins that are intimately involved with synaptic structure and function. They showed that the ERBB4 protein transcribed as a result of the CNV was mutated and unlikely to be functional.

In 2008, Xu et al. published results of an analysis of 152 individuals with schizophrenia matched against 159 controls from the same population (Xu et al., 2008). They also genotyped the biological parents of each subject, and so were able to distinguish de novo CNVs. They found 15 cases with confirmed de novo CNVs, compared with two controls, a highly significant enrichment. Two cases carried two de novo CNVs each. They then went on to assess the significance of the de novo CNVs by looking at a further cohort of 48 individuals with schizophrenia who also had an affected first- or second-degree relative. In contrast, they did not find de novo CNVs in this cohort. They also found that sporadic cases of schizophrenia were only 1.5 times more likely than unaffected controls to harbor an inherited rare CNV. They did not find any association of rare CNV events with age of onset, presence or history of learning disability, or parental age at birth. Three cases of the 22q11.2 deletion were identified in this cohort, of which two had mild facial dysmorphism.

Two much larger studies of rare copy number variation in schizophrenia were published in the same issue of *Nature* in 2008. The International Schizophrenia Consortium reported results of their analysis in 3,391 patients with schizophrenia and 3,181 ethnically matched controls (International Schizophrenia Consortium, 2008). As expected, the 22q11.2 deletion was found to be significantly associated, but the group also reported new associations with deletion CNVs at 15q13.3 and 1q21.1. The observation of a deletion at 15q13.3 was particularly salient because another group had recently associated the same region with mental retardation and seizures (Sharp et al., 2008). The region at 1q21.1 has previously been associated with schizophrenia by linkage (Brzustowicz, Hodgkinson, Chow, Honer, & Bassett, 2000). The overall burden of rare CNVs was increased 1.15-fold in cases compared to controls. Subgroups of CNVs, including those that were larger (>500 kilobase [kb] pairs), rarer in the sample (fewer than a total of six in the entire sample), and those that covered gene-coding regions were, again, particularly associated with the schizophrenia cohort. The general increase in burden of rare CNVs was reflected in the observation of large, rare deletion CNVs in other areas of the genome, which, although present in a minority of cases but no controls, failed to reach statistical significance for association (presumably because of a lack of statistical power).

In the second study, Stefansson et al., focused on 66 de novo CNV regions (i.e., regions with CNVs in an individual that were not seen in either parent) that they identified in a exploratory population sample and tested these regions for association in a sample of 1,433 cases of schizophrenia and 33,250 controls (Stefansson et al., 2008). Those regions found to have suggestive association were followed-up in a replication cohort comprised of 3,285 cases and 7,951 controls. The group identified deletions at 1q21.1, 15q11.2, and 15q13.3 as having nominal association in their first sample. All three of these deletions showed significant association in the follow-up sample. The 22q11.2 deletion was present in 8 out of 3,838 cases but, notably, absent in all controls. The authors also had access to treatment response data, family history data, and other clinical data. Clinical response to antipsychotic medication was no different from that expected normally in those with deletions at 1q21.1, 15q11.2, and 15q13.3. There was no obvious sex bias, and family history of schizophrenia in a close relative was also comparable to normal. Although only a subset of samples had data regarding cognitive function, the authors noted that three cases carrying the 1q21.1 deletion had learning disabilities and, of the eight controls who had the deletion, two had dyslexia. As with previous evidence presented for autism (Pagnamenta et al., 2010), this suggests that schizophrenia represents a disorder in which multiple deleterious genetic variants, each neither necessary nor sufficient on their own (but still associated with less severe disabilities, such as dyslexia), can combine to precipitate the clinical disorder. The formidable genetic complexity and interindividual variability of what is seen to be a similar clinical entity slowly starts to emerge with such evidence.

The role of recurrent CNVs at 16p11.2 in schizophrenia was investigated by McCarthy et al. in 2009, working on the findings of previous research implicating this region in autism (Marshall et al., 2008; Weiss et al., 2008). By this point, it was becoming clear that CNVs that segregated with specific disorders were infrequent, and this research group had previously detected a CNV in this region in two cases of childhood-onset schizophrenia (Walsh et al., 2008). They analyzed two cohorts of schizophrenia cases matched against controls. In the first cohort, they found the 16p11.2 duplication CNV in 12 out of 1,906 cases but in only 1 out of 3,971 controls. In the second cohort, they again found this duplication in 9 out of 2,645 cases but in only 1 out of 2,420 controls. This represents a 14.5-fold increased risk of schizophrenia. The deletion CNV seen in this region was not associated with schizophrenia; however, the authors performed a meta-analysis of previous reports and showed that the deletion was associated particularly with autism and developmental disorders. The duplication, in contrast, was associated with schizophrenia and bipolar disorder, as well as with autism.

This is an interesting result from a few perspectives. First, an association is shown between autism and both the deletion and the duplication, which, at first glance, appears counterintuitive. Second, the duplication is associated with schizophrenia and bipolar disorder, whereas the deletion is not. Third, some people with deletions and duplications at 16p11.2 appear to have no disorder. In reconciling these results, it is first pertinent to point out that deletion CNVs, by their nature, are more likely to be deleterious to function than are duplication CNVs. If one assumes that normal function is usually reliant on the presence of two copies of a gene, it is more likely that aberrant function will be a result of a deletion, which results in a 50 percent reduction in gene number, than of a duplication, which results in a 33.3 percent increase in gene number. As an aside, duplication events appear to be an essential driver of species evolution (Zhang, 2003). Gene duplication allows evolution of the duplicated gene copy into an analogous but different form of the original, leaving the other intact. Multiple examples of genes that share sequence similarity (or homology) with each other are present throughout the human genome, suggesting that diversification of existing genes is an evolutionary mechanism. Clearly, this cannot happen if genes are deleted (although this may, on rare occasions, be adaptive). However, these observations shed some light on why a more severe disorder such as autism or intellectual disability is only associated with the deletion CNV. Schizophrenia and bipolar disorder, being of later onset and also, usually, being treatable, can be considered less severe, and thus the association with the duplication is more likely. But this does not explain why (A) people without any disorder manifest the CNV and (B) why the duplication CNV is also seen in autism. To explain this, we must draw on the idea that a CNV event is neither necessary nor sufficient to induce a clinical disorder such as schizophrenia. It is likely that multiple deleterious variants are required, each contributing a small and probably clinically undetectable liability. Unfavorable interaction between environment, and a number of these variants in the same individual, is more likely to be associated with disorder than any one variant alone. With this in mind, points (A) and (B) are now explicable. Normal individuals who manifest the CNV may indeed have subtle problems, but not sufficient to be particularly impairing and certainly not sufficient to be diagnosed with a clinical disorder, and they are thus recruited as controls. Those with autism who manifest the duplication CNV probably also manifest many other undetected deleterious variants, perhaps interacting unfavorably with the 16p11.2 duplication. With such speculation, we perhaps gain a glimpse into the genetic complexity of these disorders, which raises important questions about the clinical utility of using microarrays in clinical diagnostics and therapeutics (discussed at the end of this essay).

In a further study, Levinson et al. described an analysis of rare CNVs in 3,945 subjects with schizophrenia or schizoaffective disorder, compared to 3,611 screened controls (Levinson et al., 2011). In this analysis, they confirmed previous associations found for deletions at 15q13.3, 1q21.1, and 22q11.2 as well as duplications at 16p11.2. They additionally identified a new association region, finding a 1.6 MB deletion at chromosome 3q29 in seven cases, but in no controls. This CNV had previously been found in 14 individuals in a sample of 14,698 individuals with mild-moderate mental retardation, and a variety of other phenotypes, including autism (Ballif et al., 2008). Overall, this paper provides strong independent support for the role of certain rare CNVs in schizophrenia.

The large CNVs already discussed in this essay usually span hundreds of thousands or millions of base pairs. Copy number variants of this size usually encompass large numbers of genes, making it difficult to pinpoint which gene or genes may be responsible for contributing to the disease. However, smaller CNVs that interrupt genes are also likely to be important since such abnormalities are likely to produce proteins that do not function normally. The gene *NRXN1* (*neurexin 1*) encodes a presynaptic neuronal cell surface protein that, along with others, mediates trans-synaptic neurotransmission by interacting with neuroligins (Sudhof, 2008). Although disruption of this gene does not appear to impair the formation of synapses, it does impair normal synaptic function. *Neurexin 1* is a highly polymorphic gene and is probably transcribed into hundreds or thousands of distinct protein variants. It is highly expressed in all areas of the brain, and mice that have had this gene removed demonstrate deficiencies in prepulse inhibition, which is also observed in cases of schizophrenia (Swerdlow et al., 2006). Copy number variation in this gene had already been demonstrated by Kirov et al. in two siblings with schizophrenia; however, an analysis of this gene within a subcohort of schizophrenia patients previously reported in 2008 found significant associations between CNVs affecting exons of this gene and schizophrenia (Kirov et al., 2007; Rujescu et al., 2009). Walsh et al. also found a deletion of *neurexin 1* affecting identical twins with early-onset schizophrenia (Walsh et al., 2008). Marshall et al. and Weiss et al., among others, have also linked variants in *neurexin 1* to autism (Feng et al., 2006; Kim et al., 2008; Marshall et al., 2008; Morrow et al., 2008; Szatmari et al., 2007; Weiss et al., 2008), and a further study has linked this gene to mental retardation (Friedman et al., 2006). The CNVs demonstrated in *neurexin 1* are not identical, variously affecting different parts of the gene. *Neurexin 1* is a complex gene, with an alpha and beta protein formed from transcription via different promoters and multiple isoforms of each alpha and beta variant produced by splicing of the messenger RNA. A disrupting CNV within *neurexin 1* is likely to disrupt the balance of splice variants produced at different sites in the central nervous system (CNS) and probably subtly affects synaptic neurotransmission, with the extent of disruption being dependent on the precise nature of the CNV involved. In a review of this gene and meta-analysis of association studies, Kirov at al. showed the CNV to be present in 17 out of 8,798 cases but in only 17 out of 42,054 controls ($p = 1.3 \times 10^{-5}$) with deletions of more than 100 kb even more strongly associated with disease ($p = 3.7 \times 10^{-6}$) (Kirov, Rujescu, et al., 2009). Here, therefore, we have quite compelling evidence of association of a single gene with disorder. But, even here, the picture is complex, with the deletion seen in normal controls, as well as in a range of cases of different neuropsychiatric disorders. Of course, in considering this gene in isolation we fail to take account of the interacting effects of other variants, which probably explains the observed variation. The neurexin proteins are intimately associated with the neuroligin proteins at the synapse (Sudhof, 2008). Deleterious variants in neuroligin genes have also been associated with autism (Jamain et al., 2003). Clearly, some variants and some combinations of variants are more likely to cause disease than others depending on the specific nature of their interaction.

B. 22q11.2 Deletion Syndromes

This region is worth special mention because its association with schizophrenia-like syndromes had been described long before anything was known about CNVs in other regions. The 22q11.2 deletions encompass between 30 and 60 genes, depending on the deletion size, most of which are expressed in brain (Maynard et al., 2003), and the region's architecture makes it, of all human chromosomal regions, the most prone to rearrangement (Saitta et al., 2004). Larger deletions are detectable using older methods such as fluorescent in-situ hybridization (FISH), and it has been known for over a decade that schizophrenia occurs in approximately 30 percent of individuals with a physical disorder—velocardio-facial syndrome—that results from a 22q11.2 deletion (Murphy, Jones, & Owen, 1999). Other disorders such as attention deficit hyperactivity disorder (ADHD), autism, and bipolar disorder have also been reported to be associated with 22q11.2 deletions and manifestations including speech and developmental delay, cardiac abnormalities, and facial and palatal deformities are frequently seen (Karayiorgou, Simon, & Gogos, 2010). The deletion is comparatively common, with estimates ranging between 1 in 3,000 and 1 in 9,000 individuals (Goodship, Cross, LiLing, & Wren, 1998; Montcel Du & Mendizabai, 1996), but these could be underestimates since the deletion may be underascertained by clinicians who, because of the variable phenotype, are not routinely testing for its presence. It is of particular interest in the study of the genetic etiology of schizophrenia because there is little evidence of elevated rates of schizophrenia in other genetic syndromes that manifest in abnormal behavior, and there is only a mild elevation in those with learning disabilities (Turner, 1989; Turk & Hill, 1995). The inference from this is that some genes within the 22q11.2 region are particularly salient in the etiology of some cases of schizophrenia. In a gene-rich area, it has not yet been possible to confidently identify which gene deletion (or combination of gene deletions) is crucial, but the region contains a number of interesting candidates including *COMT* and *PRODH*.

C. Autism

Autism appears to be one of the most heritable neuropsychiatric disorders, with heritability estimates of over 80 percent and with approximately 5 percent of patients having known genetic disorders such as fragile X syndrome, Rett syndrome, tuberous sclerosis, Potocki-Lupski syndrome and Smith-Lemli-Opitz syndrome (Freitag, 2007). Autism is now considered to comprise a spectrum of disorders, with typical autism presenting before age 3 with a core phenotype of ritualized, stereotyped, and restricted behavior and delays in the development of language and reciprocal social interaction skills. More general intellectual disability is also commonly seen. Less severe forms, such as Asperger syndrome and atypical autism, more often present with unimpaired intellectual function and speech development but with stereotyped interests and significant deficits in the development of social behavior and reciprocal social interaction remaining prominent (Bailey, Phillips, & Rutter, 1996). The prevalence of autism differs between sexes, with a strong preponderance for males. There is no medical treatment for the underlying causes of autism; however, comorbid neuropsychiatric disorders such as major depression, obsessive compulsive disorder, and schizophrenia may be seen and are treated in a broadly similar way to those seen in individuals without autism.

Approximately 5 percent of patients with autism have an identifiable genetic abnormality that is visible with the light microscope (in practice, meaning a genetic aberration of at least 3 million base pairs in length) (Vorstman et al., 2005). Previous genetic studies involving linkage analysis have implicated over 20 chromosomal regions in autism, giving an early indication of its complex genetic etiology (Klauck, 2006).

Sebat et al. (2007) used a high-density microarray to look for CNVs in autism in 165 families (118 with a single affected child and 77 with cases spanning generations of the same family), and 99 control families (Sebat et al., 2007). Within all the individual cases of autism (n = 195), they found 15 CNVs in 14 cases, mostly comprised of deletions (12 of 15). In the controls (n = 196), they only found two CNVs in two individuals, both of which were duplications. Statistically, this was a highly significant difference. Of the families with a single affected child and no affected parents, 10 percent (12 out of 118) demonstrated a CNV in the child that was not present in either parent (a de novo mutation). This contrasted with a rate of 3 percent (2 out of 77) in families with cases of autism spanning generations, and 1 percent (2 out of 196) in control families. Thus, they present early evidence that de novo CNVs may be more associated with sporadic cases of autism than with cases in which multiple family members are affected. Among the CNVs they identified, two overlapped with genomic regions already implicated in autism (15q11–13 and 16p11.2). They made the interesting observation that de novo CNVs were enriched in females in their cases. Since autism is less common in females than males (and therefore being female can be construed as protective), they postulated that the CNVs they observed in this group were more likely to be contributing to the development of autism.

In a larger subsequent study, Weiss et al. genotyped families consisting of 1,441 cases of autism along with 1,420 parents (Weiss et al., 2008). They found a significant enrichment of a de novo deletion at 16p11.2 in their autism cases, a region that had also been highlighted by Sebat et al. Interestingly, they also observed the reciprocal duplication in this area in three families within their case sample. In one family, the duplication was a de novo event affecting one of two male offspring. In the other two families, the duplication was also present in a parent. Within one of these families, it was transmitted to two of two affected children (one male, one female) and one unaffected female. Within the other family, it was transmitted to four of four affected males. They did not observe these variants in their control subjects, however referred to a sample of bipolar disorder cases in which the deletion was seen in three females. This provides evidence for the involvement of the 16p11.2 region in autism, but the observation of this CNV in a bipolar dataset suggests CNVs in this region are unlikely to segregate solely with autism; indeed, they may be seen in many other disorders (as well as in normal controls), too. Five further large duplications CNVs were observed in the 15q11–13 region of the genome, which is classically associated with Prader-Willi and Angelman syndromes. Three of these events were de novo events, one transmitted from the mother and the other was of uncertain origin. Six families carried a deletion of the gene *neurexin 1*, transmitted from the parents. The neurexin family of proteins is of particular interest in autism because of their role in supporting synaptic function. However, of these six families, the deletion was only in two members affected with autism. They also observed CNVs in other regions of the genome previously implicated with autism; however, the sample size was not large enough to provide statistically significant evidence of association. The authors commented on the large number of CNVs that did not segregate with autism, which illustrates a pressing point. Disentangling pathogenic from benign variants is extremely difficult. Some CNVs are likely to have no relevance to disease. Some may have relatively little relevance on their own, but, in combination with other variants, confer more risk than they would otherwise. Many studies in CNVs are currently not sufficiently powered or sophisticated to detect variants with low effect sizes or interaction effects.

Adding to the increasing evidence of the role of rare and de novo CNVs in autism came another genome-wide association study by Marshall et al. of CNVs in 427 families with cases of autistic spectrum disorders, compared to 500 controls (Marshall et al., 2008). They focused initially on CNVs likely to be etiological. Twenty-seven cases had CNVs that were de novo in origin. Twenty-seven unrelated cases had CNVs at 13 different loci that overlapped with each other and four cases had identical CNVs at two loci. They found CNVs in cases that have previously been associated with autism, for example, *NLGN4*, *SHANK3*, and *NRXN1*, as well as CNVs in the regions 22q11.2 and 15q11–13. Taking these CNVs, they explored whether they were present in a further set of 1,152 matched controls. They could not find any of the CNVs found in their autism cases within this control group, which gives further indirect evidence that these detected CNVs are likely to be salient in autism. Two de novo deletions of 22q11.2 were found in two brothers with autism, further emphasizing the notion that this particular CNV is likely to be underascertained in the population. Interestingly, the authors noted that the two brothers were at the opposite ends of the autistic spectrum, one having a much milder phenotype than the other. They also found 4.3 MB duplications of this region in two other families, but, in one family, this duplication was present in the normal father and not found in the affected male. Copy number variants at 16p11.2 were again highlighted, this time in two separate families, but the authors noted that unaffected individuals also carried these CNVs. Within this cohort of individuals with autism, the authors estimated that approximately 7 percent of cases with an unknown etiology harbor de novo CNVs detectable by microarrays. This is a significant number and suggests that the routine use of microarrays in clinical scenarios where autism is suspected may be appropriate.

Many of the genes implicated by the presence of CNVs in autism disorders have an interesting function in common, that of the maintenance of synaptic structure and function in excitatory neurones. In particular, genes from the cadherin family, neurexin family, the neurexin-related CNTNAP family of genes and the SHANK family of genes, all act to form and maintain the structure of excitatory synapses, and all have had rare CNVs within or across them associated with individuals with autism. Ramocki and Zoghbi have recently put forward the hypothesis that the failure of neuronal homeostasis, via a number of mechanisms, is broadly responsible for neuropsychiatric phenotypes (Ramocki & Zoghbi, 2008). However, with rare exceptions, it is likely that no single genetic abnormality, be it a CNV or otherwise, is responsible for a neuropsychiatric disorder in any particular individual. Multiple abnormalities, CNVs among them, likely come together in a single individual to precipitate a failure in neuronal development and function, with the emergence of behavioral phenotypes over the life of the individual. Focused study of individuals, especially in multiplex families (defined as families in which multiple members of a family are affected by a disorder) are giving some clues to the general etiology of a disorder such as autism.

Pagnamenta et al. performed just such a focused study on a family with a high incidence of autistic traits (Pagnamenta et al., 2009). Within a nuclear family of five were two male children affected with autism and one daughter who was unaffected with autism, but displayed poor reading speed. Neither parent exhibited autism; however, the mother, and many of her relatives, exhibited poor reading speed, and the father, along with many of his relatives, was reported to exhibit some autistic traits. The two autistic sons had previously been part of a CNV study that had detected a deletion CNV involving the genes *DOCK4* and *IMMP2L*. Within this study, the authors showed that this CNV had been inherited from the mother by both sons and the daughter, and that the deletion caused the fusion of the DOCK4 and IMMP2L genes to produce an aberrant protein that did not function as expected. This deletion CNV was found in a further six members on the maternal side of the family, all of whom had poor reading speed. Coincident to the detection of the CNV affecting *DOCK4* and *IMMP2L* was the detection of a further CNV affecting *CNTNAP5* in the sons (but not the daughter). This time, the CNV had been inherited from the father, who had been observed to have autistic-like traits. The authors postulated that the individual variants on their own were not necessarily associated with any clinical disorder but that the combination of the two, manifested in the two sons, may be contributing to the emergence of autism. Of course, this focused study does not take into account that many other genetic variants may be having an effect in this family; however, the point of principle is relatively clear. Different genetic variants, on their own, are not sufficient to cause a clinically recognizable disease state, but may come together to cause a state that is clinically recognizable, such as autism.

D. Attention Deficit Hyperactivity Disorder

Attention deficit hyperactivity disorder is a neuropsychiatric disorder seen in childhood comprising pervasive and disabling degrees of impulsivity, inattentiveness, and hyperactivity. It is overrepresented in boys. A recent meta-analysis estimated the worldwide pooled prevalence at 5.29 percent (Polanczyk, de Lima, Horta, Biederman, & Rohde, 2007). The term ADHD is largely synonymous with the term "hyperkinetic disorder," which is often used in countries outside of North America. The symptoms often continue into adulthood and are associated with substantial comorbidity. The disorder is substantially genetic, with heritability estimates from twin and adoption studies of around 76 percent (Franke, Neale, & Faraone, 2009). Linkage and GWAS of single nucleotide polymorphisms have not yielded any significant insights into the genetic architecture of the disorder.

Three studies of CNVs in ADHD have been published to date. Elia et al. compared 335 cases of ADHD and their parents with 2,026 healthy controls. Genotyping them on an Illumina microarray platform, they identified 222 inherited CNVs within the cases that were not present in any of the controls. Of these CNVs, 28 covered 22 genes previously implicated by genetic studies in various neuropsychiatric disorders including autism and schizophrenia. Eleven CNVs covered eight genes that have already had CNVs within them associated with autism and schizophrenia. When they analyzed their CNV set by known biological pathways, they found an overrepresentation of genes involved in learning, CNS development, hindbrain development, and cell adhesion. However, they found no overall difference in mean CNV count and CNV size between cases and controls, perhaps suggesting that although children with ADHD may not have a higher than normal burden of CNVs, they do have a particular burden of CNVs falling over genes associated with pathways of CNS development.

Williams et al. analyzed the rate of large, rare CNVs in 410 children with ADHD compared to 1,156 unrelated, ethnically matched controls. Some children in this study had intellectual disability (IQ <70). They discovered 57 CNVs in their case set, equivalent to 15.6 percent of cases, and 78 in the control set, equivalent to 7.5 percent. The probability of such a difference occurring by chance was 8.9×10^{-5}. When cases with intellectual disability were excluded, 12.5 percent of the cases harbored a CNV, still a statistically significant result. Thus, whereas some of the association of CNVs with ADHD in this study is due to comorbid intellectual disability, some association is also likely to be due to CNVs associating with a susceptibility to ADHD alone. It should, however, be noted that the average IQ of those cases without intellectual disability was 89, some way off the normal population average of 100. In analyzing regions that had previously been associated with autism or schizophrenia, they found that the case CNVs were significantly more likely to fall within these regions than the control CNVs, representing further evidence that the CNVs found are likely to be significant in ADHD development, but also reinforcing the notion that CNVs can confer a broadly increased risk of neuropsychiatric illness, rather than segregating by clinical disorder.

In a smaller, more focused clinical study, Lesch et al. studied 99 children and adolescents with severe ADHD (Lesch et al., 2011). Two-thirds of these children came from families with at least two members affected with ADHD, eight patients were from extended families with high numbers of individuals with ADHD, and the remainder (24 cases) were sporadic. Of interest, they excluded any case with an IQ of less than 80 in this study and any case with a diagnosis of autism or schizophrenia. They found 17 potentially syndrome-associated CNVs in their cohort, comprising four deletions and 13 duplications. Two CNVs were de novo. However, no single region was associated with the disorder.

E. Bipolar Disorder

Bipolar affective disorder is marked by mania and depression (or, less commonly, mania alone), often accompanied by disturbances in behavior and thinking. It has a lifetime prevalence of about 1 percent in both males and females. It is typically episodic, with periods of complete recovery between episodes. Evidence for a large genetic component to the disorder is substantial, with most twin studies indicating a heritability of about 80 percent (McGuffin et al., 2003). The lifetime risk for the disorder is significantly elevated in the monozygotic co-twin of a person with bipolar disorder (40–70 percent) and first-degree relatives (5–10 percent) (Craddock & Jones, 1999).

Zhang et al. described their work comparing the frequency of rare CNVs in 1,001 cases of bipolar disorder and 1,034 controls (Zhang et al., 2009). They found that samples with singleton deletions (deletion CNVs occurring only once in the entire sample) were enriched in cases versus controls (16.2 percent vs. 12.3 percent). In this study, the authors reported that some singleton CNVs covered genes already implicated in schizophrenia (*GRM7* and *LARGE*). *GRM7* encodes a metabotropic glutamate receptor and has been associated with a range of neuropsychiatric disorders. *LARGE* encodes a gene that is associated with an autosomal recessive form of muscular dystrophy that also presents with severe mental retardation (*MDC1D*). Pathway analysis of these CNVs also implicated biological pathways involved with psychological disorders, behavior, and learning. Comparisons of singleton deletion CNVs in a sample are interesting because they may represent the rarest CNVs in a population. Logically, this group is more likely to contain CNVs that are de novo and disease-causing, and thus are better candidates for more focused research on disease mechanisms.

In contrast, Grozeva et al. found little association of bipolar disorder with rare CNVs, including singleton CNVs. Indeed, they found significantly *less* deletion CNVs in the 1,697 cases of bipolar disorder that they studied from the Wellcome Trust Case Control Consortium than in 2,806 unscreened controls. It is important, however, to note that the control group in this study is quite different from that of Zhang et al., who screened their controls for an absence of neuropsychiatric morbidity. The controls in Grozeva et al., in contrast, had no such screening and instead represented an unselected population group.

At the time of writing, McQuillin et al. have published the only other study on CNVs in bipolar disorder. They compared overall CNV burden in 1,461 bipolar disorder cases to 510 controls screened for an absence of neuropsychiatric disorder. Similar to Grozeva et al., but in contrast to Zhang et al., they found no evidence to suggest an increased burden of rare CNVs, including singleton CNVs. They also found that, overall, bipolar cases tended to have a lower burden of CNVs compared to controls. Further studies in bipolar disorder are needed before any firm conclusions can be drawn.

F. Unipolar Depression

Rucker et al. have published the only study on genome-wide burden of CNVs in unipolar depressive disorder to date (Rucker et al., 2011). They studied recurrent depressive disorder of at least moderate severity, which is more likely to represent a heritable form of unipolar depressive disorder. They compared large, rare CNVs in 3,106 cases of recurrent depression, 459 controls screened for lifetime-absence of neuropsychiatric disorder, and 5,619 unscreened controls. Within these groups, they found an ordered increase in the frequency of samples with deletion CNVs, but not duplication CNVs, with the screened control sample showing the lowest frequency, cases of depression the highest frequency, and the unscreened controls falling in the middle. Further analysis showed that CNVs deleting protein coding regions of the genome were largely responsible for the association, and, within an analysis of regions previously implicated in schizophrenia, there was an overall enrichment of CNVs in cases when compared to the screened controls, although no individual area was significant. One individual was found to harbor the 22q11.2 deletion, and five individuals had CNVs at 16p11.2. Although this study highlights that part of the etiology of recurrent depression is likely to involve rare CNVs, as it is in autism and schizophrenia, it also highlighted that groups of highly functioning individuals have a particularly low frequency of deletion CNVs. The fact that a relatively uniform frequency of duplication CNVs was observed throughout the cohorts within this study may suggest that duplication CNVs are less likely to be pathogenic.

IV. Common Copy Number Variants

So far, we have only considered the impact of rare CNVs. Common CNVs account for a significant proportion of variation and are defined as occurring in more than 1 percent of the population. Although some have argued that common variants of small to moderate effect may explain all or most of the susceptibility to complex diseases (Risch & Merikangas, 1996), others have argued that both common variants of small effect and rarer variants of larger effect play their roles (Rucker & McGuffin, 2010,; Uher, 2009). Conrad and Pinto have recently published work analyzing the origins and functional impact of CNVs in the human genome. By using the observation that many common CNVs are tagged by single nucleotide polymorphisms, they were able to show that common

CNVs are unlikely to contribute to a substantial proportion of heritability in neuropsychiatric disorders (Conrad et al., 2010). Little evidence from individual GWA studies supports the role of common CNVs in the etiology of neuropsychiatric disorders, and, although future research may throw this conclusion into doubt, we do not consider common CNVs further in this essay.

V. Conclusion

In this essay, we have seen numerous studies linking rare CNVs with numerous different neuropsychiatric disorders. It is clear that there is an association, which is much cause for celebration in psychiatry. However, the etiological basis of the observed association is far from clear, and much work remains to be done. Studies in neuropsychiatric disorders are often hampered by the lack of specificity of the diagnostic criteria themselves. The human brain is the least amenable organ to study, and findings in rodent and primate brain studies do not easily translate to humans. In this context, evidence of association (of which we have much) is only weak evidence of causation (of which we have little). Some genes, such as the neurexin genes, have multiple lines of evidence pointing to their involvement; other genes have no known function in the CNS, or no known function at all. Although there is much to be excited about in neuropsychiatric genetics, is there any translatable benefit? Is there a role for the detection of CNVs in the clinic?

Although CNVs are clearly associated with human disease states, they also represent normal genetic variability within the population. Indeed, the mechanisms of CNV formation are most probably an essential driver of normal species evolution. Delineating CNVs into those that are potentially pathogenic and those that are benign is fraught with problems, especially when the clinical diagnostic boundaries are themselves blurred. Added to this, CNVs only represent one realm of variation within the genome and, in reality, are unlikely to act independently of other forms of variation. Explaining the complex effects of genetic variation to patients and their relatives is not easy. However, if a deleterious variant is detected in combination with a known clinical phenotype, then there are implications for diagnosis, genetic counseling, and family planning, as such variants may or may not be passed on to offspring, mediated largely by chance. It is here that we have seen the first truly translatable benefit from the human genome project as a whole and high-resolution microarrays in particular. Although the direct benefit to the patient (and his or her family) is simply that of knowledge, this should not in itself be underestimated. It represents the first step toward understanding the causes of neuropsychiatric disorders and better, more refined treatments.

What of the possibility of predicting the development of a neuropsychiatric disorder by screening for CNVs? We have observed in this essay how deleterious variants associated with disease are also observed in controls with no observable disease. The concept of genetic "penetrance" describes the likelihood that an individual with a genetic variant will display the disorder with which it is associated. Penetrance for the Huntington disease gene, for example, is 100 percent; all individuals with the aberrant gene develop the disorder. Vassos et al. have analyzed the penetrance of variants associated with schizophrenia occurring in the genomic regions 15q13.3, 1q21.1, 15q11.2, 17p12, 2p16.3, 16p13.1, and 16p11.2 (Vassos et al., 2010). Estimates for these variants, themselves already having been identified as significantly associated with schizophrenia in various different studies (Ingason et al., 2011; International Schizophrenia Consortium, 2008; Kirov et al., 2007; Kirov, Grozeva, et al., 2009; McCarthy et al., 2009; Rujescu et al., 2009; Stefansson et al., 2008; Vrijenhoek et al.,), ranged between 2 and 9 percent. That is, only 2–9 percent of individuals manifesting any one of these CNVs might go on to develop schizophrenia. This does not mean that an individual with a CNV in these areas might not go on to develop another neuropsychiatric disorder, and, given the strong evidence that CNVs contribute to an increased risk for a variety of disorders, the penetrance estimates for neuropsychiatric disorders in general is likely to be higher. However, the levels of penetrance seen are still much below levels seen in Mendelian disorders. This suggests that screening for deleterious CNVs in these regions is unlikely to be helpful in predicting a specific disorder, although it remains to be seen if it may be helpful in predicting a general liability to disorder. More particularly, however, if an individual without disease is found not to harbor a CNV, it will offer no reassurance that they will not go on to develop a neuropsychiatric disorder. Similarly, if an individual is found to harbor a variant, it gives no certainty that a disease state will develop. Screening, at least in this context, would appear for the moment to have little utility. If strong interactional effects are found between individual CNVs, or CNVs and other forms of genetic variation, then this picture may change; however, clinical tests for diagnosis and prediction demand high sensitivities and specificities, and this is unlikely to be developed soon.

The clinical utility of CNVs in diagnostics is rather blurred. Microarrays have already supplanted karyotyping in the routine screening of children with developmental delay (Miller et al., 2010). Kaminsky and colleagues have recently completed an analysis of clinical microarray data in children with unexplained intellectual and developmental disability from the first consortium of diagnostic laboratories using microarrays in children with developmental abnormality or delay (the International Standards for Cytogenetic Arrays consortium) (Kaminsky et al., 2011). The results show that microarrays have a significantly higher pick-up rate of potentially deleterious events than does karyotyping, although microarrays cannot detect abnormalities such as inversions or balanced translocations. Notwithstanding this, however, it is still currently difficult to classify the variants observed as either definitely or definitely not pathogenic. Some variants have strong evidence of association with disorder, but the majority are currently of unknown significance. In this situation, deletion CNVs are presumed to be more pathogenic than duplication CNVs, and de novo CNVs more likely to be etiologically significant than variants that are seen in normal parents. Ultimately, we find ourselves at a relatively early stage of knowledge development, and, although some clear associations have been found, many will remain blurred until further research is completed. This is particularly true in neuropsychiatric disorders such as autism and schizophrenia, where clear association with particular CNVs only occurs in a minority of cases. A balance needs to be struck between the need for knowledge about the extent and consequences of genetic variation and the potential risks and benefits such knowledge will bring to individual patients. Such decisions should be made on a case-by-case basis.

The clinical detection of CNVs may in the future represent an interesting area for treatment. First, the genes implicated by variant detection may produce proteins that are targets for drugs. Second, it may be possible to specifically up-regulate the transcription of genes affected by deletion CNVs, and down-regulate genes that are duplicated. Such treatments are purely hypothetical at this stage, and certainly a long way from being realized.

Copy number variants represent the first replicated evidence in the search for the genetic etiology of neuropsychiatric disorders. The evidence for association of CNVs with different neuropsychiatric diseases is particularly strong for diseases with high heritability scores, such as autism and schizophrenia; however, bipolar disorder may be a notable exception. It is important to appreciate that CNVs only contribute a small proportion of overall risk. All disorders still have large amounts of unaccounted heritability, which will likely be explained by different spheres of genetic and epigenetic variation and interaction effects. Large, rare CNVs have been most robustly identified and associated with disease. These probably represent the "low-hanging fruit" but, even these, with the exception of the 22q11.2 deletion, do not confer disease in more than a small minority of individuals who carry them. De novo CNVs, not present in either parent but present in an affected child, are particularly strong candidates for causal association between gene and disorder, although the current methodologies for detection are limited in resolution and are likely to be missing many potentially pathogenic CNVs. We may need to perform meta-analyses of studies of many different neuropsychiatric disorders to get a clearer picture of which CNVs are the most important risk factors for neuropsychiatric disorders as a whole, rather than concentrating on individual disorders (which, as we pointed out at the beginning of this essay, are not expected to neatly segregate according to their genetic etiologies anyway). Taken all together, the evidence strongly argues for a polygenic, complex model of disease of the type argued for schizophrenia by Gottesman, based on the work of Falconer, 40 years ago (Falconer, 1965; Gottesman & Shields, 1967). The effect of CNVs with smaller effect sizes, and particularly the interaction effects between different CNVs, have yet to be defined and may require novel research and analysis strategies. The next generation of genome sequencing technologies will add some clarity to this picture in the coming decade.

Neuropsychiatric disorders probably represent a broad range of interacting neurocognitive problems, themselves a result of complex interactions between genotype and environment. The discovery of robust associations with large, rare CNVs represents the first window into the genetic etiology of neuropsychiatric disorders. The overall picture is likely to be highly complex, with CNVs representing just one type of genetic variation, and genetic variation itself representing just one type of a myriad of interacting factors. Microarrays, and now next-generation sequencing technologies, have allowed vast quantities of new genetic data to be collected, and the analysis of these data has proven challenging. Nevertheless, we are now beginning to reap the rewards of the human genome project, and with this has come our first glimpse into the realm of the genetic etiology of neuropsychiatric disorders.

References

Bailey, A., Phillips, W., & Rutter, M. (1996). Autism: Towards an integration of clinical, genetic, neuropsychological, and neurobiological perspectives. *Journal of Child Psychology and Psychiatry, 37*(1), 89–126.

Ballif, B. C., Theisen, A., Coppinger, J., Gowans, G. C., Hersh, J. H., Madan-Khetarpal, S., et al. (2008). Expanding the clinical phenotype of the 3q29 microdeletion syndrome and characterization of the reciprocal microduplication. *Molecular Cytogenetics,* 1, 8.

Brzustowicz, L. M., Hodgkinson, K. A., Chow, E. W., Honer, W. G., & Bassett, A. S. (2000). Location of a major susceptibility locus for familial schizophrenia on chromosome 1q21-q22. *Science, 288*(5466), 678–682.

Craddock, N., & Jones, I. (1999). Genetics of bipolar disorder. *Journal of Medical Genetics, 36*(8), 585–594.

Conrad, D., Pinto, D., Redon, R., Feuk, L., Gokcumen, O., Zhang, Y., et al. (2010). Origins and functional impact of copy number variation in the human genome. *Nature, 464*, 704–712.

Falconer, D. (1965). The inheritance of liability to certain diseases, estimated from the incidence among relatives. *Annals of Human Genetics, 29,* 51–76.

Feng, J., Schroer, R., Yan, J., Song, W., Yang, C., Bockholt, A., et al. (2006). High frequency of neurexin 1β signal peptide structural variants in patients with autism. *Neuroscience Letters, 409*(1), 10–13.

Feuk, L., Carson, A., & Scherer, S. (2006). Structural variation in the human genome. *Nature Reviews Genetics, 7*(2), 85–97.

Franke, B., Neale, B. M., & Faraone, S. V. (2009). Genome-wide association studies in ADHD. *Human Genetics, 126*(1), 13–50.

Freitag, C. M. (2007). The genetics of autistic disorders and its clinical relevance: A review of the literature. *Molecular Psychiatry, 12*(1), 2–22.

Friedman, J. M., Baross, A., Delaney, A. D., Ally, A., Arbour, L., Asano, J., et al. (2006). Oligonucleotide microarray analysis of genomic imbalance in children with mental retardation. *American Journal of Human Genetics, 79*(3), 500–513.

Goodship, J., Cross, I., LiLing, J., & Wren, C. (1998). A population study of chromosome 22q11 deletions in infancy. *Archives of the Diseases of Childhood, 79*(4), 348–351.

Gottesman, I. I. (1991). *Schizophrenia genesis: The origins of madness.* New York: W. H. Freeman/Times Books/Henry Holt & Co.

Gottesman, I., & Shields, J. (1967). A polygenic theory of schizophrenia. *Proceedings of the National Academy of Sciences, 58*(1), 199–205.

Harris, E., & Barraclough, B. Excess mortality of mental disorder. (1998). *British Journal of Psychiatry, 173,* 11–53.

Ingason, A., Rujescu, D., Cichon, S., Sigurdsson, E., Sigmundsson, T., Pietilainen, O., et al. (2011). Copy number variations of chromosome 16p13.1 region associated with schizophrenia. *Molecular Psychiatry, 16*(1), 17–25

International Schizophrenia Consortium. (2008). Rare chromosomal deletions and duplications increase risk of schizophrenia. *Nature, 455*(7210), 237–241.

Jamain, S., Quach, H., Betancur, C., Råstam, M., Colineaux, C., Gillberg, I. C., et al. (2003). Mutations of the X-linked genes encoding neuroligins NLGN3 and NLGN4 are associated with autism. *Nature Genetics, 34*(1), 27–29.

Kaminsky, E. B., Kaul, V., Paschall, J., Church, D. M., Bunke, B., Kunig, D., et al. (2011). An evidence-based approach to establish the functional and clinical significance of copy number variants in intellectual and developmental disabilities. *Genetics in Medicine, 13*(9), 777–784.

Karayiorgou, M., Simon, T. J., & Gogos, J. A. (2010). 22q11.2 microdeletions: Linking DNA structural variation to brain dysfunction and schizophrenia. *Nature Reviews Neuroscience, 11*(6), 402–416.

Kim, H. -G., Kishikawa, S., Higgins, A. W., Seong, I. –S., Donovan, D. J., Shen, Y., et al. (2008). Disruption of neurexin 1 associated with autism spectrum disorder. *American Journal of Human Genetics, 82*(1), 199–207.

Kirov, G., Grozeva, D., Norton, N., Ivanov, D., Mantripragada, K., Holmans, P., et al. (2009). Support for the involvement of large copy number variants in the pathogenesis of schizophrenia. *Human Molecular Genetics, 18*(8), 1497–1503.

Kirov, G., Gumus, D., Chen, W., Norton, N., Georgieva, L., Sari, M., et al. (2007). Comparative genome hybridization suggests a role for NRXN1 and APBA2 in schizophrenia. *Human Molecular Genetics, 17*(3), 458–465.

Kirov, G., Rujescu, D., Ingason, A., Collier, D. A., O'Donovan, M. C., & Owen, M. J. (2009). Neurexin 1 (NRXN1) deletions in schizophrenia. *Schizophrenia Bulletin, 35*(5), 851–854.

Klauck, S. M. Genetics of autism spectrum disorder. *European Journal of Human Genetics, 14*(6), 714–720.

Lander, E., Linton, L., Birren, B., Nusbaum, C., Zody, M., Baldwin, J., et al. (2001). Initial sequencing and analysis of the human genome. *Nature, 409*(6822), 860–921.

Lesch, K.-P., Selch, S., Renner, T. J., Jacob, C., Nguyen, T. T, Hahn, T., et al. (2011). Genome-wide copy number variation analysis in attention-deficit/hyperactivity disorder: Association with neuropeptide Y gene dosage in an extended pedigree. *Molecular Psychiatry, 16*(5), 491–503.

Levinson, D. F., Duan, J., Oh, S., Wang, K., Sanders, A. R., Shi, J., et al. (2011). Copy number variants in schizophrenia: Confirmation of five previous findings and new evidence for 3q29 microdeletions and VIPR2 duplications. *American Journal of Psychiatry, 168*(3), 302–316.

Marshall, C., Noor, A., Vincent, J., Lionel, A., Feuk, L., Skaug, J., et al. (2008). Structural variation of chromosomes in autism spectrum disorder. *American Journal of Human Genetics, 82*(2), 477–488.

Maynard, T., Haskell, G., Peters, A., Sikich, L., Lieberman, J., & LaMantia, A. (2003). A comprehensive analysis of 22q11 gene expression in the developing and adult brain. *Proceedings of the National Academy of Science USA, 100*(24), 14433–14438.

McCarthy, S., Makarov, V., Kirov, G., Addington, A., McClellan, J., Yoon, S., et al. (2009). Microduplications of 16p11.2 are associated with schizophrenia. *Nature Genetics, 41*, 1223–1227

McGuffin, P., Rijsdijk, F., Andrew, M., Sham, P., Katz, R., & Cardno, A. (2003). The heritability of bipolar affective disorder and the genetic relationship to unipolar depression. *Archives of General Psychiatry, 60*(5), 497–502.

Miller, D., Adam, M., Aradhya, S., Biesecker, L., Brothman, A., Carter, N., et al. (2010). Consensus statement: Chromosomal microarray is a first-tier clinical diagnostic test for individuals with developmental disabilities or congenital anomalies. *American Journal of Human Genetics, 86*(5), 749–764.

Montcel Du, S. T., & Mendizabai, H. (1996). Prevalence of 22q11 microdeletion. *Journal of Medical Genetics, 33*, 719.

Morrow, E. M., Yoo, S. –Y., Flavell, S. W., Kim, T. –K., Lin, Y., Hill, R. S., et al. (2008). Identifying autism loci and genes by tracing recent shared ancestry. *Science, 321*(5886), 218–223.

Murphy, K., Jones, L., & Owen, M. (1999). High rates of schizophrenia in adults with velo-cardio-facial syndrome. *Archives of General Psychiatry, 56*(10), 940–945.

Pagnamenta, A. T., Bacchelli, E., de Jonge, M. V., Mirza, G., Scerri, T. S., Minopoli, F., et al. (2010). Characterization of a family with rare deletions in CNTNAP5 and DOCK4 suggests novel risk loci for autism and dyslexia. *Biological Psychiatry, 68*(4), 320–328.

Pagnamenta, A. T., Wing, K., Akha, E. S., Knight, S. J. L., Bölte, S., Schmötzer, G., Duketis, E., et al. (2009). A 15q13.3 microdeletion segregating with autism. *European Journal of Human Genetics, 17*(5), 687–692.

Perry, G. H., Ben-Dor, A., Tsalenko, A., Sampas, N., Rodriguez-Revenga, L., Tran, C. W., et al. (2008). The fine-scale and complex architecture of human copy-number variation. *American Journal of Human Genetics, 82*(3), 685–695.

Polanczyk, G., de Lima, M. S., Horta, B. L., Biederman, J., & Rohde, L. A. (2007). The worldwide prevalence of ADHD: A systematic review and metaregression analysis. *American Journal of Psychiatry, 164*(6), 942–948.

Ramocki, M., & Zoghbi, H. (2008). Failure of neuronal homeostasis results in common neuropsychiatric phenotypes. *Nature, 455*(7215), 912–918.

Risch, N., & Merikangas, K. (1996). The future of genetic studies of complex human diseases. *Science, 273*(5281), 1516–1517.

Rucker, J. J., Breen, G., Pinto, D., Pedroso, I., Lewis, C. M., & Cohen-Woods, S., et al. (2011). Genome-wide association analysis of copy number variation in recurrent depressive disorder. *Mol Psychiatry*, doi: 10.1038/mp.2011.144 [Epub ahead of print]

Rucker, J. J. H., & McGuffin, P. (2010). Polygenic heterogeneity: A complex model of genetic inheritance in psychiatric disorders. *Biological Psychiatry, 68*(4), 312–313.

Rujescu, D., Ingason, A., Cichon, S., Pietilainen, O. P. H., Barnes, M. R., Toulopoulou, T., et al. (2009). Disruption of the neurexin 1 gene is associated with schizophrenia. *Human Molecular Genetics, 18*(5), 988–996

Saitta, S. C., Harris, S. E., Gaeth, A. P., Driscoll, D. A., McDonald-McGinn, D. M., Maisenbacher, M. K., et al. (2004). Aberrant interchromosomal exchanges are the predominant cause of the 22q11.2 deletion. *Human Molecular Genetics, 13*(4), 417–428.

Schwab, S. G., & Wildenauer, D. B. (2009). Update on key previously proposed candidate genes for schizophrenia. *Current Opinion in Psychiatry, 22*(2), 147–153.

Sebat, J., Lakshmi, B., Malhotra, D., Troge, J., Lese-Martin, C., Walsh, T., et al. (2007). Strong association of de novo copy number mutations with autism. *Science, 316*(5823), 445–449.

Sharp, A., Mefford, H., Li, K., Baker, C., Skinner, C., Stevenson, R., et al. (2008). A recurrent 15q13.3 microdeletion syndrome associated with mental retardation and seizures. *Nature Genetics, 40*(3), 322–328.

Sudhof, T. (2008). Neuroligins and neurexins link synaptic function to cognitive disease. *Nature,455*(7215), 903–911.

Sullivan, P. F. (2005). The genetics of schizophrenia. *PLoS Medicine, 2*(7), e212.

Stefansson, H., Rujescu, D., Cichon, S., Pietilainen, O., Ingason, A., Steinberg, S., et al. (2008). Large recurrent microdeletions associated with schizophrenia. *Nature, 455*(7210), 232–236.

Swerdlow, N. R., Light, G. A., Cadenhead, K. S., Sprock, J., Hsieh, M. H., & Braff, D. L. (2006). Startle gating deficits in a large cohort of patients with schizophrenia: Relationship to medications, symptoms, neurocognition, and level of function. *Archives of General Psychiatry. 63*(12), 1325–1335.

Szatmari, P., Paterson, A., Zwaigenbaum, L., Roberts, W., Brian, J., Liu, X., et al. (2007). Mapping autism risk loci using genetic linkage and chromosomal rearrangements. *Nature Genetics, 39*(3), 319–328.

Turk, J., & Hill, P. (1995). Behavioural phenotypes in dysmorphic syndromes. *Clinical Dysmorphology, 4*(2), 105.

Turner, T. H. (1989). Schizophrenia and mental handicap: An historical review, with implications for further research. *Psychological Medicine, 19*(2), 301–314.

Uher, R. (2009). The role of genetic variation in the causation of mental illness: An evolution-informed framework. *Molecular Psychiatry, 14*, 1072–1082.

Vassos, E., Collier, D. A., Holden, S., Patch, C., Rujescu, D., St. Clair, D., et al. (2010). Penetrance for copy number variants associated with schizophrenia. *Human Molecular Genetics, 19*(17), 3477–3481.

Venter, J., Adams, M., Myers, E., Li, P., & Mural, R. (2001). The sequence of the human genome. *Science, 291*(5507), 1304–1351.

Vorstman, J. A. S., Staal, W. G., van Daalen, E., Van Engeland, H., Hochstenbach, P. F. R., & Franke, L. (2005). Identification of novel autism candidate regions through analysis of reported cytogenetic abnormalities associated with autism. *Molecular Psychiatry, 11*(1), 18–28.

Vrijenhoek, T., Buizervoskamp, J., Vanderstelt, I., Strengman, E., Sabatti, C., Geurstvankessel, A., et al. Recurrent CNVs disrupt three candidate genes in schizophrenia patients. *American Journal of Human Genetics, 83*(4), 504–510.

Walsh, T., McClellan, J., McCarthy, S., Addington, A., Pierce, S., Cooper, G., et al. (2008). Rare structural variants disrupt multiple genes in neurodevelopmental pathways in schizophrenia. *Science, 320*(5875), 539–543.

Weiss, L. A., Shen, Y., Korn, J. M., Arking, D. E., Miller, D. T., Fossdal, R., et al. (2008). Association between microdeletion and microduplication at 16p11.2 and autism. *New England Journal of Medicine, 358*(7), 667–675.

Wellcome Trust Case Control Consortium. (2007). Genome-wide association study of 14,000 cases of seven common diseases and 3,000 shared controls. *Nature, 447*(7145), 661–678.

Xu, B., Roos, J. L., Levy, S., van Rensburg, E. J., Gogos, J. A., & Karayiorgou, M. (2008). Strong association of de novo copy number mutations with sporadic schizophrenia. *Nature Genetics, 40*(7), 880–885

Zhang, D., Cheng, L., Qian, Y., Alliey-Rodriguez, N., Kelsoe, J., Greenwood, T., et al. (2009). Singleton deletions throughout the genome increase risk of bipolar disorder. *Molecular Psychiatry, 14*(4), 376–380.

Zhang, J. (2003). Evolution by gene duplication: An update. *Trends in Ecology & Evolution, 18*(6), 292–298.

CHAPTER

13 Molecular Foundations of the Symptoms of Major Depressive Disorder

Michael C. Chen *and* Ian H. Gotlib

Abstract

Major Depressive Disorder (MDD) is a prevalent and costly disorder with a broad range of cognitive, affective, and behavioral symptoms. Despite the absence of a clear final common molecular pathway in depression, many molecular systems have been implicated in MDD. In particular, disruptions in molecular systems like serotonin, dopamine, glutamate, and other neurotransmitters, as well as in stress hormones, cytokines, neurotrophins, and neuropeptides, may contribute to MDD. To link the symptoms of MDD with molecular dysfunction, this article examines these molecules in the context of three symptom clusters of MDD: cognitive/affective symptoms, volitional/behavioral symptoms, and homeostatic/vegetative symptoms. It examines how these molecules and their receptor, transport, and regulatory systems contribute to MDD and to the development of specific symptom clusters. It presents two possible frameworks of molecular dysfunction in MDD that encompass the interactions between vulnerability phenotypes and biochemical perturbations that may lead to the heterogeneous symptoms of this disorder.

Key Words: depression, symptoms, neurotransmitters, serotonin, dopamine, hormones, peptides, cytokines, neuromodulators

I. Introduction

Major Depressive Disorder (MDD) is one of the leading causes of disability worldwide, and by 2030 is projected to cause more disability than all other diseases, accidents, and disorders except HIV/AIDS and ischemic heart disease (Mathers & Loncar, 2006). To meet criteria for a diagnosis of MDD according to the(American Psychiatric Association's *Diagnostic and Statistical Manual of Mental Disorders DSM-IV,* 2000), individuals must have one or both of the cardinal symptoms of MDD, depressed mood and anhedonia, but may present with up to 17 additional possible symptoms across 7 broad categories of functioning, including changes in weight, appetite, sleep, and psychomotor function, fatigue, worthlessness, guilt, cognitive impairment, and suicidal thoughts or self-harm. Clearly, MDD is a heterogeneous disorder, varying in symptom profiles, chronicity, severity, and course (Rush, 2007). Because any two individuals diagnosed with MDD may have few symptoms in common, it is difficult to define a prototypical profile of this disorder. Indeed, any given individual may have different symptom profiles during different depressive episodes (Minor, Champion, & Gotlib, 2005). This heterogeneity makes it difficult to identify and elucidate specific biological profiles and molecular pathways involved in MDD. For example, are the same molecular anomalies present in a depressed individual with sadness, insomnia, weight loss, concentration difficulty, and suicidality as in a depressed individual with anhedonia, hypersomnia, weight gain, inappropriate guilt, and psychomotor retardation?

Researchers have compared MDD to other chronic, heterogeneous diseases such as diabetes mellitus (Krishnan & Nestler, 2008), which, like MDD, is projected to increase in prevalence in the coming decades (Wild, Roglic, Green, Sicree, & King, 2004). Like depression, diabetes can present with a variety of symptoms, such as polyuria, polydipsia, polyphagia, as well as with cardiovascular, digestive, and nervous system dysfunction. Unlike MDD, however, these symptoms are a product of a known molecular basis of the disease: a deficit in the production or utilization of insulin. The involvement of insulin—due to the lack of the hormone or to failure of its intended signaling—represents a final common molecular pathway (FCMP) in diabetes that unites the involvement of disparate physiological systems. This FCMP has allowed researchers to begin to use powerful animal models to understand the symptoms, risk factors, and complications of this disorder. In addition, the identification of insulin and the pancreatic β-cells involved in insulin production have led to dramatic improvements in the prevention and treatment of diabetes. At this point, however, an FCMP is absent in the study of MDD; presently, there is no neurobiological equivalent in depression of insulin from β-cells.

Over the past few decades, many molecular systems have been implicated in depression. Researchers hypothesized that monoamines and cytokines were involved in MDD because medications that target these molecules were found to have depressive side effects. For example, side effects from the antihypertensive drug reserpine led investigators to target catecholamines in MDD (see Schildkraut, 1965, for an early review of this literature); ultimately, the monoamine hypothesis posited that imbalances in such brain monoamines as serotonin, norepinephrine, and dopamine led to symptoms of depression. Similarly, depressive side effects of cytokine treatments, combined with increased levels of inflammatory markers in MDD, led to the formulation of the cytokine hypothesis of depression (Smith, 1991). Based on the observation that acute life stressors often precede a major depressive episode, researchers began to examine hormones implicated in the stress system, as well as neurotrophins that regulate neuroplasticity. Among the other molecules that have been implicated in MDD are glutamate (Paul & Skolnick, 2003), γ-aminobutyric acid (Kalueff & Nutt, 2007), norepinephrine, and other neurotransmitters, neurotrophins, peptide neuromodulators such as NPY, and hormones like oxytocin, vasopressin, and estrogen. Because of the wide array of molecular systems that are disrupted in MDD, or in which disruption leads to depression, it has been difficult for researchers and clinicians to cleanly link molecular and behavioral systems. Investigators have noted limitations with all these single-system theories of MDD; indeed, to date, there is no known single molecule that is both necessary and sufficient to explain all the varied combinations of symptoms that can be present in individuals diagnosed with MDD. Genetic studies of MDD, too, have not been definitive in determining consistent risk loci, due to heterogeneity, environmental factors, and nosological limitations (Shyn & Hamilton, 2010). Even evidence of successful antidepressant action is not sufficient to demonstrate the primacy of the targetted molecular system in the pathophysiology of MDD. In fact, some researchers have argued that anomalies in traditional molecules implicated in the pathophysiology of MDD, like monoamines or glutamate, are unlikely to be at the etiological core of depression (Krishnan & Nestler, 2010). This difficulty disentangling the causal links between symptoms profiles and molecular systems has historically led to simplifications and misconceptions of the neurobiology of depression (Lacasse & Leo, 2005).

A primary challenge for investigators is differentiating between molecules that are disrupted as a consequence of depression and molecules that, when disrupted, are involved in the etiology of MDD. Demonstrating that a particular molecule—for example, a metabolite of a neurotransmitter—is up- or down-regulated in MDD is not sufficient to claim that disruption in this molecule causes MDD. Even when molecules can be causally linked to depression, the heterogeneity of this disorder prevents researchers from linking that molecule to all forms of MDD. For example, showing that interferon alpha (IFN-) can cause MDD is not equivalent to proving that IFN- underlies all cases of the disorder. Indeed, whereas some molecular systems may be strongly associated with one particular domain of impairment (e.g., dopamine and deficits in motivation and behavior), the effects of other molecular systems, like norepinephrine, may be less specific. Adding to this complexity is the fact that MDD is a moderately heritable disorder (Levinson, 2009) in which preexisting vulnerabilities interact with stressful life events to increase risk for initial and recurrent depressive episodes in a manner similar to the kindling model of epileptogenesis (Kendler, Thornton, & Gardner, 2000). Moreover, different symptoms and symptom clusters in MDD are inherited at different rates. For example, sleep and appetite disruption have higher heritability estimates than does psychomotor retardation (Jang, Livesley, Taylor, Stein, & Moon, 2004).

Examining this heterogeneity of symptoms may provide insight concerning the molecular systems involved in MDD. Investigators have found that symptoms of depression tend to occur in discrete clusters, often depending on how they are assessed. The *DSM-IV* lists two major subtypes of MDD: (1) atypical, characterized by hypervegetative symptoms such as hypersomnia and increased appetite, and by sensitivity to rejection and reactive mood; and (2) melancholic, characterized by motivational and volitional symptoms and by mood disturbance. The World Health Organization distinguishes primary mood symptoms, encompassing cognitive and affective domains, from secondary somatic symptoms, including sleep, appetite, and anhedonia (World Health Organization, 2007). Researchers have recently used statistical methods to examine symptom clustering in MDD. Principal component analyses using the Beck Depression Inventory have yielded clusters of cognitive symptoms including sadness, hopelessness, and guilt, vegetative symptoms, and anhedonia and psychomotor symptoms (Dunn et al., 2002). A similar analysis using the Hamilton Depression Rating Scale has produced clusters of mood and cognitive symptoms, including depressed mood, hopelessness, and guilt, as well as psychomotor and anhedonia symptoms, and symptoms of sleep disturbance (Milak et al., 2005). Examining several measures of depression, Bech, Fava, Trivedi, Wisniewski, & Rush (2011) also found a separation of cognitive and vegetative symptoms, and Uher et al. (2008) further identified clusters of mood, cognitive, and vegetative symptoms. Combining the results of these studies, the symptoms of MDD can be organized into three clusters. The first contains mood and cognitive symptoms, which include the negative mood, guilt, and diminished self-worth characteristic of depression. The second cluster includes homeostatic and vegetative symptoms like insomnia or hypersomnia and changes in appetite or weight (Dozois, Dobson, & Ahnberg, 1998). The third cluster includes motivational and volitional behavioral symptoms, such as anhedonia and psychomotor retardation or agitation. These clusters are not meant to be definitive descriptions of the etiological processes of MDD, but rather, a method of organizing the heterogeneous presentation of this disorder.

In this chapter we describe several major molecular systems that have been implicated in MDD. We will not review an exhaustive list of molecules that are perturbed in MDD; instead, we will focus on those that have been linked to specific symptom profiles in depression. For example, although an extensive range of neuropeptides and neuromodulators has been implicated in MDD (Belmaker & Agam, 2008), the number of molecules that have been linked to specific symptoms of depression is much smaller. Thus, we discuss neurotransmitters systems, including serotonin and dopamine, and neuromodulatory systems, specifically cytokines, neurotrophic factors, and hormones. For each molecular system, we examine the distribution and function of these molecules in the body and brain and map the neuronal substrates that utilize these molecules. We review pharmacological interventions in humans and animals that may help to elucidate the role of these molecules in MDD. In this context, we examine where and how these molecular systems interact, with an eye toward presenting, if not an FCMP to depression, at least a framework from molecules to symptoms in the cognitive/affective, motivation/behavior, and homeostatic/vegetative domains of functioning.

II. Serotonin

The monoamine hypothesis was one of the first theories to link disturbance in molecular systems with the pathogenesis of MDD. Researchers found that reserpine, an antihypertensive drug derived from the rauwolfia serpentina plant, produced depressive side effects (Quetsch, Achor, Litin, & Faucett, 1959), especially in individuals with a history of MDD. Reserpine inhibits vesicular monoamine transporter (VMAT), inhibiting the normal packaging of monoamines into synaptic vesicles for subsequent release into the synapse, thereby depleting monoamines like dopamine, norepinephrine, and serotonin (Henry et al., 1994). Reserpine inhibits both isoforms of human VMAT (Erickson, Schafer, Bonner, Eiden, & Weihe, 1996) and may cause symptoms of MDD and exacerbate such predisposing factors as a lifetime history of MDD (Beers & Passman, 1990). This evidence led investigators to focus on the role of monoamines, like serotonin, in the pathophysiology of MDD.

Serotonin (5-HT), or 5-hydroxytryptamine, is a biogenic amine found throughout the body, notably in the gut (see McLean, Borman, & Lee, 2007 for review) but also in the central nervous system. It is synthesized in neurons from the amino acid tryptophan via hydroxylation by tryptophan hydroxylase, followed by decarboxylation by amino acid decarboxylase. Serotonin is packaged into synaptic vesicles by VMAT for release at the synapse, and then binds to 5-HT receptors (5-HTR). There are seven known subtypes in the family of 5-HT receptors, primarily G protein coupled receptors (GPCR), the effects of which are posited to be coordinated via downstream signaling and by changes in gene expression and regulation (see Hoyer, Hannon, & Martin, 2002 for review). These receptors include somatodendritic autoreceptors that regulate 5-HT release via negative feedback cycles. After release, 5-HT is transported back into the presynaptic neuron (reuptake) by monoamine transporters, particularly the serotonin transporter (SERT). There, 5-HT is degraded by monoamine oxidase (MAO). In the central nervous system of mammals, the raphe nuclei, especially the dorsal raphe nuclei (DRN), are the primary source of 5-HT. Projections from the DRN include the amygdala and other limbic regions, basal forebrain, and cerebral cortex, as well as projections to the habenula (Morris, Smith, Cowen, Friston, & Dolan, 1999). These projections situate the 5-HT system to influence multiple dimensions of behavior, cognition, and affect in MDD. Although many studies have broadly compared depressed and nondepressed individuals with respect to serotonin function, recent research has begun to examine the role of 5-HT perturbations in leading to specific symptoms of MDD. Next, we first examine the role of the 5-HT system in MDD in general and then link 5-HT function with specific symptoms and symptom clusters in depression.

Many modern antidepressants target the 5-HT system by disabling SERT, thereby inhibiting the reuptake of serotonin into the presynaptic neuron and leading to changes in serotonergic tone. These drugs, known as selective serotonin reuptake inhibitors (SSRIs), are "selective" because they have a much weaker action on other monoamine transporters than they do on SERT, although SERT itself may heterologously transport catecholamines like dopamine or norepinephrine, depending on the biochemical milieu (Larsen et al., 2011). The transport of 5-HT through other monoamine transporters such as norepinephrine transporter (NET) and dopamine transporter (DAT) represents a low-affinity pathway that may also be critical to understanding the effects of SSRIs (Daws, 2009). The effects of SSRIs on 5-HT function are controversial, as are the mechanism by which SSRIs reduce symptoms of depression. The therapeutic benefit of SSRIs has been questioned, although evidence for this challenge seems to be stronger for mild to moderate depression than for severe depressive symptoms (Fournier et al., 2010). The therapeutic delay of SSRIs (Blier, 2003) has drawn attention to the cascade of downstream neurochemical changes in the 5-HT system, rather than the immediate changes in 5-HT tone, as a mechanism for SSRI action. The acute rise in 5-HT after SSRI treatment is attenuated by 5-HT autoreceptors, particularly the somatic 5-HT_{1A} receptors that suppress subsequent 5-HT release (Hjorth et al., 2000). Chronic use of SSRIs will eventually desensitize 5-HT_{1A} autoreceptors, allowing 5-HT tone to rise once again (Lesch, Disselkamp-Tietze, & Schmidtke, 1990). Treatment with SSRIs also appears to trigger increases in neurogenesis (Malberg, Eisch, Nestler, & Duman, 2000) that have been implicated in the therapeutic efficacy of these drugs (Santarelli et al., 2003).

Over the last few decades, researchers have characterized polymorphisms in SERT that affect endogenous SERT function (Heils et al., 1996). One polymorphism in particular, which consists of a tandem repeated sequence in the upstream promoter region of SERT, has been found to be associated with depression, especially in interaction with stress. The "long" version of this allele is associated with greater transcriptional activity of the gene promoter than is the "short" version of this allele. Caspi and colleagues (2003) proposed a gene-by-environment interaction for SERT, in which individuals who carry one or two copies of the short allele and who also experience one or more major life stressors are at increased risk for developing MDD. Although evidence for this formulation has been equivocal (e.g., Risch et al., 2009; Karg, Burmeister, Shedden, & Sen, 2011), researchers have focused nonetheless on 5-HTTLPR as a human model of naturally varying 5-HT functioning. For example, a recent meta-analysis found that 5-HTTLPR moderately affects amygdala response to emotional stimuli (Munafò, Brown, & Hariri, 2008). Postmortem studies have shown reduced SERT binding but not specific 5-HTTLPR allele associations in suicide victims, suggesting that suicidality is associated with SERT dysfunction beyond 5-HTTLPR genotype (Mann et al., 2000). Furthermore, 5-HTTLPR is not a robust predictor of SSRI response in MDD (Lewis et al., 2011).

The 5-HT receptors like 5-HT_4 may underlie changes in neuronal neurogenesis, growth, or maturation in response to SSRIs (Kobayashi et al., 2010). Studies utilizing positron emission tomography (PET) imaging of the 5-HT system also show altered receptor functioning in MDD. Through competitive binding of a radiolabeled ligand, PET imaging captures 5-HT receptor availability in specific brain regions. Investigators utilizing PET imaging of currently depressed individuals, particularly depression associated with a family history of MDD, have documented decreased binding potential of 5-HT_{1A} receptors, most reliably in medial temporal lobe (Drevets et al., 2000), frontal lobe, and limbic regions (Sargent et al., 2000). Polymorphisms in the 5-HT_{2A} receptor gene, in interaction with environmental factors, are associated with depression (Jokela et al., 2007), as well as with resistance to antidepressants (Choi, Kang, Ham, Jeong, & Lee, 2005; McMahon et al., 2006). However, findings of postmortem studies of 5-HT_{2A} receptors density in cortical and subcortical areas are mixed (Stockmeier, 2003). Another receptor type, 5-HT_{2C}, may also be implicated in MDD; postmortem studies of suicide victims show altered 5-HT_{2C} receptor mRNA isoforms in prefrontal cortex resulting from post-transcriptional mRNA editing (Gurevich et al., 2002). Finally, treatment with SSRIs decreases 5-HT_{2A} receptor binding potential, although this effect is complicated by age (Meyer, Kapur, et al., 2001). Consideration of 5-HT receptors is critical because they also indirectly affect neurotransmission of other monoamine systems, particularly NE and DA.

A. Serotonin: Cognitive/affective symptoms

Of the factors derived from analyses of clinical symptoms in depression, serotonin is most often associated with negative affect and cognition. Sad/depressed mood is one of the two cardinal symptoms of MDD, but because sadness is a difficult symptom to model in animals, much of the evidence that links 5-HT with mood involves either pharmacological interventions that target 5-HT in humans or phenotyping studies of human variations in 5-HT system functioning. Depletion of the 5-HT precursor tryptophan causes sad mood in individuals who are vulnerable to depression, that is, individuals who have been depressed in the past, individuals with a family history of MDD, and individuals who are currently depressed and are being treated with antidepressants (Ruhe, Mason, & Schene, 2007).This manipulation appears to reinstate the cognitive and affective deficits that are consistent with individual phenotypes of MDD. Moreover, tryptophan depletion does not seem to affect psychomotor functioning, attention, or memory, but has robust effects on the processing of emotion (Mendelsohn, Riedel, & Sambeth, 2009). Tryptophan depletion may also interact with current mood to influence cognition (Robinson, Cools, Crockett, & Sahakian, 2010). Interestingly, this effect on emotion processing, at least in the processing of emotional faces, is associated with an increase in blood-oxygenation-level-dependent (BOLD) signal in the amygdala (Cools et al., 2005; van der Veen, Evers, Deutz, & Schmitt, 2007), a region in the medial temporal lobe that has been implicated in emotion processing (Adolphs, Tranel, Damasio, & Damasio, 1994). Other acute manipulations of 5-HT include d-Fenfluramine administration (Cleare, Murray, & O'Keane, 1998) and short-term treatment with SSRIs, which attenuates amygdala and medial temporal lobe response to negative emotions and leads to changes in emotion processing (Harmer, Mackay, Reid, Cowen, & Goodwin, 2006). In addition, 3,4 methylenedioxymethamphetamine (MDMA), a potent recreational drug with a primarily 5-HT action, potently alters affect (Baylen & Rosenberg, 2006). These studies support the role of 5-HT in the maintenance of mood state. Notably, however, 5-HT alterations such as tryptophan depletion do not induce depression in adults who are not at risk for MDD (Heninger, Delgado, & Charney, 1996), suggesting that serotonergic disruption interacts with other factors that predispose individuals to develop MDD.

Cognitive and affective symptoms that characterize depression, including cognitive biases, altered memory processing, and altered affect, may develop in concert with or even stem from alterations in SERT binding (see Disner, Beevers, Haigh, & Beck, 2011, for a review). These deficits correspond to increased binding potential of SERT in prefrontal cortex, anterior cingulate, and thalamic and striatal systems (Meyer, 2007). SERT functioning may also depend on polymorphisms that alter expression of this gene, such as 5-HTTLPR (Heils et al., 1996). Although these polymorphisms may not predict overall antidepressant efficacy in MDD (Lewis et al., 2011), 5-HTTLPR may predict blunted treatment response to SSRIs in mood and cognitive symptoms (Serretti et al., 2007). This blunted treatment response was not found for vegetative symptoms, including sleep difficulties, suggesting an interaction between 5-HTTLPR genotype and the molecular target of antidepressant in determining specific symptom response.

Dysfunction and perturbation of 5-HT in cognitive and affective functioning may also be mediated by altered 5-HT receptor functioning. Consistent with this possibility, studies using functional magnetic resonance imaging (fMRI) have found altered amygdala response to emotional stimuli in individuals who carry polymorphisms in 5-HT_{1A} receptor genes (Dannlowski et al., 2007). PET studies have found a positive association between 5-HT_{2A} receptors and predisposing risk factors, such as history of depression (Bhagwagar et al., 2006), as well as with dysfunctional cognitions (Meyer et al., 2003; Frokjaer et al., 2008). These studies show increased binding 5-HT_{2A} potential most consistently in prefrontal cortex and in insular and temporal cortices; these altered binding profiles correlate with affective and cognitive traits such as neuroticism (Takano et al., 2007) and dysfunctional attitudes (Bhagwagar et al., 2006). The localization of 5-HT_{2A} and 5-HT_{2C} receptor alterations in MDD suggest involvement of receptor dysfunction in negative cognitions (Berg, Harvey, Spampinato, & Clarke, 2008), but more evidence is needed to directly test this hypothesis. Investigators must be cautious in designing studies examining altered 5-HT, 5-HT receptor, or SERT functioning in order to avoid confounds related to acute or chronic changes in 5-HT levels due to age, antidepressant use, or other 5-HT system changes (Meyer, 2007).

B. Serotonin: Homeostatic/vegetative symptoms

Given the involvement of 5-HT in the regulation of sleep (Ursin, 2002), it is not surprising that the drugs targeting the 5-HT system perturb sleep functioning. MDD is associated with reduced rapid eye movement (REM) sleep latency and increased REM density, findings that have been explicitly linked to 5-HT functioning and adaptation of 5-HT receptors (Adrien, 2002). Almost all tricyclic, MAOI, and SSRI drugs that affect 5-HT suppress REM sleep in MDD (Winokur et al., 2001). Notably, drugs that target different molecular systems—for example, bupropion, that targets norepinephrine and dopamine reuptake—do not suppress REM sleep, although successful bupropion treatment does increase REM latency (Ott, Rao, Nuccio, Lin, & Poland, 2002). In healthy volunteers, the relation between target receptor specificity and REM effect is less clear (Rijnbeek, de Visser, Franson, Cohen, & van Gerven, 2003). SSRIs may not be as effective as tricyclic antidepressants in reducing vegetative symptoms (Uher et al., 2009); indeed, even after successful treatment with SSRIs, patients often have persistent symptoms, the most common of which is nighttime awakening insomnia, followed by anhedonia (McClintock et al., 2011) and other vegetative symptoms (Nierenberg et al., 2010). Whereas tryptophan depletion increases REM density in healthy individuals (Bhatti et al., 1998; Voderholzer et al., 1998), in depressed individuals being treated with SSRIs, tryptophan depletion disinhibits REM sleep, reconstituting the sleep symptoms of MDD (Moore et al., 1998).

Increased or decreased appetite or weight are vegetative/homeostatic symptoms of MDD. Brain 5-HT has been implicated in feeding behavior and appetite regulation (Blundell, 1986), and 5-HT has been hypothesized to underlie the relation between food cravings and symptoms of MDD (Wurtman, 1993). Dysphoric mood often precedes food craving (Hill, Weaver, & Blundell, 1991) and, more specifically, craving for carbohydrates (Christensen & Pettijohn, 2001). 5-HT is posited to influence appetite control through modulation of hypothalamic control of feeding homeostasis; drugs that increase 5-HT may reverse the preferential intake of carbohydrates (Leibowitz & Alexander, 1998). Other 5-HT targeting drugs often have weight- and appetite-related side effects; for instance, MDMA suppresses appetite via 5-HT_4 receptor signaling in the nucleus accumbens (NAcc), but not through alteration in motor activity (Francis, Kraushaar, Hunt, & Cornish, 2010). Administration of *d*-Fenfluramine, a 5-HT agonist, robustly increases satiety, an effect mediated by 5-HT_{2C} receptors (Vickers, Clifton, Dourish, & Tecott, 1999); 5-HT_{2C} knockout mice have poorly controlled feeding behavior, resulting in morbid obesity (Tecott et al., 1995). Conversely, 5-HT_{1A} agonists strongly potentiate appetite, and other 5-HT receptors have been implicated in appetite regulation (Hoyer et al., 2002). Finally, although some SSRIs, such as fluoxetine, cause weight changes, other non-SSRI antidepressants, like bupropion, appear to have similar effects (Serretti & Mandelli, 2010). Thus, further evidence is required to link 5-HT disruption specifically to appetite and weight disruption in MDD.

III. Dopamine

Dopamine (DA) is a catecholamine synthesized from L-tyrosine, first by the addition of a hydroxyl by tyrosine hydroxylase to form L-dihydroxyphenylalanine (L-DOPA), and then by the processing of L-DOPA (which, unlike DA, can cross the blood-brain barrier) into DA by dopa decarboxylase. This biosynthesis pathway includes the precursors for the production of norepinephrine and epinephrine (see section on norepinephrine). Like 5-HT, DA is packaged by VMAT for vesicular release into the synapse during neurotransmission. As a neurotransmitter ligand, DA binds to a family of G-protein coupled dopamine receptors (Dl-D5), which, like 5-HT receptors, display a remarkable functional heterogeneity. As with 5-HT, DA can be cleared from the synaptic cleft through reuptake via a specific transporter, the dopamine transporter (DAT). DA clearance from the synapse is also managed by heterologous transport via the norepinephrine transporter (NET) (Morón, Brockington, Wise, Rocha, & Hope, 2002), MAO metabolism (Wayment, Schenk, & Sorg, 2001), or enzymatic catabolysis in the synapse via catecol-*O*-methyltransferase (COMT).

In mammals, DA-producing neurons originating in the midbrain project widely to cortex and basal ganglia. The primary ascending pathways originate from DA neurons in the substantia nigra, ventral tegmental area (VTA), and retrorubral regions (Williams & Goldman-Rakic, 1998). The mesolimbic pathway originates from the VTA and projects to the NAcc and then to the amygdala, hippocampus, and bed nucleus of the stria terminalis, whereas the mesocortical pathway projects from the VTA to the prefrontal cortex. The nigrostriatal pathway originates from the pars compacta subregion of the substantia nigra and projects to the striatum, where it influences the direct and indirect striatal pathways of the basal ganglia. The anatomic distribution of DA receptors and methods of terminating DA action in the synapse play key roles in determining the behavioral, cognitive, and affective impact of perturbations in the DA system. DA receptors are distributed widely across the brain, with basal ganglia expressing Dl, D2, and D3 receptors (Meador-Woodruff et al., 1996). Dl, D2, D3, D4, and D5 receptors are all expressed at varying levels in the cortex, although Dl is notably more prevalent in the prefrontal cortex than are D2 and other receptors (Lidow, Goldman-Rakic, Gallager, & Rakic, 1991). DAT also has a heterogeneous pattern of distribution across the brain (Ciliax et al., 1999), although the role of this heterogeneity in MDD is not yet clear.

Direct evidence of DA dysfunction in MDD, while intriguing, is inconsistent (see Dunlop & Nemeroff, 2007 for review). In particular, age-related heterogeneity in DA functioning complicates many findings (Volkow et al., 2000). Plasma and cerebrospinal fluid (CSF) levels of homovanillic acid, a DA metabolite, is reduced in MDD, suggesting reduced DA turnover (Brown & Gershon, 1993; Lambert, Johansson, Agren, & Friberg, 2000). Given that similar deficits have been found in schizophrenia (Berger et al., 1980), however, DA turnover may be a nonspecific biomarker of psychopathology. Catecholamine depletion using α-methylparatyrosine depletes DA in a method analogous to tryptophan depletion and, like tryptophan depletion, reinstates depressive symptoms in vulnerable individuals (Berman et al., 1999). Using PET, investigators have identified glucose metabolism in prefrontal and limbic areas as potential markers for vulnerability to this catecholamine depletion, which itself alters metabolism in prefrontal and thalamic regions (Bremner et al., 2003). Transport of DA via DAT has also been implicated in MDD; investigators using PET have found reduced striatal binding potential of DAT in MDD (Meyer, Krüger, et al., 2001), but contradictory evidence has also been reported, including increased binding potentials (Brunswick, Amsterdam, Mozley, & Newberg, 2003) and binding potentials similar to those found in nondepressed participants (Argyelán et al., 2005). Argyelán et al. (2005) also found no association between binding altered by the DAT inhibitor bupropion and clinical efficacy, suggesting that the effects of bupropion are mediated by pharmcokinetic interactions with NET and NE levels (Horst & Preskorn, 1998). Evidence of altered striatal D2/D3 radioligand receptor binding is mixed (Hirvonen et al., 2008) and may depend on specific symptomatology. Postmortem studies of dopamine receptors in MDD have found increased D2/ D3 receptor binding potential (Klimek, Schenck, Han, Stockmeier, & Ordway, 2002) and increased D4 receptor mRNA in the amygdala (Xiang et al., 2008).

Parkinson's disorder is a disorder of movement resulting from selective lesions to DA neurons in the substantia nigra. Because MDD is prevalent in Parkinson's patients (Reijnders, Ehrt, Weber, Aarsland, & Leentjens, 2008), the specific nature of the lesion in Parkinson's disorder may reveal how disruption in DA function can lead to symptoms of MDD. In a case study, Bejjani and colleagues (Bejjani et al., 1999) administered deep brain stimulation (DBS) to the substantia nigra in a 65-year-old woman with Parkinson's disorder. This stimulation resulted in transient depressive symptoms, which were accompanied by alterations in PET-assessed metabolism. Previous vulnerability to severe depression may predict similar adverse effects after direct stimulation to the basal ganglia for treatment of Parkinson's (Voon, Kubu, Krack, Houeto, & Tröster, 2006). These studies, although not definitive in implicating DA dysfunction in the onset of MDD, are provocative.

A. *Dopamine: Motivation/behavior symptoms*

Dopaminergic circuits have been studied in reward processing but may be specifically involved in volitional, motivated behavior in response to rewarding stimuli (Salamone, Correa, Mingote, & Weber, 2005). This is consistent with the hypothesized role of DA dysfunction in MDD and, in particular, in the development of anhedonic and psychomotor symptoms. Historically, DA has been associated with motivation and motor symptoms, but not with vegetative or cognitive symptoms (Brown & Gershon, 1993). More recently, however, imaging of depressed individuals with PET and other methods has directly implicated dopamine in motivation (Dunlop & Nemeroff, 2007). Depleting dopamine with α-methylparatyrosine also reveals deficits in hedonic processing. PET studies conducted during catecholamine depletion show that metabolic changes in ventral striatum are correlated with anhedonia, whereas prefrontal cortex metabolism is related to overall levels of depressive symptoms (Hasler, Fromm, et al., 2008). This depletion also produces deficits during a reward task in remitted depressed individuals (Hasler et al., 2009). Importantly, Hasler et al. (2009) found that α-methylparatyrosine administration specifically perturbed reward processing but not general cognitive ability. These studies suggest that acute perturbation of DA pathways, interacting with predisposing risk phenotypes, produce volitional and motivational deficits. Specifically, the dopaminergic circuit projecting from VTA to NAcc underlies normal motivational processes; dysfunction in this circuit is a candidate substrate of anhedonia in MDD. For example, anhedonia as a symptom is associated with reduced NAcc volume and reduced NAcc response to reward in a monetary task (Wacker, Dillon, & Pizzagalli, 2009). Manipulations in rodent models have also implicated this pathway in depression (Nestler & Carlezon, 2006). In humans, DBS of NAcc alleviates anhedonia in depression while increasing glucose metabolism in NAcc (Schlaepfer et al., 2007).

In addition to motivational deficits in MDD that manifest as anhedonia, depressed individuals may have psychomotor agitation or retardation. Reduced CSF levels of the DA metabolite homovanillic acid are strongly associated with MDD in patients who present with psychomotor retardation (Brown & Gershon, 1993). Catecholamine depletion studies suggest a parallel to tryptophan depletion in recapitulating symptoms consistent with an individual's specific risk phenotype (Hasler et al., 2008). Remitted MDD patients undergoing catecholamine depletion report such psychomotor symptoms as agitation and restlessness in addition to core mood symptoms (Berman et al., 1999). Although it is difficult to distinguish this effect from possible side effects of the α-methylparatyrosine administered in this experiment (Engelman, Horwitz, Jéquier, & Sjoerdsma, 1968), these findings nevertheless suggest that acute DA system dysfunction is a source of psychomotor agitation in MDD. Similarly, chronically increased binding potentials of D2 and D3 receptors may represent a phenotype of motor dysfunction in MDD, rather than of MDD in general. As noted by Cannon et al. (2009) negative findings in D2/D3 receptor studies in MDD may be due to the absence of psychomotor retardation in study participants (Ebert, Feistel, Loew, & Pirner, 1996). Interestingly, caudate D1 receptor binding potential is inversely correlated with psychomotor functioning in healthy controls and with anhedonia in MDD patients (Cannon et al., 2009).

Single photon emission computerized tomography (SPECT) imaging has revealed frontal hypoperfusion, rather than basal ganglia dysfunction, in depressed individuals who are experiencing psychomotor difficulties (Narita, Odawara, Iseki, Kosaka, & Hirayasu, 2004). This finding suggests that psychomotor retardation involves more than a pure motor dysfunction alone (Loo et al., 2008). Indeed, preliminary treatment studies using transcranial magnetic stimulation on left DLPFC have found improvement in depressed individuals with psychomotor dysfunction (Baeken et al., 2010). In addition, treatment with fluoxetine aids the cognitive processes involved in psychomotor activity, alleviating psychomotor symptoms in MDD (Sabbe, van Hoof, Hulstijn, & Zitman, 1996). In contrast, fluoxetine was found to ameliorate but not eliminate deficits in a pure motor task (Sabbe, van Hoof, Hulstijn, & Zitman, 1997). Bupropion, however, improves psychomotor deficits in MDD (Fabre, Brodie, Garver, & Zung, 1983), anhedonia, and other volitional deficits (Jefferson et al., 2006). Taken together, these findings suggest a role for DA, perhaps in motor functioning, in psychomotor dysfunction beyond the cognitive aspect of psychomotor dysfunction that is amenable to SSRI treatment. Future studies determining the specific role of DA in volitional symptoms must separate the effects of drugs like bupropion on DA from the effects on other neurotransmitter systems (El Mansari, Ghanbari, Janssen, & Blier, 2008). In this context, it is noteworthy that the specific DA agonist pramipexole, when used in Parkinson's disorder, reduces anhedonia (Lemke, Brecht, Koester, & Reichmann, 2006), and investigators are currently examining the use of this agonist as an antidepressant (Corrigan, Denahan, Wright, Ragual, & Evans, 2000).

IV. Norepinephrine

Norepinephrine (NE) is a derivative of dopamine produced via interactions between DA and dopamine β-hydroxylase. Like 5-HT and DA, NE is packaged for release from vesicles into the synapse, where it binds to a family of receptors known as adrenergic receptors (based on the alternate name noradrenaline, generally used to denote NE in nonbrain tissue). Adrenergic receptors are G protein-coupled receptors and are organized into α and β subtype families. Once released, the presynaptic norepinephrine transporter (NET) controls NE reuptake, although NE can also be degraded by MAO or COMT. In the mammalian brain, NE cells originate primarily from the locus coeruleus (LC) and other NE nuclei in the brainstem and project widely to cortical and subcortical regions. The LC in turn receives input from prefrontal areas, particularly the cingulate gyrus and other regions, including the amygdala and hypothalamus. Projections from the LC are situated to affect global brain activity, and tonic activity in the LC reflects behavioral arousal. Thus, spontaneous activity in the LC decreases as sleep occurs, reaching near quiescence during REM, and increases as a function of engagement with the external environment (see Aston-Jones, Rajkowski, & Cohen, 1999, for review).

Markers of NE metabolism, such as the metabolite 4-hydroxy-3-methoxyphenylglycol (MHPG), were initially used as predictive markers in MDD, but this approach has been controversial (Blombery, Kopin, Gordon, Markey, & Ebert, 1980) and findings of follow-up studies have been mixed (see Brunello et al., 2003, for review). Because the synthesis of NE is downstream of the DA synthesis pathway, markers in this pathway can also be used to measure NE system activity. Postmortem studies have found decreased tyrosine hydroxylase in the LC of suicide victims (Biegon & Fieldust, 1992), as well as increased NE α2 receptor binding in the LC (Ordway, Widdowson, Smith, & Halaris, 1994) and in other regions (Meana, Barturen, & García-Sevilla, 1992). Polymorphisms in NET may affect risk for depression (Ryu et al., 2004; but see also Chang et al., 2007), and NET binding potential is decreased in MDD in the midcaudal LC postmortem (Klimek et al., 1997). Taken together, these findings suggest that there are chronic alterations in NE system functioning in MDD, but this theory is tempered by studies that have failed to find unequivocal NE dysfunction in MDD. For instance, investigators have found NE β1 receptors to be either increased (Mann, Stanley, McBride, & McEwen, 1986) or decreased (De Paermentier, Cheetham, Crompton, Katona, & Horton, 1991) in postmortem studies of suicide victims, with differential effects of various antidepressants likely contributing to these contradictory findings. Researchers have also found reduced NE in venoarterial blood in MDD, an anomaly that normalized following treatment with the NET inhibitor desipramine (Lambert et al., 2000). In this particular study, however, reduced turnover of DA, not of NE, was associated with clinical symptoms. Importantly, some investigators have found NE deficits with anxiety but not with MDD (Sevy, Papadimitriou, Surmont, Goldman, & Mendlewicz, 1989); differentiating these two patient populations may be critical in identifying their biological antecedents. To distinguish the NE-dependent mechanism of action from alteration in other monoamines during antidepressant treatment, it is important that investigators examine selective norepinephrine reuptake inhibitors (SNRIs) rather than tricyclic antidepressants or reuptake inhibitors that also affect DA or 5-HT, such as bupropion and venlafaxine, respectively. In this context, it is notable that the SNRI reboxetine showed poor efficacy compared to other antidepressants and placebos (Eyding et al., 2010). Thus, the effects of NE alterations in MDD may depend on interactions with other neurotransmitter and neuromodulator systems. For instance, NET knockout mice have changes not only in NE but also in 5-HT and DA-mediated behavioral responses (Haenisch & Bönisch, 2011) and neurotrophins or neuropeptides expression (Haenisch, Gilsbach, & Bönisch, 2008).

A. *Norepinephrine: Cognitive/affective symptoms*

The LC and other NE nuclei are associated with the sympathetic and stress response (Goddard et al., 2010); not surprisingly, the LC is integral to arousal associated with the affective context of a memory (Sterpenich et al., 2006). Indeed, amygdala-driven memory encoding of emotional content is blocked by the β-adrenergic antagonist propranolol, indicating that NE plays a critical role in this process (Strange & Dolan, 2004). Situated in a position to alter arousal, NE functioning is thought to be critical for the deployment of specific attention in a manner akin to sharpening signal-to-noise ratios of representations (Sara, 1985). Consistent with this role, investigators have also implicated NE dysfunction in panic disorder and other stress disorders (Charney, Woods, Goodman, & Heninger, 1987). NE may play a similar role in the dysfunctional cognitions and emotions present in MDD, including the maladaptive cognitions and responsiveness to environmental stressor that characterize this disorder. For instance, salivary 3-methoxy-phenylglycol, a NE metabolite, is increased in individuals with depressive symptoms in response to a negative mood induction (Hamer, Tanaka, Okamura, Tsuda, & Steptoe, 2007). Interestingly, the NE reuptake inhibitor reboxetine causes rapid changes in affective processing in healthy volunteers (Harmer et al., 2009), further supporting the role of NE in cognition and affect. We should be cautious in extending these results to MDD, however, because of reboxetine's ineffectiveness as an antidepressant (Eyding et al., 2010).

B. *Norepinephrine: Vegetative/homeostatic symptoms*

Firing rate changes in the LC precede sleep, wake, and REM states (Aston-Jones & Bloom, 1981), supporting a role of NE in sleep and arousal. Like SSRIs, reboxetine may alter sleep EEG stages, decreasing total REM (Kuenzel, Murck, Held, Ziegenbein, & Steiger, 2004) or REM latency (Farina et al., 2002). Similarly, nortriptyline, which is more selective for NE than for 5-HT (Sánchez & Hyttel, 1999), improves vegetative symptoms to a greater degree than does a 5-HT specific antidepressant (Uher et al., 2009). Although direct evidence linking NE dysfunction with appetite or weight changes in MDD is lacking, NE appears to be an important component of appetite regulation (Nelson & Gehlert, 2006). Drugs that target the NE system, such as the selective NE reuptake inhibitor atomoxetine, can ameliorate fatigue in particular (Papakostas, Petersen, Burns, & Fava, 2006). Reboxetine may affect psychomotor retardation (Ferguson, Mendels, & Schwart, 2002), but more research is needed to determine if this effect is specific to the psychological or motor component of retardation. NE is positioned to affect vegetative and homeostatic symptoms, but there is inadequate empirical evidence to support the claim that NE is specifically associated with an arousal deficit and vegetative symptoms in MDD.

V. Glutamate and γ-aminobutyric acid (GABA)

Glutamate and GABA form the excitatory and inhibitory backbones, respectively, of neurotransmission in the brain. Neurons that utilize glutamate, an amino acid that also functions as an excitatory neurotransmitter, are widely distributed throughout the mammalian brain, especially in the cortex. Glutamate is synthesized from glucose and is packaged into synaptic vesicles by vesicular glutamate transporters for release. Once released from the presynaptic neuron, glutamate has multiple receptors, including ionotropic AMPA and N-methyl-*D*-aspartate (NMDA), kainite receptor, and metabotropic glutamate receptor families, each composed of multiple subunits and each with variations and subtypes. The ionotropic receptors produce an excitatory depolarization; an excess of glutamate in the synapse leads to calcium influx and excitatatory neurotoxicity (Choi, 1987). This neurotoxicity underscores the role of excitatory amino-acid transporters in uptaking glutamate from the synapse. Unlike 5-HT or DA, however, the primary source of this uptake is not the presynaptic cell but rather nearby astrocytes. These cells act as a buffer for glutamate as well as conversion site of glutamate to glutamine via glutamine synthetase. Glutamine is then transported back to the presynaptic neuron, where it can be converted back to glutamate via glutaminase. This system forms a "tripartite" synapse: presynaptic neuron, postsynaptic neuron, and astrocyte acting in concert to facilitate neurotransmission (Hertz & Zielke, 2004; Machado-Vieira, Manji, & Zarate, 2009).

In contrast, GABA is synthesized from glutamate via the action of glutamic acid decarboxylase, and acts as the primary inhibitory neurotransmitter in the mammalian brain. GABA is packaged into vesicles by the GABA vesicular transporter and, upon release into the synapse, binds to either the ionotropic GABAa receptors or the metabotropic G protein-coupled GABAb receptors. After release, GABAergic neurotransmission is terminated by GABA reuptake through GABA transporters in neurons and surrounding glia. Given that GABA is a critical regulator of inhibitory activity, GABAergic deficits have been linked with excitatory neurotransmission dysregulation, leading to epilepsy; many drugs that target hyperexcitatory activity use GABA as a primary mechanism (see Petroff, 2002, for review). GABA can also be used, via the tricarboxylic acid cycle, to ultimately synthesize glutamate; consequently, glutuamate and GABA are intimately linked in both biochemical availability and neuronal functioning.

Glutamate has been implicated in a "glutamate hypothesis" of depression (Sanacora, Treccani, & Popoli, 2011). Although abnormalities in glutamate have been found in MDD, findings have been mixed. Increased levels of glutamate associated with MDD have been reported in plasma; however, glutamatergic dysfunction in a wide range of brain regions, (Yüksel & Öngür, 2010), including decreased glutamate, has also been reported in the brains of individuals diagnosed with MDD (Hasler et al., 2007) using magnetic resonance spectroscopy (MRS). The contradictory nature of these findings has been attributed to a wide range of factors, including a failure to separate bipolar from unipolar depressive disorders, not identifying sources of peripheral glutamate, and inconsistently controlling for confounding factors like medication (Sanacora, Zarate, Krystal, & Manji, 2008). Postmortem studies in MDD have also found decreased levels of glutamate transporters and increased levels of glutamate receptors (Choudary et al., 2005), as well as decreased vesicular glutamate transporter mRNA in medial temporal lobe (Uezato, Meador-Woodruff, & McCullumsmith, 2009), although the specificity of these findings to MDD rather than to general mood disturbance is unclear (Beneyto, Kristiansen, Oni-Orisan, McCullumsmith, & Meador-Woodruff, 2007). Abnormalities in GABA have also been reported in MDD, including increased GABA in plasma (Petty & Sherman, 1984), reduced GABA in CSF (Gold, Bowers, Roth, & Sweeney, 1980), and decreased GABA with increasing MDD severity in brain tissue (Honig, Bartlett, Bouras, & Bridges, 1988). MRS studies examining GABA have also found decreased GABA in occipital (Sanacora et al., 1999) and prefrontal brain regions (Hasler et al., 2007), a reduction that persists following recovery and that may be either a neurobiological scar or a vulnerability phenotype (Bhagwagar et al., 2007). Changes in a wide variety of GABAergic gene transcripts are found in prefrontal and medial temporal regions in MDD (Sequeira et al., 2009). Hetereogeneous findings have been reported regarding MDD-associated GABA receptor binding in postmortem samples. For example, $GABA_b$ receptor binding is mostly unchanged in suicide victims (Cross, Cheetham, Crompton, Katona, & Horton, 1988), but $GABA_a$ expression is altered in depressed individuals (Choudary et al., 2005).

Researchers examining genetic markers have also found associations between glutamate and GABA genes and MDD. $GABA_a$ receptor subunit genes (Horiuchi et al., 2004) have been associated with MDD, although more research is needed to dissociate these findings in MDD with other $GABA_a$ receptor subunits that have been associated with bipolar disorder (Massat et al., 2002). Associations between glutamate receptors and psychopathology have thus far focused on bipolar disorder, but some emerging findings implicate these receptors in MDD. A kainate receptor gene marker has been associated with antidepressant response (Paddock et al., 2007), whereas other glutamate receptor markers have been associated with suicidal ideation while undergoing antidepressant treatment (Laje et al., 2007). In animal models, metabotropic glutamate receptor antagonists have antidepressant effects (Chaki et al., 2004), and the NMDA receptor may play an even more critical role in mediating antidepressant response from a variety of treatments. Chronic antidepressant response, from a wide range of molecular mechanisms, may converge on long-term alterations of the NMDA receptor complex (Skolnick et al., 1996). Although SSRIs and other treatments slowly change NMDA receptor binding over time, direct targeting of this system via such NMDA receptor antagonists as ketamine leads to a fast antidepressant response (Berman et al., 2000) that may be due to disinhibition of neurotrophin expression (Autry et al., 2011).

Overall, these studies implicate altered glutamate and GABA functioning in MDD. In a study of altered GABA-related gene expression in prefrontal cortex, Klempan and colleagues (Klempan et al., 2009) point out that findings of altered glutamatergic and GABAergic functioning in MDD are most likely interrelated. Specifically, reciprocal biochemical synthesis pathways make it difficult to conclude definitively that either glutamate or GABA is dysregulated in MDD distinct from its partner. Hence, studies of glutamatergic and GABAergic functioning together often reveal concordant dysfunction, for instance, upregulation of glutamate receptors and of $GABA_a$ subunits (Choudary et al., 2005). Interactions among glutamate, GABA, and other neurotransmitter systems are ubiquitous, leading some investigators to propose that glutamate and GABA, specifically the NMDA receptor complex, represents a plausible FCMP of MDD (Marsden, 2011). The current challenge for researchers is two-fold: first, to determine whether these disruptions are specific to MDD and particular MDD symptom clusters, and second, to determine whether disrupted GABA and glutamate functioning precede, co-occur with, or are scars of MDD.

A. *Glutamate and GABA: Cognitive/affective symptoms*

Although the wide distribution of glutamatergic neurons makes it likely that this system can affect multiple symptom clusters in MDD, cognitive and affective systems may be most strongly affected by perturbations in glutamatergic functioning. Consistent with the critical role played by glutamate and its receptors in learning (Whitlock, Heynen, Shuler, & Bear, 2006), affective processes like fear conditioning are dependent on glutamatergic mechanisms, specifically trafficking of glutamate receptors such as AMPA (Rumpel, LeDoux, Zador, & Malinow, 2005). Given the wide pattern of glutamate innervation, including direct innervation of monoaminergic systems, investigators have proposed that glutamate modulates functioning in the hippocampal and prefrontal cortex circuits that are responsible for symptoms of MDD (Paul & Skolnick, 2003). It is noteworthy that the NMDA receptor antagonist ketamine leads to rapid antidepressant actions that appear to show symptom specificity. Initial studies found that ketamine ameliorated mood, helplessness, worthlessness, and suicidality symptoms, but not other symptoms of MDD (Berman et al., 2000). Resolution of other symptoms, such as sleep symptoms, may follow at necessarily later points, but notably, many of the side effects of ketamine are cognitive, perceptual, or affective disturbances. Ketamine administration rapidly resolves mood symptoms and suicidality (Price, Nock, Charney, & Mathew, 2009), but the effects have been found not to last, and relapse of MDD symptoms appears to be unfortunately common (aan het Rot, Mathew, & Charney, 2009).

Although GABA has also been implicated in anxiety disorders, and drugs targeting GABAergic transmission are used for treating anxiety (Löw et al., 2000), this system has been implicated more widely in affective and cognitive phenotypes that may be common to anxiety and MDD (Hasler & Northoff, 2011). One possible role for GABA is in self-relevant processing typical of resting-state functional circuits in the brain (Northoff et al., 2007) or in the regulation of response to negative emotional stimuli (Northoff et al., 2002). Glutamate may also play a role in modulating resting-state connectivity in regions in the salience network (Horn et al., 2010) or in the default mode network, a passive, self-referential network that shows aberrant activity in MDD (Sheline et al., 2009).

B. *Glutamate and GABA: Volitional/behavioral symptoms*

In an MRS study, Sanacora et al. (Sanacora et al., 2004) found pronounced decreased GABA concentrations and increased glutamate concentrations in individuals with the melancholic subtype of MDD. Notably, there were no correlations between GABA concentrations and severity of MDD, measured by the HAM-D. This suggests that symptoms of melancholic depression, such as volitional deficits, are associated with glutamatergic or GABAergic dysfunction. Consistent with this formulation, reduced glutamine levels in the pregenual anterior cingulate are associated with MDD, specifically anhedonia (Walter et al., 2009). These findings indicate that disruptions in GABA and glutamate functioning may be a biomarker of anhedonia in MDD, or possibly a molecular pathway to this symptom. Another link between glutamate or GABA function and anhedonia is the distribution of GABAa receptors in mood-regulating and addiction-promoting regions. Specifically, benzodiazepines, which are highly addictive, act by binding to GABAa receptors on neurons that inhibit DA neurons in the VTA pathway (Tan et al., 2010). This disinhibition of DA neurons is posited to increase VTA neuron firing susceptibility in a mechanism similar to that of other drugs of abuse, leading Luscher, Shen, & Sahir (2011) to propose that deficits in $GABA_a$ receptor subunits may mediate GABAergic anhedonia in MDD. Supporting this, $GABA_a$ receptor γ2 subunit knockdown mice produced depression-like symptoms, including latency to feed, behavioral despair, and reduced sucrose intake, behaviors analogous to symptoms of the melancholic subtype (Shen et al., 2010) Future research could profitably examine whether this mechanism underlies anhedonia in MDD or acts as a mediator of DA disruption, culminating in volitional and/or behavioral symptoms of MDD.

VI. Hypothalamic-Pituitary-Adrenal Axis

The hypothalamic-pituitary-adrenal (HPA) axis is a crucial molecular circuit for physiological responses to environmental challenges and is involved in orchestrating a wide range of neural, chemical, and behavioral adaptations through secretion of a series of hormones. In this axis, the paraventricular nuclei of the hypothalamus secrete corticotropin (adrenocorticotropic) releasing hormone (CRH) and arginine vasopressin (AVP). CRH and AVP stimulate the pituitary to produce adrenocorticotropic hormone (ACTH), also known as corticotropin, which in turn stimulates secretion of mineralocorticoids, as well as glucocorticoids such as cortisol, from the adrenal glands. Glucocorticoids regulate metabolic functioning, particularly carbohydrate metabolism (Long, Katzin, & Fry, 1940) via gluconeogenesis (Haynes, 1962). Glucocorticoids also play a critical role in modulating inflammatory response, which has led to their use as a potent anti-inflammatory drug (see Coutinho & Chapman, 2011, for review). The HPA axis, in general, is a critical regulator of homeostasis (see Chrousos, 2009, for review), and the axis itself is self-regulated via negative feedback loops. Release of glucocorticoids triggers several negative short- and long-term feedback mechanisms that depend on both neural signaling and protein synthesis at multiple sites of action that inhibit ACTH, CRH, and AVP secretion (Keller-Wood & Dallman, 1984; de Kloet, 2000; Tasker, Di, & Malcher-Lopes, 2005). HPA-axis activity varies across the day in a regulated circadian rhythm interacting with the "master clock" in the suprachiasmatic nucleus of the hypothalamus (Kalsbeek et al., 2011). The HPA axis interacts with numerous neurotransmitters to affect cognition, affect, and behavior, and dysfunction in this system, resulting in a wide range of pathologies (de Kloet, Joëls, & Holsboer, 2005).

Several factors regulate the presence of glucocorticoids, such as cortisol, in the brain. Glucocorticoids are normally bound to corticosteroid-binding globulin, although increased levels of glucocorticoid production may overwhelm this limited binding system in peripheral circulation. Access to the brain via the blood-brain barrier varies across different hormones, but cortisol in particular has relatively low permeability due to a high number of hydrogen bonds in serum (Pardridge & Mietus, 1979). Furthermore, the multidrug resistance P-glycoprotein excludes cortisol from the brain via active and semi-selective transport (Karssen et al., 2001). Despite this, cortisol is the primary glucocorticoid in primates, whereas corticosterones, which are not excluded by multidrug resistance P-glycoproteins, dominate glucocorticoid signaling in rodents. In the brain, the steroid products of the HPA-axis bind primarily to two types of receptors belonging to the nuclear receptor family: glucocorticoid receptors (GR) and mineralocorticoid receptors (MR). Cortisol binds to GR with high affinity, as does dexamethasone, a synthetic glucocorticoid that binds to pituitary GR. Corticosterone binds with high affinity to MR (see de Kloet, 2004, for review), which can also serve as a nonselective receptor for cortisol. GR and MR also have differing downstream effects; GR, for instance, can inhibit nuclear factor-B (NF-B), which is thought to underlie the immunosuppressive effects of glucocorticoids (Auphan, DiDonato, Rosette, Helmberg, & Karin, 1995). The affinities of these receptors are actively regulated, for instance, by 11-hydroxysteroid dehydrogenases (11-HSD). 11-HSD type 2 can shield MR from cortisol by converting cortisol to cortisone, which does not bind to MR (Holmes & Seckl, 2006). 11-HSD type 1 can reverse this reaction. The HPA-axis corticosteroids have widespread effects in the brain; although recent evidence has explored apoptosis-inducing (Sapolsky, Krey, & McEwen, 1985) or neurogenesis-suppressing (Fuchs & Gould, 2000) effects of glucocorticoids, a minimal level of glucocorticoid activation of MR is required to prevent cell death in dentate gyrus granule cells (Woolley, Gould, Sakai, Spencer, & McEwen, 1991).

Both basal HPA-axis activity and HPA-axis reactivity to challenge have been proposed as biomarkers of MDD. Basal HPA-axis activity can be measured as urinary, salivary, serum, or CSF levels of cortisol throughout the day, examining both a total area under the curve measure of output and rhythmic changes. HPA-axis reactivity can be measured by exposing individuals to physiological or psychological challenges. One of the most robust elicitors of an HPA-axis response is a psychosocial challenge that is either unpredictable or contains elements of social evaluation (Foley & Kirschbaum, 2010). An alternate challenge is the administration of the synthetic glucocorticoid dexamethasone; the HPA-axis responds to dexamethasone by suppressing output, and failure to suppress is indicative of overactive HPA-axis activity or impaired negative feedback. Administration of CRH has the opposite effect of stimulating ACTH release and cortisol output. Two recent meta-analyses that have examined HPA-axis activity in MDD reached different conclusions. Knorr, Vinberg, Kessing, & Wetterslev (2010) analyzed studies that assessed levels of salivary cortisol in MDD and found that depressed individuals have elevated morning cortisol and somewhat elevated evening cortisol. Knorr et al. also noted, however, a large degree of heterogeneity in studies and findings, and concluded that salivary cortisol is not a reliable marker of MDD. In a comprehensive meta-analysis of multiple HPA-axis assays, Stetler and Miller (2011) concluded that depressed individuals have elevated levels of cortisol, although again there was substantial methodological variability in the reviewed studies. Compared with their nondepressed counterparts, depressed individuals were also found to have moderately elevated ACTH, similar CRH levels, and elevated ACTH and cortisol levels during the dexamethasone suppression test. Discrepancies between these two meta-analyses may be due to the scope of the studies reviewed and to methodological differences in the included studies themselves, such as use of different fluids for assay measurement. Depressed adolescents and children also have dysregulated HPA-axis functioning, including higher basal levels of cortisol, failed suppression during dexamethasone administration, and increased reactivity and delayed recovery to psychosocial stressors (Lopez-Duran, Kovacs, & George, 2009).

Gene expression assays using postmortem tissues support the formulation that MDD is characterized by HPA-axis dysregulation. For example, Wang, Kamphuis, Huitinga, Zhou, & Swaab (2008) found increased hypothalamic expression of CRF mRNA, as well as increased AVP, MR, and sex hormones that regulate HPA-axis activity. AVP and oxytocin neuron numbers are also increased in MDD (Purba, Hoogendijk, Hofman, & Swaab, 1996), although cell counts were also found to be elevated in bipolar disorder. CRH may be a particularly strong biomarker for MDD, as postmortem cell counts of CRH-expressing neurons are increased in the paraventricular nucleus of the hyperthalamus in individuals with MDD, without any obvious lesions or hypertrophy of the actual paraventricular nucleus (Raadsheer, Hoogendijk, Stam, Tilders, & Swaab, 1994). Researchers have also documented a range of disruptions in CRH receptor expression in prefrontal regions (Merali et al., 2004). In addition, polymorphisms in CRH receptor genes have been associated with MDD, further supporting the formulation that HPA dysfunction is a contributing factor to, rather than an epiphenomenon of, MDD (Liu et al., 2006). Based on these findings, CRH receptors have been targets of pharmaceutical interventions in depression, although their efficacy extends to anxiety-related symptoms as well (Holsboer & Ising, 2008).

A. *HPA-axis hormones: Vegetative/homeostatic symptoms*

Different subtypes of depression may present with different profiles of HPA-axis functioning. Stetler and Miller (2011) reviewed several classically used subtypes: atypical depression, characterized by vegetative hypersomnia and hyperphagia, melancholic depression, characterized by hypovegetetative insomnia and hypophagia as well as by volitional deficits, and psychotic depression, characterized by hallucinations or delusions and cognitive symptoms. Although atypical depression was associated with less HPA-axis dysregulation than were other types, melancholic depression was associated with greater HPA-axis dysregulation, and MDD with psychotic features had the highest HPA-axis activity. Although stratification by symptoms is difficult to ascertain, as is the causal direction of the associations between symptom subtypes and HPA-axis dysregulation, the HPA-axis is posited to play a critical role specifically in modulating sleep (Buckley & Schatzberg, 2005) and sleep disturbances (Friess et al., 2008) in MDD. In healthy adults, cortisol administration suppresses REM sleep (Gillin, Jacobs, Fram, & Snyder, 1972), likely due to downstream effects on sleep-disrupting hormones like CRH (Held et al., 2004). In contrast, cortisol does not suppress REM sleep in depressed individuals, suggesting that interactions between the HPA-axis and other neurotransmitter systems responsible for REM sleep regulation are disrupted in MDD (Schmid et al., 2008). Indeed, initial trials with a CRH receptor antagonist found some increases in slow wave sleep in individuals with MDD (Held et al., 2004), consistent with animal models showing amelioration of stress-induced sleep disturbances using this class of drugs (Lancel, Müller-Preuss, Wigger, Landgraf, & Holsboer, 2002).

There is little empirical research examining HPA-axis hormones and appetite or weight symptoms in MDD. Similarities have been noted between MDD and anorexia in HPA-axis dysregulation, and researchers have speculated that appetite disturbance with weight loss in MDD may be associated with HPA-axis dysfunction (Kathol, Jaeckle, Lopez, & Meller, 1989). Conversely, hypoactivity of this stress system has been proposed to underlie hyperphagia and weight changes (Gold & Chrousos, 2002). It is possible that heightened HPA-axis activity, either via endogenous risk factors discussed earlier or by environmental stress, affects appetite directly or via interactions with other appetite-regulating molecules like neuropeptide Y (Adam & Epel, 2007), which itself has been implicated in MDD, anxiety, and bipolar disorder (Heilig, 2004). Given the abundance of neuropeptide modulators of appetite, it is likely that future research will be able to better delineate the relation between HPA-axis dysregulation and homeostatic dysregulation in MDD.

VII. Neurotrophins

The long-held notion that neurogenesis is nonexistent in adult mammals has been challenged in the last several decades (see Ming & Song, 2011, for review). New neurons are generated primarily from the subventricular zone and subgranular zone of the dentate gyrus of the hippocampus (Gage, 2000), with new cells having the capacity to migrate to more distal sites. Because stress is a potent negative regulator of new-cell proliferation in these neurogenesis zones (Mirescu & Gould, 2006), and stress and its associated hormones have been posited to be a precipitating factor in MDD, researchers have implicated dysregulation of adult neurogenesis as a possible neurobiological intersection for stress and MDD (Sahay & Hen, 2007). The neurotrophic theory of depression posits that disrupted neuroplasticity and neurogenesis contributes to the onset of MDD (Pittenger & Duman, 2008). For example, elevated glucocorticoid levels in MDD may lead to suppression of cell proliferation, which may be countered by antidepressants that upregulate neurogenesis (Warner-Schmidt & Duman, 2006). Indeed, SSRI treatment and amelioration of behavioral deficits in animal models of MDD may depend in part on neurogenesis (Santarelli et al., 2003); the role of hippocampal neurogenesis, however, may, in turn, depend on the molecular pathway of vulnerability (Surget et al., 2008). Thus, it is not clear if impaired neurogenesis itself is casually related to MDD, or alternatively, is necessary for antidepressant effects in MDD (Sahay & Hen, 2007).

Regulation of neurogenesis is dependent on a wide range of cellular signaling, involving glutamatergic and GABAergic input from other neurons as well as regulation from growth factors (see (Zhao, Deng, & Gage, 2008), for review). Multiple neurotrophins have been explored in the context of MDD. Insulin-like growth factor-I and fibroblast growth factor both induce neurogenesis (Aberg, Aberg, Hedbäcker, Oscarsson, & Eriksson, 2000), and VGF (Hunsberger et al., 2007) and vascular endothelial growth factor (Warner-Schmidt & Duman, 2007) have also been associated with altered neuroplasticity and antidepressant efficacy. Since BDNF expression was identified as a target of antidepressants (Nibuya, Morinobu, & Duman, 1995), this protein has been implicated not only in drug efficacy, but also as a primary deficit in MDD (see (Duman & Monteggia, 2006; Krishnan & Nestler, 2010), for review). Indeed, BDNF has received considerable attention as a potent mediator of antidepressant effects. A recently meta-analysis found that BDNF levels in serum are lower in individuals with MDD than in healthy controls (Sen, Duman, & Sanacora, 2008), with scores on severity inventories correlating with serum levels and potentially recovering after SSRI treatment (Yoshimura et al., 2007). BDNF, and its receptor TrkB, is a hypothesized pathway of the antidepressant effects of SSRIs and exercise (Li et al., 2008); conversely, BDNF knockdowns produce depression-like behavior in rodent models (Taliaz, Stall, Dar, & Zangen, 2010). Interventions that result in rapid antidepressant action, like the NMDA receptor antagonist ketamine, may act by disinhibiting the expression of BDNF (Autry et al., 2011). Convergent evidence suggests that suicide victims are characterized by both reduced TrkB (Dwivedi et al., 2003) and reduced BDNF levels (Karege, Vaudan, Schwald, Perroud, & La Harpe, 2005). Importantly, reductions in BDNF may be reversed with antidepressant medication (Chen, Dowlatshahi, MacQueen, Wang, & Young, 2001). In addition to the role of BDNF in facilitating neuroplasticity, another possible mechanism of the action of BDNF in MDD is through altered 5-HT circuitry (Pezawas et al., 2008), although more research is necessary to clarify the downstream actions of BDNF, which may include 5-HT-mediated changes in homeostatically regulated functions, such as appetite (Pelleymounter, Cullen, & Wellman, 1995). Notably, similar deficits and antidepressant interactions have been documented with other neurotrophins (Evans et al., 2004).

There have been intriguing reports of associations between BDNF polymorphisms and risk for depression (Verhagen et al., 2010), but these findings have been mixed (Duman & Monteggia, 2006). Examinations of the BDNF val66met single nucleotide polymorphism have indicated that the *val* allele is associated with neuroticism (Sen et al., 2003), as well as the cognitive-affective, but not the vegetative/homestatic, factor of depressive symptoms (Duncan, Hutchison, Carey, & Craighead, 2009). In contrast, the *met* allele has been associated with harm avoidance and comorbid MDD and anxiety (Jiang et al., 2005). These contradictions have not yet been satisfactorily addressed, and many questions remain regarding the relation between this polymorphism and MDD (Groves, 2007). Indeed, BDNF may play a role more generally in both mood and anxiety disorders. In this context, correlations have been reported between BDNF serum levels and symptoms of anxiety (Satomura et al., 2011) and psychomotor functioning (Dell'Osso et al., 2010). The actions of BDNF as a mediator between use-dependent plasticity and homeostatic processes such as sleep (Faraguna, Vyazovskiy, Nelson, Tononi, & Cirelli, 2008) also suggest a role for BDNF in vegetative or homeostatic processes in MDD.

A. Brain-derived neurotrophic factor: Volitional/behavioral symptoms

Researchers have examined BDNF in the NAcc to assess the role of this neurotrophin in volitional behavior. Infusions of BDNF have been found to chronically facilitate reward behavior, including reward-induced motor activity (Horger et al., 1999). The effect of BDNF in this pathway also mediates depression-like behavior (Eisch et al., 2003). Researchers have noted an interesting gender difference in the effect of BDNF: whereas male mice with conditional BDNF knockouts exhibit increased motor activity reminiscent of psychomotor agitation, female BDNF knockout mice show decreased sugar preference, a proposed marker of anhedonia (Monteggia et al., 2007). Indeed, other researchers have found that, under stress inductions, female BDNF knockout mice show reduced locomotor activity in addition to reduced sucrose preference (Autry et al., 2011). Different findings concerning BDNF and behavioral outcomes in animals may depend on the location of BDNF disruption and the behavioral model used (Nestler & Carlezon, 2006). More research, especially with humans, is necessary to determine the potential causal role played by BDNF in specific symptom clusters.

VIII. Cytokines

Cytokines are molecular messengers utilized by cells for communication as part of immune function. They can be broadly categorized by their putative effects on immune function: either promoting cellular response to intracellular pathogenic insults, such as viral antigens, or promoting antibody response to extracellular antigens, such as from bacterial or protozoan insults. The former group is associated with a Th1 (or type 1) pro-inflammatory response, whereas the later group is characterized by a Th2 (or type 2) anti-inflammatory response (discussion of Th17 and other regulatory responses is beyond the scope of this article). Because pro-inflammatory cytokines in particular have been implicated in MDD, we focus on them in this section. Pro-inflammatory cytokines include a range of interleukins (IL), including IL-1α, IL-1β, IL-2, IL-6, and tumor necrosis factor α (TNFα). Cytokines produced in the periphery can affect neural functioning in several ways. Cytokines can directly cross the blood-brain barrier, either by passive diffusion or via active transport by saturation of existing transporters. Cytokines can also affect peripheral nervous tissue that then influences CNS functioning; for instance, cytokines may directly influence vagal nerve firing and downstream modulation of neurotransmitter systems affecting symptoms (Bravo et al., 2011).

Like reserpine, which led to unforeseen depressive symptoms in some individuals, cytokines used for immunological effect, for instance interferon-alpha (IFN-α) for the treatment of hepatitis B (McDonald, Mann, & Thomas, 1987), or IL-2 therapy (Denicoff et al., 1987), resulted in symptoms of depression. Preexisting vulnerability phenotypes may potentiate the influence of cytokines on MDD. For instance, short-allele carriers of SERT are more likely than are long-allele carriers to develop depression during IFN-α therapy

reproductive stages in females (see Soares & Zitek, 2008, for review). Estrogen in particular has been associated with mood disorders (Osterlund, Witt, & Gustafsson, 2005). Yet despite this overall evidence, it has been difficult to link specific symptoms in MDD with estrogen changes. As with monoamines, specific negative mood symptoms developing from hormone challenges are dependent on vulnerability to MDD (Bloch et al., 2000). Furthermore, investigators have posited that estrogen and other gonadal hormones interact with 5-HT and DA systems to produce dysregulation of affective processing in premenstrual dysphoric disorder and other fluctuations in hormonal levels (Rubinow & Schmidt, 2006). Although intriguing but preliminary evidence links polymorphisms in the estrogen receptor with cognitive and affective disturbances (Sundermann, Maki, & Bishop, 2010), it is clear that further work is necessary to elucidate the nature of the relation between sex hormones and symptoms of MDD.

B. Neurotransmitters, neuropeptides, and neuromodulators: Volitional/behavioral symptoms

Orexin, melanin-concentrating hormone, melanocortin, and other hypothalamic neuropeptides all have been proposed to play a role in volitional symptoms in MDD. Neuropeptide Y (NPY) has also been reported to be altered in MDD (Heilig, 2004), and may link stress reactivity and appetitive, homeostatic, and affective symptoms. (Treadway & Zald, 2011) recently highlighted the importance of considering both the anticipatory and the consummatory aspects of anhedonia, and the importance of opioids in the latter. This includes neuropeptides that bind to opioid receptors situated in regions of the brain implicated in reward, including the NAcc. There, they may interact with prefrontal regions to code the magnitude of a reward. Thus, it is reasonable to hypothesize a role for opioids in the etiology of anhedonia. In particular, -endorphins have been implicated in MDD, specifically the volitional and behavioral symptoms (Hegadoren, O'Donnell, Lanius, Coupland, & Lacaze-Masmonteil, 2009), although evidence thus far, while promising, is not conclusive.

C. Neurotransmitters, neuropeptides, and neuromodulators: Vegetative/homeostatic symptoms

Molecular clock molecules in the suprachiasmatic nucleus responsible for maintaining the "master" clock have been implicated in the circadian component of sleep. Although unlikely to account for most cases of MDD, these genes may be dysregulated in this disorder (McClung, 2011). If the molecular clock system is implicated in MDD, it will most likely be linked to circadian rhythm dysfunction and related problems in sleep and energy. Similarly, melatonin, a circadian rhythm hormone that can be synthesized from 5-HT via the intermediary product N-acetylserotonin, has been implicated in mood disorders like seasonal affective disorder, as well as with sleep dysfunction in MDD (Srinivasan et al., 2009). The melatonin receptor agonist agomelatine has been tested as an antidepressant (Lôo, Hale, & D'haenen, 2002), and melatonin may be a promising target for treating vegetative/homeostatic symptoms. In addition, several neuropeptides have recently been investigated using both animal models and preclinical trials. Although specific symptom associations are, thus far, sparse, it is almost certain that these molecules will be the subject of future studies (Rotzinger, Lovejoy, & Tan, 2010).

X. Conclusions and Future Directions

Despite considerable progress over the last half century in our understanding of molecular aspects of MDD, several theoretical and methodological difficulties remain. One issue concerns the nature and specificity of the individual symptoms in MDD. Some symptoms, like fatigue or psychomotor retardation, may cross multiple boundaries. Fatigue, for example, may have cognitive, affective, somatic, motor, or volitional components (Stahl, 2003). Psychomotor retardation encompasses a set of psychological, motor, and motivational deficits that include positive symptoms like increased self-touching as well as negative symptoms or slowing in speech, eye movement, posture, and volitional movement (Buyukdura, McClintock, & Croarkin, 2011). Although we have presented relatively simple links between molecules and symptoms, it is important to recognize that different facets of a single symptom may have multiple molecular origins. The potentially causal order of symptoms in MDD may be the key to understanding the molecular antecedents of this disorder. For instance, the finding that insomnia precedes MDD and remains during remission (Staner, 2010) suggests that this symptom may be an underlying vulnerability factor for MDD. Disentangling these endophenotypes may inform investigators about specific neural circuits, molecular pathways, and inherited factors underlying risk for MDD. To date, no genes or loci have been associated definitively with risk for depression (Shyn & Hamilton, 2010). Although it may be possible that stronger results would be obtained by substituting endophenotypes for classical diagnoses (Gottesman & Gould, 2003), the likelihood of identifying conclusively a specific genetic cause of MDD in the near future is low. Indeed, given the heterogeneity of this disorder, it is not reasonable to expect to find clear genetic risk markers. Nevertheless, researchers may be able to elucidate how genetic, endophenotype, and environmental factors interact to precipitate the onset of MDD.

Localizing molecular dysfunction in the brain may allow researchers to focus their hypotheses and may even provide a diagnostic biomarker for MDD. Although functional magnetic resonance imaging yields remarkable spatial clarity related to neural functioning, only recently have researchers been able to combine this method with concurrent PET imaging to examine specific molecular systems. This is especially important given the disparate localizations of the molecules discussed here—from widely distributed systems like 5-HT and NE, to distributed systems with specific regions of interest, like DA in the NAcc, to localized molecules like hypothalamic peptides. Whereas the challenge for distributed systems is to determine specific neural pathways that contribute to specific symptoms, the challenge for localized molecules is to determine how they may influence wider brain function to produce symptoms of MDD. Investigators can now combine peripheral measures, such as serum cytokine levels, with brain imaging to gain a better understanding of how these systems may interact to produce the symptoms of depression. For molecules that play a large role outside the brain, as in the gut or in peripheral circulation, it is unclear how body systems interact with brain systems to produce the symptoms of depression. Few studies have examined enteric functioning and depression, and studies of molecules in the periphery and brain functioning in MDD are in their infancy. Localization of molecular perturbations outside the brain will be as important as localizing dysfunction in the brain.

Linking animal models with human phenotypes is necessary for understanding the molecular pathways in MDD. Although animal models may never be able to model complex cognitive or affective deficits like excessive guilt, there are excellent models in place for anhedonia and other symptoms of MDD. For example, studies of sucrose consumption in rodents with various experimental manipulations may be designed alongside analogous human studies to yield information about molecular mechanisms of anhedonia. Because animal models do not always produce the predicted phenotype (Haenisch & Bönisch, 2011), it may be difficult to distinguish behaviors that represent MDD compared to other psychopathologies. Yet, animal models provide valuable information about the effects of chronic versus acute disturbances, with a level of molecular specificity and detail that is difficult to obtain in human studies. Newer experimental approaches, like inducible knockouts and remote control of specific molecular circuits, can be used with behavioral measures like social defeat and withdrawal to better model MDD. Collaborations between researchers using animal models and investigators studying human clinical samples will help to ensure a match between animal and human endophenotypes.

Of the many molecules that have been examined in MDD, 5-HT, DA, and HPA-axis hormones have received the most attention to date. Recently, however, interest has expanded to gonadal hormones, neuropeptides and neuromodulators, neurotrophins, cytokines, and almost every known neurotransmitter. Because many of these molecules interact and regulate each other, it is difficult to disentangle cause from effect: if an individual with MDD has elevated cortisol, altered 5-HT receptor binding, altered catecholamine levels, and elevated IL-6, which—if any—of these is related most proximally to the etiology of MDD? Given practical limitations, most investigators have limited molecular studies of MDD to one system at a time. Moreover, human studies have been limited by the use of imperfect psychometric instruments (Bagby, Ryder, Schuller, & Marshall, 2004); similarly, animal studies lack the ability to assess several core symptoms, such as guilt or worthlessness. Unless researchers are able to pinpoint a singular molecular antecedent of MDD, which seems unlikely, we are left having to correlate molecular biomarkers with symptom clusters. By examining these symptom clusters, however, we can devise two possible frameworks going forward for studying the etiology of MDD: a divergent molecular theory and a convergent molecular theory of MDD.

The first possibility is that there is a latent molecular construct that is present in most individuals with MDD, and the divergent sequelae of this molecular foundation is responsible for disruption of secondary molecules and the various symptoms of depression. This does not preclude the involvement of multiple molecules—for instance, a disruption in both 5-HT and DA systems—but it does predict a singular pattern of molecular signatures. Many findings of altered biomarkers in MDD describe distributions of biomarker activity that, while statistically different from healthy individuals, nonetheless overlap with the comparable distribution of "normal" samples, making these biomarkers inefficient as diagnostic tools. The pattern of molecular disruption in MDD is likely to be more subtle and more complex than a simple increase or decrease in a particular molecule. This underlying phenotype may be difficult to detect, because it may manifest in multiple ways. For example, research utilizing twin populations have found that the melancholic MDD subtype is not etiologically separable from other types of MDD but rather, that they represent different severity manifestations of the same latent vulnerability (Kendler, 1997).

This divergent molecular model of MDD yields several predictions. The first is that a single molecular phenotype is necessary and sufficient to cause the various symptom clusters of depression, either directly or by disrupting other molecular systems. The second prediction stems from the consistent observation from multiple molecular systems that it is easier to generate symptoms of MDD by disturbing molecular functioning in individuals who carry a latent vulnerability to MDD than it is in nonvulnerable people (Berman et al., 1999; Ruhe et al., 2007; Leyton et al., 2000). In addition, a rich and complex literature demonstrates that life stressors are an important factor in triggering MDD (Monroe & Harkness, 2005). Thus, the divergent molecular model of MDD predicts that vulnerability factors—either familial and genetic risk factors or scars from past depressive episodes—and environmental stress factors together directly affect the central molecular pattern in MDD. One proposed example is that disruption in 5-HT signaling and 5-HT receptor densities acts as the central molecular pattern of MDD. The 5-HTTLPR polymorphism confers risk for MDD in conjunction with significant life stress like child maltreatment (Caspi et al., 2003; Karg et al., 2011; Risch et al., 2009) and directly affects stress reactivity (Gotlib, Joormann, Minor, & Hallmayer, 2008). Such 5-HT dysfunction may be sufficient to cause cognitive/affective symptoms, vegetative/homeostatic symptoms, and downstream DA disruption that lead to volitional/behavioral symptoms. Laboratory interventions like tryptophan depletion in individuals with 5-HT vulnerability may trigger this central MDD pathway (Booij, Van der Does, & Riedel, 2003). If MDD does have a central pathway, however, it is clear that current antidepressant treatments do not adequately targeting this central pathway (Rush et al., 2006).

The second possibility is that multiple molecular systems are disrupted in MDD, leading to the heterogeneous presentation of this disorder. This theory posits that separate molecular systems, each associated with specific symptoms, are disrupted in MDD. If there is sufficient disruption in multiple interacting systems, the overall symptom presentation may be consistent with the modern diagnosis of MDD. Depending on individual symptom presentation, however, there may be little overlap in the molecular disruption between different individuals with MDD. Depressed individuals may present with completely different patterns of symptoms, which may have at their basis completely different markers of molecular dysfunction. Whereas one depressed individual may have 5-HT dysfunction and elevated inflammatory markers, a second individual may have DA receptor dysfunction and elevated cortisol. Because systems interact, individuals with one molecular dysfunction, for instance 5-HT vulnerability, may then have subsequent dysfunction in other molecular systems, leading to different symptoms at different depressive episodes (Minor et al., 2005). These molecular dysfunctions and their corresponding symptoms, when summed, converge on a common profile of MDD, despite differences in underlying pathophysiology.

This second, convergent model of MDD also makes several predictions. The first is that among a common pool of individuals with a DSM-IV diagnosis of MDD, there may be multiple distinct etiologies. This possibility is troubling for researchers investigating the pathophysiology of MDD. Researchers may not find differences between depressed and healthy individuals because they are actually studying several different syndromes, albeit syndromes that appear similar on the surface. Conversely, investigators may be able to gain traction on the molecular etiology of these depressive syndromes by studying individual symptoms, regardless of DSM-IV diagnoses. Already, investigators are focusing on personality traits, like neuroticism (Takano et al., 2007), dysfunctional attitudes, and other risk factors (Bhagwagar et al.,

2006), and on specific symptoms like anhedonia (Walter et al., 2009). One difficulty with this method is that participants in these studies are likely to also have other symptoms that may preclude molecular specificity. The convergent theory of MDD also predicts that antidepressants that specifically target one molecular system may have only moderate therapeutic efficacy when administered to a sample of depressed individuals with heterogeneous molecular dysfunction. Indeed, recent meta-analyses have shown that SSRIs are less effective for mildly or moderately depressed individuals than for severely depressed persons (Fournier et al., 2010). In addition, the recent STAR*D trial found that even after flexible treatment options between SSRIs and other medications like bupropion, long-term remission rates are unsatisfactory (Rush et al., 2006). Among many possible reasons for these outcomes, the convergent theory predicts that the low rates of remission may be because medications were not tailored to pathophysiologies. In the monoamine depletion paradigm, depressive symptoms are triggered when the depleted neurotransmitter matches the molecular target of the antidepressant used by an individual, suggesting multiple but specific vulnerability pathways (Ruhe et al., 2007).

It is important to note that these theories of molecular dysfunction in MDD are not mutually exclusive. For any given individual, there may be multiple pathways of vulnerability, as in the convergent theory. Once molecular dysfunction occurs, however, interactions among systems may activate other pathways. For instance, an individual may have a 5-HT vulnerability, but interactions between 5-HT and DA in the mesolimbic pathway (Alex & Pehek, 2007) may underlie development of anhedonia. Alternatively, an individual may have several molecular dysfunctions leading to MDD, but on remission, one core symptom, such as insomnia, remains as a marker of remaining dysfunction, such as overactive NE or HPA-axis functioning. These frameworks encourage symptom-specific research that targets individuals with clinically significant impairment in a given symptom, even if they do not meet the overall DSM-IV diagnostic criteria for MDD. This approach should enable greater statistical power in human studies by minimizing the dilution that may occur by comparing healthy controls to a diverse, heterogeneous group of depressed individuals. Symptom-focused research also enables better matching between human endophenotypes and animal behavioral models, leveraging the precision of animal models more than comparisons with a heterogeneous set of symptoms in humans. The danger of this approach is studying, within a particular symptom, a mix of individuals with that symptom without a depressive vulnerability and individuals with that symptom as part of MDD. Exploring the molecular overlap between individuals with specific symptoms and individuals with the same symptoms in the context of MDD may help to determine if these markers are part of a central phenotype of MDD or piecemeal symptoms that may culminate in MDD.

The framework of final common molecular pathways, although likely impractical for MDD, may be feasible for individual symptoms of depression. As investigators move toward conceptualizing function and dysfunction in a continuous manner (Helzer, Kraemer, & Krueger, 2006; Kendler & Gardner, 1998), it will be imperative to link molecular dysfunction to individual symptoms and domains of functioning, rather than a heterogeneous label of "depression."

References

aan het Rot, M., Mathew, S. J., & Charney, D. S. (2009). Neurobiological mechanisms in major depressive disorder. *Canadian Medical Association Journal, 180*(3), 305–313. doi:10.1503/cmaj.080697

Aberg, M. A., Aberg, N. D., Hedbäcker, H., Oscarsson, J., & Eriksson, P. S. (2000). Peripheral infusion of IGF-I selectively induces neurogenesis in the adult rat hippocampus. *The Journal of Neuroscience, 20*(8), 2896–2903.

Adam, T. C., & Epel, E. S. (2007). Stress, eating and the reward system. *Physiology & Behavior, 91*(4), 449–458. doi:10.1016/j.physbeh.2007.04.011

Adolphs, R., Tranel, D., Damasio, H., & Damasio, A. (1994). Impaired recognition of emotion in facial expressions following bilateral damage to the human amygdala. *Nature, 372*(6507), 669–672. doi:10.1038/372669a0

Adrien, J. (2002). Neurobiological bases for the relation between sleep and depression. *Sleep Medicine Reviews, 6*(5), 341–351. doi:10.1053/smrv.2001.0200

Alex, K. D., & Pehek, E. A. (2007). Pharmacologic mechanisms of serotonergic regulation of dopamine neurotransmission. *Pharmacology & Therapeutics, 113*(2), 296–320. doi:10.1016/j.pharmthera.2006.08.004

American Psychiatric Association, & DSM-IV, American Psychiatric Association Task Force on. (2000). *Diagnostic and statistical manual of mental disorders: DSM-IV-TR.* American Psychiatric Association Washington, DC.

Andréasson, A., Arborelius, L., Erlanson-Albertsson, C., & Lekander, M. (2007). A putative role for cytokines in the impaired appetite in depression. *Brain, Behavior, and Immunity, 21*(2), 147–152. doi:16/j.bbi.2006.08.002

Argyelán, M., Szabó, Z., Kanyó, B., Tanács, A., Kovács, Z., Janka, Z., & Pávics, L. (2005). Dopamine transporter availability in medication free and in bupropion treated depression: A 99mTc-TRODAT-1 SPECT study. *Journal of Affective Disorders, 89*(1–3), 115–123. doi:10.1016/j.jad.2005.08.016

Aston-Jones, G, & Bloom, F. E. (1981). Activity of norepinephrine-containing locus coeruleus neurons in behaving rats anticipates fluctuations in the sleep-waking cycle. *The Journal of Neuroscience, 1*(8), 876–886.

Aston-Jones, G., Rajkowski, J., & Cohen, J. (1999). Role of locus coeruleus in attention and behavioral flexibility. *Biological Psychiatry, 46*(9), 1309–1320. doi:16/S0006-3223(99)00140-7

Auphan, N., DiDonato, J. A., Rosette, C., Helmberg, A., & Karin, M. (1995). Immunosuppression by glucocorticoids: Inhibition of NF-kappa B activity through induction of I kappa B synthesis. *Science, 270*(5234), 286–290.

Autry, A. E., Adachi, M., Nosyreva, E., Na, E. S., Los, M. F., Cheng, P., Kavalali, E. T.,... Monteggia, L. M. (2011). NMDA receptor blockade at rest triggers rapid behavioural antidepressant responses. *Nature, 475*(7354), 91–95. doi:10.1038/nature10130

Baeken, C., De Raedt, R., Santermans, L., Zeeuws, D., Vanderhasselt, M. A., Meers, M., & Vanderbruggen, N. (2010). HF-rTMS treatment decreases psychomotor retardation in medication-resistant melancholic depression. *Progress in Neuro-Psychopharmacology & Biological Psychiatry, 34*(4), 684–687. doi:10.1016/j.pnpbp.2010.03.021

Bagby, R. M., Ryder, A. G., Schuller, D. R., & Marshall, M. B. (2004). The Hamilton Depression Rating Scale: Has the gold standard become a lead weight? *The American Journal of Psychiatry, 161*(12), 2163–2177. doi:10.1176/appi.ajp.161.12.2163

Bauer, J., Hohagen, F., Gimmel, E., Bruns, F., Lis, S., Krieger, S.,... Fritsch-Montero, R. (1995). Induction of cytokine synthesis and fever suppresses REM sleep and improves mood in patients with major depression. *Biological Psychiatry, 38*(9), 611–621.

Baylen, C. A., & Rosenberg, H. (2006). A review of the acute subjective effects of MDMA/ecstasy. *Addiction, 101*(7), 933–947. doi:10.1111/j.1360-0443.2006.01423.x

Bech, P., Fava, M., Trivedi, M. H., Wisniewski, S. R., & Rush, A. J. (2011). Factor structure and dimensionality of the two depression scales in STAR*D using level 1 datasets. *Journal of Affective Disorders, 132*(3), 396–400. doi:10.1016/j.jad.2011.03.011

Beers, M. H., & Passman, L. J. (1990). Antihypertensive medications and depression. *Drugs, 40*(6), 792–799.

Bejjani, B. P., Damier, P., Arnulf, I., Thivard, L., Bonnet, A. M., Dormont, D.,... Agid, Y. (1999). Transient acute depression induced by high-frequency deep-brain stimulation. *The New England Journal of Medicine, 340*(19), 1476–1480. doi:10.1056/NEJM199905133401905

Belmaker, R. H., & Agam, G. (2008). Major depressive disorder. *The New England Journal of Medicine, 358*(1), 55–68. doi:10.1056/NEJMra073096

Beneyto, M., Kristiansen, L. V., Oni-Orisan, A., McCullumsmith, R. E., & Meador-Woodruff, J. H. (2007). Abnormal glutamate receptor expression in the medial temporal lobe in schizophrenia and mood disorders. *Neuropsychopharmacology, 32*(9), 1888–1902. doi:10.1038/sj.npp.1301312

Berg, K. A., Harvey, J. A., Spampinato, U., & Clarke, W. P. (2008). Physiological and therapeutic relevance of constitutive activity of 5-HT 2A and 5-HT 2C receptors for the treatment of depression. *Progress in brain research, 172*, 287–305. doi:10.1016/S0079-6123(08)00914-X

Berger, P. A., Faull, K. F., Kilkowski, J., Anderson, P. J., Kraemer, H., Davis, K. L., & Barchas, J. D. (1980). CSF monoamine metabolites in depression and schizophrenia. *The American Journal of Psychiatry, 137*(2), 174–180.

Berman, R. M., Cappiello, A., Anand, A., Oren, D. A., Heninger, G. R., Charney, D. S., & Krystal, J. H. (2000). Antidepressant effects of ketamine in depressed patients. *Biological Psychiatry, 47*(4), 351–354.

Berman, R. M., Narasimhan, M., Miller, H. L., Anand, A., Cappiello, A., Oren, D. A.,... Charney, D. S. (1999). Transient depressive relapse induced by catecholamine depletion: Potential phenotypic vulnerability marker? *Archives of General Psychiatry, 56*(5), 395–403.

Bhagwagar, Z., Hinz, R., Taylor, M., Fancy, S., Cowen, P., & Grasby, P. (2006). Increased 5-HT(2A) receptor binding in euthymic, medication-free patients recovered from depression: a positron emission study with [(11)C]MDL 100,907. *The American Journal of Psychiatry, 163*(9), 1580–1587. doi:10.1176/appi.ajp.163.9.1580

Bhagwagar, Z., Wylezinska, M., Jezzard, P., Evans, J., Ashworth, F., Sule, A.,... Cowen, P. J. (2007). Reduction in occipital cortex gamma-aminobutyric acid concentrations in medication-free recovered unipolar depressed and bipolar subjects. *Biological Psychiatry, 61*(6), 806–812. doi:10.1016/j.biopsych.2006.08.048

Bhatti, T., Gillin, J. C., Seifritz, E., Moore, P., Clark, C., Golshan, S.,... Kelsoe, J. (1998). Effects of a tryptophan-free amino acid drink challenge on normal human sleep electroencephalogram and mood. *Biological Psychiatry, 43*(1), 52–59. doi:10.1016/S0006-3223(97)80252-1

Biegon, A., & Fieldust, S. (1992). Reduced tyrosine hydroxylase immunoreactivity in locus coeruleus of suicide victims. *Synapse, 10*(1), 79–82. doi:10.1002/syn.890100111

Blier, P. (2003). The pharmacology of putative early-onset antidepressant strategies. *European Neuropsychopharmacology, 13*(2), 57–66.

Bloch, M., Schmidt, P. J., Danaceau, M., Murphy, J., Nieman, L., & Rubinow, D. R. (2000). Effects of gonadal steroids in women with a history of postpartum depression. *The American Journal of Psychiatry, 157*(6), 924–930.

Blombery, P. A., Kopin, I. J., Gordon, E. K., Markey, S. P., & Ebert, M. H. (1980). Conversion of MHPG to vanillylmandelic acid. Implications for the importance of urinary MHPG. *Archives of General Psychiatry, 37*(10), 1095–1098.

Blundell, J. E. (1986). Serotonin manipulations and the structure of feeding behaviour. *Appetite, 7 (Suppl)*, 39–56.

Booij, L., Van der Does, A. J. W., & Riedel, W. J. (2003). Monoamine depletion in psychiatric and healthy populations: review. *Molecular Psychiatry, 8*(12), 951–973. doi:10.1038/sj.mp.4001423

Bravo, J. A., Forsythe, P., Chew, M. V., Escaravage, E., Savignac, H. M., Dinan, T. G.,... Cryan, J. F. (2011). Ingestion of Lactobacillus strain regulates emotional behavior and central GABA receptor expression in a mouse via the vagus nerve. *Proceedings of the National Academy of Sciences of the United States of America, 108*(38), 16050–16055. doi:10.1073/pnas.1102999108

Bremner, J. D., Vythilingam, M., Ng, C. K., Vermetten, E., Nazeer, A., Oren, D. A.,... Charney, D. S.

(2003). Regional brain metabolic correlates of alpha-methylparatyrosine-induced depressive symptoms: implications for the neural circuitry of depression. *The Journal of the American Medical Association, 289*(23), 3125–3134. doi:10.1001/jama.289.23.3125

Brown, A. S., & Gershon, S. (1993). Dopamine and depression. *Journal of Neural Transmission, 91*(2–3), 75–109. doi:10.1007/BF01245227

Brunello, N., Blier, P., Judd, L. L., Mendlewicz, J., Nelson, C. J., Souery, D.,... Racagni, G. (2003). Noradrenaline in mood and anxiety disorders: Basic and clinical studies. *International Clinical Psychopharmacology, 18*(4), 191–202. doi:10.1097/01.yic.0000073880.93678.68

Brunswick, D. J., Amsterdam, J. D., Mozley, P. D., & Newberg, A. (2003). Greater availability of brain dopamine transporters in major depression shown by [99m Tc]TRODAT-1 SPECT imaging. *The American Journal of Psychiatry, 160*(10), 1836–1841.

Buckley, T. M., & Schatzberg, A. F. (2005). On the interactions of the hypothalamic-pituitary-adrenal (HPA) axis and sleep: Normal HPA axis activity and circadian rhythm, exemplary sleep disorders. *The Journal of Clinical Endocrinology and Metabolism, 90*(5), 3106–3114. doi:10.1210/jc.2004-1056

Bull, S. J., Huezo-Diaz, P., Binder, E. B., Cubells, J. F., Ranjith, G., Maddock, C.,... Pariante, C. M. (2008). Functional polymorphisms in the interleukin-6 and serotonin transporter genes, and depression and fatigue induced by interferon-[alpha] and ribavirin treatment. *Molecular Psychiatry, 14*(12), 1095–1104.

Buyukdura, J. S., McClintock, S. M., & Croarkin, P. E. (2011). Psychomotor retardation in depression: Biological underpinnings, measurement, and treatment. *Progress in Neuro-Psychopharmacology & Biological Psychiatry, 35*(2), 395–409. doi:10.1016/j.pnpbp.2010.10.019

Cannon, D. M., Klaver, J. M., Peck, S. A., Rallis-Voak, D., Erickson, K., & Drevets, W. C. (2009). Dopamine type-1 receptor binding in major depressive disorder assessed using positron emission tomography and [11C]NNC-112. *Neuropsychopharmacology, 34*(5), 1277–1287. doi:10.1038/npp.2008.194

Capuron, L, Ravaud, A., Neveu, P. J., Miller, A. H., Maes, M., & Dantzer, R. (2002). Association between decreased serum tryptophan concentrations and depressive symptoms in cancer patients undergoing cytokine therapy. *Molecular Psychiatry, 7*(5), 468–473. doi:10.1038/sj.mp.4000995

Capuron, L., Fornwalt, F. B., Knight, B. T., Harvey, P. D., Ninan, P. T., & Miller, A. H. (2009). Does cytokine-induced depression differ from idiopathic major depression in medically healthy individuals? *Journal of Affective Disorders, 119*(1–3), 181–185. doi:16/j.jad.2009.02.017

Capuron, L., Gumnick, J. F., Musselman, D. L., Lawson, D. H., Reemsnyder, A., Nemeroff, C. B., & Miller, A. H. (2002). Neurobehavioral effects of interferon-alpha in cancer patients: Phenomenology and paroxetine responsiveness of symptom dimensions. *Neuropsychopharmacology, 26*(5), 643–652. doi:10.1016/S0893-133X(01)00407-9

Capuron, L., Pagnoni, G., Demetrashvili, M. F., Lawson, D. H., Fornwalt, F. B., Woolwine, B.,... Miller, A. H. (2007). Basal ganglia hypermetabolism and symptoms of fatigue during interferon-alpha therapy. *Neuropsychopharmacology, 32*(11), 2384–2392. doi:10.1038/sj.npp.1301362

Carmichael, M. D., Davis, J. M., Murphy, E. A., Carson, J. A., Van Rooijen, N., Mayer, E., & Ghaffar, A. (2010). Role of brain macrophages on IL-1beta and fatigue following eccentric exercise-induced muscle damage. *Brain, Behavior, and Immunity, 24*(4), 564–568. doi:10.1016/j.bbi.2009.12.011

Caspi, A., Sugden, K., Moffitt, T. E., Taylor, A., Craig, I. W., Harrington, H.,... Poulton, R. (2003). Influence of life stress on depression: Moderation by a polymorphism in the 5-HTT gene. *Science, 301*(5631), 386–389. doi:10.1126/science.1083968

Chaki, S., Yoshikawa, R., Hirota, S., Shimazaki, T., Maeda, M., Kawashima, N., Nakazato, A. (2004). MGS0039: A potent and selective group II metabotropic glutamate receptor antagonist with antidepressant-like activity. *Neuropharmacology, 46*(4), 457–467. doi:10.1016/j.neuropharm.2003.10.009

Chang, C. C., Lu, R. B., Chen, C. L., Chu, C. M., Chang, H. A., Huang, C. C.,... Huang, S.-Y. (2007). Lack of association between the norepinephrine transporter gene and major depression in a Han Chinese population. *Journal of Psychiatry & Neuroscience, 32*(2), 121–128.

Charney, D. S., Woods, S., Goodman, W., & Heninger, G. (1987). Neurobiological mechanisms of panic anxiety: Biochemical and behavioral correlates of yohimbine-induced panic attacks. *The American Journal of Psychiatry, 144*(8), 1030–1036. doi:

Chen, B., Dowlatshahi, D., MacQueen, G. M., Wang, J. F., & Young, L. T. (2001). Increased hippocampal bdnf immunoreactivity in subjects treated with antidepressant medication. *Biological Psychiatry, 50*(4), 260–265. doi:16/S0006-3223(01)01083-6

Choi, D. W. (1987). Ionic dependence of glutamate neurotoxicity. *The Journal of Neuroscience, 7*(2), 369–379.

Choi, M. J., Kang, R. H., Ham, B. J., Jeong, H. Y., & Lee, M. S. (2005). Serotonin receptor 2A gene polymorphism (-1438A/G) and short-term treatment response to citalopram. *Neuropsychobiology, 52*(3), 155–162. doi:10.1159/000087847

Choudary, P. V., Molnar, M., Evans, S. J., Tomita, H., Li, J. Z., Vawter, M. P.,... Jones, E. G. (2005). Altered cortical glutamatergic and GABAergic signal transmission with glial involvement in depression. *Proceedings of the National Academy of Sciences of the United States of America, 102*(43), 15653–15658. doi:10.1073/pnas.0507901102

Christensen, L., & Pettijohn, L. (2001). Mood and carbohydrate cravings. *Appetite, 36*(2), 137–145. doi:06/appe.2001.0390

Chrousos, George P. (2009). Stress and disorders of the stress system. *Nature Reviews Endocrinology, 5*(7), 374–381. doi:10.1038/nrendo.2009.106

Ciliax, B. J., Drash, G. W., Staley, J. K., Haber, S., Mobley, C. J., Miller, G. W.,... (1999). Immunocytochemical localization of the dopamine transporter in human brain. *The Journal of Comparative Neurology, 409*(1), 38–56.

Cleare, A. J., Murray, R. M., & O'Keane, V. (1998). Assessment of serotonergic function in major depression using d-fenfluramine: Relation to clinical variables and antidepressant response. *Biological Psychiatry, 44*(7), 555–561.

Cools, R., Calder, A. J., Lawrence, A. D., Clark, L., Bullmore, E., & Robbins, T. W. (2005). Individual differences in threat sensitivity predict serotonergic modulation of amygdala response to fearful faces. *Psychopharmacology, 180*(4), 670–679. doi:10.1007/s00213-005-2215-5

cues. *Biological Psychiatry, 59*(9), 816–820. doi:16/j.biopsych.2005.10.015
Harmer, C. J., O'Sullivan, U., Favaron, E., Massey-Chase, R., Ayres, R., Reinecke, A., . . . Cowen, P. J. (2009). Effect of acute antidepressant administration on negative affective bias in depressed patients. *The American Journal of Psychiatry, 166*(10), 1178–1184. doi:10.1176/appi.ajp.2009.09020149
Hasler, G., Fromm, S., Carlson, P. J., Luckenbaugh, D. A., Waldeck, T., Geraci, M., . . . Drevets, W. C. (2008). Neural response to catecholamine depletion in unmedicated subjects with major depressive disorder in remission and healthy subjects. *Archives of General Psychiatry, 65*(5), 521–531. doi:10.1001/archpsyc.65.5.521
Hasler, G., Luckenbaugh, D. A., Snow, J., Meyers, N., Waldeck, T., Geraci, M., . . . Drevets, W. C. (2009). Reward processing after catecholamine depletion in unmedicated, remitted subjects with major depressive disorder. *Biological Psychiatry, 66*(3), 201–205. doi:10.1016/j.biopsych.2009.02.029
Hasler, G, & Northoff, G. (2011). Discovering imaging endophenotypes for major depression. *Molecular Psychiatry, 16*(6), 604–619. doi:10.1038/mp.2011.23
Hasler, G., van der Veen, J. W., Tumonis, T., Meyers, N., Shen, J., & Drevets, W. C. (2007). Reduced prefrontal glutamate/glutamine and gamma-aminobutyric acid levels in major depression determined using proton magnetic resonance spectroscopy. *Archives of General Psychiatry, 64*(2), 193–200. doi:10.1001/archpsyc.64.2.193
Haynes, R. C., Jr. (1962). Studies of an vitro effect of glucocorticoids on gluconeogenesis. *Endocrinology, 71*, 399–406.
Hegadoren, K. M., O'Donnell, T., Lanius, R., Coupland, N. J., & Lacaze-Masmonteil, N. (2009). The role of beta-endorphin in the pathophysiology of major depression. *Neuropeptides, 43*(5), 341–353. doi:10.1016/j.npep.2009.06.004
Heilig, M. (2004). The NPY system in stress, anxiety and depression. *Neuropeptides, 38*(4), 213–224. doi:10.1016/j.npep.2004.05.002
Heils, A., Teufel, A., Petri, S., Stöber, G., Riederer, P., Bengel, D., & Lesch, K. P. (1996). Allelic variation of human serotonin transporter gene expression. *Journal of Neurochemistry, 66*(6), 2621–2624.
Held, K., Künzel, H., Ising, M., Schmid, D. A., Zobel, A., Murck, H., . . . Steiger, A. (2004). Treatment with the CRH1-receptor-antagonist R121919 improves sleep-EEG in patients with depression. *Journal of Psychiatric Research, 38*(2), 129–136.
Helzer, J. E., Kraemer, H. C., & Krueger, R. F. (2006). The feasibility and need for dimensional psychiatric diagnoses. *Psychological Medicine, 36*(12), 1671–1680. doi:10.1017/S003329170600821X
Heninger, G. R., Delgado, P. L., & Charney, D. S. (1996). The revised monoamine theory of depression: A modulatory role for monoamines, based on new findings from monoamine depletion experiments in humans. *Pharmacopsychiatry, 29*(1), 2–11. doi:10.1055/s-2007-979535
Henry, J. P., Botton, D., Sagne, C., Isambert, M. F., Desnos, C., Blanchard, V., . . . Gasnier, B. (1994). Biochemistry and molecular biology of the vesicular monoamine transporter from chromaffin granules. *The Journal of Experimental Biology, 196*, 251–262.
Hertz, L., & Zielke, H. R. (2004). Astrocytic control of glutamatergic activity: astrocytes as stars of the show. *Trends in Neurosciences*, 27(12), 735–743. doi:16/j.tins.2004.10.008
Hill, A. J., Weaver, C. F. L., & Blundell, J. E. (1991). Food craving, dietary restraint and mood. *Appetite, 17*(3), 187–197. doi:16/0195-6663(91)90021-J
Hirvonen, J., Karlsson, H., Kajander, J., Markkula, J., Rasi-Hakala, H., Någren, K., . . . Hietala, J. (2008). Striatal dopamine D2 receptors in medication-naive patients with major depressive disorder as assessed with [11C]raclopride PET. *Psychopharmacology, 197*(4), 581–590. doi:10.1007/s00213-008-1088-9
Hjorth, S., Bengtsson, H. J., Kullberg, A., Carlzon, D., Peilot, H., & Auerbach, S. B. (2000). Serotonin autoreceptor function and antidepressant drug action. *Journal of Psychopharmacology, 14*(2), 177–185.
Holmes, M. C., & Seckl, J. R. (2006). The role of 11beta-hydroxysteroid dehydrogenases in the brain. Molecular and Cellular Endocrinology, 248(1–2), 9–14. doi:10.1016/j.mce.2005.12.002
Holsboer, F., & Ising, M. (2008). Central CRH system in depression and anxiety–evidence from clinical studies with CRH1 receptor antagonists. *European Journal of Pharmacology, 583*(2–3), 350–357. doi:10.1016/j.ejphar.2007.12.032
Honig, A., Bartlett, J. R., Bouras, N., & Bridges, P. K. (1988). Amino acid levels in depression: A preliminary investigation. *Journal of Psychiatric Research, 22*(3), 159–164. doi:16/0022-3956(88)90001-5
Horger, B. A., Iyasere, C. A., Berhow, M. T., Messer, C. J., Nestler, E. J., & Taylor, J. R. (1999). Enhancement of locomotor activity and conditioned reward to cocaine by brain-derived neurotrophic factor. *The Journal of Neuroscience, 19*(10), 4110–4122.
Horiuchi, Y., Nakayama, J., Ishiguro, H., Ohtsuki, T., Detera-Wadleigh, S. D., Toyota, T., . . . Arinami, T. (2004). Possible association between a haplotype of the GABA-A receptor alpha 1 subunit gene (GABRA1) and mood disorders. *Biological Psychiatry, 55*(1), 40–45.
Horn, D. I., Yu, C., Steiner, J., Buchmann, J., Kaufmann, J., Osoba, A., . . . Walter, M. (2010). Glutamatergic and resting-state functional connectivity correlates of severity in major depression—the role of pregenual anterior cingulate cortex and anterior insula. *Frontiers in Systems Neuroscience, 4*. doi:10.3389/fnsys.2010.00033
Horst, W. D., & Preskorn, S. H. (1998). Mechanisms of action and clinical characteristics of three atypical antidepressants: venlafaxine, nefazodone, bupropion. *Journal of Affective Disorders, 51*(3), 237–254.
Howren, M. B., Lamkin, D. M., & Suls, J. (2009). Associations of depression with C-reactive protein, IL-1, and IL-6: A meta-analysis. *Psychosomatic Medicine, 71*(2), 171–186. doi:10.1097/PSY.0b013e3181907c1b
Hoyer, D., Hannon, J. P., & Martin, G. R. (2002). Molecular, pharmacological and functional diversity of 5-HT receptors. *Pharmacology Biochemistry and Behavior, 71*(4), 533–554. doi:16/S0091-3057(01)00746-8
Hunsberger, J. G., Newton, S. S., Bennett, A. H., Duman, C. H., Russell, D. S., Salton, S. R., & Duman, R. S. (2007). Antidepressant actions of the exercise-regulated gene VGF. *Nature Medicine, 13*(12), 1476–1482. doi:10.1038/nm1669
Imeri, L, Bianchi, S., & Mancia, M. (1997). Muramyl dipeptide and IL-1 effects on sleep and brain temperature after

inhibition of serotonin synthesis. *The American Journal of Physiology, 273*(5 Pt 2), R1663–1668.

Imeri, L., & Opp, M. R. (2009). How (and why) the immune system makes us sleep. *Nature Reviews Neuroscience, 10*(3), 199–210. doi:10.1038/nrn2576

Irwin, M. R., & Miller, A. H. (2007). Depressive disorders and immunity: 20 years of progress and discovery. *Brain, Behavior, and Immunity, 21*(4), 374–383. doi:10.1016/j.bbi.2007.01.010

Jang, K. L., Livesley, W. J., Taylor, S., Stein, M. B., & Moon, E. C. (2004). Heritability of individual depressive symptoms. *Journal of Affective Disorders, 80*(2–3), 125–133. doi:10.1016/S0165-0327(03)00108-3

Jefferson, J. W., Rush, A. J., Nelson, J. C., VanMeter, S. A., Krishen, A., Hampton, K. D.,...Modell, J. G. (2006). Extended-release bupropion for patients with major depressive disorder presenting with symptoms of reduced energy, pleasure, and interest: Findings from a randomized, double-blind, placebo-controlled study. *The Journal of Clinical Psychiatry, 67*(6), 865–873.

Jiang, X., Xu, K., Hoberman, J., Tian, F., Marko, A. J., Waheed, J. F.,...Lipsky, R. H. (2005). BDNF variation and mood disorders: A novel functional promoter polymorphism and Val66Met are associated with anxiety but have opposing effects. *Neuropsychopharmacology, 30*(7), 1353–1361. doi:10.1038/sj.npp.1300703

Jokela, M., Keltikangas-Järvinen, L., Kivimäki, M., Puttonen, S., Elovainio, M., Rontu, R., & Lehtimäki, T. (2007). Serotonin receptor 2A gene and the influence of childhood maternal nurturance on adulthood depressive symptoms. *Archives of General Psychiatry, 64*(3), 356–360. doi:10.1001/archpsyc.64.3.356

Juengling, F. D., Ebert, D., Gut, O., Engelbrecht, M. A., Rasenack, J., Nitzsche, E. U.,...Lieb, K. (2000). Prefrontal cortical hypometabolism during low-dose interferon alpha treatment. *Psychopharmacology, 152*(4), 383–389. doi:10.1007/s002130000549

Kalsbeek, A., van der Spek, R., Lei, J., Endert, E., Buijs, R. M., & Fliers, E. (2011). Circadian rhythms in the hypothalamo-pituitary-adrenal (HPA) axis. *Molecular and Cellular Endocrinology*. doi:10.1016/j.mce.2011.06.042

Kalueff, A. V., & Nutt, D. J. (2007). Role of GABA in anxiety and depression. *Depression and Anxiety, 24*(7), 495–517. doi:10.1002/da.20262

Karege, F., Vaudan, G., Schwald, M., Perroud, N., & La Harpe, R. (2005). Neurotrophin levels in postmortem brains of suicide victims and the effects of antemortem diagnosis and psychotropic drugs. *Brain Research. Molecular Brain Research, 136*(1–2), 29–37. doi:10.1016/j.molbrainres.2004.12.020

Karg, K., Burmeister, M., Shedden, K., & Sen, S. (2011). The serotonin transporter promoter variant (5-HTTLPR), stress, and depression meta-analysis revisited: Evidence of genetic moderation. *Archives of General Psychiatry, 68*(5), 444–454. doi:10.1001/archgenpsychiatry.2010.189

Karssen, A. M., Meijer, O. C., van der Sandt, I. C., Lucassen, P. J., de Lange, E. C., de Boer, A. G., & de Kloet, E. R. (2001). Multidrug resistance P-glycoprotein hampers the access of cortisol but not of corticosterone to mouse and human brain. *Endocrinology, 142*(6), 2686–2694.

Kathol, R., Jaeckle, R., Lopez, J., & Meller, W. (1989). Pathophysiology of HPA axis abnormalities in patients with major depression: an update. *The American Journal of Psychiatry, 146*(3), 311–317.

Keller-Wood, M. E., & Dallman, M. F. (1984). Corticosteroid inhibition of ACTH secretion. *Endocrine Reviews, 5*(1), 1–24.

Kendler, K. S., & Gardner, C. O., Jr. (1998). Boundaries of major depression: an evaluation of DSM-IV criteria. *The American Journal of Psychiatry, 155*(2), 172–177.

Kendler, K. S. (1997). The diagnostic validity of melancholic major depression in a population-based sample of female twins. *Archives of General Psychiatry, 54*(4), 299–304. doi:10.1001/archpsyc.1997.01830160013002

Kendler, K. S., Thornton, L. M., & Gardner, C. O. (2000). Stressful life events and previous episodes in the etiology of major depression in women: an evaluation of the "kindling" hypothesis. *The American Journal of Psychiatry, 157*(8), 1243–1251. doi:10.1176/appi.ajp.157.8.1243

Klempan, T. A., Sequeira, A., Canetti, L., Lalovic, A., Ernst, C., ffrench-Mullen, J., & Turecki, G. (2009). Altered expression of genes involved in ATP biosynthesis and GABAergic neurotransmission in the ventral prefrontal cortex of suicides with and without major depression. *Molecular Psychiatry, 14*(2), 175–189. doi:10.1038/sj.mp.4002110

Klimek, V., Schenck, J. E., Han, H., Stockmeier, C. A., & Ordway, G. A. (2002). Dopaminergic abnormalities in amygdaloid nuclei in major depression: a postmortem study. *Biological Psychiatry, 52*(7), 740–748.

Klimek, V., Stockmeier, C., Overholser, J., Meltzer, H. Y., Kalka, S., Dilley, G., & Ordway, G. A. (1997). Reduced levels of norepinephrine transporters in the locus coeruleus in major depression. *The Journal of Neuroscience, 17*(21), 8451–8458.

Knorr, U., Vinberg, M., Kessing, L. V., & Wetterslev, J. (2010). Salivary cortisol in depressed patients versus control persons: A systematic review and meta-analysis. *Psychoneuroendocrinology, 35*(9), 1275–1286. doi:10.1016/j.psyneuen.2010.04.001

Kobayashi, K., Ikeda, Y., Sakai, A., Yamasaki, N., Haneda, E., Miyakawa, T., & Suzuki, H. (2010). Reversal of hippocampal neuronal maturation by serotonergic antidepressants. *Proceedings of the National Academy of Sciences of the United States of America, 107*(18), 8434–8439. doi:10.1073/pnas.0912690107

Krishnan, V., & Nestler, E. J. (2008). The molecular neurobiology of depression. *Nature, 455*(7215), 894–902. doi:10.1038/nature07455

Krishnan, V., & Nestler, E. J. (2010). Linking molecules to mood: New insight into the biology of depression. *The American Journal of Psychiatry, 167*(11), 1305–1320. doi:10.1176/appi.ajp.2009.10030434

Krueger, J. M., Obál, F. J., Fang, J., Kubota, T., & Taishi, P. (2001). The role of cytokines in physiological sleep regulation. *Annals of the New York Academy of Sciences, 933*, 211–221.

Kuenzel, H. E., Murck, H., Held, K., Ziegenbein, M., & Steiger, A. (2004). Reboxetine induces similar sleep-EEG changes like SSRI's in patients with depression. *Pharmacopsychiatry, 37*(5), 193–195. doi:10.1055/s-2004-827242

Lacasse, J. R., & Leo, J. (2005). Serotonin and depression: A disconnect between the advertisements and the scientific literature. *PLoS Medicine, 2*(12), e392. doi:10.1371/journal.pmed.0020392

Laje, G., Paddock, S., Manji, H., Rush, A. J., Wilson, A. F., Charney, D., & McMahon, F. J. (2007). Genetic markers

of suicidal ideation emerging during citalopram treatment of major depression. *The American Journal of Psychiatry, 164*(10), 1530–1538. doi:10.1176/appi.ajp.2007.06122018
Lambert, G., Johansson, M., Agren, H., & Friberg, P. (2000). Reduced brain norepinephrine and dopamine release in treatment-refractory depressive illness: Evidence in support of the catecholamine hypothesis of mood disorders. *Archives of General Psychiatry, 57*(8), 787–793.
Lancel, M., Müller-Preuss, P., Wigger, A., Landgraf, R., & Holsboer, F. (2002). The CRH1 receptor antagonist R121919 attenuates stress-elicited sleep disturbances in rats, particularly in those with high innate anxiety. *Journal of Psychiatric Research, 36*, 197–208. doi:10.1016/S0022-3956(02)00009-2
Larsen, M. B., Sonders, M. S., Mortensen, O. V., Larson, G. A., Zahniser, N. R., & Amara, S. G. (2011). Dopamine transport by the serotonin transporter: A mechanistically distinct mode of substrate translocation. *The Journal of Neuroscience, 31*(17), 6605–6615. doi:10.1523/JNEUROSCI.0576-11.2011
Leibowitz, S. F., & Alexander, J. T. (1998). Hypothalamic serotonin in control of eating behavior, meal size, and body weight. *Biological Psychiatry, 44*(9), 851–864.
Lemke, M. R., Brecht, H. M., Koester, J., & Reichmann, H. (2006). Effects of the dopamine agonist pramipexole on depression, anhedonia and motor functioning in Parkinson's disease. *Journal of the Neurological Sciences, 248*(1–2), 266–270. doi:16/j.jns.2006.05.024
Lesch, K. P., Disselkamp-Tietze, J., & Schmidtke, A. (1990). 5-HT1A receptor function in depression: effect of chronic amitriptyline treatment. *Journal of Neural Transmission, 80*(2), 157–161. doi:10.1007/BF01257081
Levinson, D. F. (2009). Genetics of Major Depression. In Gotlib, I. H. & Hammen, C. (Eds.), *Handbook of depression* (2nd ed., pp. 165–186). New York: Guilford Press.
Lewis, G., Mulligan, J., Wiles, N., Cowen, P., Craddock, N., Ikeda, M.,...Peters, T. J. (2011). Polymorphism of the 5-HT transporter and response to antidepressants: Randomised controlled trial. *The British journal of psychiatry: The journal of mental science, 198*(6), 464–471. doi:10.1192/bjp.bp.110.082727
Leyton, M., Ghadirian, A. M., Young, S. N., Palmour, R. M., Blier, P., Helmers, K. F., & Benkelfat, C. (2000). Depressive relapse following acute tryptophan depletion in patients with major depressive disorder. *Journal of Psychopharmacology, 14*(3), 284–287.
Li, Y., Luikart, B. W., Birnbaum, S., Chen, J., Kwon, C.-H., Kernie, S. G.,...Parada, L. F. (2008). TrkB regulates hippocampal neurogenesis and governs sensitivity to antidepressive treatment. *Neuron, 59*(3), 399–412. doi:10.1016/j.neuron.2008.06.023
Lidow, M. S., Goldman-Rakic, P. S., Gallager, D. W., & Rakic, P. (1991). Distribution of dopaminergic receptors in the primate cerebral cortex: Quantitative autoradiographic analysis using [3H]raclopride, [3H]spiperone and [3H] SCH23390. *Neuroscience, 40*(3), 657–671.
Liu, Z., Zhu, F., Wang, G., Xiao, Z., Wang, H., Tang, J.,...Li, W. (2006). Association of corticotropin-releasing hormone receptorl gene SNP and haplotype with major depression. *Neuroscience Letters, 404*(3), 358–362. doi:10.1016/j.neulet.2006.06.016
Long, C., Katzin, B., & Fry, E. G. (1940). The adrenal cortex and carbohydrate metabolism. *Endocrinology, 26*(2), 309.
Loo, C. K., Sachdev, P., Mitchell, P. B., Gandevia, S. C., Malhi, G. S., Todd, G., & Taylor, J. L. (2008). A study using transcranial magnetic stimulation to investigate motor mechanisms in psychomotor retardation in depression. *The International Journal of Neuropsychopharmacology, 11*(7), 935–946. doi:10.1017/S1461145708008821
Lôo, H., Hale, A., & D'haenen, H. (2002). Determination of the dose of agomelatine, a melatoninergic agonist and selective 5-HT(2C) antagonist, in the treatment of major depressive disorder: A placebo-controlled dose range study. *International Clinical Psychopharmacology, 17*(5), 239–247.
Lopez-Duran, N. L., Kovacs, M., & George, C. J. (2009). Hypothalamic-pituitary-adrenal axis dysregulation in depressed children and adolescents: A meta-analysis. *Psychoneuroendocrinology, 34*(9), 1272–1283. doi:10.1016/j.psyneuen.2009.03.016
Lotrich, F. E., Ferrell, R. E., Rabinovitz, M., & Pollock, B. G. (2009). Risk for depression during interferon-alpha treatment is affected by the serotonin transporter polymorphism. *Biological Psychiatry, 65*(4), 344–348. doi:10.1016/j.biopsych.2008.08.009
Löw, K., Crestani, F., Keist, R., Benke, D., Brünig, I., Benson, J. A.,...Rudolph, U. (2000). Molecular and neuronal substrate for the selective attenuation of anxiety. *Science, 290*(5489), 131–134.
Luscher, B., Shen, Q., & Sahir, N. (2011). The GABAergic deficit hypothesis of major depressive disorder. *Molecular Psychiatry, 16*(4), 383–406.
Machado-Vieira, R., Manji, H. K., & Zarate, C. A. (2009). The role of the tripartite glutamatergic synapse in the pathophysiology and therapeutics of mood disorders. *The Neuroscientist, 15*, 525–539. doi:10.1177/1073858409336093
Majer, M., Welberg, L. A. M., Capuron, L., Pagnoni, G., Raison, C. L., & Miller, A. H. (2008). IFN-alpha-induced motor slowing is associated with increased depression and fatigue in patients with chronic hepatitis C. *Brain, Behavior, and Immunity, 22*(6), 870–880. doi:10.1016/j.bbi.2007.12.009
Malberg, J. E., Eisch, A. J., Nestler, E. J., & Duman, R. S. (2000). Chronic antidepressant treatment increases neurogenesis in adult rat hippocampus. *The Journal of Neuroscience, 20*(24), 9104–9110.
Mann, J. J., Huang, Y. Y., Underwood, M. D., Kassir, S. A., Oppenheim, S., Kelly, T. M.,...Arango, V. (2000). A serotonin transporter gene promoter polymorphism (5-HTTLPR) and prefrontal cortical binding in major depression and suicide. *Archives of General Psychiatry, 57*(8), 729–738.
Mann, J. J., Stanley, M., McBride, P. A., & McEwen, B. S. (1986). Increased serotonin2 and beta-adrenergic receptor binding in the frontal cortices of suicide victims. *Archives of General Psychiatry, 43*(10), 954–959.
Marsden, W. N. (2011). Stressor-induced NMDAR dysfunction as a unifying hypothesis for the aetiology, pathogenesis and comorbidity of clinical depression. *Medical Hypotheses, 77*(4), 508–528. doi:10.1016/j.mehy.2011.06.021
Massat, I., Souery, D., Del-Favero, J., Oruc, L., Noethen, M. M., Blackwood, D.,...Mendlewicz, J. (2002). Excess of allele1 for alpha3 subunit GABA receptor gene (GABRA3) in bipolar patients: A multicentric association study. *Molecular Psychiatry, 7*(2), 201–207. doi:10.1038/sj.mp.4000953

Mathers, C. D., & Loncar, D. (2006). Projections of global mortality and burden of disease from 2002 to 2030. *PLoS Medicine, 3*(11), e442. doi:10.1371/journal.pmed.0030442

McClintock, S. M., Husain, M. M., Wisniewski, S. R., Nierenberg, A. A., Stewart, J. W., Trivedi,... Rush, A. J. (2011). Residual symptoms in depressed outpatients who respond by 50% but do not remit to antidepressant medication. *Journal of Clinical Psychopharmacology, 31*(2), 180–186. doi:10.1097/JCP.0b013e31820ebd2c

McClung, C. A. (2011). Circadian rhythms and mood regulation: Insights from pre-clinical models. *European Neuropsychopharmacology, 21 Suppl 4*, S683–693. doi:10.1016/j.euroneuro.2011.07.008

McDonald, E. M., Mann, A. H., & Thomas, H. C. (1987). Interferons as mediators of psychiatric morbidity. An investigation in a trial of recombinant alpha-interferon in hepatitis-B carriers. *Lancet, 2*(8569), 1175–1178.

McLean, P. G., Borman, R. A., & Lee, K. (2007). 5-HT in the enteric nervous system: gut function and neuropharmacology. *Trends in Neurosciences, 30*(1), 9–13. doi:16/j.tins.2006.11.002

McMahon, F. J., Buervenich, S., Charney, D., Lipsky, R., Rush, A. J., Wilson, A. F.,... Manji, H. (2006). Variation in the gene encoding the serotonin 2A receptor is associated with outcome of antidepressant treatment. *American Journal of Human Genetics, 78*(5), 804–814. doi:10.1086/503820

Meador-Woodruff, J. H., Damask, S. P., Wang, J., Haroutunian, V., Davis, K. L., & Watson, S. J. (1996). Dopamine receptor mRNA expression in human striatum and neocortex. *Neuropsychopharmacology, 15*(1), 17–29. doi:10.1016/0893-133X(95)00150-C

Meana, J. J., Barturen, F., & García-Sevilla, J. A. (1992). Alpha 2-adrenoceptors in the brain of suicide victims: Increased receptor density associated with major depression. *Biological Psychiatry, 31*(5), 471–490.

Meeusen, R., Watson, P., Hasegawa, H., Roelands, B., & Piacentini, M. F. (2006). Central fatigue: The serotonin hypothesis and beyond. *Sports Medicine, 36*(10), 881–909.

Mendelsohn, D., Riedel, W. J., & Sambeth, A. (2009). Effects of acute tryptophan depletion on memory, attention and executive functions: A systematic review. *Neuroscience and Biobehavioral Reviews, 33*(6), 926–952. doi:10.1016/j.neubiorev.2009.03.006

Merali, Z., Brennan, K., Brau, P., & Anisman, H. (2003). Dissociating anorexia and anhedonia elicited by interleukin-1beta: Antidepressant and gender effects on responding for "free chow" and "earned" sucrose intake. Psychopharmacology, 165(4), 413–418. doi:10.1007/s00213-002-1273-1

Merali, Z., Du, L., Hrdina, P., Palkovits, M., Faludi, G., Poulter, M. O., & Anisman, H. (2004). Dysregulation in the suicide brain: mRNA expression of corticotropin-releasing hormone receptors and GABA(A) receptor subunits in frontal cortical brain region. *The Journal of Neuroscience, 24*(6), 1478–1485. doi:10.1523/JNEUROSCI.4734-03.2004

Meyer, J. H. (2007). Imaging the serotonin transporter during major depressive disorder and antidepressant treatment. *Journal of Psychiatry & Neuroscience: JPN, 32*(2), 86–102.

Meyer, J. H., Kapur, S., Eisfeld, B., Brown, G. M., Houle, S., DaSilva, J.,... Kennedy, S. H. (2001). The effect of paroxetine on 5-HT(2A) receptors in depression: an [(18)F] setoperone PET imaging study. *The American Journal of Psychiatry, 158*(1), 78–85.

Meyer, J. H., Krüger, S., Wilson, A. A., Christensen, B. K., Goulding, V. S., Schaffer, A.,... Kennedy, S. H. (2001). Lower dopamine transporter binding potential in striatum during depression. *Neuroreport, 12*(18), 4121–4125.

Meyer, J. H., McMain, S., Kennedy, S. H., Korman, L., Brown, G. M., DaSilva, J. N.,... Links, P. (2003). Dysfunctional attitudes and 5-HT2 receptors during depression and self-harm. *The American Journal of Psychiatry, 160*(1), 90–99.

Milak, M. S., Parsey, R. V., Keilp, J., Oquendo, M. A., Malone, K. M., & Mann, J. J. (2005). Neuroanatomic correlates of psychopathologic components of major depressive disorder. *Archives of General Psychiatry, 62*(4), 397–408. doi:10.1001/archpsyc.62.4.397

Ming, G. L., & Song, H. (2011). Adult neurogenesis in the mammalian brain: significant answers and significant questions. *Neuron, 70*(4), 687–702. doi:16/j.neuron.2011.05.001

Minor, K. L., Champion, J. E., & Gotlib, I. H. (2005). Stability of DSM-IV criterion symptoms for major depressive disorder. *Journal of Psychiatric Research, 39*(4), 415–420. doi:10.1016/j.jpsychires.2004.11.001

Mirescu, C., & Gould, E. (2006). Stress and adult neurogenesis. *Hippocampus, 16*(3), 233–238. doi:10.1002/hipo.20155

Monroe, S. M., & Harkness, K. L. (2005). Life stress, the "kindling" hypothesis, and the recurrence of depression: considerations from a life stress perspective. *Psychological Review, 112*, 417–445. doi:10.1037/0033-295X.112.2.417

Monteggia, L. M., Luikart, B., Barrot, M., Theobold, D., Malkovska, I., Nef, S., Parada, L. F., et al. (2007). Brain-derived neurotrophic factor conditional knockouts show gender differences in depression-related behaviors. *Biological Psychiatry, 61*(2), 187–197. doi:10.1016/j.biopsych.2006.03.021

Moore, P., Gillin, J. C., Bhatti, T., DeModena, A., Seifritz, E., Clark, C.,... Kelsoe, J. (1998). Rapid tryptophan depletion, sleep electroencephalogram, and mood in men with remitted depression on serotonin reuptake inhibitors. *Archives of General Psychiatry, 55*(6), 534–539. doi:10.1001/archpsyc.55.6.534

Morón, J. A., Brockington, A., Wise, R. A., Rocha, B. A., & Hope, B. T. (2002). Dopamine uptake through the norepinephrine transporter in brain regions with low levels of the dopamine transporter: Evidence from knock-out mouse lines. *The Journal of Neuroscience, 22*(2), 389–395.

Morris, J. S., Smith, K. A., Cowen, P. J., Friston, K. J., & Dolan, R. J. (1999). Covariation of activity in habenula and dorsal raphé nuclei following tryptophan depletion. *NeuroImage, 10*(2), 163–172. doi:10.1006/nimg.1999.0455

Motivala, S. J., Sarfatti, A., Olmos, L., & Irwin, M. R. (2005). Inflammatory markers and sleep disturbance in major depression. *Psychosomatic Medicine, 67*(2), 187–194. doi:10.1097/01.psy.0000149259.72488.09

Muller, N., Schwarz, M. J., Dehning, S., Douhe, A., Cerovecki, A., Goldstein-Muller, B.,... Riedel, M. (2006). The cyclooxygenase-2 inhibitor celecoxib has therapeutic effects in major depression: results of a double-blind, randomized, placebo controlled, add-on pilot study to reboxetine. *Molecular Psychiatry, 11*(7), 680–684.

Munafò, M. R., Brown, S. M., & Hariri, A. R. (2008). Serotonin transporter (5-HTTLPR) genotype and amygdala activation: a meta-analysis. *Biological Psychiatry, 63*(9), 852–857. doi:10.1016/j.biopsych.2007.08.016

Narita, H., Odawara, T., Iseki, E., Kosaka, K., & Hirayasu, Y. (2004). Psychomotor retardation correlates with frontal hypoperfusion and the Modified Stroop Test in patients under 60-years-old with major depression. *Psychiatry and Clinical Neurosciences, 58*(4), 389–395. doi:10.1111/j.1440-1819.2004.01273.x

Nelson, D. L., & Gehlert, D. R. (2006). Central nervous system biogenic amine targets for control of appetite and energy expenditure. *Endocrine, 29*, 49–60. doi:10.1385/ENDO:29:1:49

Nestler, E. J., & Carlezon, W. A., Jr. (2006). The mesolimbic dopamine reward circuit in depression. *Biological Psychiatry, 59*(12), 1151–1159. doi:10.1016/j.biopsych.2005.09.018

Nibuya, M., Morinobu, S., & Duman, R. S. (1995). Regulation of BDNF and trkB mRNA in rat brain by chronic electroconvulsive seizure and antidepressant drug treatments. *The Journal of Neuroscience, 15*(11), 7539–7547.

Nierenberg, A. A., Husain, M. M., Trivedi, M. H., Fava, M., Warden, D., Wisniewski, S. R.,... Rush, A. J. (2010). Residual symptoms after remission of major depressive disorder with citalopram and risk of relapse: a STAR*D report. *Psychological Medicine, 40*(1), 41–50. doi:10.1017/S0033291709006011

Northoff, G., Walter, M., Schulte, R. F., Beck, J., Dydak, U., Henning, A.,... Boesiger, P. (2007). GABA concentrations in the human anterior cingulate cortex predict negative BOLD responses in fMRI. *Nature Neuroscience, 10*(12), 1515–1517. doi:10.1038/nn2001

Northoff, G., Witzel, T., Richter, A., Gessner, M., Schlagenhauf, F., Fell, J.,... Heinze, H.-J. (2002). GABA-ergic modulation of prefrontal spatio-temporal activation pattern during emotional processing: A combined fMRI/MEG study with placebo and lorazepam. *Journal of Cognitive Neuroscience, 14*(3), 348–370. doi:10.1162/089892902317361895

Ordway, G. A., Widdowson, P. S., Smith, K. S., & Halaris, A. (1994). Agonist binding to alpha 2-adrenoceptors is elevated in the locus coeruleus from victims of suicide. *Journal of Neurochemistry, 63*(2), 617–624.

Osterlund, M. K., Witt, M. R., & Gustafsson, J. A. (2005). Estrogen action in mood and neurodegenerative disorders: estrogenic compounds with selective properties-the next generation of therapeutics. *Endocrine, 28*(3), 235–242. doi:10.1385/ENDO:28:3:235

Ott, G. E., Rao, U., Nuccio, I., Lin, K. M., & Poland, R. E. (2002). Effect of bupropion-SR on REM sleep: Relationship to antidepressant response. *Psychopharmacology, 165*(1), 29–36. doi:10.1007/s00213-002-1165-4

Paddock, S., Laje, G., Charney, D., Rush, A. J., Wilson, A. F., Sorant, A. J. M.,... McMahon, F. J. (2007). Association of GRIK4 with outcome of antidepressant treatment in the STAR*D cohort. *The American Journal of Psychiatry, 164*(8), 1181–1188. doi:10.1176/appi.ajp.2007.06111790

Papakostas, G. I., Petersen, T. J., Burns, A. M., & Fava, M. (2006). Adjunctive atomoxetine for residual fatigue in major depressive disorder. *Journal of Psychiatric Research, 40*(4), 370–373. doi:10.1016/j.jpsychires.2005.04.013

Pardridge, W. M., & Mietus, L. J. (1979). Transport of steroid hormones through the rat blood-brain barrier. *Journal of Clinical Investigation, 64*(1), 145–154.

Paul, I. A., & Skolnick, P. (2003). Glutamate and depression. *Annals of the New York Academy of Sciences, 1003*(1), 250–272. doi:10.1196/annals.1300.016

Pelleymounter, M. A., Cullen, M. J., & Wellman, C. L. (1995). Characteristics of BDNF-induced weight loss. *Experimental Neurology, 131*(2), 229–238. doi:16/0014-4886(95)90045-4

Petroff, O. A. C. (2002). GABA and glutamate in the human brain. *The Neuroscientist, 8*(6), 562–573.

Petty, F., & Sherman, A. D. (1984). Plasma GABA levels in psychiatric illness. *Journal of Affective Disorders, 6*(2), 131–138. doi:16/0165-0327(84)90018-1

Pezawas, L., Meyer-Lindenberg, A., Goldman, A. L., Verchinski, B. A., Chen, G., Kolachana, B. S.,... Weinberger, D. R. (2008). Evidence of biologic epistasis between BDNF and SLC6A4 and implications for depression. *Molecular Psychiatry, 13*(7), 709–716. doi:10.1038/mp.2008.32

Pittenger, C., & Duman, R. S. (2008). Stress, depression, and neuroplasticity: a convergence of mechanisms. *Neuropsychopharmacology, 33*(1), 88–109. doi:10.1038/sj.npp.1301574

Plata-Salamán, C. R. (1998). Cytokine-induced anorexia. Behavioral, cellular, and molecular mechanisms. *Annals of the New York Academy of Sciences, 856*, 160–170.

Prather, A. A., Rabinovitz, M., Pollock, B. G., & Lotrich, F. E. (2009). Cytokine-induced depression during IFN-alpha treatment: The role of IL-6 and sleep quality. *Brain, Behavior, and Immunity, 23*(8), 1109–1116. doi:10.1016/j.bbi.2009.07.001

Price, R. B., Nock, M. K., Charney, D. S., & Mathew, S. J. (2009). Effects of intravenous ketamine on explicit and implicit measures of suicidality in treatment-resistant depression. *Biological Psychiatry, 66*(5), 522–526. doi:10.1016/j.biopsych.2009.04.029

Purba, J. S., Hoogendijk, W. J. G., Hofman, M. A., & Swaab, D. F. (1996). Increased number of vasopressin- and oxytocin-expressing neurons in the paraventricular nucleus of the hypothalamus in depression. *Archives of General Psychiatry, 53*(2), 137–143. doi:10.1001/archpsyc.1996.01830020055007

Quetsch, R. M., Achor, R. W., Litin, E. M., & Faucett, R. L. (1959). Depressive reactions in hypertensive patients: A comparison of those treated with Rauwolfia and those receiving no specific antihypertensive treatment. *Circulation, 19*(3), 366–375.

Raadsheer, F. C., Hoogendijk, W. J. G., Stam, F. C., Tilders, F. J. H., & Swaab, D. F. (1994). Increased numbers of corticotropin-releasing hormone expressing neurons in the hypothalamic paraventricular nucleus of depressed patients. *Neuroendocrinology, 60*, 436–444. doi:10.1159/000126778

Raison, C. L., Rye, D. B., Woolwine, B. J., Vogt, G. J., Bautista, B. M., Spivey, J. R., & Miller, A. H. (2010). Chronic interferon-alpha administration disrupts sleep continuity and depth in patients with hepatitis C: Association with fatigue, motor slowing, and increased evening cortisol. *Biological Psychiatry, 68*(10), 942–949. doi:10.1016/j.biopsych.2010.04.019

Reijnders, J. S. A., Ehrt, U., Weber, W. E. J., Aarsland, D., & Leentjens, A. F. G. (2008). A systematic review of prevalence studies of depression in Parkinson's disease. *Movement Disorders, 23*(2), 183–189. doi:10.1002/mds.21803

Rijnbeek, B., de Visser, S. J., Franson, K. L., Cohen, A. F., & van Gerven, J. M. A. (2003). REM sleep effects as a biomarker for the effects of antidepressants in healthy volunteers. *Journal of Psychopharmacology, 17*(2), 196–203.

Risch, N., Herrell, R., Lehner, T., Liang, K.-Y., Eaves, L., Hoh, J.,... Merikangas, K. R. (2009). Interaction between the serotonin transporter gene (5-HTTLPR), stressful life events, and risk of depression. *The Journal of the American Medical Association*, 301(23), 2462–2471. doi:10.1001/jama.2009.878

Robinson, O. J., Cools, R., Crockett, M. J., & Sahakian, B. J. (2010). Mood state moderates the role of serotonin in cognitive biases. *Journal of Psychopharmacology, 24*(4), 573–583. doi:10.1177/0269881108100257

Rotzinger, S., Lovejoy, D. A., & Tan, L. A. (2010). Behavioral effects of neuropeptides in rodent models of depression and anxiety. *Peptides, 31*(4), 736–756. doi:10.1016/j.peptides.2009.12.015

Rubinow, D. R., & Schmidt, P. J. (2006). Gonadal steroid regulation of mood: The lessons of premenstrual syndrome. *Frontiers in Neuroendocrinology, 27*(2), 210–216. doi:10.1016/j.yfrne.2006.02.003

Ruhe, H. G., Mason, N. S., & Schene, A. H. (2007). Mood is indirectly related to serotonin, norepinephrine and dopamine levels in humans: A meta-analysis of monoamine depletion studies. *Molecular Psychiatry, 12*(4), 331–359.

Rumpel, S., LeDoux, J., Zador, A., & Malinow, R. (2005). Postsynaptic receptor trafficking underlying a form of associative learning. *Science, 308*(5718), 83–88. doi:10.1126/science.1103944

Rush, A. J. (2007). The varied clinical presentations of major depressive disorder. *The Journal of Clinical Psychiatry, 68 Suppl 8*, 4–10.

Rush, A. J., Trivedi, M. H., Wisniewski, S. R., Nierenberg, A. A., Stewart, J. W., Warden, D., . . . Fava, M. (2006). Acute and longer-term outcomes in depressed outpatients requiring one or several treatment steps: A STAR*D report. *The American Journal of Psychiatry, 163*, 1905–1917. doi:10.1176/appi.ajp.163.11.1905

Ryu, S. H., Lee, S. H., Lee, H. J., Cha, J. H., Ham, B. J., Han, C. S., . . . Lee, M.-S. (2004). Association between norepinephrine transporter gene polymorphism and major depression. *Neuropsychobiology, 49*(4), 174–177. doi:10.1159/000077361

Sabbe, B., van Hoof, J., Hulstijn, W., & Zitman, F. (1996). Changes in fine motor retardation in depressed patients treated with fluoxetine. *Journal of Affective Disorders, 40*(3), 149–157. doi:16/0165-0327(96)00052-3

Sabbe, B., van Hoof, J., Hulstijn, W., & Zitman, F. (1997). Depressive retardation and treatment with fluoxetine: Assessment of the motor component. *Journal of Affective Disorders, 43*(1), 53–61. doi:16/S0165-0327(96)01409-7

Sahay, A., & Hen, R. (2007). Adult hippocampal neurogenesis in depression. *Nature Neuroscience, 10*(9), 1110–1115. doi:10.1038/nn1969

Salamone, J. D., Correa, M., Mingote, S. M., & Weber, S. M. (2005). Beyond the reward hypothesis: Alternative functions of nucleus accumbens dopamine. *Current Opinion in Pharmacology, 5*(1), 34–41. doi:10.1016/j.coph.2004.09.004

Sanacora, G., Mason, G. F., Rothman, D. L., Behar, K. L., Hyder, F., Petroff, O. A., . . . (1999). Reduced cortical gamma-aminobutyric acid levels in depressed patients determined by proton magnetic resonance spectroscopy. *Archives of General Psychiatry, 56*(11), 1043–1047.

Sanacora, G., Gueorguieva, R., Epperson, C. N., Wu, Y. T., Appel, M., Rothman, D. L., . . . Mason, G. F. (2004). Subtype-specific alterations of gamma-aminobutyric acid and glutamate in patients with major depression. *Archives of General Psychiatry, 61*(7), 705–713. doi:10.1001/archpsyc.61.7.705

Sanacora, G., Treccani, G., & Popoli, M. (2011). Towards a glutamate hypothesis of depression: An emerging frontier of neuropsychopharmacology for mood disorders. *Neuropharmacology.* doi:10.1016/j.neuropharm.2011.07.036

Sanacora, G., Zarate, C. A., Krystal, J., & Manji, H. K. (2008). Targeting the glutamatergic system to develop novel, improved therapeutics for mood disorders. *Nature Reviews Drug Discovery, 7*(5), 426–437. doi:10.1038/nrd2462

Sánchez, C., & Hyttel, J. (1999). Comparison of the effects of antidepressants and their metabolites on reuptake of biogenic amines and on receptor binding. *Cellular and Molecular Neurobiology, 19*(4), 467–489.

Santarelli, L., Saxe, M., Gross, C., Surget, A., Battaglia, F., Dulawa, S., . . . Hen, R. (2003). Requirement of hippocampal neurogenesis for the behavioral effects of antidepressants. *Science, 301*(5634), 805–809. doi:10.1126/science.1083328

Sapolsky, R., Krey, L., & McEwen, B. (1985). Prolonged glucocorticoid exposure reduces hippocampal neuron number: implications for aging. *The Journal of Neuroscience, 5*(5), 1222–1227.

Sara, S. J. (1985). Noradrenergic modulation of selective attention: its role in memory retrieval. *Annals of the New York Academy of Sciences, 444*, 178–193.

Sargent, P. A., Kjaer, K. H., Bench, C. J., Rabiner, E. A., Messa, C., Meyer, J., . . . Cowen, P. J. (2000). Brain serotonin1A receptor binding measured by positron emission tomography with [11C]WAY-100635: Effects of depression and antidepressant treatment. *Archives of General Psychiatry, 57*(2), 174–180.

Satomura, E., Baba, H., Nakano, Y., Maeshima, H., Suzuki, T., & Arai, H. (2011). Correlations between brain-derived neurotrophic factor and clinical symptoms in medicated patients with major depression. *Journal of Affective Disorders.* doi:10.1016/j.jad.2011.06.041

Schildkraut, J. J. (1965). The catecholamine hypothesis of affective disorders: A review of supporting evidence. *The American Journal of Psychiatry, 122*(5), 509–522.

Schlaepfer, T. E., Cohen, M. X., Frick, C., Kosel, M., Brodesser, D., Axmacher, N., . . . Sturm, V. (2007). Deep brain stimulation to reward circuitry alleviates anhedonia in refractory major depression. *Neuropsychopharmacology, 33*(2), 368–377.

Schmid, D. A., Brunner, H., Lauer, C. J., Uhr, M., Yassouridis, A., Holsboer, F., & Friess, E. (2008). Acute cortisol administration increases sleep depth and growth hormone release in patients with major depression. *Journal of Psychiatric Research, 42*(12), 991–999. doi:10.1016/j.jpsychires.2007.12.003

Sen, S., Duman, R., & Sanacora, G. (2008). Serum brain-derived neurotrophic factor, depression, and antidepressant medications: Meta-analyses and implications. *Biological Psychiatry, 64*(6), 527–532. doi:16/j.biopsych.2008.05.005

Sen, S., Nesse, R. M., Stoltenberg, S. F., Li, S., Gleiberman, L., Chakravarti, A., . . . Burmeister, M. (2003). A BDNF coding variant is associated with the NEO personality inventory domain neuroticism, a risk factor for depression. *Neuropsychopharmacology, 28*(2), 397–401.

Sequeira, A., Mamdani, F., Ernst, C., Vawter, M. P., Bunney, W. E., Lebel, V.,... Turecki, G. (2009). Global brain gene expression analysis links glutamatergic and GABAergic alterations to suicide and major depression. *PloS One, 4*(8), e6585. doi:10.1371/journal.pone.0006585

Serretti, A., & Mandelli, L. (2010). Antidepressants and body weight: A comprehensive review and meta-analysis. *The Journal of Clinical Psychiatry, 71*(10), 1259–1272. doi:10.4088/JCP.09r05346blu

Serretti, A., Mandelli, L., Lorenzi, C., Pirovano, A., Olgiati, P., Colombo, C., & Smeraldi, E. (2007). Serotonin transporter gene influences the time course of improvement of "core" depressive and somatic anxiety symptoms during treatment with SSRIs for recurrent mood disorders. Psychiatry Research, 149(1–3), 185–193. doi:10.1016/j.psychres.2006.03.020

Sevy, S., Papadimitriou, G. N., Surmont, D. W., Goldman, S., & Mendlewicz, J. (1989). Noradrenergic function in generalized anxiety disorder, major depressive disorder, and healthy subjects. *Biological Psychiatry, 25*(2), 141–152.

Sheline, Y. I., Barch, D. M., Price, J. L., Rundle, M. M., Vaishnavi, S. N., Snyder, A. Z.,... Raichle, M. E. (2009). The default mode network and self-referential processes in depression. *Proceedings of the National Academy of Sciences of the United States of America, 106*(6), 1942–1947. doi:10.1073/pnas.0812686106

Shen, Q., Lal, R., Luellen, B. A., Earnheart, J. C., Andrews, A. M., & Luscher, B. (2010). Gamma-Aminobutyric acid-type A receptor deficits cause hypothalamic-pituitary-adrenal axis hyperactivity and antidepressant drug sensitivity reminiscent of melancholic forms of depression. *Biological Psychiatry, 68*(6), 512–520. doi:10.1016/j.biopsych.2010.04.024

Shyn, S. I., & Hamilton, S. P. (2010). The genetics of major depression: Moving beyond the monoamine hypothesis. *The Psychiatric Clinics of North America, 33*(1), 125–140. doi:10.1016/j.psc.2009.10.004

Skolnick, P., Layer, R. T., Popik, P., Nowak, G., Paul, I. A., & Trullas, R. (1996). Adaptation of N-methyl-D-aspartate (NMDA) receptors following antidepressant treatment: Implications for the pharmacotherapy of depression. *Pharmacopsychiatry, 29*(1), 23–26. doi:10.1055/s-2007-979537

Smith, R. S. (1991). The macrophage theory of depression. *Medical Hypotheses, 35*(4), 298–306.

Soares, C. N., & Zitek, B. (2008). Reproductive hormone sensitivity and risk for depression across the female life cycle: a continuum of vulnerability? *Journal of Psychiatry & Neuroscience: JPN, 33*(4), 331–343.

Sonti, G., Ilyin, S. E., & Plata-Salamán, C. R. (1996). Neuropeptide Y blocks and reverses interleukin-1 beta-induced anorexia in rats. *Peptides, 17*(3), 517–520.

Srinivasan, V., Pandi-Perumal, S. R., Trakht, I., Spence, D. W., Hardeland, R., Poeggeler, B., & Cardinali, D. P. (2009). Pathophysiology of depression: role of sleep and the melatonergic system. *Psychiatry Research, 165*(3), 201–214. doi:10.1016/j.psychres.2007.11.020

Stahl, S. M. (2003). Deconstructing psychiatric disorders, part 2: An emerging, neurobiologically based therapeutic strategy for the modern psychopharmacologist. *The Journal of Clinical Psychiatry, 64*(10), 1145–1146.

Staner, L. (2010). Comorbidity of insomnia and depression. *Sleep Medicine Reviews, 14*(1), 35–46. doi:10.1016/j.smrv.2009.09.003

Sterpenich, V., D'Argembeau, A., Desseilles, M., Balteau, E., Albouy, G., Vandewalle, G.,... Maquet, P. (2006). The locus ceruleus is involved in the successful retrieval of emotional memories in humans. *The Journal of Neuroscience, 26*(28), 7416–7423. doi:10.1523/JNEUROSCI.1001-06.2006

Stetler, C., & Miller, G. E. (2011). Depression and hypothalamic-pituitary-adrenal activation: A quantitative summary of four decades of research. *Psychosomatic Medicine, 73*(2), 114–126. doi:10.1097/PSY.0b013e31820ad12b

Stewart, J. C., Rand, K. L., Muldoon, M. F., & Kamarck, T. W. (2009). A prospective evaluation of the directionality of the depression-inflammation relationship. *Brain, Behavior, and Immunity, 23*(7), 936–944. doi:16/j.bbi.2009.04.011

Stockmeier, C. A. (2003). Involvement of serotonin in depression: Evidence from postmortem and imaging studies of serotonin receptors and the serotonin transporter. *Journal of Psychiatric Research, 37*(5), 357–373.

Strange, B. A., & Dolan, R. J. (2004). Beta-adrenergic modulation of emotional memory-evoked human amygdala and hippocampal responses. *Proceedings of the National Academy of Sciences of the United States of America, 101*(31), 11454–11458. doi:10.1073/pnas.0404282101

Sundermann, E. E., Maki, P. M., & Bishop, J. R. (2010). A review of estrogen receptor alpha gene (ESR1) polymorphisms, mood, and cognition. *Menopause, 17*(4), 874–886. doi:10.1097/gme.0b013e3181df4a19

Surget, A., Saxe, M., Leman, S., Ibarguen-Vargas, Y., Chalon, S., Griebel, G.,... Belzung, C. (2008). Drug-dependent requirement of hippocampal neurogenesis in a model of depression and of antidepressant reversal. *Biological Psychiatry, 64*(4), 293–301. doi:10.1016/j.biopsych.2008.02.022

Takano, A., Arakawa, R., Hayashi, M., Takahashi, H., Ito, H., & Suhara, T. (2007). Relationship between neuroticism personality trait and serotonin transporter binding. *Biological Psychiatry, 62*(6), 588–592. doi:16/j.biopsych.2006.11.007

Taliaz, D., Stall, N., Dar, D. E., & Zangen, A. (2010). Knockdown of brain-derived neurotrophic factor in specific brain sites precipitates behaviors associated with depression and reduces neurogenesis. *Molecular Psychiatry, 15*(1), 80–92. doi:10.1038/mp.2009.67

Tan, K. R., Brown, M., Labouèbe, G., Yvon, C., Creton, C., Fritschy, J.-M.,... Lüscher, C. (2010). Neural bases for addictive properties of benzodiazepines. *Nature, 463*(7282), 769–774. doi:10.1038/nature08758

Tasker, J. G., Di, S., & Malcher-Lopes, R. (2005). Rapid central corticosteroid effects: Evidence for membrane glucocorticoid receptors in the brain. *Integrative and Comparative Biology, 45*(4), 665–671. doi:10.1093/icb/45.4.665

Tecott, L. H., Sun, L. M., Akana, S. F., Strack, A. M., Lowenstein, D. H., Dallman, M. F., & Julius, D. (1995). Eating disorder and epilepsy in mice lacking 5-HT2c serotonin receptors. *Nature, 374*(6522), 542–546. doi:10.1038/374542a0

Treadway, M. T., & Zald, D. H. (2011). Reconsidering anhedonia in depression: Lessons from translational neuroscience. *Neuroscience and Biobehavioral Reviews, 35*(3), 537–555. doi:10.1016/j.neubiorev.2010.06.006

Tyring, S., Gottlieb, A., Papp, K., Gordon, K., Leonardi, C., Wang, A.,... Krishnan, R. (2006). Etanercept and clinical outcomes, fatigue, and depression in psoriasis: Double-blind placebo-controlled randomised phase III trial. *Lancet, 367*(9504), 29–35. doi:10.1016/S0140-6736(05)67763-X

Uezato, A., Meador-Woodruff, J. H., & McCullumsmith, R. E. (2009). Vesicular glutamate transporter mRNA expression in the medial temporal lobe in major depressive disorder, bipolar disorder, and schizophrenia. *Bipolar Disorders, 11*(7), 711–725. doi:10.1111/j.1399-5618.2009.00752.x

Uher, R., Farmer, A., Maier, W., Rietschel, M., Hauser, J., Marusic, A., . . . Aitchison, K. J. (2008). Measuring depression: comparison and integration of three scales in the GENDEP study. *Psychological Medicine, 38*(2), 289–300. doi:10.1017/S0033291707001730

Uher, R., Maier, W., Hauser, J., Marusic, A., Schmael, C., Mors, O., . . . McGuffin, P. (2009). Differential efficacy of escitalopram and nortriptyline on dimensional measures of depression. *The British Journal of Psychiatry, 194*(3), 252–259. doi:10.1192/bjp.bp.108.057554

Ursin, R. (2002). Serotonin and sleep. *Sleep Medicine Reviews, 6*(1), 55–67. doi:53/smrv.2001.0174

van der Veen, F. M., Evers, E. A. T., Deutz, N. E. P., & Schmitt, J. A. J. (2007). Effects of acute tryptophan depletion on mood and facial emotion perception related brain activation and performance in healthy women with and without a family history of depression. *Neuropsychopharmacology, 32*(1), 216–224. doi:10.1038/sj.npp.1301212

Verhagen, M., van der Meij, A., van Deurzen, P. A. M., Janzing, J. G. E., Arias-Vásquez, A., Buitelaar, J. K., & Franke, B. (2010). Meta-analysis of the BDNF Val66Met polymorphism in major depressive disorder: effects of gender and ethnicity. *Molecular Psychiatry, 15*(3), 260–271. doi:10.1038/mp.2008.109

Vickers, S. P., Clifton, P. G., Dourish, C. T., & Tecott, L. H. (1999). Reduced satiating effect of d-fenfluramine in serotonin 5-HT(2C) receptor mutant mice. *Psychopharmacology, 143*(3), 309–314.

Voderholzer, U., Hornyak, M., Thiel, B., Huwig-Poppe, C., Kiemen, A., König, A., . . . Hohagen, F. (1998). Impact of experimentally induced serotonin deficiency by tryptophan depletion on sleep EEG in healthy subjects. *Neuropsychopharmacology, 18*(2), 112–124. doi:10.1016/S0893-133X(97)00094-8

Volkow, N. D., Logan, J., Fowler, J. S., Wang, G.-J., Gur, R. C., Wong, C., . . . Pappas, N. (2000). Association between age-related decline in brain dopamine activity and impairment in frontal and cingulate metabolism. *The American Journal of Psychiatry, 157*(1), 75–80. doi:10.1176/appi.ajp.157.1.75

Voon, V., Kubu, C., Krack, P., Houeto, J.-L., & Tröster, A. I. (2006). Deep brain stimulation: Neuropsychological and neuropsychiatric issues. *Movement Disorders, 21 Suppl 14*, S305–327. doi:10.1002/mds.20963

Wacker, J., Dillon, D. G., & Pizzagalli, D. A. (2009). The role of the nucleus accumbens and rostral anterior cingulate cortex in anhedonia: integration of resting EEG, fMRI, and volumetric techniques. *NeuroImage, 46*(1), 327–337. doi:10.1016/j.neuroimage.2009.01.058

Walter, M., Henning, A., Grimm, S., Schulte, R. F., Beck, J., Dydak, U., . . . Northoff, G. (2009). The relationship between aberrant neuronal activation in the pregenual anterior cingulate, altered glutamatergic metabolism, and anhedonia in major depression. *Archives of General Psychiatry, 66*(5), 478–486. doi:10.1001/archgenpsychiatry.2009.39

Wang, S.-S., Kamphuis, W., Huitinga, I., Zhou, J.-N., & Swaab, D. F. (2008). Gene expression analysis in the human hypothalamus in depression by laser microdissection and real-time PCR: the presence of multiple receptor imbalances. *Molecular Psychiatry, 13*(8), 786–799, 741. doi:10.1038/mp.2008.38

Warner-Schmidt, J. L., & Duman, R. S. (2006). Hippocampal neurogenesis: Opposing effects of stress and antidepressant treatment. *Hippocampus, 16*(3), 239–249. doi:10.1002/hipo.20156

Warner-Schmidt, J. L., & Duman, R. S. (2007). VEGF is an essential mediator of the neurogenic and behavioral actions of antidepressants. *Proceedings of the National Academy of Sciences of the United States of America, 104*(11), 4647–4652. doi:10.1073/pnas.0610282104

Warner-Schmidt, J. L., Vanover, K. E., Chen, E. Y., Marshall, J. J., & Greengard, P. (2011). Antidepressant effects of selective serotonin reuptake inhibitors (SSRIs) are attenuated by antiinflammatory drugs in mice and humans. *Proceedings of the National Academy of Sciences of the United States of America, 108*(22), 9262–9267. doi:10.1073/pnas.1104836108

Wayment, H. K., Schenk, J. O., & Sorg, B. A. (2001). Characterization of extracellular dopamine clearance in the medial prefrontal cortex: Role of monoamine uptake and monoamine oxidase inhibition. *The Journal of Neuroscience, 21*(1), 35–44.

Whitlock, J. R., Heynen, A. J., Shuler, M. G., & Bear, M. F. (2006). Learning induces long-term potentiation in the hippocampus. *Science, 313*(5790), 1093–1097. doi:10.1126/science.1128134

Wild, S., Roglic, G., Green, A., Sicree, R., & King, H. (2004). Global prevalence of diabetes: estimates for the year 2000 and projections for 2030. *Diabetes Care, 27*(5), 1047–1053.

Williams, S. M., & Goldman-Rakic, P. S. (1998). Widespread origin of the primate mesofrontal dopamine system. *Cerebral Cortex, 8*(4), 321–345.

Winokur, A., Gary, K. A., Rodner, S., Rae-Red, C., Fernando, A. T., & Szuba, M. P. (2001). Depression, sleep physiology, and antidepressant drugs. *Depression and Anxiety, 14*(1), 19–28. doi:10.1002/da.1043

Woolley, C. S., Gould, E., Sakai, R. R., Spencer, R. L., & McEwen, B. S. (1991). Effects of aldosterone or RU28362 treatment on adrenalectomy-induced cell death in the dentate gyrus of the adult rat. *Brain Research, 554*(1–2), 312–315.

World Health Organization. (2007). International Statistical Classification of Diseases and Related Health Problems. Retrieved from http://www.who.int/classifications/icd/en/

Wurtman, J. J. (1993). Depression and weight gain: the serotonin connection. *Journal of Affective Disorders, 29*(2–3), 183–192. doi:16/0165-0327(93)90032-F

Xiang, L., Szebeni, K., Szebeni, A., Klimek, V., Stockmeier, C. A., Karolewicz, B., . . . Ordway, G. A. (2008). Dopamine receptor gene expression in human amygdaloid nuclei: elevated D4 receptor mRNA in major depression. *Brain Research, 1207*, 214–224. doi:10.1016/j.brainres.2008.02.009

Yoshida, H., Peterfi, Z., García-García, F., Kirkpatrick, R., Yasuda, T., & Krueger, J. M. (2004). State-specific asymmetries in EEG slow wave activity induced by local application of TNFalpha. *Brain Research, 1009*(1–2), 129–136. doi:10.1016/j.brainres.2004.02.055

Yoshimura, R., Mitoma, M., Sugita, A., Hori, H., Okamoto, T., Umene, W., . . . Nakamura, J. (2007). Effects of paroxetine or milnacipran on serum brain-derived neurotrophic factor in depressed patients. *Progress in Neuro-Psychopharmacology & Biological Psychiatry, 31*(5), 1034–1037. doi:10.1016/j.pnpbp.2007.03.001

Yüksel, C., & Öngür, D. (2010). Magnetic resonance spectroscopy studies of glutamate-related abnormalities in mood disorders. *Biological Psychiatry, 68*(9), 785–794. doi:10.1016/j.biopsych.2010.06.016

Zhao, C., Deng, W., & Gage, F. H. (2008). Mechanisms and functional implications of adult neurogenesis. *Cell, 132*(4), 645–660. doi:10.1016/j.cell.2008.01.033

CHAPTER

14 Is Depression an Infectious Disease?

Turhan Canli

Abstract

In molecular psychology, the tools of molecular biology and genomics are applied to study and better understand behavior. In this chapter, the author applies this approach to present a novel perspective on major depressive disorder (MDD). The treatment approach for MDD has not changed much in half a century, and—in general—today's treatments are not more effective than earlier ones. The author suggests a new approach, one that conceptualizes MDD as an infectious disease. The author presents a set of four clues that are consistent with such a claim and lays out a molecular approach to how a search for a disease-causing agent could be conducted.

Key Words: genomics, virus, bacterial, parasite, depression

Major depressive disorder (MDD) or major depression is among the most common forms of mental illness, with a lifetime prevalence rate for a major depressive episode estimated at 16.6 percent in the United States (Kessler, Petukhova, Sampson, Zaslavsky, & Wittchen, 2012). Diagnosis of major depression requires 2 weeks of clinically significant dysphoria or anhedonia, plus the presence of five symptoms from among these behavioral domains: change in weight, change in sleep, change in psychomotor activity, fatigue, feelings of worthlessness or guilt, problems with executive function (concentration or decision making), and recurrent thoughts of death or suicide (American Psychiatric Association [APA], 2000). Given this diagnostic protocol, major depression is a highly heterogeneous disorder, challenging clinicians and researchers alike. Recurrence rates are discouraging: 50 percent of patients who have suffered a first episode of MDD and 80 percent of those who have suffered two episodes will have a recurrence at least once in their lifetime (Burcusa & Iacono, 2007). Calls-to-arms have been issued (Agid et al., 2007), but are we fighting the wrong war? In this chapter, I briefly review critiques on the current treatment approach, which suggest that a novel approach is needed. I then present four clues that are consistent with the hypothesis that MDD may be caused by an infectious agent, which could be parasitic, bacterial, or viral. I then suggest experimental approaches that could empirically test this hypothesis.

I. Critiques of the Current Pharmacological Approach

By and large, pharmacological approaches to the treatment of major depression have not changed in decades, focusing as they do on receptor-ligand interactions and the development of drugs such as selective serotonin or serotonin/norepinephrine reuptake inhibitors (SSRIs/SNRIs; Agid et al., 2007). Yet the effectiveness of current medications is not significantly different between competing products (Gartlehner et al., 2011), nor are so-called second generation drugs much better than earlier versions. Indeed, the reported effectiveness of today's drugs has actually decreased compared to three decades ago (Khan, Bhat, Kolts, Thase, & Brown, 2010). Age-cohort studies point to an increased lifetime risk and earlier onset in successive cohorts, even when controlling for confounds (Lewinsohn, Rohde, Seeley, & Fischer, 1993).

There is evidence that the published record on the efficacy of antidepressant drugs is presented as more positive than the complete dataset of published and unpublished results suggests. For example, Turner and colleagues (Turner, Matthews, Linardatos, Tell, & Rosenthal, 2008) addressed the question of publication bias in the reporting of antidepressant drug efficacy results by comparing published studies with the results from the same clinical trials registered with the Food and Drug Administration (FDA). The dataset comprised 12 antidepressant drugs from 74 studies involving 12,564 adult patients. All included studies were designed to be randomized, double-blind and placebo-controlled.

Thirty-eight of 74 registered studies (51 percent) were deemed to contain positive primary outcome results by the FDA, and all but one of these studies were published. The remaining studies were deemed by the FDA to be either negative (24 studies; these had no significant primary or secondary outcome results) or questionable (12 studies; these studies either failed or reported significant secondary, but not primary, outcome results). Of the 36 negative/questionable studies, only 3 were published as negative, 22 were not published, and 11 were published as positive by presenting positive secondary outcome results as if they were primary outcomes and de-emphasizing or omitting nonsignificant primary outcome results. Of the 12,564 patients who were registered in the FDA database, data from 3,449 (27 percent) were not published, and from 1,843 (15 percent) were published as positive when the FDA deemed the results questionable. Finally, there was a marked contrast in each of the 12 drug's effect sizes, as calculated on the basis of FDA results versus published results: published effect sizes were on average 32 percent larger than FDA-based effect sizes.

The authors acknowledged that the reasons for nonpublication are unknown and that these could reflect decisions by authors or sponsors not to submit or rejections by editors and reviewers. They also emphasized that nonsignificant results in individual trials do not necessarily reflect lacking efficacy because meta-analyses of each of the 12 drugs suggested superior efficacy relative to placebo.

The last assertion by Turner and colleagues has been challenged by Kirsch and colleagues, however (Kirsch et al., 2008). These investigators conducted a meta-analysis of both published and unpublished data on four antidepressants' effects as a function of baseline depression severity and drug efficacy. All four drugs were SSRIs (fluoxetine, venlafaxine, nefazodone, and paroxetine) and tested in short-term, double-blind, placebo-controlled, randomized control study designs. These investigators calculated drug effects as a reduction from baseline on the Hamilton Rating Scale for Depression (HRSD) and found that 31 out of 35 studies did show some drug efficacy, equivalent to a 1.8-point difference between drug and placebo HRSD change scores, which were statistically significant but below the 3-point threshold for attaining clinical significance. Clinical significance was only observed in those patients whose baseline depression severity was at the upper range of the very severely depressed spectrum (HRSD scores around 28), but, critically, this result was driven more by reduced responsiveness to placebo than by increased responsiveness to the drug treatment. Kirsch and colleagues concluded that "given these data, there seems little evidence to support the prescription of antidepressant medication to any but the most severely depressed patients" (p. 0266).

The conclusions by Kirsch and colleagues have been questioned by Fountoulakis and Moller who reanalyzed the Kirsch data and found that, at least for venlafaxine and paroxetine, clinical improvements of 3 points in the HRSD change score were reached (Fountoulakis & Moller, 2011). Yet these results still only represent very modest improvements in symptoms.

Drugs targeting other pathways have not been more effective than SSRIs. For example, a meta-analysis (Singh, Singh, & Kar, 2012) assessed the efficacy of a novel antidepressant agomelatine (marketed as Valdoxane) that is a 5-HT_{2C} receptor antagonist and a melatonergic MT_1/MT_2 receptor agonist. The meta-analysis included nine published studies and combined data from 2,390 agomelatine-treated patients compared to 689 placebo-treated and 864 SSRI-treated patient controls. The standardized mean difference to placebo was statistically significant but small (−0.26), leading the authors to question the clinical significance of the treatment. Another meta-analysis that included unpublished studies across 20 trials and 7,460 patients noted similar effect sizes (Taylor, Sparshatt, Varma, & Olofinjana, 2014). A third meta-analysis of both published and unpublished studies noted significant publication bias for positive results and concluded that a clinical effect of the drug was "unlikely" (Koesters, Guaiana, Cipriani, Becker, & Barbui, 2013). Furthermore, there was no evidence that this drug prevented relapse any more than placebo (Koesters et al., 2013; Taylor et al., 2014).

Another example is ketamine, a glutamate N-methyl-d-aspartate (NMDA) receptor antagonist that has received much attention as a fast-acting type of antidepressant since the first report (Berman et al., 2000). A recent systematic review noted the limitations of this literature, which is based on a small number of patients tested (N = 163) and shows a paucity of randomized controlled trials, active placebo conditions, and long-term outcomes (Aan Het Rot, Zarate, Charney, & Mathew, 2012). Another meta-analysis came to a similar conclusion and noted as a major weakness that the effects of ketamine are short-lived, disappearing within a matter of days to weeks (Caddy, Giaroli, White, Shergill, & Tracy, 2014).

Taken together, the literature on pharmacological interventions suggests reporting biases that amplify positive results while ignoring negative results. There is evidence that drugs may not be clinically much more effective than placebos, and, even if they are, their effects are small or may be short-lived. Given the high rate of recurrence among depressed patients, one has to conclude that, at best, these pharmaceutical interventions address symptoms rather than any root causes of the disease itself.

II. Shifting the Paradigm: Major Depression as an Infectious Disease

Given the current state of major depression treatment, I suggest a radical shift away from existing causal hypotheses. Instead of conceptualizing depression as a disorder caused by dysfunction in neurotransmitter systems, I propose to consider the possibility that major depression (and perhaps other common mental illnesses) is a form of infectious disease. The cause of the infection may be parasitic, bacterial, or viral. It may be one particular pathogen or a class of pathogens. It may be that different forms of pathogens are associated with different subtypes of depression. The pathogen may lie dormant as a latent vulnerability factor, such as a ubiquitous infectious agent that resides in neuronal cells or as a circulating retrovirus that could have been contracted at an early age without apparent symptoms. It may be passed on to offspring and among individuals sharing the same environment as part of the human *microbiome* or *virome*, or it could be a retrovirus that is embedded within the human genome, perhaps only to be assembled by multiple splicing events. It may become activated by molecular signals that are released when the host is exposed to stressful life experiences or under conditions of reduced immunity. It could explain the heterogeneity of major depression in terms of infected brain regions. To date, there is no proof for any of these suggestions, but I present a set of clues that are consistent with such a hypothesis.

III. Clues

A. Illness Behavior and Inflammatory Markers

The first clue that major depression may be an infectious disease is that depressed patients act sick. They often exhibit sickness behavior such as lethargy, loss of appetite, inability to get out of bed, and loss of interest in sex or other pleasurable activities. Characterized at the molecular level of analysis, depressed patients, relative to controls, have elevated inflammatory markers such as cytokines, chemokines, and acute-phase reactant proteins (Dowlati et al., 2010; Howren, Lamkin, & Suls, 2009; Liu, Ho, & Mak, 2012) and proinflammatory cytokines such as interleukin 1α and 1 β (IL-1α and IL-1 β), interleukin 6 (IL-6), tumor necrosis factor α (TNFα), and interferon-γ (INF-γ) that can induce sickness behavior (Dantzer, O'Connor, Freund, Johnson, & Kelley, 2008). One meta-analysis of 24 studies confirmed significantly elevated concentrations of TNFα and of IL-6 (but not any other cytokines) in patients diagnosed with MDD (N = 438 and 492, respectively) relative to controls (N = 350 and 400, respectively; Dowlati et al., 2010). Another meta-analysis of 29 studies, representing patients diagnosed with major depression (N = 901) and controls (N = 1,013), further confirmed the association between MDD and elevated TNFα and of IL-6 levels and extended it to elevated levels of soluble interleukin-2 receptor (sIL-2R; Liu et al., 2012).

Studies of postmortem brains of depressed individuals also show evidence for inflammatory markers. Tonelli and colleagues (Tonelli et al., 2008) investigated transcriptional levels of cytokines in samples obtained from suicide victims (N = 34) and controls (N = 17). The expression analyses were conducted in orbitofrontal cortex in Brodmann Area 11 (BA11), a region previously associated with suicide and suicidal ideation (Mann, 2003; Oquendo et al., 2003). Among suicide victims, these investigators observed significant increases, relative to controls, in IL-4 and IL-13 as a function of sex: IL-4 expression was significantly increased in females, whereas IL-13 was significantly increased in males. mRNA transcription levels of TNFα, which has been associated with depression (Dinan, 2009; Mikova, Yakimova, Bosmans, Kenis, & Maes, 2001), were also elevated in BA11 of female suicide victims but did not reach statistical significance.

Dean and colleagues investigated protein expression of TNFα and mRNA transcription of associated receptors in postmortem samples from MDD ($N = 10$), bipolar disorder (BD, $N = 10$) and schizophrenia (Sz, $N = 19$) patients and 30 age-matched controls (Dean et al., 2013; Dean, Tawadros, Scarr, & Gibbons, 2010). Their analyses focused on the dorsolateral prefrontal in BA46 and anterior cingulate cortex in BA24, two regions associated with cognitive control and emotion regulation (Critchley, 2005; Diener et al., 2012). They reported a 458 percent increase, relative to age-matched controls, in protein expression levels of the transmembrane TNFα (tmTNFα) in BA46 (but not BA24) in MDD patients (Dean et al., 2010). There was a double dissociation with BD patients, who showed a 158 percent increase in protein expression levels of tmTNFα in BA24 (but not BA46), whereas Sz subjects showed no significant increase in tmTNFα protein expression in either brain region (Dean et al., 2013). This follow-up study also found that mRNA transcription of TNF receptor 2 was significantly reduced in BA46 in both MDD (–51 percent) and in BD (–67 percent) patients. In contrast to these mood disordered patients, Sz patients showed an increase in TNF receptor 1 transcription in both BA24 (53 percent) and in BA46 (82 percent). Additional protein expression analyses focused on the proinflammatory marker IL-1β and did not find any changes in its expression in these brain regions. Thus, it appears that MDD, BD, and Sz all may be associated with some aspects of TNFα dysregulation in either BA24 or BA46.

Miguel-Hidalgo and colleagues compared postmortem samples from individuals diagnosed with major depression ($N = 18$) to samples from nonpsychiatric controls ($N = 19$) (Miguel-Hidalgo et al., 2011). Based on their previous finding (Miguel-Hidalgo, Nithuairisg, Stockmeier, & Rajkowska, 2007) that normal aging was associated with increased immunoreactivity in intercellular adhesion molecule-1 (ICAM-1, an inflammatory mediator in blood vessels and astrocytes) in orbitofrontal cortex in BA47, they extended the analysis to depressed individuals. Compared to controls, they noted a significant reduction in ICAM-1 immunoreactivity, which was most pronounced in those depressed individuals who had committed suicide. Although the authors noted limitations such as heterogeneity in the type and dosage of antidepressant medication and the fact that depression diagnosis was obtained postmortem, the results are consistent with their conclusion that depression is associated with changes in astrocyte-mediated immune function in the orbitofrontal cortex.

Shelton and colleagues compared genome-wide expression differences between drug-free individuals with a history of major depression and matched controls ($N = 14$ matched pairs) (Shelton et al., 2011), focusing on the anterior prefrontal cortex in BA10 as a region involved in reward processing (Rogers et al., 1999). Using a 1.5-fold expression difference as a threshold, along with analysis of molecular pathways using gene set enrichment analysis, they discovered a large set of pro- and anti-inflammatory cytokines that were differentially expressed as a function of major depression, including interleukins IL-1α, -2, -3, -5, -8, -9, -10, -12A, -13, -15, -18; IFNγ; and lymphotoxin α (LTA; TNF superfamily member 1).

Given that inflammatory markers in the periphery and in postmortem brain samples are associated with depression, is there a link to depression treatment responses? Indeed, several studies have reported improved antidepressant treatment responses in patients who received an inhibitor of proinflammatory cytokines, cyclooxygenase-2 (COX-2), in addition to their regular treatment regimen (Abbasi, Hosseini, Modabbernia, Ashrafi, & Akhondzadeh, 2012; Akhondzadeh et al., 2009; Chen, Tzeng, & Chen, 2010; Muller et al., 2006; Nery et al., 2008).

Hannestad and colleagues conducted a meta-analysis of 22 studies of MDD patients treated with psychopharmacological interventions that had measured changes in serum levels of TNFα, IL-6, and IL-1β during the course of treatment (Hannestad, DellaGioia, & Bloch, 2011). They found that although all drug treatments were associated with reduced depression symptoms, none was associated with reduced serum levels of TNFα, and only SSRIs were associated with reduced serum levels of IL-1β and IL-6 (Hannestad et al., 2011). They concluded that a reduction in depressive symptoms would not require a reduction in circulating inflammatory cytokines, but acknowledged that they may nonetheless contribute to depressive symptoms. On the other hand, one study reported that the SSRI citalopram actually increases levels of the proinflammatory cytokines TNFα and INF-γ in the frontal cortex of mice and that concurrent use of anti-inflammatory drugs during citalopram treatment actually increases treatment failure in patients (Warner-Schmidt, Vanover, Chen, Marshall, & Greengard, 2011). Thus, although inflammation may play a role in the etiology of depression (Krishnadas & Cavanagh, 2012; Raedler, 2011), the mechanisms by which classic antidepressant drugs affect inflammatory markers and how these relate to symptom reduction continues to be poorly understood.

A different approach is to consider possible antidepressant effects of drugs that directly block cytokine receptors. For example, Krugel and colleagues examined the potential antidepressant effects of a TNFα receptor blocker (etanercept, a human TNF-α receptor p75-Fc fusion protein) versus an antidepressant (imipramine) or Ringer solution in rats (Krugel, Fischer, Radicke, Sack, & Himmerich, 2013). Animals were subjected to chronic mild stress (restraint) or a nonrestraint control condition and tested in a forced swim test (FST) after 5 weeks. Compared to Ringer solution-treated controls, animals subjected to chronic mild stress and treated with etanercept exhibited reduced depression-like behavior, as measured with reduced immobile time in the FST, which was comparable to imipramine-treated animals. Antidepressant effects of etanercept have also been noted in patients with moderate to severe psoriasis in a double-blind placebo-controlled randomized clinical trial (Tyring et al., 2006) and an open-label active-comparator randomized trial in patients with rheumatoid arthritis (Bae et al., 2013). These results suggest that the direct targeting of cytokine receptors may be a useful antidepressant strategy, although even successful treatment may still only address symptoms of inflammation rather than its cause.

B. Pathogens

Inflammatory markers associated with major depression may reflect activation of immune defenses against some form of pathogen(s). Indeed, the second clue comes from known examples of pathogens—parasites, bacteria, and viruses—that have either been shown to alter host behavior on infection or that have some association with depression or other mood disorders. I do not argue that the specific examples discussed here are principal causes of major depression. Indeed, the observed associations are, or may very well be, secondary to the pathogen's primary effect on its host. Instead, I want to use these examples to illustrate possible biological pathways by which some currently undiscovered pathogen (or set of pathogens) could play a causal role in major depression.

B.1 PARASITES

There are several examples of parasites that can alter their host's behavior to generate a form of "extended phenotype." One example is the case of the colorfully named "zombie ant" (Andersen et al., 2009; Hughes et al., 2011). Worker ants that become infected by *Ophiocordyceps* fungi leave their colony and seek out an environment that is optimal for the spread of *Ophiocordyceps.* Once in this environment, the ant engages in a "death grip" bite onto a leaf. At this time, the ant's head is filled with fungal cells and, upon death of the host, which is still attached to the leaf by its mandibles, *Ophiocordyceps* grows a spore-dispersal structure from the base of the ant's head to invade the surrounding territory.

Another example is the case of the "suicidal cricket" (Thomas et al., 2002): when the cricket *Nemobius sylvestris* becomes infected by the nematomorph (hairworm) *Paragordius tricuspidatus*, it becomes more likely to jump into a pool of water, the native environment of *P. tricuspidatus.* Upon drowning of its host, *P. tricuspidatus* emerges from the host's cavity to return to its aqueous habitat. Dramatically, Thomas and colleagues describe how crickets that had been "rescued" by the researchers (i.e., extracted from the water before drowning) immediately returned to the edge of the pool to jump back in again. In rare instances, they noted that crickets that had released their worms without drowning in the process left the pool, apparently having vanquished their suicidal tendencies.

Moving from insects to rodents and humans, another example is Chagas disease, a tropical disease caused by the protozoan parasite *Trypanosoma cruzi* (Lannes-Vieira, de Araujo-Jorge, Soeiro Mde, Gadelha, & Correa-Oliveira, 2010), which has infected 16–18 million individuals in some areas of Latin America and which is exclusively carried by blood-sucking insects (Castro, de Mecca, & Bartel, 2006). During the acute phase of *T. cruzi* infection, individuals may be asymptomatic or have a fever, but life-threatening myocarditis or meningoencephalitis occurs in a subset of patients, with a death rate of 10 percent (Castro et al., 2006). During this phase, the parasite enters the central nervous system of mice and men where it can reside quietly as a chronic infection (Pittella, 2009). Mice infected with *T. cruzi* exhibit depression-like behavior, as measured in the laboratory in forced swim and tail suspension tests, and this behavior is reversed with administration of the SSRI fluoxetine and with the parasiticide drug benznidazole (Vilar-Pereira et al., 2012).

A much more widely studied parasite associated with mental pathology is *Toxoplasma gondii*, which lives in the gut of cats. When its eggs are released into the environment through feline feces, they can come into contact with rats. Rats infected with *T. gondii* exhibit altered affective behavior: rather than responding with fear to the scent of cat urine, they become attracted to it (Berdoy, Webster, & Macdonald, 2000; Vyas, Kim, Giacomini, Boothroyd, & Sapolsky, 2007). A cat that eats the fatally attracted rat will now become host to the next life cycle of *T. gondii*. Humans can become infected by coming into contact with feline feces or by consuming contaminated water or undercooked meat infected with *T. gondii* cysts. Linking the discussion of *T. gondii* infection back to the previous section on inflammation, infected hosts produce proinflammatory cytokines such as IL-6, IL-12, TNF, and IFN-γ (Hsu, Groer, & Beckie, 2014; Munoz, Liesenfeld, & Heimesaat, 2011).

Particularly relevant for this chapter on depression as an infectious disease, *T. gondii* can move beyond intestines of infected rodents to other organs, including the brain, where it accumulates in cysts in the prefrontal cortex, hippocampus, and amygdala (Berenreiterova, Flegr, Kubena, & Nemec, 2011; Vyas et al., 2007), brain regions associated with mood disorders (Drevets, Price, & Furey, 2008).

There is a growing literature on the psychological effects of *T. gondii* infection in humans, much of which initially focused on establishing a link to Sz (Torrey, Bartko, Lun, & Yolken, 2007; Torrey, Bartko, & Yolken, 2012; Torrey & Yolken, 2003; Yolken et al., 2001). Such studies are much-needed as the scale of mental illness is matched by the scale of human *T. gondii* infection: approximately 33 percent of the world's population (Montoya & Liesenfeld, 2004) and 22.5 percent of the US population older than the age of 12 (Centers for Disease Control [CDC], 2014) is believed to be infected.

Epidemiological data provide an indirect link to mood disorders. For example, neuroticism is a risk factor for depression, and the prevalence of *T. gondii* across countries correlates with aggregate neuroticism scores among those populations (Lafferty, 2006). A similar study reported a positive correlation between *T. gondii* prevalence rates and average suicide rates across 20 European nations (Lester, 2010), and reanalysis of the data highlighted this association, particularly for women of postmenopausal age (Ling, Lester, Mortensen, Langenberg, & Postolache, 2011).

More direct links have been provided by patient studies. For example, Arling and colleagues (2009) compared *T. gondii* seropositivity and antibody titers between patients diagnosed with major depression or with bipolar disorder who either had a history of suicide attempt ($N = 99$) or did not ($N = 119$) and healthy controls ($N = 39$). Patients with other types of diagnosed psychopathology, given previous links to Sz, were excluded from this study. Suicide attempters were found to have significantly higher *T. gondii* antibody titers than did mood-disordered patients without a history of suicide attempt, and there was a predictive association between antibody titers and suicide history. *T. gondii* seropositivity, on the other hand, was not significantly associated with any measures. A study based on a Turkish sample of 200 psychiatric inpatients with a history of suicide attempt and 200 controls (which included family members of the patients) reported significant differences in *T. gondii* seropositivity, with patients showing a seroprevalence rate of 41 percent versus 28 percent in controls (Yagmur, Yazar, Temel, & Cavusoglu, 2010).

The underlying molecular mechanism appears to involve dopamine (McConkey, Martin, Bristow, & Webster, 2013). Prior work has shown that total brain dopamine levels were increased by 114 percent in mice infected with the Type I *T. gondii* strain (Stibbs, 1985) and that dopamine accumulated in cyst-containing brain cells at a level that was 100-fold in vivo compared to in vitro cysts (Prandovszky et al., 2011). Treatment with the dopamine antagonist haloperidol prevents behavioral changes in infected rodents (Webster, Lamberton, Donnelly, & Torrey, 2006). *T. gondii* contains genes to synthesize tyrosine hydroxylase (Gaskell, Smith, Pinney, Westhead, & McConkey, 2009; Whitaker, McConkey, & Westhead, 2009), which converts tyrosine to L-dopa. However, because *T. gondii* does not produce dopa decarboxylase, which is needed to convert L-dopa to dopamine, this final step would have to rely on host cells containing this enzyme, specifically catecholaminergic neurons such as dopaminergic or serotonergic cells. Based on these considerations, McConkey and colleagues propose a model in which *T. gondii* may infect many brain regions but induces behavioral effects only in those regions that contain catecholaminergic neurons (McConkey et al., 2013).

Treatment for *T. gondii* is limited because of current drug toxicity, but a recent report on Bumped kinase inhibitors (BKI) demonstrated that one particular BKI, 1294, can be administered orally and is effective against toxoplasmosis in vivo (Doggett, Ojo, Fan, Maly, & Van Voorhis, 2014). To date, there are no published reports on the use of BKIs in the context of depression treatment.

B.2 BACTERIA

The role of bacteria, particularly those residing in the gastrointestinal tract, in the regulation of brain function and behavior has come into focus in recent years (Cryan & Dinan, 2012; Dinan & Cryan, 2013; Moloney, Desbonnet, Clarke, Dinan, & Cryan, 2014). The number of microorganisms in the human gastrointestinal tract exceeds the number of human cells by a factor of ten, the number of human genes by a factor of 150, and is estimated to include more than 1,000 species and more than 7,000 strains (Cryan & Dinan, 2012). The role of these microorganisms has been discussed with respect to the immune system (Bisgaard, Bonnelykke, & Stokholm, 2014; Jiao et al., 2009; Mulder et al., 2011; Olszak et al., 2012), the regulation of fat storage, nutrient absorption and gut motility (Backhed et al., 2004; Bercik, Collins, & Verdu, 2012), and overall health (Kinross, von Roon, Holmes, Darzi, & Nicholson, 2008; Moloney et al., 2014). The (bidirectional) interaction between the gastrointestinal tract and the brain has been labeled the "gut-brain axis" (Romijn, Corssmit, Havekes, & Pijl, 2008) and has been expanded to highlight the significance of microorganisms to be denoted as the "microbiome-gut-brain axis" (Rhee, Pothoulakis, & Mayer, 2009). Here, I focus on those studies that have linked this axis to affective behavior, with an emphasis on major depression.

Several animal studies have begun to link intestinal microbes with emotional behavior, beginning with the seminal discovery by Sudo and colleagues that germ-free mice exhibit enhanced hypothalamic-pituitary-adrenal (HPA) axis activity in response to psychological stress and that this response can be normalized by administration of the bacterial strain *Bifidobacteria infantis* (Sudo et al., 2004). Building on this and related findings, Desbonnet and colleagues studied the effects of *B. infantis* in a rat model of stress and depression (Desbonnet, Garrett, Clarke, Bienenstock, & Dinan, 2008). Animals treated for 14 days with *B. infantis* showed significantly reduced levels of the proinflammatory cytokines IFN-γ, TNF-α, and IL-6 following mitogen stimulation, relative to controls. These animals also showed significant increases in plasma concentrations of tryptophan, a precursor for serotonin, as well as reduced 5-hydroxyindole acetic acid (5-HIAA) in the frontal cortex and decreased dihydroxyphenylacetic acid (DOPAC) in the amygdala. On the other hand, *B. infantis* treatment did not produce any behavioral changes in a FST. Taken together, these data support an effect of *B. infantis* on both peripheral and central markers of immune function and monoaminergic activity previously associated with depression, although a behavioral marker failed to further establish a link to depressive behavior.

Later work by this group used a maternal separation (MS) model to further address possible behavioral effects of *B. infantis* (Desbonnet et al., 2010). Animals in the experimental condition were adult rats that had experienced MS as pups (3 hours separation per day from P2–P14) and that were subsequently chronically treated with *B. infantis* or with the antidepressant citalopram. MS animals without *B. infantis* or citalopram treatment exhibited reduced swim behavior and increased immobility in a FST, increased levels of IL-6 in blood, and elevated mRNA transcription levels of the corticotrophin-releasing factor gene in the amygdala. All of these measures were normalized in MS animals treated with *B. infantis*, although the effects were not as strong as in animals treated with citalopram.

The modulation of the γ-aminobutyric acid (GABA) system and emotional behavior by the *Lactobacillus rhamnosus* strain has been studied by Bravo and colleagues (Bravo et al., 2011). Adult male BALB/c mice in the experimental condition were fed broth containing *L. rhamnosus (JB-1)* for 28 days. Behavioral assessments included stress-induced hyperthermia (SIH), elevated plus maze (EPM), fear conditioning involving foot shock, and the FST. Compared to broth-only controls, *L. rhamnosus (JB-1)*-treated mice exhibited significantly more open-arm entries in the EPM, enhanced freezing behavior in the presence of the fear-conditioned stimulus, and significantly less time immobile in the FST. There was no significant difference between experimental and control animals in the SIH.

The effects of *L. rhamnosus (JB-1)* treatment were further assessed with regard to stress hormones in the periphery and GABA modulation in the brain.

In the periphery, measures of plasma corticosterone taken 30 minutes after the FST showed that *L. rhamnosus (JB-1)*-treated mice had significantly lower levels of the stress hormone than did controls. In the brain, there was evidence for differential GABA regulation as a function of *L. rhamnosus (JB-1)* treatment. Compared to controls, experimental animals showed higher levels of $GABA_{B1b}$ mRNA in the cingulate and prelimbic cortices and lower levels in hippocampus and amygdala, among other regions. Differential expression of $GABA_{A\alpha2}$ mRNA was also observed in many of these regions.

Bravo and colleagues showed that the observed behavioral and neurochemical effects were mediated by the vagus nerve because vagotomized mice did not show any differential effects as a function of *L. rhamnosus (JB-1)* treatment. Taken together, these data suggest that *L. rhamnosus (JB-1)* has some anxiolytic and antidepressant effects: it reduced unconditioned fear, as measured by open-arm entries in the EPM test, yet it also increased conditioned fear, as measured by increased freezing behavior to a fear-conditioned stimulus; it reduced depression-associated behavior, as measured by reduced immobility in the FST, which was further associated with a reduced level of peripheral corticosterone. These behavioral effects can be interpreted by vagus nerve-mediated central effects on GABA function: for example, reduced transcription levels of $GABA_{B1b}$ in amygdala and hippocampus are consistent with the antidepressant effect of $GABA_B$ receptor antagonists (Slattery, Desrayaud, & Cryan, 2005).

Human studies have investigated the link between probiotic food and mood. In a double-blind, placebo-controlled study using a general population sample, a 3-week regimen of a probiotic-containing milk drink containing *Lactobacillus casei* Shirota improved mood in those participants whose mood scores at baseline were in the lower third of the sample (Benton, Williams, & Brown, 2007). In another double-blind, placebo-controlled study of healthy volunteers, a 30-day treatment of combined *Lactobacillus helveticus* R0052 and *Bifidobacterium longum* R0175 consumption was associated with significantly reduced self-reported somatization, depression, anxiety, and hostility (Messaoudi et al., 2011).

A link between probiotic nutrition and brain activity during emotion processing was investigated by Tillisch and colleagues (Tillisch et al., 2013). Participants were healthy women without gastrointestinal or psychiatric symptoms who were placed on a 4-week regimen of a fermented milk product containing *Bifidobacterium animalis subsp. Lactis, Streptococcus thermophiles, Lactobacillus bulgaricus*, and *Lactococcus lactis subsp. Lactis*, or who were assigned to a control condition (nonfermented milk product or no intervention). At baseline and following the intervention, participants were scanned using functional magnetic resonance imaging to measure brain activation at rest and during an emotion face-processing task. Administration of the probiotics was associated with changes in midbrain connectivity at rest and with reduced activation to emotion faces in a widely distributed network of regions, including affective regions.

A mechanism by which gastrointestinal bacteria may play a role in major depression has been proposed by Maes and colleagues (Maes, Kubera, & Leunis, 2008; Maes, Kubera, Leunis, & Berk, 2012). Their "leaky gut" hypothesis states that stressors such as hypoxia, nitric oxide, oxidant stress, or cytokines can increase gastrointestinal permeability, allowing lipopolysaccharides (LPS) from gram-negative bacteria to translocate and elicit an inflammatory response that may contribute to depression. In support of this hypothesis, Maes and colleagues observed increased levels of serum concentrations of IgM and IgA against LPS of the gram-negative enterobacteria (*Hafnia alvei, Morganella morganii, Pseudomonas putida, Citrobacter koseri*, and *Klebsiella pneumoniae*) in patients with major depression relative to healthy controls (Maes et al., 2008). In a follow-up study, this group studied a larger cohort of patents with major depression ($N = 112$) and found that elevated serum concentrations of IgM and IgA against the LPS of gram-negative enterobacteria were most pronounced in patients with chronic depression but were not significantly associated with melancholia or with recurrent depression (Maes et al., 2012).

Stilling and colleagues proposed another mechanism by which bacteria in the gastrointestinal tract may affect brain and behavior, which involves epigenetic regulation (Stilling, Dinan, & Cryan, 2014). These authors speculate that microbial products from the gastrointestinal tract may activate epigenetic mechanisms in the brain, although there is no direct evidence for altered neuroepigenetic regulation from gut bacteria to date.

B.3 VIRUSES

A recent meta-analysis of 28 published case–control studies of biomarkers of microorganisms was conducted by Wang and colleagues (Wang et al., 2014). Of the 16 infectious agents included in the analysis, significant associations with depression were found for Borna disease virus (BDV), herpes simplex virus-1, varicella zoster virus, and Epstein-Barr virus (bacteria were also included in this meta-analysis: the only significant association with depression was found for *Chlamydophila trachomatis*, but, notably in light of the earlier discussion, not for *T. gondii*; however, only three *T. gondii* studies were included in this meta-analysis). Of these identified microorganisms, BDV is an interesting case for discussion because its association with depression—meta-analytic support notwithstanding—is quite controversial.

According to Wang and colleagues, BDV, a negative-stranded nonretroviral RNA virus, is the most-studied virus associated with depression, with 15 out of 28 included studies devoted to this virus (Wang et al., 2014). A study of 30 postmortem brains from a general population sample reported evidence of BDV infection in two individuals in frontal and temporal cortex, olfactory bulb, and hippocampus (Haga et al., 1997), suggesting a surprisingly high rate of frequency. However, this conclusion is questionable in light of a later study of postmortem samples from 86 psychiatric patients and 52 healthy controls, none of which showed evidence of BDV infection (Czygan et al., 1999).

Wang and colleagues found that depressed individuals had an odds ratio (OR) of 3.25 of infection, relative to healthy controls. However, ORs differed widely across studies, ranging from 0.60 to 35.0. ORs were not significantly affected by detection technique (measurements of BDV mRNA from blood or peripheral blood mononuclear cells, peripheral blood mononuclear cells [PBMCs] vs. serum antibodies), but there was significant heterogeneity across studies, explaining 46.1 percent of the OR variance.

Intriguingly, one small clinical study of 25 patients reported that treatment of depressed patients with amantadine reduced clinical depressive symptoms along with markers of BDV-infection (Dietrich & Bode, 2008), although no other replication studies can be found in the published literature. Although BDV infection has been observed in depressed patients in several studies (Amsterdam et al., 1985; Bode, Durrwald, Rantam, Ferszt, & Ludwig, 1996; Bode, Ferszt, & Czech, 1993; Bode et al., 2001; Bode, Zimmermann, Ferszt, Steinbach, & Ludwig, 1995; Fu et al., 1993; Zhang et al., 2014), other studies have failed to find any evidence for an association (Hornig et al., 2012; Iwata et al., 1998; Kim et al., 1999; Na, Tae, Song, & Kim, 2009). Among these nonreplications, the study by Hornig et al. (2012), in particular, has been hailed as setting "a gold standard for future investigations" (Oldstone, 2012) because of its use blinded serological and molecular analyses, rigor of clinical assessment, protocols for sample collection, and experimental design. However, the sample included only 80 patients diagnosed with MDD, so future studies using much larger patient cohorts would still be useful.

Given its pandemic reach, the human immunodeficiency virus (HIV) is also of interest here. It crosses the blood–brain barrier (Atwood, Berger, Kaderman, Tornatore, & Major, 1993; Chiodi, Norkrans, et al., 1988; Chiodi, Sonnerborg, et al., 1988) where it primarily targets parenchymal microglia and the perivascular microglia/macrophages, the only cells in the brain that express both the CD4 and CCR5 receptors required for HIV entry to the cell (Anthony & Bell, 2008). Later neuropathology is believed to be a secondary consequence of disruption of function in these neuron-supporting cells (Anthony & Bell, 2008).

Although depression is a common diagnosis for HIV-infected patients (Repetto et al., 2003), 10 studies concluded that the risk for depression is no greater among HIV-infected patients than it is among the general population (Ciesla & Roberts, 2001). However, when these studies were aggregated in a meta-analysis, there was a significantly higher rate of major depression among HIV-positive (160 out of N = 1,700) relative to HIV-negative (47 out of N = 896) individuals, which translated into an OR of 1.99 greater likelihood for MDD diagnosis in HIV-positive individuals relative to controls (Ciesla & Roberts, 2001). The mere association with depression, of course, does not imply a causal relationship because depression may have preceded infection or may very well represent the patient's psychological response to the diagnosis or be a side effect of drug treatment. Nonetheless, potential mechanisms by which HIV may lead to depression have been discussed by Del Guerra and colleagues (Del Guerra, Fonseca, Figueiredo, Ziff, & Konkiewitz, 2013). They suggest that psychosocial factors may interact with virally mediated pathways such as elevated cytokines, neurotoxicity affecting dopaminergic neurons, and reduced levels of monoamines and of brain-derived neurotrophic factor (BDNF).

However, there is currently no evidence linking HIV infection and depression to differential brain pathology, although one positron emission tomography (PET) study compared brain serotonergic transmission in HIV-infected patients with ($N = 9$) and without ($N = 9$) depression and healthy controls ($N = 7$) using a serotonin transporter (5-HTT)-specific ligand (Hammoud et al., 2010). After false discovery rate statistical correction, HIV-infected patients (consisting of both depressed and nondepressed individuals), relative to healthy controls, showed significantly lower 5-HTT binding potential only in the insula, suggesting possible serotonergic neuronal loss in that region. However, no significant differences among any of the groups were found as a function of depression, which may reflect the small sample size of the study population.

If the association of HIV-infection with MDD were causal, then perhaps antiretroviral treatment for HIV would reduce depressive symptoms. However, current evidence does not support this hypothesis. Springer and colleagues conducted a post hoc analysis of data from HIV-infected injection drug users (IDUs) who were enrolled in a 6-month prospective, randomized control trial of directly administered antiretroviral therapy (DAART; Springer, Chen, & Altice, 2009). Depressive symptoms were measured at baseline and at 6 months using the Center for Epidemiologic Studies of Depression Scale (CES-D). At baseline, 58 of 89 patients (63 percent) met criteria for severe depression or MDD; however, CES-D scores were significantly improved for this group, as well as for all patients, at 6 months. Regression analysis found that improved CES-D scores were associated with several factors, such as increase in CD4 count and adherence, but not with DAART. In light of my earlier discussion about the efficacy of antidepressants, improvement in CES-D scores was also not associated with use of antidepressant medication.

C. *Missing Heritability*

A third clue that major depression may be an infectious disease is its "missing heritability": that is, despite decades of twin study evidence for the heritability of depression, a direct genetic link has not yet been found. Genome-wide association studies have not identified markers that reach significance or replicate across analyses (Lewis et al., 2010; Verbeek et al., 2012). Such results have been taken to suggest that major depression is highly polygenic, with individual genes contributing only in minor ways to the phenotype. Another approach has been the study of gene-environment interactions for polymorphic candidate genes (Aguilera et al., 2009; Canli & Lesch, 2007; Canli et al., 2006; Caspi et al., 2003; Conway, Hammen, Brennan, Lind, & Najman, 2010; Eley et al., 2004; Kaufman et al., 2004; Kendler, 1998; Lesch, 2004) and epigenetics (McGowan & Kato, 2008; Oberlander et al., 2008; Philibert et al., 2007). Although these studies have begun to make advances, controversy remains (Risch et al., 2009).

In light of the previous subsection on bacteria, an entirely new perspective on the problem of missing heritability is offered by the paradigm-shifting concept of the human body as an ecosystem for microbiota and an associated "microbiome" (the combined genome of these microbiota) in the gastrointestinal tract (Human Microbiome Project, 2012*b*; Peterson et al., 2009; Walter & Ley, 2011). In addition to any dysfunction that may be caused by a "leaky gut" as discussed earlier, it is possible that the genes represented in this microbiome may contain pathogenic alleles or mutate, or that genetic variation among hosts may produce adverse interactive effects with the microbiome in some individuals. Genetic transmission in humans may not reflect direct transmission through the human germline but may instead result from the transmission of the biome because microbiota can be passed between mothers and their infants during birth or through environmental exposure (Yatsunenko et al., 2012).

Correspondingly, there is a human "virome" (Bibby, Viau, & Peccia, 2011; Foxman & Iwasaki, 2011; Wylie, Weinstock, & Storch, 2012) that harbors numerous viral species, some of which are poorly understood but may play a role in human diseases. For example, Wylie and colleagues studied the presence of viruses in children with unexplained fever (UF, febrile children), a condition that is common in children under 3 years of age, and controls (afebrile children; Wylie, Mihindukulasuriya, Sodergren, Weinstock, & Storch, 2012). They conducted deep sequencing analysis on plasma and nasopharyngeal (NP) swab samples from children age 2–36 months who presented in the emergency room with UF ($N = 73$) and age-matched children who were in the hospital for surgery but who were afebrile ($N = 103$). Across all children, viral sequences from 25 different viral genera were discovered that included several expected genera, such as adenoviruses, enteroviruses, and roseolaviruses. Febrile children had 1.5–5 times more viral sequences than did nonfebrile children and carried a larger variety of viruses. For example, none of the afebrile children plasma samples contained more than one viral genus, whereas 61 percent of febrile children plasma samples contained two to five virus genera. Some samples bore similarities to a reference genome at the level of amino acids but not at the level of nucleotides, suggesting the presence of unknown or divergent viruses. It is possible that other virome habitants may contribute to other illnesses, including major depression.

Missing heritability related to depression may not only be explained by the human biome or virome, but also by overlooked elements of the human genome itself, such as those regions of the human genome that represent insertions of originally exogenous, nonhuman sequences. One example for such a process is horizontal gene transfer, which is the exchange of genetic material between nonmating species (Gilbert, Schaack, Pace, Brindley, & Feschotte, 2010). Using horizontal gene transfer, parasites might be able to enter the human genome. Another example by which exogenous gene sequences may insert themselves into the human genome is through human genome domestication of retroviruses, whose sequences actually make up 8 percent of the human genome (Lander et al., 2001). Some retroviral insertions into the human genome appear to have been protected from mutational degeneration because they serve some function that is useful to the host, a process known as domestication of exogenous retroviruses (Kurdyukov et al., 2001). For instance, retroviral envelope (*env*) genes play a critical role in the transfer of nutrition from the mother to the embryo through the placenta, and several such genes have been discovered, with the oldest example, the *Syncytin-Car1* gene, being domesticated 60 million years ago (Cornelis et al., 2012). It is possible that other domesticated retrograde viruses play useful biological roles in their human hosts but may render a subset of human hosts vulnerable to major depression. Alternatively, depressive behavior itself may be considered adaptive in some contexts (Allen & Badcock, 2003; Andrews & Thomson, 2009; Gilbert, 2006; Nesse, 2000).

In light of the previous discussion of BDV, it is striking that gene segments of this virus have recently been discovered in the human germline and may have entered vertebrate genomes 40 million years ago (Belyi, Levine, & Skalka, 2010). Compared across species, the presence of genomic DNA sequences that seem derived from viral BDV RNA correlates positively with species' resistance to the disease caused by BDV; this advantage could be eroded in some genetically vulnerable individuals who might then exhibit cognitive and affective disorders.

Taken together, the conceptualization of a human biome or virome potentially contributing to depression vastly expands the number of genomes to be investigated for links to the illness. Horizontal DNA transfer or domestication of retroviruses into the human genome may also represent an overlooked area of genetic contributions to major depression. Each of these processes may then further interact with (1) environmental factors, such as stressful life events; (2) endogenous human genetic variants; (3) gene epistatic interactions; (4) epigenetic regulatory mechanisms; or (5) aging-related or other dynamic hormonal changes to render some individuals more susceptible to major depression while imbuing others with resilience. Missing heritability of depression may lurk in any of these unexplored spaces.

D. Depression Contagion

The final clue hinting at depression as an infectious disease is particularly speculative and refers to "depression contagion." Surprisingly, analysis of data from the Framingham Heart Study, a longitudinal study of more than 30,000 individuals over a period of more than 30 years, found that depressive symptoms in one person were significantly associated with those of their friends and neighbors, up to three degrees of separation (Rosenquist, Fowler, & Christakis, 2011). These data suggest an environmental influence in spreading depression, presumably through social communication.

An interesting perspective on the argument for environmental spread of depression is provided by one study highlighting the spread of human gut biomes through environmental exposure (Yatsunenko et al., 2012). The study analyzed the bacterial content of fecal samples from 531 individuals from 151 families across three different regions (Amazonas of Venezuela, rural Malawi, and US metropolitan regions) and included mono- and dizygotic twins, siblings, and parents. This sample allowed for analysis of vertical transmission of the gut microbiome across generations and within shared environments. The authors found that the phylogenetic architecture of the fecal microbiota in children and teenagers was similar among monozygotic twins as it was among dizygotic twins, which replicated a previous study by this group in adult twins (Turnbaugh et al., 2009), suggesting low heritability of the microbiome. However, there was strong evidence for environmental transmission, as fecal microbiota were similar between teenage twins and both their parents, and microbiota of parents were significantly more similar to each other than were those of individuals from different families. These results led the authors to the conclusion that gut microbial content may be shaped by common environmental exposures. It is possible that the environmental spread of gut microbiota may, in part, contribute to the environmental spread of depression.

IV. The Road Ahead

The foregoing discussion laid out four clues that are consistent with the hypothesis that major depression is an infectious disease. I presented examples of parasites, bacteria, and viruses that illustrate the possible biological pathways by which a currently unknown pathogen (or set of pathogens) could cause MDD. The hunt for such a pathogen would need to start with a large sample of carefully clinically characterized patients with MDD and matched healthy controls. Analyses should be conducted on multiple and different samples from each individual, such as fecal, blood, and other bodily fluids; postmortem brain tissue from donors with clinically diagnosed MDD and nondepressed controls should also be included. Brain samples should be obtained from regions that have robust associations with depression, such as amygdala, insula, anterior cingulate cortex, thalamus, and striatum (Diener et al., 2012; Hamilton et al., 2012). Data could be compared with existing available resources, such as the National Institutes of Health (NIH) Human Microbiome Project (Peterson et al., 2009).

Experimental approaches should be modeled after gold-standard study protocols, bioinformatics approaches, and quality controls that have already been established and continue to be refined (Human Microbiome Project, 2012*a*; Kuczynski et al., 2012). Examples of a methodologically rigorous approach to patient–control comparisons are studies of concordant or discordant twin pairs and their parents, as have been conducted in the study of obesity (Turnbaugh et al., 2009) and malnutrition (Yatsunenko et al., 2012) or studies using blinded serological and molecular analyses, rigorous protocols for clinical assessment, sample collection, and experimental design (Hornig et al., 2012). If gut microbiota were discovered that are strongly associated with MDD, then a next step could be the behavioral and neurological study of gnotobiotic mice colonized with these microbiomes. Eventual human treatment approaches could include the transfer of bacterial colonies.

A focus on viral associations with major depression could utilize methods that were developed to detect particular classes of pathogens, such as those designed to discover adenoviruses, bunyaviruses, coronaviruses, enteroviruses, flaviviruses, herpesviruses, paramyxoviruses, and parechoviruses (Barber et al., 2012). Methods have also been developed to use genomic sequencing approaches to discover hitherto unknown viruses (Allen et al., 2011; Barzon et al., 2013; Cook et al., 2013; Kreuze et al., 2009; Tang & Chiu, 2010; Zeigler Allen et al., 2013). Of course, any discovery of a pathogen playing a causal role in major depression would simply be a starting point for a long journey toward developing new treatments. But rather than settling for symptom reduction, these new approaches could represent aspirations for discovering a genuine cure. Hopelessness, so much a feature of depression, may be replaced with hope for patients and their families, after all.

Acknowledgments

I thank Christopher Gill, M.D., for his critical feedback on a draft of this chapter.

References

Aan Het Rot, M., Zarate, C. A., Jr., Charney, D. S., & Mathew, S. J. (2012). Ketamine for depression: Where do we go from here? *Biological Psychiatry, 72*(7), 537–547. doi: 10.1016/j.biopsych.2012.05.003

Abbasi, S. H., Hosseini, F., Modabbernia, A., Ashrafi, M., & Akhondzadeh, S. (2012). Effect of celecoxib add-on treatment on symptoms and serum IL-6 concentrations in patients with major depressive disorder: Randomized double-blind placebo-controlled study. *Journal of Affective Disorders, 141*(2–3), 308–314. doi: 10.1016/j.jad.2012.03.033

Agid, Y., Buzsaki, G., Diamond, D. M., Frackowiak, R., Giedd, J., Girault, J. A., . . . Weinberger, D. (2007). How can drug discovery for psychiatric disorders be improved? *Nature Reviews. Drug Discovery, 6*(3), 189–201. doi: Nrd2217 [pii] 10.1038/nrd2217

Aguilera, M., Arias, B., Wichers, M., Barrantes-Vidal, N., Moya, J., Villa, H., . . . Fananas, L. (2009). Early adversity and 5-HTT/BDNF genes: New evidence of gene-environment interactions on depressive symptoms in a general population. *Psychological Medicine, 39*(9), 1425–1431. doi: 10.1017/S0033291709005248. Epub 2009 Feb 12.

Akhondzadeh, S., Jafari, S., Raisi, F., Nasehi, A. A., Ghoreishi, A., Salehi, B., . . . Kamalipour, A. (2009). Clinical trial of adjunctive celecoxib treatment in patients with major depression: A double blind and placebo controlled trial. *Depression and Anxiety, 26*(7), 607–611.

Allen, N. B., & Badcock, P. B. (2003). The social risk hypothesis of depressed mood: Evolutionary, psychosocial, and neurobiological perspectives. *Psychological Bulletin, 129*(6), 887–913. doi: 10.1037/0033-2909.129.6.887

Allen, L. Z., Ishoey, T., Novotny, M. A., McLean, J. S., Lasken, R. S., & Williamson, S. J. (2011). Single virus genomics: A new tool for virus discovery. *PLoS One, 6*(3), e17722. doi: 10.1371/journal.pone.0017722

American Psychiatric Association (APA). (2000). *Diagnostic and statistical manual of mental disorders* (4th ed., text rev., *DSM-IV-TR).* Washington, DC: Author.

Amsterdam, J. D., Winokur, A., Dyson, W., Herzog, S., Gonzalez, F., Rott, R., & Koprowski, H. (1985). Borna disease virus. A possible etiologic factor in human affective disorders? *Archives of General Psychiatry, 42*(11), 1093–1096.

Andersen, S. B., Gerritsma, S., Yusah, K. M., Mayntz, D., Hywel-Jones, N. L., Billen, J., ... Hughes, D. P. (2009). The life of a dead ant: The expression of an adaptive extended phenotype. *American Naturalist, 174*(3), 424–433. doi: 10.1086/603640

Andrews, P. W., & Thomson, J. A., Jr. (2009). The bright side of being blue: Depression as an adaptation for analyzing complex problems. *Psychological Review, 116*(3), 620–654. doi: 10.1037/a0016242

Anthony, I. C., & Bell, J. E. (2008). The neuropathology of HIV/AIDS. *International Review of Psychiatry, 20*(1), 15–24. doi: 10.1080/09540260701862037

Arling, T. A., Yolken, R. H., Lapidus, M., Langenberg, P., Dickerson, F. B., Zimmerman, S. A., ... Postolache, T. T. (2009). Toxoplasma gondii antibody titers and history of suicide attempts in patients with recurrent mood disorders. *The Journal of nervous and mental disease, 197*(12), 905–908. doi: 10.1097/NMD.0b013e3181c29a23

Atwood, W. J., Berger, J. R., Kaderman, R., Tornatore, C. S., & Major, E. O. (1993). Human immunodeficiency virus type 1 infection of the brain. *Clinical Microbiology Reviews, 6*(4), 339–366.

Backhed, F., Ding, H., Wang, T., Hooper, L. V., Koh, G. Y., Nagy, A., ... Gordon, J. I. (2004). The gut microbiota as an environmental factor that regulates fat storage. *Proceedings of the National Academy of Sciences of the United States of America, 101*(44), 15718–15723.

Bae, S. C., Gun, S. C., Mok, C. C., Khandker, R., Nab, H. W., Koenig, A. S., ... Singh, A. (2013). Improved health outcomes with etanercept versus usual DMARD therapy in an Asian population with established rheumatoid arthritis. *BMC Musculoskeletal Disorders, 14*, 13. doi: 10.1186/1471-2474-14-13

Barber, R. M., Porter, B. F., Li, Q., May, M., Claiborne, M. K., Allison, A. B., ... Schatzberg, S. J. (2012). Broadly reactive polymerase chain reaction for pathogen detection in canine granulomatous meningoencephalomyelitis and necrotizing meningoencephalitis. *Journal of Veterinary Internal Medicine, 26*(4), 962–968. doi: 10.1111/j.1939-1676.2012.00954.x

Barzon, L., Lavezzo, E., Costanzi, G., Franchin, E., Toppo, S., & Palu, G. (2013). Next-generation sequencing technologies in diagnostic virology. *Journal of Clinical Virology, 58*(2), 346–350. doi: 10.1016/j.jcv.2013.03.003

Belyi, V. A., Levine, A. J., & Skalka, A. M. (2010). Unexpected inheritance: Multiple integrations of ancient bornavirus and ebolavirus/marburgvirus sequences in vertebrate genomes. *PLoS Pathogens, 6*(7), e1001030. doi: 10.1371/journal.ppat.1001030

Benton, D., Williams, C., & Brown, A. (2007). Impact of consuming a milk drink containing a probiotic on mood and cognition. *European Journal of Clinical Nutrition, 61*(3), 355–361. doi: 10.1038/sj.ejcn.1602546

Bercik, P., Collins, S. M., & Verdu, E. F. (2012). Microbes and the gut-brain axis. *Neurogastroenterology and Motility, 24*(5), 405–413. doi: 10.1111/j.1365-2982.2012.01906.x

Berdoy, M., Webster, J. P., & Macdonald, D. W. (2000). Fatal attraction in rats infected with Toxoplasma gondii. *Proceedings Biological Sciences/The Royal Society, 267*(1452), 1591–1594.

Berenreiterova, M., Flegr, J., Kubena, A. A., & Nemec, P. (2011). The distribution of Toxoplasma gondii cysts in the brain of a mouse with latent toxoplasmosis: Implications for the behavioral manipulation hypothesis. *PLoS One, 6*(12), e28925.

Berman, R. M., Cappiello, A., Anand, A., Oren, D. A., Heninger, G. R., Charney, D. S., & Krystal, J. H. (2000). Antidepressant effects of ketamine in depressed patients. *Biological Psychiatry, 47*(4), 351–354.

Bibby, K., Viau, E., & Peccia, J. (2011). Viral metagenome analysis to guide human pathogen monitoring in environmental samples. *Letters in Applied Microbiology, 52*(4), 386–392. doi: 10.1111/j.1472-765X.2011.03014.x

Bisgaard, H., Bonnelykke, K., & Stokholm, J. (2014). Immune-mediated diseases and microbial exposure in early life. *Clinical and Experimental Allergy, 44*(4), 475–481. doi: 10.1111/cea.12291

Bode, L., Durrwald, R., Rantam, F. A., Ferszt, R., & Ludwig, H. (1996). First isolates of infectious human Borna disease virus from patients with mood disorders. *Molecular Psychiatry, 1*(3), 200–212.

Bode, L., Ferszt, R., & Czech, G. (1993). Borna disease virus infection and affective disorders in man. *Archives of Virology, Suppl., 7*, 159–167.

Bode, L., Reckwald, P., Severus, W. E., Stoyloff, R., Ferszt, R., Dietrich, D. E., & Ludwig, H. (2001). Borna disease virus-specific circulating immune complexes, antigenemia, and free antibodies—the key marker triplet determining infection and prevailing in severe mood disorders. *Molecular Psychiatry, 6*(4), 481–491. doi: 10.1038/sj.mp.4000909

Bode, L., Zimmermann, W., Ferszt, R., Steinbach, F., & Ludwig, H. (1995). Borna disease virus genome transcribed and expressed in psychiatric patients. *Nature Medicine, 1*(3), 232–236.

Bravo, J. A., Forsythe, P., Chew, M. V., Escaravage, E., Savignac, H. M., Dinan, T. G., ... Cryan, J. F. (2011). Ingestion of Lactobacillus strain regulates emotional behavior and central GABA receptor expression in a mouse via the vagus nerve. *Proceedings of the National Academy of Sciences USA, 108*(38), 16050–16055. doi: 10.1073/pnas.1102999108

Burcusa, S. L., & Iacono, W. G. (2007). Risk for recurrence in depression. *Clinical Psychological Review, 27*(8), 959–985. doi: S0272-7358(07)00058-X [pii] 10.1016/j.cpr.2007.02.005

Caddy, C., Giaroli, G., White, T. P., Shergill, S. S., & Tracy, D. K. (2014). Ketamine as the prototype glutamatergic antidepressant: Pharmacodynamic actions, and a systematic review and meta-analysis of efficacy. *Therapeutic Advances in Psychopharmacology, 4*(2), 75–99. doi: 10.1177/2045125313507739

Canli, T., & Lesch, K. -P. (2007). Long story short: The serotonin transporter in emotion regulation and social cognition. *Nature Neuroscience, 10*(9), 1103–1109.

Canli, T., Qiu, M., Omura, K., Congdon, E., Haas, B. W., Amin, Z.,... Lesch, K. P. (2006). Neural correlates of epigenesis. *Proceedings of the National Academy of Sciences of the United States of America, 103*(43), 16033–16038.

Caspi, A., Sugden, K., Moffitt, T. E., Taylor, A., Craig, I. W., Harrington, H.,... Poulton, R. (2003). Influence of life stress on depression: Moderation by a polymorphism in the 5-HTT gene. *Science, 301*(18 Jul 2003), 386–389.

Castro, J. A., de Mecca, M. M., & Bartel, L. C. (2006). Toxic side effects of drugs used to treat Chagas' disease (American trypanosomiasis). *Human and Experimental Toxicology, 25*(8), 471–479.

Centers for Disease Control (CDC). (2014). *Parasites—Toxoplasmosis (Toxoplasma infection).* Retrieved April 14, 2014, from http://www.cdc.gov/parasites/toxoplasmosis/

Chen, C. -Y., Tzeng, N. -S., & Chen, Y. -C. (2010). Maintenance therapy of celecoxib for major depression with mimicking neuropsychological dysfunction. *General Hospital Psychiatry, 32*(6), 647.e647–e649.

Chiodi, F., Norkrans, G., Hagberg, L., Sonnerborg, A., Gaines, H., Froland, S.,... Vandvik, B. (1988). Human immunodeficiency virus infection of the brain. II. Detection of intrathecally synthesized antibodies by enzyme linked immunosorbent assay and imprint immunofixation. *Journal of the Neurological Sciences, 87*(1), 37–48.

Chiodi, F., Sonnerborg, A., Albert, J., Gaines, H., Norkrans, G., Hagberg, L.,... Fenyo, E. M. (1988). Human immunodeficiency virus infection of the brain. I. Virus isolation and detection of HIV specific antibodies in the cerebrospinal fluid of patients with varying clinical conditions. *Journal of the Neurological Sciences, 85*(3), 245–257.

Ciesla, J. A., & Roberts, J. E. (2001). Meta-analysis of the relationship between HIV infection and risk for depressive disorders. *American Journal of Psychiatry, 158*(5), 725–730.

Conway, C. C., Hammen, C., Brennan, P. A., Lind, P. A., & Najman, J. M. (2010). Interaction of chronic stress with serotonin transporter and catechol-O-methyltransferase polymorphisms in predicting youth depression. *Depression and Anxiety, 27*(8), 737–745.

Cook, S., Chung, B. Y., Bass, D., Moureau, G., Tang, S., McAlister, E.,... Firth, A. E. (2013). Novel virus discovery and genome reconstruction from field RNA samples reveals highly divergent viruses in dipteran hosts. *PLoS One, 8*(11), e80720. doi: 10.1371/journal.pone.0080720

Cornelis, G., Heidmann, O., Bernard-Stoecklin, S., Reynaud, K., Veron, G., Mulot, B.,... Heidmann, T. (2012). Ancestral capture of syncytin-Car1, a fusogenic endogenous retroviral envelope gene involved in placentation and conserved in Carnivora. *Proceedings of the National Academy of Sciences USA, 109*(7), E432–E441. doi: 10.1073/pnas.1115346109

Critchley, H. D. (2005). Neural mechanisms of autonomic, affective, and cognitive integration. *Journal of Comparative Neurology, 493*(1), 154–166. doi: 10.1002/cne.20749

Cryan, J. F., & Dinan, T. G. (2012). Mind-altering microorganisms: The impact of the gut microbiota on brain and behaviour. *Nature Reviews Neuroscience, 13*(10), 701–712. doi: 10.1038/nrn3346

Czygan, M., Hallensleben, W., Hofer, M., Pollak, S., Sauder, C., Bilzer, T.,... Lieb, K. (1999). Borna disease virus in human brains with a rare form of hippocampal degeneration but not in brains of patients with common neuropsychiatric disorders. *Journal of Infectious Diseases, 180*(5), 1695–1699. doi: 10.1086/315068

Dantzer, R., O'Connor, J. C., Freund, G. G., Johnson, R. W., & Kelley, K. W. (2008). From inflammation to sickness and depression: When the immune system subjugates the brain. *Nature Reviews Neuroscience, 9*(1), 46–56. doi: 10.1038/nrn2297

Dean, B., Gibbons, A. S., Tawadros, N., Brooks, L., Everall, I. P., & Scarr, E. (2013). Different changes in cortical tumor necrosis factor-alpha-related pathways in schizophrenia and mood disorders. *Molecular Psychiatry, 18*(7), 767–773. doi: 10.1038/mp.2012.95

Dean, B., Tawadros, N., Scarr, E., & Gibbons, A. S. (2010). Regionally-specific changes in levels of tumour necrosis factor in the dorsolateral prefrontal cortex obtained postmortem from subjects with major depressive disorder. *Journal of Affective Disorders, 120*(1–3), 245–248. doi: 10.1016/j.jad.2009.04.027

Del Guerra, F. B., Fonseca, J. L., Figueiredo, V. M., Ziff, E. B., & Konkiewitz, E. C. (2013). Human immunodeficiency virus-associated depression: Contributions of immuno-inflammatory, monoaminergic, neurodegenerative, and neurotrophic pathways. *Journal of Neurovirology, 19*(4), 314–327. doi: 10.1007/s13365-013-0177-7

Desbonnet, L., Garrett, L., Clarke, G., Bienenstock, J., & Dinan, T. G. (2008). The probiotic Bifidobacteria infantis: An assessment of potential antidepressant properties in the rat. *Journal of Psychiatric Research, 43*(2), 164–174. doi: 10.1016/j.jpsychires.2008.03.009

Desbonnet, L., Garrett, L., Clarke, G., Kiely, B., Cryan, J. F., & Dinan, T. G. (2010). Effects of the probiotic Bifidobacterium infantis in the maternal separation model of depression. *Neuroscience, 170*(4), 1179–1188. doi: 10.1016/j.neuroscience.2010.08.005

Diener, C., Kuehner, C., Brusniak, W., Ubl, B., Wessa, M., & Flor, H. (2012). A meta-analysis of neurofunctional imaging studies of emotion and cognition in major depression. *Neuroimage, 61*(3), 677 685. doi: S1053-8119(12)00386-2 [pii] 10.1016/j.neuroimage.2012.04.005

Dietrich, D. E., & Bode, L. (2008). Human Borna disease virus-infection and its therapy in affective disorders. *Acta Pathologica, Microbiologica, et Immunologica Scandinavica, Suppl.*(124), 61–65.

Dinan, T. G. (2009). Inflammatory markers in depression. *Current Opinion in Psychiatry, 22*(1), 32–36. doi: 10.1097/YCO.0b013e328315a561

Dinan, T. G., & Cryan, J. F. (2013). Melancholic microbes: A link between gut microbiota and depression? *Neurogastroenterology and Motility, 25*(9), 713–719. doi: 10.1111/nmo.12198

Doggett, J. S., Ojo, K. K., Fan, E., Maly, D. J., & Van Voorhis, W. C. (2014). Bumped kinase inhibitor 1294 treats established Toxoplasma gondii infection. *Antimicrobial Agents and Chemotherapy.* doi: 10.1128/AAC.01823-13

Dowlati, Y., Herrmann, N., Swardfager, W., Liu, H., Sham, L., Reim, E. K., & Lanctot, K. L. (2010). A meta-analysis of cytokines in major depression. *Biological Psychiatry, 67*(5), 446–457. doi: S0006-3223(09)01229-3 [pii] 10.1016/j.biopsych.2009.09.033

Drevets, W. C., Price, J. L., & Furey, M. L. (2008). Brain structural and functional abnormalities in mood disorders: Implications for neurocircuitry models of depression. *Brain Structure and Function, 213*(1–2), 93–118.

Eley, T. C., Sugden, K., Corsico, A., Gregory, A. M., Sham, P., McGuffin, P.,... Craig, I. W. (2004). Gene-environment interaction analysis of serotonin system markers with adolescent depression. *Molecular Psychiatry, 9*(10), 908–915.

Fountoulakis, K. N., & Moller, H. -J. (2011). Efficacy of antidepressants: A re-analysis and re-interpretation of the Kirsch data. *International Journal of Neuropsychopharmacology (CINP), 14*(3), 405–412.

Foxman, E. F., & Iwasaki, A. (2011). Genome-virome interactions: Examining the role of common viral infections in complex disease. *Nature Reviews Microbiology, 9*(4), 254–264. doi: Nrmicro2541 [pii] 10.1038/nrmicro2541

Fu, Z. F., Amsterdam, J. D., Kao, M., Shankar, V., Koprowski, H., & Dietzschold, B. (1993). Detection of Borna disease virus-reactive antibodies from patients with affective disorders by western immunoblot technique. *Journal of Affective Disorders, 27*(1), 61–68.

Gartlehner, G., Hansen, R. A., Morgan, L. C., Thaler, K., Lux, L., Van Noord, M.,... Lohr, K. N. (2011). Comparative benefits and harms of second-generation antidepressants for treating major depressive disorder: An updated meta-analysis. *Annals of Internal Medicine, 155*(11), 772–785. doi: 155/11/772 [pii] 10.1059/0003-4819-155-11-201112060-00009

Gaskell, E. A., Smith, J. E., Pinney, J. W., Westhead, D. R., & McConkey, G. A. (2009). A unique dual activity amino acid hydroxylase in Toxoplasma gondii. *PLoS One, 4*(3), e4801.

Gilbert, C., Schaack, S., Pace, J. K., 2nd, Brindley, P. J., & Feschotte, C. (2010). A role for host-parasite interactions in the horizontal transfer of transposons across phyla. *Nature, 464*(7293), 1347–1350. doi: 10.1038/nature08939

Gilbert, P. (2006). Evolution and depression: Issues and implications. *Psychological Medicine, 36*(3), 287–297. doi: 10.1017/S0033291705006112

Haga, S., Yoshimura, M., Motoi, Y., Arima, K., Aizawa, T., Ikuta, K.,... Ikeda, K. (1997). Detection of Borna disease virus genome in normal human brain tissue. *Brain Research, 770*(1–2), 307–309.

Hamilton, J. P., Etkin, A., Furman, D. J., Lemus, M. G., Johnson, R. F., & Gotlib, I. H. (2012). Functional neuroimaging of major depressive disorder: A meta-analysis and new integration of base line activation and neural response data. *American Journal of Psychiatry, 169*(7), 693–703. doi: 10.1176/appi.ajp.2012.11071105

Hammoud, D. A., Endres, C. J., Hammond, E., Uzuner, O., Brown, A., Nath, A.,... Pomper, M. G. (2010). Imaging serotonergic transmission with [11C]DASB-PET in depressed and non-depressed patients infected with HIV. *Neuroimage, 49*(3), 2588–2595. doi: 10.1016/j.neuroimage.2009.10.037

Hannestad, J., DellaGioia, N., & Bloch, M. (2011). The effect of antidepressant medication treatment on serum levels of inflammatory cytokines: A meta-analysis. *Neuropsychopharmacology, 36*(12), 2452–2459. doi: Npp2011132 [pii] 10.1038/npp.2011.132

Hornig, M., Briese, T., Licinio, J., Khabbaz, R. F., Altshuler, L. L., Potkin, S. G.,... Lipkin, W. I. (2012). Absence of evidence for bornavirus infection in schizophrenia, bipolar disorder and major depressive disorder. *Molecular Psychiatry, 17*(5), 486–493. doi: Mp2011179 [pii] 10.1038/mp.2011.179

Howren, M. B., Lamkin, D. M., & Suls, J. (2009). Associations of depression with C-reactive protein, IL-1, and IL-6: A meta-analysis. *Psychosomatic Medicine, 71*(2), 171–186. doi: 10.1097/PSY.0b013e3181907c1b

Hsu, P. C., Groer, M., & Beckie, T. (2014). New findings: Depression, suicide, and Toxoplasma gondii infection. *Journal of the American Association of Nurse Practitioners.* doi: 10.1002/2327-6924.12129

Hughes, D. P., Andersen, S. B., Hywel-Jones, N. L., Himaman, W., Billen, J., & Boomsma, J. J. (2011). Behavioral mechanisms and morphological symptoms of zombie ants dying from fungal infection. *BMC Ecology, 11*, 13. doi: 10.1186/1472-6785-11-13

Human Microbiome Project Consortium. (2012*a*). A framework for human microbiome research. *Nature, 486*(7402), 215–221. doi: 10.1038/nature11209

Human Microbiome Project Consortium. (2012*b*). Structure, function and diversity of the healthy human microbiome. *Nature, 486*(7402), 207–214. doi: 10.1038/nature11234

Iwata, Y., Takahashi, K., Peng, X., Fukuda, K., Ohno, K., Ogawa, T.,... Shigeta, S. (1998). Detection and sequence analysis of borna disease virus p24 RNA from peripheral blood mononuclear cells of patients with mood disorders or schizophrenia and of blood donors. *Journal of Virology, 72*(12), 10044–10049.

Jiao, L., Han, X., Wang, S., Fan, Y., Yang, M., Qiu, H., & Yang, X. (2009). Imprinted DC mediate the immune-educating effect of early-life microbial exposure. *European Journal of Immunology, 39*(2), 469–480. doi: 10.1002/eji.200838367

Kaufman, J., Yang, B. Z., Douglas-Palumberi, H., Houshyar, S., Lipschitz, D., Krystal, J. H., & Gelernter, J. (2004). Social supports and serotonin transporter gene moderate depression in maltreated children. *Proceedings of the National Academy of Sciences USA, 101*(49), 17316–17321. Epub 12004 Nov 17324.

Kendler, K. S. (1998). Major depression and the environment: A psychiatric genetic perspective. *Pharmacopsychiatry, 31*, 5–9.

Kessler, R. C., Petukhova, M., Sampson, N. A., Zaslavsky, A. M., & Wittchen, H. -U. (2012). Twelve-month and lifetime prevalence and lifetime morbid risk of anxiety and mood disorders in the United States. *International Journal of Methods in Psychiatric Research, 21*(3), 169–184. doi: 10.1002/mpr.1359

Khan, A., Bhat, A., Kolts, R., Thase, M. E., & Brown, W. (2010). Why has the antidepressant-placebo difference in antidepressant clinical trials diminished over the past three decades? *CNS Neuroscience and Therapeutics, 16*(4), 217–226. doi: CNS151 [pii] 10.1111/j.1755-5949.2010.00151.x

Kim, Y. K., Kim, S. H., Choi, S. H., Ko, Y. H., Kim, L., Lee, M. S.,... Song, J. W. (1999). Failure to demonstrate Borna disease virus genome in peripheral blood mononuclear cells from psychiatric patients in Korea. *Journal of Neurovirology, 5*(2), 196–199.

Kinross, J. M., von Roon, A. C., Holmes, E., Darzi, A., & Nicholson, J. K. (2008). The human gut microbiome: Implications for future health care. *Current Gastroenterology Reports, 10*(4), 396–403.

Kirsch, I., Deacon, B. J., Huedo-Medina, T. B., Scoboria, A., Moore, T. J., & Johnson, B. T. (2008). Initial severity and antidepressant benefits: A meta-analysis of data submitted to the Food and Drug Administration. *PLoS Medicine, 5*(2), e45. doi: 07-PLME-RA-0042 [pii] 10.1371/journal.pmed.0050045

Koesters, M., Guaiana, G., Cipriani, A., Becker, T., & Barbui, C. (2013). Agomelatine efficacy and acceptability revisited: Systematic review and meta-analysis of published and unpublished randomised trials. *British Journal of Psychiatry, 203*(3), 179–187. doi: 10.1192/bjp.bp.112.120196

Kreuze, J. F., Perez, A., Untiveros, M., Quispe, D., Fuentes, S., Barker, I., & Simon, R. (2009). Complete viral genome sequence and discovery of novel viruses by deep sequencing of small RNAs: A generic method for diagnosis, discovery and sequencing of viruses. *Virology, 388*(1), 1–7. doi: 10.1016/j.virol.2009.03.024

Krishnadas, R., & Cavanagh, J. (2012). Depression: An inflammatory illness? *Journal of Neurology Neurosurgery and Psychiatry, 83*(5), 495–502. doi: Jnnp-2011-301779 [pii] 10.1136/jnnp-2011-301779

Krugel, U., Fischer, J., Radicke, S., Sack, U., & Himmerich, H. (2013). Antidepressant effects of TNF-alpha blockade in an animal model of depression. *Journal of Psychiatric Research, 47*(5), 611–616. doi: 10.1016/j.jpsychires.2013.01.007

Kuczynski, J., Lauber, C. L., Walters, W. A., Parfrey, L. W., Clemente, J. C., Gevers, D., & Knight, R. (2012). Experimental and analytical tools for studying the human microbiome. *Nature Reviews Genetics, 13*(1), 47–58. doi: Nrg3129 [pii] 10.1038/nrg3129

Kurdyukov, S. G., Lebedev, Y. B., Artamonova, II, Gorodentseva, T. N., Batrak, A. V., Mamedov, I. Z., . . . Sverdlov, E. D. (2001). Full-sized HERV-K (HML-2) human endogenous retroviral LTR sequences on human chromosome 21: Map locations and evolutionary history. *Gene, 273*(1), 51–61.

Lafferty, K. D. (2006). Can the common brain parasite, Toxoplasma gondii, influence human culture? *Proceedings. Biological Sciences/The Royal Society, 273*(1602), 2749–2755. doi: R8663292G0L526V6 [pii] 10.1098/rspb.2006.3641

Lander, E. S., Linton, L. M., Birren, B., Nusbaum, C., Zody, M. C., Baldwin, J., . . . International Human Genome Sequencing Consortium. (2001). Initial sequencing and analysis of the human genome. *Nature, 409*(6822), 860–921. doi: 10.1038/35057062

Lannes-Vieira, J., de Araujo-Jorge, T. C., Soeiro Mde, N., Gadelha, P., & Correa-Oliveira, R. (2010). The centennial of the discovery of Chagas disease: Facing the current challenges. *PLoS Neglected Tropical Diseases, 4*(6), e645. doi: 10.1371/journal.pntd.0000645

Lesch, K. P. (2004). Gene-environment interaction and the genetics of depression. *Journal of Psychiatry and Neuroscience, 29*(3), 174–184.

Lester, D. (2010). Brain parasites and suicide. *Psychological Reports, 107*(2), 424.

Lewinsohn, P. M., Rohde, P., Seeley, J. R., & Fischer, S. A. (1993). Age-cohort changes in the lifetime occurrence of depression and other mental disorders. *Journal of Abnormal Psychology, 102*(1), 110–120.

Lewis, C. M., Ng, M. Y., Butler, A. W., Cohen-Woods, S., Uher, R., Pirlo, K., . . . McGuffin, P. (2010). Genome-wide association study of major recurrent depression in the U.K. population. *American Journal of Psychiatry, 167*(8), 949–957. doi: 10.1176/appi.ajp.2010.09091380

Ling, V. J., Lester, D., Mortensen, P. B., Langenberg, P. W., & Postolache, T. T. (2011). Toxoplasma gondii seropositivity and suicide rates in women. *Journal of Nervous and Mental Disease, 199*(7), 440–444. doi: 10.1097/NMD.0b013e318221416e

Liu, Y., Ho, R. C., & Mak, A. (2012). Interleukin (IL)-6, tumour necrosis factor alpha (TNF-alpha) and soluble interleukin-2 receptors (sIL-2R) are elevated in patients with major depressive disorder: A meta-analysis and meta-regression. *Journal of Affective Disorders, 139*(3), 230–239. doi: S0165-0327(11)00460-5 [pii] 10.1016/j.jad.2011.08.003

Maes, M., Kubera, M., & Leunis, J. C. (2008). The gut-brain barrier in major depression: Intestinal mucosal dysfunction with an increased translocation of LPS from gram negative enterobacteria (leaky gut) plays a role in the inflammatory pathophysiology of depression. *Neuro Endocrinology Letters, 29*(1), 117–124.

Maes, M., Kubera, M., Leunis, J. C., & Berk, M. (2012). Increased IgA and IgM responses against gut commensals in chronic depression: Further evidence for increased bacterial translocation or leaky gut. *Journal of Affective Disorders, 141*(1), 55–62. doi: 10.1016/j.jad.2012.02.023

Mann, J. J. (2003). Neurobiology of suicidal behaviour. *Nature Reviews Neuroscience, 4*(10), 819–828. doi: 10.1038/nrn1220

McConkey, G. A., Martin, H. L., Bristow, G. C., & Webster, J. P. (2013). Toxoplasma gondii infection and behaviour—location, location, location? *Journal of Experimental Biology, 216*(Pt. 1), 113–119.

McGowan, P. O., & Kato, T. (2008). Epigenetics in mood disorders. *Environmental Health and Preventive Medicine, 13*(1), 16–24. doi: 10.1007/s12199-007-0002-0

Messaoudi, M., Lalonde, R., Violle, N., Javelot, H., Desor, D., Nejdi, A., . . . Cazaubiel, J. M. (2011). Assessment of psychotropic-like properties of a probiotic formulation (Lactobacillus helveticus R0052 and Bifidobacterium longum R0175) in rats and human subjects. *British Journal of Nutrition, 105*(5), 755–764. doi: 10.1017/S0007114510004319

Miguel-Hidalgo, J. J., Nithuairisg, S., Stockmeier, C., & Rajkowska, G. (2007). Distribution of ICAM-1 immunoreactivity during aging in the human orbitofrontal cortex. *Brain Behavior and Immunity, 21*(1), 100–111. doi: 10.1016/j.bbi.2006.05.001

Miguel-Hidalgo, J. J., Overholser, J. C., Jurjus, G. J., Meltzer, H. Y., Dieter, L., Konick, L., . . . Rajkowska, G. (2011). Vascular and extravascular immunoreactivity for intercellular adhesion molecule 1 in the orbitofrontal cortex of subjects with major depression: Age-dependent changes. *Journal of Affective Disorders, 132*(3), 422–431. doi: 10.1016/j.jad.2011.03.052

Mikova, O., Yakimova, R., Bosmans, E., Kenis, G., & Maes, M. (2001). Increased serum tumor necrosis factor alpha concentrations in major depression and multiple sclerosis. *European Neuropsychopharmacology, 11*(3), 203–208.

Moloney, R. D., Desbonnet, L., Clarke, G., Dinan, T. G., & Cryan, J. F. (2014). The microbiome: Stress, health and disease. *Mammalian Genome, 25*(1–2), 49–74. doi: 10.1007/s00335-013-9488-5

Montoya, J. G., & Liesenfeld, O. (2004). Toxoplasmosis. *Lancet, 363*(9425), 1965–1976. doi: 10.1016/S0140-6736(04)16412-X

Mulder, I. E., Schmidt, B., Lewis, M., Delday, M., Stokes, C. R., Bailey, M., . . . Kelly, D. (2011). Restricting microbial exposure in early life negates the immune benefits associated with gut colonization in environments of high microbial diversity. *PLoS One, 6*(12), e28279. doi: 10.1371/journal.pone.0028279

Muller, N., Schwarz, M. J., Dehning, S., Douhe, A., Cerovecki, A., Goldstein-Muller, B.,... Riedel, M. (2006). The cyclooxygenase-2 inhibitor celecoxib has therapeutic effects in major depression: Results of a double-blind, randomized, placebo controlled, add-on pilot study to reboxetine. *Molecular Psychiatry, 11*(7), 680–684.

Munoz, M., Liesenfeld, O., & Heimesaat, M. M. (2011). Immunology of Toxoplasma gondii. *Immunological Reviews, 240*(1), 269–285.

Na, K. S., Tae, S. H., Song, J. W., & Kim, Y. K. (2009). Failure to detect borna disease virus antibody and RNA from peripheral blood mononuclear cells of psychiatric patients. *Psychiatry Investigation, 6*(4), 306–312. doi: 10.4306/pi.2009.6.4.306

Nery, F. G., Monkul, E. S., Hatch, J. P., Fonseca, M., Zunta-Soares, G. B., Frey, B. N.,... Soares, J. C. (2008). Celecoxib as an adjunct in the treatment of depressive or mixed episodes of bipolar disorder: A double-blind, randomized, placebo-controlled study. *Human Psychopharmacology, 23*(2), 87–94.

Nesse, R. M. (2000). Is depression an adaptation? *Archives of General Psychiatry, 57*(1), 14–20.

Oberlander, T. F., Weinberg, J., Papsdorf, M., Grunau, R., Misri, S., & Devlin, A. M. (2008). Prenatal exposure to maternal depression, neonatal methylation of human glucocorticoid receptor gene (NR3C1) and infant cortisol stress responses. *Epigenetics, 3*(2), 97–106.

Oldstone, M. B. (2012). The game's afoot: Seeking viruses that cause chronic and degenerative neurologic and psychiatric disorder. *Molecular Psychiatry, 17*(5), 472–473. doi: 10.1038/mp.2012.5

Olszak, T., An, D., Zeissig, S., Vera, M. P., Richter, J., Franke, A.,... Blumberg, R. S. (2012). Microbial exposure during early life has persistent effects on natural killer T cell function. *Science, 336*(6080), 489–493. doi: 10.1126/science.1219328

Oquendo, M. A., Placidi, G. P., Malone, K. M., Campbell, C., Keilp, J., Brodsky, B.,... Mann, J. J. (2003). Positron emission tomography of regional brain metabolic responses to a serotonergic challenge and lethality of suicide attempts in major depression. *Archives of General Psychiatry, 60*(1), 14–22.

Peterson, J., Garges, S., Giovanni, M., McInnes, P., Wang, L., Schloss, J. A.,... Guyer, M. (2009). The NIH Human Microbiome Project. *Genome Research, 19*(12), 2317–2323. doi: Gr.096651.109 [pii] 10.1101/gr.096651.109

Philibert, R. A., Sandhu, H., Hollenbeck, N., Gunter, T., Adams, W., & Madan, A. (2007). The relationship of 5HTT (SLC6A4) methylation and genotype on mRNA expression and liability to major depression and alcohol dependence in subjects from the Iowa Adoption Studies. *American Journal of Medical Genetics. Part B, Neuropsychiatric Genetics, 147B*(5), 543–549.

Pittella, J. E. H. (2009). Central nervous system involvement in Chagas disease: A hundred-year-old history. *Transactions of the Royal Society of Tropical Medicine and Hygiene, 103*(10), 973–978.

Prandovszky, E., Gaskell, E., Martin, H., Dubey, J. P., Webster, J. P., & McConkey, G. A. (2011). The neurotropic parasite Toxoplasma gondii increases dopamine metabolism. *PLoS One, 6*(9), e23866.

Raedler, T. J. (2011). Inflammatory mechanisms in major depressive disorder. *Current Opinion in Psychiatry, 24*(6), 519–525. doi: 10.1097/YCO.0b013e32834b9db6

Repetto, M. J., Evans, D. L., Cruess, D. G., Gettes, D. R., Douglas, S. D., & Petitto, J. M. (2003). Neuropsychopharmacologic treatment of depression and other neuropsychiatric disorders in HIV-infected individuals. *CNS Spectrums, 8*(1), 59–63.

Rhee, S. H., Pothoulakis, C., & Mayer, E. A. (2009). Principles and clinical implications of the brain-gut-enteric microbiota axis. *Nature Reviews Gastroenterological Hepatology, 6*(5), 306–314. doi: 10.1038/nrgastro.2009.35

Risch, N., Herrell, R., Lehner, T., Liang, K. Y., Eaves, L., Hoh, J.,... Merikangas, K. R. (2009). Interaction between the serotonin transporter gene (5-HTTLPR), stressful life events, and risk of depression: A meta-analysis. *Journal of the American Medical Association, 301*(23), 2462–2471. doi: 301/23/2462 [pii] 10.1001/jama.2009.878

Rogers, R. D., Owen, A. M., Middleton, H. C., Williams, E. J., Pickard, J. D., Sahakian, B. J., & Robbins, T. W. (1999). Choosing between small, likely rewards and large, unlikely rewards activates inferior and orbital prefrontal cortex. *Journal of Neuroscience, 19*(20), 9029–9038.

Romijn, J. A., Corssmit, E. P., Havekes, L. M., & Pijl, H. (2008). Gut-brain axis. *Current Opinion in Clinical Nutrition and Metabolic Care, 11*(4), 518–521. doi: 10.1097/MCO.0b013e328302c9b0

Rosenquist, J. N., Fowler, J. H., & Christakis, N. A. (2011). Social network determinants of depression. *Molecular Psychiatry, 16*(3), 273–281. doi: Mp201013 [pii] 10.1038/mp.2010.13

Shelton, R. C., Claiborne, J., Sidoryk-Wegrzynowicz, M., Reddy, R., Aschner, M., Lewis, D. A., & Mirnics, K. (2011). Altered expression of genes involved in inflammation and apoptosis in frontal cortex in major depression. *Molecular Psychiatry, 16*(7), 751–762. doi: 10.1038/mp.2010.52

Singh, S. P., Singh, V., & Kar, N. (2012). Efficacy of agomelatine in major depressive disorder: Meta-analysis and appraisal. *International Journal of Neuropsychopharmacology, 15*(3), 417–428. doi: 10.1017/S1461145711001301

Slattery, D. A., Desrayaud, S., & Cryan, J. F. (2005). GABAB receptor antagonist-mediated antidepressant-like behavior is serotonin-dependent. *Journal of Pharmacology and Experimental Therapeutics, 312*(1), 290–296. doi: 10.1124/jpet.104.073536

Springer, S. A., Chen, S., & Altice, F. (2009). Depression and symptomatic response among HIV-infected drug users enrolled in a randomized controlled trial of directly administered antiretroviral therapy. *AIDS Care, 21*(8), 976–983. doi: 10.1080/09540120802657555

Stibbs, H. H. (1985). Changes in brain concentrations of catecholamines and indoleamines in Toxoplasma gondii infected mice. *Annals of Tropical Medicine and Parasitology, 79*(2), 153–157.

Stilling, R. M., Dinan, T. G., & Cryan, J. F. (2014). Microbial genes, brain & behaviour—epigenetic regulation of the gut-brain axis. *Genes Brain Behavior, 13*(1), 69–86. doi: 10.1111/gbb.12109

Sudo, N., Chida, Y., Aiba, Y., Sonoda, J., Oyama, N., Yu, X. N.,... Koga, Y. (2004). Postnatal microbial colonization programs the hypothalamic-pituitary-adrenal system for stress response in mice. *Journal of Physiology, 558*(Pt. 1), 263–275. doi: 10.1113/jphysiol.2004.063388

Tang, P., & Chiu, C. (2010). Metagenomics for the discovery of novel human viruses. *Future Microbiology, 5*(2), 177–189. doi: 10.2217/fmb.09.120

Taylor, D., Sparshatt, A., Varma, S., & Olofinjana, O. (2014). Antidepressant efficacy of agomelatine: Meta-analysis of published and unpublished studies. *British Medical Journal, 348*, g1888. doi: 10.1136/bmj.g1888

Thomas, F., Schmidt-Rhaesa, A., Martin, G., Manu, C., Durand, P., & Renaud, F. (2002). Do hairworms (Nematomorpha) manipulate the water seeking behaviour of their terrestrial hosts? *Journal of Evolutionary Biology, 15*(3), 356–361.

Tillisch, K., Labus, J., Kilpatrick, L., Jiang, Z., Stains, J., Ebrat, B., . . . Mayer, E. A. (2013). Consumption of fermented milk product with probiotic modulates brain activity. *Gastroenterology, 144*(7), 1394–1401, 1401 e1391–1394. doi: 10.1053/j.gastro.2013.02.043

Tonelli, L. H., Stiller, J., Rujescu, D., Giegling, I., Schneider, B., Maurer, K., . . . Postolache, T. T. (2008). Elevated cytokine expression in the orbitofrontal cortex of victims of suicide. *Acta Psychiatrica Scandinavica, 117*(3), 198–206. doi: 10.1111/j.1600-0447.2007.01128.x

Torrey, E. F., Bartko, J. J., Lun, Z. R., & Yolken, R. H. (2007). Antibodies to Toxoplasma gondii in patients with schizophrenia: A meta-analysis. *Schizophrenia Bulletin, 33*(3), 729–736. doi: Sbl050 [pii] 10.1093/schbul/sbl050

Torrey, E. F., Bartko, J. J., & Yolken, R. H. (2012). Toxoplasma gondii and other risk factors for schizophrenia: An update. *Schizophrenia Bulletin, 38*(3), 642–647. doi: Sbs043 [pii] 10.1093/schbul/sbs043

Torrey, E. F., & Yolken, R. H. (2003). Toxoplasma gondii and schizophrenia. *Emerging Infectious Diseases, 9*(11), 1375–1380. doi: 10.3201/eid0911.030143

Turnbaugh, P. J., Hamady, M., Yatsunenko, T., Cantarel, B. L., Duncan, A., Ley, R. E., . . . Gordon, J. I. (2009). A core gut microbiome in obese and lean twins. *Nature, 457*(7228), 480–484. doi: Nature07540 [pii] 10.1038/nature07540

Turner, E. H., Matthews, A. M., Linardatos, E., Tell, R. A., & Rosenthal, R. (2008). Selective publication of antidepressant trials and its influence on apparent efficacy. *New England Journal of Medicine, 358*(3), 252–260. doi: 358/3/252 [pii] 10.1056/NEJMsa065779

Tyring, S., Gottlieb, A., Papp, K., Gordon, K., Leonardi, C., Wang, A., . . . Krishnan, R. (2006). Etanercept and clinical outcomes, fatigue, and depression in psoriasis: Double-blind placebo-controlled randomised phase III trial. *Lancet, 367*(9504), 29–35. doi: 10.1016/S0140-6736(05)67763-X

Verbeek, E. C., Bakker, I. M., Bevova, M. R., Bochdanovits, Z., Rizzu, P., Sondervan, D., . . . Heutink, P. (2012). A fine-mapping study of 7 top scoring genes from a GWAS for major depressive disorder. *PLoS One, 7*(5), e37384. doi: 10.1371/journal.pone.0037384

Vilar-Pereira, G., Silva, A. A., Pereira, I. R., Silva, R. R., Moreira, O. C., de Almeida, L. R., . . . Lannes-Vieira, J. (2012). Trypanosoma cruzi-induced depressive-like behavior is independent of meningoencephalitis but responsive to parasiticide and TNF-targeted therapeutic interventions. *Brain Behavior and Immunity, 26*(7), 1136–1149. doi: 10.1016/j.bbi.2012.07.009

Vyas, A., Kim, S. K., Giacomini, N., Boothroyd, J. C., & Sapolsky, R. M. (2007). Behavioral changes induced by Toxoplasma infection of rodents are highly specific to aversion of cat odors. *Proceedings of the National Academy of Sciences USA, 104*(15), 6442–6447. doi: 10.1073/pnas.0608310104

Walter, J., & Ley, R. (2011). The human gut microbiome: Ecology and recent evolutionary changes. *Annual Review of Microbiology, 65*, 411–429. doi: 10.1146/annurev-micro-090110-102830

Wang, X., Zhang, L., Lei, Y., Liu, X., Zhou, X., Liu, Y., . . . Xie, P. (2014). Meta-analysis of infectious agents and depression. *Scientific Reports, 4*, 4530. doi: 10.1038/srep04530

Warner-Schmidt, J. L., Vanover, K. E., Chen, E. Y., Marshall, J. J., & Greengard, P. (2011). Antidepressant effects of selective serotonin reuptake inhibitors (SSRIs) are attenuated by antiinflammatory drugs in mice and humans. *Proceedings of the National Academy of Sciences USA, 108*(22), 9262–9267. doi: 10.1073/pnas.1104836108

Webster, J. P., Lamberton, P. H., Donnelly, C. A., & Torrey, E. F. (2006). Parasites as causative agents of human affective disorders? The impact of anti-psychotic, mood-stabilizer and anti-parasite medication on Toxoplasma gondii's ability to alter host behaviour. *Proceedings. Biological Sciences/The Royal Society, 273*(1589), 1023–1030. doi: F314V9210P112451 [pii] 10.1098/rspb.2005.3413

Whitaker, J. W., McConkey, G. A., & Westhead, D. R. (2009). The transferome of metabolic genes explored: Analysis of the horizontal transfer of enzyme encoding genes in unicellular eukaryotes. *Genome Biology, 10*(4), R36.

Wylie, K. M., Mihindukulasuriya, K. A., Sodergren, E., Weinstock, G. M., & Storch, G. A. (2012). Sequence analysis of the human virome in febrile and afebrile children. *PLoS One, 7*(6), e27735. doi: 10.1371/journal.pone.0027735 PONE-D-11-15957 [pii]

Wylie, K. M., Weinstock, G. M., & Storch, G. A. (2012). Emerging view of the human virome. *Translational Research, 160*(4), 283–290. doi: S1931-5244(12)00128-4 [pii] 10.1016/j.trsl.2012.03.006

Yagmur, F., Yazar, S., Temel, H. O., & Cavusoglu, M. (2010). May Toxoplasma gondii increase suicide attempt-preliminary results in Turkish subjects? *Forensic Science International, 199*(1–3), 15–17. doi: 10.1016/j.forsciint.2010.02.020

Yatsunenko, T., Rey, F. E., Manary, M. J., Trehan, I., Dominguez-Bello, M. G., Contreras, M., . . . Gordon, J. I. (2012). Human gut microbiome viewed across age and geography. *Nature, 486*(7402), 222–227. doi: http://www.nature.com/nature/journal/v486/n7402/abs/nature11053.htmlsupplementary-information

Yolken, R. H., Bachmann, S., Ruslanova, I., Lillehoj, E., Ford, G., Torrey, E. F., & Schroeder, J. (2001). Antibodies to Toxoplasma gondii in individuals with first-episode schizophrenia. *Clinical Infectious Diseases, 32*(5), 842–844. doi: CID000667 [pii] 10.1086/319221

Zeigler Allen, L., Ishoey, T., Novotny, M. A., McLean, J. S., Lasken, R. S., & Williamson, S. J. (2013). Isolation and genome analysis of single virions using "single virus genomics". *Journal of Visualized Experiments, 75*, e3899. doi: 10.3791/3899

Zhang, L., Xu, M. M., Zeng, L., Liu, S., Liu, X., Wang, X., . . . Xie, P. (2014). Evidence for Borna disease virus infection in neuropsychiatric patients in three western China provinces. *European Journal of Clinical Microbiology and Infectious Diseases, 33*(4), 621–627. doi: 10.1007/s10096-013-1996-4

CHAPTER

15 Consequences of Early-Life Experiences on Cognition and Emotion: A Role for Nutrition and Epigenetic Mechanisms

Eva F. G. Naninck, Paul J. Lucassen *and* Aniko Korosi

Abstract

Perinatal experiences during a critical developmental period program brain structure and function "for life", thereby determining vulnerability to psychopathology and cognition in adulthood. Although these functional consequences are associated with alterations in HPA-axis activity and hippocampal structure and function, the underlying mechanisms remain unclear. The parent-offspring relationship (i.e., sensory and nutritional inputs by the mother) is key in mediating these lasting effects. This chapter discusses how early-life events, for example, the amount of maternal care, stress, and nutrition, can affect emotional and cognitive functions later in life. Interestingly, effects of perinatal malnutrition resemble the perinatal stress-induced long-term deficits. Because stress and nutrition are closely interrelated, it proposes that altered stress hormones and changes in specific key nutrients during critical developmental periods act synergistically to program brain structure and function, possibly via epigenetic mechanisms. Understanding how the adult brain is shaped by early experiences is essential to develop behavioural and nutritional preventive therapy.

Key Words: early-life, HPA-axis, hippocampus, plasticity, neurogenesis, epigenetics, maternal care, nutrition

I. Introduction

Perinatal life is a period of unique sensitivity during which specific experiences and environmental conditions can have lasting effects on the susceptibility to develop mental disorders later in life (Gluckman, Hanson, & Beedle, 2007; Loman, Gunnar, 2010; Lupien, McEwen, Gunnar, & Heim, 2009). The active process, whereby a stimulus or insult, given or occurring during a critical period, has irreversible long-term effects on the organism, is called programming. During the critical developmental period in early-life (EL), species-specific parental stimulation is the main determinant of the quality of the EL environment. Several crucial components are embedded in the parent-offspring relationships (e.g., sensorimotor, nutritional). The loss or alteration of any of these components results in a deregulation of the structural and functional changes that occur during development, and can result in disturbances in adulthood.

Indeed, there is clinical evidence for programming of mental health "for life" by prenatal as well as by early postnatal events. Prenatal events, for example, maternal depression (Gutteling, de Weerth & Buitelaar, 2005; O'Connor et al., 2005), severe stress during pregnancy caused by war (van Os & Selten, 1998), earthquake (Watson, Mednick, Huttunen & Wang, 1999), exposure to the World Trade Center attacks on 9/11 in New York City (Berkowitz et al., 2003; Yehuda et al., 2005), famine (Roseboom, Painter, van Abeelen, Veenendaal & de Rooij, 2011) or prenatal alcohol exposure (Willoughby, Sheard, Nash & Rovet, 2008) have all been associated with higher incidence of psychopathologies and an increased activity of the neuroendocrine stress system (but see: Selten, van der Graaf, van Duursen, Gispen-de Wied & Kahn, 1999), cognitive impairments and lasting abnormalities in hippocampal development. Similarly, during postnatal development, adverse EL experiences, for example, the loss of a parent, social deprivation, physical and/or sexual abuse, or rearing by a drug or alcohol-abusive mother, are associated with increased vulnerability to psychopathology (Maselko, Kubzansky, Lipsitt & Buka, 2011; Stevens et al., 2008) and impaired cognitive function later in life (Chugani et al., 2001; Kaplan et al., 2001; Mueller et al., 2010; Nelson et al., 2007). Interestingly, these changes are associated with abnormal brain activity in limbic areas (Chugani et al., 2001) and diminished hippocampal volume (Teicher, Anderson, & Polcari, 2012).

In contrast, positive EL experiences (e.g., excellent maternal care) have been associated with resilience to affective disorders (Gourion et al., 2008; Maselko et al., 2011) and larger hippocampal volume (Luby et al., 2012). In addition, children exposed to prenatal and/or early postnatal malnutrition also exhibit cognitive deficits in adolescence and adulthood (Benton, 2010; de Rooij, Wouters, Yonker, Painter & Roseboom, 2010; de Souza, Fernandes & Tavares do Carmo, 2011; Galler, Ramsey, Solimano, Lowell & Mason, 1983; Klein et al., 1976; Laus, Duarte Manhas Ferreira Vales, Braga Costa & Sousa Almeida, 2011; Lucas, 1998; Walker, Grantham-McGregor, Powell & Chang, 2000), (but see de Groot et al., 2011) and enhanced risk of developing psychiatric disorders such as depression (Brown, van Os, Driessens, Hoek & Susser, 2000; Costello, Worthman, Erkanli & Angold, 2007). Considering that, in most cases, the aforementioned environments cannot be avoided, it is fundamental to understand the mechanisms underlying the lasting effects of early-life stress (ELS) on brain functioning and disease susceptibility in order to develop appropriate therapeutic and preventive intervention.

In the literature, two key hypotheses lay the basis for most clinical and animal research aiming to unravel the mechanisms underlying programming by EL experience. The first one is formulated by Barker as "*developmental origins of health and disease*" (DOHaD) (Barker, 2004; 2007; Gluckman et al., 2007). It proposes that the EL environment, from the beginning of gestation until the beginning of puberty, can set the stage for adult pathology. The second hypothesis is referred to as the *predictive adaptive response* hypothesis and proposes that the EL environment is perceived as the anticipated future environment. Therefore, in response to changes in the EL environment, the progeny will adapt to improve its chance of survival (Gluckman & Hanson, 2004) specifically within that predicted environment. However, disparities between the predicted and the actual environment may result in disease. Throughout this chapter we will discuss the validity of these hypotheses.

The conclusions that can be drawn from human epidemiological research are complex and limited because, in clinical studies, it remains difficult to prove causality between EL experiences and long-term outcomes. For example, retrospective reports are likely biased by recall, and prospective longitudinal studies might be influenced by the genetic background and uncontrollable environmental factors. Therefore, animal models are key in unravelling the precise molecular mechanisms underlying lasting effects of EL experience on mental health. These models allow for prospective longitudinal studies, manipulation of single environmental factors, and control of genetic make-up (Nestler & Hyman, 2010).

Because the most commonly used models are based on rats and mice, we will largely review studies based on rodent models, next to evidence from relevant clinical studies. In this chapter, we will further explore in great detail how the various key elements of the EL environment, including maternal care, stress, and nutrition, can program emotional and cognitive function and affect susceptibility to psychopathology and cognitive decline later in life.

II. Early-life experience and stress-related emotional disorders

Over the past 50 years, there has been a wealth of clinical evidence for the bidirectional programming of emotional health by early-life (EL) experience (Gutteling et al., 2005; Maselko et al., 2011; O'Connor et al., 2005; Stevens et al., 2008; van Os & Selten, 1998; Watson et al., 1999; Willoughby et al., 2008; Yehuda et al., 2005). Programming of the hypothalamic-pituitary-adrenal (HPA) axis, which controls the stress response, is thought to be instrumental in this and to underlie the lasting influence of EL experience on neuroendocrine and behavioral responses to stress and thereby on emotional health (Heim, Plotsky & Nemeroff, 2004). Upon exposure to stress, the HPA axis is activated, leading to the initial release of corticotrophin-releasing hormone (CRH) from the hypothalamic paraventricular nucleus (PVN), stimulation of pituitary ACTH secretion into the blood, and the subsequent release of glucocorticoids from the adrenal glands: corticosterone (CORT) in rodents and cortisol in humans. Negative feedback takes place when glucocorticoids bind to glucocorticoid receptors (GRs) in hippocampus, PVN, prefrontal cortex, and pituitary and, thereby, inhibits the release of CRH and ACTH (Spencer, Miller, Stein & McEwen, 1991; Swanson & Simmons, 1989).

Basal stress hormone levels and the magnitude of the stress response are tightly regulated and influenced by levels of both hippocampal GR and hypothalamic CRH (Joels & Baram, 2009; Walker & Dallman, 1993). In rodents, the development of hypothalamic neuronal circuits takes place during a critical time-window encompassing the first 2–3 postnatal weeks of rodent life (Morgane, Mokler, & Galler, 2002). Because these circuits are involved in regulating the neuroendocrine stress axis, disturbance in the chronology of cellular and molecular events during this particular period may induce irreversible and long-lasting changes in HPA-axis tone. Indeed, experimentally induced positive or negative modulation of the EL environment alters basal and stress induced HPA-axis activity in adulthood in experimental animals (Avishai-Eliner, Gilles, Eghbal-Ahmadi, Bar-El & Baram, 2001; Brunson, Eghbal-Ahmadi, Bender, Chen & Baram, 2001; Hess, 1969; Korosi et al., 2010; Levine & Lewis, 1959; Meaney et al., 1996). Next, we will discuss these effects in detail and gain insight into the possible underlying mechanisms.

A. *Stress hormone regulation early in life*

Interestingly, HPA-axis activity early in life has been described to be different from its activity in adulthood. In fact, the first 2 weeks of rodent postnatal life are often referred to as the stress-hyporesponsive period because of experiments showing that neither (a) stressors that are effective in the adult, nor (b) direct stimulation with ACTH, can significantly elevate plasma CORT levels in 2–13-day-old pups (Sapolsky & Meaney, 1986; Walker, Perrin, Vale & Rivier, 1986). However, exposure of pups to age-appropriate physiological and psychological stressors, such as cold (Yi & Baram, 1994) or prolonged (e.g., 3 hours or longer) or repeated maternal separation (MS), does result in short-term increases in HPA-axis activity (Avishai-Eliner, Yi, Newth & Baram, 1995; Dent, Okimoto, Smith & Levine, 2000; Plotsky & Meaney, 1993; Suchecki, Mozaffarian, Gross, Rosenfeld & Levine, 1993). In addition, prolonged maternal separation (MS) augments the CORT response to application of exogenous ACTH in pups (Stanton, Gutierrez & Levine, 1988).

Therefore, during the first two weeks of rodent life, when the pups are dependent on the dam for nutrition, warmth, and sensory stimulation, it appears that manipulations, particularly those that disturb any component of the dam-pup relationship, are very effective in inducing early-life stress (ELS) conditions (see next section). Thus, the stress-hyporesponsive period per se does not completely describe the full extent of stress responsiveness during the early postnatal phase. In the following sections, we will consider how adverse or positive experiences during the pre- and postnatal developmental period can affect HPA-axis activity long term.

B. *Lasting consequences of early-life stress exposure on the set point of the neuroendocrine stress axis and on emotional outcomes*

Epidemiological studies have reported that *prenatal* stress is associated with an increased risk to various psychopathologies and a higher anxiety later in life (see Glover, 2011). Indeed, high maternal cortisol and psychosocial stress during pregnancy have been found to alter HPA-activity in infants (Davis, Glynn, Waffarn & Sandman, 2011), pre-adolescent children (Gutteling et al., 2005), and in young adults (Entringer, Kumsta, Hellhammer, Wadhwa & Wüst, 2009). The effects of prenatal factors on HPA-axis activity and emotion of the offspring have been extensively studied in rodent models as well. In these models, the intrauterine environment is experimentally manipulated by exposure of the pregnant dam to alcohol, toxins, or stress, for example. Typically, prenatal stress is induced by exposure of the pregnant dam to single or repeated restraint (Weinstock, 2001) or to psychological stress (Abe et al., 2007). Prenatal repeated restraint stress is associated with elevated basal CORT levels and reduced GR expression in the hippocampus and PFC of the adult male offspring (Green et al., 2011; but see Szuran, Pliska, Pokorny, & Welzl, 2000). In addition to alterations in HPA-activity, prenatal stress has been found to increase behavioural measures of anxiety in adulthood (Abe et al., 2007; Zuena et al., 2008).

During *postnatal* life, the most potent stressors consist of disrupting the dam-pup interaction by single prolonged separation for 24 hours (maternal deprivation; MD) or repeated daily separations of the dam and pups for 3–5 hours (maternal separation; MS). These manipulations have been found to modulate HPA-axis activity and stress responsiveness acutely and long term (Oomen et al., 2011; Plotsky & Meaney, 1993; Workel, Oitzl, Fluttert, Lesscher, Karssen & de Kloet, 2001). Following MD, basal expression of CRH mRNA in the PVN was reported to be unchanged (Avishai-Eliner et al., 1995; Dent et al., 2000) or reduced (Schmidt Enthoven, et al., 2004). This was accompanied by reduced CRH receptor 2 ($CRFR_2$) mRNA expression in the ventromedial hypothalamus (Eghbal-Ahmadi, Hatalski, Avishai-Eliner & Baram, 1997) on postnatal day (P) 8. ACTH and CORT response to a stressor, such as cold exposure or restraint, were markedly increased in prepubertal animals (Avishai-Eliner et al., 1995; Dent et al., 2000). Effects of MD on HPA-axis activity in response to stress are still present in adulthood, but the direction of the effect largely depends on the age of the animal and on the testing conditions (Workel, et al., 2001).

Similarly, repeated daily MS during the first 2 postnatal weeks augments the hormonal stress response in rodents throughout life and is accompanied by a decreased GR binding in both hippocampus and hypothalamus (Plotsky & Meaney, 1993) (Chen et al., 2012) but see (Lehmann et al., 2002; Pryce, Bettschen, & Feldon, 2001), and leads to increased fearfulness in tests of novelty and anxiety (Caldji et al., 1998; Ladd et al., 2000; Wigger & Neumann, 1999) and to depressive-like behaviour at an adult age (Huot, Plotsky, Lenox, & McNamara, 2002; Ladd et al., 2000). Interestingly, also the expression level of several genes that are essential regulators of the neuroendocrine and behavioural stress response are persistently altered in recurrently separated rats. They include: increased CRH expression in the amygdala, locus ceruleus (LC), and parabrachial nucleus; increased stress-induced CRH hnRNA expression in the PVN (Chen et al., 2012); and increased CRH receptor 1 ($CRFR_1$) mRNA expression levels in the LC and raphe nucleus (Ladd, Owens & Nemeroff, 1996). Furthermore, levels of arginine vasopressin (AVP) in the PVN and bed nucleus of the stria terminalis (BnST) have been reported to be either increased or decreased by MS (Desbonnet, Garrett, Daly, McDermott & Dinan, 2008; Murgatroyd et al., 2009; Veenema & Neumann, 2009).

So far, we have described how MS/MD affects the HPA-axis and emotional function later in life. However, as these models induce acute or recurrent stress, they lack important aspects of ELS as present in humans, in which stress is typically chronic and the mother/caregiver is present, but unable to give her child the appropriate care (Kendall-Tackett, 2007; Koenen, Moffitt, Caspi, Taylor & Purcell, 2003; Whipple & Webster-Stratton, 1991). As unpredictable and fragmented maternal behaviour is a hallmark of the human neglect/abuse situation (Gaudin, Polansky, Kilpatrick & Shilton, 1996; Whipple & Webster-Stratton, 1991), an animal model of *chronic* ELS was developed to recapitulate these important elements of the human condition (Avishai-Eliner et al., 2001; Ivy, Brunson, Sandman & Baram, 2008; Rice, Sandman, Lenjavi & Baram, 2008). In this animal model, the amount of nesting and bedding material is limited during the first postnatal week, which induces chronic stress in the dam and results in erratic, fragmented maternal care (i.e., shortened bouts of nurturing behaviour and frequent shifts between behaviours). The fragmentation of maternal care subsequently causes chronic ELS in the pups, as evidenced by their elevated basal plasma CORT levels, a transient weight gain reduction by the end of the one-week stress period (at P9), increased adrenal weight, and reduced CRH receptor binding in the pituitary. These physiological changes are accompanied by reductions in CRH and GR mRNA expression in the PVN and in hippocampal $CRFR_1$ expression and GR in the frontal cortex, whereas hypothalamic AVP levels are unaffected (Avishai-Eliner et al., 2001; Gilles, Schultz & Baram, 1996). Although the ELS induced reduction in hypothalamic CRH was found to be transient in rats (Avishai-Eliner et al., 2001), it was still present in adult mice (Rice et al., 2008).

In summary, these studies indicate that acute, recurrent, and chronic stress early in life have lasting consequences on stress responsiveness and emotionality in adulthood associated with specific molecular changes.

C. Programming of stress responsiveness by enriched early-life environment

In contrast to the lasting effects of adverse EL experience, there is evidence that enriched EL experiences lead to resilience to stress-related pathologies at an adult age (Gourion et al., 2008; Maselko et al., 2011). As a result, several animal models of enriched EL environments have been developed and studied in order to understand how such resilience to adult mental disorders is conferred. These models consist of increased sensory stimulation by the mother. Next to natural variation in maternal care provided by the dam to the pup (Liu et al., 1997; Liu, Diorio, Day, Francis & Meaney, 2000), an experimental paradigm used to create an "enriched" EL environment is "handling." Typically, handling consists of brief (15 minute), recurrent, daily separations of the dam from her pups starting at P2 and lasting for at least a week (Fenoglio, Chen, & Baram, 2006b; Weaver et al., 2001), or even up to 3 weeks (Bhatnagar & Meaney, 1995; Hess, 1969; Plotsky & Meaney, 1993). Such brief separations of dam and pup are not considered to be stressful because, in nature, the dam is regularly away from the nest for periods of 20–30 minutes (Rosenblatt, 1975) and daily handling augments the maternally derived sensory input to the pups as they are extensively groomed and licked upon reunion with the dam (Fenoglio, Chen, & Baram, 2006; Korosi et al., 2010). Handled rats (H) at weaning exhibit lower plasma CORT concentrations in response to mild stress (exposure to novel stimuli) when compared to nonhandled (NH) controls (Levine, 1967). In contrast, elevations in plasma CORT following application of an electric shock are more rapid and initially higher in H than in NH rats (Levine, 1962) but show a more rapid return to basal levels (Denenberg, Brumaghim, Haltmeyer, & Zarrow, 1967).

Thus, H rats cope differently with stressful stimuli of varying intensities than controls (but see Ader, 1970). Although they perceive and respond to mild stressors, they recover faster from severe stressors (Chen et al., 2010). In addition, handling leads to resilience to depressive-like behavior (Meaney et al., 1991). The "handled phenotype" is thought to arise from lasting changes in the expression levels of key regulators of the HPA-axis. Indeed this hormonal and behavioral phenotype is accompanied by reduced CRH mRNA in the PVN, increased expression of hippocampal GR (Plotsky & Meaney, 1993), reduced levels of AVP in the median eminence (Viau, Sharma, Plotsky, & Meaney, 1993) and either decreased or increased AVP levels in the PVN (Todeschin et al., 2009; Viau et al., 1993).

These handling evoked changes in HPA-axis activity do not occur all at once but develop sequentially. A reduction in *Crh* gene expression in the PVN can already be observed at P9, at the end of a week of daily handling (Avishai-Eliner et al., 2001; Fenoglio, 2005). This event is followed by decreases in the hormonal responses to stress by P23 and in an increased hippocampal GR expression between P23–45. Next to the early onset of reduction in *Crh* expression, the key role of CRH for EL programming of the HPA-axis is further strengthened as pharmacological blockade of the $CRFR_1$ in NH rats during P10–17, is sufficient to upregulate hippocampal GR persistently and to confer the behavioral phenotype of improved cognitive functions seen in adult H rats (Fenoglio, 2005; Korosi & Baram, 2009).

The question of how maternal care reaches the PVN and mediates changes in *Crh* expression was addressed by using the immediate-early gene Fos (Fenoglio, Chen, & Baram, 2006). The increase in maternal nurturing behavior after reunion of H pups with the dam elicited a transient activation of neurons (expressing Fos) in the BnST, in the central nucleus of the amygdala (ACe) and in the thalamic paraventricular nucleus (PVT) (Fenoglio, Chen, & Baram, 2006). ACe and BnST are afferent to the PVN and are known to augment CRH expression (Akana & Dallman, 1997; Choi et al., 2007; Feldman, Conforti, Itzik, & Weidenfeld, 1994), while PVT is a region with major inhibitory output onto the ACe. It is, therefore, plausible to assume that the handling-induced activation of PVT neurons changes their activity (e.g., firing rate, neurotransmitter release), and alters the activity of the BnST and the ACe, and, thereby, the entire afferent input onto the PVN. Accordingly, a drastic but temporary reduction in the number and function of excitatory synapses on CRH-expressing neurons (without a change in inhibitory synapses) has been demonstrated in H pups immediately after the handling period (Korosi, Shanabrough, et al., 2010). Considering that PVN CRH levels are persistently reduced in H offspring, this "rewiring" of the hypothalamus is thought to underlie the initiation of the changes in CRH expression found in H rats, but probably other molecular mechanisms are responsible for maintaining these reduced levels of expression later in life. Later, these mechanisms will be discussed in more detail.

Strikingly, although positive and negative EL experiences result in different phenotypes, both handling and chronic ELS result in reduced CRH mRNA levels on P9. Yet, how can this CRH reduction on P9 be associated with different phenotypes later in life? The chronic ELS-evoked CRH reduction is present immediately after the stress period (at P9) in both rats (Avishai-Eliner et al., 2001) and mice (Rice et al., 2008), although this reduction is transient only in rats and still apparent in adult mice. The reduced CRH expression comes in parallel with indices of chronic stress, such as: elevated basal CORT levels, increased adrenal weight, decreased body weight, decreased GR mRNA in the hypothalamus and frontal cortex, and reduced CRH receptor binding in the pituitary (Avishai-Eliner et al., 2001). It is, therefore, likely that an impaired glucocorticoid negative feedback is responsible for the reduction in CRH mRNA after chronic ELS. In addition, prolonged enhanced CRH release, without an elevation in compensatory synthesis, may result in depletion of hypothalamic CRH mRNA. In contrast to this chronic ELS scenario, the handling–evoked CRH reduction in the PVN is not associated with changes in basal CORT, ACTH, or hypothalamic GR expression. Therefore, CRH repression induced by handling, followed by an ameliorated neuroendocrine stress response and increased hippocampal GR expression, is thought to be independent from glucocorticoid negative feedback.

In summary, both epidemiological and animal studies provide strong evidence for a causal relationship between the quality of EL events and the programming of the stress response and emotional function later in life. We will further discuss, in detail, the effects of EL experience on hippocampal structure and cognitive functions (see later), and address in the last part of this chapter; (a) the molecular mechanisms that may underlie the lasting changes after ELS, and (b) whether the lasting effects of adverse EL experiences can be prevented or (partially) reversed, and (c) whether the effects of positive EL experiences can be induced to improve emotional health.

III. Programming of cognitive functions, hippocampal structure, and synaptic plasticity by early-life experience

Earlier we have discussed the effects of early-life (EL) experiences on the long-term programming of HPA-axis responsiveness and on emotional function. As mentioned in the introduction, however, EL experience programs cognitive functions as well. Indeed, in humans, stressful EL experiences have been associated with impaired cognitive function later in life (Kaplan et al., 2001; Mueller et al., 2010; Nelson et al., 2007). This has also been confirmed in animal studies (Aisa, Tordera, Lasheras, Del Río & Ramírez, 2007; Brunson, Kristen, et al., 2005; Ivy et al., 2010; Oitzl, Workel, Fluttert, Frösch, & de Kloet, 2000; Oomen et al., 2010). In order to understand the biological mechanisms responsible for the lasting effects of EL experiences on cognition, much research has focused on the hippocampus. This limbic brain structure is of particular interest because it is not only important in regulating the stress response (Snyder, Soumier, Brewer, Pickel, & Cameron, 2011) but also crucial for learning, memory, and cognition (Squire, 1992; Squire, 2004). Therefore, it is important to realize that the EL experience-induced structural and functional changes that we will discuss in the following sections might not only be relevant for cognitive functions per se, but may also have implications for the lasting effects on emotional function as described earlier. Hippocampal volume appears to be bidirectionally affected by positive and negative EL experiences (Bremner et al., 1997; Luby et al., 2012; Teicher et al., 2012; Vythilingam et al., 2002). Although maternal support in early childhood is associated with larger hippocampal volume in young adolescents (Luby et al., 2012), adverse EL experiences (e.g., childhood abuse or maltreatment) correlates with reduced hippocampal volume in adults (Bremner et al., 1997; Teicher et al., 2012; Vythilingam et al., 2002). In addition, alterations in hippocampal volume have been associated with several psychopathologies, including major depression (Geuze, Vermetten, & Bremner, 2005). In the following section, we will briefly introduce hippocampal anatomy and plasticity before focusing on animal studies that provide evidence for an association between EL environment and cognitive functions. We will further explore whether alterations in hippocampal volume and integrity could be the basis for the lasting alterations in cognitive functions induced by EL environment.

A. The hippocampal formation and its forms of neuronal and synaptic plasticity

The hippocampal formation consists of: the hippocampus proper, including CA1, CA2, and CA3 (CA: cornu ammonis, Ammon's horn); the dentate gyrus (DG, which surrounds the hilus); the subiculum; and the entorhinal cortex (EC). The main hippocampal information processing unit is the trisynaptic circuit, which consists of connections between the EC and DG via performant path axons, and from there to CA3 neurons via their mossy fiber axons, and from CA3 to the CA1 subregion via Schaffer collaterals (for an extensive review see van Strien, Cappaert, & Witter,

2009). Converging evidence indicates that the hippocampus is functionally segmented along its dorso-ventral (or septotemporal) axis (Fanselow & Dong, 2010; Moser & Moser, 1998). Although the dorsal hippocampus has a prominent role in declarative memory (including spatial memory) (Moser, Moser, & Andersen, 1993), the ventral (or temporal) part of the hippocampus has been implicated in the regulation of stress, anxiety, and emotional memory, and affection (Bannerman et al., 2004; Fanselow & Dong, 2010; Trivedi & Coover, 2004).

Interestingly, the hippocampus is a brain region with an incredibly high degree of structural plasticity throughout life; it undergoes continuous dynamic changes in neuronal connectivity (e.g., synapse formation/elimination and dendritic extension/retraction). These can be assessed either morphologically (e.g., by characterization of dendritic complexity or arborization and spine shape/density) or functionally by recording, for example, long-term potentiation (LTP; Whitlock, Heynen, Shuler, & Bear, 2006). In addition, the hippocampus exhibits the ability to produce new neurons during adulthood. This fundamental, new form of structural plasticity is termed adult neurogenesis (AN) and comprises the multistep process in which stem cells and neuronal progenitor cells present in the adult brain undergo proliferation, migration, and subsequent differentiation into fully functional neurons that integrate into the existing circuitry (Kempermann, Wiskott, & Gage, 2004). It has been suggested that these newly generated neurons make relatively large contributions to hippocampal functioning and correlative changes in their number or survival have been described in relation to several hippocampal tasks or properties (Marin-Burgin, Mongiat, Pardi, & Schinder, 2012; Schmidt-Hieber, Jonas, & Bischofberger, 2004).

During adulthood, several hormonal and environmental conditions modify levels of AN. Factors that stimulate AN are exposure to an enriched environment (Kempermann, Kuhn & Gage, 1997), physical activity (Brown & Kuhn, 2003; van Praag, Kempermann & Gage, 1999), treatment with antidepressant drugs like selective serotonin reuptake inhibitors (SSRIs) and monoamine oxidase inhibitors (MAOIs) (Duman, Nakagawa & Malberg, 2001; Perera, Park & Nemirovskaya, 2008), and hippocampus-dependent learning (Gould, Beylin, Tanapat, Reeves & Shors, 1999). In contrast, ageing (Heine, Maslam, Joels, & Lucassen, 2004) and exposure to acute or chronic stress are potent inhibitors of AN (Cameron & Gould, 1994; Heine, Maslam, Zareno, Joels & Lucassen, 2004; Lucassen et al., 2010). Because the hippocampus is richly endowed with mineralocorticoid and glucocorticoid receptors (MR and GR) (Reul & de Kloet, 1985) and $CRFR_1$ (Chen, 2004; Chen, Brunson, Muller, Cariaga & Baram, 2000; Van Pett et al., 2000), it is not surprising that stress hormones significantly influence hippocampal structure and function (Joels, Pu, Wiegert, Oitzl, & Krugers, 2006). In humans, deregulation of hippocampal neurogenesis has been associated with cognitive impairment, addiction, and mood disorders (Castrén, 2005; Czéh & Lucassen, 2007). Similar aspects have been proposed in animal studies (e.g., Shors et al., 2001; Snyder et al., 2011). In general, abnormalities or reductions in the level of neuronal plasticity and synaptic function within the hippocampus are associated with impaired hippocampus-dependent learning. For instance, reduced AN (Shors et al., 2001), impaired LTP (Ivy et al., 2010), and reduced dendritic complexity (Bagot et al., 2009; Chen et al., 2010) have all been found to hamper learning and memory in hippocampus-dependent tasks.

In addition to factors during adulthood, EL experience has lasting effects on these forms of plasticity as well (Fenoglio, Brunson, & Baram, 2006; Korosi & Baram, 2010; Korosi et al., 2010; Korosi et al., 2011; McClelland, Korosi, Cope, Ivy & Baram, 2011;), thereby possibly affecting emotional and cognitive functions in later life. Functional deficiencies associated with a disturbed programming of the hippocampus include: heightened emotionality (Abe et al., 2007), heightened anxiety (Huot, Thrivikraman, Meaney & Plotsky, 2001), altered HPA-axis responsiveness (Brunson, Kramar, et al., 2005; Huot et al., 2002) and diminished cognitive capacity in learning and memory tasks (Aisa et al., 2007; Brunson, Kramar, et al., 2005; Ivy et al., 2010; Oitzl et al., 2000; Oomen et al., 2010). So, when stress is experienced during EL, its effects on hippocampal structure and function are persistent throughout life and have notable consequences for cognition and disease susceptibility, whereas when stress is experienced during adulthood, its effects are often reversible, after appropriate recovery periods or drug treatment (Heine, Maslam, Zareno, et al., 2004; Oomen, Mayer, de Kloet, Joels, & Lucassen, 2007; Sousa, Lukoyanov, Madeira, Almeida, & Paula-Barbosa, 2000).

The reason that hippocampal structure and function is highly susceptible to the EL environment might be partly due to the fact that hippocampal development is not complete at the time of birth (Rice & Barone, 2000). In rodents, formation of the DG is initiated around embryonic day 18 by migration of granule cell progenitors from the dentate neuroepithelium (Pleasure, Collins, & Lowenstein, 2000) and continues during the first two postnatal weeks. Therefore, it is plausible to assume that EL experience might interrupt or corrupt the functional and structural maturation of the hippocampal network in an irreversible manner. In the following section, we will discuss how both adverse and positive EL experiences can provoke permanent changes at various levels of hippocampal neuronal plasticity and discuss its implications on cognitive functions.

B. Early-life experiences alter hippocampus-dependent cognitive functions

EL experience, in addition to emotional functions, also affects cognition and hippocampus-dependent spatial learning and memory in rodents. *Prenatal* stress has been reported to induce spatial learning and memory deficits (Lemaire, Koehl, Le Moal, & Abrous, 2000; Wu et al., 2007; Zuena et al., 2008). For example, repeated restraint of the pregnant dam for 45 minutes, 3 times daily during the last week of gestation is found to impair spatial learning and memory in the offspring (Lemaire et al., 2000). Acquisition of spatial information in the Morris water maze (MWM; a spatial navigation task for rodents) was also impaired in adult rats exposed to *postnatal* stress, such as MD (Oitzl et al., 2000; Oomen et al., 2010) or repeated MS (Aisa et al., 2007; Huot et al., 2002). In addition, adult rats (Brunson et al., 2005a; Ivy et al., 2010) and mice (Rice et al., 2008) exposed to chronic ELS exhibited impaired performance in the MWM and in the novel-object recognition task (a task based on spontaneous exploratory behaviour of unfamiliar objects to test declarative memory in rodents). Interestingly, the cognitive deficits seem to progress with ageing as they are present in old (12 months old) but not in young adult (4–5 months old) rats, whereas the cognitive dysfunction in mice can already be detected at 4 months of age (Rice et al., 2008). Finally, neonatal handling is found to improve cognitive performance in adult offspring (Fenoglio, 2005; Fenoglio, Brunson, & Baram, 2006; Kosten, Lee, & Kim, 2007).

Thus, hippocampus-dependent cognitive functions are affected bidirectionally and permanently by EL experiences. In order to determine the underlying biological mechanisms, the effects of EL experiences on hippocampal structure and synaptic physiology can be addressed at various levels of complexity: synaptic connectivity, dendritic morphology, and levels of AN. The effects of EL experience on these forms of plasticity will be discussed in the next section.

C. Early-life experiences alter hippocampal structure

C.1 PRENATAL STRESS

Impaired spatial learning in 4–5-week-old male rats exposed to prenatal stress (footshock, prenatal restraint, psychological stress, or variable stress during gestation) has been associated with suppression of hippocampal LTP and facilitation of long-term depression (LTD) in CA1 (Yaka, Salomon, Matzner, & Marta Weinstock, 2007; Yang, Han, Cao, Li, & Xu, 2006). These changes were associated with alterations in spine density and dendritic length in DG, CA1, and CA3 (Weinstock, 2011), and with reduced levels of neuronal proliferation and survival in the DG (Lucassen, Bosch, Jousma, Krömer, Andrew, Seckl, & Neumann, 2009; Lemaire et al., 2000; Odagiri et al., 2008: Zuena et al., 2008) starting as early as P1 (Van den Hove et al., 2006) and P10 (Kawamura, Jihuan Chen, Takahashi, Ichitani, & Nakahara, 2006) and lasting up to 22 months of age (Lemaire et al., 2000), whereas differentiation was mostly unaffected by these prenatal EL experiences. Interestingly, learning could not upregulate neurogenesis in prenatally stressed rats (Lemaire et al., 2000), suggesting that the hippocampus of rats exposed to prenatal stress exhibits reduced plasticity to factors known otherwise to stimulate such dynamic changes.

C.2 POSTNATAL STRESS

Lasting changes in hippocampal neuronal plasticity and synaptic integrity have also been reported after exposure to postnatal stress in rodents. Adults rats subjected to chronic ELS during the first postnatal week (Brunson, Kramar, et al., 2005; Ivy et al., 2010) or low levels of maternal care (Bagot et al., 2009; Champagne et al., 2008) exhibit impaired LTP in CA1 and CA3, as well as aberrant growth of mossy fibers in CA3 and dendritic atrophy in CA1. In addition, MS and MD in early postnatal life have been shown to alter dendritic complexity (Oomen et al., 2011), reduce the number of granule cells in the DG (Oomen et al., 2011), the number of hippocampal glia (Leventopoulos et al., 2007), and mossy fiber density (Huot et al., 2002) in the offspring, notably in a sex-dependent manner. ELS affects levels of AN as well. Neuronal survival was decreased and apoptosis was increased in offspring of low-caring mothers versus offspring of high-caring mothers (Bredy, Grant, Champagne, & Meaney, 2003; Weaver, Grant, & Meaney, 2002). In addition, repeated MS leads to transiently increased (Nair et al., 2007) and lastingly decreased levels of proliferation (Mirescu, Peters, & Gould, 2004; Oomen et al., 2010), without affecting neuronal survival (Greisen, Altar, Bolwig, Whitehead, & Wörtwein, 2005; Mirescu et al., 2004) in the DG of the offspring. Similarly, MD is found to transiently increased numbers of immature (DCX-positive) neurons in rats at 3 weeks of age (Oomen et al., 2009), ultimately leading to reduced proliferation throughout the full rostro-caudal axis of the DG, and reduced differentiation in the caudal part of the DG at 10 weeks of age (Oomen et al., 2009). The evidence just presented suggests that the ELS-induced reduced neurogenic capacity observed later in life might be due to an increase in neurogenesis during the postnatal phase that might result in depletion of the neurogenic pool.

In contrast, an enriched EL environment, either attributable to natural variation in maternal care (Liu et al., 1997), or experimentally induced by early handling, induces the opposite effect when compared to the effects of MS/MD. Neonatal handling induces a direct increase in hippocampal proliferation associated with increased number of neurons and astrocytes in pyramidal cell layers of CA1, CA2, and CA3 both directly after the 10-day handling procedure as well as at P90 (Winkelmann-Duarte et al., 2011). Interestingly, although adult offspring of low-caring mothers or maternally deprived rats showed a reduced synaptic plasticity when studied under basal conditions, they exhibited improved learning and memory, and enhanced hippocampal synaptic plasticity under conditions that mimic stress (Bagot et al., 2009; Champagne et al., 2008; Oomen et al., 2010). In fact, contextual learning was enhanced and LTP in DG and CA1 was facilitated in the presence of CORT (simulating stressful conditions) both in maternally deprived rats and in offspring of low-caring mothers. These data, in line with the adaptive programming hypothesis, suggest that adverse EL events might prepare the organism to respond optimally to stressful contexts encountered later in life.

In summary, the studies described have shown that EL experiences during both pre- and postnatal development can bidirectionally alter hippocampal neuronal plasticity and synaptic integrity. This strongly supports the possibility that these structural changes might be involved in affected cognitive function.

IV. Molecules and mechanisms underlying early programming

In the previous sections, we reviewed evidence from human and animal studies demonstrating that early-life (EL) experience programs the stress response and cognitive functions, affects neuronal structure and synaptic functions, and persistently alters the expression levels of key genes (see also Figure 15.1).

Programming occurs in a bidirectional manner and is persistent throughout life. However, the determining factors of this EL environment and the mechanisms mediating these lasting consequences are yet unclear. Understanding the neurobiological basis of the enduring consequences of this programming is fundamental for promoting healthy human neurological function and preventing stress-related cognitive and affective disorders.

In this section, we will address this question at different levels of complexity and discuss (a) the active role of maternal care, (b) the role of circulating stress hormones and stress-related neuropeptides, and finally (c) address the possibility that epigenetic mechanisms are responsible for the lasting effects of EL experience on altered gene expression and, thereby, on brain structure and function.

A. Does maternal care actively modulate the lasting effects of early-life experience?

Considering that the mother-infant relationship is the main determinant of the quality of the EL environment, it is reasonable to assume that maternal care including tactile stimulation (e.g., ano-genital licking essential to induce urination and defecation in the pups), and provision of nutrition, are key factors in mediating the lasting effects of EL experience on brain structure and function of the offspring. In the section 'The Role of Perinatal Nutrition in Adult Emotional and Cognitive Functions' we will discuss in detail the role of nutrition in EL programming, whereas here we will focus on the proposed function of active maternal care in mediating the lasting effects of EL experience. The available evidence that EL manipulations, including maternal separation (MS) or

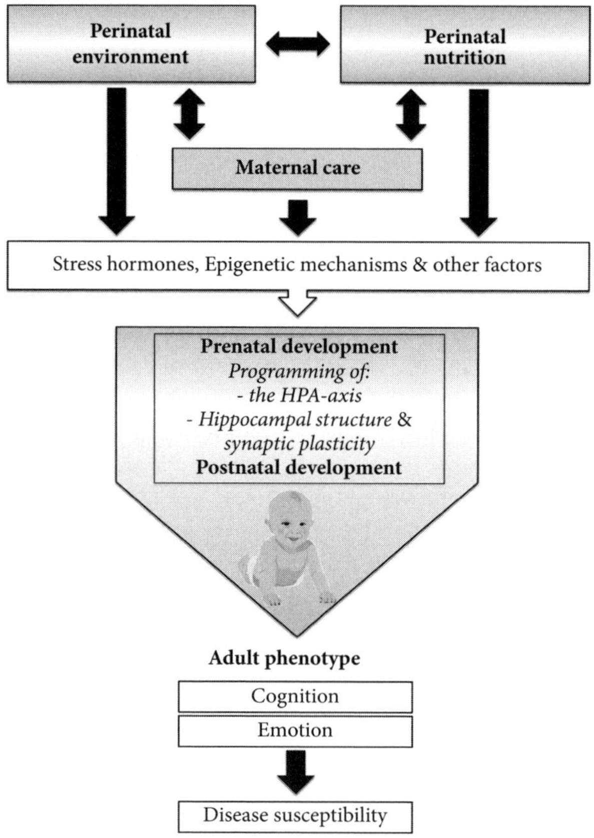

Fig. 15.1 Schematic representation of the pathways by which early-life experiences affect cognitive and emotional outcomes in adulthood and thereby determine disease susceptibility. During prenatal development (*in orange*) and postnatal development, various factors can lastingly influence the biology of the organism in a bidirectional manner. This scheme depicts the complex interaction among these factors, which are environment (stressful/enriched), nutrition, and maternal care. By affecting, amongst other things, stress hormone levels and epigenetic mechanisms, these factors can alter programming of the HPA-axis and programming of the hippocampal structure and synaptic plasticity.

deprivation (MD), chronic ELS and handling, alter the quality and/or quantity of maternal care (Brown, Smotherman & Levine, 1977; Brunson, Kramar, et al., 2005; Fenoglio, Brunson, & Baram, 2006; Korosi & Baram, 2010; Macri, Mason, & Wurbel, 2004; Pryce et al., 2001) strongly suggests that mother-infant interaction is crucial in programming brain and behavior. However, these studies do not directly address whether maternal care, under normal conditions, is actively involved in these regulations.

In support of this notion, it has been demonstrated that natural variation in maternal care (Bagot et al., 2009; Champagne et al., 2008; Liu et al., 1997; 2000) and individual within-litter variation in the amount of active care received (van Hasselt et al., 2012) leads to differences in stress response and cognitive functions in later life. These changes are associated with alterations in the endocrine system and in brain structure and plasticity as well. In line with this evidence from animal studies, in preterm and term neonates, moderate touch in the form of massage, kangaroo care (skin-to–skin contact between mother and infant) or other forms of sensory stimulation have been shown to lead to beneficial consequences, including reduction of pain responsiveness (Cignacco et al., 2007) and reduced reactivity to stress (Feldman, Singer, & Zagoory, 2010).

B. Role of glucocorticoids and stress-related neuropeptides in mediating the lasting effects of early-life experience

EL experience permanently alters, in a bidirectional way, the circulating levels of stress hormone (ACTH and corticosterone) and the expression of stress-related genes including CRH (Chen et al., 2012) GR (Weaver et al., 2004) AVP (Murgatroyd et al., 2009) and brain derived neurotrophic factor (BDNF) (Roth, Lubin, Funk, & Sweatt, 2009). If these persistently altered hormone levels and genes are indeed responsible for the functional consequences discussed earlier, then pharmacologically inhibiting or inducing such changes, or blocking the action of these peptides on their receptors, should prevent or reverse (in case of adverse EL environment) or induce (in case of enriched EL environment) these functional consequences. So far, the two most investigated candidates for the lasting effects of EL experience on stress response and cognitive functions are the CORT-GR/MR and the CRH-$CRFR_1$ systems.

B.1 GLUCOCORTICOIDS

Although EL experience has lasting consequences on the circulating level of the stress hormone corticosterone (CORT) (Chen et al., 2012; Ivy et al., 2008; Rice et al., 2008), as well as on GR and MR expression levels (Weaver et al., 2004), it is not clear whether these alterations are responsible for the lasting effects on brain structure and function. Even though CORT acting via its receptors is a logical candidate, there is some controversy in the literature about this possibility. First, CORT alteration is not consistently observed throughout models (Brunson, Kramar, et al., 2005; Lajud, Roque, Cajero, Gutiérrez-Ospina & Torner, 2011; Mirescu et al., 2004). Secondly, although neonatal dexamethasone treatment has lasting effects on spatial learning and hippocampal synaptic plasticity (Kamphuis et al., 2003), permanently reducing CORT levels by adrenalectomy at P10 does not alter subsequent levels of adult neurogenesis (Brunson, Baram, & Bender, 2005). The paucity of data points to the need for further research in this area; however, the contradictory data from the existing studies suggest that other factors may also contribute to the mechanisms by which EL experience programs brain structure and function.

B.2 CORTICOTROPIN RELEASING HORMONE (CRH)

There is ample evidence demonstrating that CRH expression is persistently altered by EL experience. Chronic ELS (Brunson, Kramar, et al., 2005; Ivy et al., 2010; Rice et al., 2008), MS (Chen et al., 2012), handling (Fenoglio, Brunson, & Baram, 2006; Korosi et al., 2010), or natural variation in maternal care (Liu et al., 1997) all lead to lifelong alterations of CRH expression in the PVN and chronic ELS increased CRH expression in the CA1 and CA3 regions of the hippocampus (Ivy et al., 2010). Consistent with the notion that CRH alterations are critically involved in the lasting effects of EL experience, chronic exposure to CRH alone alters hippocampal structure in a similar fashion as chronic ELS does (Brunson, Kramar, et al., 2005). Moreover, CRH repression is the first alteration in the sequence of changes induced by handling (Avishai-Eliner et al., 2001). Therefore, CRF receptor antagonists are potential candidates for reversing the deleterious effects of ELS or for promoting resilience to affective disorders.

Indeed, partially blocking CRF receptor type 1 ($CRFR_1$) in control rats during early development (P10–P17) was sufficient to confer the resilient phenotype that is characteristic of rats that received increased amounts of maternal care, including an up-regulation of hippocampal GR mRNA and improved cognitive functions (Fenoglio, 2005). Similarly, selective blocking of $CRFR_1$ during the week after chronic ELS (P10-P17) prevented LTP impairment and dendritic atrophy in CA1, and preserved hippocampus-dependent cognitive functioning (Ivy et al., 2010). In addition, conditional $CRFR_1$ knockout mice subjected to chronic ELS exhibited restored LTP and spine density in CA3 and enhanced LTP and spine density in CA1, in parallel with restored cognitive function (Wang et al., 2011).

In line with the concept of $CRFR_1$ as a molecular target for influencing emotional and cognitive function, it is interesting that the presence of specific combinations of single nucleotide polymorphisms in the *Crhr1* gene are protective against depressive symptoms in individuals when they are maltreated in childhood (Bradley et al., 2008; Tyrka et al., 2009). Therefore, in addition to obvious social and behavioural intervention strategies, pharmacological interventions targeting CRH-CRF receptor signaling may promote enhanced resilience to human stress-related disorders associated with EL experience.

Although up to now we discussed the actions of glucocorticoid and CRH as separate entities, they are obviously interrelated factors and, thus, their action might converge on the same cell and contribute in an additive manner to the consequences of EL experience. Indeed, administration early in life of CRH (Brunson, Kramar, et al., 2005) or glucocorticoids (Kamphuis et al., 2003) results in similar cognitive deficits later in life. Little is known though about the effects of CRH or CORT administration in EL on expression of GR/MR or CRFR, respectively, and the interaction between these two systems, therefore, warrants more research.

C. Do epigenetic mechanisms mediate the early-life experience induced lasting changes in gene expression?

A next question in understanding the consequences of early stress exposure relates to its molecular basis. An increasing number of reports support the possibility that epigenetic mechanisms are implicated in mediating the persistent effects of EL experience on gene expression (Chen et al., 2012; Murgatroyd et al., 2009; Roth et al., 2009; Weaver et al., 2004). Epigenetic modifications consist of alteration in the chromatin, modifications of the DNA (such as DNA methylation), post-translational modifications of histone proteins and regulation by noncoding RNAs. Alterations in chromatin structure determine whether a region of DNA is open and transcriptionally active (euchromatin), or whether it is condensed and largely transcriptionally inactive (heterochromatin). In contrast to the genome, the epigenome is dynamic, thereby allowing the organism to adapt to the environment. Some epigenetic modifications (e.g., changes in DNA methylation) are typically more stable than others and are, therefore, better candidates to mediate the lasting effects of EL experiences on the brain.

Indeed, the permanent increase in hippocampal GR expression induced by high levels of maternal care is associated with decreased DNA methylation and increased histone acetylation and transcription factor binding at the promoter region of the GR (Weaver et al., 2004; but see Daniels et al., 2009). Accordingly, translational research has found lower hippocampal GR expression and increased GR promoter DNA-methylation in suicide victims with a history of childhood abuse and/or neglect (McGowan et al., 2009).

Similarly, the ELS-induced persistent increase in levels of AVP is associated with enduring DNA hypomethylation at a key regulatory region of the *Avp* gene (Murgatroyd et al., 2009). In addition, reduced expression of brain-derived neurotrophic factor (BDNF) in the prefrontal cortex induced by ELS is accompanied by increased levels of DNA methylation at its IV exon promoter (Roth et al., 2009). Finally, the MD-induced increase in CRH expression in response to stress is accompanied by decreased methylation at the cyclic AMP-responsive element (a region critical for CRH transcriptional activation) (Chen et al., 2012). Accordingly, handling induced reduction in hypothalamic CRH expression is associated with persistently elevated levels of neuron restrictive silencing factor (NRSF) (Korosi et al., 2010). NRSF is a transcriptional repressor that functions by recruiting epigenetic machinery (Naruse, Aoki, Kojima & Mori, 1999; Roopra et al., 2000; Seth & Majzoub, 2001), supporting the notion that the persistent alteration of CRH expression by EL experience is mediated by epigenetic mechanisms.

Taken together, these studies indicate that environmental factors in EL can trigger the epigenetic machinery to persistently repress or activate transcription of a particular gene. So far we have discussed studies that focussed on epigenetic modifications at specific genes of interest (e.g., *Gr, Crh, Avp* and *Bdnf*). However, environmental factors in EL might also change the epigenome more globally. Indeed, there is evidence that the epigenetic response to maternal care is coordinated in clusters across broad genomic areas, suggesting that the epigenetic response to EL experience involves not only single candidate genes but also more global changes across the genome (McGowan et al., 2011). In addition, MS in mice changed global levels of histone deacetylases (HDACs), further supporting the idea of global changes across the epigenome (Levine, Worrell, Zimnisky & Schmauss, 2011). Accordingly, in human studies, small but statistically significant differences in whole-genome DNA methylation levels have been reported between institutionalized children and children that were raised by their biological parents, including specific genes known to be involved in the biosynthesis of hormones and neurotransmitters and in the control of their receptor activity (Naumova et al., 2011).

In summary, these studies indicate that the mechanisms underlying the lasting consequences of EL experience encompass different levels of complexity, spanning from sensory stimuli to epigenetic regulation of gene expression. These mechanisms most likely interact with each other and overlap in their effect; therefore, preventive and therapeutic interventions should take into account these various levels of complexity.

V. The role of perinatal nutrition in adult emotional and cognitive functions

In the previous sections we have discussed the consequences of negative and positive early-life (EL) experience on vulnerability to psychiatric disorders and cognitive decline later in life. Consistent with the notion that the mother-offspring interaction, and in particular maternal tactile stimulation, is important for brain development and later function, the amount of active maternal care received by rodent pups is closely linked to phenotypic outcome. Because the mother-offspring interaction consists of several components, it is likely that, next to tactile stimulation, additional features of the dam-pup interaction such as *nutrition* play an important role during the sensitive developmental period as well. Several studies indicate that the quantity and content of EL nutrition represents one of the principal nongenetic factors that is directly related to formation, growth, and functional organization of the organism later in life. A nutritional insult during the early period of life will cause irreversible morphological, neurophysiological, and functional damage that adversely influence brain development. Indeed, clinical evidence indicates that children exposed to prenatal and/or early postnatal malnutrition not only exhibit cognitive deficits in adolescence and adulthood (Benton, 2010; de Rooij et al., 2010; de Souza et al., 2011; Galler et al., 1983; Klein et al., 1976; Laus et al., 2011; Lucas, 1998; Walker et al., 2000; but see de Groot et al., 2011), but also an enhanced risk of developing psychiatric disorders like depression (Brown et al., 2000; Costello et al., 2007).

Unlike starvation, undernutrition and malnutrition are seen throughout the entire world, including the western society. The term *undernutrition* refers to chronic caloric insufficiency, when all nutrients required by the species are available in the diet but in insufficient amounts. The term *malnutrition* refers to an imbalanced diet in which one or more essential nutrients are absent (or present but in the wrong proportion). Although many nutrients are important for neuronal cell growth and development, some appear to have greater effects during the perinatal period, such as: protein, iron, zinc, selenium, iodine, folate, vitamin A, choline and long-chain polyunsaturated fatty acids. Obviously, the severity, timing, and duration of the nutritional insult will further determine the effects on emotional and cognitive development.

Several animal models have been established to investigate the mechanisms by which EL nutrition induces persistent changes in emotional and cognitive function later in life, a concept often referred to as *nutritional programming*. Because the offspring's diet early in life is exclusively dependent on maternal milk, and because its composition is determined by maternal nutrition, most animal models used to study how mal- or undernutrition in EL affects brain function are based on altering maternal nutrition. These models include: restriction of maternal protein intake during pregnancy and/or lactation (PR), global dietary restriction of 50% (FR50), and overnutrition with high fat diet (HF). Indeed, in adult animals exposed to perinatal malnutrition, deficits in learning and memory (Castro, Tracy & Rudy, 1989; Fukuda, Françolin-Silva, & Almeida, 2002; Jordan, Cane & Howells, 1981; Laus et al., 2011; Martínez et al., 2009; Tonkiss & J R Galler, 1990; Valadares, Fukuda, Françolin-Silva, Hernandes, & Almeida, 2010) but see (Bedi, 1992; Campbell & Bedi, 1989; Huang, 2003) and enhanced responsiveness to stress (Bertram, Trowern, Copin, Jackson, & Whorwood, 2001) have been documented.

In the next sections we will discuss how nutritional interventions during the early life period affect emotional and cognitive function later in life and next, we will discuss the possible mechanisms that might underlie these persistent consequences. Considering the tight relationship between nutrition, stress (Pijlman, Wolterink & Van Ree, 2003), and quality of maternal care (Wiener, Smotherman & Levine, 1976), we will consider the possible role of EL nutrition in mediating the long-term consequences of ELS in the offspring, next to the direct effects of EL nutrition (by the presence and/or absence of specific nutrients in the maternal diet).

A. Effects of perinatal malnutrition on stress and emotional function

Epidemiological evidence suggests that perinatal malnutrition leads to enhanced risk of developing psychiatric disorders such as depression (Brown et al., 2000; Costello et al., 2007). In order to understand how nutrition during critical developmental periods can affect the vulnerability to develop psychiatric disorders, we will first consider how nutrition interferes with the development of the brain regions involved in HPA-axis activity and consider the reciprocal influence that metabolic signals, nutrients, and the HPA-axis have on each other.

As mentioned earlier, development of the hypothalamic circuits takes place during critical phases in early postnatal life (Morgane et al., 2002). Indeed, perinatal maternal undernutrition and malnutrition lead to increased proliferation in the median eminence, PVN, and arcuate nucleus of the hypothalamus until P15 (Coupé et al., 2009) and to an increased number of neurons in the ventromedial hypothalamic nucleus and PVN at P20 (Plagemann et al., 2000). The hypothalamus is not only involved in the regulation of the HPA-axis but also important for mediating food intake and energy balance. This relationship between food intake, energy balance, and activity of the HPA-axis is well known as several studies reported alterations in basal HPA-axis activity and stress responsiveness in genetically obese rats (Duclos, Timofeeva, Michel, & Denis Richard, 2005), in rodents fed HF diet (Tannenbaum et al., 1997) or subjected to perinatal FR (Lesage, Blondeau, Grino, Bréant, & Dupouy, 2001; Lingas & Matthews, 2001; Sebaai, Lesage, Breton, Vieau, & Deloof, 2004). Alternatively, chronic stress conditions have been correlated with changes in food intake (Pijlman et al., 2003). Thus, food intake and activity of the HPA-axis are closely interrelated with many overlaying neuronal pathways and structures that can respond to (and integrate) both nutritional and stressful stimuli.

The HPA-axis is sensitive to modulation by metabolic signals including: leptin, insulin, glucose, and ghrelin. Leptin and glucose, elevated in a state of positive energy balance, reduce ACTH and CORT secretion (Benedict et al., 2009; Huang, Rivest & Richard, 1998). In contrast, plasma ghrelin, which is present at high concentrations under fasting conditions, stimulates the HPA-axis via CRH and AVP (Wren et al., 2002). This is also true early in life (Schmidt et al., 2006; Walker, 2010) when food intake and nutrition of the progeny are tightly dependent on the presence of the mother, the quality of her care, and the maternal diet. Indeed, Schmidt and colleagues demonstrated that metabolic signals are also crucial in programming the HPA axis in response to MS, next to quality and quantity of maternal care and circulating stress hormones (Schmidt et al., 2006). In fact, MS that is associated with decreased CRH in the PVN and an increase in CORT, also leads to reduced plasma glucose and leptin levels and increased ghrelin levels in the offspring. Pharmacologically blocking the reduction of glucose, or the increase in ghrelin, attenuated the HPA-axis response to MS. This suggests that metabolic signals play an important role in triggering the HPA-axis response of the neonate to MS (Schmidt et al., 2006). Importantly, the concentration of circulating leptin in the offspring is furthermore modified both by the availability as well as the composition of the maternal milk.

Maternal milk is rich in fat, which is required for growth and brain development during EL (D'Asti et al., 2010; Walker et al., 2008; Walker, 2010). Feeding mothers HF diet from gestational day 14 (G14) throughout lactation increased maternal milk fatty acid and leptin content. This, in turn, lastingly increases plasma fat and leptin levels in the offspring (Walker et al., 2008). These metabolic changes are associated with a blunted hormonal stress response on P10 (Trottier et al., 1998). This effect on HPA-axis activity could be direct, possibly due to the elevations in circulating leptin levels (Oates, Woodside, & Walker, 2000; Walker, 2010), or could be indirect, possibly due to differences in maternal diet that might have affected maternal behavior (Wiener et al., 1976). It is tempting to speculate in this respect that the increased levels of leptin that are observed in pups may change the odor of their urine, possibly stimulating increased licking and grooming by the mother.

Thus, the naturally elevated intake of fat, as well as the high exposure to leptin in pups via maternal milk, is important for the development of adequate brain function and an optimal stress responsiveness. This suggests that the dam expresses a regulatory influence on the activity of the HPA-axis of her offspring that is, at least partly, mediated through her dietary intake of fat (Walker, 2005). Similar to fat content, protein restriction (PR) (Bertram et al., 2001) as well as general undernutrition (Dutriez-Casteloot et al., 2008) during gestation and/or lactation, affect the HPA-axis of the progeny during development and adulthood. Next to the direct effects of food restriction on the quality of maternal milk (and, therefore, on the offspring's diet), these nutritional restrictions increase the circulating stress hormones and decrease placental 11β-hydroxy-steroid dehydrogenase type 2 (11β-HSD2) gene expression of the mother. This enzyme protects the foetus from high levels of maternal CORT by catalyzing the inactivation of glucocorticoids to 11-ketoforms. These alterations may thus lead to fetal overexposure to maternal CORT, which, in turn, might be responsible for some of the effects of maternal food restriction throughout the perinatal period, exemplifying again the tight relationship between stress, nutrition, and metabolic signals.

B. *Effects of perinatal malnutrition on cognitive function*

In addition to an increased vulnerability to develop psychiatric disorders, children that are exposed to prenatal and/or early postnatal malnutrition also exhibit cognitive deficits (Benton, 2010; de Rooij et al., 2010; de Souza et al., 2011; Galler et al., 1983; Klein et al., 1976; Laus et al.,

2011; Lucas, 1998; Walker et al., 2000). Whereas some animal studies reported cognitive deficits (Barnes, Pond, Kwong & Reid, 1966; Bhate et al., 2008; Castro et al., 1989; Fukuda et al., 2002; Jordan et al., 1981; Tonkiss & Galler, 1990; Beard et al., 2006; Felt & Lozoff, 1996; McEchron, Cheng, Liu, Connor, & Gilmartin, 2005), others reported no differences in learning between the offspring of well-nourished and malnourished dams (Campbell & Bedi, 1989; Tonkiss, Shultz, & Galler, 1994; Wolf, Almli, Finger, Ryan, & Morgane, 1986). These discrepancies probably depend on the fact that some forms of memory are more vulnerable to EL nutritional stress than others. Indeed, adult rats malnourished during EL exhibit cognitive deficits in reference memory tasks (such as the distal version of the MWM and the object recognition test), but not in working memory tasks (measured in the proximal cue version of the MWM and the water T-maze) (Castro & Rudy, 1987; Fukuda et al., 2002; Goodlett, Valentino, Morgane, & Resnick, 1986; Ranade et al., 2008; Valadares et al., 2010). In addition, fetal and neonatal iron deficiency results in long-term deficits in trace recognition memory, procedural memory, and spatial navigation (Beard et al., 2006; Felt & Lozoff, 1996; McEchron et al., 2005; Wolf et al., 1986)

As discussed earlier, the hippocampus is crucial for cognitive functions. Given its continued postnatal development and its high degree of structural plasticity (see section "The Hippocampal Formation and Its Forms of Neuronal and Synaptic Plasticity), this brain region is particularly vulnerable to malnutrition during the early life period. Nutritional manipulations during perinatal development induce decreased hippocampal width (Katz, Davies, & Dobbing, 1982), age-specific alterations in hippocampal neurogenesis (Coupé et al., 2009; King, 2004; Lewis, Patel, & Balázs, 1979; Bedi, 1991; Debassio, Kemper, Tonkiss, & Galler, 1996; King, 2004; Matos, Orozco-Solís, Souza, Castro, & Bolaños-Jiménez, 2011), reduced granular cell size, dendritic complexity and synaptic spine density of the DG (Diáz-Cintra et al., 1991), decreased soma size, dendritic branches and spine density of CA3 pyramidal cells (Diáz-Cintra, García-Ruiz, Corkidi, & Cintra, 1994) and reduced dendritic arborization of CA1 pyramidal cells (Jorgenson, Wobken, & Georgieff, 2003). These structural changes are associated with enhanced interneuron-mediated inhibition in the DG of the hippocampus (Blatt, Chen, Rosene, Volicer, & Galler, 1994; Mokler, Bronzino, Galler, & Morgane, 1999), and deficits in the induction and maintenance of LTP in malnourished animals (Bronzino, Austin La France, Morgane, & Galler, 1996; Morgane et al., 1993). Taken together, these data indicate that synapses in the malnourished hippocampus are less capable of supporting plasticity, and that alterations in the hippocampal circuitry during development could account for learning and memory deficits induced by perinatal malnutrition.

In summary, perinatal malnutrition has persistent consequences on brain development and function as demonstrated by the lasting effects on stress responsiveness and cognitive functions. Interestingly, the effects of perinatal malnutrition on stress responsiveness and cognitive function at adulthood resembles several of the long-term deficits induced by perinatal stress. Both insults result in cognitive deficits and an increased susceptibility to psychiatric disorders such as anxiety and depression (Brown et al., 2000; Costello et al., 2007; Heim, Shugart, Craighead, & Nemeroff, 2010). One possibility is that both instances depend on alterations in maternal behavior and thereby altered HPA-axis activity (Fig. 15.1).

In line with this hypothesis, maternal undernutrition results in altered maternal behavior (Wiener et al., 1976) and in high plasma levels of CORT in adult offspring along with reduced GR expression (Bertram et al., 2001; Lesage et al., 2001; Núñez et al., 2008), suggesting that the deleterious effects of maternal nutrient restriction on brain structure and function could also result from differential sensory input from the mother and enhanced persistent HPA-axis activity.

Thus, HPA-axis dysfunction could be the common etiologic factor in the programming effect of perinatal stress and early malnutrition. However, the opposite possibility should be considered as well. During stress, regulation of appetite and metabolism are altered, thereby affecting the intake of essential macro- and micronutrients. Thus, next to the regulation of brain structure and function by sensory stimuli from the mother and stress hormones, the lack of key nutrients might also affect the brain directly, both in the case of restricted nutrition as well as in the case of perinatal stress. In the next section, we will discuss how the lack of key nutrients might affect the brain permanently, and we will explore the possibility that epigenetic mechanisms underlie these persistent effects.

VI. Potential molecular mechanisms underlying the effects of malnutrition on brain structure and function

Next to effects of nutrients at the structural level, EL nutrition has the capacity to alter chromatin structure and gene expression as well, as demonstrated both in human and animal studies (Canani et al., 2011; Heijmans et al., 2008; McKay et al., 2011; Steegers-Theunissen et al., 2009; Wiedmeier, Joss-Moore, Lane & Neu, 2011). Thus, different molecular and epigenetic mechanisms may (partially) explain how and which dietary factors during critical developmental periods might program adult disease susceptibility. For example, protein malnutrition results in reduced neuronal DNA and RNA content and altered fatty acid profile (Bass, Netsky, & Young, 1970; Benítez-Bribiesca, la Rosa-Alvarez, & Mansilla-Olivares, 1999; Jones & Dyson, 1981). This could ultimately result in fewer neurons, reduced protein synthesis and hypomyelination and thereby in reduced synapse numbers and dendritic arborisation.

The mechanism by which EL nutrition modulates the epigenome is unclear but may involve alterations either in the supply of methyl donors, essential for DNA methylation, in the activities of DNMTs and/or through the activities of specific transcription factors. As to supply of dietary methyl donors, folate, vitamin B6/B12, methionine, choline and betaine have been extensively studied for their effect on DNA methylation. Choline availability during fetal brain development induces epigenetic changes modulating the expression of genes involved in hippocampal neuronal differentiation (Craciunescu et al., 2004; Craciunescu, Albright, Mar, Song, & Zeisel, 2003; Craciunescu, Johnson, & Zeisel, 2010) and maternal choline deficiency is shown to alter development and neurogenesis in the fetal mouse hippocampus (Albright, Mar, Craciunescu, Song, & Zeisel, 2005). However, whether lack of dietary methyl donor during the critical developmental period persistently affects brain structure and function and whether this involves epigenetic mechanisms remains to be determined.

Furthermore, perturbations in maternal diet can alter DNMT expression. Hyperglycaemia and hyperinsulinaemia have been reported to alter the 1-carbon (homocysteine) metabolism and to enhance DNMT activity (Chiang et al., 1996). Pregnant rats fed a PR diet show increased blood homocysteine concentration in early gestation (Petrie, Duthie, Rees, & McConnell, 2002), which is associated with a reduction in DNMT1 expression and inhibited binding of DNMT1 at the GR promoter in the liver of the offspring (Lillycrop et al., 2007). Because DNMT1 expression is negatively regulated by homocysteine (James, Melnyk, Pogribna, Pogribny, & Caudill, 2002), yet increased by folic acid (Lillycrop et al., 2007), modulation of DNMT1 expression by differences in 1-carbon metabolism could well provide a link between components in the maternal diet and epigenetic regulation of gene expression in the fetus.

A. Can dietary intervention reverse the effects of perinatal malnutrition?

Despite the apparent stability of methylation marks, as demonstrated in various animal studies, alterations in DNA methylation induced by maternal diet or differential nurturing behavior can be prevented and actually reversed by interventions in postnatal life. Although relatively little research has been done in this area and the reversibility of the effects of maternal malnutrition on brain structure and function by dietary intervention has not been proven yet, it has been shown that both folate and glycine supplementation to the maternal diet could reverse the effects of a low-protein diet in pregnancy on the blood pressure and vascular function of the offspring (Brawley et al., 2004), and it even prevented epigenetic changes (Burdge, Lillycrop, & Jackson, 2009). In addition, the phenotype and genotype of the offspring from PR-diet-fed dams was altered by folate supplementation during the juvenile-pubertal period (Burdge, Lillycrop, Phillips, et al., 2009). Finally, central infusion of l-methionine (i.e., the essential amminoacid that serves as a methyldonor group for DNA and histone methylation) for 7 days at P90, changed the methylation status of DNA previously programmed by differential maternal-nurturing behavior (Weaver, 2005). Taking these data together, nutrition seems to be a very promising candidate to prevent and/or reverse at least some of the lasting consequences of EL experience on adult brain structure and function.

VII. Final conclusion

Throughout this chapter we have presented ample evidence demonstrating that the adult phenotype is not only determined by gene X environment interaction, but rather by a three-way interaction of gene X environment X development. The environment during critical developmental periods (pre- and postnatal) affects gene expression in a lasting manner. This contributes to permanent alterations in brain structure and function, ultimately determining the cognitive and emotional outcomes in adulthood in response to specific adult challenges and experiences. Indeed, positive and negative early-life (EL) experiences affect cognitive and emotional health later in life in an interrelated and bidirectional manner. During these critical periods in EL, many important developmental processes take place including programming of the HPA-axis and development of brain structures important for emotion and cognition. These processes are tightly regulated by maternal care (sensory stimuli), exposure to hormones and neuropeptides (e.g., CORT and CRH), the presence and/or absence of required macro- and micronutrients and epigenetic mechanisms (Fig. 15.1). Therefore, any events that alter these components of the EL environment will interfere with these processes. In this chapter we have proposed several mechanisms responsible for the lasting effects of early-life experiences on emotion and cognition.

Much evidence points toward a crucial role for epigenetic mechanisms in mediating the effects of EL experiences. However, what triggers such epigenetic modifications is not clear as yet. Next to maternal care, stress hormones, and neuropeptides, nutrition seems to play an important role as well. These components all influence each other (Figure 15.1); for example EL nutrition is not only critical for growth and (brain) development, but also crucial for proper functioning of the epigenetic machinery. In addition, stressful EL experiences alter the regulation of appetite and metabolism and might thereby reduce nutrient availability; Finally undernutrition leads to an enhanced HPA-axis activity. Because of these interactions, and because these environmental cues often occur simultaneously, so far it remains difficult to dissect the separate contributions of the EL environment and nutrition on (epigenetic) programming of brain structure and function.

Understanding the mechanisms by which nutrition and other environmental cues influence epigenetic regulation of genes and the periods of susceptibility and stability of the induced changes are critical for the identification of individuals at risk and for the development of novel intervention strategies. Noninvasive interventions, targeted at maternal nutrition and maternal care, are relatively easy to implement and might have a significant and lasting effect on the health outcome of the offspring.

VIII. Future directions

The previously discussed evidence leads us to pose some questions to enable the development of therapeutic and preventive interventions to improve human health. As presented here, there has been great progress in our knowledge of the molecules and mechanisms involved in the lasting effects of EL stress. However, to be able to design interventions aiming to reverse the consequences of adverse EL experiences, we will need to determine if there is a critical time-window during which these effects can be reversed and, if so, how long this period is. In addition, as several of the effects of EL experience are brain-region specific, it is necessary to understand the mechanisms that underlie this brain specificity in order to develop targeted therapies. Finally, we described that alterations in the EL environment not only affect the quality and quantity of maternally mediated sensory stimuli but also of the nutrition provided. Therefore, we propose that it is crucial to consider the combined effects of early-life stress and early-life nutrition. Identification of the exact relationship between early-life stress and food intake/metabolism/nutrient availability will facilitate the development of specific nutritional interventions, which represent a relatively inexpensive and noninvasive therapy.

Acknowledgments

PJL is supported by the Dutch Brain foundation, NWO PRIOMED and ISAO.

References

Abe, H., Hidaka, N., Kawagoe, C., Odagiri, K., Watanabe, Y., Ikeda, T., . . ., Ishida, Y. (2007). Prenatal psychological stress causes higher emotionality, depression-like behavior, and elevated activity in the hypothalamo-pituitary-adrenal axis. *Neuroscience Research*, *59*(2), 145–151.

Ader, R. (1970). The effects of early experience on the adrenocortical response to different magnitudes of stimulation. *Physiology & Behavior*, *5*(8), 837–839.

Aisa, B., Tordera, R., Lasheras, B., Del Río, J., & Ramírez, M. J. (2007). Cognitive impairment associated to HPA axis hyperactivity after maternal separation in rats. *Psychoneuroendocrinology*, *32*(3), 256–266.

Akana, S. & Dallman, M. (1997). Chronic cold in adrenalectomized, corticosterone (B)-treated rats: Facilitated corticotropin responses to acute restraint emerge as B increases. *Endocrinology*, *138*(8), 3249–3258.

Albright, C., Mar, M., Craciunescu, C., Song, J., & Zeisel, S. (2005). Maternal dietary choline availability alters the balance of netrin-1 and DCC neuronal migration proteins in fetal mouse brain hippocampus. *Developmental Brain Research*, *159*(2), 149–154.

Avishai-Eliner, S., Gilles, E., Eghbal-Ahmadi, M., Bar-El, Y., & Baram, T. Z. (2001). Altered regulation of gene and protein expression of hypothalamic-pituitary-adrenal axis

components in an immature rat model of chronic stress. *Journal of Neuroendocrinology, 13*(9), 799–807.
Avishai-Eliner, S., Yi, S. J., Newth, C. J., & Baram, T. Z. (1995). Effects of maternal and sibling deprivation on basal and stress induced hypothalamic-pituitary-adrenal components in the infant rat. *Neuroscience Letters, 192*(1), 49–52.
Bagot, R. C., van Hasselt, F. N., Champagne, D. L., Meaney, M. J., Krugers, H. J., & Joels, M. (2009). Maternal care determines rapid effects of stress mediators on synaptic plasticity in adult rat hippocampal dentate gyrus. *Neurobiology of Learning and Memory, 92*(3), 292–300.
Bannerman, D. M., Rawlins, J. N. P., McHugh, S. B., Deacon, R. M. J., Yee, B. K., Bast, T.,…, Feldon, J. (2004). Regional dissociations within the hippocampus—memory and anxiety. *Neuroscience and biobehavioral reviews, 28*(3), 273–283.
Barker, D. J. P. (2004). Developmental origins of adult health and disease. *Journal of Epidemiology and Community Health, 58*(2), 114–115.
Barker, D. J. P. (2007). The origins of the developmental origins theory. *Journal of Internal Medicine, 261*(5), 412–417.
Barnes, R. H., Pond, W. G., Kwong, E., & Reid, I. (1966). Effect of severe protein-calorie malnutrition in the baby pig upon relative utilization of different dietary proteins. *Journal of Nutrition, 89*(3), 355–364.
Bass, N. H., Netsky, M. G., & Young, E. (1970). Effects of neonatal malnutrition on developing cerebrum I. Microchemical and histologic study of cellular differentiation in the rat. *Archives of Neurology, 23*(4), 289–302.
Beard, J. L., Felt, B., Schallert, T., Burhans, M., Connor, J. R., & Georgieff, M. K. (2006). Moderate iron deficiency in infancy: Biology and behavior in young rats. *Behavioural Brain Research, 170*(2), 224–232.
Bedi, K. S. (1991). Early-life undernutrition causes deficits in rat dentate gyrus granule cell number. *Experientia, 47*(10), 1073–1074.
Bedi, K. S. (1992). Spatial learning ability of rats undernourished during early postnatal life. *Physiology & Behavior, 51*(5), 1001–1007.
Benedict, C., Kern, W., Schmid, S. M., Schultes, B., Born, J., & Hallschmid, M. (2009). Early morning rise in hypothalamic-pituitary-adrenal activity: a role for maintaining the brain's energy balance. *Psychoneuroendocrinology, 34*(3), 455–462.
Benítez-Bribiesca, L., De la Rosa-Alvarez, I., & Mansilla-Olivares, A. (1999). Dendritic spine pathology in infants with severe protein-calorie malnutrition. *Pediatrics, 104*(2)
Benton, D. (2010). The influence of dietary status on the cognitive performance of children. *Molecular Nutrition & Food Research, 54*(4), 457–470.
Berkowitz, G. S., Wolff, M. S., Janevic, T. M., Holzman, I. R., Yehuda, R., & Landrigan, P. J. (2003). The World Trade Center disaster and intrauterine growth restriction *JAMA: Journal of the American Medical Association, 290*(5), 595–596.
Bertram, C., Trowern, A. R., Copin, N., Jackson, A. A., & Whorwood, C. B. (2001). The maternal diet during pregnancy programs altered expression of the glucocorticoid receptor and type 2 11beta-hydroxysteroid dehydrogenase: potential molecular mechanisms underlying the programming of hypertension in utero. *Endocrinology, 142*(7), 2841–2853.
Bhate, V., Deshpande, S., Bhat, D., Joshi, N., Ladkat, R., Watve, S.,…, Yajnik, C. (2008). Vitamin B12 status of pregnant Indian women and cognitive function in their 9-year-old children. *Food and Nutrition Bulletin, 29*(4), 249–254.
Bhatnagar, S., & Meaney, M. (1995). Hypothalamic-pituitary-adrenal function in chronic intermittently cold-stressed neonatally handled and non handled rats. *Journal of Neuroendocrinology, 7*(2), 97–108.
Blatt, G. J., Chen, J. C., Rosene, D. L., Volicer, L., & Galler, J. R. (1994). Prenatal protein malnutrition effects on the serotonergic system in the hippocampal formation: an immunocytochemical, ligand binding, and neurochemical study. *Brain Research Bulletin, 34*(5), 507–518.
Bradley, R., Binder, E., Epstein, M., Tang, Y., Nair, H., Liu, W.,…, Ressler, K.J. (2008). Influence of child abuse on adult depression: moderation by the corticotropin-releasing hormone receptor gene. *Archives of General Psychiatry, 65*(2), 190–200.
Brawley, L., Torrens, C., Anthony, F. W., Itoh, S., Wheeler, T., Jackson, A. A.,…, Hanson, M.A. (2004). Glycine rectifies vascular dysfunction induced by dietary protein imbalance during pregnancy. *The Journal of physiology, 554*(2), 497–504.
Bredy, T. W., Grant, R. J., Champagne, D. L. & Meaney, M. J. (2003). Maternal care influences neuronal survival in the hippocampus of the rat. *The European Journal of Neuroscience, 18*(10), 2903–2909.
Bremner, J. D., Randall, P., Vermetten, E., Staib, L., Bronen, R. A., Mazure, C.,…, Charney, D.S. (1997). Magnetic resonance imaging-based measurement of hippocampal volume in posttraumatic stress disorder related to childhood physical and sexual abuse-a preliminary report. *Biological Psychiatry, 41*(1), 23–32.
Bronzino, J. D., Austin La France, R. J., Morgane, P. J., & Galler, J. R. (1996). Diet-induced alterations in the ontogeny of long-term potentiation. *Hippocampus, 6*(2), 109–117.
Brown, A. S., van Os, J., Driessens, C., Hoek, H. W., & Susser, E. S. (2000). Further evidence of relation between prenatal famine and major affective disorder. *American Journal of Psychiatry, 157*(2), 190–195.
Brown, C. P., Smotherman, W. P. & Levine, S. (1977). Interaction-induced reduction in differential maternal responsiveness: an effect of cue-reduction or behavior. *Developmental Psychobiology, 10*(3), 273–280.
Brown, J., & Kuhn, C. C. (2003). Enriched environment and physical activity stimulate hippocampal but not olfactory bulb neurogenesis *European Journal of Neuroscience, 17*(10), 2042–6
Brunson, K. L, Kramár, E., Lin, B., Chen, Y., Colgin, L. L., Yanagihara, T. K.,…, Baram, T.Z. (2005). Mechanisms of late-onset cognitive decline after early-life stress. *The Journal of Neuroscience, 25*(41), 9328–9338.
Brunson, K., Eghbal-Ahmadi, M., Bender, R., Chen, Y., & Baram, T. Z. (2001). Long-term, progressive hippocampal cell loss and dysfunction induced by early-life administration of corticotropin-releasing hormone reproduce the effects of early-life stress. *Proceedings of the National Academy of Sciences of the United States of America, 98*(15), 8856–8861.
Brunson, K. L, Baram, T. Z., & Bender, R. A. (2005). Hippocampal neurogenesis is not enhanced by lifelong reduction of glucocorticoid levels. *Hippocampus, 15*(4), 491–501.

Burdge, G. C., Lillycrop, K. A. & Jackson, A. A. (2009). Nutrition in early life, and risk of cancer and metabolic disease: alternative endings in an epigenetic tale. *British Journal of Nutrition, 101*(5), 619–630.

Burdge, G. C., Lillycrop, K. A., Phillips, E. S., Slater-Jefferies, J. L., Jackson, A. A. & Hanson, M. A. (2009). Folic acid supplementation during the juvenile-pubertal period in rats modifies the phenotype and epigenotype induced by prenatal nutrition. *Journal of Nutrition, 139*(6), 1054–1060.

Caldji, C., Tannenbaum, B., Sharma, S., Francis, D., Plotsky, P., & Meaney, M. J. (1998). Maternal care during infancy regulates the development of neural systems mediating the expression of fearfulness in the rat. *Proceedings of the National Academy of Sciences of the United States of America, 95*(9), 5335–5340.

Cameron, H. A., & Gould, E. (1994). Adult neurogenesis is regulated by adrenal steroids in the dentate gyrus. *Neuroscience, 61*(2), 203–209.

Campbell, L. F. & Bedi, K. S. (1989). The effects of undernutrition during early life on spatial learning *Physiology & Behavior, 45*(5), 883–890.

Canani, R. B., Di Costanzo, M., Leone, L., Bedogni, G., Brambilla, P., Cianfarani, S.,..., Agostoni, C. (2011). Epigenetic mechanisms elicited by nutrition in early life. *Nutrition Research Reviews, 24*(02), 198–205.

Castrén, E. (2005). Is mood chemistry. *Nature Reviews Neuroscience, 6*(3), 241–246.

Castro, C. A., & Rudy, J. W. (1987). Early-life malnutrition selectively retards the development of distal- but not proximal-cue navigation. *Developmental Psychobiology, 20*(5), 521–537.

Castro, C. A., Tracy, M., & Rudy, J. W. (1989). Early-life undernutrition impairs the development of the learning and short-term memory processes mediating performance in a conditional-spatial discrimination task. *Behavioural Brain Research, 32*(3), 255–264.

Champagne, D. L., Bagot, R. C., van Hasselt, F., Ramakers, G., Meaney, M. J., de Kloet,..., Krugers, H. (2008). Maternal care and hippocampal plasticity: evidence for experience-dependent structural plasticity, altered synaptic functioning, and differential responsiveness to glucocorticoids and stress. *The Journal of Neuroscience, 28*(23), 6037–6045.

Chen, J., Evans, A. N., Liu, Y., Honda, M., Saavedra, J. M., & Aguilera, G. (2012). Maternal deprivation in rats is associated with corticotropin releasing hormone (Crh) promoter hypomethylation and enhances Crh transcriptional responses to stress in adulthood. *Journal of Neuroendocrinology. 24*(7), 1055–1064.

Chen, Y. (2004). Hippocampal corticotropin releasing hormone: Pre- and postsynaptic location and release by stress. *Neuroscience, 126*(3), 533–540.

Chen, Y, Brunson, K., Muller, M., Cariaga, W. & Baram, T. Z. (2000). Immunocytochemical distribution of corticotropin-releasing hormone receptor type-1 (CRF(1))-like immunoreactivity in the mouse brain: Light microscopy analysis using an antibody directed against the C-terminus. *Journal of Comparative Neurology, 420*(3), 305–323.

Chen, Y, Rex, C., Rice, C., Dube, C., Gall, C., Lynch, G., & Baram, T.Z. (2010). Correlated memory defects and hippocampal dendritic spine loss after acute stress involve corticotropin-releasing hormone signaling. *Proceedings of the National Academy of Sciences of the United States of America, 107*(29), 13123–13128.

Chiang, P. K., Gordon, R. K., Tal, J., Zeng, G. C., Doctor, B. P., Pardhasaradhi, K., & McCann, P. P. (1996). S-Adenosylmethionine and methylation *The FASEB Journal, 10*(4), 471–480.

Choi, D. C., Evanson, N. K., Furay, A. R., Ulrich-Lai, Y. M., Ostrander, M. M., & Herman, J. P. (2007). The anteroventral bed nucleus of the stria terminalis differentially regulates hypothalamic-pituitary-adrenocortical axis responses to acute and chronic stress. *Endocrinology, 149*(2), 818–826.

Chugani, H. T., Behen, M. E., Muzik, O., Juhász, C., Nagy, F., & Chugani, D. C. (2001). Local brain functional activity following early deprivation: a study of postinstitutionalized Romanian orphans. *NeuroImage, 14*(6), 1290–1301.

Cignacco, E., Hamers, J. P. H., Stoffel, L., van Lingen, R. A., Gessler, P., McDougall, J., & Nelle, M. (2007). The efficacy of non-pharmacological interventions in the management of procedural pain in preterm and term neonates, a systematic literature review. *European Journal of Pain, 11*(2), 139–152.

Costello, E. J., Worthman, C., Erkanli, A., & Angold, A. (2007). Prediction from low birth weight to female adolescent depression: a test of competing hypotheses. *Archives of General Psychiatry, 64*(3), 338–344.

Coupé, B., Dutriez-Casteloot, I., Breton, C., Lefèvre, F., Mairesse, J., Dickes-Coopman, A.,..., Vieau, D. (2009). Perinatal undernutrition modifies cell proliferation and brain-derived neurotrophic factor levels during critical time-windows for hypothalamic and hippocampal development in the male rat. *Journal of Neuroendocrinology, 21*(1), 40–48.

Craciunescu, C. N., Albright, C. D., Mar, M., Song, J., & Zeisel, S. H. (2003). Choline availability during embryonic development alters progenitor cell mitosis in developing mouse hippocampus. *The Journal of Nutrition, 133*(11), 3614–3618.

Craciunescu, C. N., Brown, E. C., Mar, M., Albright, C. D., Nadeau, M. R., & Zeisel, S. H. (2004). Folic acid deficiency during late gestation decreases progenitor cell proliferation and increases apoptosis in fetal mouse brain. *The Journal of Nutrition, 134*(1), 162–166.

Craciunescu, C. N., Johnson, A. R., & Zeisel, S. H. (2010). Dietary choline reverses some, but not all, effects of folate deficiency on neurogenesis and apoptosis in fetal mouse brain. *The Journal of Nutrition, 140*(6), 1162–1166.

Czéh, B., & Lucassen, P. J. (2007). What causes the hippocampal volume decrease in depression? Are neurogenesis, glial changes and apoptosis implicated? *European Archives of Psychiatry and Clinical Neuroscience, 257*(5), 250–260.

Daniels, W. M. U., Fairbairn, L. R., van Tilburg, G., McEvoy, C. R. E., Zigmond, M. J., Russell, V., A. & Stein, D. J. (2009). Maternal separation alters nerve growth factor and corticosterone levels but not the DNA methylation status of the exon *1*(7) glucocorticoid receptor promoter region. *Metabolic Brain Disease, 24*(4), 615–627.

D'Asti, E., Long, H., Tremblay-Mercier, J., Grajzer, M., Cunnane, S. C., Di Marzo, V., & Walker, C. (2010). Maternal dietary fat determines metabolic profile and the magnitude of endocannabinoid inhibition of the stress response in neonatal rat offspring. *Endocrinology, 151*(4), 1685–1694.

on pregnancy, postpartum, and breastfeeding. *Trauma Violence Abuse, 8*(3), 344–353.

King, R. (2004). Effects of prenatal protein malnutrition and acute postnatal stress on granule cell genesis in the fascia dentata of neonatal and juvenile rats. *Developmental Brain Research, 150*(1), 9–15.

Klein, R. E., Arenales, P., Delgado, H., Engle, P. L., Guzmán, G., Irwin, M.,..., Yarbrough, C. (1976). Effects of maternal nutrition on fetal growth and infant development. *Bulletin of the Pan American Health Organization, 10*(4), 301–306.

Koenen, K., Moffitt, T., Caspi, A., Taylor, A., & Purcell, S. (2003). Domestic violence is associated with environmental suppression of IQ in young children. *Development and Psychopathology, 15*(2), 297–311.

Korosi, A. & Baram, T. Z. (2009). The pathways from mother's love to baby's future. *Frontiers in Behavioral Neuroscience, 3*, 27.

Korosi, A. & Baram, T. Z. (2010). Plasticity of the stress response early in life: mechanisms and significance. *Developmental Psychobiology, 52*(7), 661–670.

Korosi, A, Naninck, E. F. G., Oomen, C. A., Schouten, M., Krugers, H., Fitzsimons, C., & Lucassen, P. J. (2011). Early-life stress mediated modulation of adult neurogenesis and behavior. *Behavioural Brain Research, 227*(2), 400–409.

Korosi, A., Shanabrough, M., McClelland, S., Liu, Z., Borok, E., Gao, X.,..., Baram, T.Z. (2010). Early-life experience reduces excitation to stress-responsive hypothalamic neurons and reprograms the expression of corticotropin-releasing hormone. *Journal of Neuroscience, 30*(2), 703–713.

Kosten, T. A., Lee, H. J., & Kim, J. J. (2007). Neonatal handling alters learning in adult male and female rats in a task-specific manner. *Brain Research, 1154*, 144–153.

Ladd, C.O., Huot, R. L., Thrivikraman, K. V., Nemeroff, C.B., Meaney, M.J., & Plotsky, P. M. (2000). Long-term behavioral and neuroendocrine adaptations to adverse early experience. *Progress in Brain Research, 122*, 81–103.

Ladd, C. O., Owens, M. J., & Nemeroff, C. B. (1996). Persistent changes in corticotropin-releasing factor neuronal systems induced by maternal deprivation. *Endocrinology, 137*(4), 1212–1218.

Lajud, N., Roque, A., Cajero, M., Gutiérrez-Ospina, G., & Torner, L. (2011). Periodic maternal separation decreases hippocampal neurogenesis without affecting basal corticosterone during the stress hyporesponsive period, but alters HPA axis and coping behavior in adulthood. *Psychoneuroendocrinology*, 1–11.

Laus, M. F., Duarte Manhas Ferreira Vales, L., Braga Costa, T. M., & Sousa Almeida, S. (2011). Early postnatal protein-calorie malnutrition and cognition: A review of human and animal studies. *International Journal of Environmental Research and Public Health, 8*(2), 590–612.

Lehmann, J., Pryce, C., Jongen-Relo, A., Stohr, T., Pothuizen, H., & Feldon, J. (2002). Comparison of maternal separation and early handling in terms of their neurobehavioral effects in aged rats. *Neurobiology of Aging, 23*(3), 457–466.

Lemaire, V., Koehl, M., Le Moal, M., & Abrous, D. N. (2000). Prenatal stress produces learning deficits associated with an inhibition of neurogenesis in the hippocampus. *Proceedings of the National Academy of Sciences of the United States of America, 97*(20), 11032–11037.

Lesage, J, Blondeau, B., Grino, M., Bréant, B., & Dupouy, J. P. (2001). Maternal undernutrition during late gestation induces fetal overexposure to glucocorticoids and intrauterine growth retardation, and disturbs the hypothalamo-pituitary adrenal axis in the newborn rat. *Endocrinology, 142*(5), 1692–1702.

Leventopoulos, M., Rüedi-Bettschen, D., Knuesel, I., Feldon, J., Pryce, C. R., & Opacka-Juffry, J. (2007). Long-term effects of early life deprivation on brain glia in Fischer rats. *Brain Research, 1142*, 119–126.

Levine, A., Worrell, T. R., Zimnisky, R., & Schmauss, C. (2011). Early life stress triggers sustained changes in histone deacetylase expression and histone H4 modifications that alter responsiveness to adolescent antidepressant treatment. *Neurobiology of Disease, 45*(1), 488–498.

Levine, S. (1962). Plasma-free corticosteroid response to electric shock in rats stimulated in infancy. *Science, 135*(3506), 795–796.

Levine, S. (1967). Maternal and environmental influences on the adrenocortical response to stress in weanling rats. *Science, 156*(772), 258–260.

Levine, S., & Lewis, G. (1959). The relative importance of experimenter contact in an effect produced by extra-stimulation in infancy. *Journal of Comparative Physiological Psychology, 52*(3), 368–369.

Lewis, P. D., Patel, A. J., & Balázs, R. (1979). Effect of undernutrition on cell generation in the rat hippocampus. *Brain Research, 168*(1), 186–189.

Lillycrop, K. A., Slater-Jefferies, J. L., Hanson, M. A., Godfrey, K. M., Jackson, A. A., & Burdge, G. C. (2007). Induction of altered epigenetic regulation of the hepatic glucocorticoid receptor in the offspring of rats fed a protein-restricted diet during pregnancy suggests that reduced DNA methyltransferase-1 expression is involved in impaired DNA methylation and changes in histone modifications. *British Journal of Nutrition, 97*(06), 1064.

Lingas, R. I., & Matthews, S. G. (2001). A short period of maternal nutrient restriction in late gestation modifies pituitary-adrenal function in adult guinea pig offspring *Neuroendocrinology, 73*(5), 302–311.

Liu, D., Diorio, J., Tannenbaum, B., Caldji, C., Francis, D., Freedman, A.,..., Meaney, M.J. (1997). Maternal care, hippocampal glucocorticoid receptors, and hypothalamic-pituitary-adrenal responses to stress. *Science, 277*(5332), 1659–1662.

Liu, D., Diorio, J., Day, J., Francis, D., & Meaney, M.J. (2000). Maternal care, hippocampal synaptogenesis and cognitive development in rats. *Nature Neuroscience, 3*(8), 799–806.

Loman, M. M., & Gunnar, M. R., (2010). Early experience and the development of stress reactivity and regulation in children. *Neuroscience and biobehavioral reviews, 34*(6), 867–876.

Luby, J. L., Barch, D. M., Belden, A., Gaffrey, M. S., Tillman, R., Babb, C.,..., Botteron, K.N. (2012). Maternal support in early childhood predicts larger hippocampal volumes at school age. *Proceedings of the National Academy of Sciences.*

Lucas, A. (1998). Programming by early nutrition: an experimental approach. *The Journal of Nutrition, 128*(2 Suppl), 401S–406S.

Lucassen, P. J., Meerlo, P., Naylor, A. S., van Dam, A. M., Dayer, A. G., Fuchs, E.,..., Czéh, B. (2010). Regulation of adult neurogenesis by stress, sleep disruption, exercise and inflammation: Implications for depression and antidepressant action. *European Neuropsychopharmacology, 20*(1), 1–17.

Lucassen, P. J., Bosch, O. J., Jousma, E., Krömer, S. A., Andrew, R., Seckl, J. R., & Neumann, I. D. (2009). Prenatal stress reduces postnatal neurogenesis in rats selectively bred for high, but not low, anxiety: Possible key role of placental 11beta-hydroxysteroid dehydrogenase type 2. *European Journal of Neuroscience, 29*(1), 97–103.

Lupien, S. J., McEwen, B. S., Gunnar, M. R., & Heim, C. (2009). Effects of stress throughout the lifespan on the brain, behaviour and cognition. *Nature Reviews. Neuroscience, 10*(6), 434–445.

Macri, S., Mason, G. J., & Wurbel, H. (2004). Dissociation in the effects of neonatal maternal separations on maternal care and the offspring's HPA and fear responses in rats. *European Journal of Neuroscience, 20*(4), 1017–1024.

Marin-Burgin, A., Mongiat, L. A., Pardi, M. B., & Schinder, A. F. (2012). Unique processing during a period of high excitation/inhibition balance in adult-born neurons. *Science, 335*(6073), 1238–1242.

Martínez, Y., Díaz-Cintra, S., León-Jacinto, U., Aguilar-Vázquez, A., Medina, A. C., Quirarte, G. L., & Prado-Alcalá, R. A. (2009). Effects of postnatal malnutrition and senescence on learning, long-term memory, and extinction in the rat. *Behavioural Brain Research, 203*(1), 48–53.

Maselko, J., Kubzansky, L., Lipsitt, L., & Buka, S. L. (2011). Mother's affection at 8 months predicts emotional distress in adulthood. *Journal of Epidemiology and Community Health, 65*(7), 621–625.

Matos, R. J. B., Orozco-Solís, R., Souza, S. L. D., Castro, R. M. D., & Bolaños-Jiménez, F. (2011). Nutrient restriction during early life reduces cell proliferation in the hippocampus at adulthood but does impairs the neuronal differentiation process of the new generated cells. *Neuroscience, 196*, 16–24.

McClelland, S., Korosi, A., Cope, J., Ivy, A., & Baram, T. Z. (2011). Emerging roles of epigenetic mechanisms in the enduring effects of early-life stress and experience on learning and memory. *Neurobiology of Learning and Memory, 96*(1), 79–88.

McEchron, M. D., Cheng, A. Y., Liu, H., Connor, J. R., & Gilmartin, M. R. (2005). Perinatal nutritional iron deficiency permanently impairs hippocampus-dependent trace fear conditioning in rats. *Nutritional Neuroscience, 8*(3), 195–206.

McGowan, P. O., Sasaki, A., D'Alessio, A. C., Dymov, S., Labonté, B., Szyf, M.,..., Meaney, M. J. (2009). Epigenetic regulation of the glucocorticoid receptor in human brain associates with childhood abuse. *Nature Neuroscience, 12*(3), 342–348.

McGowan, P. O., Suderman, M., Sasaki, A., Huang, T. C. T., Hallett, M., Meaney, M. J. & Szyf, M. (2011). Broad epigenetic signature of maternal care in the brain of adult rats. *PLoS ONE, 6*(2), e14739.

McKay, J. A., Xie, L., Harris, S., Wong, Y. K., Ford, D., & Mathers, J. C. (2011). Blood as a surrogate marker for tissue-specific DNA methylation and changes due to folate depletion in post-partum female mice. *Molecular Nutrition & Food Research, 55*(7), 1026–1035.

Meaney, M.J., Viau, V., Bhatnagar, S., Betito, K., Iny, L.J., O'Donnell, D., & Mitchell, J. B. (1991). Cellular mechanisms underlying the development and expression of individual differences in the hypothalamic-pituitary-adrenal stress response. *Journal of Steroid Biochemistry and Molecular Biology, 39*(2), 265–274.

Meaney, M., Diorio, J., Francis, D., Widdowson, J., LaPlante, P., Caldji, C.,..., Plotsky, P.M. (1996). Early environmental regulation of forebrain glucocorticoid receptor gene expression: Implications for adrenocortical responses to stress. *Developmental Neuroscience, 18*(1–2), 49–72.

Mirescu, C., Peters, J. D., & Gould, E. (2004). Early life experience alters response of adult neurogenesis to stress. *Nature Neuroscience, 7*(8), 841–846.

Mokler, D. J., Bronzino, J. D., Galler, J. R., & Morgane, P. J. (1999). The effects of median raphé electrical stimulation on serotonin release in the dorsal hippocampal formation of prenatally protein malnourished rats. *Brain Research, 838*(1–2), 95–103.

Morgane, P. J., Austin-LaFrance, R., Bronzino, J., Tonkiss, J., Diáz-Cintra, S., Cintra, L.,..., Galler, J.R. (1993). Prenatal malnutrition and development of the brain. *Neuroscience and Biobehavioral Reviews, 17*(1), 91–128.

Morgane, P. J, Mokler, D. J., & Galler, J. R. (2002). Effects of prenatal protein malnutrition on the hippocampal formation. *Neuroscience and Biobehavioral Reviews, 26*(4), 471–483.

Moser, E., Moser, M. B., & Andersen, P. (1993). Spatial learning impairment parallels the magnitude of dorsal hippocampal lesions, but is hardly present following ventral lesions. *The Journal of Neuroscience, 13*(9), 3916–3925.

Moser, M. B. & Moser, E. I. (1998). Functional differentiation in the hippocampus. *Hippocampus, 8*(6), 608–619.

Mueller, S. C., Maheu, F. S., Dozier, M., Peloso, E., Mandell, D., Leibenluft, E.,..., Ernst, M. (2010). Early-life stress is associated with impairment in cognitive control in adolescence: an fMRI study. *Neuropsychologia, 48*(10), 3037–3044.

Murgatroyd, C., Patchev, A. V., Wu, Y., Micale, V., Bockmühl, Y., Fischer, D.,..., Spengler, D. (2009). Dynamic DNA methylation programs persistent adverse effects of early-life stress. *Nature Neuroscience, 12*(12), 1559–1566.

Nair, A., Vadodaria, K. C., Banerjee, S. B., Benekareddy, M., Dias, B. G., Duman, R. S., & Vaidya, V. A. (2007). Stressor-specific regulation of distinct brain-derived neurotrophic factor transcripts and cyclic AMP response element-binding protein expression in the postnatal and adult rat hippocampus. *Neuropsychopharmacology, 32*(7), 1504–1519.

Naruse, Y., Aoki, T., Kojima, T., & Mori, N. (1999). Neural restrictive silencer factor recruits mSin3 and histone deacetylase complex to repress neuron-specific target genes. *Proceedings of the National Academy of Sciences of the United States of America, 96*(24), 13691–13696.

Naumova, O. Y., Lee, M., Koposov, R., Szyf, M., Dozier, M., & Grigorenko, E. L. (2011). Differential patterns of whole-genome DNA methylation in institutionalized children and children raised by their biological parents. *Development and psychopathology, 24*(01), 143–155.

Nelson, C. A., Zeanah, C. H., Fox, N. A., Marshall, P. J., Smyke, A. T., & Guthrie, D. (2007). Cognitive recovery in socially deprived young children: The Bucharest Early Intervention Project. *Science, 318*(5858), 1937–1940.

Nestler, E. J., & Hyman, S. E. (2010). Animal models of neuropsychiatric disorders. *Nature Neuroscience, 13*(10), 1161–1169.

Núñez, H., Ruiz, S., Soto-Moyano, R., Navarrete, M., Valladares, L., White, A., & Pérez, H. (2008). Fetal undernutrition induces overexpression of CRH mRNA and CRH protein in hypothalamus and increases CRH and corticosterone in plasma during postnatal life in the rat. *Neuroscience Letters, 448*(1), 115–119.

Oates, M., Woodside, B., & Walker, C. (2000). Chronic leptin administration in developing rats reduces stress responsiveness partly through changes in maternal behavior. *Hormones and Behavior, 37*(4), 366–376.

O'Connor, T. G., Ben-Shlomo, Y., Heron, J., Golding, J., Adams, D., & Glover, V. (2005). Prenatal anxiety predicts individual differences in cortisol in pre-adolescent children. *Biological Psychiatry 58*(3), 211–217.

Odagiri, K., Abe, H., Kawagoe, C., Takeda, R., Ikeda, T., Matsuo, H.,. . ., Ishida, Y. (2008). Psychological prenatal stress reduced the number of BrdU immunopositive cells in the dorsal hippocampus without affecting the open field behavior of male and female rats at one month of age. *Neuroscience letters, 446*(1), 25–29.

Oitzl, M. S., Workel, J. O., Fluttert, M., Frösch, F., & de Kloet, E. R. (2000). Maternal deprivation affects behaviour from youth to senescence: amplification of individual differences in spatial learning and memory in senescent Brown Norway rats. *The European Journal of Neuroscience, 12*(10), 3771–3780.

Oomen, C. A., Girardi, C. E. N., Cahyadi, R., Verbeek, E. C., Krugers, H., Joels, M., & Lucassen, P. J. (2009). Opposite effects of early maternal deprivation on neurogenesis in male versus female rats. *PLoS ONE, 4*(1), e3675.

Oomen, C. A., Mayer, J. L., de Kloet, E. R., Joels, M., & Lucassen, P. J. (2007). Brief treatment with the glucocorticoid receptor antagonist mifepristone normalizes the reduction in neurogenesis after chronic stress. *European Journal of Neuroscience, 26*(12), 3395–3401.

Oomen, C. A., Soeters, H., Audureau, N., Vermunt, L., van Hasselt, F. N., Manders, E. M. M.,. . ., Krugers, H. (2010). Severe early life stress hampers spatial learning and neurogenesis, but improves hippocampal synaptic plasticity and emotional learning under high-stress conditions in adulthood. *The Journal of Neuroscience, 30*(19), 6635–6645.

Oomen, C. A., Soeters, H., Audureau, N., Vermunt, L., van Hasselt, F. N., Manders, E. M. M.,. . ., Lucassen, P.J. (2011). Early maternal deprivation affects dentate gyrus structure and emotional learning in adult female rats. *Psychopharmacology, 214*(1), 249–260.

Perera, T. D., Park, S., & Nemirovskaya, Y. (2008). Cognitive role of neurogenesis in depression and antidepressant treatment. *Neuroscientist, 14*(4), 326–338.

Petrie, L., Duthie, S. J., Rees, W. D., & McConnell, J. M. L. (2002). Serum concentrations of homocysteine are elevated during early pregnancy in rodent models of fetal programming. *British Journal of Nutrition, 88*(5), 471–477.

Pijlman, F. T. A., Wolterink, G., & Van Ree, J. M. (2003). Physical and emotional stress have differential effects on preference for saccharine and open field behaviour in rats. *Behavioural Brain Research, 139*(1–2), 131–138.

Plagemann, A., Harder, T., Rake, A., Melchior, K., Rohde, W. & Dörner, G. (2000). Hypothalamic nuclei are malformed in weanling offspring of low protein malnourished rat dams. *Journal of Nutrition, 130*(10), 2582–2589.

Pleasure, S. J., Collins, A. E., & Lowenstein, D. H. (2000). Unique expression patterns of cell fate molecules delineate sequential stages of dentate gyrus development. *The Journal of Neuroscience, 20*(16), 6095–6105.

Plotsky, P. M., & Meaney, M. J., (1993). Early, postnatal experience alters hypothalamic corticotropin-releasing factor (CRF) mRNA, median eminence CRF content and stress-induced release in adult rats. *Brain Research. Molecular Brain Research, 18*(3), 195–200.

Pryce, C. R., Bettschen, D., & Feldon, J. (2001). Comparison of the effects of early handling and early deprivation on maternal care in the rat. *Developmental Psychobiology, 38*(4), 239–251.

Ranade, S. C., Rose, A., Rao, M., Gallego, J., Gressens, P., & Mani, S. (2008). Different types of nutritional deficiencies affect different domains of spatial memory function checked in a radial arm maze. *Neuroscience, 152*(4), 859–866.

Reul, J. & de Kloet, E. (1985). Two receptor systems for corticosterone in rat brain: microdistribution and differential occupation. *Endocrinology, 117*(6), 2505–2511.

Rice, C. J., Sandman, C. A., Lenjavi, M. R., & Baram, T. Z. (2008). A novel mouse model for acute and long-lasting consequences of early life stress. *Endocrinology, 149*(10), 4892–4900.

Rice, D., & Barone, S. (2000). Critical periods of vulnerability for the developing nervous system: evidence from humans and animal models. *Environmental Health Perspectives, 108 (Suppl 3)*, 511–533.

Roopra, A., Sharling, L., Wood, I., Briggs, T., Bachfischer, U., Paquette, A. & Buckley, N. (2000). Transcriptional repression by neuron-restrictive silencer factor is mediated via the Sin3-histone deacetylase complex. *Molecular and Cell Biology, 20*(6), 2147–2157.

Roseboom, T. J., Painter, R. C., van Abeelen, A. F. M., Veenendaal, M. V. E., & de Rooij, S. R. (2011). Hungry in the womb: What are the consequences? Lessons from the Dutch famine. *Maturitas, 70*(2), 141–145.

Rosenblatt, J. S. (1975). Prepartum and postpartum regulation of maternal behaviour in the rat. *Ciba Foundation Symposium*, (33), 17–37.

Roth, T. L., Lubin, F. D., Funk, A. J., & Sweatt, J. D. (2009). Lasting epigenetic influence of early-life adversity on the BDNF gene. *Biological Psychiatry, 65*(9), 760–769.

Sapolsky, R. M., & Meaney, M. J. (1986). Maturation of the adrenocortical stress response: Neuroendocrine control mechanisms and the stress hyporesponsive period. *Brain Research, 396*(1), 64–76.

Schmidt, M., Enthoven, L., van Woezik, J. H. G., Levine, S., de Kloet, E. R., & Oitzl, M. S. (2004). The dynamics of the hypothalamic-pituitary-adrenal axis during maternal deprivation. *Journal of Neuroendocrinology, 16*(1), 52–57.

Schmidt, M., Levine, S., Alam, S., Harbich, D., Sterlemann, V., Ganea, K.,. . ., Müller, M.B. (2006). Metabolic signals modulate hypothalamic-pituitary-adrenal axis activation during maternal separation of the neonatal mouse. *Journal of Neuroendocrinology, 18*(11), 865–874.

Schmidt-Hieber, C., Jonas, P., & Bischofberger, J. (2004). Enhanced synaptic plasticity in newly generated granule cells of the adult hippocampus. *Nature, 429*(6988), 184–187.

Sebaai, N., Lesage, J., Breton, C., Vieau, D. & Deloof, S. (2004). Perinatal food deprivation induces marked alterations of the hypothalamo-pituitary-adrenal axis in 8-month-old male rats both under basal conditions and

after a dehydration period. *Neuroendocrinology, 79*(4), 163–173.
Selten, J. P., van der Graaf, Y., van Duursen, R., Gispen-de Wied, C. C., & Kahn, R. S. (1999). Psychotic illness after prenatal exposure to the 1953 Dutch Flood Disaster. *Schizophrenia Research, 35*(3), 243–245.
Seth, K. A. & Majzoub, J. A. (2001). Repressor element silencing transcription factor/neuron-restrictive silencing factor (REST/NRSF) can act as an enhancer as well as a repressor of corticotropin-releasing hormone gene transcription. *Journal of Biological Chemistry, 276*(17), 13917–13923.
Shors, T. J., Miesegaes, G., Beylin, A., Zhao, M., Rydel, T., & Gould, E. (2001). Neurogenesis in the adult is involved in the formation of trace memories. *Nature, 410*(6826), 372–376.
Snyder, J. S., Soumier, A., Brewer, M., Pickel, J., & Cameron, H. A. (2011). Adult hippocampal neurogenesis buffers stress responses and depressive behaviour. *Nature, 476*(7361), 458–461.
Sousa, N., Lukoyanov, N. V., Madeira, M. D., Almeida, O. F., & Paula-Barbosa, M. M. (2000). Reorganization of the morphology of hippocampal neurites and synapses after stress-induced damage correlates with behavioral improvement. *Neuroscience, 97*(2), 253–266.
Spencer, R., Miller, A., Stein, M., & McEwen, B. (1991). Corticosterone regulation of type I and type II adrenal steroid receptors in brain, pituitary, and immune tissue. *Brain Research, 549*(2), 236–246.
Squire, L. R. (1992). Memory and the hippocampus: A synthesis from findings with rats, monkeys, and humans. *Psychological Review, 99*(2), 195–231.
Squire, L. R. (2004). Memory systems of the brain: A brief history and current perspective *Neurobiology of Learning and Memory, 82*(3), 171–177.
Stanton, M., Gutierrez, Y., & Levine, S. (1988). Maternal deprivation potentiates pituitary-adrenal stress responses in infant rats. *Behavioral Neuroscience, 102*(5), 692–700.
Steegers-Theunissen, R. P., Obermann-Borst, S. A., Kremer, D., Lindemans, J., Siebel, C., Steegers, E. A.,. . ., Heijmans, B.T. (2009). Periconceptional maternal folic acid use of 400 μg per day is related to increased methylation of the IGF2 gene in the very young child. *PLoS ONE, 4*(11), e7845.
Stevens, S. E., Sonuga-Barke, E. J. S., Kreppner, J. M., Beckett, C., Castle, J., Colvert, E.,. . ., Rutter, M. (2008). Inattention/overactivity following early severe institutional deprivation: presentation and associations in early adolescence. *Journal of Abnormal Child Psychology, 36*(3), 385–398.
Suchecki, D., Mozaffarian, D., Gross, G., Rosenfeld, P., & Levine, S. (1993). Effects of maternal deprivation on the ACTH stress response in the infant rat. *Neuroendocrinology, 57*(2), 204–212.
Swanson, L., & Simmons, D. (1989). Differential steroid hormone and neural influences on peptide mRNA levels in CRH cells of the paraventricular nucleus: a hybridization histochemical study in the rat. *Journal of Comparative Neurology, 285*(4), 413–435.
Szuran, T. F., Pliska, V., Pokorny, J., & Welzl, H. (2000). Prenatal stress in rats: Effects on plasma corticosterone, hippocampal glucocorticoid receptors, and maze performance. *Physiology & Behavior, 71*(3–4), 353–362.
Tannenbaum, B. M., Brindley, D. N., Tannenbaum, G. S., Dallman, M. F., McArthur, M. D. & Meaney, M. J. (1997). High-fat feeding alters both basal and stress-induced hypothalamic-pituitary-adrenal activity in the rat. *American Journal of Physiology, 273*(6 Pt 1), E1168–77.
Teicher, M. H., Anderson, C. M., & Polcari, A. (2012). Childhood maltreatment is associated with reduced volume in the hippocampal subfields CA3, dentate gyrus, and subiculum *Proceedings of the National Academy of Sciences, 109*(9), 563–572.
Todeschin, A. S., Winkelmann-Duarte, E. C., Jacob, M. H. V., Aranda, B. C. C., Jacobs, S., Fernandes, M. C.,. . ., Lucion, A.B. (2009). Effects of neonatal handling on social memory, social interaction, and number of oxytocin and vasopressin neurons in rats. *Hormones and Behavior, 56*(1), 93–100.
Tonkiss, J., & Galler, J. R. (1990). Prenatal protein malnutrition and working memory performance in adult rats. *Behavioural Brain Research, 40*(2), 95–107.
Tonkiss, J., Shultz, P., & Galler, J R. (1994). An analysis of spatial navigation in prenatally protein malnourished rats *Physiology & Behavior, 55*(2), 217–224.
Trivedi, M. A. & Coover, G. D. (2004). Lesions of the ventral hippocampus, but not the dorsal hippocampus, impair conditioned fear expression and inhibitory avoidance on the elevated T-maze. *Neurobiology of Learning and Memory, 81*(3), 172–184.
Trottier, G., Koski, K. G., Brun, T., Toufexis, D. J., Richard, D., & Walker, C. D. (1998). Increased fat intake during lactation modifies hypothalamic-pituitary-adrenal responsiveness in developing rat pups: a possible role for leptin. *Endocrinology, 139*(9), 3704–3711.
Tyrka, A. R., Price, L. H., Gelernter, J., Schepker, C., Anderson, G. M. & Carpenter, L. L. (2009). Interaction of childhood maltreatment with the corticotropin-releasing hormone receptor gene: effects on hypothalamic-pituitary-adrenal axis reactivity. *Biological Psychiatry, 66*(7), 681–685.
Valadares, C. T., Fukuda, M. T. H., Françolin-Silva, A. L., Hernandes, A. S.,. & Almeida, S. S. (2010). Effects of postnatal protein malnutrition on learning and memory procedures. *Nutritional Neuroscience, 13*(6), 274–282.
Van den Hove, D. L. A., Steinbusch, H. W. M., Scheepens, A., Van de Berg, W. D. J., Kooiman, L. A. M., Boosten, B. J. G.,. . ., Blanco, C.E. (2006). Prenatal stress and neonatal rat brain development. *Neuroscience, 137*(1), 145–155.
van Hasselt, F. N., Cornelisse, S., Zhang, T. Y., Meaney, M. J., Velzing, E. H., Krugers, H. J., & Joels, M. (2012). Adult hippocampal glucocorticoid receptor expression and dentate synaptic plasticity correlate with maternal care received by individuals early in life. *Hippocampus, 22*(2), 255–266.
van Os, J., & Selten, J. P. (1998). Prenatal exposure to maternal stress and subsequent schizophrenia. The May 1940 invasion of The Netherlands. *The British Journal of Psychiatry, 172*, 324–326.
Van Pett, K., Viau, V., Bittencourt, J. C., Chan, R. K., Li, H. Y., Arias, C.,. . ., Sawchenko, P. E. (2000). Distribution of mRNAs encoding CRF receptors in brain and pituitary of rat and mouse. *The Journal of Comparative Neurology, 428*(2), 191–212.
van Praag, H., Kempermann, G., & Gage, F. H. (1999). Running increases cell proliferation and neurogenesis in the adult mouse dentate gyrus. *Nature Neuroscience, 2*(3), 266–270.
van Strien, N. M., Cappaert, N. L. M., & Witter, M. P. (2009). The anatomy of memory: An interactive overview of the

parahippocampal-hippocampal network. *Nature Reviews Neuroscience, 10*(4), 272–282.
Veenema, A. H., & Neumann, I. D. (2009). Maternal separation enhances offensive play-fighting, basal corticosterone and hypothalamic vasopressin mRNA expression in juvenile male rats. *Psychoneuroendocrinology, 34*(3), 463–467.
Viau, V., Sharma, S., Plotsky, P. M., & Meaney, M. J. (1993). Increased plasma ACTH responses to stress in nonhandled compared with handled rats require basal levels of corticosterone and are associated with increased levels of ACTH secretagogues in the median eminence. *Journal of Neuroscience, 13*(3), 1097–1105.
Vythilingam, M., Heim, C., Newport, J., Miller, A. H., Anderson, E., Bronen, R.,..., Bremner, J.D. (2002). Childhood trauma associated with smaller hippocampal volume in women with major depression. *The American Journal of Psychiatry, 159*(12), 2072–2080.
Walker, C. (2005). Nutritional aspects modulating brain development and the responses to stress in early neonatal life. *Progress in Neuro-Psychopharmacology and Biological Psychiatry, 29*(8), 1249–1263.
Walker, C., & Dallman, M. (1993). Neonatal facilitation of stress-induced adrenocorticotropin secretion by prior stress: evidence for increased central drive to the pituitary. *Endocrinology, 132*(3), 1101–1107.
Walker, C. D. (2010). Maternal touch and feed as critical regulators of behavioral and stress responses in the offspring. *Developmental Psychobiology, 52*(7), 638–650.
Walker, C. D., Naef, L., D'Asti, E., Long, H., Xu, Z., Moreau, A., & Azeddine, B. (2008). Perinatal maternal fat intake affects metabolism and hippocampal function in the offspring: A potential role for leptin. *Annals of the New York Academy of Sciences, 1144*, 189–202.
Walker, C. D., Perrin, M., Vale, W. & Rivier, C. (1986). Ontogeny of the stress response in the rat: role of the pituitary and the hypothalamus *Endocrinology, 118*(4), 1445–1451.
Walker, S. P., Grantham-McGregor, S. M., Powell, C. A., & Chang, S. M. (2000). Effects of growth restriction in early childhood on growth, IQ, and cognition at age 11 to 12 years and the benefits of nutritional supplementation and psychosocial stimulation. *Journal of Pediatrics, 137*(1), 0036–0041.
Wang, X.D., Rammes, G., Kraev, I., Wolf, M., Liebl, C., Scharf, S. H.,..., Schmidt, M.V. (2011). Forebrain CRF1 modulates early-life stress-programmed cognitive deficits. *The Journal of Neuroscience, 31*(38), 13625–13634.
Watson, J. B., Mednick, S. A., Huttunen, M., & Wang, X. (1999). Prenatal teratogens and the development of adult mental illness. *Development and Psychopathology, 11*(3), 457–466.
Weaver, I. C. G. (2005). Reversal of maternal programming of stress responses in adult offspring through methyl supplementation: altering epigenetic marking later in life. *Journal of Neuroscience, 25*(47), 11045–11054.
Weaver, I. C. G, Cervoni, N., Champagne, F. A., D'Alessio, A. C., Sharma, S., Seckl, J. R.,..., Meaney, M.J. (2004). Epigenetic programming by maternal behavior. *Nature Neuroscience, 7*(8), 847–854.
Weaver, I. C. G., Grant, R. J., & Meaney, M. J. (2002). Maternal behavior regulates long-term hippocampal expression of BAX and apoptosis in the offspring *Journal of Neurochemistry, 82*(4), 998–1002.
Weaver, I., La Plante, P., Weaver, S., Parent, A., Sharma, S., Diorio, J.,..., Meaney, M.J. (2001). Early environmental regulation of hippocampal glucocorticoid receptor gene expression: characterization of intracellular mediators and potential genomic target sites. *Molecular and Cellular Endocrinology., 185*(1–2), 205–218.
Weinstock, M. (2001). Alterations induced by gestational stress in brain morphology and behaviour of the offspring. *Progress in Neurobiology, 65*(5), 427–451.
Weinstock, M. (2011). Sex-dependent changes induced by prenatal stress in cortical and hippocampal morphology and behaviour in rats: an update. *Stress, 14*(6), 604–613.
Whipple, E. & Webster-Stratton, C. (1991). The role of parental stress in physically abusive families. *Child Abuse and Neglect, 15*(3), 279–291.
Whitlock, J. R., Heynen, A. J., Shuler, M. G., & Bear, M. F. (2006). Learning induces long-term potentiation in the hippocampus. *Science, 313*(5790), 1093–1097.
Wiedmeier, J. E., Joss-Moore, L. A., Lane, R. H., & Neu, J. (2011). Early postnatal nutrition and programming of the preterm neonate. *Nutrition Reviews, 69*(2), 76–82.
Wiener, S. G., Smotherman, W. P., & Levine, S. (1976). Influence of maternal malnutrition on pituitary-adrenal responsiveness to offspring. *Physiology & Behavior, 17*(6), 897–901.
Wigger, A., & Neumann, I. D. (1999). Periodic maternal deprivation induces gender-dependent alterations in behavioral and neuroendocrine responses to emotional stress in adult rats. *Physiology & Behavior, 66*(2), 293–302.
Willoughby, K. A., Sheard, E. D., Nash, K., & Rovet, J. (2008). Effects of prenatal alcohol exposure on hippocampal volume, verbal learning, and verbal and spatial recall in late childhood. *Journal of the International Neuropsychological Society: JINS, 14*(6), 1022–1033.
Winkelmann-Duarte, E. C., Padilha-Hoffmann, C. B., Martins, D. F., Schuh, A. F. S., Fernandes, M. C.,..., Lucion, A.B. (2011). Early-life environmental intervention may increase the number of neurons, astrocytes, and cellular proliferation in the hippocampus of rats. *Experimental Brain Research, 215*(2), 163–172.
Wolf, C., Almli, C. R., Finger, S., Ryan, S., & Morgane, P. J. (1986). Behavioral effects of severe and moderate early malnutrition. *Physiology & Behavior, 38*(5), 725–730.
Workel, J. O., Oitzl, M. S., Fluttert, M., Lesscher, H., Karssen, A., & de Kloet, E. R. (2001). Differential and age-dependent effects of maternal deprivation on the hypothalamic-pituitary-adrenal axis of brown Norway rats from youth to senescence. *Journal of Neuroendocrinology, 13*(7), 569–580.
Wren, A. M., Small, C. J., Fribbens, C. V., Neary, N. M., Ward, H. L., Seal, L. J.,..., Bloom, S.R. (2002). The hypothalamic mechanisms of the hypophysiotropic action of ghrelin. *Neuroendocrinology, 76*(5), 316–324.
Wu, J., Song, T., Li, Y., He, K., Ge, L., & Wang, L. (2007). Prenatal restraint stress impairs learning and memory and hippocampal PKCbeta1 expression and translocation in offspring rats. *Brain Research, 1141*, 205–213.
Yaka, R., Salomon, S., Matzner, H., & Weinstock, M. (2007). Effect of varied gestational stress on acquisition of spatial memory, hippocampal LTP and synaptic

proteins in juvenile male rats. *Behavioural Brain Research, 179*(1), 126–132.

Yang, J., Han, H., Cao, J., Li, L., & Xu, L. (2006). Prenatal stress modifies hippocampal synaptic plasticity and spatial learning in young rat offspring. *Hippocampus, 16*(5), 431–436.

Yehuda, R., Engel, S. M., Brand, S. R., Seckl, J., Marcus, S. M., & Berkowitz, G. S. (2005). Transgenerational effects of posttraumatic stress disorder in babies of mothers exposed to the World Trade Center attacks during pregnancy. *Journal of Clinical Endocrinology and Metabolism, 90*(7), 4115–4118.

Yi, S. J., & Baram, T.Z. (1994). Corticotropin-releasing hormone mediates the response to cold stress in the neonatal rat without compensatory enhancement of the peptide & aposis gene expression. *Endocrinology, 135*(6), 2364–2368.

Zuena, A. R., Mairesse, J., Casolini, P., Cinque, C., Alemà, G. S., Morley-Fletcher, S., . . ., Maccari, S. (2008). Prenatal restraint stress generates two distinct behavioral and neurochemical profiles in male and female rats. *PLoS ONE, 3*(5), e2170.

CHAPTER 16

Genetic Tools in the Erasure of Emotional Memory

Derya Sargin, Chen Yan *and* Sheena A. Josselyn

Abstract

Fear is an important emotion; remembering fearful events/places/stimuli is key for survival. However, dysregulation of fear may underlie the etiology of several psychiatric diseases. Inappropriate storage and/or recall of fearful events can lead to maladaptive fear behaviors and physiological responses that contribute to emotional disorders. Much research has provided insights into the neural processes mediating the formation of fear memories. In addition, some new research has begun to provide insights into how fear memories may be weakened. A more thorough understanding of the molecular, cellular, and circuit basis of the formation and storage of fear memories may one day provide insights into how we can rid ourselves of aberrant fear memories associated with psychopathological responses.

Key Words: fear memory, amygdala, fear extinction, HSV, CREB, depression

I. Fear memory

Fear has evolved as a survival mechanism, enabling animals to recognize and respond appropriately to threatening circumstances (Fanselow & Lester, 1988). Learning to fear threatening situations and to act appropriately (freezing, attacking or escaping) (LeDoux, 1996; Ohman & Mineka, 2001) enables individuals to cope with dangerous contingencies that could potentially decrease the probability of having offspring and transmitting their genes to future generations.

Despite the significant role of fear learning and memory in mediating survival, too much fear may be maladaptive. Overgeneralization of fear may be detrimental to mental health and has been proposed to lead to the development of pathological conditions. For instance, roughly 10% of individuals who experience a traumatic event will go on to develop a fear- or anxiety-related disorder, including panic disorder, post-traumatic stress disorder (PTSD), or depression (Kessler, Chiu, Demler, Merikangas, & Walters, 2005; Kessler, Sonnega, Bromet, Hughes, & Nelson, 1995). Current treatments for fear and anxiety disorders primarily focus on behavioral and cognitive approaches, such as exposure and cognitive processing therapies. These therapies may ultimately lead to reduction of symptoms (Bomyea & Lang, 2012) but they are far from ideal. Therefore, a more thorough understanding of how fear is normally acquired, stored, and expressed may provide novel therapeutic targets for the treatment of these disorders.

Classical fear conditioning has been used extensively to delineate the neural systems mediating fear learning in the lab. In these experiments, a previously neutral stimulus (e.g., light or tone; conditioned stimulus [CS]) that is paired with an aversive stimulus (e.g., footshock; unconditioned stimulus [US]) acquires the ability to elicit conditioned fear responses. However, the expression of fear acquired under one set of circumstances may be inappropriate or even disadvantageous in another set of circumstances in which the threat is no longer dangerous. Therefore, many experimental studies have begun to probe how fear may be reduced.

Fear may be reduced or inhibited in several ways. For example, extinction training, in which a stimulus that was previously paired with shock is presented repeatedly in the absence of the shock (CS-no shock), gradually reduces conditioned fear responses (Pavlov, 1927). Conditioned inhibition of fear, in which a stimulus signals the absence of shock, is another procedure that reduces conditioned fear responses (Rescorla, 1969). Perhaps the ultimate way of decreasing fear would be to "erase" the actual fear memory. In this chapter we provide an overview of experimental models and tools that are currently being used to understand the molecular and cellular mechanisms underlying fear memory erasure.

A. Neural circuitry of fear

Evidence from genetic, biochemical, electrophysiological, and behavioral studies from many labs, including those of Mike Davis, Michael Fanselow, Joe LeDoux, and Steven Maren, established that the amygdala, particularly the lateral nucleus of the amygdala (LA), is required for the formation of conditioned fear memories (Davis, 1992; Fanselow & Gale, 2003; J. E. LeDoux, 2000; Maren & Quirk, 2004; Rumpel, LeDoux, Zador, & Malinow, 2005) (but see McGaugh, 2002). The LA serves as the sensory gateway into the amygdala because it receives strong inputs from sensory systems of the brain including the visual, auditory, somatosensory, taste, and olfactory inputs from thalamus and cortex (reviewed in LeDoux, 2007). The intra- and internuclear projections that are thought to underlie fear learning in the amygdala mainly follow a principal pathway that runs from the LA to the basolateral nucleus (BLA) and from the BLA to the central nucleus (CEA). The neurons of the LA that receive the CS input and send projections to the BLA consist mainly of glutamatergic projection neurons and, to a lesser extent, GABAergic inhibitory interneurons. The BLA projects to the CEA, which is an important output region because it connects to the brainstem and to hypothalamic areas that control the expression of emotional and physiological reactions, such as the freezing response. As the CEA projects to the brainstem, the BLA sends connections to the striatum, which plays a role in governing motor responses. Inhibitory intercalated (ITC) interneurons of the LA form another relay site between the LA, BLA, and the CEA. This GABAergic cell mass also receives connections from the sensory cortex and gates the information flow between the BLA and the CEA (reviewed in Ehrlich, et al., 2009; J. LeDoux, 2007; Pape & Pare, 2010). A simplified principal circuitry thought to underlie fear learning is shown in Figure 16.1.

Further evidence for the importance of the amygdala in fear has been shown by studies in humans. Neuroimaging studies show that the amygdala is activated when the subject is fear conditioned or presented with fearful facial expressions. Furthermore, the degree of amygdala activation correlates with the emotional intensity (LaBar, Gatenby, Gore, LeDoux, & Phelps, 1998; Morris, et al., 1996). Moreover, individuals with amygdala damage have impairments in recognizing and expressing fearful emotions. For instance, patient SM046, who has bilateral and selective amygdala damage due to a rare genetic disorder, Urbach-Wiethe disease (Adolphs, Tranel, Damasio, & Damasio, 1994; Adolphs, Tranel, Damasio, & Damasio, 1995), is able to discern the correct emotion presented in most facial expressions but is unable to recognize fearful emotional expression. In addition, removal of the amygdala (to control epileptic seizures) is also associated with impaired recognition of fearful face expressions (Adolphs, Tranel, & Damasio, 2001) or changes in the perception of aversive words (Anderson & Phelps, 2001). Therefore, in both humans and rodents, the amygdala is particularly important in fear and fear memory.

The formation of long-term memory (LTM) requires gene expression and protein synthesis. The formation of long-term fear memory is no exception, and Pavlovian fear conditioning induces distinct patterns of gene activation in the LA. For instance, fear conditioning leads to an increase in mRNA and protein levels of immediate early genes (IEGs), including *c-fos* and *zif268/egr-1*, in the amygdala (Campeau, et al., 1991; Malkani & Rosen, 2000; Radulovic, Kammermeier, & Spiess, 1998; Rosen, Fanselow, Young, Sitcoske, & Maren, 1998). Moreover, the formation of fear memory is impaired by manipulations that block LA NMDA receptor function (Bauer, Schafe, & LeDoux, 2002; Campeau, Miserendino, & Davis, 1992; Fanselow & Kim, 1994; Fendt, 2001; Gewirtz & Davis, 1997; Goosens & Maren, 2003; Lee, Choi, Brown, & Kim, 2001; Maren, Aharonov, Stote, & Fanselow, 1996; Miserendino, Sananes, Melia, & Davis, 1990; Rodrigues, Schafe, & LeDoux, 2001),

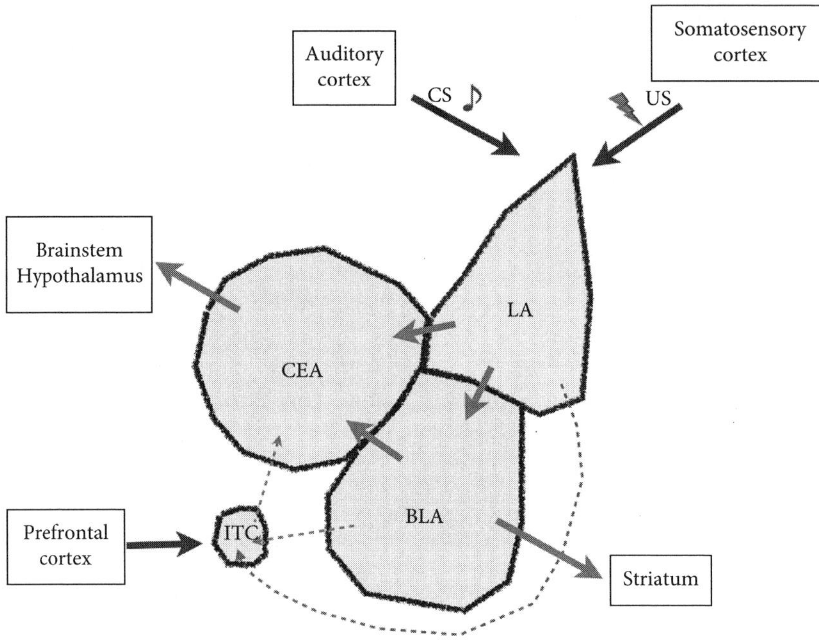

Fig. 16.1 *legend*: A simplified circuitry of fear learning. CS; conditioned stimulus, US; unconditioned stimulus, amygdala areas: LA; lateral nucleus, BLA; basolateral nucleus, CEA; central nucleus, ITC; intercalated interneurons.

L-type voltage-gated calcium channel (LVGCC) function (Bauer, et al., 2002), or RNA (Bailey, Kim, Sun, Thompson, & Helmstetter, 1999; Duvarci, Nader, & LeDoux, 2008) and protein synthesis (Maren, Ferrario, Corcoran, Desmond, & Frey, 2003; Nader, Schafe, & Le Doux, 2000; Schafe & LeDoux, 2000). Together these studies suggest that synaptic plasticity in LA is important for the formation of fear memories.

B. Synaptic processing in LA and fear memory

The notion that learning induces synaptic modifications in the brain was initially proposed by Hebb (1949) and has been reformulated as the synaptic plasticity and memory hypothesis in more recent years. The core of this hypothesis is that synaptic changes between neurons occur in response to learning, and that these synaptic changes are the mechanisms by which memory traces are formed and stored in the brain (Martin, Grimwood, & Morris, 2000). Behavioral training produces long-lasting changes in neurons in the form of long-term potentiation (LTP), which leads to the increase of synaptic strength and the long-term maintenance of memories in discrete neuronal networks. It has been hypothesized that LTP is a molecular and cellular correlate of memory formation.

Fear conditioning has been shown to enhance CS-evoked potentiation of LA synapses (McKernan & Shinnick-Gallagher, 1997; Rogan, Staubli, & LeDoux, 1997). Early studies found enhanced α-amino-3-hydroxy-5-methyl-4-isoxazolepropionic acid receptor (AMPAR)-mediated synaptic transmission in the LA neurons of fear conditioned rats compared to those of naive or unpaired animals, indicating that AMPAR responses are potentiated and LTP is enhanced in the amygdala by fear learning (McKernan & Shinnick-Gallagher, 1997). Likewise, paired training led to an increase in the slope and amplitude of the auditory CS-evoked field potentials (CS-EPs) in the LA of freely behaving rats, reminiscent of the effects of LTP induction. Unpaired training did not lead to the same effect. Enhanced synaptic response induced by fear training was accompanied by increased freezing response (Rogan, et al., 1997).

Another early in vivo study, carried out by Rosenkranz and Grace showed that changes in LA synaptic efficacy are correlated with fear learning (Rosenkranz & Grace, 2002). Using in vivo intracellular recordings of LA neurons during Pavlovian fear conditioning in rats, they demonstrated that repeated pairing of an odor (CS) with a foot-shock (US) led to an enhancement in the amplitude of CS-evoked excitatory postsynaptic potentials (EPSPs) in LA neurons. In these experiments, repeated unpaired odor presentation led to a gradual decrease in odor-evoked depolarizing responses of LA neurons. In contrast, when an odor was paired with a train of foot-shocks, enhanced post-synaptic response of LA neurons was observed (Rosenkranz & Grace, 2002). These pioneering studies suggested that fear conditioning elicits LTP-like processes that underlie fear learning.

Further evidence of the involvement of LA synapses in the fear circuit came from experiments performed by Kwon and Choi, and their use of intracranial stimulation of thalamic afferents to the LA of rats (Kwon & Choi, 2009). Bipolar stimulating electrodes were implanted in the medial geniculate nucleus (MGm) of the thalamus, and stimulation of these electrodes substituted for the CS. A recording electrode was implanted into the LA to measure the synaptic responses of LA neurons after auditory fear conditioning. Fear conditioning involved pairing a train of stimulation (CS) with a footshock (US). The MGm-paired group showed significant freezing to the CS and a higher level of CS-evoked field potentials in the LA when compared with the unpaired group (Kwon & Choi, 2009). Taken together, activity-dependent synaptic modifications in LA neurons are the cellular correlates of the fear memory trace that is formed and stored following Pavlovian fear conditioning.

II. Extinction of fear memory

In Pavlovian fear conditioning, if the CS is presented repeatedly in the absence of the US, expression of the conditioned response (CR) gradually decreases. It is thought that a novel association between the CS and no-US is formed. The extinction of a fear memory was first described by Pavlov in 1927 (Pavlov, 1927). In extinction training, it is thought that the original memory for CS-US association is maintained, but that CS presentations alone inhibit the output of this association and subsequently prevent the expression of the conditioned fear. Failure to fully extinguish fear responses may underlie anxiety disorders, and these disorders may result from a dysfunction of the neural mechanisms underlying extinction learning (Rauch, Shin, & Phelps, 2006).

The neural circuits and mechanisms involved in extinction learning have been extensively studied (Myers & Davis, 2007; Quirk & Mueller, 2008). Given the significant role of the amygdala in acquisition and consolidation of fear memory, it is not surprising that the amygdala has been identified as a crucial component of extinction learning. Studies using intra-amygdala infusions of NMDAR antagonists have shown that extinction learning is dependent on NMDARs in the BLA (Falls, Miserendino, & Davis, 1992; Sotres-Bayon, Bush, & LeDoux, 2007; Walker, Ressler, Lu, & Davis, 2002). Similarly, a blockade of glutamate receptors (Kim, Lee, Park, Song, et al., 2007) or an infusion of a LVGCC antagonist (Cain, Blouin, & Barad, 2002) into the BLA of mice impaired extinction learning. Inhibition of MAPK/ERK signaling in the BLA prior to extinction training also prevented acquisition of fear extinction (Herry, Trifilieff, Micheau, Luthi, & Mons, 2006). The cannabinoid system has also been implicated in the regulation of extinction learning in LA. Cannabinoid receptor 1 (CB1) deficient mice were severely impaired in short- and long-term extinction of auditory fear conditioning, which was mimicked by systemic injections of a CB1 antagonist (Marsicano, et al., 2002). Overall, these data indicate that LA plays a major role in fear extinction learning.

Fear extinction is regulated by the activation of the inhibitory intercalated (ITC) interneurons of the LA, which receive glutamatergic afferents from the infralimbic (IL) region of the ventromedial prefrontal cortex (vmPFC) (McDonald, 1992). ITC neurons are situated between the BLA and CEA and relay inhibitory information to the CEA, which is the main output nucleus of the amygdala (Pare & Smith, 1993). Using an immunotoxin that only targets cells expressing high levels of μ-opioid receptors (μORs), such as ITC cells, Likhtik and colleagues have recently shown that selective ITC lesions result in extinction deficits. Their results further emphasize the role of the ITC neurons as the significant component of fear extinction learning (Likhtik, Popa, Apergis-Schoute, Fidacaro, & Pare, 2008). By acting as the inhibitory interface between the BLA and CEA, the ITC neurons maintain the regulatory response necessary for the inhibition of CS-induced conditioned fear.

Because the vmPFC relays critical information to the ITC, the involvement of the vmPFC in fear extinction learning has been extensively studied. Some of the important studies demonstrating the requirement of the vmPFC for consolidation of extinction were carried out by Greg Quirk and colleagues (Burgos-Robles, Vidal-Gonzalez, Santini, & Quirk, 2007). Rats with vmPFC lesions were able to extinguish fear to the CS presented without the US within one session; however they were impaired in extinction recall when tested 24 hours later (Quirk, Russo, Barron, & Lebron, 2000). The same group studied the vmPFC response in extinction further by infusing the competitive NMDAR antagonist CPP into the vmPFC of fear-conditioned rats prior to or immediately after extinction training. In both cases, inhibition of NMDAR activity in the vmPFC resulted in a high freezing response, indicating poor recall of extinction memory (Burgos-Robles, et al., 2007).

Although these studies provided strong evidence for the importance of the vmPFC-ITC pathway for the recall of fear extinction memory, microstimulation of different parts of vmPFC, however, yields different fear behavior, suggesting heterogeneity within the vmPFC in fear regulation (Vidal-Gonzalez, Vidal-Gonzalez, Rauch, & Quirk, 2006). This is supported by a recent study showing a dissociation between the infralimbic (IL) and prelimbic (PrL) region of vmPFC in modulating conditioned fear (Sierra-Mercado, Padilla-Coreano, & Quirk, 2011). Sierra-Mercado and colleagues injected a small dose of $GABA_A$ receptor agonist muscimol to inhibit bilateral IL or PrL in fear-conditioned rats immediately before extinction training. Rats with IL inhibition showed higher freezing during extinction training and testing; however, inactivating PrL led to lower level freezing during training but normal extinction memory expression during drug-free testing. These results indicate that IL is necessary for the formation of extinction memory, and PrL is important for fear memory expression but not extinction learning (Sierra-Mercado et al., 2011).

Heterogeneity of the fear pathway has also been observed downstream from vmPFC. Luthi and colleagues studied the behavior of individual neurons within the mPFC and BLA network in response to fear memory extinction (Herry, et al., 2008). They implanted recording electrodes in the amygdala of mice and trained them with two distinct auditory CS with different frequencies. One CS (CS^+) was paired with the footshock (US), whereas the other was unpaired (CS^-). Mice showed increased freezing only to the CS^+ 24 hours after conditioning. Mice were exposed to a series of CS^+ presentations without the US during the extinction training, which caused the level of freezing to the CS^+ to decrease to the level induced by presentation of the CS^-. The study identified two distinct classes of neurons in the BLA; "fear neurons" with a selective increase in CS^+-evoked spike firing during and after fear training and "extinction neurons" with increased firing only during extinction training. Therefore, CS-US pairing evoked enhanced firing in fear neurons when mice were exposed to the CS alone, whereas a different population of neurons, the extinction neurons, responded to the CS undergoing extinction. Upon fear renewal by moving the mice from the extinction context to the conditioning context, activity was switched from extinction to fear neurons. Inactivation of the BLA neurons, by infusion of muscimol, inhibited both extinction learning and fear renewal, emphasizing the necessity of these neurons for the observed behavior. Fear and extinction neurons exhibited a differential connectivity pattern: Fear neurons were shown to receive inputs from the hippocampus and project to the mPFC, whereas extinction neurons were reciprocally connected with the mPFC (Herry, et al., 2008).

A recent study by Lee and colleagues suggests the switch that toggles fear learning or extinction may lie in the mediodorsal thalamus (mdTh) nucleus (Lee, et al., 2011). This brain structure has significant connections to both BLA and mPFC, and, interestingly, contain neurons that fires in two distinct modes: a phospholipase C β4 (PLCβ4) - mGluR1 dependent tonic firing, and a low voltage-activated T-type Ca^{2+} channel (LVCC) dependent burst firing. They have shown that PLCβ4 knockout mice, or wild-type (WT) mice with PLCβ4 pathway inhibited by shRNA injection or pharmacological manipulations, have normal fear learning but impaired fear extinction (Lee, et al., 2011). This behavior of PLCβ4 knockout mice is correlated with a higher baseline burst firing in mdTh and lower tonic firing during fear extinction. On the other hand, pharmacologically inhibiting LVCC in mdTh enhances fear extinction in both WT and PLCβ4 knockout mice, but impairs fear acquisition. Furthermore, direct electrophysiological manipulation that induces tonic firing in mdTh enhances fear extinction, whereas inducing burst firing suppresses extinction (Lee et al., 2011). The results suggest that the firing pattern of mdTh neurons, which is controlled by their intrinsic membrane property, play an important role in switching fear learning and fear extinction. Taken together, these studies shed light on some of the important networks that mediate fear extinction learning.

Where the amygdala governs the major processing of fear-conditioned responses, the hippocampus is involved in the regulation of contextual fear conditioning (Anagnostaras, Maren, & Fanselow, 1999; Kennedy & Shapiro, 2004; Sanders,

Wiltgen, & Fanselow, 2003) as well as extinction of fear learning (Corcoran, Desmond, Frey, & Maren, 2005; Corcoran & Maren, 2001, 2004). The inhibitory response of the ITC neurons in the LA is activated by the vmPFC. The vmPFC receives inputs from several brain regions including cortex, thalamus, and hippocampus (Myers-Schulz & Koenigs, 2012). Among these areas, the hippocampus plays a major role in the context-specific encoding and retrieval of fear extinction. Expression of memory strongly depends on the context in which memory has been retrieved, and the hippocampus is a key region involved in the regulation of memory retrieval based on the contextual cues. The involvement of hippocampus in extinction learning has first been shown in rats with impaired reinstatement of conditional fear after permanent hippocampal lesions (Wilson, Brooks, & Bouton, 1995). The role of hippocampus in context-specific extinction has further been supported by studies involving electrolytic hippocampal lesions in rats which exhibited impaired freezing during a context extinction test (Ji & Maren, 2005; Maren & Fanselow, 1997). Moreover, inactivation of the dorsal hippocampus using $GABA_A$ receptor agonist muscimol, blocked context-specific expression of fear memory extinction (Corcoran & Maren, 2001, 2004). Schimanski et al. showed that extinction of contextual fear was impaired in mice that had a reduced hippocampal commissure (9XCA/Wah strain) (Schimanski, Wahlsten, & Nguyen, 2002). A variety of complex signaling cascade activation in the dorsal hippocampus, including PKA, MEK/ERK (Fischer, et al., 2007; Szapiro, Vianna, McGaugh, Medina, & Izquierdo, 2003), BDNF (Heldt, Stanek, Chhatwal, & Ressler, 2007), and Cdk5 (Sananbenesi, et al., 2007) signaling pathways have been shown to regulate extinction of contextual fear.

Taken together, extinction is a type of inhibitory learning that leads to a reduction in the original fear response and thus serves as a significant treatment for a variety of anxiety disorders. More effective fear- and anxiety-treatment strategies could be developed with a better understanding of the neural circuitry between the vmPFC, BLA, and hippocampus that regulate fear learning and excitation.

III. Extinction or forgetting?

Extinction training is thought to suppress, but not destroy, the conditioned fear memory (Rescorla, 2001). Whereas extinction training might be useful in inhibiting a fear memory, the extinguished fear may spontaneously recover over time (Baum, 1988), be reinstated after being re-exposed to the original context in which the US was delivered (Rescorla & Heth, 1975), or be renewed by a change to the context in which extinction took place (Bouton & Bolles, 1979). In this way, extinction is distinguished from forgetting, as the fear memory remains intact. This provides a challenge for therapeutic interventions utilizing extinction-based therapies for treatment of fear disorders. Although pharmacological studies using drugs aimed at manipulating NMDAR (Ledgerwood, Richardson, & Cranney, 2003, 2005; Walker, et al., 2002), adrenergic (Bernardi & Lattal, 2010; Do-Monte, Allensworth, & Carobrez, 2010), or cannabinoid (Lutz, 2007) signaling have proven to be useful tools to enhancing extinction learning, the manipulation and efficacy of such drugs in a clinical setting remain elusive.

Several recent studies have provided contrary evidence to the general concept of extinction as inhibitory learning. These studies center around the notion that fear conditioning induces LTP in amygdala synapses and increases the expression of cell-surface AMPARs (Rumpel, et al., 2005; Yeh, Mao, Lin, & Gean, 2006). The studies posit that extinction is merely the reversal of training-elicited synaptic efficacy in LA, represented as depotentiation of conditioning-induced LTP. Mao et al. showed that fear-conditioning-induced increases in the surface expression of glutamate receptor subunit 1 (GluR1) of AMPARs in the LA of rats could be reversed when rats were given an extinction-training paradigm 1 hour after conditioning. Rats also demonstrated an inhibition in startle potentiation (Mao, Hsiao, & Gean, 2006). However, when extinction training was conducted 24 hours after conditioning, it did not reverse the increase in GluR1 expression, although it still led to a reduction in startle potentiation. If low frequency stimulation (LFS) is applied 1 hour after tetanic stimulation (TS), conditioning induced LTP in the amygdala is erased and depotentiation is augmented (Mao, et al., 2006). Likewise, surface GluR1 expression is enhanced in the presence of D-cycolserine (DCS), a positive modulator of NMDARs that normally facilitates extinction-induced reduction of fear responses (Mao, et al., 2006). Similarly, synaptic potentiation induced by fear conditioning in rats was reversed by extinction training through depotentiation of thalamic afferents to LA. This effect was suggested to occur via a mechanism that involved activation of NMDAR and enhanced the expression of surface AMPARs (Kim, Lee, Park, Hong, et al., 2007). Consistent with these results, Myers et al. proposed a model for memory erasure where a memory is susceptible to erasure when extinction is initiated within a specific time interval (e.g., 1 hour) after conditioning. Inhibitory learning is thought to occur after longer post-training intervals (Myers, Ressler, & Davis, 2006).

Clem and Huganir further tested the concept of extinction as the inhibitory learning by examining Ca^{2+}-permeable AMPARs in the mouse LA (Clem & Huganir, 2010). They first showed that auditory fear conditioning led to an enhancement in the GluR2 lacking, Ca^{2+}-permeable AMPAR (CP AMPAR)-mediated excitatory postsynaptic currents at thalamic synapses to LA. This increase peaked at 24 hours and persisted for 7 days after fear learning, suggesting that it might be critical for the stability of fear memory. Increased CP-AMPAR response at the synapse correlated with enhanced long-term depression (LTD) due to augmented capacity for synaptic removal of these receptors. Reactivation of fear memory performed by CS exposure in the training context 30 minutes before extinction led to no fear response 2 or 7 days after fear conditioning, indicative of fear memory erasure. Retrieval before extinction significantly reduced AMPA-mediated current and LTD response, suggesting a decrease in potentiated transmission mediated by fear conditioning. These results were explained by retrieval-induced removal of CP-AMPARs from LA synapses when memory reactivation occurred within a critical time window before CP-AMPARs return to their basal levels in the synapse (Clem & Huganir, 2010).

Memories can also be lost during a process called reconsolidation. In the reconsolidation process, a previously acquired fear memory becomes labile upon reactivation during retrieval and requires additional protein synthesis to persist further (Misanin, Miller, & Lewis, 1968; Nader, et al., 2000; Tronson, Wiseman, Olausson, & Taylor, 2006). Several studies have shown that pharmacological manipulation (i.e., protein synthesis inhibitors) within the reconsolidation window interferes with memory restorage and might lead to the loss of the consolidated memory (Alberini, 2005; Duvarci & Nader, 2004; Hupbach, Gomez, Hardt, & Nadel, 2007; Nader, et al., 2000). The reconsolidation window was shown to be within 6 hours after memory reactivation in rats (Nader, et al., 2000) and might serve as the time required to integrate retrieved information and to re-stabilize the memory. Monfils et al. combined reconsolidation and extinction learning in a single paradigm that preserved the reduction of the learned fear memory (Monfils, Cowansage, Klann, & LeDoux, 2009). These researchers reactivated the fear memory in conditioned rats by exposing them to a single CS (retrieval) prior to extinction training. Rats that received extinction training within the reconsolidation window (10 minutes and 1 hour) did not exhibit spontaneous recovery, renewal, or reinstatement of fear. Therefore, extinction training within a critical time window during which the memory becomes labile prevented the return of fear (Monfils, et al., 2009).

The discovery that long-term memories become fragile within a critical time window and thus could be manipulated, gave rise to the initiation of clinical trials that made use of pharmacological manipulations aimed at treating PTSD by targeting this critical time period. The first clinical trials initiated by Roger Pitman focused on a β-adrenergic blocker propranolol that had previously been used to treat high blood pressure (Lithell, Pollare, & Vessby, 1992). Studies on rodents had shown that preconditioned responses could be weakened when propranolol was administered in animals after re-presentation of the CS (Debiec & Ledoux, 2004; Przybyslawsky, Roullet, & Sara, 1999). Therefore, Pitman and colleagues decided to investigate whether re-activating the memories of PTSD patients would induce this critical time window during which propranolol could be used to reduce the traumatic memories (Brunet, et al., 2008). 19 PTSD patients were recruited for the study and asked to describe the details of the traumatic events that caused their PTSD for re-activation of traumatic memories. The patients were then given either propanolol or placebo right after recall. Physiological reactions of the patients were recorded a week later as they listened to their traumatic scripts recited by a researcher. Overall, stress responses of the propranolol-treated patients were significantly smaller compared to those who had received placebo. This study held promise by showing for the first time that propranolol treatment immediately after retrieval of a post-traumatic memory reduced the physiological responses associated with the recall of that memory (Brunet, et al., 2008). In order to find out whether the effect of propranolol was long-lasting, the same group conducted a follow-up study in which propranolol was administered once a week for 6 weeks and PTSD scores were measured over a 6-month period. The decrease in PTSD scores over this follow-up period was highly significant. About 71% of the participants no longer met the criteria for PTSD at the end of the study (Brunet, et al., 2011). The study, however, was criticized for not including a control group.

Tronel and Alberini tested the role of the glucocorticoid system during reconsolidation of reactivated traumatic memories in rats. They injected the glucocorticoid receptor antagonist RU 38486 into BLA following reactivation of inhibitory avoidance (IA) memory. They showed that IA memory was disrupted upon blocking the glucocorticoid system in BLA right after recall, and that disruption in memory was persistent (Tronel & Alberini, 2007). As the glucocorticoid system was proven to be important in fear memory manipulation, a more recent study focused on the natural stress hormone, hydrocortisone, and hypothesized that development of PTSD might render individuals with an inability to produce sufficient cortisol and thus make them more susceptible to trauma (Zohar, et al., 2011). A single high-dose hydrocortisone treatment in patients within 6 hours after the trauma exposure proved to be protective against subsequent development of PTSD. The group also studied the morphological consequences of high-dose steroid treatment in dentate gyrus neurons of stressed rats. Stress-induced decreases in dendritic growth and spine density was exacerbated by high-dose hydrocortisone treatment in rats, offering a potential mechanism for the protective effects of cortisol in patients (Zohar, et al., 2011). Additional studies have also proposed a potential treatment role for morphine (Holbrook, Galarneau, Dye, Quinn, & Dougherty, 2010) and ketamine (McGhee, Maani, Garza, Gaylord, & Black, 2008) in PTSD.

Consequently, reversal of the critical mechanisms that underlie the expression of fear, such as depotentiation of synaptic strength within the amygdala shortly after training-induced potentiation of synapses, or interfering with the reactivation of the acquired memories within a critical time window might erase conditioned fear association and thus lead to permanent elimination of the fear response. However, one of the most difficult aspects of using fear memory elimination in a clinical setting will be specifically targeting maladaptive memories.

IV. Genetic/molecular manipulations to study fear memory erasure

In addition to the extensive search for manipulations to facilitate extinction learning, recent studies have also examined whether it is possible to "erase" a fear memory.

A. Genetic and protein manipulation

Cao et al. used a chemical-genetic approach to study the effect of α-Ca^{2+}/calmodulin-dependent protein kinase (αCAMKII) activity on retrieval of fear memories (Cao, et al., 2008). αCAMKII is a protein known to be a crucial component of memory consolidation (reviewed in Lisman, Schulman, & Cline, 2002). Cao and colleagues used transgenic mice in their study; the mice carried a modified αCAMKII with a F89G mutation (αCAMKII-F89G) that renders the protein accessible to an inhibitor that is normally too large to fit into the unmodified kinases. αCAMKII-F89G was expressed under the control of forebrain specific αCAMKII promoter (Wang, et al., 2003). Activity of αCAMKII-F89G could be inhibited by intraperitoneal (i.p.) injections of a chemically synthesized inhibitor, 1-Naphthylmethyl (NM)-PP1 that enters the brain within 3–5 minutes and suppresses αCAMKII-F89G activity within 10 minutes postinjection (Cao, et al., 2008; Wang, et al., 2003). The researchers showed that αCAMKII-F89G mice were impaired in cued- and contextual-fear conditioning, and that this effect could be reversed by NM-PP1 injection. Mice with increased αCAMKII-F89G activity at the time of recall 1 hour or 30 days after being trained in contextual or auditory fear conditioning had significant deficits in fear memory retrieval. However, these deficits could be overcome by inhibiting the activity of αCAMKII-F89G at the time of recall at either time points. Moreover, memory erasure was specific to the memory being retrieved during αCAMKII-F89G expression. Mice were impaired for the retrieval of contextual memory in the presence of αCAMKII-F89G activity; however, cued fear memory remained intact upon αCAMKII-F89G inhibition (Cao, et al., 2008). These results showed that elevated αCAMKII activity during retrieval of a specific fear memory led to a permanent degradation of the memory trace being actively retrieved. Consequently, both existing and newly formed memories could rapidly be erased in a regulated manner while leaving the rest intact (Cao, et al., 2008).

Several recent studies of LTM formation have focused on the protein kinase C isoform M zeta (PKMzeta). This enzyme is of particular importance in the maintenance of LTM (reviewed in Sacktor, 2008) and has been shown to be a significant component of the mechanisms underlying auditory fear memory formation in the amygdala (Serrano, et al., 2008). Karim Nader and colleagues studied the effect of inhibition of PKMzeta on the disruption of established fear memories (Migues, et al., 2010). Infusion of a cell-permeable selective PKMzeta inhibitor, PKM pseudosubstrate inhibitory peptide (ZIP), into the BLA of rats 1 day or 7 days after auditory fear training resulted in a significant impairment in conditioned fear memory. When retrained, the animals could reestablish auditory fear memory, showing that the ability of the BLA to acquire and establish fear memories was not abolished upon ZIP infusion. They also showed that PKMzeta inactivation by ZIP is mediated by the internalization of cell-surface GluR2 containing AMPARs, which led to a decrease of these receptors in postsynaptic sites. Co-infusion of a $GluR2_{3Y}$ peptide, which blocks AMPAR internalization, prevented the decrease of GluR2 receptors at the synapse and rescued the fear memory deficits in rats. Thus, a compound that interfered with AMPAR trafficking in BLA neurons was sufficient to disrupt established auditory fear memory in rats (Migues, et al., 2010). A recent study by Parsons and Davis questioned whether similar treatment with the ZIP peptide produced a permanent erasure of fear memory (Parsons & Davis, 2011).

B. Viral manipulation

B.1 IMPORTANCE OF CREB FOR FEAR MEMORY

Studies in our lab, as well as many other labs, have focused on a transcription factor cyclic adenosine-3′–5′-monophosphate (cAMP) response element binding protein (CREB) that has long been known for its critical role in memory formation (reviewed in Benito & Barco, 2010; Josselyn & Nguyen, 2005; Won & Silva, 2008). Genetic mouse models developed to study the role of CREB in the brain have proven to be powerful tools in providing insight into the role of CREB in memory. One of the first genetic mouse models to implicate the role of CREB in memory was developed by Gunter Schutz and colleagues. The mouse was made by inserting a neomycin resistant gene cassette into the exon 2 of *creb1* gene, which led to the targeted disruption of α and δ isoforms of CREB (Bourtchuladze, et al., 1994; Hummler, et al., 1994). Although these mice show an overexpression of the CREB β isoform (Blendy, Kaestner, Schmid, Gass, & Schutz, 1996), they have impairments in memory formation and LTP (Bourtchuladze, et al., 1994). CREB null mutant mice, deficient in all CREB isoforms, die prematurely due to respiratory deficits, making it necessary to utilize conditional knockout strategies to further study the role of CREB in memory formation of the adult mouse brain. To study the specific role of CREB in fear memory, Alcino Silva and colleagues generated a mutant mouse model in which CREB function was temporally and reversibly inhibited (Kida, et al., 2002). In this model, the CREB α isoform carrying a S133A mutation ($CREB^{S133A}$), which blocks the interaction between CREB and its co-activator CREB binding protein (CBP), was fused to a mutated ligand binding domain of the human estrogen receptor (LBD^{G521R}). The activity of the LBD^{G521R} is regulated by 4-hydroxytamoxifen (TAM), as opposed to estrogen; LBD^{G521R}—$CREB^{S133A}$ (CREB repressor, $CREB^{IR}$) fusion protein is inactive in the absence of TAM. However, when TAM is administered, activated $CREB^{IR}$ inhibits CRE-dependent gene transcription by competing with endogenous CREB. Spatial control of $CREB^{IR}$ was achieved by expressing it under a CAMKIIα promoter, which restricts transgene expression to principal excitatory neurons of the forebrain including cortex, hippocampus, amygdala and striatum (Mayford, et al., 1996). $CREB^{IR}$ mice that received TAM before fear training were impaired in both contextual and auditory fear memory tested 24 hours after learning. Moreover, stabilization of a reactivated fear memory was tested by repressing CREB activity 6 hours before a 90-second re-exposure to the training context or to the tone that was previously paired with shock. During re-exposure, mice did not receive shock. Memory was tested 24 hours after re-exposure. Mice with $CREB^{IR}$ activity before re-exposure showed impaired LTM, demonstrating the importance of CREB for the stability of fear memories (Kida, et al., 2002).

B.2 NEURONAL COMPETITION FOR INTEGRATION INTO A FEAR MEMORY TRACE (H3)

There are various types of viral delivery systems that can be used to alter gene expression in the brain (table 16.1). To manipulate CREB levels in the brain, our lab made use of Herpes Simplex Virus (HSV) vectors, which are replication-defective viral vectors suitable for efficient gene transduction

Table 16.1 Viral vectors commonly used for in vivo gene delivery

	Herpes Simplex Virus (HSV)	Adeno-associated virus (AAV)	Alphavirus Semliki Forest virus, Sindbis Virus	Adenovirus	Simple Retrovirus (Moloney murine leukemia virus, gammaretrovirus)	Complex Retrovirus (Lentivirus, HIV-1)
Viral genome	ds DNA	ss DNA	ss RNA	ds DNA	ss RNA	ss RNA
Target tissue	Neurotropic	Neurotropic	Neurotropic	Infects both neurons and glia	Infects dividing cells (not suitable for neurons)	Infects both neurons and glia
Maintenance in the host cell	Episomal	Integrates into the host genome	Does not integrate into the host genome	Episomal	Integrates into the host genome	Integrates into the host genome (episomal lentiviral vectors have been developed)
Transgene capacity	Large insert size (up to 150kb)	Insert size 〈5kb	Insert size 〈4kb	Large insert size (up to 30kb)	Insert size up to 8kb	Insert size up to 8kb
Duration of transgene expression	Expression 24–72h post-infection; lasts only a few days	Slow but long-term expression	Rapid (1d post-infection) but transient expression	Slow but long-term expression	Long-term expression	Long-term expression
Toxicity	Requirement of a helper virus during packaging might induce cytotoxicity	Very low cytotoxicity; vector of choice for human gene therapy	High levels of cytotoxicity; not suitable for long-term studies	Cytotoxicity due to immune-genicity and helper virus contamination	Cytotoxicity due to integration	Cytotoxicity due to integration

ss; single-stranded, ds; double-stranded, kb; kilobase

into neurons (reviewed in Carlezon, Nestler, & Neve, 2000; Cuchet, Potel, Thomas, & Epstein, 2007; Manservigi, Argnani, & Marconi, 2010). The ability of these nontoxic viral vectors to accommodate large pieces of foreign DNA (up to 150 kb) make them potentially useful tools to investigate functions of genes within specific sites of CNS, and allowing viral vectors to be potentially utilized for gene-therapy strategies. The advantage of the use of viral vectors over genetically modified mouse models is that viral vectors require a relatively shorter time for gene delivery and expression, and there is less cost associated with developing new genetic models. Moreover, the onset of gene expression with viral vectors is more easily and precisely controlled, because the gene delivery can be performed at any time during the life of an animal. Therefore, HSV viral vectors carrying specific genes of interest can be stereotaxically delivered into discrete brain regions at critical time windows of behavioral testing, allowing for genetic manipulation to be restricted both in time and location.

In searching for the crucial components of the fear memory trace induced by auditory fear conditioning, we carried out experiments in which we microinjected HSV vectors encoding CREB fused to GFP (or solely GFP to serve as the control) into the LA of wild-type (WT) mice (Han, et al., 2007). The initial experiments focused on manipulating CREB levels to show the importance of CREB signaling for selecting distinct neurons to be involved in a given memory trace. CREB activity in LA neurons was augmented after auditory fear training with increased phosphorylated CREB (pCREB) levels detectable in the LA of fear-conditioned mice when compared to the homecage controls. This observation indicated that CREB activity is important in selecting distinct populations of neurons to be included in the fear memory trace. Infusing HSV-CREB vector into the LA of CREB deficient (CREB$^{\alpha\delta-/-}$) mice, and enhancing CREB levels in about 18% of LA neurons, was sufficient to rescue the auditory fear memory deficits. Moreover, CREB-enhanced auditory fear memory in WT mice that were trained with a low shock intensity (0.4 mA), which does not normally induce fear memory formation. However, it still remained unclear how manipulating only a small subset of LA neurons could be sufficient to enhance fear memory.

To study the contribution of CREB activity to a given memory trace, we used fluorescent in situ hybridization (catFISH) to detect subcellular distribution of an immediate early gene *Arc* (activity-regulated cytoskeleton associated protein; Arg3.1). *Arc* has previously shown to be rapidly synthesized in response to neuronal activity (Guzowski, McNaughton, Barnes, & Worley, 1999). Auditory fear conditioning and fear memory testing activate a significant proportion of LA neurons that can be identified by nuclear-localization of *Arc* RNA within 5–15 minutes of activity. *Arc* RNA translocates to the cytoplasm within 30 minutes of activity, making it possible to follow the time course of neuronal activation right after training or testing. We infused the HSV-CREB vector into the LA of WT mice prior to auditory fear conditioning, tested fear memory 24 hours after training and removed the brain 5 minutes after the memory test to detect *Arc* RNA levels in LA. Neurons infected with the CREB vector were 3 times more likely to be *Arc* positive (*Arc+*) when compared to their noninfected neighbors. On the other hand, neurons infected with the control vector (HSV-GFP) were no more likely to be *Arc+* than their uninfected neighbors. Similarly, neurons infected with HSV-CREB were 10 times more likely to be activated in CREB$^{\alpha\delta-/-}$ mice. We obtained similar results when we tested *Arc* levels in neurons activated by fear memory training instead of testing (Han, et al., 2007). These experiments provided the first hint that CREB infusion does not lead to an overall activity enhancement within the LA and that neurons infected with the CREB overexpressing vector were specifically selected to participate in a fear memory trace.

In a similar experiment, we decreased the CREB function within a subset of LA neurons by infusing them with the dominant negative CREB (CREB^{S133A}) vector. Fear memory was normal in mice infected with the HSV- CREB^{S133A} vector, since the remaining, uninfected neurons, with intact CREB function, were selectively recruited into the fear memory trace and were sufficient to encode fear memory. As expected, the probability of detecting *Arc+* neuronal nuclei was 12 times lower in the infected neurons with low levels of CREB compared to their noninfected neighbors (Han, et al., 2007). Overall, we showed that competition occurs between neurons in the formation of fear memory. Furthermore, when a neuron is activated, the amount of CREB within a given neuron determines its ability to be selected for the memory trace.

B.3 FEAR MEMORY ERASURE BY SELECTIVE ABLATION OF CREB NEURONS

To further identify the subset of LA neurons that are crucial components of the auditory fear memory, we investigated the effect of ablating those neurons that are preferentially recruited into the fear memory trace after training or testing. We hypothesized that once the selected neurons within a given memory trace are disrupted, this should have detrimental effects on fear memory formation. To do this, we combined HSV vector delivery with an inducible diphtheria-toxin (DT) strategy. We made use of Cre recombinase (cre)-inducible DT receptor (DTR) transgenic mice (iDTR) that were produced by Ari Weissman and colleagues (Buch, et al., 2005). DT is a potent toxin that can cross the blood-brain barrier and bind to its specific receptor (DTR) expressed on the cell surface. Once binding occurs, the toxin is internalized by receptor mediated endocytosis and is translocated into the cytosol where it blocks protein synthesis and leads to apoptosis. iDTR mice were developed by overexpressing a simian DTR whose expression is initially silenced by a loxP flanked STOP cassette that can be excised by cre recombinase leading to the constitutive expression of DTR. We generated a HSV vector that co-expressed CREB and cre recombinase (HSV-CREB-cre). Since mouse cells do not normally express a functional DTR, infusion of HSV-CREB-cre vector in the LA of iDTR mice only activates DTR in the neurons with increased CREB expression. Infusion of DT will, therefore, induce apoptosis only in those neurons with high CREB levels, the ones that were preferentially recruited into the fear memory trace. Control-cre (Cntrl-cre) vector, which expressed GFP instead of CREB, was microinjected into the LA and led to apoptosis in a similar proportion of nonactive neurons upon DT injection (Han, et al., 2009).

To detect the neurons activated by fear memory testing, we followed a similar protocol as in the previous experimental paradigm. We infused CREB-cre or Cntrl-cre vectors into the LA of WT mice before fear training, carried out the memory test 24 hours after infusion and removed the brains 5 min after testing to detect *Arc* levels in the LA. As shown before, *Arc* RNA was preferentially localized to the nucleus of LA neurons that were infused with the CREB-cre vector, those overexpressing high levels of CREB (Han, et al., 2009).

The next step was to see whether ablation of CREB expressing selectively activated neurons would erase fear memory. If LA neurons with high levels of CREB are preferentially chosen over their neighbors to be included in the fear memory trace, then deleting these neurons would ideally lead to a complete removal of the fear memory. To test this, we infused CREB-cre or Cntrl-cre vector into the LA of iDTR transgenic mice before weak auditory fear training (1 tone-shock pairing), performed fear memory testing 24 hours after training, administered DT to induce apoptosis in the neurons expressing CREB cre or Cntrl cre vectors, and repeated the memory test after DT administration. As expected, CREB-cre infusion led to enhancement in the first fear memory test, performed 24 hours after weak training. DT administration produced substantial apoptosis that was selective to the LA neurons infected with the cre vectors. Ablation of neurons overexpressing CREB was sufficient to abolish the enhancement in fear memory (Han, et al., 2009).

Strong auditory fear training (2 tone-shock pairings) induces a strong fear memory that cannot be further enhanced by CREB overexpression in LA. However, similar to the memory induced after weak training, neurons with high levels of CREB are still selectively recruited to be a part of the fear memory trace, compared to those with low CREB expression after strong training. Therefore, ablation of only the CREB overexpressing neurons, but not those with the Cntrl-cre vector, led to the loss of fear memory induced by strong training (Han, et al., 2009). Altogether, these results showed that both weak and robust memory was completely erased after the loss of CREB overexpressing neurons that were previously recruited to the fear memory trace. Deletion of those with endogenous CREB levels did not influence memory formation, although apoptosis was induced in a similar subset and number of LA neurons. Therefore, the mechanisms underlying the memory loss were highly specific.

Memory erasure upon deletion of CREB overexpressing neurons was also permanent, because no recovery was observed when the mice were tested as long as 12 days after training. Moreover memory loss could not be due to extinction since mice infused with the Cntrl-cre vector still showed a robust and stable memory over repeated testing. Although fear memory was lost after deletion of CREB overexpressing neurons, mice were able to acquire novel fear memory when they were retrained. Furthermore, fear memory acquired before infusion was not lost after the ablation of neurons with increased CREB levels, emphasizing the importance of CREB-induced competitive selection at the time of memory encoding (Han, et al., 2009). These findings were especially intriguing in the sense that not only was enhancing CREB in a small subset of LA neurons sufficient to enhance fear memory, but deleting these neurons was capable of producing a specific, selective, and persistent memory loss.

B.4 FEAR MEMORY ERASURE BY SILENCING CREB NEURONS

Zhou and colleagues further studied the role of CREB in mediating competitive neuronal selection in fear memory by GFP-tagged CREB viral (HSV-CREB) infusions in the LA of WT mice (Zhou, et al., 2009). They first demonstrated that CREB infusion increased the excitability of LA neurons, making them more likely to be activated during fear memory formation. Based on their altered synaptic properties, neurons with higher CREB levels and higher excitability were preferentially selected to contribute to the fear memory expression, which they termed "memory allocation." They used a functional approach to silence CREB overexpressing neurons involved in the fear memory response by co-expressing a *Drosophila* allatostatin receptor (AlstR). Upon binding to its ligand allatostatin, AlstR activates G-protein coupled inwardly rectifying K^+ (GIRK) channels, leading to hyperpolarization of the membrane and inhibition of the synaptic response. They confirmed that allatostatin administration altered membrane properties of only GFP positive (thus CREB overexpressing) LA neurons while leaving their noninfected neighbors unaffected, thereby selectively inactivating only those neurons with high CREB levels. Next, they asked whether silencing CREB overexpressing neurons in LA would affect fear memory. To test this, they co-infused HSV-CREB and -AlstR viruses into LA prior to auditory fear conditioning and tested fear memory 24 hours later. Inactivation of CREB neurons during training upon allatostatin administration reduced CREB-enhanced fear memory to control levels. Interestingly, administration of allatostatin during testing also disrupted freezing. However when administered both during training and testing, allatostatin did not result in inhibition of the fear response. Therefore, CREB overexpressing neurons must be active during training to be selectively recruited to the fear memory circuitry. In the absence of this selection, uninfected cells with endogenous CREB levels are sufficient to outcompete their silenced neighbors to initiate the fear response. Once again, fear memory disruption was specific because similar treatments did not affect memory for a different amygdala-dependent behavioral paradigm, conditioned taste-aversion task. Moreover, when auditory fear memory was acquired before viral infusions, memory was not affected by allatostatin treatment. However, memory for a different amygdala-dependent behavioral paradigm, conditioned taste-aversion task, was impaired when acquired after viral infusions, highlighting the behavioral specificity of CREB effects in LA (Zhou, et al., 2009).

V. Conclusion

Undoubtedly, fear memories are an essential component of our survival strategy. However, as the recurrent collection of all the traumatic events we face in life lead to long-lasting memories, the psychological burden can become intolerable over time. Thanks to the considerable amount of animal research, today we have significant information about how fearful memories are encoded and sustained. The advent of elegant experimental approaches developed in recent years has provided a growing number of techniques to suppress and ultimately erase unwanted fears. Genetics research aimed at understanding the mechanisms of fear memory formation and erasure holds great potential for the treatment and perhaps the cure of emotional disorders.

VI. Future directions

With the advancement of genetic tools to manipulate discrete neuronal populations in specific areas of the brain, we are now able to study the mechanisms underlying the appearance and removal of memory traces. Genetically modified mouse models and viral manipulations acting to either overexpress or knockout specific genes in different brain regions, have proven to be useful in revealing the functions of specific genes in the nervous system. However, these research techniques for fear memory erasure currently cannot provide precise temporal control of activity over a specific subpopulation of neurons. A new tool developed by researchers such as Georg Nagel and Karl Deisseroth (Boyden, Zhang, Bamberg, Nagel, & Deisseroth, 2005; Nagel, et al., 2003; Zhang, et al., 2007) takes advantage of a class of molecules called channelrhodopsins, which are ion channels gated by visible light. Neurons overexpressing channelrhodopsins can be noninvasively activated or inhibited at millisecond temporal resolution under an external light source. Combined with viral mediated gene delivery, researchers were able to control the activity of specific subpopulation of neurons (reviewed in Fenno, Yizhar, & Deisseroth, 2011). Although only few studies have used channelrhodopsins to investigate fear memory erasure, it has great potential to reveal circuitry mechanisms for fear memory formation.

Promising effects of optogenetics in animal models (Berdyyeva & Reynolds, 2009; Lobo, Nestler, & Covington, 2012; Witten, et al., 2011) have raised the question whether the use of similar techniques for treatment of neuropsychiatric disorders in humans is feasible. Optogenetics, combined with viral mediated gene delivery, would prove useful in targeting specific neurons disrupted in certain diseases in a temporally controlled manner. This would not only target the specific mechanisms underlying the disease in question but would also be essential in preventing unwanted side effects. Instead of infusing a drug to antagonize or induce general brain circuitry, optogenetic approaches would allow manipulation of only a specific circuitry in the affected area and thus might overcome the inconclusive results obtained in prior clinical trials. The preceding year has successfully shown that optogenetic methods could be utilized in nonhuman primates (reviewed in Han, 2012). The use of viral vectors however, for targeted delivery of specific genes in humans is still debatable. To date, viral mediated gene therapy has been utilized in retinal (Carter, 2005) and neurodenegerative disorders (reviewed in Weinberg, Samulski, & McCown, 2013). Adeno-associated viral (AAV) vectors are especially useful tools for in vivo gene transfer due to their persistence in an episomal form and lack of immunogenicity in humans. Based on these favorable characteristics, the use of AAV vectors has substantially increased in the recent years, with more than 80 clinical trials being initiated (gene therapy clinical trial database; http://www.wiley.com/legacy/wileychi/genmed/clinical/). Although promising, these clinical trials are still in their infancy.

Another critical debate in manipulation of optogenetic techniques for human disease includes the invasive nature of the methodology. Other invasive techniques such as deep brain stimulation (DBS), however, have currently been utilized for treatment of Parkinson's disease (Okun, 2012) and depression (Anderson, et al., 2012; Kammer & Spitzer, 2012) in humans without adverse side effects. DBS, induced by placement of electrodes in affected brain areas, leads to electrical stimulation of the affected circuits potentially restoring the balance between the affected areas. Because of the heterogeneity of the areas in which electrodes are placed, however, the approach does not have cell-type specificity. Thus, optogenetic methods might also be utilized to drive a more targeted and cell-specific form of DBS (Gradinaru, Mogri, Thompson, Henderson, & Deisseroth, 2009).

In summary, development of novel techniques such as optogenetics offers great opportunities for basic research. Utilization of optogenetic methods for treatment of human disease still face unpredictable risks and challenges; however, rapidly advancing technology holds great promise for future therapeutic approaches of neurological diseases.

References

Adolphs, R., Tranel, D., & Damasio, H. (2001). Emotion recognition from faces and prosody following temporal lobectomy. *Neuropsychology, 15*(3), 396–404.

Adolphs, R., Tranel, D., Damasio, H., & Damasio, A. (1994). Impaired recognition of emotion in facial expressions following bilateral damage to the human amygdala. *Nature, 372*(6507), 669–672.

Adolphs, R., Tranel, D., Damasio, H., & Damasio, A. R. (1995). Fear and the human amygdala. *Journal of Neuroscience, 15*(9), 5879–5891.

Alberini, C. M. (2005). Mechanisms of memory stabilization: are consolidation and reconsolidation similar or distinct processes? *Trends in Neuroscience, 28*(1), 51–56.

Anagnostaras, S. G., Maren, S., & Fanselow, M. S. (1999). Temporally graded retrograde amnesia of contextual fear after hippocampal damage in rats: within-subjects examination. *Journal of Neuroscience, 19*(3), 1106–1114.

Anderson, A. K., & Phelps, E. A. (2001). Lesions of the human amygdala impair enhanced perception of emotionally salient events. *Nature, 411*(6835), 305–309.

Anderson, R. J., Frye, M. A., Abulseoud, O. A., Lee, K. H., McGillivray, J. A., Berk, M., & Tye, S. J. (2012). Deep brain stimulation for treatment-resistant depression: efficacy, safety and mechanisms of action. *Neuroscience and Biobehavioral Reviews, 36*(8), 1920–1933.

Bailey, D. J., Kim, J. J., Sun, W., Thompson, R. F., & Helmstetter, F. J. (1999). Acquisition of fear conditioning in rats requires the synthesis of mRNA in the amygdala. *Behavioral Neuroscience, 113*(2), 276–282.

Bauer, E. P., Schafe, G. E., & LeDoux, J. E. (2002). NMDA receptors and L-type voltage-gated calcium channels contribute to long-term potentiation and different components of fear memory formation in the lateral amygdala. *Journal of Neuroscience, 22*(12), 5239–5249.

Baum, M. (1988). Spontaneous recovery from the effects of flooding (exposure) in animals. *Behavioral Research and Therapy, 26*(2), 185–186.

Benito, E., & Barco, A. (2010). CREB's control of intrinsic and synaptic plasticity: implications for CREB-dependent memory models. *Trends in Neuroscience, 33*(5), 230–240.

Berdyyeva, T. K., & Reynolds, J. H. (2009). The dawning of primate optogenetics. *Neuron, 62*(2), 159–160.
Bernardi, R. E., & Lattal, K. M. (2010). A role for alpha-adrenergic receptors in extinction of conditioned fear and cocaine conditioned place preference. *Behavioral Neuroscience, 124*(2), 204–210.
Blendy, J. A., Kaestner, K. H., Schmid, W., Gass, P., & Schutz, G. (1996). Targeting of the CREB gene leads to up-regulation of a novel CREB mRNA isoform. *EMBO Journal, 15*(5), 1098–1106.
Bomyea, J., & Lang, A. J. (2012). Emerging interventions for PTSD: Future directions for clinical care and research. *Neuropharmacology, 62*(2), 607–616.
Bourtchuladze, R., Frenguelli, B., Blendy, J., Cioffi, D., Schutz, G., & Silva, A. J. (1994). Deficient long-term memory in mice with a targeted mutation of the cAMP-responsive element-binding protein. *Cell, 79*(1), 59–68.
Bouton, M. E., & Bolles, R. C. (1979). Role of conditioned contextual stimuli in reinstatement of extinguished fear. *Journal of Experimental Psychology: Animal Behavior Processes, 5*(4), 368–378.
Boyden, E. S., Zhang, F., Bamberg, E., Nagel, G., & Deisseroth, K. (2005). Millisecond-timescale, genetically targeted optical control of neural activity. *Nature Neuroscience, 8*(9), 1263–1268.
Brunet, A., Orr, S. P., Tremblay, J., Robertson, K., Nader, K., & Pitman, R. K. (2008). Effect of post-retrieval propranolol on psychophysiologic responding during subsequent script-driven traumatic imagery in post-traumatic stress disorder. *Journal of Psychiatric Research, 42*(6), 503–506.
Brunet, A., Poundja, J., Tremblay, J., Bui, E., Thomas, E., Orr, S. P., . . . Pitman, R. K. (2011). Trauma reactivation under the influence of propranolol decreases posttraumatic stress symptoms and disorder: 3 open-label trials. *Journal of Clinical Psychopharmacology, 31*(4), 547–550.
Buch, T., Heppner, F. L., Tertilt, C., Heinen, T. J., Kremer, M., Wunderlich, F. T., . . . Waisman, A. (2005). A Cre-inducible diphtheria toxin receptor mediates cell lineage ablation after toxin administration. *Nature Methods, 2*(6), 419–426.
Burgos-Robles, A., Vidal-Gonzalez, I., Santini, E., & Quirk, G. J. (2007). Consolidation of fear extinction requires NMDA receptor-dependent bursting in the ventromedial prefrontal cortex. *Neuron, 53*(6), 871–880.
Cain, C. K., Blouin, A. M., & Barad, M. (2002). L-type voltage-gated calcium channels are required for extinction, but not for acquisition or expression, of conditional fear in mice. *Journal of Neuroscience, 22*(20), 9113–9121.
Campeau, S., Hayward, M. D., Hope, B. T., Rosen, J. B., Nestler, E. J., & Davis, M. (1991). Induction of the c-fos proto-oncogene in rat amygdala during unconditioned and conditioned fear. *Brain Research, 565*(2), 349–352.
Campeau, S., Miserendino, M. J., & Davis, M. (1992). Intra-amygdala infusion of the N-methyl-D-aspartate receptor antagonist AP5 blocks acquisition but not expression of fear-potentiated startle to an auditory conditioned stimulus. *Behavioral Neuroscience, 106*(3), 569–574.
Cao, X., Wang, H., Mei, B., An, S., Yin, L., Wang, L. P., & Tsien, J. Z. (2008). Inducible and selective erasure of memories in the mouse brain via chemical-genetic manipulation. *Neuron, 60*(2), 353–366.
Carlezon, W. A., Jr., Nestler, E. J., & Neve, R. L. (2000). Herpes simplex virus-mediated gene transfer as a tool for neuropsychiatric research. *Critical Reviews in Neurobiology, 14*(1), 47–67.
Carter, B. J. (2005). Adeno-associated virus vectors in clinical trials. *Human Gene Therapy, 16*(5), 541–550.
Clem, R. L., & Huganir, R. L. (2010). Calcium-permeable AMPA receptor dynamics mediate fear memory erasure. *Science, 330*(6007), 1108–1112.
Corcoran, K. A., Desmond, T. J., Frey, K. A., & Maren, S. (2005). Hippocampal inactivation disrupts the acquisition and contextual encoding of fear extinction. *Journal of Neuroscience, 25*(39), 8978–8987.
Corcoran, K. A., & Maren, S. (2001). Hippocampal inactivation disrupts contextual retrieval of fear memory after extinction. *Journal of Neuroscience, 21*(5), 1720–1726.
Corcoran, K. A., & Maren, S. (2004). Factors regulating the effects of hippocampal inactivation on renewal of conditional fear after extinction. *Learning & Memory, 11*(5), 598–603.
Cuchet, D., Potel, C., Thomas, J., & Epstein, A. L. (2007). HSV-1 amplicon vectors: a promising and versatile tool for gene delivery. *Expert Opinions in Biological Therapy, 7*(7), 975–995.
Davis, M. (1992). The role of the amygdala in fear and anxiety. *Annual Review of Neuroscience, 15*, 353–375.
Debiec, J., & Ledoux, J. E. (2004). Disruption of reconsolidation but not consolidation of auditory fear conditioning by noradrenergic blockade in the amygdala. *Neuroscience, 129*(2), 267–272.
Do-Monte, F. H., Allensworth, M., & Carobrez, A. P. (2010). Impairment of contextual conditioned fear extinction after microinjection of alpha-1-adrenergic blocker prazosin into the medial prefrontal cortex. *Behavioural Brain Research, 211*(1), 89–95.
Duvarci, S., & Nader, K. (2004). Characterization of fear memory reconsolidation. *Journal of Neuroscience, 24*(42), 9269–9275.
Duvarci, S., Nader, K., & LeDoux, J. E. (2008). De novo mRNA synthesis is required for both consolidation and reconsolidation of fear memories in the amygdala. *Learning & Memory, 15*(10), 747–755.
Ehrlich, I., Humeau, Y., Grenier, F., Ciocchi, S., Herry, C., & Luthi, A. (2009). Amygdala inhibitory circuits and the control of fear memory. *Neuron, 62*(6), 757–771.
Falls, W. A., Miserendino, M. J., & Davis, M. (1992). Extinction of fear-potentiated startle: blockade by infusion of an NMDA antagonist into the amygdala. *Journal of Neuroscience, 12*(3), 854–863.
Fanselow, M. S., & Gale, G. D. (2003). The amygdala, fear, and memory. *Annals of the New York Academy of Science*, 985, 125–134.
Fanselow, M. S., & Kim, J. J. (1994). Acquisition of contextual Pavlovian fear conditioning is blocked by application of an NMDA receptor antagonist D,L-2-amino-5-phosphonovaleric acid to the basolateral amygdala. *Behavioral Neuroscience, 108*(1), 210–212.
Fanselow, M. S., & Lester, L.S. (1988). A functional behavioristic approach to aversively motivated behavior: predatory imminence as a determinant of the topography of defensive behavior. In R. C. Bolles and M. D. Beecher (Eds.), *Evolution and Learning* (pp. 185–211). Hillsdale, NJ: Erlbaum.
Fendt, M. (2001). Injections of the NMDA receptor antagonist aminophosphonopentanoic acid into the lateral nucleus of

the amygdala block the expression of fear-potentiated startle and freezing. *Journal of Neuroscience, 21*(11), 4111–4115.
Fenno, L., Yizhar, O., & Deisseroth, K. (2011). The development and application of optogenetics. *Annual Review of Neuroscience, 34*, 389–412.
Fischer, A., Radulovic, M., Schrick, C., Sananbenesi, F., Godovac-Zimmermann, J., & Radulovic, J. (2007). Hippocampal Mek/Erk signaling mediates extinction of contextual freezing behavior. *Neurobiology of Learning & Memory, 87*(1), 149–158.
Gewirtz, J. C., & Davis, M. (1997). Second-order fear conditioning prevented by blocking NMDA receptors in amygdala. *Nature, 388*(6641), 471–474.
Goosens, K. A., & Maren, S. (2003). Pretraining NMDA receptor blockade in the basolateral complex, but not the central nucleus, of the amygdala prevents savings of conditional fear. *Behavioral Neuroscience, 117*(4), 738–750.
Gradinaru, V., Mogri, M., Thompson, K. R., Henderson, J. M., & Deisseroth, K. (2009). Optical deconstruction of parkinsonian neural circuitry. *Science, 324*(5925), 354–359.
Guzowski, J. F., McNaughton, B. L., Barnes, C. A., & Worley, P. F. (1999). Environment-specific expression of the immediate-early gene Arc in hippocampal neuronal ensembles. *Nature Neuroscience, 2*(12), 1120–1124.
Han, J. H., Kushner, S. A., Yiu, A. P., Cole, C. J., Matynia, A., Brown, R. A.,...Josselyn, S. A. (2007). Neuronal competition and selection during memory formation. *Science, 316*(5823), 457–460.
Han, J. H., Kushner, S. A., Yiu, A. P., Hsiang, H. L., Buch, T., Waisman, A.,...Josselyn, S. A. (2009). Selective erasure of a fear memory. *Science, 323*(5920), 1492–1496.
Han, X. (2012). Optogenetics in the nonhuman primate. *Progress in Brain Research, 196*, 215–233.
Hebb, D. O. (1949). *The Organization of Behavior.* New York: Wiley.
Heldt, S. A., Stanek, L., Chhatwal, J. P., & Ressler, K. J. (2007). Hippocampus-specific deletion of BDNF in adult mice impairs spatial memory and extinction of aversive memories. *Molecular Psychiatry, 12*(7), 656–670.
Herry, C., Ciocchi, S., Senn, V., Demmou, L., Muller, C., & Luthi, A. (2008). Switching on and off fear by distinct neuronal circuits. *Nature, 454*(7204), 600–606.
Herry, C., Trifilieff, P., Micheau, J., Luthi, A., & Mons, N. (2006). Extinction of auditory fear conditioning requires MAPK/ERK activation in the basolateral amygdala. *European Journal of Neuroscience, 24*(1), 261–269.
Holbrook, T. L., Galarneau, M. R., Dye, J. L., Quinn, K., & Dougherty, A. L. (2010). Morphine use after combat injury in Iraq and post-traumatic stress disorder. *New England Journal of Medicine, 362*(2), 110–117.
Hummler, E., Cole, T. J., Blendy, J. A., Ganss, R., Aguzzi, A., Schmid, W.,...Schütz, G. (1994). Targeted mutation of the CREB gene: compensation within the CREB/ATF family of transcription factors. *Proceedings of the National Academy of Science of the United States of America, 91*(12), 5647–5651.
Hupbach, A., Gomez, R., Hardt, O., & Nadel, L. (2007). Reconsolidation of episodic memories: a subtle reminder triggers integration of new information. *Learning & Memory, 14*(1–2), 47–53.
Ji, J., & Maren, S. (2005). Electrolytic lesions of the dorsal hippocampus disrupt renewal of conditional fear after extinction. *Learning & Memory, 12*(3), 270–276.
Josselyn, S. A., & Nguyen, P. V. (2005). CREB, synapses and memory disorders: past progress and future challenges. *Current Drug Targets–CNS and Neurological Disorders, 4*(5), 481–497.
Kammer, T., & Spitzer, M. (2012). Brain stimulation in psychiatry: methods and magnets, patients and parameters. *Current Opinions in Psychiatry, 25*(6), 535–541.
Kennedy, P. J., & Shapiro, M. L. (2004). Retrieving memories via internal context requires the hippocampus. *Journal of Neuroscience, 24*(31), 6979–6985.
Kessler, R. C., Chiu, W. T., Demler, O., Merikangas, K. R., & Walters, E. E. (2005). Prevalence, severity, and comorbidity of 12-month DSM-IV disorders in the National Comorbidity Survey Replication. *Archives of General Psychiatry, 62*(6), 617–627.
Kessler, R. C., Sonnega, A., Bromet, E., Hughes, M., & Nelson, C. B. (1995). Posttraumatic stress disorder in the National Comorbidity Survey. *Archives of General Psychiatry, 52*(12), 1048–1060.
Kida, S., Josselyn, S. A., Pena de Ortiz, S., Kogan, J. H., Chevere, I., Masushige, S., & Silva, A. J. (2002). CREB required for the stability of new and reactivated fear memories. *Nature Neuroscience, 5*(4), 348–355.
Kim, J., Lee, S., Park, H., Song, B., Hong, I., Geum, D.,...Choi, S. (2007). Blockade of amygdala metabotropic glutamate receptor subtype 1 impairs fear extinction. *Biochemical and Biophysical Research Communications, 355*(1), 188–193.
Kim, J., Lee, S., Park, K., Hong, I., Song, B., Son, G.,...Choi, S. (2007). Amygdala depotentiation and fear extinction. *Proceedings of the National Academy of Science of the United States of America, 104*(52), 20955–20960.
Kwon, J. T., & Choi, J. S. (2009). Cornering the fear engram: long-term synaptic changes in the lateral nucleus of the amygdala after fear conditioning. *Journal of Neuroscience, 29*(31), 9700–9703.
LaBar, K. S., Gatenby, J. C., Gore, J. C., LeDoux, J. E., & Phelps, E. A. (1998). Human amygdala activation during conditioned fear acquisition and extinction: a mixed-trial fMRI study. *Neuron, 20*(5), 937–945.
Ledgerwood, L., Richardson, R., & Cranney, J. (2003). Effects of D-cycloserine on extinction of conditioned freezing. *Behavioral Neuroscience, 117*(2), 341–349.
Ledgerwood, L., Richardson, R., & Cranney, J. (2005). D-cycloserine facilitates extinction of learned fear: effects on reacquisition and generalized extinction. *Biological Psychiatry, 57*(8), 841–847.
LeDoux, J. (1996). Emotional networks and motor control: a fearful view. *Progress in Brain Research, 107*, 437–446.
LeDoux, J. (2007). The amygdala. *Current Biology, 17*(20), R868–874.
LeDoux, J. E. (2000). Emotion circuits in the brain. *Annual Review of Neuroscience, 23*, 155–184.
Lee, H. J., Choi, J. S., Brown, T. H., & Kim, J. J. (2001). Amygdalar nmda receptors are critical for the expression of multiple conditioned fear responses. *Journal of Neuroscience, 21*(11), 4116–4124.
Lee, S., Ahmed, T., Kim, H., Choi, S., Kim, D. S., Kim, S. J.,...Shin, H. S. (2011). Bidirectional modulation of fear extinction by mediodorsal thalamic firing in mice. *Nature Neuroscience, 15*(2), 308–314.
Likhtik, E., Popa, D., Apergis-Schoute, J., Fidacaro, G. A., & Pare, D. (2008). Amygdala intercalated neurons

are required for expression of fear extinction. *Nature, 454*(7204), 642–645.

Lisman, J., Schulman, H., & Cline, H. (2002). The molecular basis of CaMKII function in synaptic and behavioural memory. *Nature Reviews Neuroscience, 3*(3), 175–190.

Lithell, H., Pollare, T., & Vessby, B. (1992). Metabolic effects of pindolol and propranolol in a double-blind cross-over study in hypertensive patients. *Blood Pressure, 1*(2), 92–101.

Lobo, M. K., Nestler, E. J., & Covington, H. E., III. (2012). Potential utility of optogenetics in the study of depression. *Biological Psychiatry, 71*(12), 1068–1074.

Lutz, B. (2007). The endocannabinoid system and extinction learning. *Molecular Neurobiology, 36*(1), 92–101.

Malkani, S., & Rosen, J. B. (2000). Specific induction of early growth response gene 1 in the lateral nucleus of the amygdala following contextual fear conditioning in rats. *Neuroscience, 97*(4), 693–702.

Manservigi, R., Argnani, R., & Marconi, P. (2010). HSV Recombinant Vectors for Gene Therapy. *Open Virology Journal, 4*, 123–156.

Mao, S. C., Hsiao, Y. H., & Gean, P. W. (2006). Extinction training in conjunction with a partial agonist of the glycine site on the NMDA receptor erases memory trace. *Journal of Neuroscience, 26*(35), 8892–8899.

Maren, S. (2003). The amygdala, synaptic plasticity, and fear memory. *Annals of the New York Academy of Science, 985*, 106–113.

Maren, S., Aharonov, G., Stote, D. L., & Fanselow, M. S. (1996). N-methyl-D-aspartate receptors in the basolateral amygdala are required for both acquisition and expression of conditional fear in rats. *Behavioral Neuroscience, 110*(6), 1365–1374.

Maren, S., & Fanselow, M. S. (1997). Electrolytic lesions of the fimbria/fornix, dorsal hippocampus, or entorhinal cortex produce anterograde deficits in contextual fear conditioning in rats. *Neurobiology of Learning & Memory, 67*(2), 142–149.

Maren, S., Ferrario, C. R., Corcoran, K. A., Desmond, T. J., & Frey, K. A. (2003). Protein synthesis in the amygdala, but not the auditory thalamus, is required for consolidation of Pavlovian fear conditioning in rats. *European Journal of Neuroscience, 18*(11), 3080–3088.

Maren, S., & Quirk, G. J. (2004). Neuronal signalling of fear memory.*Nature Reviews Neuroscience, 5*(11), 844–852.

Marsicano, G., Wotjak, C. T., Azad, S. C., Bisogno, T., Rammes, G., Cascio, M. G.,... Lutz, B. (2002). The endogenous cannabinoid system controls extinction of aversive memories. *Nature, 418*(6897), 530–534.

Martin, S. J., Grimwood, P. D., & Morris, R. G. (2000). Synaptic plasticity and memory: an evaluation of the hypothesis. *Annual Review of Neuroscience, 23*, 649–711.

Mayford, M., Bach, M. E., Huang, Y. Y., Wang, L., Hawkins, R. D., & Kandel, E. R. (1996). Control of memory formation through regulated expression of a CaMKII transgene. *Science, 274*(5293), 1678–1683.

McDonald, A. J. (1992). Projection neurons of the basolateral amygdala: a correlative Golgi and retrograde tract tracing study. *Brain Research Bulletin, 28*(2), 179–185.

McGaugh, J. L. (2002). Memory consolidation and the amygdala: a systems perspective. *Trends in Neuroscience, 25*(9), 456.

McGhee, L. L., Maani, C. V., Garza, T. H., Gaylord, K. M., & Black, I. H. (2008). The correlation between ketamine and posttraumatic stress disorder in burned service members. *Journal of Trauma, 64(2 Suppl)*, S195–198; Discussion S197–198.

McKernan, M. G., & Shinnick-Gallagher, P. (1997). Fear conditioning induces a lasting potentiation of synaptic currents in vitro. *Nature, 390*(6660), 607–611.

Migues, P. V., Hardt, O., Wu, D. C., Gamache, K., Sacktor, T. C., Wang, Y. T., & Nader, K. (2010). PKMzeta maintains memories by regulating GluR2-dependent AMPA receptor trafficking. *Nature Neuroscience, 13*(5), 630–634.

Misanin, J. R., Miller, R. R., & Lewis, D. J. (1968). Retrograde amnesia produced by electroconvulsive shock after reactivation of a consolidated memory trace. *Science, 160*(3827), 554–555.

Miserendino, M. J., Sananes, C. B., Melia, K. R., & Davis, M. (1990). Blocking of acquisition but not expression of conditioned fear-potentiated startle by NMDA antagonists in the amygdala. *Nature, 345*(6277), 716–718.

Monfils, M. H., Cowansage, K. K., Klann, E., & LeDoux, J. E. (2009). Extinction-reconsolidation boundaries: key to persistent attenuation of fear memories. *Science, 324*(5929), 951–955.

Morris, J. S., Frith, C. D., Perrett, D. I., Rowland, D., Young, A. W., Calder, A. J., & Dolan R. J. (1996). A differential neural response in the human amygdala to fearful and happy facial expressions. *Nature, 383*(6603), 812–815.

Myers-Schulz, B., & Koenigs, M. (2012). Functional anatomy of ventromedial prefrontal cortex: implications for mood and anxiety disorders. *Molecular Psychiatry, 17*(2), 132–141.

Myers, K. M., & Davis, M. (2007). Mechanisms of fear extinction. *Molecular Psychiatry, 12*(2), 120–150.

Myers, K. M., Ressler, K. J., & Davis, M. (2006). Different mechanisms of fear extinction dependent on length of time since fear acquisition. *Learning & Memory, 13*(2), 216–223.

Nader, K., Schafe, G. E., & Le Doux, J. E. (2000). Fear memories require protein synthesis in the amygdala for reconsolidation after retrieval. *Nature, 406*(6797), 722–726.

Nagel, G., Szellas, T., Huhn, W., Kateriya, S., Adeishvili, N., Berthold, P.,... Bamberg, E. (2003). Channelrhodopsin-2, a directly light-gated cation-selective membrane channel. *Proceedings of the National Academy of Science of the United States of America, 100*(24), 13940–13945.

Ohman, A., & Mineka, S. (2001). Fears, phobias, and preparedness: toward an evolved module of fear and fear learning. *Psychological Review, 108*(3), 483–522.

Okun, M. S. (2012). Deep-brain stimulation for Parkinson's disease. *New England Journal of Medicine, 367*(16), 1529–1538.

Pape, H. C., & Pare, D. (2010). Plastic synaptic networks of the amygdala for the acquisition, expression, and extinction of conditioned fear. *Physiological Reviews, 90*(2), 419–463.

Pare, D., & Smith, Y. (1993). The intercalated cell masses project to the central and medial nuclei of the amygdala in cats. *Neuroscience, 57*(4), 1077–1090.

Parsons, R. G., & Davis, M. (2011). Temporary disruption of fear-potentiated startle following PKMzeta inhibition in the amygdala. *Nature Neuroscience, 14*(3), 295–296.

Pavlov, I. P. (1927). *Conditioned reflexes, an investigation of the physiological activity of the cerebral cortex.* London: Oxford University Press.

Przybyslawski, J., Roullet, P., & Sara, S. J. (1999). Attenuation of emotional and nonemotional memories after their reactivation: role of beta adrenergic receptors. *Journal of Neuroscience, 19*(15), 6623–6628.

Quirk, G. J., & Mueller, D. (2008). Neural mechanisms of extinction learning and retrieval. *Neuropsychopharmacology, 33*(1), 56–72.

Quirk, G. J., Russo, G. K., Barron, J. L., & Lebron, K. (2000). The role of ventromedial prefrontal cortex in the recovery of extinguished fear. *Journal of Neuroscience, 20*(16), 6225–6231.

Radulovic, J., Kammermeier, J., & Spiess, J. (1998). Relationship between fos production and classical fear conditioning: effects of novelty, latent inhibition, and unconditioned stimulus preexposure. *Journal of Neuroscience, 18*(18), 7452–7461.

Rauch, S. L., Shin, L. M., & Phelps, E. A. (2006). Neurocircuitry models of posttraumatic stress disorder and extinction: human neuroimaging research—past, present, and future. *Biological Psychiatry, 60*(4), 376–382.

Rescorla, R. A. (1969). Pavlovian conditioned inhibition. *Psychological Bulletin, 72*(2), 77–94.

Rescorla, R. A. (2001). Retraining of extinguished Pavlovian stimuli. *Journal of Experimental Psychology:Animal Behavior Processes, 27*(2), 115–124.

Rescorla, R. A., & Heth, C. D. (1975). Reinstatement of fear to an extinguished conditioned stimulus. *Journal of Experimental Psychology: Animal Behavior Processes, 1*(1), 88–96.

Rodrigues, S. M., Schafe, G. E., & LeDoux, J. E. (2001). Intra-amygdala blockade of the NR2B subunit of the NMDA receptor disrupts the acquisition but not the expression of fear conditioning. *Journal of Neuroscience, 21*(17), 6889–6896.

Rogan, M. T., Staubli, U. V., & LeDoux, J. E. (1997). Fear conditioning induces associative long-term potentiation in the amygdala. *Nature, 390*(6660), 604–607.

Rosen, J. B., Fanselow, M. S., Young, S. L., Sitcoske, M., & Maren, S. (1998). Immediate-early gene expression in the amygdala following footshock stress and contextual fear conditioning. *Brain Research, 796*(1–2), 132–142.

Rosenkranz, J. A., & Grace, A. A. (2002). Dopamine-mediated modulation of odour-evoked amygdala potentials during pavlovian conditioning. *Nature, 417*(6886), 282–287.

Rumpel, S., LeDoux, J., Zador, A., & Malinow, R. (2005). Postsynaptic receptor trafficking underlying a form of associative learning. *Science, 308*(5718), 83–88.

Sacktor, T. C. (2008). PKMzeta, LTP maintenance, and the dynamic molecular biology of memory storage. *Progress in Brain Research, 169*, 27–40.

Sananbenesi, F., Fischer, A., Wang, X., Schrick, C., Neve, R., Radulovic, J., & Tsai, L. H. (2007). A hippocampal Cdk5 pathway regulates extinction of contextual fear. *Nature Neuroscience, 10*(8), 1012–1019.

Sanders, M. J., Wiltgen, B. J., & Fanselow, M. S. (2003). The place of the hippocampus in fear conditioning. *European Journal of Pharmacology, 463*(1–3), 217–223.

Schafe, G. E., & LeDoux, J. E. (2000). Memory consolidation of auditory pavlovian fear conditioning requires protein synthesis and protein kinase A in the amygdala. *Journal of Neuroscience, 20*(18), RC96.

Schimanski, L. A., Wahlsten, D., & Nguyen, P. V. (2002). Selective modification of short-term hippocampal synaptic plasticity and impaired memory extinction in mice with a congenitally reduced hippocampal commissure. *Journal of Neuroscience, 22*(18), 8277–8286.

Serrano, P., Friedman, E. L., Kenney, J., Taubenfeld, S. M., Zimmerman, J. M., Hanna, J.,... Fenton, A. A. (2008). PKMzeta maintains spatial, instrumental, and classically conditioned long-term memories. *PLoS Biology, 6*(12), 2698–2706.

Sierra-Mercado, D., Padilla-Coreano, N., & Quirk, G. J. (2011). Dissociable roles of prelimbic and infralimbic cortices, ventral hippocampus, and basolateral amygdala in the expression and extinction of conditioned fear. *Neuropsychopharmacology, 36*(2), 529–538.

Sotres-Bayon, F., Bush, D. E., & LeDoux, J. E. (2007). Acquisition of fear extinction requires activation of NR2B-containing NMDA receptors in the lateral amygdala. *Neuropsychopharmacology, 32*(9), 1929–1940.

Szapiro, G., Vianna, M. R., McGaugh, J. L., Medina, J. H., & Izquierdo, I. (2003). The role of NMDA glutamate receptors, PKA, MAPK, and CAMKII in the hippocampus in extinction of conditioned fear. *Hippocampus, 13*(1), 53–58.

Tronel, S., & Alberini, C. M. (2007). Persistent disruption of a traumatic memory by postretrieval inactivation of glucocorticoid receptors in the amygdala. *Biological Psychiatry, 62*(1), 33–39.

Tronson, N. C., Wiseman, S. L., Olausson, P., & Taylor, J. R. (2006). Bidirectional behavioral plasticity of memory reconsolidation depends on amygdalar protein kinase A. *Nature Neuroscience, 9*(2), 167–169.

Vidal-Gonzalez, I., Vidal-Gonzalez, B., Rauch, S. L., & Quirk, G. J. (2006). Microstimulation reveals opposing influences of prelimbic and infralimbic cortex on the expression of conditioned fear. *Learning & Memory, 13*(6), 728–733.

Walker, D. L., Ressler, K. J., Lu, K. T., & Davis, M. (2002). Facilitation of conditioned fear extinction by systemic administration or intra-amygdala infusions of D-cycloserine as assessed with fear-potentiated startle in rats. *Journal of Neuroscience, 22*(6), 2343–2351.

Wang, H., Shimizu, E., Tang, Y. P., Cho, M., Kyin, M., Zuo, W.,... Tsien J. Z. (2003). Inducible protein knockout reveals temporal requirement of CaMKII reactivation for memory consolidation in the brain. *Proceedings of the National Academy of Science of the United States of America, 100*(7), 4287–4292.

Weinberg, M. S., Samulski, R. J., & McCown, T. J. (2013). Adeno-associated virus (AAV) gene therapy for neurological disease. *Neuropharmacology, 69*, 82–88.

Wilson, A., Brooks, D. C., & Bouton, M. E. (1995). The role of the rat hippocampal system in several effects of context in extinction. *Behavioral Neuroscience, 109*(5), 828–836.

Witten, I. B., Steinberg, E. E., Lee, S. Y., Davidson, T. J., Zalocusky, K. A., Brodsky, M.,... Deisseroth, K. (2011). Recombinase-driver rat lines: tools, techniques, and optogenetic application to dopamine-mediated reinforcement. *Neuron, 72*(5), 721–733.

Won, J., & Silva, A. J. (2008). Molecular and cellular mechanisms of memory allocation in neuronetworks. *Neurobiology of Learning & Memory, 89*(3), 285–292.

Yeh, S. H., Mao, S. C., Lin, H. C., & Gean, P. W. (2006). Synaptic expression of glutamate receptor after encoding of fear memory in the rat amygdala. *Molecular Pharmacology, 69*(1), 299–308.

Zhang, F., Wang, L. P., Brauner, M., Liewald, J. F., Kay, K., Watzke, N.,... Deisseroth, K. (2007). Multimodal

fast optical interrogation of neural circuitry. *Nature, 446*(7136), 633–639.
Zhou, Y., Won, J., Karlsson, M. G., Zhou, M., Rogerson, T., Balaji, J., . . . Silva, A. J. (2009). CREB regulates excitability and the allocation of memory to subsets of neurons in the amygdala. *Nature Neuroscience, 12*(11), 1438–1443.
Zohar, J., Yahalom, H., Kozlovsky, N., Cwikel-Hamzany, S., Matar, M. A., Kaplan, Z., . . . Cohen, H. (2011). High dose hydrocortisone immediately after trauma may alter the trajectory of PTSD: interplay between clinical and animal studies. *European Neuropsychopharmacology, 21*(11), 796–809.

CHAPTER

17 Genetics of Autonomic Nervous System Activity

Eco de Geus, Rene van Lien, Melanie Neijts, *and* Gonneke Willemsen

Abstract

Large individual differences in the activity of the autonomic nervous system (ANS) play a key role in risk for cardiovascular disease. This chapter presents an overview of the measurement strategies that can be used to study ANS activity in samples that are sufficiently large to allow genetic analyses. Heart rate variability, in particular, respiratory sinus arrhythmia (RSA) is identified as the measure of choice to index parasympathetic activity, whereas preejection period (PEP) is the measure of choice to index sympathetic activity. Twin studies have demonstrated significant genetic contributions to resting levels of both RSA (heritability estimates range from 25 to 71 percent) and PEP (heritability estimates range from 48 to 74 percent) and the genetic variance in these traits seems to further increase under conditions of psychological stress. Identifying the genetic variants that influence parasympathetic and sympathetic activity may increase our understanding of the role of the ANS in cardiovascular disease.

Key Words: sympathetic, parasympathetic, heart rate variability, respiratory sinus arrhythmia, preejection period, twins, heritability

Cardiovascular disease (CVD) is one of the main causes of death in Westernized countries. The etiology of CVD is complex, with many different factors (demographical, lifestyle, psychological, and genetic) contributing to an increased risk of CVD development (Brotman, Golden, & Wittstein, 2007; Brotman, Walker, Lauer, & O'Brien, 2005). The physiological risk factors that form the final common pathway to CVD include the metabolic syndrome, with hypertension, hyperlipidemia, hyperglycemia, and android obesity as core features (Bayturan et al., 2010); inflammation (Danesh et al., 2008); coagulation/fibrinolysis imbalance (Libby & Theroux, 2005); reduced heart rate variability (Dekker et al., 1997; 2000); and increased heart rate (Fox et al., 2007). Strikingly, activity of the autonomic nervous system (ANS) is associated with all of these physiological risk factors (Charkoudian & Rabbitts, 2009; Lambert & Lambert, 2011; Malpas, 2010; Straub, Wiest, Strauch, Harle, & Scholmerich, 2006; Task Force of the European Society of Cardiology the North American Society of Pacing, 1996; Tracey, 2009; von Kanel, Mills, Fainman, & Dimsdale, 2001). Because the ANS is very sensitive to psychosocial stress, it plays a key role in almost all models in biobehavioral medicine that try to account for the well-known role of social (Karasek et al., 1988; Rosengren et al., 2004; Siegrist, Peter, Junge, Cremer, & Seidel, 1990) and psychological (Nicholson, Kuper, & Hemingway, 2006) sources of chronic stress in hypertension, diabetes, and cardiac disease.

There are large individual differences in the activity of the ANS in the basal resting state (Berntson, Cacioppo, & Quigley, 1994*b*; Berntson,

Norman, Hawkley, & Cacioppo, 2008; Cacioppo et al., 1994; Grossman & Kollai, 1993; Light, Kothandapani, & Allen, 1998; Salomon, Matthews, & Allen, 2000). These differences in ANS activity are further amplified in response to brief laboratory stressors (de Geus, Kupper, Boomsma, & Sneider, 2007; Houtveen, Rietveld, & de Geus, 2002; Lucini, Norbiato, Clerici, & Pagani, 2002; Wang et al., 2009) as well as prolonged psychosocial stress (Riese, van Doornen, Houtman, & de Geus, 2000; Vrijkotte, van Doornen, & de Geus, 2004). This essay reviews the contribution of genetic factors to the individual differences in ANS activity at rest and during stress. We first present a short overview of the ANS and a detailed review of the measurement strategies used to study it, with a focus on measures that can be used in population-based samples that are sufficiently large to allow genetic analyses. Next, we review the twin studies on the heritability of these ANS measures and present a list of candidate genes that have already been found to influence ANS activity.

I. The Autonomic Nervous System

The term "autonomic nervous system" was coined by Langley in 1898. Based on anatomical and functional criteria, he divided the ANS into three separate branches: the sympathetic nervous system (SNS) including the adrenal medulla, the parasympathetic nervous system (PNS), and the enteric nervous system, a collection of neurons embedded within the walls of the entire gastrointestinal tract that control gastrointestinal motility and secretions. In more recent use of the term, the enteric system is discarded and ANS is usually synonymous with the sympathetic and parasympathetic branches.

The sympathetic branch is best known for its key role in the "fight-or-flight" response. Activity of the SNS causes, among other things, an increase in heart rate, contractility, blood pressure, breathing rate, bronchodilation, sweat production, epinephrine secretion, and a redistribution of blood flow favoring the muscles. The PNS, on the other hand, promotes the maintenance of the body by acquiring energy from food and getting rid of wastes. The PNS is therefore often labeled as the "rest and digest" branch of the ANS. Its activity causes slowing of the heart, constriction of the pupils, stimulation of the gut and salivary glands, and other responses that help restore energy. Many organs are innervated by both the sympathetic and the parasympathetic branches of the ANS, and an increase in the activity of these branches typically exerts opposing actions. However, some organs are not dually innervated (e.g., sweat glands) and, even for dually innervated organs, the autonomic branches may have synergistic rather than opposing effects (e.g., salivary glands).

The main function of the ANS is coordinating bodily functions to ensure homeostasis and performing adaptive responses when faced with changes in the external and internal environment, such as those due to physical activity, posture change, food consumption, or hemorrhage. In addition, the ANS is capable of substantial heterostatic action; it can prepare the body for anticipated threats to homeostasis even in the absence of actual changes in bodily activity. The best known example is the anticipatory response that prepares the body for physical activity in response to a vast range of stressors that can be purely symbolic in nature and are often not followed by actual physical activity (fight-or-flight) or changes in internal environment (e.g., through blood loss or infection). This response is called the *physiological stress response.*

In humans, subjective experience of stress can be sufficient to trigger the physiological stress response. Subjective experience of stress typically occurs when there is an imbalance between perceived threats/demands and perceived abilities/resources. In the brain, the perception of internal (thoughts) or external (environmental events) threats by neocortical areas leads to the activation of limbic areas, in particular the amygdala (Lovallo, 2005). The amygdala, in turn, projects to paraventricular and other hypothalamic nuclei as well as to a network of neurons in the rostral ventrolateral medulla (RVLM) and the nucleus of the solitary tract (NTS) that initiates changes in the activity of sympathetic neurons in the intermediolateral (IML) column and in activity of the parasympathetic neurons in the *n.* ambiguus.

A. Parasympathetic Nervous System Activity

The preganglionic fibers of the PNS leave from the cell bodies of the motor nuclei of cranial nerves (CN) III, VII, IX, and X in the brainstem and from the second, third, and fourth sacral segments of the spinal cord. The vagus nerve (CN X) carries fibers to the heart and lungs (as well as to other organs) and is the primary source of parasympathetic innervation of these organs. Many efferent fibers in the vagus originate in the *n.* ambiguus. The preganglionic axons terminate in parasympathetic ganglia, which lie within or very close to the organs innervated by the short postganglionic neurons. The preganglionic neurons employ acetylcholine (ACh) as the primary neurotransmitter, which binds to a nicotinic receptor subtype on the post-ganglionic neurons in the ganglia. Postganglionic parasympathetic fibers also employ ACh as a primary neurotransmitter, but the receptor subtypes on the target organ are commonly muscarinic. For instance, the parasympathetic postganglionic receptors in the sinoatrial (SA) node of the heart are type 2 muscarinic (M2) and their activation reduces heart rate.

B. Sympathetic Nervous System Activity

The preganglionic fibers from neurons in the IML column leave the central nervous system from the thoracic and lumbar regions of the spinal cord. They synapse onto a chain of sympathetic ganglia that lie close to the spinal cord, known as the *sympathetic trunk*. The preganglionic neurons from the IML column to the sympathetic ganglia employ ACh as the primary neurotransmitter. The post-ganglionic neurons from the sympathetic ganglia to the organs employ norepinephrine as the primary neurotransmitter, which can act on α_1- adrenergic (e.g., in arterioles) or β_1- and β_2-adrenergic receptors (e.g., on the heart). Stimulation of the α_1-adrenergic receptors causes vasoconstriction by acting on the smooth muscles in the medial layer of the blood vessels. Stimulation of the cardiac β-adrenergic receptors by norepinephrine released from the cardiac sympathetic nerves (*nn.* accelerantes) increases the pacemaker frequency of the SA node (i.e., heart rate), as well as contractility of the ventricles. Together, vasoconstriction and increased cardiac performance account for the increase in blood pressure seen during increased sympathetic activity.

A first exception to the use of norepinephrine as the final effector in the SNS is found in the sympathetic innervation of eccrine sweat glands, which is cholinergic rather than adrenergic. A second exception is a set of preganglionic neurons that end in a special ganglion, namely the adrenal medulla. On activation by preganglionic neurons, the adrenal medulla releases a small amount of norepinephrine into the bloodstream, but most of the released norepinephrine is converted to epinephrine, which is excreted in much larger amounts than norepinephrine (5:1). Circulating epinephrine preferentially binds to β_2- receptors in the vessels and on the heart, causing vasodilatation (mostly in muscle tissue) and increases in heart rate and contractility.

II. Measurements of Autonomic Nervous System Activity

Many studies of the ANS have focused on the fight-or-flight response, which is often characterized by reciprocal increases in SNS activity and decreases in PNS activity. Such a pattern gives rise to increases in heart rate and blood pressure, and heart rate and blood pressure reactivity are still among the most used variables to indicate changes in ANS activity. Laboratory studies generally involve the measurement of heart rate and blood pressure during one or more rest periods and during mental and physical challenges, with each period often lasting no more than 5–15 minutes. Such studies provide valuable information on the mechanisms underlying ANS responses to stress and have been instrumental in establishing the existence of stable individual differences in the physiological stress response. However, these individual differences in ANS responses to standardized laboratory situations do not seem to transfer to responses to actual real-life stressors because the association between laboratory and ambulatory measurements is moderate at best (Gerin, Rosofsky, Pieper, & Pickering, 1994; Kamarck & Lovallo, 2003; van Doornen, Knol, Willemsen, & de Geus, 1994).

It is possible that the psychological and physiological processes induced by laboratory stress are only a poor reflection of the actual processes in everyday real-life stress situations. Perhaps as a consequence, the predictive value of ANS responses to laboratory challenges for future CVD is low, with the response to a challenge hardly contributing to the prediction of disease when basal levels have been taken into account (Barnett, Spence, Manuck, & Jennings, 1997; Carroll et al., 1998; Coresh, Kleg, Mead, Liang, & Whelton, 1992). As an alternative to bringing "everyday situations to the laboratory," researchers have increasingly tried to bring the "laboratory to everyday situations." This is done by using miniaturized versions of the recording equipment to perform prolonged ambulatory monitoring in naturalistic settings (Fahrenberg, Myrtek, Pawlik, & Perrez, 2007; Houtveen & de Geus, 2009; Wilhelm & Grossman, 2010). The expectation is that ambulatory measurement of physiological levels and reactivity in the natural environment will lead to better prediction of morbidity and mortality. For blood pressure, the added value of the ambulatory approach has already been demonstrated (Mallion, Baguet, Siche, Tremel, & de Gaudemaris, 1999; Palatini & Jullius, 2004;

Table 17.1 Measurement strategies of autonomic nervous system activity in humans.

Technique	Invasiveness	Principle	Measure	References
PARASYMPATHETIC				
Parasympathetic microneurography	Very High	Direct measurement of action potentials in the parasympathetic nerves converted to bursts with a fixed time constant integrator.	Vagal (burst count/time)	Jewett, 1964; Kunze, 1972; Cerati & Schwartz, 1991
Microdialysis	Very High	Measurement of acetylcholine (ACh) concentrations in the dialysate samples, for instance in the sino-atrial (SA) node, using high-performance liquid chromatography.	[Ach]	Shimizu et al., 2009
Pharmacological blockade	Moderate	When the heart rate measured in the unblocked state is subtracted from the heart rate measured during full muscarinic blockade, the change in heart rate (ΔHR) yields a measure of vagal activity.	ΔHR (bpm)	Cacioppo et al., 1994; Berntson et al., 1994; Martinmaki et al., 2006
Respiratory Sinus Arrhythmia	Noninvasive	Heart rate variability due to respiratory gating of tonic vagal effects on the SA node scales with the height of that activity.	RMSSD (ms) pvRSA (ms) HF (ms^2)	Katona & Jih, 1975; Akselrod et al., 1981; Cerutti et al., 2001; Sztajzel, 2004
Heart Rate Variability (TP, ULF, VLF, LF)	Noninvasive	Same principle as RSA above, but lower frequencies in heart rate variability are additionally influenced by modulation of the sympathetic effects on the SA node.	SDNN (ms)/TP (ms^2) SDANN/ULF (ms^2) VLF (ms^2) LF (ms^2)	Task Force of the European Society of Cardiology the North American Society of Pacing, 1996
Baroreflex sensitivity (BRS)	High (if intra-arterial BP) Noninvasive (if Finapres)	The BRS is computed as the mean slope of the regression line relating beat-to-beat changes in SBP to changes in IBI. As the sympathetic contribution to fast changes in the IBI is minimal, this reflects mainly cardiac vagal activity.	BRS (ms/mmHg)	La Rovere et al., 2008; DiRienzo et al., 2009
SYMPATHETIC				
Sympathetic microneurography	High	Direct measurement of action potentials by a tungsten microelectrode inserted into the fascicle of a sympathetic nerve in the skin or muscle (m. peroneus). The voltage is rectified and integrated with a fixed time constant (100 ms).	MSNA (burst count/time) SSNA (burst count/time)	Hagbarth & Vallbo, 1968; Wallin, 1984; Wallin, 2004

(*Continued*)

Table 17.1 (Continued)

Technique	Invasiveness	Principle	Measure	References
Regional norepinephrine spillover	Very High	During constant-rate infusion of radiolabeled NE, and with regional catheterization, the organ-specific rate of spillover of NE to plasma can be determined based on the degree of dilution of infused radioactive-labeled NE by endogenous-released NE.	Total NE spillover (ng/min) Organ NE spillover (ng/min)	Esler et al., 1988; Grassi & Esler, 1999; Esler & Kaye, 2000; Eisenhofer, 2005;
Pharmacological blockade	Moderate	When the response measured in the unblocked state is subtracted from the response measured during blockade, this yields a measure of sympathetic activity to the organ. For instance, (1) the heart rate decrease (ΔHR) that is induced by blockade of β-adrenergic receptors, (2) the blood pressure increase that is induced by blockade of α-adrenergic receptors.	ΔHR (bpm) ΔDBP (bpm)	Julius et al., 1971; Cacioppo et al., 1994; Berntson et al., 1994
Plasma catecholamines	Low (if venous) High(if arterial)	Measurements of concentrations of NE or E in arterial or venous blood using high-performance liquid chromatography. Caveat: intraneuronal vesicular storage/leakage and synaptic reuptake contribute in unknown ways to plasma NE concentrations and (changes in) tissue clearance strongly co-determine the level of both catecholamines.	[NE] pg/mL [E] pg/mL	Esler et al., 1990; Hjemdahl, 1990; Goldstein et al., 2003
Urinary catecholamines	Noninvasive	Measurement of (24-h) urinary excretion of NE or E (relative to creatinine). Caveat: same as for plasma levels, and the kidneys themselves also produce NE.	[NE] (pg/mg creatinine) [E] (pg/mg creatinine)	Esler et al., 1990; Hjemdahl, 1990; Goldstein et al., 2003
Skin conductance	Noninvasive	Sympathetic activity of skin nerves increases the activity of the sweat glands, which in turn yields a measurable change in the conductance of an applied current across the skin.	SCL (μS) nsSCRs (counts/time)	Fowles, 1986; Boucsein, 1992; Dawson et al., 2000
Salivary α-amylase activity	Noninvasive	NE release from sympathetic nerve terminals near adrenoceptors in the acinar cells of the saliva glands cause an immediate increase in the protein-to-fluid ratio of many	sAA (U/mL)	Nater & Rohleder, 2009

(Continued)

Table 17.1 (Continued)

Technique	Invasiveness	Principle	Measure	References
		salivary proteins, including α-amylase. Caveat: PNS activity also increases α-amylase secretion.		
LF/HF ratio	Noninvasive	The LF/HF ratio of spectral power of the heart rate in the lower frequencies centered around 0.1 Hz (LF) divided by the power in the higher frequencies centered around the respiratory frequency (HF). Also expressed as LF/(LF+HF) = LFnu. Caveat: validity is very controversial.	LF/HF (dimensionless) LFnu (dimensionless)	Pagani & Malliani, 2000
Systolic Time Intervals	Noninvasive	From the thorax impedance cardiogram, the preejection period is derived as the time interval between the onset of ventricular depolarization and the opening of the semilunar valves. This interval shortens with increased sympathetic drive to the left ventricle. Caveat: posture should be controlled.	PEP (ms)	Sherwood et al., 1990

Verdecchia, 2000; Verdecchia, Schillaci, Reboldi, Franklin, & Porcellati, 2001).

Although a great deal has been learned from studies focusing on heart rate and blood pressure as indicators of ANS activity, a disadvantage of these variables is that they represent an unknown mix of sympathetic and parasympathetic effects. It has been shown that the classical reciprocal pattern of sympathetic activation with parasympathetic deactivation describes only a limited part of the total autonomic space (Berntson, Cacioppo, & Quigley, 1991). Different patterns of co-activation, reciprocal activation and co-inhibition are found across individuals performing the same task or within individuals performing different tasks. For example, dental phobia patients engaged in a stressful mental arithmetic task showed an increase in their SNS activity with decreased PNS activity; but, when exposed to phobic stimuli, the same subjects showed increased SNS activity with *increased* PNS activity (Bosch, de Geus, Veerman, & Amerongen, 2000). Most important, health outcomes of sympathetic hyperreactivity need not be the same as those of parasympathetic hyperreactivity. Hyperactivity of the SNS has been mostly associated with an increased risk for hypertension, the metabolic syndrome, and left ventricular failure (Brotman et al., 2007; Esler, 2010; Esler et al., 2008; Lambert, Straznicky, Lambert, Dixon, & Schlaich, 2010), whereas loss of PNS activity causes a reduction in the electrical stability of the heart (Schwartz et al., 2003; Vanoli et al., 1991) and may play a key role in the proinflammatory state (Rosas-Ballina & Tracey, 2009; Tracey, 2009).

Because heart rate and blood pressure do not reveal the underlying pattern of ANS activity, studies in the past two decades began indexing sympathetic and parasympathetic activity separately. Next, we review the various measures of SNS and PNS activity currently in use (see Table 17.1 for an overview).

A. *Parasympathetic Activity*

The ideal way to assess parasympathetic activity is the direct measurement of action potentials in the parasympathetic nerves (Cerati & Schwartz, 1991; Jewett, 1964; Kunze, 1972). For cardiac vagal activity, an alternative was developed that assesses the changes in (ACh) concentration in the SA node by microdialysis (Shimizu et al., 2009). Unfortunately, both these "golden standard" measures are too invasive to be used in research with humans. A theoretical alternative would be to measure the spillover of ACh to plasma by venipuncture, but this is not feasible because of the rapid and extensive clearance of the transmitter in the synaptic space by acetylcholinesterase.

Human studies of parasympathetic activity have therefore focused on the *effects* of parasympathetic activity on the innervated organs rather than on activity per se. For instance, to assess cardiac parasympathetic activity to the SA node, the heart rate change can be measured in response to pharmacological blockade of the muscarinic receptors (Berntson et al., 1994*a*; Cacioppo et al., 1994; Martinmaki, Rusko, Kooistra, Kettunen, & Saalasti, 2006). High doses of M2-antagonists, like atropine, effectively remove all parasympathetic effects on the heart. When the heart rate measured in the unblocked state is subtracted from the heart rate measured during full blockade, subjects with higher parasympathetic activity will show a larger increase in heart rate during blockade than will subjects with lower activity. The increase in heart rate is not a perfect measure of parasympathetic activity, since the SNS and PNS are known to interact (Mizuno et al., 2008). High levels of SNS activity will act to reduce the release of ACh through the action of the α_2-autoreceptors on the terminal buttons of vagal axons. However, these interactive effects can be dealt with using a dual blockade strategy (Berntson et al., 1994*a*; Berntson et al., 1994*b*; Cacioppo et al., 1994).

A disadvantage of pharmacological blockade studies is that they are confined to well-controlled (hospital) settings and are not readily amenable to be used in larger samples or in recordings in naturalistic settings. Fortunately, reliable noninvasive estimation of parasympathetic cardiac effects is possible by measuring time- or frequency-domain indices of heart rate variability in the respiratory frequency range, also called *respiratory sinus arrhythmia* (RSA). Respiratory sinus arrhythmia is the difference in heart rate during the inspiration and expiration phases of the respiratory cycle. Respiratory sinus arrhythmia is generated when tonic firing of motor neurons in the *n.* ambiguus and sympathetic nuclei is modulated by phasic inhibition and excitation coupled to the respiratory cycle. This modulation is caused by connections between the nuclei that control the respiratory generator in the pre-Bötzinger and Bötzinger complexes and the vagal and sympathetic motor neurons, which lie in close proximity in the brainstem (Rekling & Feldman, 1998). This respiration–ANS coupling yields an oscillatory pattern in the release of norepinephrine and ACh in the SA node, such that ACh levels increase during expiration and decrease during inspiration, whereas norepinephrine shows the reverse pattern of increases during inspiration and decreases during expiration. The effect of this respiratory "gating" (Eckberg, 2003) is that the heart rate increases during inspiration and decreases during expiration. This has an advantageous effect on the efficiency of oxygen exchange in the lungs, at least in the resting state (Yasuma & Hayano, 2004).

Fortuitously, the effect of the respiratory-related changes in vagal activity on heart rate variability is much more prominent than the effect of the respiratory-related changes in sympathetic activity. This is due to the differential filter characteristics of the muscarinergic and adrenergic receptors (Berntson, Cacioppo, & Quigley, 1993). The muscarinergic receptor rapidly opens potassium channels after ACh release, and closure of calcium channels with parallel changes in the hyperpolarization of the SA node cells occur within hundreds of milliseconds. The β-receptors, that first need to activate protein kinases before channels for sodium and calcium are opened, act much slower and influence the speed of depolarization of the SA cells only on the scale of seconds. This causes the high-frequency changes due to respiration to be filtered from the sympathetic effects on the heart. In keeping, RSA shows relatively little sensitivity to sympathetic blockade but is affected in a dose–response way by muscarinergic blockers in humans (Martinmaki et al., 2006) or vagal cooling in animals (Katona & Jih, 1975). This has led to the use of RSA as a proxy for vagal cardiac activity (Berntson et al., 1997; Task Force of the European Society of Cardiology the North American Society of Pacing, 1996), although it is acknowledged that large changes in sensitivity of chemoreceptor and baroreceptor reflexes (Berntson et al., 1997; Berntson et al., 1993; Houtveen et al., 2002) or respiratory behavior (Grossman & Kollai, 1993; Grossman, Wilhelm, & Spoerle, 2004; Ritz & Dahme, 2006) are important confounders.

Respiratory sinus arrhythmia can be derived from the interbeat interval (IBI) time series obtained from the R waves in the electrocardiogram (ECG) by taking the root mean square of differences (RMSSD) between successive IBIs (Sztajzel, 2004; Task Force of the European Society of Cardiology the North American Society of Pacing, 1996). When the respiratory signal is co-registered with the ECG, RSA can also be derived by peak-valley estimation (pvRSA). Estimates of pvRSA are obtained by subtracting the shortest IBI during heart rate acceleration in the inspiration phase from the longest IBI during heart rate deceleration in the expiration phase (Katona & Jih, 1975). Respiratory sinus arrhythmia can also be derived in the frequency domain by Fourier (Akselrod et al., 1981) or wavelet analysis (Houtveen & Molenaar, 2001; Pichot et al., 1999; Wiklund, Aka, & Niklasson, 1997) or by autoregressive model estimation (Cerutti, Bianchi, & Mainardi, 2001). These analyses describe the periodic oscillations of the heart rate signal decomposed at different frequencies and amplitudes and provide information on the amount of their relative contribution to the variance (also termed *power*) in the heart rate. Power in the respiratory frequency range of 0.15–0.40 Hz (HF power) can be used to index RSA.

In standardized laboratory recordings, as well as in ambulatory settings, the different time- and frequency-domain measures of RSA (e.g., RMSSD, HF power, pvRSA) were highly correlated, with all *r*'s being greater than 0.80 (Hayano et al., 1991*a*; Houtveen & Molenaar, 2001; Grossman, van Beek, & Wientjes, 1990; Penttila et al., 2001), and this high intercorrelation of the various RSA measures proved stable across a wide range of values for respiration and heart rates (Goedhart, Kupper, Willemsen, Boomsma, & de Geus, 2007). An important feature of the RSA measures is that they can be reliably measured under naturalistic conditions with the use of ambulatory monitoring (de Geus, Willemsen, Klaver, & van Doornen, 1995; Wilhelm, Roth, & Sackner, 2003). For the average 24-hour levels of RMSSD and HF power, high test–retest correlations ($.63 < r < .90$) were found after 3 to 65 days in both healthy individuals and cardiac patients (Bigger, Jr. et al., 1992; Hohnloser, Klingenheben, Zabel, Schroder, & Just, 1992; Kleiger et al., 1991; Sinnreich, Kark, Friedlander, Sapoznikov, & Luria, 1998; Stein, Rich, Rottman, & Kleiger, 1995). Good long-term temporal stability ($.58 < r < .76$) for 24-hour levels of pvRSA, HF, and RMSSD has been shown over a period of 7 months (Pitzalis et al., 1996) to 3.4 years (Goedhart et al., 2007).

Respiratory sinus arrhythmia is only one component of the total variability in the heart rate. Other heart rate variability measures include the power in the low-frequency (LF, 0.04–0.15 Hz), very-low-frequency (VLF, 0.003–0.04 Hz), and ultra-low-frequency (ULF, < 0.003 Hz) bands (Sztajzel, 2004; Task Force of the European Society of Cardiology the North American Society of Pacing, 1996). Together with RSA, these measures have received a lot of attention in medicine because lowered levels of heart rate variability predict adverse cardiovascular events, including atrial fibrillation, myocardial infarction, congestive heart failure, and coronary artery disease (Bigger et al., 1990; Bigger, Jr. et al., 1992; Bigger, Jr., Fleiss, Rolnitzky, & Steinman, 1993; Dekker et al., 1997; 2000; Hayano et al., 1991*b*; Kleiger, Miller, Bigger, & Moss, 1987; Lombardi et al., 1987; Nolan et al., 1998; Saul et al., 1988; Singer et al., 1988; Tsuji et al., 1996; Vikman et al., 2003). Heart rate variability in the LF band arises from so-called Mayer waves, which are periodic oscillations in arterial blood pressure around the 0.1 Hz frequency (Julien, 2006). Through the action of the baroreceptors, these periodic changes in blood pressure are met by parallel changes in vagal activity and sympathetic vascular tone to keep blood pressure constant. This gives rise to a "10-s rhythm" in the heart rate that can be detected by spectral analysis because the power in the LF band heart rate variability in the ULF band has been hypothesized to reflect circadian rhythms in bradykinin, renin, or angiotensin release, or thermoregulatory effects on peripheral vasomotor tone. However, in ambulatory recordings, the ULF, which is highly correlated to the standard deviation of the average heart rate across all 5-minute segments of an entire recording (SDANN), seems to arise predominantly from changes in behavior from passive to physically active and vice versa, particularly around the transitions to and from sleep (Roach, Wilson, Ritchie, & Sheldon, 2004). This may also be true of the standard deviation of the heart rate (SDNN), which reflects all of these same components. The SDNN can also be indexed by the total power (TP), which is almost identical to the sum of ULF + VLF + LF + HF.

A final measure that has been used to index cardiac vagal activity, particularly in the context of increased risk for cardiac disease, is the sensitivity of the baroreflex (BRS, La Rovere et al., 2003; La Rovere, Bigger, Marcus, Mortara, & Schwartz, 1998). The baroreflex loop counteracts deviations in blood pressure from an ongoing setpoint by changing sympathetic outflow to the blood vessels to affect peripheral resistance and by changing sympathetic and vagal activity to the SA node to affect cardiac output (DiRienzo, Parati, Radaelli, & Castiglioni, 2009; La Rovere, Pinna, & Raczak, 2008). Deviation of blood pressure from the setpoint is detected by stretch receptors (baroreceptors), mainly located on the wall of the aorta and the carotid arteries. These receptors interface with the sympathetic and parasympathetic motor neurons of the *n.* ambiguus and the *n.* tractus solitarius to generate the required ANS responses. To allow blood pressure to increase in emergency situations, the sensitivity of the baroreflex is kept variable; substantial within-subject variation in the BRS over time is found, as well as large between-subject differences (DiRienzo et al., 2009). Classical approaches to measure BRS have manipulated baroreceptor firing through pharmacological agents that affect blood pressure and by lower body negative pressure, forced expiration against resistance (the Valsalva maneuver), or a neck suction/pressure cuff over the carotid baroreceptors. Changes in mean arterial blood pressure are then regressed on the changes in heart rate to obtain the cardiac BRS, which is defined as the change in heart rate induced by a fixed rise in blood pressure. These methods assess the integrated effects of baroreflex-induced changes in sympathetic vascular activity and sympathetic and vagal cardiac activity.

To selectively index the cardiac *vagal* component of the baroreflex loop, two methods are available. Both need the continuous time series of IBIs as well as the beat-to-beat changes in blood pressure that can be obtained from intra-arterial pressure recordings or noninvasively from the "Finapres" vascular unloading technique to determine finger arterial pressure (Langewouters, Settels, Roelandt, & Wesseling, 1998). In the sequence method, all occurrences of three or more consecutive beats are identified with progressive increases/decreases in systolic blood pressure of at least 1 mm Hg that are followed by progressive increases/decreases in the IBI. The BRS is computed as the mean slope of the regression line relating changes in systolic blood pressure to changes in IBI (Steptoe & Sawada, 1989). In the spectral method, the BRS can be obtained by calculating the strength of the linear coupling (coherence) between the beat-to-beat fluctuations in the systolic blood pressure and the IBI time series in the low-frequency (0.04–0.15 Hz) band (Robbe et al., 1987). Because the effects of cardiac sympathetic activity are again too slow to follow the rapid beat-to-beat changes in blood pressure, changes in the cardiac BRS are mainly thought to reflect changes in cardiac vagal activity. A depressed BRS (<3 ms/mm Hg) was shown to increase the risk for cardiac mortality by 2.8 times, independent of left heart failure and mostly by increasing sudden death (La Rovere et al., 1998).

B. *Sympathetic Activity*

The golden standards to measure SNS activity in humans are the direct microneurographic recording of action potentials from superficial nerves innervating the skeletal muscle (MSNA) or the skin (SSNA) (Hagbarth, Hallin, Wallin, Torebjork, & Hongell, 1972; Hagbarth & Vallbo, 1968; Svedenhag, Wallin, Sundlof, & Henriksson, 1984; Wallin, 1984; 2004) or the measurement of spillover of the postganglionic neurotransmitter norepinephrine using radioactive tracers (Eisenhofer, 2005; Esler et al., 1988; Esler & Kaye, 2000; Kingwell et al., 1994). The advantage of norepinephrine spillover is that it can be measured on an organ-to-organ basis, which allows separate measurement of, for instance, renal, lung, or cardiac sympathetic activity (Eisenhofer, 2005). This is important because the notion of a single emergency SNS response that affects all organs to the same extent has proven untenable. In some circumstances, like exercise, the SNS acts as a "unitary system," but, in many other situations, it is capable of differentiated regulation of its activity to separate organs to a substantial degree (Hjemdahl, Freyschuss, Juhlin-Dannefelt, & Linde, 1984).

Much less invasive measurements of norepinephrine and/or its metabolites in antecubital venous blood are possible by venipuncture or by assessing the excretion of norepinephrine and/or metabolites in urine. However, concerns have been raised about differences in intraneuronal vesicular storage and leakage, reuptake, extraneuronal clearance, and urinary filtration/secretion that may (severely) distort the relation between actual SNS activity and plasma and urine norepinephrine concentrations (Eisenhofer, Kopin, & Goldstein, 2004; Esler et al., 1990; Goldstein, Eisenhofer, & Kopin, 2003). Only a very small proportion of norepinephrine released from sympathetic nerves reaches the bloodstream, and the final plasma levels depend on an unknown mixture of changes in the rate of release, as well as on the rate of removal by tissues. Plasma epinephrine levels reflect neural outflow to the adrenal medulla rather well

(Goldstein et al., 2003), but the substantial clearance of adrenaline by adrenoceptors in the tissues, which further increases during vasoconstriction, causes antecubital plasma levels to be much lower than the arterial levels (Hjemdahl, 1990). Nonetheless, venous epinephrine levels show systematic increases to psychosocial stress that seem to selectively enhance adrenomedullary activation over general sympathetic nerve activation.

As was true for the PNS, most human studies of parasympathetic activity have focused on the *effects* of sympathetic activity on the innervated organs rather than on activity per se. Sympathetic nervous system effects can be measured using pharmacological blockade of either α-receptors (e.g., phentolamine) or β-receptors (e.g., propranolol) or even specific β_1- (e.g., metoprolol) or β_2-adrenergic (ICI 118–551) receptors. This has been extensively done for the assessment of cardiac and vascular sympathetic activity. Cardiac sympathetic activity, for instance, is estimated as the difference between heart period in the unblocked state and during complete blockade of cardiac sympathetic effects (Berntson et al., 1994*a*; Berntson et al., 1994*b*; Cacioppo et al., 1994; Lewis, Nylander, Gad, & Areskog, 1980; Shi, Stevens, Foresman, Stern, & Raven, 1995).

Skin SNS effects can be noninvasively measured by the activity of the eccrine sweat glands. Acetylcholine release from the preganglionic sympathetic nerves increases the activity of the sweat glands, which in turn increases the conductance of an applied current across the skin (Foster & Weiner, 1970; Fowles, 1986). Because sweat glands are at the highest density in palmar and plantar regions, approximately 400/mm^2, most researchers measure skin conductance at these sites. Electrodermal activity incorporates both slow tonic shifts in basal skin conductance level (SCL) and more rapid phasic transient events, that is, skin conductance responses (SCRs), which are also referred to as *galvanic skin responses* (Boucsein, 1992; Dawson, Schell, & Filion, 2000; Fowles, 1986). The frequency of the nonspecific SCRs (nsSCRs) is termed electrodermal lability (Mundy-Castle & McKiever, 1953). Both SCL and nsSCRs have been shown to be influenced by emotional stress (Boucsein, 1992; Critchley, 2002; Schell, Dawson, & Filion, 1988). The test–retest reliability coefficients over time periods encompassing 1 day to a year for SCL levels ranged from 0.40 to 0.85 and from 0.40 to 0.76 for nsSCRs (Freixa i Baque, 1982; Iacono et al., 1984; Schell, Dawson, Nuechterlein, Subotnik, & Ventura, 2002; Schell et al., 1988; Vossel & Zimmer, 1990).

The salivary glands are also innervated by the ANS, and salivary α-amylase concentration (sAA) has been suggested to be a noninvasive marker for SNS activity (Nater & Rohleder, 2009). Salivary amylase is a digestive enzyme that breaks down insoluble starch into soluble maltose and dextrin. It comprises approximately 30 percent of total protein secretion from the parotid glands, submandibular glands, and sublingual glands, as well as from minor glands in the submucosa underlying the lips, cheeks, and palate (Humphrey & Williamson, 2001). Various studies have revealed that conditions known to evoke sympathetic activation uniformly increase sAA secretion, including stressful academic examination (Bosch, de Geus, Ring, Nieuw Amerongen, & Stowell, 2004; Chatterton, Jr., Vogelsong, Lu, Ellman, & Hudgens, 1996), stressful computer games (Skosnik, Chatterton, Swisher, & Park, 2000; Takai et al., 2004), watching a stressful video (Bosch et al., 2003), mental arithmetic test (Noto, Sato, Kudo, Kurata, & Hirota, 2005), cold pressor test (van Stegeren et al., 2008), and the Trier Social Stress Test (Nater et al., 2006; Nater et al., 2005; Rohleder, Nater, Wolf, Ehlert, & Kirschbaum, 2004). Moreover, administration of the β-adrenergic antagonists reduces sAA concentration in unstimulated whole mouth saliva (Nederfors & Dahlöf, 1992) and attenuates the stress-induced increases in sAA concentration (van Stegeren, Rohleder, Everaerd, & Wolf, 2006).

Unfortunately, most studies used an inadequate methodology that was validated for cortisol research but is not suitable for sAA measurements (Bosch, Veerman, de Geus, & Proctor, 2011). Even more problematic is that the salivary glands are innervated by both branches of the ANS, not just the sympathetic branch (Proctor & Carpenter, 2001). Parasympathetic activity can influence sAA concentrations: (1) via α-amylase release from glands that are solely or mainly parasympathetically innervated (e.g., the palate and sublingual glands); (2) via synergistic sympathetic–parasympathetic interactions whereby parasympathetic activity amplifies sympathetic effects; and (3) via the effects of (parasympathetically-mediated) salivary flow rate (Bosch et al., 2011). In fact, Bosch et al. (2003) showed that a passive-coping stressor that evoked parasympathetic activation (viewing a surgical video), as measured by increases in salivary flow and RSA, also evoked a strong sAA release, which was much larger than the release during a stressor that elicited a dominant cardiac sympathetic activation (a time-pace memory task). The important role of the PNS in sAA and protein secretion invalidates the use of sAA concentration as an exclusive read-out of sympathetic activity.

Pagani and coworkers have advanced the notion that a single ratio, the power of the heart period time series in the LF band divided by the power in the HF band, may capture changes in the ratio of sympathetic to vagus nerve traffic to the heart (Malliani et al., 1991; Montano et al., 1994; Pagani et al., 1986; Pagani et al., 1997; Pagani & Malliani, 2000). The idea behind the LF/HF ratio is that, during sympathetic activation, LF and HF power both decrease as does the TP, but the decrease in LF power relative to the decrease in total power is less strong than that in HF power, whereas the reverse occurs during vagal activation, where the increase in LF power relative to the increase in total power is less strong than that in HF power. Expressing the spectral components in absolute units prevents the appreciation of this fractional redistribution of the power across the LF and HF bands. This information is regained when LF and HF are expressed as a ratio, or when LF and HF power are measured in normalized units (LFnu), which represent the relative value of each power component in proportion to the TP minus the VLF/ULF components (Burr, 2007).

The predictive power of LF power for CVD is beyond question (Bigger, Jr. et al., 1993; Dekker et al., 2000; Tsuji et al., 1996). However, the use of the normalized LF power as a potential measure of cardiac sympathetic control is the subject of controversy (Eckberg, 1997; Goedhart, Willemsen, Houtveen, Boomsma, & de Geus, 2008). The strongest concern about the validity of the LF/HF ratio is provided by those studies that directly compare it against invasive measures of sympathetic activity, such as peroneal muscle nerve activity or cardiac norepinephrine spillover. Although some studies did report a correlation of these measures with the LF/HF ratio (Pagani et al., 1997), most studies did not find this correlation across a range of clinical contexts (Grassi & Esler, 1999; Kingwell et al., 1994).

Together with electrodermal activity, measurement of cardiac contractility is currently the preferred noninvasive method to measure SNS activity. Contractility is influenced only by the sympathetic branch of the ANS because there is an abundance of functional β-adrenergic receptors on the human ventricle but no ACh receptors. Activation of the β-receptors exerts inotropic effects on the cardiac muscle through the opening of calcium channels in the membrane, as well as of the T tubules of the muscle fibers. The calcium influx increases contractile force and contraction speed of the ventricle. This increased contractility is reflected in a larger ejection fraction of the left ventricle. The ejection fraction reflects the ratio between the stroke volume and the end diastolic volume. The ejection fraction can be obtained from recordings of end diastolic and end systolic volumes (the difference equals the stroke volume) by (contrast) echocardiography or magnetic resonance imaging (Malm, Frigstad, Sagberg, Larsson, & Skjaerpe, 2004).

Contractility is also reflected in a more rapid start of the ejection phase after the onset of ventricular depolarization, a time interval referred to as the preejection period (PEP). The PEP can be measured by using impedance cardiography, in which a HF alternating current is introduced across the thorax by electrodes at the level of the neck and belly (Sherwood et al., 1990). Electrodes at the level of the top and bottom of the sternum measure the changes in the impedance of the enclosed thorax column (dZ). The first derivative of the pulsatile changes in transthoracic impedance (dZ/dt) is called the *impedance cardiogram* (ICG), and it reflects the momentary changes in aortic blood flow during the systolic phase. From the combined ECG and ICG, the PEP can be derived as the time interval between the onset of ventricular depolarization (QRST-onset) and the opening of the semilunar valves (sharp upstroke in the dZ/dt). Within-subject changes in PEP reliably index changes in β-adrenergic drive to the left ventricle, as was shown in laboratory studies that employed manipulations known to increase cardiac sympathetic activity such as epinephrine infusion, amyl nitrite inhalation, mental stress, and exercise. These manipulations systematically decrease PEP (de Geus et al., 2007; Houtveen, Groot, & de Geus, 2005; Krzeminski et al., 2000; Mezzacappa, Kelsey, & Katkin, 1999; Miyamoto et al., 1983; Nelesen, Shaw, Ziegler, & Dimsdale, 1999; Newlin & Levenson, 1979; Schachinger, Weinbacher, Kiss, Ritz, & Langewitz, 2001; Sherwood et al., 1986; Smith et al., 1989; Svedenhag et al., 1986; Richter & Gendolla, 2009). In addition, pharmacological blockade of cardiac sympathetic effects results in the expected prolongation of the PEP (Berntson et al., 1994*a*; Cacioppo et al., 1994; Harris, Schoenfeld, & Weissler, 1967; Schachinger et al., 2001; Winzer et al., 1999), whereas PEP is hardly affected by blockade of cardiac vagal effects (Berntson et al., 1994*a*; Cacioppo et al., 1994; Martinsson, Larso, & Hjemdahl, 1987).

Since the PEP can be reliably obtained noninvasively from only 5–7 spot electrodes, a number of ambulatory devices are available that allow recording of the PEP in naturalistic settings (Cybulski, 2000; Nakonezny et al., 2001; Sherwood, McFetridge, & Hutcheson, 1998; Willemsen, de Geus, Klaver, van Doornen, & Carroll, 1996). The only caveat in using PEP as an index of changes in cardiac sympathetic activity is its sensitivity to preload and afterload effects. Cardiac contractility can increase independently from sympathetic effects when the stretch of the myocardial muscle fibers increases through the Frank-Starling mechanism. Thus, increases in end diastolic volume (preload) can shorten the PEP in the absence of increased sympathetic activity, leading to the erroneous suggestion of *increased* cardiac sympathetic effects. The reverse problem occurs when the pressure in the aorta is increased (afterload) in the presence of unchanged sympathetic activity. Because it takes longer for the aortic valves to open, the PEP becomes longer, erroneously suggesting *decreased* cardiac sympathetic effects. Postural changes have a major effect on preload and afterload and indeed lead to paradoxical responses of the PEP. Head-up tilting from supine to upright systematically prolongs the PEP (Frey & Kenney, 1979; Lewis, Rittgers, Forester, & Boudoulas, 1977; Ovadia, Gear, Thoele, & Marcus, 1995), and longer PEPs have also been demonstrated when subjects go from supine to sitting to standing (Houtveen et al., 2005; Sherwood & Turner, 1993; Waldstein, Neumann, & Merrill, 1998). Clearly, posture needs to be taken into account when using PEP as a measure of sympathetic activity.

When posture is controlled, the PEP has been shown to be a stable individual characteristic. In the laboratory, test–retest correlations between 0.45 to 0.88 have been found for baseline and stress-task levels of PEP across retest intervals ranging from 28 days to 3 years (Burleson et al., 2003; Matthews, Salomon, Kenyon, & Allen, 2002; Willemsen et al., 1998). For ambulatory 24-hour PEP, high stability ($.67 < r < .93$) has been found across a few days (Vrijkotte et al., 2004), as well as across a much longer period of 2–5 years (Goedhart et al., 2006). It is important to note that the between-subject differences in PEP reflect the extent to which subjects differ in the degree of sympathetic *effects* on their cardiac contractility. These effects are likely to be highly correlated with differences in sympathetic activity, but the correlation is not perfect. Inotropic responses to norepinephrine and circulating epinephrine will be modulated by individual differences in the effectiveness of the cardiac β_1- and β_2-adrenergic receptors. Density, affinity, and distribution of these receptors may show large individual differences (Liggett, 1995; Liggett et al., 2006). These individual differences in receptor status may, for instance, lead to a paradoxically long PEP in patients with high levels of cardiac sympathetic nerve activity who have very low ventricular β-receptor densities. In healthy subjects, however, a high interindividual correlation (.82) was found between PEP levels and cardiac sympathetic effects as assessed through dual blockade (Cacioppo et al., 1994).

III. Genetics of Autonomic Nervous System Activity

To determine the genetic contribution to the individual differences in ANS measures, we can resort to family studies. The essence of all family studies is that they relate the degree of genetic resemblance to the degree of trait resemblance. From basic biometrical principles, we can compute the estimated genetic covariance between two individuals based on the degree of their genetic relatedness (Falconer & Mackay, 1996; Lynch & Walsh, 1997). If two individuals are genetically unrelated, their expected genetic covariance is zero. It is important to note that this does not mean that they cannot show trait resemblance, but that such resemblance must be due to factors other than genetic ones (i.e., environmental factors). Things change when the covariance is computed between individuals who *do* have a genetic relationship. Relatives, for instance, parent and offspring, will more often have identical variants at a DNA locus than will unrelated individuals. If genetic factors contribute to the trait, the covariance between family members will vary systematically with the degree of genetic relatedness.

A general formula for the genetic part of the covariance between a trait measured in two relatives is:

$$\text{Covariance}\,(P_1, P_2) = \mathbf{u} * V_A + \mathbf{r} * V_D$$

where P_1 is a trait measured in relative 1, P_2 the same trait measured in relative 2, V_A the additive genetic variance caused by all loci contributing to the trait, V_D the dominant genetic variance caused by all loci contributing to the trait, **u** the coefficient of relationship representing the correlation between the relatives for loci acting additive, and **r** the coefficient of dominance. A locus acts additively if both variants at a locus contribute equally to trait variance, whereas in dominance, one variant exerts a (much) stronger effect than the other variant. For parent–offspring pairs, u= ½ and r =0, for full siblings u = ½ and r = ¼, for uncles (aunts) with the nephews (nieces) u = ¼ and r = 0, for grandparent and grandchild u = ¼ and r = 0, and for single first cousins u = ⅛ and r = 0. When the covariance between family members in a trait is measured across many different family relationships (e.g., different values for **u** and **r**), this yields a set of equations that can be algebraically solved to estimate V_A and V_D. This is the essence of the use of pedigrees (parent–offspring, grandparent–grandchild, uncle–niece, siblings, etc.) studies to estimate the broad heritability of a trait, which is simply the genetic variance (V_A and V_D) in the trait divided by the total variance (V_P) in the trait.

Studying family resemblance is a powerful tool to detect genetic contribution to a trait but, as already noted by Sir Francis Galton (1869), a major shortcoming of this type of family studies is that the degree of genetic relatedness can be confounded with the degree of shared family environment. Full siblings, for instance, do not only share more genes than nieces and nephews, they are also typically reared in the same household. The shared family environment includes potentially important factors like parental socioeconomic status (SES), neighborhood, school, sports club, family diet, and parental attitudes and general rearing style. The solution to separating the genetic and shared environmental effects on the trait resemblance of family members has been found in what Galton called "a unique experiment of nature": monozygotic (MZ) and dizygotic (DZ) twin pregnancies.

Monozygotic twinning occurs when, for reasons that are still incompletely understood, a fertilized egg divides before it nestles in the uterus. Monozygotic twins are usually said to inherit identical genetic material. This is not entirely correct. Genetic imprinting patterns can be found to differ in MZ twins as is illustrated, for instance, in an MZ twin pair discordant for Beckwith-Wiedemann syndrome due to differential imprinting (Martin, Boomsma, & Machin, 1997). A number of other occurrences can make the genetic identity of MZ twins less than 100 percent. Since these occurrences are all rare, the assumption that MZs have 100 percent genetic identity is quite defensible, particularly since deviation from perfect identity will lead to an *under*estimation of genetic effects (i.e., the assumption is a conservative one).

If more than one egg is released from the ovaries during a menstrual cycle, and each egg is fertilized by a separate sperm, the result is a nonidentical twin also known as a DZ or fraternal twin. Dizygotic twinning rates have risen in the last decades in most countries because of artificial reproduction techniques and the higher age at which mothers get their first child, which may be paired to higher levels of follicle stimulating hormone. Genetically, DZ twins do not differ from singleton brother–brother, sister–sister, or brother–sister pairs; that is, they share on average 50 percent of their genetic material (Hoekstra et al., 2004). Opposite-sex (DOS) twins are always DZ twins.

In a twin study, four possible factors and their interactions and correlations are assumed to contribute to the total variance in a trait: unique environmental factors (E), shared environmental factors (C), additive genetic factors (A), and dominant genetic factors (D). Shared environmental factors and additive and dominant genetic factors can cause twin resemblance, whereas the extent to which twins do not resemble each other is ascribed to the unique (or nonshared) environmental factors. These include all unique experiences such as differential jobs or lifestyle, accidents, or other life events, and, in childhood, differential treatment by the parents and nonshared peers. Twin researchers typically use structural equation modeling (SEM) to estimate the relative contribution of the A, D, C, and E factors to the individual differences in the trait. In SEM, the relationships between several latent unobserved variables (e.g., genetic and environmental factors) and observed variables (e.g., ANS measures) are summarized by a series of equations. Additional equations can specify the correlation between the latent genetic and environmental factors based on the known genetic relationship. For instance, the latent additive and dominant genetic factors influencing an ANS measure are correlated unity in MZ twins, but only 0.5 and 0.25, respectively, in DZ twins.

It is possible to derive the expected variance-covariance matrix implied by the total set of equations through the use of covariance algebra. Using maximum likelihood estimation, the fit of the expected covariance/variance matrix to the actual observed covariance/variance matrix is iteratively tested in a sample of hundreds or thousands of twins over a range of possible values for the path coefficients. From the best fitting model, the estimates for the path coefficients (e.g., a, c, d, and e) are used to estimate the relative contribution of the latent factors A, C, D, and E to the total variance in, for instance, an ANS measure. Heritability of the ANS measure, defined as the relative proportion of the total variance explained by genetic factors, is obtained as the ratio of a2 + d2/(a2 + d2 + e2 + c2). The heritability can also be expressed as a percentage by multiplying this ratio by 100.

Using the classic twin design, we can only estimate three components of variance at the same time (either A, C, and E, or A, D, and E). One solution is to add parents or offspring of twins to create an extended twin-family design (Keller et al., 2009). If such family data are not available, one needs to make the assumption that either C or D is absent. This can be inferred in twin studies by first inspecting the MZ and DZ correlations to see whether dominance or shared environmental effects are actually likely to play a role. The presence of dominance can be recognized because it yields DZ correlations that are much lower than half the MZ correlation. In contrast, the presence of shared environmental effects yields DZ correlations that are much higher than half the MZ correlations. If the MZ correlation is about twice as high as the DZ correlation an AE model will fit best.

A. Heritability of Indicators of Autonomic Nervous System Activity

Tables 17.2 and 17.3 summarize the results from twin studies on the heritability of the previously described measures of ANS activity. Only studies with at least 50 twin pairs were included. Table 17.2 focuses on studies that have estimated the heritability of cardiac vagal activity using the different measures of RSA. The table excludes a study on HF power in 5-month-old infants, in which HF power was only weakly influenced by genetic factors (11 percent), and unique environmental factors explained most of the HF variance, particularly in boys (Dubreuil et al., 2003). Because myelinization of vagal fibers is still incomplete up until age 2 (Loeliger, Tolcos, Leditschke, Campbell, & Rees, 2000), it is uncertain whether the relation between vagal activity and RSA in such young children is comparable to that in adolescents and adults. For completeness Table 17.2 also includes the heritability estimates for the BRS and total heart rate variability, either measured as the standard deviation of all IBIs (SDNN) or the TP of the IBI spectrum, and for the VLF and ULF power. Table 17.3 focuses on studies that have estimated the heritability of measures of SNS activity that, to date, only include MSNA, plasma catecholamines, SCL/nsSCRs, and PEP. The table excludes a study on nine MZ twin pairs that used the golden standard MSNA approach (Wallin, Kunimoto, & Sellgren, 1993). This study reported that the MZ twin resemblance was as good as the average test–retest reliability of the method.

B. Heritability of Measures of Cardiac Vagal Activity

In laboratory twin studies that record RSA in quiet resting conditions, a significant but modest genetic contribution has been systematically reported (Busjahn et al., 1998; de Geus et al., 2007; Kupper et al., 2005; Riese et al., 2006; Riese et al., 2007; Su et al., 2009; Tank et al., 2001; Tuvblad, Isen, Baker, Raine, Lozano, & Jacobson, 2010; Uusitalo et al., 2007; Wang, Thayer, Treiber, & Snieder, 2005; Wang et al., 2009; Zhang et al., 2007). Heritability estimates at rest for pvRSA range from 25 to 55 percent, for RMSSD from 36 to 71 percent, and for HF from 37 to 63 percent. Heritability estimates for BRS ranged from 22 to 55 percent. Heritability estimates for European Americans and African Americans were very similar (Wang et al., 2005), and none of the studies reported a sex difference in heritability or evidence for different genes being expressed in males and females.

The difference in heritability estimates across studies and measures probably does not reflect a truly different genetic architecture. Instead, it may simply reflect the large age range across the studies (mean age 9 to mean age 51) and the different duration of the resting condition and the posture of the subject (sitting, supine). One study that used multiple measures in the same set of subjects found a very high genetic correlation among SDNN, RMSSD, and HF, indicating that these three RSA measures reflect largely the same underlying genetic factor (Wang et al., 2005).

Strikingly, heritability of ambulatory values of RSA tends to be higher than the values at rest. Two studies that measured pvRSA, HF, or BRS at rest and during a series of mental stressors reported increased genetic variance in these measures under stress (de Geus et al., 2007; Riese et al., 2006; Wang et al., 2009). This suggests that genetic influences on cardiac vagal activity become more pronounced when the subject is challenged by mentally and emotionally “engaging” conditions.

Table 17.2 Studies reporting heritability of cardiac parasympathetic nervous system activity.

Reference	Number of subjects	Age	Sex	Protocol	Measure	Heritability (%)	Best fitting model
Boomsma et al., 1990	MZ = 140 DZ = 180	16.7	m/f	Laboratory (8-minute rest – quiet sitting)	pvRSA	25	AE
Snieder et al., 1997	MZ = 182 DZ = 234	44.2	m/f	Laboratory (8-minute rest – quiet sitting)	pvRSA	31	AE
Busjahn et al., 1998	MZ = 190 DZ = 92	33		Ambulatory 30-minute recording	SDNN	60	ADE
					RMSSD	65	ADE
					HF	39	AE
Tank et al., 2001	MZ = 176 DZ =122	33	m/f	Laboratory (10-minute rest – semi-supine)	BRS	43	AE
Kupper et al., 2004	MZ = 218 DZ = 301 Sibs = 253	31	m/f	Ambulatory 24-hour recording, analysis restricted to sitting activities and sleep	SDNN SDNN SDNN SDNN	35 (morning) 36 (afternoon) 47 (evening) 43 (night)	AE AE AE AE
					RMSSD RMSSD RMSSD RMSSD	41 (morning) 48 (afternoon) 48 (evening) 40 (night)	AE AE AE AE
Kupper et al., 2005	MZ = 222 DZ = 305 Sibs = 253	31	m/f	Ambulatory 24-hour recording, analysis restricted to sitting activities and sleep	pvRSA pvRSA pvRSA pvRSA	40 (morning) 49 (afternoon) 55 (evening) 54 (night)	AE AE AE AE
Wang et al., 2005	MZ = 104 DZ = 102	15.2	m/f	Laboratory (10-minute rest – supine)	SDNN	66	AE
					RMSSD	71	AE
					HF	63	AE
					LF	45	AE
					LF/HF	32	AE

(Continued)

Table 17.2 (Continued)

Reference	Number of subjects	Age	Sex	Protocol	Measure	Heritability (%)	Best fitting model
De Geus et al., 2007	Re-analysis of Boomsma 1990 dataset	16.7	m/f	Laboratory (8-minute rest – quiet sitting)	pvRSA	31	AE
				Laboratory (mental stress)	pvRSA	54	AE
				Laboratory (reactivity)	pVRSA	ns	AE
	Re-analysis of Snieder 1997 dataset	44.2	m/f	Laboratory (3-minute rest – quiet sitting)	pvRSA	32	AE
				Laboratory (mental stress)	pvRSA	44	AE
				Laboratory (reactivity)	pvRSA	ns	AE
Riese et al., 2006	MZ =148 DZ = 102	23.2	f	Laboratory (5-minute rest – quiet sitting)	BRS	22	AE
				Laboratory (mental stress)	BRS	42	AE
Riese et al., 2007	MZ = 115 DZ = 91	22.8	f	Laboratory (rest + stress in a single latent factor)	BRS	53	AE
					TP	51	AE
Zhang et al., 2007	MZ = 336 DZ = 106	15–84	m/f	Laboratory (5-minute rest – sitting)	BRS	55	AE
					TP	23	AE
Uusitalo et al., 2007	MZ = 208 (104 chronic diseased) DZ = 296 (173 chronic diseased)	51.5	m	Laboratory (5-minute rest – supine)	RMSSD	36	ADE
					LF	28	AE
					HF	37	ADE
					LF/HF	28	AE
Wang et al., 2009	MZ = 282 DZ = 372 Sibs = 81	17.8	m/f	Laboratory (15-minute rest – supine)	RMSSD	48	AE
				Laboratory (mental stress)	RMSSD	58	AE
				Laboratory (reactivity)	RMSSD	18	AE

(Continued)

Table 17.2 (Continued)

Reference	Number of subjects	Age	Sex	Protocol	Measure	Heritability (%)	Best fitting model
				Laboratory (15 minute rest – supine)	HF	50	AE
				Laboratory (mental stress)	HF	58	AE
				Laboratory (reactivity)	HF	49	AE
Su et al., 2010	198 MZ = 121 DZ = 77	55.0	m	Ambulatory 24-hour recording, physical activity restricted to light walking	TP	63	AE
					ULF	59	AE
					VLF	57	AE
					LF	43	AE
					HF	56	AE
Tuvblad et al., 2010	MZ = 482 DZ = 361	9.6	m/f	Laboratory (3-minute rest – quiet sitting)	pvRSA	39	AE

MZ, monozygotic; DZ, dizygotic, Sibs, singleton siblings of the twins added in an extended twin design.

C. Heritability of Measures of Sympathetic Nervous System Activity

Table 17.3 summarizes the twin studies that have estimated the heritability of (putative) measures of SNS activity. The studies on plasma and urinary norepinephrine levels reported heritability estimates of 42 to 70 percent for plasma and 46 to 76 percent for urinary levels. For epinephrine, heritability estimates of 64 to 74 percent were found for plasma and 65 to 68 percent for urinary levels (Jedrusik et al., 2004; Rao et al., 2007*b*; Rao et al., 2007*a*; Rao et al., 2010; Williams, Puddey, Beilin, & Vandongen, 1993; Zhang et al., 2006; Zhang et al., 2009; Zhang et al., 2010*b*; Zhang et al., 2004).

For electrodermal lability, heritability estimates between 30 and 43 percent have been found (Crider, Kremen, Xian, Jacobson, Waterman, Eisen, Tsuang, Lyons, 2004; Lykken, Iacono, Haroian, McGue, & Bouchard, 1988), and, for skin conductance level, heritability was 26 percent in males and 34 percent in females (Tuvblad et al., 2010).

We could not find twin studies addressing the ejection fraction, but one study used the sibling correlation in echocardiographic recordings to estimate heritability. This suggested that 40 percent of the variance in contractility was due to genetic factors (Fox et al., 2010). In our own studies in the Netherlands Twin Registry, we used the PEP to index cardiac contractility (de Geus et al., 2007; Kupper, Willemsen, Boomsma, & de Geus, 2006). We found substantial heritability of the PEP in laboratory and ambulatory settings, which varied from 48 percent in nighttime recordings to 74 percent during mental stress. In an adolescent sample, PEP reactivity to a mental stressor was also heritable (54 percent). This converges with the finding of significant genetic contribution to the stress reactivity of heart rate and blood pressure, the variables that strongly depend on cardiac SNS reactivity. Mental stress-induced increases in heart rate and systolic and diastolic blood pressure were found to be 26 percent and 43 percent (Wu, Snieder, & de Geus, 2010), respectively.

In the middle-aged groups, but not in the adolescents, the DOS correlation for PEP was lower than the correlation in same-sex DZ pairs (de Geus et al.,

Table 17.3 Studies reporting heritability of measures of sympathetic nervous system activity.

Authors	Number of subjects	Age	Sex	Protocol	Measure	Heritability (%)	Best fitting model
Williams et al., 2003	MZ + DZ = 196	17–65	m/f	Venipuncture in supine rest	Plasma [NE]	57	AE
					Plasma [E]	74/64[a]	AE
Crider et al., 2004	MZ = 325 DZ = 291	47.8	M	4 + 3 minutes rest – quiet sitting	nsSCR	43	AE
Jedrusik et al., 2004	MZ = 78 DZ = 74	34.5	m/f	Venipuncture in rest	Plasma [NE]	42	AE
					Urinary [NE]	76	AE
					Plasma [E]	69	AE
					Urinary [E]	65	AE
O' Connor group (data republished in multiple papers; here Rao et al. 2006 was used)	MZ = 238 DZ = 106	15–84	m/f	Venipuncture in rest	Plasma [NE]	70	AE
					Urinary [NE]	46	AE
					Plasma [E]	67	AE
					Urinary [E]	68	AE
Kupper et al., 2006	MZ = 218 DZ = 301 Sibs = 253	31	m/f	24 hour recording, analysis restricted to sitting activities & sleep	PEP PEP PEP PEP	62 (morning) 62 (afternoon) 55 (evening) 48 (sleep)	AE AE AE AE
De Geus et al., 2007	MZ = 140 DZ = 180	16.7	m/f	Laboratory (8 minute rest – quiet sitting) Laboratory (mental stress) Laboratory (reactivity)	PEP PEP PEP	70 74 54	AE AE AE
De Geus et al., 2007	MZ = 182 DZ = 234	44.2	m/f	Laboratory (3 minute rest – quiet sitting) Laboratory (mental stress) Laboratory (reactivity)	PEP PEP PEP	64 56 Ns	AE AE AE
Tuvblad et al., 2010	MZ = 512 DZ = 484	9.6	m/f	Laboratory (3 minute rest – quiet sitting)	SCL	26/34[a]	AE
					nsSCR	30	AE

NE, norepinephrine; E, epinephrine; MZ, monozygotic; DZ, dizygotic; Sibs, singleton siblings of the twins added in an extended twin design.

[a]Different estimate in males/females

2007; Kupper et al., 2006). This indicates that different genes play a role in individual differences in cardiac sympathetic activity in adult men and women. The most likely explanation for this sex difference is an interaction between adrenoceptor signaling and the male and female sex hormones. Several studies have shown the presence of such interaction. Testosterone regulates gene expression of the major calcium regulatory proteins in isolated ventricular myocytes (Golden, Marsh, & Jiang, 2004; Golden, Marsh, Jiang, & Moulden, 2005). A role for female sex hormones is supported by several studies showing that estrogen inhibits β_1- adrenergic receptor activation on the heart (Thawornkaiwong, Preawni, & Wattanapermpool, 2003).

D. Which Genetic Variants Cause the Heritability of SNS and PNS Activity?

Although twin studies show that individual differences in basal ANS activity and its responses to stress are significantly influenced by genetic factors, these factors are modeled as latent factors, and the actual gene networks that harbor the DNA variants underlying the heritability of the ANS are left unspecified. Identification of these genes and their functional variants is an important next step because they could help elucidate the biological pathways through which ANS activity contributes to CVD risk.

One of the main gene finding strategies is a candidate gene association study. Such a study tests whether a particular variant in a candidate gene and a trait co-occur above chance level, given the frequency of the variant and the distribution of the trait in the population (McCaffery, Snieder, Dong, & de Geus, 2007). The selection of genes is often based on the known biological pathways involved in the trait of interest. For the PNS, genes involved in biosynthesis, transport, and breakdown of ACh seem obvious candidates, as are the genes for the muscarinic receptors. For the SNS, likely candidates are genes controlling catecholamine synthesis and metabolism, neuronal norepinephrine reuptake, and adrenergic receptor function. The variants to be typed within the candidate gene—for example, single nucleotide polymorphisms (SNPs), repeat polymorphisms, or insertion/deletion polymorphisms—are prioritized by their location within coding, promoter, or splice regions, or, if known, their functional effects on the gene product or on gene expression.

Table 17.4 presents a sample of the genetic variants that have been found, at least once, to be significantly associated with an ANS measure in human subjects (Baccarelli et al., 2008; Boccardi et al., 2010; Busjahn et al., 1998; Ellis, Beevers, Hixon, & McGeary, 2011; Kurnik et al., 2007; Masuo et al., 2005; Matsunaga et al., 2005; Matsunaga et al., 2007; Matsunaga et al., 2009; Matsunaga et al., 2010; Milan et al., 2005; Neumann et al., 2005; Neumann et al., 2006; Neumann et al., 2009; Neumeister et al., 2005; Newton-Cheh et al., 2007; Nishikino et al., 2006; Park et al., 2006; Probst-Hensch et al., 2008; Rao et al., 2008; Shihara et al., 1999; Shihara et al., 2001; Stolarz et al., 2004; Suzuki et al., 2003; Thayer et al., 2003; Tingley et al., 2007; Viola, James, Archer, & Dijk, 2008; Yang et al., 2010; Yasuda et al., 2004; Ylitalo et al., 2000; Zhang et al., 2010*a*). Such a list is, by its nature, unfinished, because ongoing progress in the understanding of the biology of the ANS will increase the list, whereas failure to replicate typically does not lead to removal of the gene as a "candidate" (because it still meets the criterion "associated at least once"). If the more rigorous demands were made that (1) an association needs to be detected in a sample that was a priori sufficiently powered and (2) the association needs to be replicated exactly (same SNP, same variant, same direction of effect, same ANS variable) in independent samples (Sullivan, 2007), then none of the candidates on the list would survive. This is probably too rigid, and at least two genes are very likely to be truly involved in ANS activity based on experimental confirmation in animal studies; namely, the genes encoding the G-protein-regulated phosducin and the dual-specific A kinase-anchoring protein 2 (Tingley et al., 2007; Beetz et al., 2009, respectively).

Mice with a targeted deletion of Pdc displayed elevated catecholamine turnover, and their postganglionic sympathetic neurons showed prolonged action potential firing after stimulation with ACh and increased firing frequencies during membrane depolarization (Beetz et al., 2009). Interestingly, Pdc –/– mice displayed exaggerated increases in blood pressure in response to stress, and, in two different human populations, several SNPs in the *PDC* gene were associated with stress-induced blood pressure reactivity. These findings suggest that phosducin is an important modulator of sympathetic activity. For the PNS, subjects with a variant in the dual-specific A kinase-anchoring protein 2 (*AKAP10*) gene were found to have lower SDNN in a small patient sample (Tingley et al., 2007). In a much larger sample of healthy middle-aged adults, the same variant was again found to be associated with SDNN and also with RMSSD.

Table 17.4 Studies reporting association of candidate genes with measures of autonomic nervous system activity.

Reference	ANS measure	Genetic variant	Finding	Remarks
Busjahn et al., 1998	HF RMSSD SDNN	Angiotensin-converting enzyme (*ACE*)	The DD variant of an insertion deletion polymorphism was associated with increased RSA at rest.	Association opposite to that found in African Americans by Thayer et al.
Ylitalo et al., 2000	BRS	Aldosterone synthase gene (*CYP11B2*)	BRS became incrementally lower with more C-alleles in the promoter C-344T SNP.	Findings were stronger in women than in men.
Shihara et al., 2001	VLF	Uncoupling protein 1 (*UCP1*) β3 adrenergic receptor (*β3AR*)	GG homozygotes of the *UCP1* promoter A-3826G SNP had lower VLF at supine rest.	The inhibitory effect of *UCP1* on VLF was observed only with occurrence of the Trp64Arg variant of the *β3AR* gene.
Suzuki et al., 2003	LF VLF	α-adrenergic receptor type 2B (*ADRA2B*)	Short/Short homozygotes of a three-amino acid deletion polymorphism had significantly greater LF and VLF than Long/Long homozygotes.	
Thayer et al., 2003	HF	Angiotensin-converting enzyme (*ACE*)	The II variant of an insertion deletion polymorphism was associated with increased RSA at rest.	Association opposite to that found in European Americans by Busjahn et al.
Stolarz et al., 2004	LF/HF LF HF	Aldosterone synthase (*CYP11B2*)	Supine LF and LF/HF increased whereas HF decreased with the number of T-alleles at the *CYP11B2* C-344T promoter SNP.	Association only seen in subjects with sodium excretion > 190 mmol.
Stolarz et al., 2004	LF/HF LF HF	Type-1-angiotensin receptor (*AT1R*)	Orthostatic changes in LF, HF, and LF/HF were blunted in C-allele carriers of the *AT1R* A1166C SNP.	Association only seen in subjects with sodium excretion > 190 mmol.
Yasuda et al., 2004	HF	α-Subunit Gs-protein (*GNAS1*)	T-allele carriers at the *GNAS1* T393C SNP had lower supine HF, but not standing HF.	
Zhang et al., 2004	Plasma [E] Urinary [NE]	Tyrosine Hydroxylase (*TH*)	Increased number of the (TCAT)10i allele of a repeat polymorphism was associated with higher levels of plasma E and urinary NE.	

(*Continued*)

Table 17.4 (Continued)

Reference	ANS measure	Genetic variant	Finding	Remarks
Matsunaga et al., 2005	LF/HF	G-Protein β_3 Subunit (*GNB3*)	CC homozygotes for the *GNB3* C825T SNP had lower LF/HF and higher HF/TP ratio's than CT/TT genotypes while standing.	
Milan et al., 2005	BRS	Bradykinin β_2 receptor gene(*B2R*)	The BRS increased with the number of T-alleles at the *B2R* C258T SNP.	B2R genotype was a strong independent predictor of BRS, accounting for 12% of its variation.
Neumann et al., 2005, 2006.	HF LF LF/HF	Acetylcholine transporter (*CHT1, SLC5A7*)	G-allele homozygotes for a *CHT1* non-coding 3′ UTR SNP (rs333229) had lower HF power and higher LF power and higher LF/HF ratio.	Same SNP is also associated to depressed mood and brain activity in regions of the "central autonomic network."
Neumeister et al., 2005	Total body NE spillover	α-Adrenergic receptor type 2C (*ADRA2C*)	At rest, homozygotes for the Del322–325 polymorphism had higher total body NE spillover than heterozygotes or noncarriers.	The same In-frame deletion of *ADRA2C* increases the risk of congestive heart failure.
Kurnik et al., 2006	plasma [NE]	α-Adrenergic receptor type 2A (*ADRA2A*)	Two uncommon variants (G>C at -1903 and C>G at -1607, identified in 3 black subjects) were associated with higher plasma NE levels.	
Nishikino et al., 2006	LF	Type-1-angiotensin receptor (*AT1R*)	Higher LF in C-allele carriers of the *AT1R* A1166C SNP.	
Park et al., 2006	HF	Hemochromatosis gene (*HFE*)	C282Y(rs1800562) and H63D (rs1799945) major alleles were associated with lower HF.	Association was conditional on high level of particulate air pollution.
Matsunaga et al., 2007	HF LF/HF LF	β-Adrenergic receptor type 1 (*ADRB1*) β-Adrenergic receptor type 2 (*ADRB2*)	At supine rest, *ADRB1* Arg16 homozygotes had lower LF/HF and higher HF than the Gly16 allele carriers. *ADRB2* Glu27 allele carriers had higher LF than Gln27 homozygotes.	
Newton-Cheh et al., 2007	SDNN	α-Adrenergic receptor type 1A (*ADRA1A*)	6 SNPs in the gene reached significance at classical p <.05	As the SNPs were part of a genome-wide association scan, the significance level may have been too liberal.

(Continued)

Table 17.4 (Continued)

Reference	ANS measure	Genetic variant	Finding	Remarks
	SDNN/TP	α-Adrenergic receptor type 1B (*ADRA1B*)	1 SNP(rs2195926) in the gene reached significance at classical p <.05	As the SNP was part of a genome-wide association scan, the significance level may have been too liberal.
	TP	α-Adrenergic receptor type 2B (*ADRA2B*)	1 SNP(rs9325124) in the gene reached significance at classical p <.05	As the SNP was part of a genome-wide association scan, the significance level may have been too liberal.
Tingley et al., 2007	SDNN	α-Kinase anchoring protein 10 (*AKAP10*)	Homozygotes for the G-allele (Val) at the *AKAP10* I646V SNP (rs203462) have lower SDNN.	In mice *AKAP10* was shown to modulate the sensitivity of cardiac cells to cholinergic stimulation.
Baccarelli et al., 2008	SDNN	Methylene-tetrahydrofolate reductase (*MTHFR*)	T-allele carriers of the *MTHFR* C677T SNP (rs1801133) had lower SDNN than CC homozygotes.	
Rao et al., 2008	plasma [NE]	Tyrosine hydroxylase (*TH*)	TH promoter haplotype #2 (TGGG) increased NE excretion during stress.	
Matsunaga et al., 2009	HF LF/HF	Uncoupling proteins 2 & 3 (*UCP2, UCP3*)	At supine rest the II variant of the UCP2 45-bp insertion/deletion polymorphism was associated higher LF/HF ratio. Carriers of the T-allele of the *UCP3* S55 C/T SNP had lower LF/HF ratio and higher HF than CC homozygotes.	The II variant of the UCP2 insertion/deletion polymorphism was also associated with higher blood pressure.
Neumann et al., 2009	SDNN RMSSD	α-Kinase anchoring protein 10 (*AKAP10*)	G-allele (Val) carriers at the *AKAP10* I646V SNP (rs203462) have lower SDNN and RMSSD.	SDNN exactly replicates Tingley et al., 2007
Rana et al., 2009	plasma [E]	Nicotinic acetylcholine receptor (*CHRNA*)	*CHRNA3* K95K (G) allele (rs3743075) was associated with higher E levels.	
Su et al., 2009	ULF	C-reactive protein (*CRP*)	Subjects homozygous for the T-allele at *CRP* SNP rs1205 had higher ULF than CC homozygotes.	The SNP was significantly associated with both CRP and ULF and explained 11% of their genetic covariance.

(*Continued*)

Table 17.4 (Continued)

Reference	ANS measure	Genetic variant	Finding	Remarks
Viola et al., 2009	SDNN LFnu	Circadian clock gene PERIOD3 (*PER3*)	Homozygotes for the longer allele (5/5) in a variable number tandem repeat polymorphism in the coding region of *PER3* had higher LFnu and lower SDNN compared to 4/4 homozygotes, in particular during baseline sleep.	The *PER3* VNTR polymorphism also affected slow wave sleep duration.
Boccardi et al., 2010		Transcription factor 7-like 2 (*TCF7L2*)	TT homozygotes of a *TCF7L2* G/T SNP (rs12233572) had higher LF/HF ratio during glucose ingestion compared to heterozygotes and CC homozygotes.	No effects of the *TCF7L2* SNP were found on baseline LF/HF.
Ellis et al., 2010	HF	Serotonin transporter (5-HTT) repeat polymorphism	The short variant of the 5-HTTLPR was associated with lower HF at rest.	
Matsunaga et al., 2010	HF LF VLF TP	Estrogen receptor α gene (*ESR1*)	Haplotype analysis based on the PvuII and XbaI polymorphisms showed that *ESR1* haplotype 2 (PvuII C allele and XbaI A allele) had lower TP, VLF, LF, and HF at supine rest.	Carriers of the ESR1 haplotype 2 also had a higher systolic and mean arterial blood pressure.
Probst-Hensch et al., 2010	TP LF	Glutathione S-transferase (*GST*)	Participants missing both copies of the *GSTT1* gene had lower TP and LF.	
Yang et al., 2010	HF RMSSD LF/HF	Brain-derived neurotrophic factor (*BDNF*)	Met/Met homozygotes at the *BDNF* Val66Met polymorphism had lower HF/RMSSD and higher LF/HF ratio compared to the Val/Val homozygotes.	

NE, norepinephrine; E, epinephrine

The importance of this gene for cardiac vagal activity was further confirmed by experimental disruption of the gene in mouse embryonic stem cells that were differentiated into cardiac cells (Tingley et al., 2007). Mutation of mouse *Akap10* increased the sensitivity of cultured cardiac cells to cholinergic signals, and this result was reproduced in living mice.

Apart from these two genes, the yield of the candidate gene approach has been modest. This may reflect a general disadvantage of an approach that capitalizes entirely on existing biological disease models. Many of the hundreds or thousands of genes relevant to the disease may not have been properly annotated, or they may reside in pathways that have not been linked to the disease before. Increasingly, gene finding attempts in large-scale samples have used the more agnostic strategy of a genome-wide association study (GWAS) that makes no a priori assumptions on the biological pathways involved (Manolio & Collins, 2009). Genome-wide association studies exploit the availability of the human genome sequence and the rapid development across the past decades of affordable, high-accuracy, high-throughput technologies to genotype from SNPs scattered across the entire genome. Although a single subject will deviate in around 4 million SNPs from the major variant in the base population, a subset of 500,000 to 1 million well-chosen tagging SNPs can capture the whole genome of that subject. Single-nucleotide polymorphisms that lie in proximity on the same chromosome are often transmitted together across generations as a block of SNP genotypes (also called a *haplotype block*). Recombination among the SNPs in the haplotype block is infrequent, and combining the tag-SNPs with the detailed inventory of the haplotype structure of the world's main populations in the HapMap/1000 genomes projects allows the imputation of the unmeasured SNPs with high precision. The final set of (imputed) SNPs can be used to test genetic association to ANS parameters on a genome-wide scale.

At the time of writing, only one GWAS had been published that focused specifically on the ANS (Newton-Cheh et al., 2007). In the Framingham Heart Study, 548 subjects were typed on the Affymetrix 100K SNP chip. No genome-wide association test results for SDNN/TP nor for the LF/HF ratio met the stern p-value criterion of $5 \times 10^{e-8}$ commonly applied in GWAS to deal with the huge multiple testing burden when half a million SNPs are tested. Taken the current experience of the GWA field, this study was a priori underpowered to detect a significant effect of a single SNP, which will typically contribute no more than 0.1–0.01 percent to the genetic variance in the trait. To detect significant SNP contribution in GWAS, analyses on tens to hundreds of thousands of subjects are needed for which whole-genome SNP data and ANS traits are available. No single research group can mount such numbers, and large GWA consortia have been formed worldwide that exchange their association results in an across-study meta-analysis. These meta-analyses have been hugely successful (http://www.genome.gov/gwastudies/) for many traits, including cardiovascular risk factors (Demirkan et al., 2011; Dehghan et al., 2011; Eijgelsheim et al., 2010; Dupuis et al., 2010; Wain et al., 2011). In 2011, the Genetic Variability in Heart Rate Variability (VgHRV) consortium was set up to perform meta-analyses on RSA and other heart rate variability traits.

IV. Conclusion

A number of valid and reliable measurement strategies can be used to study PNS and SNS activity in samples that are sufficiently large to allow genetic analyses. Respiratory sinus arrhythmia seems the measure of choice to index parasympathetic activity in such large-scale studies and PEP the measure of choice to index sympathetic activity. These measures are noninvasive and can be obtained in controlled laboratory situations, as well as in ambulatory recordings, without large discomfort for the participants.

Twin studies demonstrate significant genetic contributions to RSA (heritability estimates range between 25 and 71 percent) and to a second indicator of cardiac vagal activity, BRS (heritability between 22 and 53 percent). For the SNS, diverse indicators of activity to different organs converge on an even more substantial contribution of genetic factors. Heritability of plasma and urinary norepinephrine varied between 42 and 76 percent, heritability of epinephrine levels between 64 and 74 percent, heritability of nsSCR/SCL between 26 and 43 percent, and heritability of PEP between 48 and 74 percent. For PEP and RSA, the genetic variance in these variables was shown to increase during stress, either because of the emergence of newly expressed genetic variation or through the amplification of existing genetic influences in the resting state.

V. Future Directions

Gene finding is a major direction for future research in this area. Identification of genetic variants and their genomic mechanisms will not only advance our understanding of the biology of the ANS, but it will also allow us to address the causality in the association between ANS activity and CVD. As alluded to earlier, the ANS plays a key role in almost all models in biobehavioral medicine that try to account for the well-known role of social (Karasek et al., 1988; Rosengren et al., 2004; Siegrist et al., 1990) and psychological (Nicholson et al., 2006) sources of chronic stress in hypertension, diabetes, and cardiac disease. Differences in the activity of the ANS between individuals have been robustly associated with differences in most of the physiological risk factors for CVD. The direction of causation of this association, however, remains to be established. The dominant hypothesis is that increases in SNS activity and decreases in PNS activity exert a causal effect on the risk factors (Bayturan et al., 2010; Charkoudian & Rabbitts, 2009; Danesh et al., 2008; Danesh et al., 2004; Esler, 2010; Lambert & Lambert, 2011; Libby & Theroux, 2005; Malpas, 2010). The idea that the ANS has causal effects is supported by arguments of biological plausibility and by prospective epidemiological evidence, but actual proof of causality beyond reasonable doubt is still missing. Two alternative explanations for an association between ANS activity and CVD risk need to be taken into account. The first is that of reverse causality, in which an increase in the CVD risk factors itself causes compensatory changes in ANS activity. The increase in cardiac sympathetic activity in heart failure patients may be one such example (Charkoudian & Rabbitts, 2009). The second alternative is a set of underlying genetic factors that not only influence SNS or PNS activity but also independently influence the processes leading to CVD. This phenomenon, in which low level biological variation has effects on multiple complex traits at the organ level, is called *genetic pleiotropy*. If present in a time-lagged form—that is, if genetic effects on the ANS precede effects of the same genetic variants on CVD—this phenomenon can cause longitudinal correlations that mimic a causal effect of the ANS.

A powerful approach to test causality is *Mendelian randomization*. In this approach, genetic variants that cause variation in ANS activity are identified, after which the effects of these variants on CVD are established. If such effects are significant, they make a strong case for causal effects (e.g., of ANS on CVD) because "randomization" of genotypes during meiosis takes place well before the disease could cause compensatory changes in ANS activity. The reverse is also true: the absence of a genetic association between variants for ANS and CVD falsifies causality because, under the causal model, each factor influencing ANS *must* also influence CVD. The Mendelian randomization approach requires large samples, particularly if the causal effect of ANS on CVD risk is only modest. It also requires knowledge of two sets of genetic variants: the variants that influence ANS, as well as the variants that influence CVD risk factors. The latter set is now identified with increasing success by GWA studies (Demirkan et al., 2011; Dehghan et al., 2011; Dupuis et al., 2010; Eijgelsheim et al., 2010; Wain et al., 2011). Gene finding for ANS has not kept pace so far and should be a high priority.

In addition to being heritable itself, ANS activity can modulate the expression of other genetic risk factors through its influence on regulatory DNA elements that control gene expression. Genes in the proinflammatory response provide a clear example. The proinflammatory response is activated by proinflammatory cytokines tumor necrosis factor (TNF)α and interleukin (IL)-1. TNFα is a key initiator of the IL-6 response that, in turn, stimulates the acute-phase response, reflected in the synthesis of fibrinogen and C-reactive protein (CRP). These two acute-phase reactants are widely accepted biomarkers of CVD risk (Danesh et al., 2004; Kannel, Wolf, Castelli, & D'Agostino, 1987; Ridker, Hennekens, Buring, & Rifai, 2000; Wilhelmsen et al., 1984), and IL-6 itself also strongly indexes disease risk. In 17 population-based prospective studies of IL-6 and clinical coronary outcomes (i.e., infarction or coronary death), a combined odds ratio of 1.61 (95 percent confidence interval [CI] 1.42–1.83) per 2 standard deviations (SD) increase was found for IL-6 (Danesh et al., 2008).

The expression of proinflammatory cytokines like TNFα and IL-6 is regulated by, among others, the regulatory transcription factors nuclear factor-κB (NF-κB) and GATA1. Exogenous and endogenous molecular products of infection and injury interact with membrane-bound receptors on immune cells to induce increased expression of proinflammatory cytokine genes. This, in turn, activates a PNS-controlled anti-inflammatory pathway. Activity of the *n.* ambiguus to the celiac ganglion causes the splenic nerve to release ACh in the spleen which, through a nicotinic receptor (α7nAChR), causes the suppression of NF-κB activation and the inhibition of *IL6* gene expression (Rosas-Ballina & Tracey, 2009; Tracey, 2009). In contrast, SNS activation is a powerful inducer of *IL6* gene expression via β-adrenergic activation of the GATA1 transcription factor (Cole et al., 2010). The effects of SNS activation on IL6 induction depend on a genetic variant in the *IL6* gene, giving rise to a gene-by-SNS interactive effect: G-variant carriers of a SNP (rs1800795) in GATA1-binding motif in the *IL6* promoter region showed much stronger induction of *IL6* than C-variant carriers. IL-6, in short, provides a clear example of how SNS and PNS activity can modulate the expression of genetic variation in a CVD risk factor. A similar modulatory effect of the ANS is likely to be found on many other genetic risk factors and, given its sensitivity to the environment, ANS activity may be a major source of G×E interaction in CVD.

The questions "Which genetic variants have an effect on ANS activity?," "Are the effects of SNS and PNS activity on CVD causal?," and "How does SNS and PNS activity modulate the genetic risk for CVD?" can be fully addressed in the near future using current genomic technologies. Answers to these burning questions will deepen our understanding of the complex interplay between genetic and psychosocial risk factors in the onset and progression of CVD.

References

Akselrod, S., Gordon, D., Ubel, F. A., Shannon, D. C., Berger, A. C., & Cohen, R. J. (1981). Power spectrum analysis of heart rate fluctuation: A quantitative probe of beat-to-beat cardiovascular control. *Science, 213*, 220–222.

Baccarelli, A., Cassano, P. A., Litonjua, A., Park, S. K., Suh, H., Sparrow, D. et al. (2008). Cardiac autonomic dysfunction: Effects from particulate air pollution and protection by dietary methyl nutrients and metabolic polymorphisms. *Circulation, 117*, 1802–1809.

Barnett, P. A., Spence, J. D., Manuck, S. B., & Jennings, J. R. (1997). Psychological stress and the progression of carotid artery disease. *Journal of Hypertension, 15*, 49–55.

Bayturan, O., Tuzcu, E. M., Lavoie, A., Hu, T., Wolski, K., Schoenhagen, P. et al. (2010). The metabolic syndrome, its component risk factors, and progression of coronary atherosclerosis. *Archives of Internal Medicine, 170*, 478–484.

Beetz, N., Harrison, M. D., Brede, M., Zong, X., Urbanski, M. J., Sietmann, A. et al. (2009). Phosducin influences sympathetic activity and prevents stress-induced hypertension in humans and mice. *Journal of Clinical Investigation, 119*, 3597–3612.

Berntson, G. G., Bigger, J. T., Jr., Eckberg, D. L., Grossman, P., Kaufmann, P. G., Malik, M. et al. (1997). Heart rate variability: Origins, methods, and interpretive caveats. *Psychophysiology, 34*, 623–648.

Berntson, G. G., Cacioppo, J. T., Binkley, P. F., Uchino, B. N., Quigley, K. S., & Fieldstone, A. (1994*a*). Autonomic cardiac control. III. Psychological stress and cardiac response in autonomic space as revealed by pharmacological blockades. *Psychophysiology, 31*, 599–608.

Berntson, G. G., Cacioppo, J. T., & Quigley, K. S. (1991). Autonomic determinism: The modes of autonomic control, the doctrine of autonomic space, and the laws of autonomic constraint. *Psychological Reviews, 98*, 459–487.

Berntson, G. G., Cacioppo, J. T., & Quigley, K. S. (1993). Respiratory sinus arrhythmia: Autonomic origins, physiological mechanisms, and psychophysiological implications. *Psychophysiology, 30*, 183–196.

Berntson, G. G., Cacioppo, J. T., & Quigley, K. S. (1994*b*). Autonomic cardiac control. I. Estimation and validation from pharmacological blockades. *Psychophysiology, 31*, 572–585.

Berntson, G. G., Norman, G. J., Hawkley, L. C., & Cacioppo, J. T. (2008). Cardiac autonomic balance versus cardiac regulatory capacity. *Psychophysiology, 45*, 643–652.

Bigger, J. T., Jr., Fleiss, J. L., Rolnitzky, L. M., & Steinman, R. C. (1993). Frequency domain measures of heart period variability to assess risk late after myocardial infarction. *Journal of the American College of Cardiology, 21*, 729–736.

Bigger, J. T., Jr., Fleiss, J. L., Steinman, R. C., Rolnitzky, L. M., Kleiger, R. E., & Rottman, J. N. (1992). Correlations among time and frequency domain measures of heart period variability two weeks after acute myocardial infarction. *American Journal of Cardiology, 69*, 891–898.

Bigger, J. T., Hoover, A. C., Steinman, R. C., Rolnitzky, L. M., Fleiss, J. L., & Multicenter Study of Silent Myocardial ischemia investigators (1990). Autonomic nervous system activity during myocardial ischemia in man estimated by power spectral analysis of heart period variability. *American Journal of Cardiology, August 15*, 497–498.

Boccardi, V., Ambrosino, I., Papa, M., Fiore, D., Rizzo, M. R., Paolisso, G. et al. (2010). Potential role of TCF7L2 gene variants on cardiac sympathetic/parasympathetic activity. *European Journal of Human Genetics, 18*, 1333–1338.

Bosch, J. A., de Geus, E. E., Ring, C., Nieuw Amerongen, A. V., & Stowell, J. R. (2004). Academic examinations and immunity: Academic stress or examination stress? *Psychosomatic Medicine, 66*, 625–626.

Bosch, J. A., de Geus, E. J. C., Veerman, E. C. I., & Amerongen, A. V. N. (2000). Innate secretory immunity in response to laboratory stressors that evoke distinct patterns of cardiac autonomic activity. *Psychosomatic Medicine, 62*, 1276.

Bosch, J. A., Turkenburg, M., Nazmi, K., Veerman, E. C., de Geus, E. J., & Nieuw Amerongen, A. V. (2003). Stress as a determinant of saliva-mediated adherence and coadherence of oral and nonoral microorganisms. *Psychosomatic Medicine, 65*, 604–612.

Bosch, J. A., Veerman, E. C., de Geus, E. J., & Proctor, G. B. (2011). alpha-Amylase as a reliable and convenient measure of sympathetic activity: Don't start salivating just yet! *Psychoneuroendocrinology, 36*, 449–453.

Boucsein, W. (1992). *Electrodermal Activity*. New York: Plenum Press.

Brotman, D. J., Golden, S. H., & Wittstein, I. S. (2007). The cardiovascular toll of stress. *Lancet, 370*, 1089–1100.

Brotman, D. J., Walker, E., Lauer, M. S., & O'Brien, R. G. (2005). In search of fewer independent risk factors. *Archives of Internal Medicine, 165*, 138–145.

Burleson, M. H., Poehlmann, K. M., Hawkley, L. C., Ernst, J. M., Berntson, G. G., Malarkey, W. B. et al. (2003). Neuroendocrine and cardiovascular reactivity to stress in mid-aged and older women: Long-term temporal consistency of individual differences. *Psychophysiology, 40*, 358–369.

Burr, R. L. (2007). Interpretation of normalized spectral heart rate variability indices in sleep research: A critical review. *Sleep, 30*, 913–919.

Busjahn, A., Voss, A., Knoblauch, H., Knoblauch, M., Jeschke, E., Wessel, N. et al. (1998). Angiotensin-converting enzyme and angiotensinogen gene polymorphisms and heart rate variability in twins. *American Journal of Cardiology, 81*, 755–760.

Cacioppo, J. T., Berntson, G. G., Binkley, P. F., Quigley, K. S., Uchino, B. N., & Fieldstone, A. (1994). Autonomic cardiac control. II. Noninvasive indices and basal response as revealed by autonomic blockades. *Psychophysiology, 31*, 586–598.

Carroll, D., Smith, G. D., Willemsen, G., Sheffield, D., Sweetnam, P. M., Gallacher, J. E. J. et al. (1998). Blood pressure reactions to the cold pressor test and the prediction of ischaemic heart disease: Data from the Caerphilly Study. *Journal of Epidemiology and Community Health, 52*, 528–529.

Cerati, D., & Schwartz, P. J. (1991). Single cardiac vagal fiber activity, acute myocardial ischemia, and risk for sudden death. *Circulation Research, 69*, 1389–1401.

Cerutti, S., Bianchi, A. M., & Mainardi, L. T. (2001). Advanced spectral methods for detecting dynamic behaviour. *Autonomic Neuroscience, 90*, 3–12.

Charkoudian, N., & Rabbitts, J. A. (2009). Sympathetic neural mechanisms in human cardiovascular health and disease. *Mayo Clinic Proceedings, 84*, 822–830.

Chatterton, R. T., Jr., Vogelsong, K. M., Lu, Y. C., Ellman, A. B., & Hudgens, G. A. (1996). Salivary alpha-amylase as a measure of endogenous adrenergic activity. *Clinical Physiology, 16*, 433–448.

Cole, S. W., Arevalo, J. M., Takahashi, R., Sloan, E. K., Lutgendorf, S. K., Sood, A. K. et al. (2010). Computational identification of gene-social environment interaction at the human IL6 locus. *Proceedings of the National Academy of Sciences of the United States of America, 107*, 5681–5686.

Coresh, J., Klag, M. J., Mead, L. A., Liang, K. Y., & Whelton, P. K. (1992). Vascular Reactivity in Young-Adults and Cardiovascular-Disease – A Prospective-Study. *Hypertension, 19*, 218–223.

Crider, A,, Kremen, W.S., Xian, H., Jacobson, K.C., Waterman, B., Eisen, S.A., Tsuang, M.T., Lyons, M.J. (2004). Stability, consistency, and heritability of electrodermal response lability in middle-aged male twins. Psychophysiology, *41(4)*, 501–509

Critchley, H. D. (2002). Electrodermal Responses: What Happens in the Brain. *The Neuroscientist, 8*, 132–142.

Cybulski, G. (2000). Ambulatory impedance cardiography: New possibilities. *Journal of Applied Physiology, 88*, 1509–1510.

Danesh, J., Kaptoge, S., Mann, A. G., Sarwar, N., Wood, A., Angleman, S. B. et al. (2008). Long-term interleukin-6 levels and subsequent risk of coronary heart disease: Two new prospective studies and a systematic review. *PLoS Medicine, 5*, e78.

Danesh, J., Wheeler, J. G., Hirschfield, G. M., Eda, S., Eiriksdottir, G., Rumley, A. et al. (2004). C-reactive protein and other circulating markers of inflammation in the prediction of coronary heart disease. *The New England Journal of Medicine, 350*, 1387–1397.

Dawson, M. E., Schell, A. M., & Filion, D. L. (2000). The Electrodermal System. In J. T. Cacioppo, L. G. Tassinary, & G. G. Berntson (Eds.), *Handbook of Psychophysiology* (pp. 200–223). New York: Cambridge University Press.

de Geus, E. J., Willemsen, G. H., Klaver, C. H., & van Doornen, L. J. (1995). Ambulatory measurement of respiratory sinus arrhythmia and respiration rate. *Biological Psychology, 41*, 205–227.

de Geus, E. J. C., Kupper, N., Boomsma, D. I., & Snieder, H. (2007). Bivariate Genetic Modeling of Cardiovascular Stress Reactivity: Does Stress Uncover Genetic Variance? *Psychosomatic Medicine, 69*, 356–364.

Dehghan, A., Dupuis, J., Barbalic, M., Bis, J. C., Eiriksdottir, G., Lu, C. et al. (2011). Meta-analysis of genome-wide association studies in >80 000 subjects identifies multiple loci for C-reactive protein levels. *Circulation, 123*, 731–738.

Dekker, J. M., Crow, R. S., Folsom, A. R., Hannan, P. J., Liao, D., Swenne, C. A. et al. (2000). Low Heart Rate Variability in a 2-Minute Rhythm Strip Predicts Risk of Coronary Heart Disease and Mortality From Several Causes: The ARIC Study. *Circulation, 102*, 1239–1244.

Dekker, J. M., Schouten, E. G., Klootwijk, P., Pool, J., Swenne, C. A., & Kromhout, D. (1997). Heart rate variability from short electrocardiographic recordings predicts mortality from all causes in middle-aged and elderly men. The Zutphen Study. *American Journal of Epidemiology, 145*, 899–908.

Demirkan, A., Amin, N., Isaacs, A., Jarvelin, M. R., Whitfield, J. B., Wichmann, H. E. et al. (2011). Genetic architecture of circulating lipid levels. *European Journal of Human Genetics, 19*, 813–819.

DiRienzo M., Parati, G., Radaelli, A., & Castiglioni, P. (2009). Baroreflex contribution to blood pressure and heart rate oscillations: Time scales, time-variant characteristics and nonlinearities. *Philosophical Transactions. Series A, Mathematical, Physical, and Engineering Sciences, 367*, 1301–1318.

Dubreuil, E., Ditto, B., Dionne, G., Pihl, R. O., Tremblay, R. E., Boivin, M. et al. (2003). Familiality of heart rate and cardiac-related autonomic activity in five-month-old twins: The Quebec newborn twins study. *Psychophysiology, 40*, 849–862.

Dupuis, J., Langenberg, C., Prokopenko, I., Saxena, R., Soranzo, N., Jackson, A. U. et al. (2010). New genetic loci implicated in fasting glucose homeostasis and their impact on type 2 diabetes risk. *Nature Genetics, 42*, 105–116.

Eckberg, D. L. (2003). The human respiratory gate. *Journal of Physiology, 548*, 339–352.

Eckberg, D. L. (1997). Sympathovagal Balance: A Critical Appraisal. *Circulation, 96*, 3224–3232.

Eijgelsheim, M., Newton-Cheh, C., Sotoodehnia, N., de Bakker, P. I., Muller, M., Morrison, A. C. et al. (2010). Genome-wide association analysis identifies multiple loci related to resting heart rate. *Human Molecular Genetics, 19*, 3885–3894.

Eisenhofer, G. (2005). Sympathetic nerve function--assessment by radioisotope dilution analysis. *Clinical Autonomic Research, 15*, 264–283.

Eisenhofer, G., Kopin, I. J., & Goldstein, D. S. (2004). Catecholamine metabolism: A contemporary view with implications for physiology and medicine. *Pharmacological Reviews, 56*, 331–349.

Ellis, A. J., Beevers, C. G., Hixon, J. G., & McGeary, J. E. (2011). Serotonin transporter promoter region (5-HTTLPR) polymorphism predicts resting respiratory sinus arrhythmia. *Psychophysiology, 48*, 923–926.

Esler, M. (2010). The 2009 Carl Ludwig Lecture: Pathophysiology of the human sympathetic nervous system in cardiovascular diseases: The transition from mechanisms to medical management. *Journal of Applied Physiology, 108*, 227–237.

Esler, M., Eikelis, N., Schlaich, M., Lambert, G., Alvarenga, M., Dawood, T. et al. (2008). Chronic mental stress is a cause of essential hypertension: Presence of biological markers of stress. *Clinical and Experimental Pharmacology and Physiology, 35*, 498–502.

Esler, M., Jennings, G., Korner, P., Willett, I., Dudley, F., Hasking, G. et al. (1988). Assessment of human sympathetic nervous system activity from measurements of norepinephrine turnover. *Hypertension, 11*, 3–20.

Esler, M., Jennings, G., Lambert, G., Meredith, I., Horne, M., & Eisenhofer, G. (1990). Overflow of Catecholamine Neurotransmitters to the Circulation – Source, Fate, and Functions. *Physiological Reviews, 70*, 963–985.

Esler, M., & Kaye, D. (2000). Measurement of sympathetic nervous system activity in heart failure: The role of norepinephrine kinetics. *Heart Failure Review, 5*, 17–25.

Fahrenberg, J., Myrtek, M., Pawlik, K., & Perrez, M. (2007). Ambulatory assessment-monitoring behavior in daily life settings – A Behavioral-Scientific challenge for psychology. *European Journal of Psychological Assessment, 23*, 206–213.

Falconer, D. S., & Mackay, T. F. C. (1996). *Introduction to quantitative genetics*. (4 ed.) Essex: Pearson Education Limited.

Foster, K. G., & Weiner, J. S. (1970). Effects of Cholinergic and Adrenergic Blocking Agents on Activity of Eccrine Sweat Glands. *Journal of Physiology-London, 210*, 883–895.

Fowles, D. C. (1986). The Eccrine System and Electrodermal Activity. In M. G. H. Coles, E. Donchin, & S. W. Porges (Eds.), *Psychophysiology: Systems Processes and Applications*. (pp. 51–96). New York: The Guilford Press.

Fox, E. R., Klos, K. L., Penman, A. D., Blair, G. J., Blossom, B. D., Arnett, D. et al. (2010). Heritability and genetic linkage of left ventricular mass, systolic and diastolic function in hypertensive African Americans (From the GENOA Study). *American Journal of Hypertension, 23*, 870–875.

Fox, K., Borer, J. S., Camm, A. J., Danchin, N., Ferrari, R., Sendon, J. L. L. et al. (2007). Resting heart rate in cardiovascular disease. *Journal of the American College of Cardiology, 50*, 823–830.

Freixa i Baque, E. (1982). Reliability of electrodermal measures: A compilation. *Biological Psychology, 14*, 219–229.

Frey, M. A. B., & Kenney, R. A. (1979). Systolic-Time Intervals During Combined Hand Cooling and Head-Up Tilt. *Aviation Space and Environmental Medicine, 50*, 218–222.

Galton, F. R. S. (1869). *Hereditary genius*. London: MacMillan & Co.

Gerin, W., Rosofsky, M., Pieper, C., & Pickering, T. G. (1994). A test of generalizability of cardiovascular reactivity using a controlled ambulatory procedure. *Psychosomatic Medicine, 56*, 360–368.

Goedhart, A. D., Kupper, N., Willemsen, G., Boomsma, D. I., & de Geus, E. J. (2006). Temporal stability of ambulatory stroke volume and cardiac output measured by impedance cardiography. *Biological Psychology, 72*, 110–117.

Goedhart, A. D., van der Sluis, S. S., Houtveen, J. H., Willemsen, G., & de Geus, E. J. (2007). Comparison of time and frequency domain measures of RSA in ambulatory recordings. *Psychophysiology, 44*, 203–215.

Goedhart, A. D., Willemsen, G., Houtveen, J. H., Boomsma, D. I., & de Geus, E. J. (2008). Comparing low frequency heart rate variability and preejection period: Two sides of a different coin. *Psychophysiology, 45*, 1086–1090.

Golden, K. L., Marsh, J. D., & Jiang, Y. (2004). Testosterone regulates mRNA levels of calcium regulatory proteins in cardiac myocytes. *Hormone and Metabolic Research, 36*, 197–202.

Golden, K. L., Marsh, J. D., Jiang, Y., & Moulden, J. (2005). Acute actions of testosterone on contractile function of isolated rat ventricular myocytes. *European Journal of Endocrinology, 152*, 479–483.

Goldstein, D. S., Eisenhofer, G., & Kopin, I. J. (2003). Sources and significance of plasma levels of catechols and their metabolites in humans. *Journal of Pharmacology and Experimental Therapeutics, 305*, 800–811.

Grassi, G., & Esler, M. (1999). How to assess sympathetic activity in humans. *Journal of Hypertension, 17*, 719–734.

Grossman, P., & Kollai, M. (1993). Respiratory sinus arrhythmia, cardiac vagal tone, and respiration: Within- and between-individual relations. *Psychophysiology, 30*, 486–495.

Grossman, P., van Beek, J., & Wientjes, C. (1990). A comparison of three quantification methods for estimation of respiratory sinus arrhythmia. *Psychophysiology, 27*, 702–714.

Grossman, P., Wilhelm, F. H., & Spoerle, M. (2004). Respiratory sinus arrhythmia, cardiac vagal control, and daily activity. *AJP – Heart and Circulatory Physiology, 287*, H728–H734.

Hagbarth, K. E., Hallin, R. G., Wallin, B. G., Torebjork, H., & Hongell, A. (1972). General Characteristics of Sympathetic Activity in Human Skin Nerves. *Acta Physiologica Scandinavica, 84*, 164–&.

Hagbarth, K. E., & Vallbo, A. B. (1968). Pulse and respiratory grouping of sympathetic impulses in human muscle-nerves. *Acta Physiologica Scandinavica, 74*, 96–108.

Harris, W. S., Schoenfeld, C. D., & Weissler, A. M. (1967). Effects of adrenergic receptor activation and blockade on the systolic preejection period, heart rate, and arterial pressure in man. *Journal of Clinical Investigation, 46*, 1704–1714.

Hayano, J., Skakibara, Y., Yamada, A., Yamada, M., Mukai, S., Fujinami, T. et al. (1991*a*). Accuracy of assessment of cardiac vagal tone by heart rate variability in normal subjects. *American Journal of Cardiology, 67*, 199–204.

Hayano, J., Yamada, A., Mukai, S., Skakibara, Y., Yamada, M., Ohte, N. et al. (1991*b*). Severity of coronary atherosclerosis correlates with the respiratory component of heart rate variability. *American Heart Journal, 121*, 1070–1079.

Hjemdahl, P. (1990). Physiology of the autonomic nervous system as related to cardio vascular function: Implications for stress research. In D. G. Byrne & R. H. Roseman (Eds.), *Anxiety and the heart* (pp. 95–158). New York: Hemisphere Publishing.

Hjemdahl, P., Freyschuss, U., Juhlin-Dannfelt, A., & Linde, B. (1984). differentiated sympathetic activation during mental stress evoked by the Stroop test. *Acta Physiologica Scandinavica, 527*, 25–29.

Hoekstra, C., Meijer, P., Kluft, C., Heutink, P., Smit, G., de Geus, E. et al. (2004). Genetics of dizygotic twinning: A feasibility study for a biobank. *Twin Research and Human Genetics, 7*, 556–563.

Hohnloser, S. H., Klingenheben, T., Zabel, M., Schroder, F., & Just, H. (1992). Intraindividual reproducibility of heart rate variability. *Pacing and Clinical Electrophysiology, 15*, 2211–2214.

Houtveen, J. H., & de Geus E. J. C. (2009). Noninvasive psychophysiological ambulatory recordings: Study design and data analysis strategies. *European Psychologist, 12(2)*, 132–141.

Houtveen, J. H., Groot, P. F., & de Geus, E. J. (2005). Effects of variation in posture and respiration on RSA and preejection period. *Psychophysiology, 42*, 713–719.

Houtveen, J. H., & Molenaar, P. C. (2001). Comparison between the Fourier and Wavelet methods of spectral analysis applied to stationary and nonstationary heart period data. *Psychophysiology, 38*, 729–735.

Houtveen, J. H., Rietveld, S., & de Geus, E. J. (2002). Contribution of tonic vagal modulation of heart rate, central respiratory drive, respiratory depth, and respiratory frequency to respiratory sinus arrhythmia during mental stress and physical exercise. *Psychophysiology, 39*, 427–436.

Humphrey, S. P., & Williamson, R. T. (2001). A review of saliva: Normal composition, flow, and function. *Journal of Prosthetic Dentistry, 85*, 162–169.

Iacono, W. G., Peloquin, L. J., Lykken, D. T., Haroian, K. P., Valentine, R. H., & Tuason, V. B. (1984). Electrodermal Activity in Euthymic Patients with Affective-Disorders – One-Year Retest Stability and the Effects of Stimulus-Intensity and Significance. *Journal of Abnormal Psychology, 93*, 304–311.

Jedrusik, P., Januszewicz, A., Busjahn, A., Wocial, B., Ignatowska-Switalska, H., Strelau, J. et al. (2004). The effects of genetic factors on selected indicators of the activity of the sympathoadrenal system and the renin-angiotensin-aldosterone system in twins. *Kardiol Pol, 61*, 423–429.

Jewett, D. L. (1964). Activity of single efferent fibers in the cervical vagus nerve of the dog, with special reference to possible cardio-inhibitory fibres. *Journal of Physiology, 175*, 321–357.

Julien, C. (2006). The enigma of Mayer waves: Facts and models. *Cardiovascular Research, 70*, 12–21.

Kamarck, T. W., & Lovallo, W. R. (2003). Cardiovascular Reactivity to Psychological Challenge: Conceptual and Measurement Considerations. *Psychosomatic Medicine, 65*, 9–21.

Kannel, W. B., Wolf, P. A., Castelli, W. P., & D'Agostino, R. (1987). Fibrinogen and risk of cardiovascular disease. *JAMA, 258*, 1183–1186.

Karasek, R., Theorell, T., Schwartz, J. E., Pieper, C., Schnall, P. L., & Michela, J. L. (1988). Job characteristics in relation to the prevalence of myocardial infarction in the US: Health Examination Survey (HES) and Health and Nutrition Examination Survey (HANES). *American Journal of Public Health, 78*, 910–919.

Katona, P. G., & Jih, R. (1975). Respiratory sinus arrhythmia: A non-invasive measure of parasympathetic cardiac control. *Journal of Applied Physiology, 39*, 801–805.

Keller, M. C., Medland, S. E., Duncan, L. E., Hatemi, P. K., Neale, M. C., Maes, H. H. et al. (2009). Modeling extended twin family data I: Description of the Cascade model. *Twin Research and Human Genetics, 12*, 8–18.

Kingwell, B. A., Thompson, J. M., Kaye, D. M., Mcpherson, G. A., Jennings, G. L., & Esler, M. D. (1994). Heart-Rate Spectral-Analysis, Cardiac Norepinephrine Spillover, and Muscle Sympathetic-Nerve Activity During Human Sympathetic Nervous Activation and Failure. *Circulation, 90*, 234–240.

Kleiger, R. E., Bigger, J. T., Bosner, M. S., Chung, M. K., Cook, J. R., Rolnitzky, L. M. et al. (1991). Stability over time of variables measuring heart rate variability in normal subjects. *American Journal of Cardiology, 68*, 626–630.

Kleiger, R. E., Miller, J. P., Bigger, J. T., & Moss, A. J. (1987). Decreased heart rate variability and its association with mortality after myocardial infarction. *American Journal of Cardiology, 59*, 256–262.

Krzeminski, K., Kruk, B., Nazar, K., Ziemba, A. W., Cybulski, G., & Niewiadomski, W. (2000). Cardiovascular, metabolic and plasma catecholamine responses to passive and active exercises. *Journal of Physiology and Pharmacology, 51*, 267–278.

Kunze, D. L. (1972). Reflex discharge patterns of cardiac vagal efferent fibres. *Journal of Physiology, 222*, 1–15.

Kupper, N., Willemsen, G., Boomsma, D. I., & de Geus, E. J. (2006). Heritability of indices for cardiac contractility in ambulatory recordings. *Journal of Cardiovascular Electrophysiology, 17*, 877–883.

Kupper, N., Willemsen, G., Posthuma, D., De Boer, P., Boomsma, D. I., & de Geus, E. J. C. (2005). A genetic analysis of ambulatory cardiorespiratory coupling. *Psychophysiology, 42*, 202–212.

Kurnik, D., Muszkat, M., Friedman, E. A., Sofowora, G. G., Diedrich, A., Xie, H. G. et al. (2007). Effect of the alpha2C-adrenoreceptor deletion322–325 variant on sympathetic activity and cardiovascular measures in healthy subjects. *Journal of Hypertension, 25*, 763–771.

La Rovere, M. T., Bigger, J. T., Marcus, F. I., Mortara, A., & Schwartz, P. J. (1998). Baroreflex sensitivity and heart-rate variability in prediction of total cardiac mortality after myocardial infarction. *Lancet, 351*, 478–484.

La Rovere, M. T., Pinna, G. D., & Raczak, G. (2008). Baroreflex sensitivity: Measurement and clinical implications. *Annals of Noninvasive Electrocardiology, 13*, 191–207.

La Rovere, M. T., Pinna, G. D., Maestri, R., Mortara, A., Capomolla, S., Febo, O. et al. (2003). Short-Term Heart Rate Variability Strongly Predicts Sudden Cardiac Death in Chronic Heart Failure Patients. *Circulation, 107*, 565–570.

Lambert, E. A., & Lambert, G. W. (2011). Stress and its role in sympathetic nervous system activation in hypertension and the metabolic syndrome. *Current Hypertension Reports, 13*, 244–248.

Lambert, G. W., Straznicky, N. E., Lambert, E. A., Dixon, J. B., & Schlaich, M. P. (2010). Sympathetic nervous activation in obesity and the metabolic syndrome--causes, consequences and therapeutic implications. *Pharmacology & Therapeutics, 126*, 159–172.

Langewouters, G. J., Settels, J. J., Roelandt, R., & Wesseling, K. H. (1998). Why use Finapres or Portapres rather than intra-arterial or intermittent non-invasive techniques of blood pressure measurement? *Journal of Medical Engineering & Technology, 22*, 37–43.

Lewis, R. P., Rittgers, S. E., Forester, W. F., & Boudoulas, H. (1977). Critical-Review of Systolic-Time Intervals. *Circulation, 56*, 146–158.

Lewis, S. F., Nylander, E., Gad, P., & Areskog, N. H. (1980). Non-Autonomic Component in Bradycardia of Endurance Trained Men at Rest and During Exercise. *Acta Physiologica Scandinavica, 109*, 297–305.

Libby, P., & Theroux, P. (2005). Pathophysiology of coronary artery disease. *Circulation, 111*, 3481–3488.

Liggett, S. (1995). Functional properties of human beta(2)-adrenergic receptor polymorphisms. *News in Physiological Sciences, 10*, 265–273.

Liggett, S. B., Mialet-Perez, J., Thaneemit-Chen, S., Weber, S. A., Greene, S. M., Hodne, D. et al. (2006). A polymorphism within a conserved beta1-adrenergic receptor motif alters cardiac function and {beta}-blocker response in human heart failure. *Proceedings of the National Academy of Sciences of the United States of America, 103*(30): 11288–11293.

Light, K. C., Kothandapani, R. V., & Allen, M. T. (1998). Enhanced cardiovascular and catecholamine responses in

women with depressive symptoms. *International Journal of Psychophysiology, 28*, 157–166.
Loeliger, M., Tolcos, M., Leditschke, J., Campbell, P., & Rees, S. (2000). Tracing cranial nerve pathways (glossopharyngeal, vagus, and hypoglossal) in SIDS and control infants: A DiI study. *Journal of Neuropathology and Experimental Neurology, 59*, 822–829.
Lombardi, F., Sandrone, G., Pernpruner, S., Sala, R., Garimoldi, M., Cerutti, S. et al. (1987). Heart rate variability as an index of sympathovagal interaction after acute myocardial infarction. *American Journal of Cardiology, 60*, 1239–1245.
Lovallo, W. R. (2005). *Stress and health: Biological and psychological interactions* (2nd ed). Thousand Oaks, CA: Sage.
Lucini, D., Norbiato, G., Clerici, M., & Pagani, M. (2002). Hemodynamic and Autonomic Adjustments to Real Life Stress Conditions in Humans. *Hypertension, 39*, 184–188.
Lykken, D.T., Iacono, W.G., Haroian, K., McGue, M., Bouchard, T.J. Jr. (1988). Habituation of the skin conductance response to strong stimuli: a twin study. *Psychophysiology, 25(1)*, 4–15.
Lynch, M., & Walsh, B. (1997). *Genetics and analysis of quantitative traits*. Sunderland, MA: Sinauer Associates.
Malliani, A., Pagani, M., Lombardi, F., Furlan, R., Guzzetti, S., & Cerutti, S. (1991). Spectral analysis to assess increased sympathetic tone in arterial hypertension. *Hypertension, 14*, III36–III43.
Mallion, J. M., Baguet, J. P., Siche, J. P., Tremel, F., & de Gaudemaris, R. (1999). Clinical value of ambulatory blood pressure monitoring. *Journal of Hypertension, 17*, 585–595.
Malm, S., Frigstad, S., Sagberg, E., Larsson, H., & Skjaerpe, T. (2004). Accurate and reproducible measurement of left ventricular volume and ejection fraction by contrast echocardiography: A comparison with magnetic resonance imaging. *Journal of the American College of Cardiology, 44*, 1030–1035.
Malpas, S. C. (2010). Sympathetic nervous system overactivity and its role in the development of cardiovascular disease. *Physiological Reviews, 90*, 513–557.
Manolio, T. A., & Collins, F. S. (2009). The HapMap and genome-wide association studies in diagnosis and therapy. *Annual Review of Medicine, 60*, 443–456.
Martin, N., Boomsma, D., & Machin, G. (1997). A twin-pronged attack on complex traits. *Nature Genetics, 17*, 387–392.
Martinmaki, K., Rusko, H., Kooistra, L., Kettunen, J., & Saalasti, S. (2006). Intraindividual validation of heart rate variability indexes to measure vagal effects on hearts. *AJP – Heart and Circulatory Physiology, 290*, H640–H647.
Martinsson, A., Larsson, K., & Hjemdahl, P. (1987). Studies in vivo and in vitro of terbutaline-induced beta-adrenoceptor desensitization in healthy subjects. *Clinical Science, 72*, 47–54.
Masuo, K., Katsuya, T., Fu, Y., Rakugi, H., Ogihara, T., & Tuck, M. L. (2005). Beta2- and beta3-adrenergic receptor polymorphisms are related to the onset of weight gain and blood pressure elevation over 5 years. *Circulation, 111*, 3429–3434.
Matsunaga, T., Gu, N., Yamazaki, H., Adachi, T., Yasuda, K., Moritani, T. et al. (2010). Association of estrogen receptor-alpha gene polymorphisms with cardiac autonomic nervous activity in healthy young Japanese males. *Clinica Chimica Acta, 411*, 505–509.
Matsunaga, T., Gu, N., Yamazaki, H., Tsuda, M., Adachi, T., Yasuda, K. et al. (2009). Association of UCP2 and UCP3 polymorphisms with heart rate variability in Japanese men. *Journal of Hypertension, 27*, 305–313.
Matsunaga, T., Nagasumi, K., Yamamura, T., Gu, N., Nishikino, M., Ueda, Y. et al. (2005). Association of C825T polymorphism of G protein beta3 subunit with the autonomic nervous system in young healthy Japanese individuals. *American Journal of Hypertension, 18*, 523–529.
Matsunaga, T., Yasuda, K., Adachi, T., Gu, N., Yamamura, T., Moritani, T. et al. (2007). Association of beta-adrenoceptor polymorphisms with cardiac autonomic modulation in Japanese males. *American Heart Journal, 154*, 759–766.
Matthews, K. A., Salomon, K., Kenyon, K., & Allen, M. T. (2002). Stability of children's and adolescents' hemodynamic responses to psychological challenge: A three-year longitudinal study of a multiethnic cohort of boys and girls. *Psychophysiology, 39*, 826–834.
McCaffery, J. M., Snieder, H., Dong, Y., & de Geus, E. (2007). Genetics in psychosomatic medicine: Research designs and statistical approaches. *Psychosomatic Medicine, 69*, 206–216.
Mezzacappa, E. S., Kelsey, R. M., & Katkin, E. S. (1999). The effects of epinephrine administration on impedance cardiographic measures of cardiovascular function. *International Journal of Psychophysiology, 31*, 189–196.
Milan, A., Mulatero, P., Williams, T. A., Carra, R., Schiavone, D., Martuzzi, R. et al. (2005). Bradykinin B2 receptor gene (-58T/C) polymorphism influences baroreflex sensitivity in never-treated hypertensive patients. *Journal of Hypertension, 23*, 63–69.
Miyamoto, Y., Higuchi, J., Abe, Y., Hiura, T., Nakazono, Y., & Mikami, T. (1983). Dynamics of cardiac output and systolic time intervals in supine and upright exercise. *Journal of Applied Physiology, 55*, 1674–1681.
Mizuno, M., Kamiya, A., Kawada, T., Miyamoto, T., Shimizu, S., Shishido, T. et al. (2008). Accentuated antagonism in vagal heart rate control mediated through muscarinic potassium channels. *Journal of Physiology Science, 58*, 381–388.
Montano, N., Ruscone, T. G., Porta, A., Lombardi, F., Pagani, M., & Malliani, A. (1994). Power Spectrum Analysis of Heart-Rate-Variability to Assess the Changes in Sympathovagal Balance During Graded Orthostatic Tilt. *Circulation, 90*, 1826–1831.
Mundy-Castle, A. C., & McKiever, B. I. (1953). The Psychophysiological Significance of the Galvanic Skin Response. *Journal of Experimental Psychology, 46*, 15–24.
Nakonezny, P. A., Kowalewski, R. B., Ernst, J. M., Hawkley, L. C., Lozano, D. L., Litvack, D. A. et al. (2001). New ambulatory impedance cardiograph validated against the Minnesota Impedance Cardiograph. *Psychophysiology, 38*, 465–473.
Nater, U. M., La Marca, R., Florin, L., Moses, A., Langhans, W., Koller, M. M. et al. (2006). Stress-induced changes in human salivary alpha-amylase activity — associations with adrenergic activity. *Psychoneuroendocrinology, 31*, 49–58.
Nater, U. M., & Rohleder, N. (2009). Salivary alpha-amylase as a non-invasive biomarker for the sympathetic nervous system: Current state of research. *Psychoneuroendocrinology, 34*, 486–496.

Nater, U. M., Rohleder, N., Gaab, J., Berger, S., Jud, A., Kirschbaum, C. et al. (2005). Human salivary alpha-amylase reactivity in a psychosocial stress paradigm. *International Journal of Psychophysiology, 55*, 333–342.

Nederfors, T., Dahlöf, C. (1992). Effects of the beta-adrenoceptor antagonists atenolol and propranolol on human whole saliva flow rate and composition. *Archives of Oral Biology, 37(7)*, 579–84.

Nelesen, R. A., Shaw, R., Ziegler, M. G., & Dimsdale, J. E. (1999). Impedance cardiography-derived hemodynamic responses during baroreceptor testing with amyl nitrite and phenylephrine: A validity and reliability study. *Psychophysiology, 36*, 105–108.

Neumann, S. A., Brown, S. M., Ferrell, R. E., Flory, J. D., Manuck, S. B., & Hariri, A. R. (2006). Human choline transporter gene variation is associated with corticolimbic reactivity and autonomic-cholinergic function. *Biological Psychiatry, 60*, 1155–1162.

Neumann, S. A., Lawrence, E. C., Jennings, J. R., Ferrell, R. E., & Manuck, S. B. (2005). Heart rate variability is associated with polymorphic variation in the choline transporter gene. *Psychosomatic Medicine, 67*, 168–171.

Neumann, S. A., Tingley, W. G., Conklin, B. R., Shrader, C. J., Peet, E., Muldoon, M. F. et al. (2009). AKAP10 (I646V) functional polymorphism predicts heart rate and heart rate variability in apparently healthy, middle-aged European-Americans. *Psychophysiology, 46*, 466–472.

Neumeister, A., Charney, D. S., Belfer, I., Geraci, M., Holmes, C., Sharabi, Y. et al. (2005). Sympathoneural and adrenomedullary functional effects of alpha2C-adrenoreceptor gene polymorphism in healthy humans. *Pharmacogenetics Genomics, 15*, 143–149.

Newlin, D. B., & Levenson, R. W. (1979). Pre-ejection period: Measuring beta-adrenergic influences upon the heart. *Psychophysiology, 16*, 546–553.

Newton-Cheh, C., Guo, C. Y., Wang, T. J., O'Donnell, C. J., Levy, D., & Larson, M. G. (2007). Genome-wide association study of electrocardiographic and heart rate variability traits: The Framingham Heart Study. *BMC Medical Genetics, 8*.

Nicholson, A., Kuper, H., & Hemingway, H. (2006). Depression as an aetiologic and prognostic factor in coronary heart disease: A meta-analysis of 6362 events among 146 538 participants in 54 observational studies. *European Heart Journal, 27*, 2763–2774.

Nishikino, M., Matsunaga, T., Yasuda, K., Adachi, T., Moritani, T., Tsujimoto, G. et al. (2006). Genetic variation in the renin-angiotensin system and autonomic nervous system function in young healthy Japanese subjects. *Journal of Clinical Endocrinology & Metabolism, 91*, 4676–4681.

Nolan, J., Batin, P. D., Andrews, R., Lindsay, S. J., Brooksby, P., Mullen, M. et al. (1998). Prospective Study of Heart Rate Variability and Mortality in Chronic Heart Failure: Results of the United Kingdom Heart Failure Evaluation and Assessment of Risk Trial (UK-Heart). *Circulation, 98*, 1510–1516.

Noto, Y., Sato, T., Kudo, M., Kurata, K., & Hirota, K. (2005). The relationship between salivary biomarkers and state-trait anxiety inventory score under mental arithmetic stress: A pilot study. *Anesthesia & Analgesia., 101*, 1873–1876.

Ovadia, M., Gear, K., Thoele, D., & Marcus, F. I. (1995). Accelerometer Systolic-Time Intervals As Fast-Response Sensors of Upright Posture in the Young. *Circulation, 92*, 1849–1859.

Pagani, M., Lombardi, F., Guzzetti, S., Rimoldi, O., Furlan, R., Pizzinelli, P. et al. (1986). Power spectral analysis of heart rate and arterial pressure variabilities as a marker of sympatho-vagal interaction in man and conscious dog. *Circulation Research., 59*, 178–193.

Pagani, M., & Malliani, A. (2000). Interpreting oscillations of muscle sympathetic nerve activity and heart rate variability. *Journal of Hypertension, 18*, 1709–1719.

Pagani, M., Montano, N., Porta, A., Malliani, A., Abboud, F. M., Birkett, C. et al. (1997). Relationship between spectral components of cardiovascular variabilities and direct measures of muscle sympathetic nerve activity in humans. *Circulation, 95*, 1441–1448.

Palatini, P., & Jullius, S. (2004). Elevated heart rate: A major risk factor for cardiovascular disease. *Clinical and Experimental Hypertension, 26*, 637–644.

Park, S. K., O'Neill, M. S., Wright, R. O., Hu, H., Vokonas, P. S., Sparrow, D. et al. (2006). HFE genotype, particulate air pollution, and heart rate variability - A gene-environment interaction. *Circulation, 114*, 2798–2805.

Penttila, J., Helminen, A., Jartti, T., Kuusela, T., Huikuri, H. V., Tulppo, M. P. et al. (2001). Time domain, geometrical and frequency domain analysis of cardiac vagal outflow: Effects of various respiratory patterns. *Clinical Physiology, 21*, 365–376.

Pichot, V., Gaspoz, J. M., Molliex, S., Antoniadis, A., Busso, T., Roche, F. et al. (1999). Wavelet transform to quantify heart rate variability and to assess its instantaneous changes. *Journal of Applied Physiology, 86*, 1081–1091.

Pitzalis, M. V., Mastropasqua, F., Massari, F., Forleo, C., Di Maggio, M., Passantino, A. et al. (1996). Short- and long-term reproducibility of time and frequency domain heart rate variability measurements in normal subjects. *Cardiovascular Research, 32*, 226–233.

Probst-Hensch, N. M., Imboden, M., Felber, D. D., Barthelemy, J. C., Ackermann-Liebrich, U., Berger, W. et al. (2008). Glutathione S-transferase polymorphisms, passive smoking, obesity, and heart rate variability in nonsmokers. *Environmental Health Perspectives, 116*, 1494–1499.

Proctor, G. B., & Carpenter, G. H. (2001). Chewing stimulates secretion of human salivary secretory immunoglobulin A. *Journal of Dental Research, 80*, 909–913.

Rao, F., Wessel, J., Wen, G., Zhang, L., Rana, B. K., Kennedy, B. P. et al. (2007a). Renal albumin excretion: Twin studies identify influences of heredity, environment, and adrenergic pathway polymorphism. *Hypertension, 49*, 1015–1031.

Rao, F., Zhang, K., Zhang, L., Rana, B. K., Wessel, J., Fung, M. M. et al. (2010). Human tyrosine hydroxylase natural allelic variation: Influence on autonomic function and hypertension. *Cell and Molecular Neurobiology, 30*, 1391–1394.

Rao, F., Zhang, L., Wessel, J., Zhang, K., Wen, G., Kennedy, B. P. et al. (2007b). Tyrosine hydroxylase, the rate-limiting enzyme in catecholamine biosynthesis: Discovery of common human genetic variants governing transcription, autonomic activity, and blood pressure in vivo. *Circulation, 116*, 993–1006.

Rao, F., Zhang, L., Wessel, J., Zhang, K., Wen, G., Kennedy, B. P. et al. (2008). Adrenergic polymorphism and the human stress response. *Annals of the New York Academy of Sciences, 1148*, 282–296.

Rekling, J. C., & Feldman, J. L. (1998). PreBotzinger complex and pacemaker neurons: Hypothesized site and kernel for respiratory rhythm generation. *Annual Review of Physiology, 60*, 385–405.
Richter, M., & Gendolla, G. H. (2009). The heart contracts to reward: Monetary incentives and preejection period. *Psychophysiology, 46*, 451–457.
Ridker, P. M., Hennekens, C. H., Buring, J. E., & Rifai, N. (2000). C-reactive protein and other markers of inflammation in the prediction of cardiovascular disease in women. *The New England Journal of Medicine., 342*, 836–843.
Riese, H., Rijsdijk, F. V., Ormel, J., van Roon, A. M., Neeleman, J., & Rosmalen, J. G. (2006). Genetic influences on baroreflex sensitivity during rest and mental stress. *Journal of Hypertension, 24*, 1779–1786.
Riese, H., Rosmalen, J. G., Ormel, J., van Roon, A. M., Oldehinkel, A. J., & Rijsdijk, F. V. (2007). The genetic relationship between neuroticism and autonomic function in female twins. *Psychological Medicine, 37*, 257–267.
Riese, H., van Doornen, L. J., Houtman, I. L., & de Geus, E. J. (2000). Job strain and risk indicators for cardiovascular disease in young female nurses. *Health Psychology, 19*, 429–440.
Ritz, T., & Dahme, B. (2006). Implementation and interpretation of respiratory sinus arrhythmia measures in psychosomatic medicine: Practice against better evidence? *Psychosomatic Medicine, 68*, 617–627.
Roach, D., Wilson, W., Ritchie, D., & Sheldon, R. (2004). Dissection of long-range heart rate variability: Controlled induction of prognostic measures by activity in the laboratory. *Journal of the American College of Cardiology, 43*, 2271–2277.
Robbe, H. W. J., Mulder, L. J. M., Rueddel, H., Langewitz, W. A., Veldman, J. B. P., & Mulder, G. (1987). Assessment of baroreceptor reflex sensitivity by means of spectral analysis. *Hypertension, 10*, 538–543.
Rohleder, N., Nater, U. M., Wolf, J. M., Ehlert, U., & Kirschbaum, C. (2004). Psychosocial stress-induced activation of salivary alpha-amylase: An indicator of sympathetic activity? *Annals of the New York Academy of Sciences, 1032*, 258–263.
Rosas-Ballina, M., & Tracey, K. J. (2009). Cholinergic control of inflammation. *Journal of Internal Medicine, 265*, 663–679.
Rosengren, A., Hawken, S., Ounpuu, S., Sliwa, K., Zubaid, M., Almahmeed, W. A. et al. (2004). Association of psychosocial risk factors with risk of acute myocardial infarction in 11119 cases and 13648 controls from 52 countries (the INTERHEART study): Case-control study. *Lancet, 364*, 953–962.
Salomon, K., Matthews, K. A., & Allen, M. T. (2000). Patterns of sympathetic and parasympathetic reactivity in a sample of children and adolescents. *Psychophysiology, 37*, 842–849.
Saul, J. P., Arai, Y., Berger, R. D., Lilly, N. S., Collucci, W. S., & Cohen, R. J. (1988). Assessment of autonomic regulation in chronic congestive heart failure by heart rate spectral analysis. *American Journal of Cardiology, 61*, 1291–1299.
Schachinger, H., Weinbacher, M., Kiss, A., Ritz, R., & Langewitz, W. (2001). Cardiovascular Indices of Peripheral and Central Sympathetic Activation. *Psychosomatic Medicine, 63*, 788–796.
Schell, A. M., Dawson, M. E., & Filion, D. L. (1988). Psychophysiological Correlates of Electrodermal Lability. *Psychophysiology, 25*, 619–632.
Schell, A. M., Dawson, M. E., Nuechterlein, K. H., Subotnik, K. L., & Ventura, J. (2002). The temporal stability of electrodermal variables over a one-year period in patients with recent-onset schizophrenia and in normal subjects. *Psychophysiology, 39*, 124–132.
Schwartz, A. R., Gerin, W., Davidson, K. W., Pickering, T. G., Brosschot, J. F., Thayer, J. F. et al. (2003). Toward a Causal Model of Cardiovascular Responses to Stress and the Development of Cardiovascular Disease. *Psychosomatic Medicine, 65*, 22–35.
Sherwood, A., Allen, M. T., Fahrenberg, J., Kelsey, R. M., Lovallo, W. R., & van Doornen, L. J. (1990). Methodological guidelines for impedance cardiography. *Psychophysiology, 27*, 1–23.
Sherwood, A., Allen, M. T., Obrist, P. A., & Langer, A. W. (1986). Evaluation of beta-adrenergic influences on cardiovascular and metabolic adjustments to physical and psychological stress. *Psychophysiology, 23*, 89–105.
Sherwood, A., McFetridge, J., & Hutcheson, J. S. (1998). Ambulatory impedance cardiography: A feasibility study. *Journal of Applied Physiology, 85*, 2365–2369.
Sherwood, A., & Turner, J. R. (1993). Postural stability of hemodynamic responses during mental challenge. *Psychophysiology, 30*, 237–244.
Shi, X. R., Stevens, G. H. J., Foresman, B. H., Stern, S. A., & Raven, P. B. (1995). Autonomic Nervous-System Control of the Heart – Endurance Exercise Training. *Medicine and Science in Sports and Exercise, 27*, 1406–1413.
Shihara, N., Yasuda, K., Moritani, T., Ue, H., Adachi, T., Tanaka, H. et al. (1999). The association between Trp64Arg polymorphism of the beta3-adrenergic receptor and autonomic nervous system activity. *Journal of Clinical Endocrinology & Metabolism, 84*, 1623–1627.
Shihara, N., Yasuda, K., Moritani, T., Ue, H., Uno, M., Adachi, T. et al. (2001). Synergistic effect of polymorphisms of uncoupling protein 1 and beta3-adrenergic receptor genes on autonomic nervous system activity. *International Journal of Obesity, 25*, 761–766.
Shimizu, S., Akiyama, T., Kawada, T., Shishido, T., Yamazaki, T., Kamiya, A. et al. (2009). In vivo direct monitoring of vagal acetylcholine release to the sinoatrial node. *Autonomic Neuroscience, 148*, 44–49.
Siegrist, J., Peter, R., Junge, A., Cremer, P., & Seidel, D. (1990). Low status control, high effort at work and ischaemic heart disease: Prospective evidence from blue-collar men. *Social Science & Medicine, 31*, 1127.
Singer, D. H., Martin, G. J., Magid, N., Weiss, J. S., Schaad, J. W., Kehoe, R. et al. (1988). Low heart rate variability and sudden cardiac death. *Journal of Electrocardiology*, S46–S55.
Sinnreich, R., Kark, J. D., Friedlander, Y., Sapoznikov, D., & Luria, M. H. (1998). Five minute recordings of heart rate variability for population studies: Repeatability and age-sex characteristics. *Heart, 80*, 156–162.
Skosnik, P. D., Chatterton, R. T., Jr., Swisher, T., & Park, S. (2000). Modulation of attentional inhibition by norepinephrine and cortisol after psychological stress. *International Journal of Psychophysiology, 36*, 59–68.
Smith, J. J., Muzi, M., Barney, J. A., Ceschi, J., Hayes, J., & Ebert, T. J. (1989). Impedance-derived cardiac

indices in supine and upright exercise. *Annals of Biomedical Engineering, 17,* 507–515.
Stein, P. K., Rich, M. W., Rottman, J. N., & Kleiger, R. E. (1995). Stability of index of heart rate variability in patients with congestive heart failure. *American Heart Journal, 129,* 975–981.
Steptoe, A., & Sawada, Y. (1989). Assessment of baroreceptor reflex during mental stress and relaxation. *Psychophysiology, 26,* 140–147.
Stolarz, K., Staessen, J. A., Kawecka-Jaszcz, K., Brand, E., Bianchi, G., Kuznetsova, T. et al. (2004). Genetic variation in CYP11B2 and AT1R influences heart rate variability conditional on sodium excretion. *Hypertension, 44,* 156–162.
Straub, R. H., Wiest, R., Strauch, U. G., Harle, P., & Scholmerich, J. (2006). The role of the sympathetic nervous system in intestinal inflammation. *Gut, 55,* 1640–1649.
Su, S., Lampert, R., Zhao, J., Bremner, J. D., Miller, A., Snieder, H. et al. (2009). Pleiotropy of C-reactive protein gene polymorphisms with C-reactive protein levels and heart rate variability in healthy male twins. *American Journal of Cardiology, 104,* 1748–1754.
Sullivan, P. F. (2007). Spurious genetic associations. *Biological Psychiatry, 61,* 1121–1126.
Suzuki, N., Matsunaga, T., Nagasumi, K., Yamamura, T., Shihara, N., Moritani, T. et al. (2003). Alpha(2B)-adrenergic receptor deletion polymorphism associates with autonomic nervous system activity in young healthy Japanese. *Journal of Clinical Endocrinology & Metabolism, 88,* 1184–1187.
Svedenhag, J., Martinsson, A., Ekblom, B., & Hjemdahl, P. (1986). Altered cardiovascular responsiveness to adrenaline in endurance-trained subjects. *Acta Physiologica Scandinavia, 126,* 539–550.
Svedenhag, J., Wallin, B. G., Sundlof, G., & Henriksson, J. (1984). Skeletal-Muscle Sympathetic Activity at Rest in Trained and Untrained Subjects. *Acta Physiologica Scandinavica, 120,* 499–504.
Sztajzel, J. (2004). Heart rate variability: A noninvasive electrocardiographic method to measure the autonomic nervous system. *Swiss Medical Weekly., 134,* 514–522.
Takai, N., Yamaguchi, M., Aragaki, T., Eto, K., Uchihashi, K., & Nishikawa, Y. (2004). Effect of psychological stress on the salivary cortisol and amylase levels in healthy young adults. *Archives of Oral Biology, 49,* 963–968.
Tank, J., Jordan, J., Diedrich, A., Stoffels, M., Franke, G., Faulhaber, H. D. et al. (2001). Genetic influences on baroreflex function in normal twins. *Hypertension, 37,* 907–910.
Task Force of the European Society of Cardiology the North American Society of Pacing (1996). Heart Rate Variability: Standards of Measurement, Physiological Interpretation, and Clinical Use. *Circulation, 93,* 1043–1065.
Thawornkaiwong, A., Preawnim, S., & Wattanapermpool, J. (2003). Upregulation of beta 1-adrenergic receptors in ovariectomized rat hearts. *Life Sciences, 72,* 1813–1824.
Thayer, J. F., Merritt, M. M., Sollers, J. J., Zonderman, A. B., Evans, M. K., Yie, S. et al. (2003). Effect of angiotensin-converting enzyme insertion/deletion polymorphism DD genotype on high-frequency heart rate variability in African Americans. *American Journal of Cardiology, 92,* 1487–1490.
Tingley, W. G., Pawlikowska, L., Zaroff, J. G., Kim, T., Nguyen, T., Young, S. G. et al. (2007). Gene-trapped mouse embryonic stem cell-derived cardiac myocytes and human genetics implicate AKAP10 in heart rhythm regulation. *Proceedings of the National Academy of Sciences of the United States of America, 104,* 8461–8466.
Tracey, K. J. (2009). Reflex control of immunity. *Nature Reviews Immunology, 9,* 418–428.
Tsuji, H., Larson, M. G., Venditti, F. J., Jr., Manders, E. S., Evans, J. C., Feldman, C. L. et al. (1996). Impact of reduced heart rate variability on risk for cardiac events. The Framingham Heart Study. *Circulation, 94,* 2050–2055.
Tuvblad, C., Isen, J., Baker, L. A., Raine, A., Lozano, D. I., & Jacobson, K. C. (2010). The genetic and environmental etiology of sympathetic and parasympathetic activity in children. *Behavior Genetics, 40,* 452–466.
Uusitalo, A. L., Vanninen, E., Levalahti, E., Battie, M. C., Videman, T., & Kaprio, J. (2007). Role of genetic and environmental influences on heart rate variability in middle-aged men. *American Journal of Physiology: Heart and Circulatory Physiology, 293,* H1013–H1022.
van Doornen, L. J. P., Knol, D. L., Willemsen, G., & de Geus, E. J. C. (1994). The relationship between stress reactivity in the laboratory and in real-life: Is reliability the limiting factor? *Journal of Psychophysiology, 8,* xx.
van Stegeren, A., Rohleder, N., Everaerd, W., & Wolf, O. T. (2006). Salivary alpha amylase as marker for adrenergic activity during stress: Effect of betablockade. *Psychoneuroendocrinology, 31,* 137–141.
van Stegeren, A. H., Wolf, O. T., & Kindt, M. (2008). Salivary alpha amylase and cortisol responses to different stress tasks: Impact of sex. *International Journal of Psychophysiology, 69,* 33–40.
Vanoli, E., De Ferrari, G. M., Stramba-Badiale, M., Hull, S. S., Jr., Foreman, R. D., & Schwartz, P. J. (1991). Vagal stimulation and prevention of sudden death in conscious dogs with a healed myocardial infarction. *Circulation Research, 68,* 1471–1481.
Verdecchia, P. (2000). Prognostic value of ambulatory blood pressure – Current evidence and clinical implications. *Hypertension, 35,* 844–851.
Verdecchia, P., Schillaci, G., Reboldi, G., Franklin, S. S., & Porcellati, C. (2001). Ambulatory monitoring for prediction of cardiac and cerebral events. *Blood Pressure Monitoring, 6,* 211–215.
Vikman, S., Makikallio, T. H., Yli-Mayry, S., Nurmi, M., Airaksinen, K. E., & Huikuri, H. V. (2003). Heart rate variability and recurrence of atrial fibrillation after electrical cardioversion. *Annals of Medicine, 35,* 36–42.
Viola, A. U., James, L. M., Archer, S. N., & Dijk, D. J. (2008). PER3 polymorphism and cardiac autonomic control: Effects of sleep debt and circadian phase. *American Journal of Physiology: Heart and Circulatory Physiology, 295,* H2156–H2163.
von Kanel, R., Mills, P. J., Fainman, C., & Dimsdale, J. E. (2001). Effects of psychological stress and psychiatric disorders on blood coagulation and fibrinolysis: A biobehavioral pathway to coronary artery disease? *Psychosomatic Medicine, 63,* 531–544.
Vossel, G., & Zimmer, H. (1990). Psychometric Properties of Nonspecific Electrodermal Response-Frequency for A Sample of Male-Students. *International Journal of Psychophysiology, 10,* 69–73.

Vrijkotte, T. G. M., van Doornen, L. J. P., & de Geus, E. J. C. (2004). Overcommitment to Work Is Associated With Changes in Cardiac Sympathetic Regulation. *Psychosomatic Medicine, 66*, 656–663.

Wain, L. V., Verwoert, G. C., O'Reilly, P. F., Shi, G., Johnson, T., Johnson, A. D. et al. (2011). Genome-wide association study identifies six new loci influencing pulse pressure and mean arterial pressure. *Nature Genetics, 43*, 1005–1011.

Waldstein, S. R., Neumann, S. A., & Merrill, J. A. (1998). Postural effects on hemodynamic response to interpersonal interaction. *Biological Psychology, 48*, 57–67.

Wallin, B. G. (1984). Sympathetic activity in human extremity nerves and its relationship to plasma norepinephrine. In M. G. Ziegler & C. R. Lake (Eds.), *Norepinephrine* (Baltimore: Williams & Wilkins.

Wallin, B. G. (2004). Microneurographic assessment of sympathetic nerve traffic. *Supplements to Clinical Neurophysiology, 57*, 345–351.

Wallin, B. G., Kunimoto, M. M., & Sellgren, J. (1993). Possible genetic influence on the strength of human muscle nerve sympathetic activity at rest. *Hypertension, 22*, 282–284.

Wang, X., Ding, X., Su, S., Li, Z., Riese, H., Thayer, J. F. et al. (2009). Genetic influences on heart rate variability at rest and during stress. *Psychophysiology, 46*, 458–465.

Wang, X., Thayer, J. F., Treiber, F., & Snieder, H. (2005). Ethnic differences and heritability of heart rate variability in African- and European American youth. *American Journal of Cardiology, 96*, 1166–1172.

Wiklund, U., Akay, M., & Niklasson, U. (1997). Short-term analysis of heart-rate variability by adapted wavelet transforms. *Engineering in Medicine and Biology Magazine, IEEE, 16*, 113–8, 138.

Wilhelm, F. H., & Grossman, P. (2010). Emotions beyond the laboratory: Theoretical fundaments, study design, and analytic strategies for advanced ambulatory assessment. *Biological Psychology*.

Wilhelm, F. H., Roth, W. T., & Sackner, M. A. (2003). The lifeShirt. An advanced system for ambulatory measurement of respiratory and cardiac function. *Behavior Modification, 27*, 671–691.

Wilhelmsen, L., Svärdsudd, K., Korsan-Bengtsen, K., Larsson, B., Welin, L., & Tibblin, G. (1984). Fibrinogen as a risk factor for stroke and myocardial infarction. *The New England Journal of Medicine, 311*, 501–505.

Willemsen, G., Ring, C., Carroll, D., Evans, P., Clow, A., & Hucklebridge, F. (1998). Secretory immunoglobulin A and cardiovascular reactions to mental arithmetic and cold pressor. *Psychophysiology, 35*, 252–259.

Willemsen, G. H., de Geus, E. J., Klaver, C. H., van Doornen, L. J., & Carroll, D. (1996). Ambulatory monitoring of the impedance cardiogram. *Psychophysiology, 33*, 184–193.

Williams, P. D., Puddey, I. B., Beilin, L. J., & Vandongen, R. (1993). Genetic influences on plasma catecholamines in human twins. *Journal of Clinical Endocrinology & Metabolism, 77*, 794–799.

Winzer, A., Ring, C., Carroll, D., Willemsen, G., Drayson, M., & Kendall, M. (1999). Secretory immunoglobulin A and cardiovascular reactions to mental arithmetic, cold pressor, and exercise: Effects of beta-adrenergic blockade. *Psychophysiology, 36*, 591–601.

Wu, T., Snieder, H., & de Geus, E. (2010). Genetic influences on cardiovascular stress reactivity. *Neuroscience and Biobehavioral Review, 35*, 58–68.

Yang, A. C., Chen, T. J., Tsai, S. J., Hong, C. J., Kuo, C. H., Yang, C. H. et al. (2010). BDNF Val66Met polymorphism alters sympathovagal balance in healthy subjects. *American Journal of Medical Genetics: B Neuropsychiatric Genetics, 153B*, 1024–1030.

Yasuda, K., Matsunaga, T., Moritani, T., Nishikino, M., Gu, N., Yoshinaga, M. et al. (2004). T393C polymorphism of GNASI associated with the autonomic nervous system in young, healthy Japanese subjects. *Clinical and Experimental Pharmacology and Physiology, 31*, 597–601.

Yasuma, F., & Hayano, J. (2004). Respiratory sinus arrhythmia: Why does the heartbeat synchronize with respiratory rhythm? *Chest, 125*, 683–690.

Ylitalo, A., Airaksinen, K. E., Hautanen, A., Kupari, M., Carson, M., Virolainen, J. et al. (2000). Baroreflex sensitivity and variants of the renin angiotensin system genes. *Journal of the American College of Cardiology, 35*, 194–200.

Zhang, K., Rao, F., Rana, B. K., Gayen, J. R., Calegari, F., King, A. et al. (2009). Autonomic function in hypertension; role of genetic variation at the catecholamine storage vesicle protein chromogranin B. *Circulation: Cardiovascular Genetics, 2*, 46–56.

Zhang, K., Rao, F., Wang, L., Rana, B. K., Ghosh, S., Mahata, M. et al. (2010*a*). Common functional genetic variants in catecholamine storage vesicle protein promoter motifs interact to trigger systemic hypertension. *Journal of the American College of Cardiology, 55*, 1463–1475.

Zhang, K., Rao, F., Wen, G., Salem, R. M., Vaingankar, S., Mahata, M. et al. (2006). Catecholamine storage vesicles and the metabolic syndrome: The role of the chromogranin A fragment pancreastatin. *Diabetes, Obesity and Metabolism, 8*, 621–633.

Zhang, K., Zhang, L., Rao, F., Brar, B., Rodriguez-Flores, J. L., Taupenot, L. et al. (2010*b*). Human tyrosine hydroxylase natural genetic variation: Delineation of functional transcriptional control motifs disrupted in the proximal promoter. *Circulation: Cardiovascular Genetics, 3*, 187–198.

Zhang, L., Rao, F., Wessel, J., Kennedy, B. P., Rana, B. K., Taupenot, L. et al. (2004). Functional allelic heterogeneity and pleiotropy of a repeat polymorphism in tyrosine hydroxylase: Prediction of catecholamines and response to stress in twins. *Physiologic Genomics, 19*, 277–291.

Zhang, L., Rao, F., Zhang, K., Khandrika, S., Das, M., Vaingankar, S. M. et al. (2007). Discovery of common human genetic variants of GTP cyclohydrolase 1 (GCH1) governing nitric oxide, autonomic activity, and cardiovascular risk. *Journal of Clinical Investigation, 117*, 2658–2671.

CHAPTER

18 The Genetic Basis of Positive Emotionality

Martin Reuter *and* Christian Montag

Abstract

The chapter addresses the molecular genetic basis of the personality trait positive emotionality (PE). Beginning with historical aspects of heritability estimation and personality assessment, the main portion of this chapter discusses the molecular genetics basis of PE, which is investigated far less frequently than that of negative emotionality (NE). The studies reviewed focus on individual differences in PE in healthy subjects and include only those studies that assess PE by broadly accepted personality inventories. The review is concentrated on dopaminergic and serotonergic genes because these genes show the most association with PE.

Key Words: positive emotionality, genetics, heritability estimation, personality assessment, negative emotionality, dopaminergic gene, serotonergic gene, genes

Individual differences in personality have always sparked human interest. The implicit knowledge that humans differ with respect to temperament and character dimensions and that these differences exert consequences on social success is widely accepted. We often ask ourselves: "Why am I not as extraverted as my friend?" or "Why am I so anxious?" Although everyone has made such observations with regard to individual differences in personality, explaining them is an entirely different endeavor.

Simple examples like these make it clear that personality dimensions such as extraversion or neuroticism (so-called traits, which are more stable) have to be distinguished from immediate emotions (emotional states). Personality traits are defined as the habitual disposition to respond to a certain class of environmental stimuli with a specific class of reaction patterns. For example, a person characterized by high positive emotionality (PE) is sensitive toward stimuli predicting reward or novelty and is likely to exhibit approach behavior, whereas a person marked by negative emotionality (NE) is sensitive to threatening stimuli and often shows flight or avoidance behavior. Although personality is the proclivity for a given behavioral response set, it does not preclude a high PE person from being anxious or sad under certain circumstances. In spite of their inability to predict behavioral responses in all situations, personality traits are useful in predicting behavior under defined environmental conditions.

I. Historical Aspects

The debate behind the reasons for personality differences is very old and has been addressed in science as the classic nature–nurture debate (Krubitzer & Kahn, 2003). The nature–nurture debate aims to disentangle genetic and environmental influences on phenotypes such as personality or intelligence (Angoff, 1988). The birth of behavioral genetics as a scientific discipline provided an impetus for systematic research on this topic. In 1869, a half-cousin of Charles Darwin named Francis Galton published a book entitled *Hereditary Genius* that contained the first pioneering work in quantitative genetics (Galton, 1869). It was also Galton who spearheaded the main techniques in quantitative genetics: namely, family, twin, and adoption studies that are still applied in the twenty-first century, in contrast to his ill-conceived promotion of eugenics, which has rightfully been long abandoned.

Today, no one doubts the presence of a strong genetic influence on personality. Numerous twin studies have reported heritability estimates for personality traits of up to 0.50, depending on the conceptualization of personality (the questionnaires used) and the methods applied (for an overview, see Plomin, DeFries, McClearn, & McGuffin, 2008). Due to these strong genetic effects on personality, scientists are interested in identifying the genes that form the basis for this heritability. The revolutionary invention of the polymerase chain reaction (PCR) by Kary Mullis in 1983 provided the starting point for molecular genetics research; a few years later, this technique was applied to the study of personality. In the following two decades, the field was dominated by the so-called *candidate gene approach*, wherein the scientific agenda was completely theory driven. Based on pharmacological, lesion, and positron emission tomography (PET) studies demonstrating the significance of certain neurotransmitters for behavior, polymorphisms on related genes were investigated. Subsequently, the realization that single genes could not explain the total variance in complex behaviors led to the development of high throughput genotyping devices. The newest developments in molecular genetics introduced genome-wide association studies (GWAS) that enable simultaneous genotyping of 3 million single-nucleotide polymorphisms (SNPs; variation of a single letter [nucleotide] of the genetic code that takes place in at least 1 percent of the population) (Wang, Jia, Wolfinger, Chen, & Zhao, 2011). In a GWAS analysis, the specific primers (oligonucleotides that mark the region surrounding the polymorphism) for each of the 3 million loci are placed on a microarray chip. Results from the first GWAS on personality will be presented here, and an outlook on future directions will be given at the end of this chapter.

II. Assessment of Positive Emotionality

Despite the early merits of Francis Galton in nineteenth-century quantitative methods in personality research, the twentieth century was mainly dominated by the factor analytic era. Scientists used the lexical approach to identify major personality dimensions based on all adjectives that are useful in describing an individual's personality. The statistical method applied was factor analysis. Today, the so-called "Big Five" personality theory by Costa and McCrae (1985) is the dominant theory in the field, aiming to cover the entire personality spectrum by means of five traits. The proposed dimensions are: Openness (to Experience), Conscientiousness, Extraversion, Agreeableness, and Neuroticism. A controversy arose whether personality can be best described by three, five, six, or more personality factors. This (in our view futile) dispute widely neglected the reasons behind individual differences in temperament and character, a question that is probably of greater relevance for each individual and one that was already addressed years ago by Francis Galton. Biologically oriented personality researchers are exceptions to this school of thought because they search for the biological underpinnings of personality with the conviction that nature plays an important role in the genesis of personality. The first, pure biologically oriented personality theory was the *reinforcement sensitivity theory* (RST) by Jeffrey Gray (Gray, 1970). Gray, being a student of Hans-Juergen Eysenck, who was interested in biological influences but also used factor analytic methods (Eysenck, 1990), started his work on animals by applying lesion studies and pharmacological challenge tests in rodents. His focus was on drugs influencing fear and anxiety, such as alcohol and barbiturates. The merit of Gray's work was the identification of neurotransmitter systems and neuroanatomical structures related to the fundamental behavioral systems, called the behavioral approach system (BAS), behavioral inhibition system (BIS), and the fight–flight system (FFS). His findings were later extrapolated to humans, and his theory was substantially revised in 2000 (Gray & McNaughton, 2000). Reinforcement sensitivity theory provided the starting point for molecular genetics research because the neurotransmitter systems postulated to be involved in personality are influenced by genes. Thus, there are specific genes that code for and enable the synthesis of receptors, transporters, and enzymes related to the activity of the neurotransmitter systems. The theory-driven candidate gene approach outlined earlier marked the first steps in personality genetics and is still used today, despite the ability to conduct hypothesis-free GWAS. The second and most popular biological personality theory was proposed by Robert Cloninger. Cloninger's sociobiological theory of personality claimed three factors of personality in its initial version: Novelty Seeking, Reward Dependence, and Harm Avoidance (Cloninger, Svrakic, & Przybeck, 1993). These dimensions were assessed by means of the Tridimensional Personality Questionnaire (TPQ; Cloninger, Przybeck, Svrakic, & Wetzel, 1991). Later, a fourth dimension of Persistence was extracted from the broader trait of Reward Dependence. The newest conceptualization is manifest in the Temperament and Character Inventory (TCI; Cloninger, Przybeck, Svrakic, & Wetzel, 1994), wherein the four former traits (now referred to as temperaments) were complemented by three character dimensions: Self-Directedness, Cooperativeness, and Self-Transcendence. According to Cloninger's theory, biological factors should exert a stronger influence on temperaments, whereas characters should be more influenced by environmental factors. However, twin studies also have demonstrated a strong genetic influence on the character dimensions (Gillespie, Cloninger, Heath, & Martin, 2003). The newest biologically oriented personality theory available is based on Jaak Panksepp's work on affective neuroscience. Although Panksepp's work was influenced by Gray and Cloninger, he studied the biological underpinnings of animal emotions in an incomparably elaborate way. He used pharmacological challenge studies, electrical brain stimulation, and lesion studies to identify the neuronal circuits of the basic emotion systems in animals (Panksepp, 1998). Today, his work is adopted to investigate individual differences in human personality traits and emotions and to explain the biological basis of neuropsychiatric disorders (e.g., Felten, Montag, Markett, Walter, & Reuter, 2011; Leboyer et al., 1994; Reuter, Weber, Fiebach, Elger, & Montag, 2009). Applied to human work, the six traits of Panksepp's theory—Seek, Play, Care, Fear, Sadness, and Anger—are assessed along with a Spirituality scale by the Affective Neuroscience Personality Scales (ANPS; Davis, Panksepp, & Normansell, 2003).

III. Positive Emotionality

The current literature contains a myriad of personality theories and related questionnaires, such that it is difficult to develop an overview. At first glance, it becomes apparent that there is a significant overlap between some of these theories, creating the impression that, to some extent, the same concepts are labeled by different names. When students criticize this and ask for a system to clarify the bulk of theories, we answer that each personality theory postulates at least one dimension representing positive emotionality (PE) but that most theories have several dimensions reflecting negative emotionality (NE). This imbalance between the number of dimensions representing PE and NE is mirrored by the number of basic emotions. The work of Ekman and others has shown that we have only one basic positive emotion, happiness, but numerous negative basic emotions (e.g., anger, fear, sadness, disgust) (Ekman, 1992). Personality dimensions representing PE include, for example, Extraversion, Novelty Seeking, Seek, and the behavioral approach system (BAS) (depending on the respective theory), whereas Neuroticism, Fear, Anger, and the fight–flight system (FFS) constitute NE. A quick peek into personality psychology literature unequivocally reveals a strong imbalance between the number of publications on NE and the number of publications on PE. The latter is far more underrepresented. The reason for this is based not only on the disproportionate number of personality dimensions, but also grounded in the relevance to psychiatry. It is widely accepted by psychiatrists that individual differences in personality within the normal range have the same biological basis as psychopathologies, marking the extreme end of the same continuum. For example, anxiety disorders constitute the extreme form of neuroticism. For this reason, psychiatry research hopes to profit from research on personality and, indeed, there is plenty of evidence supporting the continuum model. One of the most prominent findings is that the serotonin transporter gene promoter polymorphism (5-HTTLPR) is related to neuroticism, as well as to the affective disorders of anxiety and depression (Brown & Harris, 2008; Domschke & Dannlowski, 2010; Lesch et al., 1996). It is understandable that clinical research is more likely to invest money in research projects in which a direct benefit to patient health is expected. However, PE is also of relevance to psychiatry. The search for the biological basis of PE is grounded in a prominent theory stemming from addiction research, the *theory of the final common pathway of reward* (Spanagel & Weiss, 1999). This theory postulates that mesolimbic dopaminergic neurons serve as a final common neural pathway for mediating reinforcement processes. In other words, each behavior or action leading to positive feelings (a salary increase, eating a delicious meal, lottery winnings, having sex, or, in the extreme, the consumption of drugs) is related to an increase in extracellular dopamine levels. This theory had tremendous impact on addiction research and was supported by the fact that most drugs causing addiction (e.g., amphetamines, cocaine, and nicotine) are directly or indirectly related to the dopamine system. There are only few exceptions (e.g., benzodiazepines) that contradict the theory of a common final pathway of reward. However, dopamine is still at the focus of drug research and therefore it is no surprise that this neurotransmitter has strongly influenced research on the biological basis of PE.

IV. Dopamine and Positive Emotionality

Before molecular genetics techniques became available for biologically oriented personality research, pharmacological challenge tests and endocrinological studies were the gold standards in human research. These techniques were successfully previously applied in animal work along with lesion studies, which were, of course, not applicable in humans for ethical reasons. In humans, dopaminergic drugs were either administered or the metabolites of neurotransmitters or enzymes related to their metabolism were assessed (Depue & Collins, 1999; Depue, Luciana, Arbisi, Collins, & Leon, 1994; Reuter, Netter, Toll, & Hennig, 2002; Zuckerman, 1985). These findings support the link between PE and the dopaminergic system and were the starting point for molecular genetics studies. In this section, we provide an overview of the studies that aimed to relate dopaminergic genes to PE.

A. COMT

Catechol-O-methyltransferase (COMT) is an enzyme that catabolizes catecholamines in the synaptic cleft, predominately in the prefrontal cortex. The COMT gene is located on the q11 band of human chromosome 22 and consists of 27,000 bases and six exons. In exon 4 (codon 158), an SNP (rs4680) occurs wherein a G→A transition results in three- to four-fold reduction in COMT enzyme activity by encoding the synthesis of the amino acid methionine (MET) instead of valine (VAL). Carriers of the Val/Val genotype have the highest levels of COMT activity, whereas carriers of the Met/Met genotype have the lowest levels of COMT activity, and heterozygotes (Val/Met genotype) fall in between (Lachman et al., 1996). The strong functional effect of rs4680 on dopamine metabolism caused by a single base exchange, along with countless positive associations with all sorts of phenotypes, have transformed this SNP into one of the most prominent polymorphisms in the literature. One of the most essential and replicated findings of rs4680 is the positive effect of the Met allele on cognitive functioning. Better performance on a diverse battery of cognitive tasks covering all sorts of executive control processes was associated with the decreased enzyme activity of the Met allele and resultant higher prefrontal dopamine levels (Meyer-Lindenberg et al., 2005; Tunbridge, Harrison, & Weinberger, 2006).

However, the COMT Val158Met polymorphism was also demonstrated to influence the processing of emotions and therefore personality. Before providing a brief review of the association between COMT Val158Met and personality, it is prudent to point out that there are numerous reasons for the observed inconsistencies in genetic association study results. First, the conceptualization of personality (i.e., the questionnaire measures used) has a strong influence on the findings. Second, it is well established that allele frequencies often differ across ethnic groups. For example, the evolutionarily younger Met allele is as frequent as the ancestral Val allele in Caucasian populations. However, in Asian populations, the allele frequency of the minor Met allele is about 0.25. Therefore, ethnic diversity across studies may have an influence on the results of association studies. Third, the sample size is important. Small samples are more likely to be influenced by stratification errors and so have an increased risk of drawing a sample that is not representative of the respective population.

Reuter and Hennig (2005) reported that Caucasian carriers of the Val/Val genotype (referred to as Met–) had significantly higher Extraversion scores as measured by the NEO-Five Factor Inventory (FFI) than did carriers of at least one Met allele (genotypes Met/Met or Val/Met; referred to as Met+). This finding was supported by Hamidovic et al. (2010) using the Multidimensional Personality Questionnaire (MPQ, Tellegen, 1982). Also, Stein et al. (2005) and Hoth et al. (2006) reported significantly lower NEO Extraversion scores in participants (mostly Caucasians) with the Met/Met genotype, corroborating the hypothesis that the Val allele is related to PE.

All personality theories postulate orthogonality (i.e., statistical independence, or lack of correlation) between their personality factors. However, empirical evidence shows that this is often not the case. For example, Extraversion and Neuroticism correlate up to −0.50 (Reuter, 2008), suggesting that scoring high on PE decreases the likelihood of also scoring high on NE. This leads us to formulate the "yin-yang hypothesis" of personality, which states that PE and NE represent different sides of the same coin. Based on this theory, it is hypothesized that, whereas the Val allele is associated with high PE, it is also related to low NE, and whereas the Met allele is associated with low PE, it is also related to high NE. Molecular genetics studies support this hypothesis: Enoch et al. (2003) and Hashimoto et al. (2007) reported highest Harm Avoidance and Stein et al. (2005) reported highest Neuroticism scores in carriers of the Met/Met genotype. According to the inverse relationship between NE and PE just hypothesized, Stein et al. (2005) not only reported highest Neuroticism, but also lowest Extraversion scores in Met/Met carriers. Surprisingly, the results were homogenous across different ethnic groups. However, there are also null findings with respect to COMT Val158Met and personality in the literature (e.g., Sheldrick et al., 2008).

B. DAT

The dopamine transporter (DAT) plays a major role in terminating dopamine neurotransmission by an active reuptake of dopamine into the presynapse. Two polymorphisms, one in the promoter region (the SNP -67 A/T) and another in intron 8 (a 40 bp variable number of tandem repeats polymorphism, VNTRs) on the DAT gene (SLC6A3) were reported to affect DAT expression. In a sample of 654 healthy Japanese participants, Shibuya et al. (2009) observed that female participants without the A allele (associated with high DAT activity) had lower scores on Self-Directedness and Cooperativeness than did those with an A allele. In males, no effect on personality could be observed. In both gender groups, the VNTR in intron 8 was not related to personality. Recently, Kazantseva et al. (2011) reported the association between a SNP located in the 3′ untranslated region (UTR) of the DAT gene (rs27072) and Persistence in a Russian sample. Samochowiec et al. (2001) found lower scores on the RD4 subscale of Reward Dependence in carriers of the 9/9 genotype in the 40 bp VNTR when compared to subjects with the 10/10 genotype. However, the sample size in this Polish study (N = 127) was rather small. Hünnerkopf et al. (2007) reported a gene–gene interaction (a so-called epistatic effect) between the DAT VNTR and the brain derived neurotrophic factor (BDNF) Val66Met polymorphism on NEO-PI-R Neuroticism and Harm Avoidance measured by the TPQ. Among individuals with at least one copy of the DAT 9-repeat allele, carriers of the BDNF Met allele exhibited significantly lower Neuroticism and Harm Avoidance scores than did noncarriers of the 9-repeat allele. This genetic interaction was not related directly to PE but could constitute a resilience factor against NE. Another epistatic effect was recently observed by our own group (Felten et al., 2011). Carriers of the 9/9 genotype of DAT who also had the Val/Val genotype of COMT Val158Met showed significantly reduced Sadness scores on the ANPS questionnaire. The Met allele of COMT also showed a significant main effect on sadness supporting the above-mentioned results, suggesting a positive association between the Val allele and PE, and a negative association between the Val allele and NE. Additional evidence for the role of DAT in PE comes from a PET study. Laine et al. (2001) found a positive correlation between Novelty Seeking scores and DAT density in the brain in a sample of alcoholics. This is in line with Cloninger's personality theory, postulating that type I alcoholics exhibit elevated Novelty Seeking scores and that the dopaminergic system forms the biological basis for this trait (Cloninger et al., 1993).

C. DRD2

The D_2 receptor is the most investigated of the five dopaminergic receptors (D_1–D_5). This interest in the D_2 receptor is based on pharmacological studies demonstrating its prominent role in reward and psychosis (Seeman & Seeman, 1988; Volkow, Fowler, & Wang, 1999). All traditional neuroleptic drugs administered in the therapy of schizophrenia were D_2 receptor antagonists before other receptor subtypes were identified as useful therapeutic targets (Seeman, 2010). The D_2 receptor was especially related to positive symptoms in schizophrenia characterized by an excess of dopamine availability. In addition, psychostimulants exert their reinforcing mechanisms by increasing synaptic dopamine levels. For these reasons, the DRD2 gene became one of the first candidate genes for drug addiction and, later, for PE. The first molecular genetic studies investigating the effect of DRD2 on personality and drug addiction were published in the 1990s. The focus was on a functional SNP, the DRD2 Taq Ia (also known as DRD2/ANKK1 Taq Ia), in which the minor A1 allele was related to a 30–40 percent reduction in D_2 receptor density when compared to the major A2 allele (Ritchie & Noble, 2003). At the time, study sample sizes were extremely small (less than 100 cases and controls) and would not meet the standards of today's scientific practice. However, without this pioneering work, new advances in molecular genetics would have been impossible. For example, Noble and colleagues reported higher Novelty Seeking scores in a group of boys carrying the minor allele in three gene loci of DRD2, including the Taq Ia locus (Noble et al., 1998). The finding that the A1+ allele (genotypes A1/A1 and A1/A2) of DRD2 Taq Ia is related to higher scores in PE and lower scores in NE in comparison to A1− subjects (genotype A2/A2) could be replicated. Higher Extraversion scores in A1+ carriers were reported by Smillie et al. (2010), higher Self-Directedness scores were found in A1+ carriers in a sample of children of alcoholics (Antolin et al., 2009), and lower Neuroticism and higher BAS Reward-Responsiveness scores were observed in A1+ carriers (Kazantseva et al., 2011; Lee, Ham, Cho, Lee, & Shim, 2007; Wacker, Reuter, Hennig, & Stemmler, 2005). Nyman et al. (2009) found an association between two DRD2 5′ end SNPs and Novelty Seeking in a Finnish population-based birth cohort study. However, negative findings are also present in the DRD2 gene literature (e.g., Cruz et al., 1995; Hibino, Tochigi, Otowa, Kato, & Sasaki, 2006).

D. DRD1, DRD3, and DRD5

Given the prominent role of the dopaminergic system in PE and reward, the DRD1, DRD3, and DRD5 genes were investigated in relation to personality. The results were mainly discouraging, although not all polymorphisms in these genes have been evaluated at this time. In an extensive Finnish population-based birth cohort study including genotypes of more than 1,400 participants, 23 HapMap tag SNPs and resulting haplotypes located on DRD1–DRD5 were analyzed for associations with Cloninger's TCI personality questionnaire (Nyman et al., 2009). There were no significant associations for any of the SNPs in DRD1, DRD3, or DRD5. In line with this, Heck and colleagues investigated 15 SNPs in the DRD3 gene but found no association with Cloninger's temperament dimensions in two independent Caucasian samples (Heck et al., 2009).

E. DRD4

The DRD4 gene is located on chromosome 11 p5.5 and shows a much-studied VNTR polymorphism in exon 3. This polymorphism is characterized by the repetition of a 48 bp sequence (2 to 7 repeats) in which the 2-, 4-, and 7-repeat alleles are the most frequent variants in most populations (Gelernter, Kennedy, van Tol, Civelli, & Kidd, 1992). In 1996, two pioneering studies reported the first association between the DRD4 48 bp VNTR and Novelty Seeking (Benjamin et al., 1996; Ebstein et al., 1996). Carriers of the 7-repeat allele had significantly higher Novelty Seeking scores than did those without it. Several follow-up studies in Caucasian samples were able to replicate this finding (Benjamin et al., 2000; Noble et al., 1998; Poston et al., 1998; Serretti et al., 2006; Strobel, Wehr, Michel, & Brocke, 1999), yet many other studies could not (Gebhardt et al., 2000; Gelernter et al., 1997; Herbst, Zonderman, McCrae, & Costa, 2000; Hill, Zezza, Wipprecht, Locke, & Neiswanger, 1999; Jönsson et al., 1997; 1998; Keltikangas-Järvinen et al., 2003; Kühn et al., 1999; Malhotra et al., 1996; Persson et al., 2000; Pogue-Geile, Ferrell, Deka, Debski, & Manuck, 1998; Strobel et al., 2002; 2003; Szekely et al., 2004). Two meta-analyses have since failed to confirm the initial association between the 7-repeat allele and higher Novelty Seeking scores (Munafò, Yalcin, Willis-Owen, & Flint, 2008; Schinka, Busch, & Robichaux-Keene, 2002). The findings are also heterogeneous in studies from Asian populations, in which participants are classified as short or long allele carriers due to the fact that the 7-repeat allele is a rare variant. Tomitaka et al. (1999) and Lee et al. (2003*a*) reported an association between the long allele and higher Novelty Seeking, whereas other studies were unable to replicate this finding (Kim, Kim, Kim, & Lee, 2006; Lee et al., 2003*b*; Mitsuyasu et al., 2001; Ono et al., 1997; Tochigi et al., 2006). In a Japanese sample, Tsuchimine et al. (2009) reported an association between the DRD4 48 bp VNTR in exon 3 and Novelty Seeking; however, this association was present only when contrasting the 5/5 genotype with the 2/2 or 4/4 genotypes.

F. DARPP-32

Thus far, attempts to identify the genetic basis of personality have mainly focused on genes coding for receptors, transporters, and enzymes involved in the metabolism and clearance of dopamine. However, it is increasingly more apparent that the signal transduction of neurotransmitters must also be investigated. Recent findings have supported this hypothesis by highlighting the importance of DARPP-32 (dopamine- and cAMP-regulated phosphorprotein, 32 kDa), a key regulatory molecule in the dopaminergic signaling pathway that functions by influencing phosphorylation cascades. DARPP-32 is abundantly expressed in GABAergic medium-sized spiny neurons of the neostriatum, where it integrates dopaminergic and glutamatergic neurotransmission (Ouimet, Miller, Hemmings, Walaas, & Greengard, 1984), as well as in the cerebellar cortex at the level of the Purkinje cell layer (Brene et al., 1994). DARPP-32 is encoded by the PPP1R1B gene, and some SNPs in this gene have been successfully related to personality. Li et al. (2011) reported a negative heterosis effect (the phenotype is weaker when expressed in heterozygous than in homozygous individuals) for rs12601930 C/T on Novelty Seeking in a healthy Han Chinese sample. In the same study, two additional polymorphisms in the PPP1R1B gene, rs879606 and rs3764352, were associated with the NE-related personality dimension of Harm Avoidance. Two years earlier, a positive association was found between the DARPP-32 gene polymorphism rs907094 and trait Anger in Caucasian subjects using the ANPS, a self-report measure of personality based on Panksepp's personality theory (Reuter et al., 2009). Carriers of the rs907094 T-allele (genotypes TT and CT) had significantly higher Anger scores than did individuals with the CC genotype. However, the CC genotype was not related to PE.

V. Serotonin and Positive Emotionality

Traditionally, the serotonergic system is considered to be the neurotransmitter system at the core of NE (but not PE) research (for examples, see Buckholtz and Meyer-Lindenberg's chapter in this volume on a discussion of serotonergic signaling and human aggression and violence). A deficit in serotonin (5-HT) (aside from a deficit in noradrenaline [NA]) was (and still is) assumed to be the major cause of affective disorders culminating in the monoamine hypothesis of depression (Schildkraut, 1974). Restoration of 5-HT levels in the central nervous system by means of selective serotonin reuptake inhibitors (SSRIs) has turned out to be one of the most effective pharmaceutical regimens in treating patients with affective disorders (Taylor, Freemantle, Geddes, & Bhagwagar, 2006). SSRIs increase the availability of 5-HT in the synaptic cleft by blocking reuptake into the presynaptic cleft. A polymorphism in the SLC6A4 gene encoding the serotonin transporter (5-HTT) also known as the serotonin transporter-linked polymorphic region (5-HTTLPR), is the most investigated genetic variation in the field of NE and psychopathology. In 1996, Klaus-Peter Lesch and colleagues described this 43 bp insertion/deletion polymorphism and revealed an association between the short s allele and Neuroticism. Moreover, they demonstrated that the short s allele of 5-HTTLPR is associated with a three-fold reduction in mRNA levels. To date, hundreds of studies demonstrate an association of the s allele with various phenotypes, mostly related to NE. The association between the s allele of 5-HTTLPR and NE-related personality traits such as Neuroticism and Harm Avoidance was often replicated and at the focus of some prominent meta-analyses, with partly controversial results (Munafò et al., 2009; Schinka et al., 2004; Sen, Burmeister, & Ghosh, 2004). The benefits of molecular genetics research for the treatment of psychiatric disorders was demonstrated by studying the SSRI response dependent on the 5-HTTLPR (Smeraldi et al., 1998). Although the findings are controversial, a majority of studies show that ll-homozygotes respond better to SSRI therapy than do carriers of at least one s allele (genotypes ss or sl) (Illi et al., 2011). This conclusion is further supported by a recent meta-analysis (Porcelli, Fabbri, & Serretti, 2012). According to the earlier mentioned yin-yang hypothesis of personality, one can assume that the serotonergic gene alleles that are negatively related to NE are indeed positively associated with PE.

A. 5-HTTLPR

As mentioned earlier, the prominent 5-HTTLPR was a stronger candidate polymorphism for NE rather than for PE. However, most studies aiming to relate the 5-HTTLPR to NE traits administered a complete personality questionnaire to their participants, resulting in unexpected correlations to PE traits. For example, no main effect of the 5-HTTLPR was found on any of the Big Five traits in a representative Swedish adolescent population sample; however, an epistatic effect between 5-HTTLPR and the functional BDNF Val66Met SNP was detected (Hiio et al., 2011). Brain derived neurotrophic factor is essential for brain plasticity, and BDNF levels influence neuronal sprouting, cell survival, and apoptosis. The BDNF gene has become a candidate gene for NE and is also related to hippocampal volume (Hajek, Kopecek, & Höschl, 2011; Montag, Weber, Fliessbach, Elger, & Reuter, 2009). Carriers of both the Met allele in BDNF Val66Met, along with the 5-HTTLPR s/s genotype had by far the lowest Conscientiousness scores when compared to the remaining allelic configurations. Conscientiousness, which is marked by a need for achievement, diligence, ambition, and orderliness, is the best personality predictor of job performance (Dudley, Orvis, Lebiecki, & Cortina, 2006). In an Estonian longitudinal population-based study investigating children at ages 9 and 15, ll genotype carriers had significantly higher scores in the PE-related personality traits of Openness, Agreeableness, and Conscientiousness when compared with s allele carriers (Harro et al., 2009). Lower Agreeableness scores in a US sample (Greenberg et al., 2000) and lower Cooperativeness in a Japanese sample were also associated with the s allele (Kumakiri et al., 1999). Along the same lines, the l allele of 5-HTTLPR was associated with elevated Extraversion scores in a US sample (Gillihan, Farah, Sankoorikal, Breland, & Brodkin, 2007). On the contrary, s allele carriers had significantly higher Openness scores in another US Caucasian sample (Stoltenberg et al., 2002). Comparison of subjects homozygous for the s allele with subjects with at least one l allele revealed significantly higher Persistence scores in the ss group in a Korean sample (Kim et al., 2005). In a combined Italian sample consisting of abstinent subjects who had never experimented with psychotropic drugs and subjects who had experimented with illegal drugs, ss carriers had higher Novelty Seeking scores than did ll carriers (Gerra et al., 2005).

B. 5-HT2A

The above-mentioned genetic association study by Heck et al. (2009) not only investigated dopaminergic genes but also serotonergic genes for a possible relationship to personality. Across two independent samples, an intronic SNP, rs927544, located in the 5-HT2A gene was associated with the personality trait of Novelty Seeking. Homozygous carriers of the T or the C allele had significantly higher Novelty Seeking scores than heterozygous CT carriers, and strongest effects were observed for the Extravagance subscale. Seretti and colleagues investigated four SNPs (rs643627, rs594242, rs6311, and rs6313) in the 5-HT2A gene with respect to their effects on TCI personality dimensions using healthy and pathological samples from Germany and Italy (Serretti et al., 2007). Overall, only marginal gene effects of 5-HT2A on personality were observable: rs594242 showed an association with Self-Directedness in the German sample, whereas rs6313 was marginally associated with Novelty Seeking in the Italian mood disorder sample. In an experimental approach, the A allele of the 5-HT2A -1438 A/G SNP was related to impulsive behavior in a go/no-go task in Japanese participants (Nomura et al., 2006). Impulsivity is an important subfacet of Novelty Seeking and had been part of Eysenck's Extraversion trait. In a Korean sample, there was a reported association between the GG genotype and higher Self-Directedness as measured by the TCI (Ham et al., 2004). Our own group showed an association between the 5-HT2A T102C SNP—which is in linkage with the functional -1438 A/G SNP—with the Absorption trait of the Tellegen Absorption Scale (Ott, Reuter, Hennig, & Vaitl, 2005). Participants with the T/T genotype of the T102C polymorphism (whose 5-HT2A receptor shows a stronger binding potential), had significantly higher Absorption scores. Although Absorption is not a classic PE trait, there is evidence that it is related to positive symptoms observed in schizophrenia. Absorption represents a disposition to experience altered states of consciousness characterized by intensively focused attention. It is correlated with hypnotic susceptibility and includes phenomena ranging from vivid perceptions and imaginations to mystical experiences. Recently, Absorption has been shown to modulate the reward circuitry of the brain while listening to music (Montag, Reuter, & Axmacher, 2011). In a Japanese study, no associations between five SNPs in the 5-HT2A gene, including T102C and -1438 A/G, and TCI-assessed personality were reported (Kusumi et al., 2002). However, the sample size of this study ($N = 133$) was small. Using a different personality inventory, the Karolinska Scales of Personality, no association was observed between these 5-HT2A SNPs and personality in a Swedish sample (Jönsson et al., 2001).

VI. Genome-Wide Association Studies

Given the small number of GWAS in personality research, we deviate from our aim to concentrate exclusively on GWAS reporting associations with PE and give a short overview of all personality GWAS published thus far.

Shifman et al. (2008) selected 2,054 extreme scorers from a large genomic data bank consisting of 88,142 English participants for their GWAS. They were able to show an association between a haplotype consisting of three SNPs on the PDE4D gene (rs702543, rs17782374, rs35277) and Neuroticism. Moreover, they also demonstrated a modest association between SNP rs10514901, which is located in a copy number variation region on chromosome 17q21.3, and Neuroticism.

In 2008, using a genome-wide association study approach, van den Oord and colleagues identified the MAM domain containing the glycosylphosphatidylinositol anchor 2 gene (MDGA 2 or MAMDC1) as a new candidate gene for Neuroticism (van den Oord et al., 2008). MDGA2 is a recently described gene that is expressed in a variety of human tissues, including the nervous system, and is proposed to be involved in regulating neuronal migration and axonal guidance (Díaz-Lopéz et al., 2005). The GWAS finding by van den Oord et al. (2008) could be replicated by Heck et al. (2011) who conducted fine mapping of the MDGA2 gene by surveying 100 SNPs. In addition, SNP rs2416054, also showed a strong association with the Harm Avoidance subscale HA4 "Fatigability and Asthenia" of Cloninger's TPQ that remained significant after correction for multiple testing. In the same year, Gillespie et al. (2008) were able to show several—although not statistically significant—associations between gene loci on diverse chromosomes with personality dimensions of the Junior Eysenck Personality questionnaire in 1,280 Australian twins and siblings.

Calboli et al. (2010) conducted the largest GWAS on Neuroticism thus far. They investigated more than 1.6 million SNPs in 2,235 British participants. The gene NKAIN2, which has not been previously reported in the context of association studies, showed the strongest association with Neuroticism. Another gene, GPC6 showed evidence suggestive of an interaction with age on Neuroticism. Although the authors found support for one previously reported association involving the phosphodiesterase 4D gene (PDE4D) (see Shifman et al., 2008, mentioned earlier), they failed to replicate other recent reports.

Terracciano et al. (2010) reported associations with all Big Five scales in their GWAS. They found associations between Neuroticism and SNAP25 (rs362584), Extraversion with BDNF and two cadherin genes (CDH13 and CDH2), Openness with CNTNAP2 (rs10251794), Agreeableness with CLOCK (rs6832769), and Conscientiousness with DYRK1A (rs2835731). The effect sizes were extremely small and explained less than 1 percent of the variance. Unfortunately, Terracciano and colleagues failed to replicate these findings in their follow-up studies.

Verweij et al. (2010) conducted a GWAS study on more than 5,000 Australian participants of European ancestry by investigating more than 1.2 million SNPs. They did not find any significant associations with Cloninger's personality traits after correcting for multiple testing. This is a pessimistic finding, given the fact that Verweij and colleagues tested a large sample with more than 90 percent power to detect SNPs that explain only 1 percent of the variance in personality. The authors conclude that individual common genetic variants of this size or greater do not contribute to personality trait variation. This pessimistic view is shared by others who are also of the opinion that psychiatric disorders are likely to be caused by the composite of many mildly harmful but rare polymorphisms, impossible to detect with conventional GWAS (Keller & Miller, 2006).

The only GWAS that deliberately investigated PE was conducted by Bartels et al. (2010). The trait under investigation was subjective Happiness, as assessed by the Subjective Happiness Scale (Lyubomirsky & Lepper, 1999). In a large sample of adolescent twins and their siblings, subjective Happiness was found to be part of the overall construct of Subjective Well-Being (SWB), and the broad-sense heritability of Happiness was estimated at 40 percent (Bartels & Boomsma, 2009). The genome scan investigated about 3,700 microsatellite markers at 1 cM resolution. This is a rather small amount of gene loci investigated when compared to traditional GWAS using SNPs. The Dutch group identified two loci, one on the end of the long arm of chromosome 19 (marker D19s254 at 110 cM) and one on the short arm of chromosome 1 (marker D1s534 at 153 cM) that showed at least a small association with Happiness.

VII. Conclusion

This chapter aims to provide an overview of the genetic basis of PE. Following a short introduction contrasting positive emotions as states from the trait PE, a short summary of the history of personality genetics is given. Also, the conceptualization of personality has a long tradition, leading to two clusters of personality theories—namely, the factorial and the biological theories—whereby only the latter aims to explain the causes behind individual differences in personality. Molecular genetics research on PE was largely inspired by quantitative genetic studies, impressively demonstrating the strong heritability of personality. The aim to identify the genes forming the basis of this heritability began with the candidate gene approach. Whereas the 5-HTTLPR was shown to be a robust predictor of NE and related psychopathologies, the findings for PE are very scarce and heterogeneous. Meta-analyses are the gold standards that bring clarity to this blurred picture, but even the few available NE meta-analyses have yielded conflicting results (Munafò et al., 2009; Schinka, Letsch, & Crawford, 2004; Sen et al., 2004), and a meta-analysis investigating the association between DRD4 and PE found negative results (Munafò et al., 2008). As in genetic association studies, meta-analyses have their methodological pitfalls related to inclusion criteria, ethical heterogeneity, and confounding influences of age and sex, as well as to heterogeneity in the assessment of the phenotype under investigation. Without conducting one's own meta-analysis, it is difficult to assess the potential contribution of the cited association studies toward identification of gene loci relevant to PE. However, the Val allele of COMT Val158 Met and the A1+ allele of DRD2 seem to be good candidates. With respect to the Val allele and PE, David Goldman introduced the "Warrior-Worrier" Model (Goldman, Oroszi, & Ducci, 2005). The Val allele associated with PE should reflect the warrior, whereas the Met allele, despite providing an advantage in cognitive functioning, should represent the worrier due to an association with NE.

The literature review also suggests that alleles associated with NE traits are often good candidates for association with low levels of PE; for example, the Val allele of COMT is associated with PE, whereas the Met allele is associated with NE (Enoch et al., 2003; Reuter & Hennig, 2005; Stein et al., 2005), the s allele of 5-HTTLPR is associated with NE and the LL genotype is associated with PE (e.g., Greenberg et al., 2000; Harro et al., 2009). This questions many personality theories claiming orthogonality of PE and NE.

Today, the candidate gene approach receives competition from GWAS studies. However, GWAS studies are extremely expensive, and positive findings have to also be replicated in independent samples. Correction for multiple testing is a further issue, as the exploratory approach of testing 3 million polymorphisms at a time harbors the risks of false-positive results. In addition, GWAS are unable to reveal the effects of genetic epistasis. However, in the end, even if GWAS generate reliable markers, these gene loci are again nothing other than candidates lacking proof of functionality, a topic addressed in the next paragraph that will show that we need a paradigm shift from industrialized high-throughput genotyping to cumbersome work at the lab bench.

A. *Demonstrating Functionality*

Without the intention of being exhaustive, the present review makes it clear that there is substantial literature reporting positive associations between PE traits and numerous polymorphisms. However, these studies are of a correlational nature, lacking proof of causality. We demonstrate this through the following example. Serretti et al. (2007) reported that the SNP rs594242 in the 5-HT2A gene is related to Self-Directedness. rs594242 is located in an intronic region within the 5-HT2A gene. For this reason, the SNP does not cause an amino-acid exchange, and so functionality might stem from alteration in the binding site for enhancer or repressor elements resulting in altered gene expression. If this is not the case, the reported association with PE can potentially be a false-positive finding, or it is possible that the locus is in linkage with another SNP that is the true source of altered 5-HT2A receptor functionality. Thus far, functionality (i.e., an effect on brain metabolism) could only be demonstrated for a small number of polymorphisms. Among these are the DRD2 Taq Ia, the COMT Val158Met, and the 5-HTTLPR. For the vast majority of polymorphisms, the evidence for functionality is missing. Demonstrating functionality is an extremely time-consuming and demanding process. This is the reason for the given disproportionality between the number of identified SNPs (a great merit of the Human Hapmap Project; www.hapmap.org) and the number of SNPs whose functionality has been tested. There are several ways to demonstrate SNP functionality in humans. It is possible to relate neurotransmitter availability or receptor density/affinity to an allelic variation through PET studies (imaging studies that use radioactive receptor ligands). Another way to test for functionality is to search for a potential association between mRNA expression in brain tissue and genetic variants. This is only possible if human cellular tissue is available (postmortem or gained through therapeutic surgery in epilepsy or cancer patients). The drawback of using tissue from living patients is the likelihood of obtaining necrotic tissue with altered cell metabolism, thereby confounding the effects of mRNA expression. The gold standard is to demonstrate functionality through cloning, wherein the human DNA sequence surrounding the SNP is inserted into a bacterial luciferase reporter plasmid and mRNA expression is measured through luciferase assays. A- SNP's influence on protein level is assessed, for example, through Western blot analysis. More indirect methods include investigating the association of allelic variants with neurotransmitter metabolites or hormones measured in the periphery (blood/saliva). Our own group used this method in the context of molecular genetics PE research. In a genetic association study (Reuter, Schmitz, Corr, & Hennig, 2006), we tested two nonsynonymous SNPs found in dopaminergic candidate genes—the previously mentioned DRD2 Taq Ia and the COMT Val158Met—for their effect on PE as defined by Gray's behavioral approach system (BAS). The rationale for choosing these SNPs was as follows. It is clear that the activity of the dopaminergic system is influenced by several processes: The synthesis of dopamine, its catabolism, its reuptake, and the density of its receptors. By investigating at least two of these processes simultaneously, the catabolism of dopamine (COMT Val 158Met) and the receptor density (DRD2 Taq Ia), a deeper understanding is expected. Results showed that neither DRD2 Taq Ia nor COMT Val158Met alone were related to PE, but the epistasis (interaction) effect of both gene loci was indeed related to PE. Carriers of the Val+/A1+ or the Val−/A1− configurations had significantly higher PE scores than the rest of the participants. The question arose as to what this epistasis effect might tell us about the relationship between dopamine and PE. Are the allelic configurations Val+/A1+ and Val−/A1− related to high or low dopamine levels? We attempted to answer this question by means of an endocrinological approach. Due to the fact that dopamine is the primary inhibiting molecule of the pituitary hormone prolactin (PRL), it is possible to infer dopamine levels from PRL blood levels. Results showed that the earlier mentioned allelic configurations associated with high PE are related to low PRL levels, indicating high central dopamine levels. Therefore, PE seems to be related to high dopamine levels, as theoretically hypothesized (Cloninger et al., 1993), and not to low dopamine levels, as inferred by Eysenck's personality theory (Eysenck, 1990) that postulated low cortex arousability in extraverts.

B. *Future Perspectives*

Quantitative geneticists have claimed that genetic variance in personality traits is almost exclusively additive. However, this is difficult to accept, given that neurotransmitter systems interact in the brain, and epistasis effects have been reported (e.g., Hünnerkopf et al., 2007; Montag et al., 2010; Reuter et al., 2006). GWAS conducted thus far concentrate on the identification of the genetic contribution of single loci, thereby neglecting gene–gene interactions. Finding these interactions will be a major challenge for computational bioinformatics, along with handling the large resulting datasets. A further direction will be the proof of functionality, as discussed in the preceding paragraph. Until the functional impact of a gene variant on brain biochemistry is demonstrated, these studies are only correlational and provide no causal explanation for individual differences in personality. In our opinion, the investigation of epigenetic factors will give us new insights into gene–environment interactions (Ballestar, 2011). For a long time, it was inconceivable that the environment could influence our genes. On a descriptive level, a very prominent study by Caspi et al. (2003) supported the idea of gene–environment interaction by showing that the s allele of 5-HTTLPR is only a risk factor for depression in the presence of critical life events. We now assume that this process is mediated by epigenetic effects. Environmental stimuli can influence the methylation and histone acetylation status of our genes, thereby rendering them active or inactive by influencing gene transcription.

We will be eager to see what the future holds and are optimistic due to the substantial progress made in molecular genetics over the course of the past few decades.

References

Angoff, W. H. (1988). The nature-nurture debate, aptitudes, and group differences. *American Psychologist, 43*(9), 713–720.

Antolin, T., Berman, S. M., Conner, B. T., Ozkaragoz, T. Z., Sheen, C. L., Ritchie, T. L., & Noble, E. P. (2009). D2 dopamine receptor (DRD2) gene, P300, and personality in children of alcoholics. *Psychiatry Research, 166*(2–3), 91–101. Epub 2009 Mar 10.

Ballestar, E. (2011). An introduction to epigenetics. *Advances in Experimental Medicine and Biology, 711*, 1–11.

Bartels, M., & Boomsma, D. I. (2009). Born to be happy? The etiology of Subjective Wellbeing. *Behavior Genetics, 39*, 605–615.

Bartels, M., Saviouk, V., de Moor, M. H., Willemsen, G., van Beijsterveldt, T. C., Hottenga, J. J., de Geus, E. J., & Boomsma, D. I. (2010). Heritability and genome-wide linkage scan of subjective happiness. *Twin Research and Human Genetics, 13*(2), 135–142.

Benjamin, J., Li, L., Patterson, C., Greenberg, B. D., Murphy, D. L., & Hamer, D. H. (1996). Population and familial association between the D4 dopamine receptor gene and

measures of Novelty Seeking. *Nature Genetics, 12*(1), 81–84.
Benjamin, J., Osher, Y., Kotler, M., Gritsenko, I., Nemanov, L., Belmaker, R. H., et al. (2000). Association between tridimensional personality questionnaire (TPQ) traits and three functional polymorphisms: Dopamine receptor D4 (DRD4), serotonin transporter promoter region (5-HTTLPR) and catechol O-methyltransferase (COMT). *Molecular Psychiatry, 5*, 96–100.
Brene, S., Lindefors, N., Ehrlich, M., Taubes, T., Horiuchi, A., Kopp, J., et al. (1994). Expression of mRNAs encoding ARPP-16/19, ARPP-21, and DARPP-32 in human brain tissue. *Journal of Neuroscience, 14*, 985–998.
Brown, G. W., & Harris, T. O. (2008). Depression and the serotonin transporter 5-HTTLPR polymorphism: A review and a hypothesis concerning gene-environment interaction. *Journal of Affective Disorders, 111*(1), 1–12.
Calboli, F. C., Tozzi, F., Galwey, N. W., Antoniades, A., Mooser, V., Preisig, M., et al. (2010). A genome-wide association study of neuroticism in a population-based sample. *PLoS One, 5*(7), e11504.
Caspi, A., Sugden, K., Moffitt, T. E., Taylor, A., Craig, I. W., Harrington, et al. (2003). Influence of life stress on depression: Moderation by a polymorphism in the 5-HTT gene. *Science, 301*(5631), 386–389.
Cloninger, C. R., Przybeck, T. R., & Svrakic, D. M. (1991). The Tridimensional Personality Questionnaire: U. S. normative data. *Psychological Reports, 69*(3 Pt 1), 1047–1057.
Cloninger, C. R., Przybeck, T. R., Svrakic, D. M., & Wetzel, R. D. (1994). *The temperament and character inventory (TCI): A guide to its development and use*. St. Louis: Washington University Center for Psychobiology of Personality.
Cloninger, C. R., Svrakic, D. M., & Przybeck, T. R. (1993). A psychobiological model of temperament and character. *Archives of General Psychiatry, 50*(12), 975–990.
Costa, P. T. Jr., & McCrae, R. R. (1985). *The NEO personality inventor*. Odessa, FL: Psychological Assessment Resources.
Cruz, C., Camarena, B., Mejía, J. M., Páez, F., Eroza, V., Ramón, De La Fuente, J., et al. (1995). The dopamine D2 receptor gene TaqI A1 polymorphism and alcoholism in a Mexican population. *Archives of Medical Research, 26*(4), 421–426.
Davis, K. L., Panksepp, J., & Normansell, L. (2003). The affective neuroscience personality scales: Normative data and implications. *Neuro-Psychoanalysis, 5*(1), 57–69.
Depue, R. A., & Collins, P. F. (1999). Neurobiology of the structure of personality: Dopamine, facilitation of incentive motivation, and extraversion. *Behavioral and Brain Sciences, 22*(3), 491–517; discussion 518–569.
Depue, R. A., Luciana, M., Arbisi, P., Collins, P., & Leon, A. (1994). Dopamine and the structure of personality: Relation of agonist-induced dopamine activity to positive emotionality. *Journal of Personality and Social Psychology, 67*(3), 485–498.
Díaz-López, A., Rivas CIniesta, P., Morán, A., García-Aranda, C., Megías, D., Sánchez-Pernaute, A., Torres, A., et al. (2005). Characterization of MDGA1, a novel human glycosylphosphatidylinositol-anchored protein localized in lipid rafts. *Experimental Cell Research, 307*(1), 91–99.
Domschke, K., & Dannlowski, U. (2010). Imaging genetics of anxiety disorders. *Neuroimage, 53*(3), 822–831. Epub 2009 Nov 26.
Dudley, N. M., Orvis, K. A., Lebiecki, J. E., & Cortina, J. M. (2006). A meta-analytic investigation of conscientiousness in the prediction of job performance: examining the intercorrelations and the incremental validity of narrow traits. *Journal of Applied Psychology, 91*(1), 40–57.
Ebstein, R. P., Novick, O., Umansky, R., Priel, B., Osher, Y., Blaine, D., et al. (1996). Dopamine D4 receptor (D4DR) exon III polymorphism associated with the human personality trait of novelty seeking. *Nature Genetics, 12*(1), 78–80.
Enoch, M. A., Schuckit, M. A., Johnson, B. A., & Goldman, D. (2003). Genetics of alcoholism using intermediate phenotypes. *Alcoholism: Clinical and Experimental Research, 27*(2), 169–176.
Ekman, P. (1992). Are there basic emotions? *Psychological Review, 99*, 550–553.
Eysenck, H. J. (1990). Biological dimensions of personality. In L. A. Pervin (Ed.), *Handbook of personality: Theory and research* (pp. 244–276). New York: Guilford.
Felten, A., Montag, C., Markett, S., Walter, N. T., & Reuter, M. (2011). Genetically determined dopamine availability predicts disposition for depression. *Brain and Behavior*, 1, 109–118.
Galton, F. (1869). Hereditary genius: An inquiry into its laws and consequences. London: Macmillan & Co.
Gebhardt, C., Leisch, F., Schüssler, P., Fuchs, K., Stompe, T., Sieghart, W., et al. (2000). Non-association of dopamine D4 and D2 receptor genes with personality in healthy individuals. *Psychiatric Genetics, 10*, 131–137.
Gerra, G., Garofano, L., Castaldini, L., Rovetto, F., Zaimovic, A., Moi, G., et al. (2005). Serotonin transporter promoter polymorphism genotype is associated with temperament, personality traits and illegal drugs use among adolescents. *Journal of Neural Transmission, 112*(10), 1397–1410.
Gelernter, J., Kennedy, J. L., van Tol, H. H., Civelli, O., & Kidd, K. K. (1992). The D4 dopamine receptor (DRD4) maps to distal 11p close to HRAS. *Genomics, 13*(1), 208–210.
Gelernter, J., Kranzler, H., Coccaro, E., Siever, L., New, A., & Mulgrew, C. L. (1997). D4 dopamine-receptor (DRD4) alleles and novelty seeking in substance-dependent, personality-disorder, and control subjects. *American Journal of Human Genetics, 61*, 1144–1152.
Gillespie, N. A., Cloninger, C. R., Heath, A. C., & Martin, N. G. (2003). The genetic and environmental relationship between Cloninger's dimensions of temperament and character. *Personality and Individual Differences, 35*, 1931–1946.
Gillespie, N. A., Zhu, G., Evans, D. M., Medland, S. E., Wright, M. J., & Martin, N. G. (2008). A genome-wide scan for Eysenckian personality dimensions in adolescent twin sibships: Psychoticism, extraversion, neuroticism, and lie. *Journal of Personality, 76*(6), 1415–1446.
Gillihan, S. J., Farah, M. J., Sankoorikal, G. M., Breland, J., & Brodkin, E. S. (2007). Association between serotonin transporter genotype and extraversion. *Psychiatric Genetics, 17*(6), 351–354.
Goldman, D., Oroszi, G., & Ducci, F. (2005). The genetics of addictions: Uncovering the genes. *Nature Reviews Genetics, 6*(7), 521–532.
Gray, J. A. (1970). The psychophysiological basis of introversion-extraversion. *Behaviour Research and Therapy, 8*, 249–266.

Gray, J. A., & McNaughton, N. (2000). *The neuropsychology of anxiety: An enquiry into the functions of the septo-hippocampal system* (2 ed.). (Oxford Psychology Series No. 33). Oxford: Oxford University Press.

Greenberg, B. D., Li, Q., Lucas, F. R., Hu, S., Sirota, L. A., Benjamin, J., Lesch, K. P., et al. (2000). Association between the serotonin transporter promoter polymorphism and personality traits in a primarily female population sample. *American Journal of Medical Genetics, 96*(2), 202–216.

Hajek, T., Kopecek, M., & Höschl, C. (2011). Reduced hippocampal volumes in healthy carriers of brain-derived neurotrophic factor Val66Met polymorphism: Meta-analysis. *World Journal of Biological Psychiatry*, [Epub ahead of print] PubMed PMID: 21722019.

Ham, B. J., Kim, Y. H., Choi, M. J., Cha, J. H., Choi, Y. K., & Lee, M. S. (2004). Serotonergic genes and personality traits in the Korean population. *Neuroscience Letters, 354*(1), 2–5.

Hamidovic, A., Dlugos, A., Palmer, A. A., de Wit, H. (2010). Catechol-O-methyltransferase val158met genotype modulates sustained attention in both the drug-free state and in response to amphetamine. *Psychiatric Genetics, 20*(3), 85–92.

Harro, J., Merenäkk, L., Nordquist, N., Konstabel, K., Comasco, E., & Oreland, L. (2009). Personality and the serotonin transporter gene: Associations in a longitudinal population-based study. *Biological Psychology, 81*(1), 9–13.

Hashimoto, R., Noguchi, H., Hori, H., Ohi, K., Yasuda, Y., Takeda, M., & Kunugi, H. (2007). A possible association between the Val158Met polymorphism of the catechol-O-methyl transferase gene and the personality trait of harm avoidance in Japanese healthy subjects. *Neuroscience Letters, 428*(1), 17–20.

Heck, A., Lieb, R., Ellgas, A., Pfister, H., Lucae, S., Roeske, D., et al. (2009). Investigation of 17 candidate genes for personality traits confirms effects of the HTR2A gene on novelty seeking. *Genes, Brain and Behavior, 8*(4), 464–472.

Heck, A., Pfister, H., Czamara, D., Müller-Myhsok, B., Pütz, B., Lucae, S., et al. (2011). Evidence for associations between MDGA2 polymorphisms and harm avoidance—replication and extension of a genome-wide association finding. *Psychiatric Genetics*, [Epub ahead of print] PubMed PMID: 21399569.

Herbst, J. H., Zonderman, A. B., McCrae, R. R., & Costa, P. T., Jr. (2000). Do the dimensions of the temperament and character inventory map a simple genetic architecture? Evidence from molecular genetics and factor analysis. *American Journal of Psychiatry, 157*, 1285–1290.

Hibino, H., Tochigi, M., Otowa, T., Kato, N., & Sasaki, T. (2006). No association of DRD2, DRD3, and tyrosine hydroxylase gene polymorphisms with personality traits in the Japanese population. *Behavioral and Brain Functions, 2*, 32.

Hiio, K., Merenäkk, L., Nordquist, N., Parik, J., Oreland, L., Veidebaum, T., & Harro, J. (2011). Effects of serotonin transporter promoter and BDNF Val66Met genotype on personality traits in a population representative sample of adolescents. *Psychiatric Genetics, 21*(5), 261–264. doi: 10.1097/YPG.0b013e32834371e8

Hill, S. Y., Zezza, N., Wipprecht, G., Locke, J., & Neiswanger, K. (1999). Personality traits and dopamine receptors (D2 and D4): Linkage studies in families of alcoholics. *American Journal of Medical Genetics, 88*, 634–641.

Hoth, K. F., Paul, R. H., Williams, L. M., Dobson-Stone, C., Todd, E., Schofield, P. R., et al. (2006). Associations between the COMT Val/Met polymorphism, early life stress, and personality among healthy adults. *Neuropsychiatric Disease and Treatment, 2*(2), 219–225.

Hünnerkopf, R., Strobel, A., Gutknecht, L., Brocke, B., & Lesch, K. P. (2007). Interaction between BDNF Val66Met and dopamine transporter gene variation influences anxiety-related traits. *Neuropsychopharmacology, 32*(12), 2552–2560.

Illi, A., Poutanen, O., Setälä-Soikkeli, E., Kampman, O., Viikki, M., Huhtala, H., et al. (2011). Is 5-HTTLPR linked to the response of selective serotonin reuptake inhibitors in MDD? *European Archives of Psychiatry and Clinical Neuroscience, 261*(2), 95–102.

Jönsson, E. G., Nöthen, M. M., Gustavsson, J. P., Berggård, C., Bunzel, R., Forslund, K., et al. (2001). No association between serotonin 2A receptor gene variants and personality traits. *Psychiatric Genetics, 11*(1), 11–17.

Jönsson, E. G., Nöthen, M. M., Gustavsson, J. P., Neidt, H., Brené, S., Tylec, A., et al. (1997). Lack of evidence for allelic association between personality traits and the dopamine D4 receptor gene polymorphisms. *American Journal of Psychiatry, 154*(5), 697–699.

Jönsson, E. G., Nöthen, M. M., Gustavsson, J. P., Neidt, H., Forslund, K., Mattila-Evenden, M., et al. (1998). Lack of association between dopamine D4 receptor gene and personality traits. *Psychological Medicine, 28*, 985–989.

Kazantseva, A., Gaysina, D., Malykh, S., & Khusnutdinova, E. (2011). The role of dopamine transporter (SLC6A3) and dopamine D2 receptor/ankyrin repeat and kinase domain containing 1 (DRD2/ANKK1) gene polymorphisms in personality traits. *Progress in Neuropsychopharmacology and Biological Psychiatry, 35*(4), 1033–1040.

Keller, M. C., & Miller, G. (2006). Resolving the paradox of common, harmful, heritable mental disorders: Which evolutionary genetic models work best? *Behavioral and Brain Sciences, 29*(4), 385–404.

Keltikangas-Järvinen, L., Elovainio, M., Kivimäki, M., Lichtermann, D., Ekelund, J., & Peltonen, L. (2003). Association between the type 4 dopamine receptor gene polymorphism and novelty seeking. *Psychosomatic Medicine, 65*, 471–476.

Kim, S. J., Kim, Y. S., Choi, N. K., Hong, H. J., Lee, H. S., & Kim, C. H. (2005). Serotonin transporter gene polymorphism and personality traits in a Korean population. *Neuropsychobiology, 51*(4), 243–247. Epub 2005 May 18.

Kim, S. J., Kim, Y. S., Kim, C. H., & Lee, H. S. (2006). Lack of association between polymorphisms of the dopamine receptor D4 and dopamine transporter genes and personality traits in a Korean population. *Yonsei Medical Journal, 47*, 787–792 B.

Kumakiri, C., Kodama, K., Shimizu, E., Yamanouchi, N., Okada, S., Noda, S., et al. (1999). Study of the association between the serotonin transporter gene regulatory region polymorphism and personality traits in a Japanese population. *Neuroscience Letters, 263*(2–3), 205–207.

Krubitzer, L., & Kahn, D. M. (2003). Nature versus nurture revisited: An old idea with a new twist. *Progress in Neurobiology, 70*(1), 33–52.

Kühn, K. U., Meyer, K., Nöthen, M. M., Gänsicke, M., Papassotiropoulos, A., & Maier, W. (1999). Allelic variants of dopamine receptor D4 (DRD4) and serotonin

receptor 5HT2c (HTR2c) and temperament factors: replication tests. *American Journal of Medical Genetics, 88*, 168–172.
Kusumi, I., Suzuki, K., Sasaki, Y., Kameda, K., Sasaki, T., & Koyama, T. (2002). Serotonin 5-HT(2A) receptor gene polymorphism, 5-HT(2A) receptor function and personality traits in healthy subjects: a negative study. *Journal of Affective Disorders, 68*(2–3), 235–241.
Lachman, H. M., Papolos, D. F., Saito, T., Yu, Y. M., Szumlanski, C. L., & Weinshilboum, R. M. (1996). Human catechol-O-methyltransferase pharmacogenetics: Description of a functional polymorphism and its potential application to neuropsychiatric disorders. *Pharmacogenetics, 6*, 243–250.
Laine, T. P., Ahonen, A., Räsänen, P., & Tiihonen, J. (2001). Dopamine transporter density and novelty seeking among alcoholics. *Journal of Addictive Diseases, 20*(4), 91–96.
Leboyer, M., Bouvard, M. P., Recasens, C., Philippe, A., Guilloud-Bataille, M., Bondoux, D., et al. (1994). Difference between plasma N-and C-terminally directed beta-endorphin immunoreactivity in infantile autism. *American Journal of Psychiatry, 151*(12), 1797–1801.
Lee, H. J., Lee, H. S., Kim, Y. K., Kim, L., Lee, M. S., Jung, I. K., et al. (2003a). D2 and D4 dopamine receptor gene polymorphisms and personality traits in a young Korean population. *American Journal of Medical Genetics, B: NeuroPsychiatric Genetics, 121*, 44–49.
Lee, H. J., Lee, H. S., Kim, Y. K., Kim, S. H., Kim, L., Lee, M. S., et al. (2003b). Allelic variants interaction of dopamine receptor D4 polymorphism correlate with personality traits in young Korean female population. *American Journal of Medical Genetics, B: NeuroPsychiatric Genetics, 118B*, 76–80.
Lee, S. H., Ham, B. J., Cho, Y. H., Lee, S. M., & Shim, S. H. (2007). Association study of dopamine receptor D2 TaqI A polymorphism and reward-related personality traits in healthy Korean young females. *Neuropsychobiology, 56*(2–3), 146–151.
Lesch, K. P., Bengel, D., Heils, A., Sabol, S. Z., Greenberg, B. D., Petri, S., Benjamin, J., Müller, C. R., Hamer, D. H., & Murphy, D. L. (1996). Association of anxiety-related traits with a polymorphism in the serotonin transporter gene regulatory region. *Science, 274*(5292), 1527–1531.
Li, J., Ma, H., Zhou, H., Huang, Y., Wu, L., Li, J., & Zhu, G. (2011). Association between DARPP-32 gene polymorphism and personality traits in healthy Chinese-Han subjects. *Journal of Molecular Neuroscience, 44*(1), 48–52.
Lyubomirsky, S., & Lepper, H. S. (1999). A measure of subjective happiness: Preliminary reliability and construct validation. *Social Indicators Research, 46*, 137–155.
Malhotra, A. K., Virkkunen, M., Rooney, W., Eggert, M., Linnoila, M., & Goldman, D. (1996). The association between the dopamine D4 receptor (D4DR) 16 amino acid repeat polymorphism and novelty seeking. *Molecular Psychiatry, 1*, 388–391.
Meyer-Lindenberg, A., Kohn, P. D., Kolachana, B., Kippenhan, S., McInerney-Leo, A., Nussbaum, R., et al. (2005). Midbrain dopamine and prefrontal function in humans: Interaction and modulation by COMT genotype. *Nature Neuroscience, 8*(5), 594–596.
Mitsuyasu, H., Hirata, N., Sakai, Y., Shibata, H., Takeda, Y., Ninomiya, H., et al. (2001). Association analysis of polymorphisms in the upstream region of the human dopamine D4 receptor gene (DRD4) with schizophrenia and personality traits. *Journal of Human Genetics, 46*, 26–31.
Montag, C., Markett, S., Basten, U., Stelzel, C., Fiebach, C., Canli, T. & Reuter, M. (2010). Epistasis of the DRD2/ANKK1 Taq Ia and the BDNF Val66Met polymorphism impacts novelty seeking and harm avoidance. *Neuropsychopharmacology, 35*, 1860–1867.
Montag, C., Reuter, M., & Axmacher, N. (2011). How one's favorite song activates the reward circuitry of the brain: Personality matters! *Behavioural Brain Research, 225*, 511–514.
Montag, C., Weber, B., Fliessbach, K., Elger, C., & Reuter, M. (2009). The BDNF Val66Met polymorphism impacts parahippocampal and amygdala volume in healthy humans: Incremental support for a genetic risk factor for depression. *Psychological Medicine, 39*, 1831–1839.
Munafò, M. R., Freimer, N. B., Ng, W., Ophoff, R., Veijola, J., Miettunen, J., et al. (2009). 5-HTTLPR genotype and anxiety-related personality traits: a meta-analysis and new data. *American Journal of Medical Genetics, B: NeuroPsychiatric Genetics, 150B*(2), 271–281.
Munafò, M. R., Yalcin, B., Willis-Owen, S. A., & Flint, J. (2008). Association of the dopamine D4 receptor (DRD4) gene and approach-related personality traits: Meta-analysis and new data. *Biological Psychiatry, 63*, 197–206.
Noble, E. P., Ozkaragoz, T. Z., Ritchie, T. L., Zhang, X., Belin, T. R., & Sparkes, R. S. (1998). D2 and D4 dopamine receptor polymorphisms and personality. *American Journal of Medical Genetics, 81*, 257–267.
Nomura, M., Kusumi, I., Kaneko, M., Masui, T., Daiguji, M., Ueno, T., et al. (2006). Involvement of a polymorphism in the 5-HT2A receptor gene in impulsive behavior. *Psychopharmacology* (*Berl*), *187*(1), 30–35.
Nyman, E. S., Loukola, A., Varilo, T., Ekelund, J., Veijola, J., Joukamaa, M., et al. (2009). Impact of the dopamine receptor gene family on temperament traits in a population-based birth cohort. *American Journal of Medical Genetics, B: NeuroPsychiatric Genetics, 150B*(6), 854 865.
Ono, Y., Manki, H., Yoshimura, K., Muramatsu, T., Mizushima, H., Higuchi, S., et al. (1997). Association between dopamine D4 receptor (D4DR) exon III polymorphism and novelty seeking in Japanese subjects. *American Journal of Medical Genetics, 74*, 501–503.
Ouimet, C. C., Miller, P. E., Hemmings, H. C. J., Walaas, S. I., & Greengard, P. (1984). DARPP-32, a dopamine- and adenosine 39:59-monophosphate-regulated phosphoprotein enriched in dopamine-innervated brain regions. III. Immunocytochemical localization. *Journal of Neuroscience, 4*, 111–124.
Ott, U., Reuter, M., Hennig, J., & Vaitl, D. (2005). Evidence for a common biological basis of the Absorption trait, hallucinogen effects, and positive symptoms: Epistasis between 5-HT2a and COMT polymorphisms. *American Journal of Medical Genetics, B: NeuroPsychiatric Genetics, 137B*(1), 29–32.
Panksepp, J. (1998). *Affective neuroscience: The foundations of human and animal motions* (Series in Affective Science). New York: Oxford University Press.
Persson, M. L., Wasserman, D., Geijer, T., Frisch, A., Rockah, R., Michaelovsky, E., et al. (2000). Dopamine D4 receptor gene polymorphism and personality traits in healthy volunteers. *European Archives of Psychiatry and Clinical Neuroscience, 250*, 203–206.

Plomin, R., DeFries, J. C., McClearn, G. E., & McGuffin, P. (2008). *Behavioral genetics* (5th ed.). New York: Worth Publishers.
Pogue-Geile, M., Ferrell, R., Deka, R., Debski, T., & Manuck, S. (1998). Human novelty-seeking personality traits and dopamine D4 receptor polymorphisms: A twin and genetic association study. *American Journal of Medical Genetics, 81*, 44–48.
Porcelli, S., Fabbri, C., & Serretti, A. (2012). Meta-analysis of serotonin transporter gene promoter polymorphism (5-HTTLPR) association with antidepressant efficacy. *European Neuropsychopharmacology, 22*(4), 239–258.
Poston II, W. S., Ericsson, M., Linder, J., Haddock, C. K., Hanis, C. L., Nilsson, T., et al. (1998). D4 dopamine receptor gene exon III polymorphism and obesity risk. *Eating and Weight Disorders, 3*, 71–77.
Reuter, M. (2008). Genetics and neuroimaging studies. In P. Corr (Ed.), *The reinforcement sensitivity theory of personality* (pp. 317–344). Cambridge: Cambridge University Press.
Reuter, M., & Hennig, J. (2005). Association of the functional catechol-O-methyltransferase VAL158MET polymorphism with the personality trait of extraversion. *Neuroreport, 16*(10), 1135–1138.
Reuter, M., Netter, P., Toll, C., & Hennig, J. (2002). Dopamine agonist and antagonist responders as related to types of nicotine craving and facets of extraversion. *Progress in Neuropsychopharmacology and Biological Psychiatry, 26*(5), 845–853.
Reuter, M., Schmitz, A., Corr, P., & Hennig, J. (2006). Molecular genetics support Gray's personality theory: The interaction of COMT and DRD2 polymorphisms predicts the behavioural approach system. *International Journal of Neuropsychopharmacology, 9*(2), 155–166.
Reuter, M., Weber, B., Fiebach, C. J., Elger, C., & Montag, C. (2009). The biological basis of anger: associations with the gene coding for DARPP-32 (PPP1R1B) and with amygdala volume. *Behavioral Brain Research, 202*(2), 179–183.
Ritchie, T., & Noble, E. P. (2003). Association of seven polymorphisms of the D2 dopamine receptor gene with brain receptor-binding characteristics. *Neurochemical Research, 28*, 73–82.
Samochowiec, J., Rybakowski, F., Czerski, P., Zakrzewska, M., Stepień, G., Pełka-Wysiecka, J., et al. (2001). Polymorphisms in the dopamine, serotonin, and norepinephrine transporter genes and their relationship to temperamental dimensions measured by the Temperament and Character Inventory in healthy volunteers. *Neuropsychobiology, 43*(4), 248–253.
Schildkraut, J. J. (1974). Biogenic amines and affective disorders. *Annual Review of Medicine, 25*(0), 333–348.
Schinka, J. A., Busch, R. M., & Robichaux-Keene, N. (2004). A meta-analysis of the association between the serotonin transporter gene polymorphism (5-HTTLPR) and trait anxiety. *Molecular Psychiatry, 9*(2), 197–202.
Seeman, P. (2010). Dopamine D2 receptors as treatment targets in schizophrenia. *Clinical Schizophrenia and Related Psychoses, 4*(1), 56–73.
Seeman, M. V., & Seeman, P. (1988). Psychosis and positron tomography. *Canadian Journal of Psychiatry, 33*(4), 299–306.
Sen, S., Burmeister, M., & Ghosh, D. (2004). Meta-analysis of the association between a serotonin transporter promoter polymorphism (5-HTTLPR) and anxiety-related personality traits. *American Journal of Medical Genetics, B NeuroPsychiatric Genetics, 127B*(1), 85–89.
Serretti, A., Mandelli, L., Lorenzi, C., Landoni, S., Calati, R., Insacco, C., & Cloninger, C. R. (2006). Temperament and character in mood disorders: influence of DRD4, SERTPR, TPH and MAO-A polymorphisms. *Neuropsychobiology, 53*(1), 9–16.
Serretti, A., Calati, R., Giegling, I., Hartmann, A. M., Möller, H. J., Colombo, C., & Rujescu, D. (2007). 5-HT2A SNPs and the Temperament and Character Inventory. *Progress in Neuropsychopharmacology and Biological Psychiatry, 31*(6), 1275–1281.
Serretti, A., Mandelli, L., Lorenzi, C., Landoni, S., Calati, R., Insacco, C., et al. (2006). Temperament and character in mood disorders: Influence of DRD4, SERTPR, TPH and MAO-A polymorphisms. *Neuropsychobiology, 53*, 9–16.
Sheldrick, A. J., Krug, A., Markov, V., Leube, D., Michel, T. M., Zerres, K., et al. (2008). Effect of COMT val158met genotype on cognition and personality. *European Psychiatry, 23*(6), 385–389.
Shibuya, N., Kamata, M., Suzuki, A., Matsumoto, Y., Goto, K., & Otani, K. (2009). The -67 A/T promoter polymorphism in the dopamine transporter gene affects personality traits of Japanese healthy females. *Behavioral Brain Research, 203*(1), 23–26.
Shifman, S., Bhomra, A., Smiley, S., Wray, N. R., James, M. R., Martin, N. G., et al. (2008). A whole genome association study of neuroticism using DNA pooling. *Molecular Psychiatry, 13*(3), 302–312. Epub 2007 Jul 31.
Schinka, J. A., Letsch, E. A., & Crawford, F. C. (2002). DRD4 and novelty seeking: Results of meta-analyses. *American Journal of Medical Genetics, 114*, 643–648.
Shifman, S., Bhomra, A., Smiley, S., Wray, N. R., James, M. R., et al. (2008). A whole genome association study of neuroticism using DNA pooling. *Molecular Psychiatry, 13*, 302–312.
Smeraldi, E., Zanardi, R., Benedetti, F., Di Bella, D., Perez, J., & Catalano, M. (1998). Polymorphism within the promoter of the serotonin transporter gene and antidepressant efficacy of fluvoxamine. *Molecular Psychiatry, 3*, 508–511.
Smillie, L. D., Cooper, A. J., Proitsi, P., Powell, J. F., & Pickering, A. D. (2010). Variation in DRD2 dopamine gene predicts Extraverted personality. *Neuroscience Letters, 468*(3), 234–237.
Spanagel, R., & Weiss, F. (1999). The dopamine hypothesis of reward: Past and current status. *Trends in Neuroscience, 22*(11), 521–527.
Stein, M. B., Fallin, M. D., Schork, N. J., & Gelernter, J. (2005). COMT polymorphisms and anxiety-related personality traits. *Neuropsychopharmacology, 30*(11), 2092–2102.
Stoltenberg, S. F., Twitchell, G. R., Hanna, G. L., Cook, E. H., Fitzgerald, H. E., Zucker, R. A., & Little, K. Y. (2002). Serotonin transporter promoter polymorphism, peripheral indexes of serotonin function, and personality measures in families with alcoholism. *American Journal of Medical Genetics, 114*(2), 230–234.
Strobel, A., Lesch, K. P., Hohenberger, K., Jatzke, S., Gutzeit, H. O., Anacker, K., et al. (2002). No association between dopamine D4 receptor gene exon III and -521C/T polymorphism and novelty seeking. *Molecular Psychiatry, 7*, 537–538.
Strobel, A., Spinath, F. M., Angleitner, A., Riemann, R., & Lesch, K. P. (2003). Lack of association between

polymorphisms of the dopamine D4 receptor gene and personality. *Neuropsychobiology, 47*, 52–56.
Strobel, A., Wehr, A., Michel, A., & Brocke, B. (1999). Association between the dopamineD4 receptor (DRD4) exon III polymorphism and measures of novelty seeking in a German population. *Molecular Psychiatry, 4*, 378–384.
Szekely, A., Ronai, Z., Nemoda, Z., Kolmann, G., Gervai, J., & Sasvari-Szekely, M. (2004). Human personality dimensions of persistence and harm avoidance associated with DRD4 and 5-HTTLPR polymorphisms. *American Journal of Medical Genetics, B NeuroPsychiatric Genetics, 126B*, 106–110.
Taylor, M. J., Freemantle, N., Geddes, J. R., & Bhagwagar, Z. (2006). Early onset of selective serotonin reuptake inhibitor antidepressant action: systematic review and meta-analysis. *Archives of General Psychiatry*, 2006 Nov, *63*(11):1217–23.
Tellegen A. (1982). *Multidimensional personality questionnaire manual.* Minneapolis, MN: University of Minnesota Press.
Terracciano, A., Sanna, S., Uda, M., Deiana, B., Usala, G., Busonero, F., et al. (2010). Genome-wide association scan for five major dimensions of personality. *Molecular Psychiatry 15*(6), 647–656.
Tochigi, M., Hibino, H., Otowa, T., Kato, C., Marui, T., Ohtani, T, et al. (2006). Association between dopamine D4 receptor (DRD4) exon III polymorphism and neuroticism in the Japanese population. *Neuroscience Letters, 398*, 333–336.
Tomitaka, M., Tomitaka, S., Otuka, Y., Kim, K., Matuki, H., Sakamoto, K., et al. (1999). Association between novelty seeking and dopamine receptor D4 (DRD4) exon III polymorphism in Japanese subjects. *American Journal of Medical Genetics, 88*, 469–471.
Tsuchimine, S., Yasui-Furukori, N., Kaneda, A., Saito, M., Sugawara, N., & Kaneko, S. (2009). Minor genetic variants of the dopamine D4 receptor (DRD4) polymorphism are associated with novelty seeking in healthy Japanese subjects. *Progress in Neuropsychopharmacology and Biological Psychiatry, 33*(7), 1232–1235.
Tunbridge, E. M., Harrison, P. J., & Weinberger, D. R. (2006). Catechol-o-methyltransferase, cognition, and psychosis: Val158Met and beyond. *Biological Psychiatry, 60*(2), 141–151.
van den Oord, E. J., Kuo, P. H., Hartmann, A. M., Webb, B. T., Möller, H. J., Hettema, J. M., et al. (2008). Genomewide association analysis followed by a replication study implicates a novel candidate gene for neuroticism. *Archives of General Psychiatry, 65*(9), 1062–1071.
Verweij, K. J., Zietsch, B. P., Medland, S. E., Gordon, S. D., Benyamin, B., Nyholt, D. R., et al. (2010). A genome-wide association study of Cloninger's temperament scales: Implications for the evolutionary genetics of personality. *Biological Psychology, 85*(2), 306–317.
Volkow, N. D., Fowler, J. S., & Wang, G. J. (1999). Imaging studies on the role of dopamine in cocaine reinforcement and addiction in humans. *Journal of Psychopharmacology, 13*(4), 337–345.
Wacker, J., Reuter, M., Hennig, J., & Stemmler, G. (2005). Sexually dimorphic link between dopamine D2 receptor gene and neuroticism-anxiety. *Neuroreport, 16*(6), 611–614.
Wang, L., Jia, P., Wolfinger, R. D., Chen, X., & Zhao, Z. (2011). Gene set analysis of genome-wide association studies: Methodological issues and perspectives. *Genomics, 98*(1), 1–8. Epub 2011 Apr 30.
Zuckerman, M. (1985). Sensation seeking, mania, and monoamines. *Neuropsychobiology, 13*(3), 121–128.

PART 3

Science and Society

likely to change substantially. Genetics may affect the adjudication of individual cases and sentencing decisions, but it is unlikely to lead to major changes.

The basic questions for the law are always the same. What is the relevance of this causal information to legal doctrine and practice? Does the evidence seem to suggest the need for radical, fundamental restructuring or reform of law or for more limited, discrete reforms, if any, of particular doctrine or practices? Advocates often tend to make overclaims based on science (Morse, 2006) or on unjustified genetic essentialism (Dar-Nimrod & Heine, 2011), and we must not make broad claims about human nature based on small statistical associations (Landis & Insel, 2008); experience indicates, however, that most of what we learn suggests no major overhaul of law.

Human behavior is generally the product of immensely complex factors, including biological, psychological, and sociological variables. Human beings are fundamentally biological machines who always interact with their environments (Noe, 2009). In particular, genetics is surely part of the explanation of many and perhaps most behaviors. The genetic explanation will tend to be extremely complex and seldom will it be nonexistent (Turkheimer, 2011). Although this is true, human beings are a special type of causal end product of the interaction between biology and environment. Our unique capacities, such as our ability to use fully developed propositional language and the ability to deliberate about what we have reason to do, have enormous moral and legal implications.

It is important to stress that responsibility is a distinguishable issue from prophylactic or rehabilitative concerns. Even if particular scientific knowledge has no bearing on responsibility, it may well have implications for the ex ante prevention of undesirable behavior, the ex post prevention of recidivism, the control of predictably dangerous people, and the promotion of desirable behavior.

Criminal responsibility and its consequences, such as whether and how much to punish a wrongdoer, pose normative moral, political, and, ultimately, legal questions. They address how we should live together. Empirical discoveries about the causes of behavior can make profound contributions to debates about what we have reason to do, but they do not by themselves dictate any normative conclusion. On the ultimate issue of how we should live, science must fall silent because how we should live is a matter of practical and not theoretical reason. Scientific discoveries may indicate what constraints on human arrangements exist, but the content of those arrangements cannot be straightforwardly derived from those discoveries (Buchanan, 2011). The issue is normative and not factual.

Finally, and with few exceptions, retribution—giving people their just deserts—is a necessary precondition for blame and punishment in the American criminal justice system, and it plays a role in all developed Western legal systems. Most criminal justice theorists justify punishment on mixed retributive and consequential grounds. Consequential concerns do matter to the definition of crimes and to the appropriate sentence to be imposed, but American and Western criminal justice systems agree that no one should be punished unless they deserve it and no more than they deserve. Retribution is a well-recognized justification for the state infliction of pain on its citizens, and it must be distinguished from revenge, with which it is often confused. Revenge is the primitive desire to hurt those who have hurt you. It usually involves anger and often includes psychological catharsis. Retribution, in contrast, is a theory of justice. There are many different accounts of retribution, but all agree that it is a good in itself to give people what they morally and legally deserve. It is not a primitive and prescientific form of human response, but a considered theory of justice. Blaming and punishing people for wronging others because they deserve such a response is the mirror image of praising and rewarding people for helping others because they deserve such a response. Accepting a retributive theory of punishment does not entail whether or how much criminals should be punished. One could be permissive or obligatory about whether retribution demands punishment, and a retributivist can be harsh or tender about how much punishment is deserved. The only basic commitment is that people should not be punished unless they deserve it and no more than they deserve.

II. The Law's Psychology

The law—a reification in itself—has not explicitly adopted any psychology or concept of the person. What follows is a rational reconstruction of the law's implicit psychology and concept of the person and why these are necessary if law is to make coherent sense and to serve a useful function in society. The next section demonstrates that criminal responsibility is an instantiation of the law's psychology and flows naturally from it.

A. *The Law's Psychology*

Lawyers take the criminal law's implicit psychology for granted because there is seldom any need to identify or question it. Genetics and other scientific explanations of behavior may appear to call the law's psychology into question, however, so it is crucial consciously to recognize it and to understand what would be entailed if it were undermined.

Criminal law presupposes the "folk psychological" view of the person and behavior. This psychological theory causally explains behavior in part by mental states such as desires, beliefs, intentions, willings, and plans. Biological, other psychological, and sociological variables also play a role, but folk psychology considers mental states fundamental to a full explanation of human action. Human behavior cannot be adequately understood if mental state causation is completely excluded or eliminated. Lawyers, philosophers, and scientists do, of course, argue about the definitions of mental states and about theories of mind and action, but that does not undermine the general claim that mental states are fundamental to law. Indeed, the arguments and evidence disputants use to convince others itself presupposes the folk psychological view of the person. Brains don't convince each other; people do.

For example, the folk psychological explanation for why you are reading this essay is, roughly, that you desire to understand the relation of genetics to criminal responsibility and sentencing, you believe that reading the essay will help fulfill that desire, and thus you formed the intention to read it. This is a "practical" explanation rather than a deductive syllogism.

Folk psychology does not presuppose the truth of free will (which will be discussed further later), it is not dualist (although it and ordinary speech sound that way), it is perfectly consistent with the truth of determinism, and it presupposes no particular moral or political view. It does not presuppose that all mental states are necessarily conscious or that people go through a conscious decision-making process each time that they act. It allows for "thoughtless," automatic, and habitual actions, and for nonconscious intentions. For example, consider the behavior of putting on your undershorts in the morning. You probably have not considered which leg to put through first since you learned to dress yourself. There is no conscious decision-making process and choice; the behavior is utterly automatic and habitual. This makes sense because there is no reason to give any thought to which leg goes first. Absolutely nothing turns on this. Suppose, however, that you had good reason to put the other leg first, say, a threat to kill you if you put the usual leg through first. You would then bring this automatic behavior under the control of reason. This example is an illustration of folk psychology's presupposition that human action will at least be rationalizable by mental state explanations or that it will be responsive to reasons, including incentives, under the right conditions.

The definition of folk psychology used in this essay does not depend on the truth of any particular bit of folk wisdom about how people are motivated, feel, or act. Any of these bits, such as that experiencing disrespect often produces anger, might turn out to be wrong after further empirical investigation. The definition insists only that human action is, in part, causally explained by mental states.

B. *The Inevitability of Folk Psychology*

Brief reflection should indicate that the law's psychology must be a folk psychological theory, a view of the person as a conscious (and potentially self-conscious) creature who forms and acts on intentions that are the product of the person's other mental states, such as desires, beliefs, willings, and plans. Law is primarily an action-guiding system of rules and standards (Sher, 2006) that could not guide people ex ante and ex post unless people could use rules as premises in their reasoning about how they should behave. Otherwise, law as an action-guiding, normative system of rules would be useless and perhaps incoherent. As the philosopher, John Searle wrote (2002),

> Once we have the possibility of explaining particular forms of human behavior as following rules, we have a very rich explanatory apparatus that differs dramatically from the explanatory apparatus of the natural sciences. When we say we are following rules, we are accepting the notion of mental causation and the attendant notions of rationality and existence of norms...
>
> ...The content of the rule does not just describe what is happening, but plays a part in *making it happen* (p. 35).

Legal rules are not simply mechanistic causes that produce "reflex" compliance, although they can certainly help to inculcate law-abiding "habits." They operate within the domain of folk psychology. Agents are meant to and can only use these rules as potential reasons for action as they decide about what they should do. Legal rules are thus action-guiding primarily because they provide an agent with good moral or prudential reasons for forbearance or action.

Unless people are capable of understanding and then using legal rules to guide their conduct, law would be powerless to affect human behavior (Shapiro, 2000). Law can directly and indirectly affect the world we inhabit only by its influence on human beings who can potentially use legal rules to guide conduct. For example, no "instinct" governs how fast a person drives on the open highway. Among the variables that explain the speed at which a person drives, the posted speed limit and the belief in the probability of suffering the consequences for exceeding it surely play a large role in the driver's choice of speed. The law thus guides action.

Human behavior can be modified by means other than influencing deliberation, and human beings do not always deliberate before they act. Nonetheless, the law presupposes folk psychology, even when we most habitually follow the legal rules. All citizens constantly act in the "shadow of the law," especially when criminal conduct is at stake.

The legal view of the person does not hold that people must always reason or consistently behave rationally according to some preordained, normative notion of rationality. Rather, the law's view is that people are capable of acting for reasons and are capable of minimal rationality according to predominantly conventional, socially constructed standards. The type of rationality the law requires is the ordinary person's commonsense view of rationality, not the technical notion that might be acceptable within the disciplines of economics, philosophy, psychology, computer science, and the like.

Virtually everything for which agents deserve to be praised, blamed, rewarded, or punished is the product of mental causation and, in principle, responsive to reason. I do not mean to imply dualism here. I am simply accepting the folk psychological view that mental states—which are fully produced by and realizable in the brain—play a genuinely causal role in explaining human behavior. Machines may cause harm, but they cannot do wrong, and they cannot violate expectations about how people ought to live together. Machines do not deserve praise, blame, reward, punishment, concern, or respect because they exist or because of the results they cause. Only people, intentional agents with the potential to act, can violate expectations of what they owe each other, and only people can do wrong.

Many scientists and some philosophers of mind and action consider folk psychology to be a primitive or prescientific view of human behavior. No one, however, has even remotely suggested a replacement psychology for the law that would conceivably be practical, and law is an intensely practical enterprise. For the foreseeable future, then, the law will be based on the folk psychological model of the person and behavior, and this essay will proceed on that premise. Until and unless scientific discoveries convince us that our view of ourselves is radically wrong—and nothing science has discovered begins to support this claim (Morse, 2008)—the basic explanatory apparatus of folk psychology will remain central. The folk psychological theory of personhood that the law implicitly adopts seems secure. As the eminent philosopher of mind, Jerry Fodor, has written (1987):

> [I]f commonsense intentional psychology were really to collapse, that would be, beyond comparison, the greatest intellectual catastrophe in the history of our species; if we're that wrong about the mind, then that's the wrongest we've ever been about anything. The collapse of the supernatural, for example, doesn't compare.... Nothing except, perhaps, our commonsense physics...comes as near our cognitive core as intentional explanation does. We'll be in deep, deep trouble if we have to give it up.... But be of good cheer; everything is going to be all right. *(p. xii)*

It is vital that we not lose sight of the folk psychological model lest we fall into confusion when various claims based on genetics or other causal variables are made. Any genetics data or evidence must always be relevant to the law's folk psychological criteria. If genetics is to have any justifiable influence on legal decisions about criminal responsibility and sentencing, it must be almost entirely through this framework.

III. The Criteria for Criminal Responsibility

The criteria for responsibility are all folk psychological. The prosecution must first prove the "elements" of the crime, which is simply the legal term for the criteria for criminal conduct. These elements are composed primarily of acts and mental states. All are infused with mental states. With exceptions that need not detain us, all crimes include a "voluntary" act requirement, which is defined, roughly, as an *intentional* bodily movement (or an omission in cases in which the person has a duty to act) done while the agent is in a reasonably integrated state of consciousness. Although the meaning of an intentional bodily movement is seldom specified, the best definition is a bodily movement that, in principle, can be understood according to the person's mental states. One can almost always ask of any action, "Why did you do that?", and expect some explicit or implicit mental explanation. If there is none even implicitly possible, it is probable that the agent's bodily movement was not an act at all. For example, reflexes and neuromuscular spasms involve bodily movements, but they are not actions because they were not intentional bodily movements and generally cannot respond directly to reasons or incentives.

With few exceptions that are themselves controversial, all crimes also require a culpable further mental state, such as purpose, knowledge, or conscious awareness that one is risking a prohibited harm (a mental state lawyers call "recklessness"). Some crimes are also defined with the mental state of negligence, which is defined as an unreasonable failure to be aware of the risk of a prohibited harm. Negligence appears to be the absence of a mental state. This is a controversial issue among legal scholars, but the best explanation for criminalizing negligence is that the failure to pay attention when the agent was creating a substantial and unjustifiable level of risk is itself a type of culpable omission.

To make this discussion more concrete, consider the following standard definition of murder mentioned in the opening hypothetical about the homicidal armed robber: killing another human being with the intent to cause death. Any intentional killing conduct—e.g., shooting, stabbing, strangling, poisoning, bludgeoning—is sufficient to meet the act requirement as long as the agent was in a state of reasonably integrated consciousness when engaging in the killing conduct. Further, the agent must engage in the conduct with the intent to cause death. If the agent lacked that intent and, say, just risked killing the victim, then the agent will not be guilty of intentional murder, but may be guilty of some other homicide crime that requires a different mental state.

Even if the prosecution is able to prove all the elements of crime, the defendant may still avoid criminal liability by successfully establishing what is called an affirmative defense of justification or excuse. In cases of justification, conduct that would otherwise be criminal is right or at least permissible under the circumstances. Self-defense is a classic example. Intentionally killing another human being is ordinarily murder, but an agent may be justified in killing if the other person is threatening the agent with wrongful, deadly force. In such cases, intentional killing is considered right or at least permissible. Note that in cases of justification, the agent is a fully responsible person.

In contrast, cases of excuse involve wrongful action performed by an agent who is not responsible. Roughly speaking, lack of rational capacity, external compulsion (e.g., acting in response to a "do-it-or-else" threat of death or grievous bodily injury), and, more controversially, lack of the capacity to control oneself (sometimes referred to as an internal compulsion or control problem) are the basic grounds for criminal law's doctrinal excuses. Note that in the cases of lack of rational capacity or lack of the capacity to control oneself, responsibility requires only the possession of the general capacity at the time in question, even if the agent did not exercise the capacity on that occasion. For example, acting irrationally, arationally, and foolishly are common even among people with the greatest capacity for rational conduct. Failure to exercise a capacity does not necessarily mean that one lacks that capacity. For the law, if the person is capable of exercising the capacity for rationality or self-control if there is good reason to do so—as there always is when important interests of potential victims are at stake—then the person may be held responsible even if he or she failed to exercise that capacity.

If the person lacks the relevant capacity, an excuse is warranted. Legal insanity is a classic example. Suppose a person delusionally believes that he is about to be killed by a secret agent and kills the suspected secret agent in what the person believes is self-defense. The person has wrongfully killed intentionally, but he would be excused because he was not a rational agent when he killed. In a very real sense, the killer did not know what he was doing.

In short, criminal responsibility is established if the prosecution can prove all the elements of the crime charged and the defendant cannot establish an affirmative defense. Criminal responsibility is defeated if the prosecution cannot prove all the elements of the crime charged or if the defendant can establish an affirmative defense. The elements and affirmative defenses all involve mental states. Of course, the person's mental state is influenced by biological, psychological, and sociological variables, including genetics, and knowledge of these variables may help determine what the person's mental state was. Nevertheless, the law is ultimately concerned with the person as an acting agent who has acted for reasons. The final explanatory pathway for law is always folk psychological. Any relevant data from genetics or other sciences must be translated into the law's folk psychological criteria (Morse, 2011).

We turn to the relation of genetics to criminal law's responsibility criteria in section V. Before doing so, however, this essay briefly discusses some false starts and distractions about responsibility.

IV. Distractions and False Starts in Criminal Responsibility Analysis

Here, we consider the following distractions that have bedeviled attempts to understand the relation between scientific explanations for criminal behavior and criminal responsibility: the free will debate, the belief that causation is per se an excusing condition, the belief that causation by abnormal variables is per se an excusing condition, and the belief that causation is the equivalent of compulsion. Many of the arguments in this section probably will be unfamiliar to many scientists, but they are crucial to proper understanding and therefore deserve careful consideration.

A. The Non-Problem of Free Will in Criminal Law

There is a problem about free will, but not in criminal law (Morse, 2007). Free will, as the term is used in the philosophical debate about free will and responsibility, is not a criterion for any legal doctrine. Criminal law criteria involve questions genuinely related to responsibility, including issues concerning consciousness, the formation of mental states such as intention and knowledge, the capacity for rationality, and compulsion, but they never address the presence or absence of free will. People sometimes use "free will" loosely to refer to genuine responsibility doctrines, but this simply distracts attention from the real issues and perpetuates confusion.

The philosophical problem of free will is metaphysical and often clouds clear thinking about the foundation for criminal responsibility. Specialists in the philosophy of free will and responsibility often distinguish between freedom of action, the freedom to do as one chooses, and freedom of the will, the freedom to choose what one would prefer to choose (Kane, 2006). This essay subsumes both under the locution "freedom of the will" or "free will."

Roughly, the notion of free will used in the debate refers to whether an agent has the ability to cause his or her own behavior uncaused by anything else. In a phrase, the buck stops entirely with the agent. This ability is sometimes called *contra-causal freedom*, *agent origination*, *metaphysical libertarianism*, and other like phrases. Only a small number of philosophers adhere to this view, which has been termed a "panicky" metaphysics (P. F. Strawson, 1982 p. 80), because it is so implausible (Bok, 1998).

Even if this type of free will is not a criterion for any criminal law doctrine, many people nonetheless believe that this type of power or ability is a foundational assumption for legal responsibility and for justifying the fair imposition of blame and punishment. Thus, if people do not possess this godlike power, then doctrines and practices relating to responsibility may be entirely incoherent. But, as we shall see, contra-causal freedom is not a necessary support for current responsibility doctrines and practices.

Most philosophers and, I speculate, virtually all scientists, believe that the universe is deterministic or universally caused, or nearly so, especially above the subatomic level. There is no uncontroversial definition of determinism, and we will never be able to confirm that it is true or not. As a working definition, however, let us assume, roughly, that all events have causes that operate according to the physical laws of the universe and that were themselves caused by those same laws operating on prior states of the universe in a continuous thread of causation going back to the first state. Even if this is too strong, the universe seems so sufficiently regular and lawful that rationality demands that we assume that universal causation is approximately correct. Philosopher G. Strawson (1989) terms this assumption the "reality constraint."

It is important to understand that, for the determinist, biological causes, including interactions of biology with the environment, pose no more or less challenge to responsibility than nonbiological or social causes. As a conceptual and empirical matter, we do not necessarily have more control over psychological or social causal variables than over biological causal variables. More important, in a world of universal causation or determinism, causal mechanisms are indistinguishable in this respect, and biological causation creates no greater threat to our life hopes than does psychological or social causation (Richards, 2000). For purposes of the metaphysical free will debate, a cause is just a cause, whether it is biological, psychological, sociological, or astrological.

If determinism is true, the people we are and the actions we perform have been caused by a chain of causation over which we mostly had no rational control and for which we could not possibly be responsible. If we do not have contra-causal freedom, how can it be rational and fair for criminal law to hold anyone accountable for anything, including blaming and punishing people because they allegedly deserve it?

Those who believe that responsibility is not compatible with determinism are called "incompatibilists" and adopt different conclusions depending on their view of determinism. "Libertarian" incompatibilists believe that determinism is not true for most actions because we have metaphysical libertarian freedom and, therefore, we are responsible. "Hard determinist" incompatibilists believe that determinism is true, deny that we have contra-causal freedom, and conclude that responsibility is impossible. "Compatibilists" believe that determinism is true, deny that contra-causal freedom is necessary for responsibility, and hold that responsibility is possible under the right conditions.

No analysis of this problem could conceivably persuade everyone. There are no decisive, analytically incontrovertible arguments to resolve the metaphysical question of the relation among determinism, libertarian free will, and responsibility. Indeed, the debate is so fraught that even theorists who adopt the same general approach to the metaphysical challenge substantially disagree. Nevertheless, the view one adopts has profound consequences for legal (and moral) theory and practice.

Let us begin with hard determinist incompatibilism. (I have already rejected libertarianism as empirically implausible. The rest of the discussion will therefore focus only on hard determinist incompatibilism, which is a coherent position held by many.) Incompatibilism does not try either to explain or justify our responsibility concepts and practices. It simply assumes that genuine responsibility is metaphysically unjustified. For example, a central incompatibilist argument is that people can be responsible only if they could have acted other than they did, but if determinism is true, they could not have acted other than they did. This is sometimes called the "principle of alternate possibilities." It has generated endless disputes between incompatibilists, who believe it is flatly inconsistent with responsibility, and compatibilists, who believe that it is not inconsistent with responsibility (Wallace, 1994). Based on this principle and similar arguments, the incompatibilist claims that even if an internally coherent account of responsibility and related practices can be given, it will be a superficial basis for responsibility, which is only an illusion (Smilansky, 2000).

Incompatibilism based on any level of scientific cause, including genetics, thus provides an external critique of responsibility. To see why, remember that causal determinism "goes all the way down." It applies to all people, to all events. Thus, if determinism is true and is inconsistent with responsibility, then no one can ever be really responsible for anything, and responsibility attributions cannot properly justify further action. But Western theories of morality and the law do hold some people responsible and excuse others, and the law responds accordingly. And, when we do excuse, it is not because there has been a little local determinism at work. For example, young children are not considered fully responsible because they are incapable of recognizing and properly weighing the right reasons for action and forbearance, not because they are determined creatures but adults are not. Determinism does not loosen its grip on us as we age.

The question, then, is whether as rational agents we must swallow our pride, accept incompatibilism because it is so self-evidently true, and somehow transform the legal system accordingly into a system that abandons desert and relies on a prediction and prevention model of social control that is untethered from considerations of genuine responsibility. Such systems have been proposed (e.g., Wootton, 1963), but they have been criticized for their harsh and potentially inhumane implications that profoundly threaten liberty and dignity (Hart, 1968; Lewis, 1953). Once again, until scientific explanations, whether from genetics or others, convince us that we are not acting agents, such a system is exceptionally unlikely to gain assent.

Compatibilists, who agree with hard determinist incompatibilists that determinism is true, have three basic answers to the incompatibilist challenge. First, they claim that responsibility attribution and related practices are human activities constructed by us for good reason and that they need not conform to any ultimate metaphysical facts about "ultimate" responsibility. Indeed, some compatibilists deny that conforming to ultimate metaphysical facts is even a coherent goal in this context. Second, compatibilism holds that our positive doctrines of responsibility are fully consistent with determinism. Third, compatibilists believe that our responsibility doctrines and practices are normatively desirable and consistent with moral, legal, and political theories that we firmly embrace. The first claim is theoretical; the third is primarily normative. There are very powerful arguments for the first and third claims (Lenman, 2006; Morse, 2004). For our current purpose of determining whether criminal law has a free will problem, the second claim is the most important.

Let us begin with the most general responsibility and excusing conditions. Recall that the capacity for rationality is the primary responsibility criterion, and its lack is the primary excusing condition. Now, it is simply a fact about human beings that they have different capacities for rationality in general and in specific contexts. Once again, for example, young children in general have less rational capacity than do adults. It is also true that rationality differences differentially affect agents' capacity to grasp and to be guided by good reason. Differences in rational capacity and its effects are real even if determinism is true. Compulsion is also an excusing condition, but it is simply another fact about human beings that some people act in response to external or internal hard choice threats to which persons of reasonable firmness might yield, and most people most of the time are not in such situations when they act. This is true even if determinism is true and even if people could not have acted otherwise.

For examples, consider again the specific doctrines of criminal responsibility. Assume that the defendant has caused a prohibited harm. Recall that responsibility requires that the defendant's behavior was an action and performed with a requisite mental state. Now, it is simply true that some bodily movements are intentional and performed in a state of reasonably integrated consciousness and some are not. It is also true that some defendants possess the requisite mental state and some do not. The truth of determinism does not mean that actions are indistinguishable from nonactions or that people do not have different mental states when they act. These facts are true and make a perfectly rational legal difference, even if determinism is true. Determinism is fully consistent with making distinctions among defendants about whether the elements of the crime charged can be proven.

Now consider the defense of legal insanity, which was briefly addressed earlier. Some people with mental disorder do not know right from wrong; others do. Once again, legally differentiating these cases makes perfect sense according to dominant retributive and consequential theories of punishment. A causal account, including from genetics, in an appropriate case, can explain how these variations were caused to occur, but it does not mean that they do not exist. Determinism is fully consistent with both the presence and absence of affirmative defenses. In sum, the legal criteria used to identify which defendants are criminally responsible map onto real behavioral differences that justify differential legal responses.

A causal determinist account would become inconsistent with our responsibility practices only if our scientific investigations convinced us that we are not the types of creatures the law takes us to be—conscious and intentional creatures who act for reasons. If it is true, for example, that we are all automata, then no one is acting and no one can be responsible for action. I have variously termed this the "No Action Thesis" or the "Victims of Neuronal Circumstances" thesis (Morse, 2003*b*; 2008; 2011). Unlike the claimed inconsistency between determinism and responsibility, which is a metaphysical question, this critique is empirical and, in principle, capable of resolution. The conclusion that we are essentially automatons would once again provide an external critique of responsibility and leave no rational room for legal decision making about genuine responsibility. Although some scientists are gesturing in this direction (Wegner, 2002), there is little in current science that suggests that most people most of the time are not conscious and intentional creatures who act for reasons or whose behavior can be guided by reason and incentives (Morse, 2008; 2011).

Compatibilism is consistent with our criminal responsibility doctrines and practices, and there is no convincing theoretical reason to reject it. All participants in the criminal justice system, including scientists who contribute to legal policy making and decisions in individual cases, have good reason to embrace compatibilism. Scientists can comfortably continue to play a crucial role in assisting the promotion of more rational criminal justice without being distracted by the irrelevant issue of free will.

B. Causation per se Does Not Excuse: The "Fundamental Psycholegal Error"

The most persistent confusion about our actual doctrines and practices concerning responsibility, which I have termed the "fundamental psycholegal error"(Morse, 1994), is the mistaken belief that causation, especially by an abnormal cause, is per se an excusing condition. In brief, this error relies on the same argument the incompatibilist makes, but without recognizing that it provides an external critique that must deny the possibility of any responsibility. If the truth of determinism or universal causation is an excusing condition, it applies not simply in any particular legal context, such as guilt or sentencing proceedings. It applies everywhere and always.

In a causally deterministic universe, all phenomena, including human actions, are fully caused. If causation were per se an excusing condition, no one could ever be responsible for anything. Thus, causation cannot be an excusing condition in law and morals, both of which hold some people responsible and excuse others. Although this is a simple and straightforward analytic point, the error persists (e.g., Kaye, 2005).

If the causal chain were different, the ensuing action would be different. The question for law is not whether behavior was caused. It is whether the legal criteria have been satisfied. For purposes of assessing responsibility, it does not matter whether the cause of the behavior in question is biological, psychological, sociological, or some combination of the three. Adducing a genetic or neurophysiological cause does no more work than adducing an environmental or interactive cause. The question is always whether the legal criteria for non-responsibility are met, however the behavior in question may have been caused. A person who is mentally disordered and does not know right from wrong will be excused from criminal responsibility whether her rationality impairment was primarily a product of faulty genetics, a neurotransmitter defect, bad parenting, social stress, the alignment of the planets, or some combination of the above. The most important question for criminal law is whether the legal excusing condition was present, not how it was caused. Causal knowledge, if sufficiently precise, may help establish whether or the likelihood that the legal criterion in question was satisfied, but the person will be excused if the excusing condition is present, even if we have no idea how the condition was caused. Actions speak louder than SNPs or neuroimages, however, because, as we have seen, the criteria for responsibility are all behavioral (acts and mental states). If the defendant's behavior clearly meets or does not meet the legal criterion in question, we must believe the behavior, even if the biological evidence seems inconsistent.

For example, in *Roper v. Simmons* (2005), the United States Supreme Court addressed the question of whether adolescents who committed capital murder when they were 16 or 17 years old should be categorically excluded from being sentenced to death. Advocates for abolition for this group of murderers argued that the demonstrated lack of complete myelination of the cortical neurons of the adolescent brain was reason to believe that 16- and 17-year-old murderers were insufficiently responsible to deserve capital punishment. Rigorous behavioral studies had already confirmed the average differences in rational capacity between adolescents and adults. The moral and constitutional implications of the data may be controversial, but the data are not. At most, the neuroscientific evidence provided a partial causal explanation of why the observed behavioral differences exist and thus some further evidence of the validity of the behavioral differences. The neuroscience was thus of only limited and indirect relevance to responsibility assessment, which is based on behavioral criteria concerning rationality. Diminished responsibility follows from diminished rationality, however the latter is caused.

Finally, it follows logically that if full causation is not per se an excusing condition, then "partial causation" also does not partially or fully excuse the agent. Most of the time, we possess only imperfect, partial understanding of the causes of behavior. It is important to remember, however, that not possessing knowledge of the complete causal account of a person's behavior does not mean that a complete causal account does not exist. Indeed, the notion that only some phenomena are caused or determined, but others are not, is incoherent. If there is causal closure—that is, if this is a universally caused or deterministic universe—then all phenomena are caused, whether or not we have knowledge of those causes.

In any case, discovering a partial normal or abnormal cause for behavior does not partially or completely excuse the agent unless that cause produces a genuine mitigating or excusing condition. For example, various causes we discover may, in part, explain why an agent's rationality is fully or partially impaired, but then it is the impairment of rationality, not causation, that is doing the excusing work.

C. *Abnormal Causation Does Not Excuse Per Se*

Abnormal causation, say, by mental disorder, also does not excuse per se, but excuses only if it produces a genuine excusing condition, such as lack of capacity to appreciate the criminality of one's actions. For example, a person suffering from mental disorder that plays a causal role in the sufferer's behavior may nonetheless retain sufficient capacity for rationality to be held fully responsible. A clinically hypomanic mugger, for example, may be especially energetic, mentally acute, and confident when he mugs. Indeed, but for the clinical condition, the mugger may not have mugged, but there is no question about the agent's criminal responsibility in this case. The mugger is sufficiently rational to be held fully responsible.

D. *Causation Is Not Compulsion*

Causation is also not the equivalent of compulsion, even if some type of normal or abnormal causal variable played a role in explaining the criminal behavior in question. All behaviors are lawfully caused, but the laws of nature are not coercive (Hume, 1978; Scanlon, 2008). Not all behavior is the product of the external or internal coercive conditions that meet moral and legal criteria for compulsion. If causation were the equivalent of compulsion, everyone would always be compelled and excused. For example, just because a person is predisposed to antisocial activity by a G×E interaction, it does not mean that the person was compelled to act. For another example, a delusional belief or a hallucination may produce irrational reasons for action, but irrational reasons are not per se more compelling than rational reasons. A person who delusionally believes in the need to use deadly self-defense is no more compelled to act than a nondelusional agent with the same honest belief. The former may be excused because he is irrational, but compulsion plays no role in such cases. By the same logic, discovering part of the causation of behavior does not mean that the behavior was compelled to that degree. Causation is not per se compulsion and "partial causation" is not per se partial or complete compulsion.

In conclusion, genetics will not affect responsibility analysis simply because it is a cause of criminal behavior, even if it produces an immensely predisposing cause. Genetics will be relevant only if it helps prove or disprove the existence of actual criminal law criteria. And, again, actions speak louder than SNPs. Let us therefore next examine the specific relevance of genetics to those criteria.

V. Genetics and Criminal Responsibility

The question of the relation of genetics to criminal responsibility reduces to whether genetic evidence or data cast doubt on the elements of crimes, such as action and mental states, or help to establish (or cast doubt about) the existence of a complete or partial affirmative defense. Genetic causation, no matter how powerful its explanatory role may be, will only negate responsibility if it prevents a defendant from meeting the responsibility criteria of the criminal law.

Using the example of the armed robber and the G×E interaction of childhood abuse and a genetic MAOA deficiency with which this essay opened, let us consider how this G×E affects responsibility. Although I will concentrate on this specific example, the argument is fully generalizable to any genetic condition that might causally contribute to criminal behavior.

The first criterion for responsibility is the "act" requirement. The defendant's bodily movements that appear to have violated the law must have been intentional and performed in a state of reasonably integrated consciousness. In other words, the agent must have acted. There is no evidence in the G×E literature under consideration to suggest that the agents thereby predisposed to criminal behavior are not acting when they commit crimes. Their bodily movements are not the equivalent of reflexes or spasms, and they are not performed in a dissociative state, such as sleepwalking. Unless some other explanation for lack of action is forthcoming, our armed robber's behavior is clearly action even if G×E played a causal role.

A similar analysis applies to whether the defendant possessed the mental state, the *mens rea*, such as intention, knowledge, recklessness, or negligence, when the crime was committed. Again, nothing in the literature of the G×E interaction in question suggests that this causal variable prevents people from forming culpable mental states. Our armed robber, for example, surely had both the intent to steal using a weapon as a threat, and, ultimately, the intent to kill.

It is possible, however, that the effect of the interaction on a defendant's mental capacities may interfere with the formation of some culpable mental states. For example, in some states, intentional killings that are performed after relatively cool, rational deliberation (so-called premeditated murder) are considered especially heinous and punished accordingly. Although I know of no such data, suppose it could be demonstrated that G×E made it extremely difficult to plan coolly and rationally for those subject to it. In that case, the G×E evidence would help the law decide if an intentional killer in fact premeditated. Notice, however, that it is lack of premeditation, not G×E itself, which might explain why the killer is not guilty of the aggravated degree of intentional homicide. If there were clear behavioral evidence that the defendant did premeditate or was capable of doing so despite the G×E—and, of course, there will be variation among people with this G×E—then the defendant would be guilty of aggravated homicide despite the evidence that G×E tends generally to make it difficult to premeditate.

Another example would be recklessness. Recall that the criminal law's definition of recklessness is conscious awareness that one's conduct is creating a high risk of a prohibited harm. It is a subjective mental state. Suppose good evidence suggested that people with G×E are extremely poor estimators of the consequences of their conduct. If so, the G×E evidence would be relevant and probative about whether a defendant actually was aware of the risk created. And again, if the behavioral evidence suggested that the defendant was actually aware of the risk despite the G×E, say by adverting to it in comments to accomplices, then G×E data would be trumped.

Consider negligence as a final example. Negligence is the failure to be aware that one's conduct is creating a high risk of a prohibited harm in situations in which a reasonable person should have been aware of the risk. It is considered an "objective" mental state because we are comparing the defendant to a hypothetical reasonable person. Suppose good evidence suggests that G×E makes it very difficult for those subject to it to behave as reasonable people should. Although there is a good argument for excusing the presence of negligence in that case (Hart, 1968), the law is unforgiving about negligence. Everyone is held to the standard of the reasonable person, including those people who may find it supremely difficult to meet that standard through no fault of their own. Any mitigation based on the defendant's deficiencies would only be considered at sentencing, which is addressed in this essay's next section.

In short, G×E is not likely to have much if any effect on the formation of the act and mental state elements of the crime charged.

G×E is most likely to be relevant to the generic excusing conditions: lack of rational capacity and lack of the capacity to control oneself. G×E is not likely to be relevant to criminal action compelled by "do-it-or-else" threats using death or grievous bodily harm as the "or-else" because these threats must be external. Lack of rational and lack of control capacities have technical legal doctrinal criteria, but, for our purposes, it is sufficient to use the generic justifications for the doctrinal excuses, such as infancy, legal insanity, and duress.

Before we can answer whether G×E is relevant to rational and control capacities, it is necessary to explore what we mean by these capacities in a bit more detail. There is no consensual definition of rationality in any of the relevant disciplines, such as psychology, psychiatry, economics, and philosophy, that study this issue. Likewise, there is no uncontroversial legal definition of rationality or of what kind and how much is required for responsibility in various legal contexts. Rationality for the law must be understood according to some contingent, normative notion both of rationality and of how much capability is required. For example, legal responsibility might require the capability of understanding the reason for an applicable rule, as well as understanding the rule's narrow behavior command and the consequence for failure to comply. These are matters of moral, political and, ultimately, legal judgment, about which reasonable people can and do differ. These are normative issues about intentional behavior guided by reasons.

If one examines the various legal responsibility doctrines that implicitly address rationality defects, however, one can infer that the law's general definition is a congeries of abilities that closely track an ordinary person's commonsense definition. For example, the agent must be able to get the facts right, to know what he is doing, to be able to respond reasonably to reasons and incentives in the context in question, and the like. For example, one criterion for legal insanity is that the agent does not know what he is doing. Deciding whether such criteria are met requires an implicit, commonsensical folk psychological definition of rationality of the sort we all use every day to evaluate our own conduct and the conduct of others.

It is much harder to provide a folk psychological account for the lack of capacity to control oneself that is independent of a rationality defect. People commonly use locutions like, "I can't help myself," or "I lost control of myself" and similar expressions, but what do they mean? When people "lose control" and act badly, they are surely acting, but what is the folk psychological process that suggests loss of control? Various models have been used to try to inject content into the process, including a collapse of control problems into rationality defects (Morse, 2002), but none has seemed conclusive. Moreover, at present, we have no way of distinguishing action that a person cannot control from action that he or she simply did not control. Such practical difficulties in part accounted for why both the American Bar Association and the American Psychiatric Association recommended abolishing "control" criteria for legal insanity in the middle 1980s.

The definitional problems persist. A recent, influential legal example is the Supreme Court's decision in *Kansas v. Crane* (2002), in which a criterion of "serious difficulty" controlling oneself was constitutionally required to be proven before a state could civilly commit a so-called mentally abnormal sexually violent predator. There was a withering dissent about the difficulty of making this conclusory judgment, and the courts that have since tried to interpret this requirement have been unable to do more than simply repeat the "serious difficulty" formula without further operationalization.

A current example from psychiatry is the conclusion that the drug seeking and using of addicts is "compulsive." It is question begging to say the addict cannot control seeking and using because those behaviors are signs of a disease. Actions are different from the purely mechanistic signs and symptoms of most diseases, such as fevers or metastases. The lack of ability to control action must be demonstrated independently (Fingarette & Hasse, 1979). The psychiatric conclusion about compulsion is not based on such an independent, operationalized measure. It is based on the commonsense conclusion that people who persist in behavior that often creates ruinous medical, psychological, fiscal, personal, and legal problems and who report feelings such as craving must not have good control over their drug-related behavior. Although there is reason to question this conclusion, especially in its strongest form (Heyman, 2009), no folk psychological process has been specified even if we do accept it.

In short, adequately defining control capacity is a problem for criminal law that science has not yet helped the law solve. Nonetheless, the ensuing discussion will assume for the sake of argument that we can make sense of control problems. Like the rationality criterion for responsibility, the degree of control the law will require is a normative matter that can vary from one legal context to another.

To be relevant to legal excuses, G×E data will have to be translated into the law's folk psychological definitions of rational capacity and control capacity. The G×E studies in question were not clinical or thick phenomenological descriptions of the high-risk subjects with both MAOA deficiency and childhood abuse. For the sake of general argument, however, let us assume a number of folk psychological variables that might in part account for the high rates of offending among these subjects. Nothing turns on whether these are the correct variables because the argument that will be made is general and would apply to whatever variables are doing the explanatory work. Consequently, let us simply assume that the subjects may have been highly sensation-seeking, impulsive, or suffering from poor judgment.

These types of folk psychological variables are relevant to commonsense notions of rational and control capacities. Moreover, science can help operationalize and measure such variables. For example, the amicus ("friend of the court") briefs of the American Psychological Association in the *Roper* case, and in both *Graham v. Florida* (2010) and *Miller v. Alabama* (2012), the Supreme Court cases holding, respectively, that it is unconstitutional to sentence juvenile offenders convicted of nonhomicide crimes to life in prison without the possibility of parole (JLWOP) and that it is unconstitutional to impose on juveniles a mandatory sentence of life without the possibility of parole for homicide crimes, are replete with studies of the decision-making abilities and other psychological variables that are relevant to whether adolescents as a class are on average less responsible than adults because they have less rational capacity. Whether the differences between adolescents and adults are sufficiently large in quantity and quality to warrant differential treatment is, of course, a normative moral and legal question that science cannot answer, but the data are surely relevant to the legal decision that must be made.

The same type of analysis applies to G×E offenders, including our armed robber. Assuming that the folk psychological process that G×E produces adversely affects rational and control capacities, the question will be whether the adverse effects are sufficiently large to warrant excuse or mitigation for criminal conduct. Treated as a general matter of legal policy, a question would be whether the effects are so marked that we should mitigate the culpability of G×E offenders as a class and forego individualized evaluation on a case-by-case basis. The law generally disfavors such general as opposed to individualized decision making, although it has been willing to prohibit the capital punishment of murderers from the reasonably well-delineated classes of people with retardation (*Atkins v. Virginia*, 2002) and of killers who were 16 or 17 years old at the time of the crime (*Roper v. Simmons*, 2005).

Assuming that G×E does have psychological effects that bear on responsibility, it is virtually certain that G×E offenders would have their culpability assessed individually through doctrinal excuses or at sentencing. Recall that the offender's actual behavior will be more probative than the group data. If the offender's history and conduct at the time of the crime indicate no substantial defects, the group data will be of little avail. Actions speak louder than G×E. If the offender's capacities are unclear, however, the group data might help.

Unfortunately for G×E offenders, no general excusing condition seems to apply. The insanity defense may seem like the strongest opportunity, but G×E offenders will not qualify for the insanity defense unless they also suffer from a major mental disorder that causes them to lose touch with reality. This is required by many jurisdictions, and, in practice, suffering from gross loss of contact with reality is necessary to succeed with an insanity defense even if the legal rule does not specify that the disorder must have psychotic features. No other general excusing condition is even remotely applicable. Criminal law has few mitigating doctrines that are considered at trial to which the folk psychological processes that G×E produces would apply. I have proposed that criminal law should adopt a generic "partial responsibility" doctrine based on diminished rationality that would be considered at trial and that would mitigate the offender's degree of conviction and punishment if the claim were successful (Morse, 2003*a*). Alas, no jurisdiction has adopted or is even considering this proposal. Again, such factors may be considered at sentencing, which the next section considers.

The arguments that I have been making about the G×E that Caspi et al. discovered are fully generalizable to any future genetics discoveries. The evidence would have to be relevant to the criminal law's folk psychological criteria for whether the defendant acted, whether he possessed the mental state required by the definition of the crime, and whether an excusing condition, such as lack of rational or control capacity, is established. Whether genetics played a causal role in explaining the criminal behavior is legally irrelevant unless one falls prey to the "fundamental psycholegal error" of thinking that causation is an excuse. This conclusion applies even in the unlikely event that *every* person subject to a particular genetic condition commits a crime. A genuine excusing or mitigating condition would still have to be established to defeat the allegation of criminal responsibility. After all, in a lawful causal world, all human behavior is fully explained by the causal background that produced it. If causes were excuses, no one would be responsible. That is not the legal and moral world we inhabit, and it is not likely to be.

A final point about the relevance of genetics data to criminal responsibility criteria concerns what I call the "clear-cut problem" (Morse, 2011). Identifying genetic markers for legally relevant behaviors or capacities will require that those behaviors and capacities are already well-characterized and clear. Nonetheless, in virtually all cases there will be overlap between the experimental and control subjects even if there is a robust statistical difference between the groups. If the behavior is clear, however, we do not need biological or other evidence to confirm it. What if the behavior is not clear, however? Nonbehavioral evidence would be very useful to help make a legal judgment, but it is in just those cases that genetic and other biological evidence is likely to be less sensitive because the overlaps between the curves are likely to be very large in such behaviorally unclear cases. Until our genetic and other biological evidence becomes much more precise, it will be least useful when we need it most. Future scientific discoveries may ameliorate the clear-cut problem, but for now it is a severe hindrance to the rational use of genetics evidence to aid guilt determinations.

In short, genetics will seldom play much role in guilt determinations, but it may play a more extensive role in sentencing and parole decisions, to which we now turn.

VI. Genetics and Sentencing

Questions concerning mitigation for diminished culpability and aggravation for potential future dangerousness are primarily the province of sentencing decisions. The question at sentencing is how genetics evidence would be relevant to sentencing criteria. This section begins by considering sentencing practices generally, and then turns to how genetics might be relevant.

In most American jurisdictions, there is a range of permissible sentences for each crime, and the sentencing judge has virtually complete discretion to impose any sentence within that range or to place the defendant on probation (unless the crime carries a mandatory minimum sentence). A minority of jurisdictions, including the federal system, have guidelines that constrain judicial discretion, but even in such constrained systems, judges have some discretion about the sentence to be imposed. Thus, sentencing judges can consider mitigating and aggravating factors not considered at trial to adjust the offender's sentence up or down within the statutorily permitted range of sentence for the crime. Sentencing criteria in noncapital cases are under-theorized and often not specified, thus leaving judges unguided discretion. Capital punishment must be decided by a jury (*Ring v. Arizona*, 2002), and there are typically statutory aggravating and mitigating factors the jury must consider. In addition, beginning with *Lockett v. Ohio* (1978), the Supreme Court made clear in many decisions that the defendant facing capital punishment may present virtually any evidence that could conceivably be mitigating, even if the statutory mitigating factors do not include the factor the evidence supports.

Consider mitigation first. Sentencing practices are often considerably less clear than the criteria for guilt. For example, the erroneous "causal theory of excuse" appears to be taken into account for mitigation. The rationale seems to be that understanding the causes of the defendant's behavior somehow reduces responsibility per se. The Italian judges' reduction of the defendants' sentences in part on the ground that the defendants were predisposed to violence was implicitly adopting this rationale. Conversely, the evidence introduced at sentencing may suggest that mitigation is warranted because the defendant suffered from substantial rationality or control deficits, even if they were not sufficiently substantial to rise to the level of a full legal excuse. Such a claim would be entirely supported by the retributive and perhaps consequential justifications for punishment that we firmly embrace. Even if the defendant was criminally responsible, she may deserve a lesser sentence if she was not fully rational at the time of the crime.

An important aggravating sentencing factor is the predicted future dangerous conduct of the defendant, which is considered not only at sentencing itself, but also for parole decisions. In capital sentencing statutes, future dangerousness is often expressed by criteria based on past behavior, such as prior convictions, and sometimes it is expressed directly. This ground for aggravation or for denial of parole is purely consequential and forward-looking—the protection of the public (in this case, most usually other inmates and correctional officials). Of course, by the same logic, defendants who are less predictably dangerous should receive lesser sentences on consequential grounds.

The major practical issue is determining how accurately we can predict future dangerous conduct and how much contribution to such accuracy genetic evidence might contribute. Current law accepts predictions based on weak evidence (e.g., *Barefoot v. Estelle*, 1983), but well-validated prediction factors have the potential to make prediction decisions more rational and just. Other aggravating factors include the defendant's failure to show remorse and committing the crime in a particularly dangerous or cruel manner (which is a culpability factor as well as an indication of dangerousness).

Note one final feature of using prediction of future dangerousness as an aggravating factor. It appears to deny the offender's agency and dignity by suggesting that the offender cannot be guided adequately by reason (Duff, 2007). This may be justified by consequential justifications, and there is no problem concerning retribution as long as the sentence remains within the statutorily authorized range, but it is nonetheless an undesirable aspect of prediction practices concerning responsible agents.

Now let us consider the relation of genetics to this brief account of sentencing, returning to our armed robber for final consideration. There is little dispute that people with his G×E are highly predisposed to committing criminal or otherwise antisocial acts. To the extent that the judge or jury in a capital case accepts the "causal theory of mitigation," the sentence may be reduced despite any clear rationale for doing so. Another theory might also be lurking to support mitigation, although it depends on only the abuse part of the interaction. The rationale is that the hard life suffered by the defendant has been "payment in advance" (Klein, 1990) or sufficient previous suffering that should reduce the amount of suffering that should be imposed now. This theory has no legal basis, but it may play a role psychologically. In any case, note that this theory has little to do with the G×E specifically. It is possible, however, that the same psychological characteristics G×E produces that predispose the armed robber to crime are rationality or control defects. If so, there will be good theoretical reason to mitigate the armed robber's sentence.

Now let us turn to how genetics might be relevant to sentencing aggravation, parole, or commitment decisions. The very same evidence of G×E predisposing the armed robber to future criminal behavior would certainly be considered a risk factor for future criminality and thus would support a longer sentence and denial of parole. The issue would be the practical one of accurately assessing how much G×E predisposes to future criminal behavior. Also, if this G×E were linked not only to future dangerousness, but also to lack of remorse, then it might be further confirmation of behavioral indications that this defendant lacked remorse.

In brief, G×E evidence can be a knife that cuts both ways, supporting both mitigation and aggravation.

VII. G×E for Criminal Justice Issues Other Than Responsibility and Sentencing

Understanding the causes of criminal behavior may be vitally useful, within and without the criminal justice system, to questions concerning rehabilitation and prevention. For example, outside the criminal justice system, it may be useful for establishing policies and programs that will reduce the risk of antisocial behavior. Fully discussing these uses raises complex issues and would require an essay in itself, but I will gesture at them in this section.

For many reasons, including issues of retributive and distributive justice and civil liberties, rehabilitation is no longer considered a prime goal of the criminal justice system, and it is of diminished importance within the delinquency jurisdiction of juvenile justice when juveniles who commit crimes are tried as adults. Nonetheless, if causal knowledge could help create effective, cost–benefit justified rehabilitation methods for some prisoners, there would be impetus to use them because recidivism imposes major costs on our society. Whether a particular genetic condition that predisposes people to criminal offending is amenable to specific rehabilitation methods derived from that causal knowledge is, of course, an open empirical question that good research can help answer. Whether an apparently effective intervention is cost–benefit justified is, of course, a normative moral, political, and legal question that science cannot decide.

Causal genetics knowledge may also be the key to prophylactic policies and programs that would be established outside the criminal justice system. Yet again, whether specific policies and programs would be effective and cost–benefit justified are open empirical and normative questions. There are risks associated with identifying classes of people and individuals as "at risk" for criminal behavior. Labeling effects, interventions that are unnecessary and often counterproductive for many recipients, and privacy issues are examples of the potential negative effects of such policies and programs. Let us hope, however, that advancing causal knowledge, whether from genetics or other fields, does point the way to successful preventive intervention and that the potential negative effects could be minimized.

Acknowledgments

This essay is a substantially revised and adapted version of a previously published chapter, Morse, S. J. (2011). Gene-environment interactions, criminal responsibility and sentencing, in K. A. Dodge & M. Rutter, M. (Eds.), *Gene-environment interactions in developmental psychopathology* (pp. 207–234). New York/London: Guilford Press. The author thanks Ed Greenlee for his invaluable help.

References

Aspinwall, L. G., Brown, T. R., & Tabery, J. (2012). The double-edged sword: Does biomechanism increase or decrease judges' sentencing of psychopaths? *Science, 337*, 846–849.

Atkins v. Virginia, 536 U.S. 304 (2002).

Barefoot v. Estelle, 463 U.S. 880 (1983).

Bernet W., Venecak-Jones, C., Farahany, N., & Montgomery, S. (2007). Bad nature, bad nurture, and testimony regarding MAOA and SLC6A4 genotyping at murder trials. *Journal of Forensic Science, 52*, 1–10.

Bok, H. (1998). *Freedom and responsibility.* Princeton, NJ: Princeton University Press.

Buchanan, A. (2011). *Beyond Humanity? The Ethics of Biomedical Enhancement.* Oxford: Oxford University Press.

Caspi, A., McClay, J., Moffit, T., Mill, J., Martin, J., Craig, I., Taylor, A., & Poulton, R. (2002). Role of genotype in the cycle of violence in maltreated children. *Science, 297*, 851–854.

Dar-Nimrod, I., & Heine, S. J. (2011). Genetic essentialism: On the deceptive determinism of DNA. *Psychological Bulletin, 137*, 800–818.

Denno, D. (2009). Behavioral genetics evidence in criminal cases: 1994–2007. In N. Farahany (Ed.), *The impact of behavioral sciences on criminal law* (pp. 317–354). New York: Oxford University Press.

Duff, A. (2007). *Answering for crime: Responsibility and liability in the criminal law*. Oxford: Hart Publishing.

Feresin, E. (2009). Lighter sentence for murderer with "bad genes." Retrieved November 21, 2009, from www.nature.com/news/2009/091030/full/news.2009.1050.html

Feresin, E. (2011). Italian court reduces murder sentence based on neuroimaging data. Retrieved September 1, 2011, from http://blogs.nature.com/news/2011/09/italian_court_reduces_murder_s.html#more

Fingarette, H., & Hasse, A. (1979). *Mental disabilities and criminal responsibility*. Berkeley: University of California Press.

Fodor, J. (1987). *Psychosemantics: The problem of meaning in the philosophy of mind*. Cambridge: MIT Press.

Graham v. Florida, 560 US 1 (2010).

Hart, H. L. A. (1968). *Punishment and responsibility: Essays in the philosophy of law*. New York: Oxford University Press.

Heyman, G. (2009). *Addiction: A disorder of choice*. Cambridge, MA: Harvard University Press.

Hume, D. (1978). *A treatise on human nature*. (P. Nidditch, Ed.). Oxford: Oxford University Press.

Kane, R. (2006). *A contemporary introduction to free will*. New York: Oxford University Press.

Kansas v. Crane, 534 U.S. 407 (2002).

Kaye, A. (2005). Resurrecting the causal theory of the excuses. *Nebraska Law Review, 83*, 1116–1177.

Klein, M. (1990). *Determinism, blameworthiness and deprivation*. Oxford: Oxford University Press.

Landis, S., & Insel, T. R. (2008). The "neuro" in neurogenetics. *Science, 322*, 821.

Lenman, J. (2006). Compatibilism and contractualism: The possibility of moral responsibility. *Ethics, 117*, 7–31.

Lewis, C. S. (1953). The humanitarian theory of punishment. *Res Judicatae, 6*, 224–230.

Lockett v. Ohio, 438 U.S. 586 (1978).

Miller v. Alabama, 132 S. Ct. 2455, 567 U.S. (2012).

Morse, S. (1994). Culpability and control. *University of Pennsylvania Law Review, 142*, 1587–1660.

Morse, S. (2002). Uncontrollable urges and irrational people. *Virginia Law Review, 88*, 1025–1078.

Morse, S. (2003*a*). Diminished rationality, diminished responsibility. *Ohio State Journal of Criminal Law, 1*, 289–308.

Morse, S. (2003*b*). Inevitable mens rea. *Harvard Journal of Law and Public Policy, 27*, 51–64.

Morse, S. (2004). Reason, results and criminal responsibility. *Illinois Law Review*, 2004, 363–444.

Morse, S. (2006). Brain overclaim syndrome and criminal responsibility: A diagnostic note. *Ohio State Journal of Criminal Law, 3*, 397–412.

Morse, S. (2007). The non-problem of free will in forensic psychiatry and psychology. *Behavioral Sciences & the Law, 25*, 203–220.

Morse, S. (2008). Determinism and the death of folk psychology: Two challenges to responsibility from neuroscience. *Minnesota Journal of Law, Science, & Technology, 9*, 1–35.

Morse, S. (2011). Lost in translation?: An essay on law and neuroscience. In M. Freeman (Ed.), *Law and neuroscience* (pp. 529–562). Oxford: Oxford University Press.

Noe, A. (2009). *Out of our heads: Why you are not your brain, and other lessons from the biology of consciousness*. New York: Hill and Wang.

Richards, J. R. (2000). *Human nature after Darwin: A philosophical introduction*. London: Routledge.

Ring v. Arizona, 536 U.S. 584 (2002).

Roper v. Simmons, 543 U.S. 551 (2005).

Scanlon, T. (2008). *Moral dimensions: Permissibility, meaning, blame*. Cambridge: Belknap Press.

Special Section. (2008). Genetics of behavior. *Science, 322*, 891–914.

Searle, J. (2002). End of the revolution. *New York Review of Books, 49*, 33–35.

Shapiro, S. (2000). Law, morality, and the guidance of conduct. *Legal Theory, 6*, 127–170.

Sher, G. (2006). *In praise of blame*. Oxford: Oxford University Press.

Smilansky, S. (2000). *Free will and illusion*. Oxford: Oxford University Press.

Strawson, G. (1989). Consciousness, free will and the unimportance of determinism. *Inquiry, 32*, 3–27.

Strawson, P. F. (1982). *Freedom and resentment*. In G. Watson (Ed.), *Free will* (pp. 59–80). Oxford: Oxford University Press.

Turkheimer, E. (2011). Genetics and human agency: Comment on Dar-Nimrod & Heine (2011). *Psychological Bulletin, 137*, 825–828.

Wallace, R. J. (1994). *Responsibility and the moral sentiments*. Cambridge, MA: Harvard University Press.

Wegner, D. (2002). *The illusion of conscious will*. Cambridge, MA: MIT Press.

Wootton, B. (1963). *Crime and the criminal law: Reflection of a magistrate and social scientist*. London: Stevens & Sons.

CHAPTER

20 Neurogenethics

Turhan Canli

Abstract

Ethical inquiry has followed advances in biology for decades, and different fields within biology have given rise to overlapping yet distinct areas of ethical inquiry. Genethics focuses on the ethics of genetics. Neuroethics focuses on the ethics of neuroscience. The author suggests that developments in molecular psychology, in which the tools of molecular biology are applied to study behavior, bring a new confluence of factors to generate a set of new questions, unique to the combination of neuroscience and genetics: neurogenethics—the ethics of neurogenetics. The questions are unique and novel because they emerge when genetic techniques are applied to the brain: to reprogram neural circuits and psychological processes, better predict behavior, personalize mental health treatment, and understand the Self. This chapter outlines the problem space and discusses specific examples of future neurogenethics research areas: viral-mediated gene therapy directed at the brain, optogenetics, imaging genetics, therapygenetics, and direct-to-consumer genomics.

Key Words: bioethics, brain, genethics, neuroethics, optogenetics

Bioethics can trace its roots all the way to the Hippocratic Oath, yet the field is constantly evolving, reflecting advances in biology. McWhirter argued that the critical questions remain unchanged: concerns about striking a balance between the rights of the individual versus community and autonomy versus paternalism (McWhirter, 2012), although she welcomed a reexamination of these questions with each new wave of advances in biology.

Others have identified "proprietary" issues belonging to certain subfields, specifically when delineating "genethics," the ethics of genetics, from "neuroethics," the ethics of neuroscience. For example, while acknowledging commonalities between genethics and neuroethics with regard to such issues as disease prediction, privacy, and identity, Illes and Racine argued that the interpretation of noninvasive functional neuroimaging data in particular is bound by cultural and anthropological factors that go beyond genethics (Illes & Racine, 2005). In a similar vein, Roskies (2007) discussed the positioning of neuroethics relative to genethics. She identified as common issues those related to consent and access to information, information misuse, distributive justice, health predictions, and the boundaries between pathology and normality. She also assigned unique issues to each field. For genethics, she named the implications of changes to the germ line as one among others. For neuroethics, she discussed three unique issues in depth: consciousness; decision making, control and free will; and moral cognition.

I suggest that the work presented in this book and in the larger peer-reviewed literature leads to the next stage in the evolution of biologically based ethics: *neurogenethics*, the ethics of neurogenetics. I see neurogenethics as an offspring of its two parent fields, in that it certainly shares some commonalities with each, but will inevitably—as it matures—acquire its own unique identity. I attempt to delineate some of these unique elements in the next section.

I. Conceptualizing the Problem Space

What kind of questions could or should be within the purview of neurogenethics? One approach is to organize the field based on the technological drivers and concomitant discoveries that have catalyzed it. I suggest conceptualizing the problem space for neurogenethics in three dimensions: the first dimension is scale, ranging from the molecular level of analysis to brains, behavior, and social structures. The second dimension captures brain processes associated with psychological functions, such as cognition, emotion, personality, social and cultural processes, development and aging, and issues related to consciousness and mind in both healthy and pathological states. The third dimension represents ethical, legal, and social implications (ELSI) and stretches across ethics considerations, ranging from procedural/technical/administrative issues such as safety and efficacy and informed consent to higher level concerns such as identity, empowerment, and personhood, or regulatory aspects of neurogenethics (see Figure 20.1).

In this section, I briefly introduce the scientific database, represented by the first two dimensions, that serves as the basis for neurogenethics considerations, which I will then introduce in subsequent sections.

I begin at the molecular level of analysis. Here, much work has focused on the mechanisms underlying cognitive processes in animal models, often using genetically modified mice. Several examples of this approach are represented by chapters in this volume, such as those presenting work on gene-by-environment (G×E) mouse models for mood disorders (Razzoli, Bartolomucci, and Carola), social bonding and social cognition (Barrett and Young), and memory (Sargin, Yan, and Josselyn).

Another large amount of work at the molecular level of analysis has focused on epigenetic and other gene regulatory mechanisms. These have become the focus of intense interest ever since the surprise

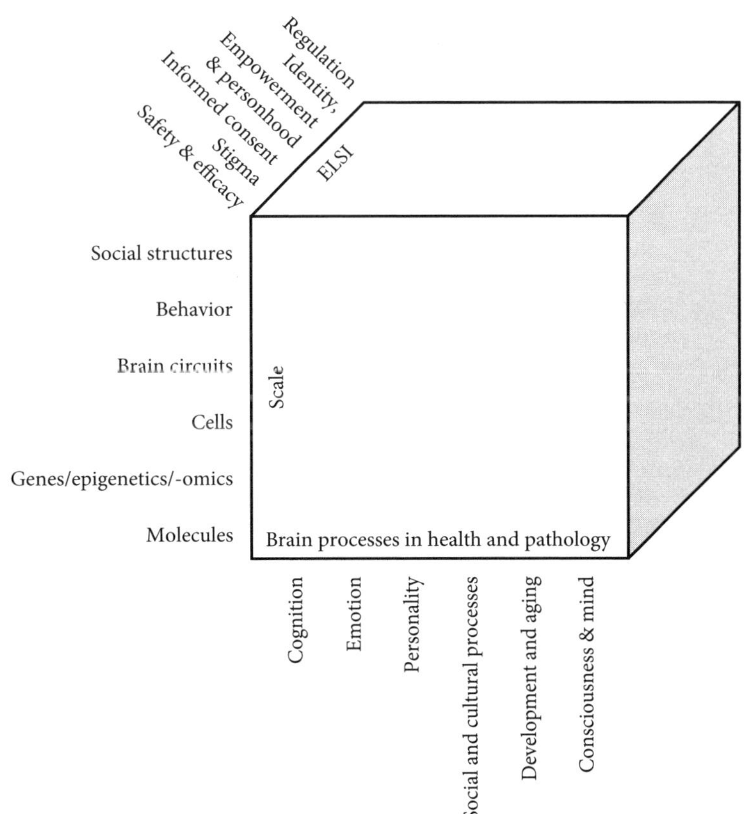

Fig. 20.1 The problem space of neurogenethics.

Illustrated here are the three dimensions that map out the problem space of neurogenetics: scale, brain processes, and ethical, legal, and social implications (ELSI).

finding from the Human Genome Project that there were far fewer protein-coding genes in the human genome than expected. As Mattick and colleagues have shown (Taft, Pheasant, & Mattick, 2007), whereas the number of protein-coding genes is relatively stable across species of increasing organismal complexity, the amount of non-protein-coding genomic regions (which make up 98 percent of the human genome) is positively correlated with organismal complexity. These non-protein-coding regions are believed to play a major role in the regulation of gene expression. In this volume, several chapters have focused on gene regulatory mechanisms, for example the role of nutrition and epigenetic mechanisms in programming neural systems of cognition and emotion (Naninck, Lucassen and Korosi), neuromodulation of social behavior in insects (Simpson and Stevenson), and the social regulation of gene expression in insects (Simpson and Stevenson) and in the African cichlid fish *Astatotilapia burtoni* (Maruska and Fernald).

Straddling the molecular and cellular levels of analysis are molecular approaches designed to repair defective genes or modify neuronal function. One example of this approach is gene therapy. For example, viral-mediated gene therapy has been successfully used to restore hearing in mice with a critical gene deletion (Akil et al., 2012). Other projects are under way that attempt to use a similar approach to replace or repair genetic defects in the brain, and clinical applications in the human brain are ongoing. Another example of this approach is optogenetics, which allows for the precise temporal and spatial activation or inactivation of genetically defined cells (Zhang, Aravanis, Adamantidis, de Lecea, & Deisseroth, 2007). This level of precision has generated a host of breakthrough discoveries in the past few years, illuminating functions in the prefrontal cortex (Warden et al., 2012), nucleus accumbens (Witten et al., 2010), ventral tegmental area (Chaudhury et al., 2013), and hippocampus (Kheirbek et al., 2013; Liu et al., 2012; Ramirez et al., 2013) as they relate to emotion, cognition, social processes, and neurodegenerative illnesses (Tye & Deisseroth, 2012).

Moving from the molecular/cellular level of analysis to genes associated with brain and behavior, it was the pioneering work of Lesch and colleagues (Lesch et al., 1996), who first discovered the link between a common variant of the serotonin transporter gene and trait neuroticism, which eventually gave rise to imaging genetics. In parallel, two other groups discovered a link between a common variant of the dopamine D_4 receptor gene (*DRD4*) and trait novelty seeking (Benjamin et al., 1996; Ebstein et al., 1996), although much of the following work has focused on the link between *DRD4* genetic variation and attention deficit hyperactivity disorder (ADHD; Faraone, Doyle, Mick, & Biederman, 2001; Gizer, Ficks, & Waldman, 2009; Kebir, Tabbane, Sengupta, & Joober, 2009; Wu, Xiao, Sun, Zou, & Zhu, 2012). In this volume, a number of imaging genetics studies have been discussed in chapters on aggression (Buckholtz and Meyer-Lindenberg), social decision making (Ebstein, Zhong, Chark, Lai, and Chew), affiliation (Westberg and Walum), and traits (Hyde, Bogdan, and Hariri).

I (Canli, 2005) and others (Tairyan & Illes, 2009) have discussed the neuroethics of imaging genetics before, but it is worth revisiting this subject in the next section, in light of more recent studies. Imaging genetics studies have addressed a wide range of psychopathologies, such as schizophrenia (Meyer-Lindenberg, 2010), ADHD (Durston, 2010), mood disorders (Scharinger, Rabl, Sitte, & Pezawas, 2010), and anxiety disorders (Domschke & Dannlowski, 2010), as well as mental processes in cognitive aging (Mattay, Goldberg, Sambataro, & Weinberger, 2008) and in social cognition (Zink & Meyer-Lindenberg, 2012) (also see chapters from Meyer-Lindenberg and from Westberg in this volume).

Whereas earlier imaging studies have focused on individual candidate genes, Meyer-Lindenberg suggests that future imaging genetics studies will be designed around genome-wide genetic variants and other levels of gene expression (Meyer-Lindenberg, 2012). Indeed, imaging genetics has already been applied to validate novel gene candidates obtained from genome-wide association data (Potkin et al., 2009). Imaging genetics is also expanding to discover associations between brain activity and other molecular markers relevant to gene regulation. For example, Ursini and colleagues have linked methylation of the methyltransferase (COMT) Val allele to prefrontal activation during a working memory task (Ursini et al., 2011). My group has discovered links between telomere length and brain functional (Ferri, Izzi, & Canli, 2014) and brain resting activation (Izzi, Ferri, & Canli, 2014), respectively.

Moving from brain levels of analysis to the individual, genetic information is beginning to predict patient responses to therapy (Eley et al., 2012) and shape individuals' sense of self (Tutton & Prainsack, 2011). In the case of the former example, current work focuses on single candidate genes, whereas in the case of the latter, information is increasingly processed at the level of the entire genome. Indeed, genome-wide association studies (GWAS) have discovered new genetic leads for a host of physical diseases (Manolio, Brooks, & Collins, 2008), although GWAS of psychiatric illnesses have not lived up to early expectations (Bondy, 2011). One reason for the shortcomings of psychiatric GWAS could be sample-sourcing: most GWAS are based on samples that can easily be obtained, such as from buccal cells or blood draws. These may not be the most suitable sources of tissue for studies of psychiatric illness if the genetic association rests on patterns of gene expression rather than on variation within DNA sequences per se. Indeed, as Hardy and Singleton (2009) noted: "One observation that has taken many observers by surprise is that most loci that have been discovered through genome-wide association analysis do not map to amino acid changes in proteins. Indeed, many of the loci do not even map to recognizable protein open reading frames but rather may act in the RNA world by altering either transcriptional or translational efficiency. They are thus predicted to affect gene expression." Thus, GWAS of psychiatric illness should ideally be conducted in brain tissue from pertinent regions, which requires access to postmortem or otherwise excised brain tissue (e.g., brain tumors).

Genome science has been brought to the individual through the emergence of direct-to-consumer (DTC) companies that may have a profound impact on people's (and perhaps even peoples') understanding of their identity. These companies can provide genomic information relevant for brain function, personality traits, and risk for psychopathology in addition to other types of health-related or genealogical information. Furthermore, the online platform provided by DTC companies such as 23andMe also aggregates consumers into large communities that can drive research discovery and may even have a broader societal impact that some have termed "biological citizenship" (Rose & Novas, 2005).

With the problem space mapped out, I next explore specific examples at different levels of scale by first laying some empirical groundwork before turning to a discussion of the ethics considerations.

II. Viral-Mediated Gene Therapy Directed at the Brain

A. Empirical Studies

A recent study by Akil and colleagues used virally mediated gene therapy to restore hearing in mice that were genetically engineered to be congenitally deaf (Akil et al., 2012). About half of all cases of human congenital deafness have genetic causes (Di Domenico et al., 2011), and current treatments remain wanting (Kral & O'Donoghue, 2010). Akil and colleagues used a mouse model of hereditary deafness in which the gene encoding the vesicular glutamate transporter-3 (VGLUT3) was missing in a VGLUT3 knockout mouse. In mice, VGLUT3 is expressed in hair cells of the inner ear, and its lack of release causes loss of synaptic transmission and deafness (Seal et al., 2008). In humans, a similar mutation of the gene encoding VGLUT3 is also associated with hearing loss (Ruel et al., 2008). Akil and colleagues injected an adeno-associated virus type 1 (AAV1) containing the VLGUT3 gene into the cochlea. This procedure resulted in protein expression of VGLUT3 in inner hair cells exclusively, which is remarkable because the intermediate gene product, VGLUT3 mRNA, was more widely expressed. This suggests an unknown regulatory posttranscriptional mechanism that ensures that VGLUT3 is only functional in the cells that would normally require it. The virally inserted VGLUT3 gene was not only associated with protein expression in inner hair cells, but also with restored hearing in treated animals.

Indeed, recombinant AAV (rAAV) vectors have been used in brain-directed gene therapy. These rAAV vectors are generally efficacious and safe, eliciting no inflammation in the brain (Bowers, Breakefield, & Sena-Esteves, 2011; High & Aubourg, 2011; Klein et al., 2002; McCown, 2011; Tenenbaum, Lehtonen, & Monahan, 2003; Weinberg, Samulski, & McCown, 2013). However, different vectors can produce different immune responses: AAV1 and -2 have been shown to produce higher antibody titres than did AAV5, -8, and -9 (Boutin et al., 2010). Administration of these vectors has been invasive, usually requiring injection of the vector directly into the brain. Less invasive or noninvasive methods are therefore of great preclinical interest. For example, self-complementary (sc) rAAV9 vectors were successfully used to cross the blood–brain barrier through intravenous injection (Duque et al., 2009): this technique produced transgene expression in rat and cat spinal cord motor neurons that lasted for 5 months. Compared to the common single-stranded types, these double-stranded sc rAAV vectors are 20–100 times more efficient (Gray et al., 2011). However, several factors limit their utility (Iida, Takino, Miyauchi, Shimazaki, & Muramatsu, 2013): for example, these vectors can only package small genes; they may produce a larger immune response than conventional vectors; and in mice and non-human primates, this delivery system targeted not neurons but astrocytes (Bevan et al., 2011; Foust et al., 2010; Samaranch et al., 2012). One group developed a novel vector by mutating AAV9 to produce the tyrosine-mutant AAV9/3 vector, which was delivered to the brain of anesthetized mice by intracardiac injection (Iida et al., 2013). This vector achieved robust gene delivery in the brain that could be further targeted by selection of different gene promoters: in its unmodified form, AAV9/3 produced gene expression primarily in glial cells; when modified with a neuron-specific synapsin I promoter, gene expression was observed in the neurons of frontal cortex, hippocampus, and amygdala 6 weeks after the procedure; when modified with a cerebellar Purkinje cell-specific L7 promoter, gene expression was observed selectively in Purkinje cells.

Chtarto and colleagues described several clinical studies of human patients with neurological diseases using AAV vectors (Chtarto et al., 2013). For example, they highlighted current work in Parkinson disease (PD), which is caused by degeneration of nigrostriatal dopaminergic neurons. Two phase I safety gene therapy studies were designed to alleviate PD symptoms. In the first study, which was directed at addressing the lack of dopamine in the putamen, ten PD patients received injections of an rAAV2 vector expressing aromatic acid decarboxylase (AADC; which converts L-dopa into dopamine) into the putamen, and this procedure was shown to express AADC over a period of 4 years without safety or tolerability side effects (Eberling et al., 2008; Mittermeyer et al., 2012). However, even at the highest dose, a slow deterioration in PD symptoms was observed following the initial first-year improvement (Mittermeyer et al., 2012). In the second study, which was directed at inhibiting overactive motor neurons in the subthalamic nucleus (STN), 12 PD patients received injections of an rAAV2 vector expressing glutamic acid decarboxylase (GAD; which synthesizes γ-aminoglutaric acid [GABA], an inhibitory neurotransmitter) into the STN. Ten of these patients showed symptom improvement 12 months later (Kaplitt et al., 2007), and a follow-up study of 45 patients using a more rigorous double-blind, sham-surgery controlled design (after removing some patients from statistical analysis for various reasons, 16 patients receiving the vector and 21 patients receiving sham-surgery were included in the final analysis) replicated these results (LeWitt et al., 2011). A different approach was taken by another group, which aimed to address the cause of the disease by restoring DA neurons through expression of glial cell-derived neurotrophic factor (GDNF) or neurturin (NTN) through injection of an AAV2-neurturin vector into the putamen and substantia nigra (Marks et al., 2008). This work has produced some symptom improvements 18 months post treatment (Marks et al., 2010) and has now progressed to a phase II clinical trial (Bartus et al., 2013).

Charta and colleagues review efforts toward the next generation of AAV vectors (Chtarto et al., 2013). These include efforts to design vectors whose transgene expression is regulated by exogenous drugs, so that the vector-inserted gene can be expressed or repressed at will, vectors that target specific cell types or neuronal subtypes, and vectors with central nervous system-limited expression, as well as efforts to improve efficacy, increase gene size, and minimize invasiveness of vector delivery.

B. Ethics Considerations

Turning to the ethical implications of this technology, Lowenstein commented on the ethics of informed consent in gene therapy trials and addressed some of these comments specifically at PD gene therapy trials (Lowenstein, 2008). For example, he noted that some safety considerations also produced useful scientific information: early PD trials (Eberling et al., 2008; Marks et al., 2008) involved unilateral injection of the vector out of safety concerns, which allowed for comparison of the treated versus nontreated hemisphere within patients. In other ways, different ethical concerns may be in conflict with one another: Lowenstein noted that ethical considerations of patient selection allowed only for the recruitment of those PD patients for whom all other conventional treatments had been exhausted without halting disease progression. Yet, from a therapeutic point of view, one would assume that patients at the early stages of disease progression would be most likely to benefit from treatment, but these patients were denied the opportunity for such treatment. As Lowenstein argued "[t]he ethical principles used in these clinical trials may have been sound when considered in isolation, but the price paid is a delay in initiating treatment of those patient populations who could best benefit from gene therapy" (p. 3).

Twenty-nine patients from the previously discussed double-blind, sham-surgery controlled

PD gene therapy study by LeWitt and colleagues (2011) also participated in a qualitative interview study about their expectations (Kim et al., 2013). The interview study was designed to differentiate between patients' "understanding" versus "appreciation" of the randomization process based on a distinction introduced by Appelbaum and Grisso (1988), in which the former term refers to an intellectual understanding of the meaning of information presented in the consenting procedure, and the latter term refers to the ability to apply that information to one's own situation. This distinction has been drawn on by studies that documented cases of research participants who may have understood the randomization process in general terms, but who were unable to apply the implications of this procedure to their own individual chances of receiving experimental treatment (Appelbaum, Roth, & Lidz, 1982). This inability has been labeled "unrealistic optimism" that the odds of receiving beneficial treatment cannot be statistically accurate (Jansen et al., 2011). In the study by Kim and colleagues, most of the responders (22/26, or 83 percent, with missing data from 3 responders) understood the randomization procedure. Of these, a slim majority (12/22, 55 percent) also appreciated correctly how the randomization procedure applied to their own chances of being in the treatment condition. The remaining 45 percent did not answer the appreciation question in quantitative terms (the randomization procedure was one-to-one, and correct quantitative answers would be "50 percent" or "50/50") but instead provided verbal comments. Detailed content analysis of responders' answers strongly suggested that these individuals did, in fact, have an appreciation of their odds but introjected into their statements additional sentiments that appealed to hope or prayer as a matter of conversational context. Kim and colleagues concluded that

> [P]erhaps we need to be more hesitant about labeling research participants as "irrational," "unrealistic," and harboring "misconceptions," unless such conclusions are drawn from carefully designed studies that take into account the relevant social and conversational contexts of their responses to questions posed to them. Such studies may take the form of careful conversational analyses or experiments that test specific hypotheses. There is no doubt that some individuals' expectations about participating in research are indeed misconceptions; but researchers assessing expectations cannot simply assume that individuals are merely vehicles for knowledge and information transfer. We risk stigmatizing prospective and actual research participants not because of their vulnerabilities, but because we measure their behavior and speech in inadequate ways. (p. 8)

III. Optogenetics

A. *Empirical Studies*

The toolset of optogenetics was developed by Boyden, Deisseroth, and colleagues (Boyden, Zhang, Bamberg, Nagel, & Deisseroth, 2005). The technical aspects of this groundbreaking technology have been reviewed elsewhere (Bernstein & Boyden, 2011; Kramer, Mourot, & Adesnik, 2013; Packer, Roska, & Hausser, 2013; Tye & Deisseroth, 2012; Williams & Deisseroth, 2013; Zhang et al., 2007). Very briefly put, the approach is based on using viral vectors (such as lentivirus or AAV, discussed in the previous section on gene therapy) to insert light-sensitive molecules called *opsins* into cells of the central nervous system that can be genetically defined in precise ways. Once in place, these molecules can be activated by specific wavelengths of light to open channels in the cell's membrane to excite or inhibit cell signaling. This technology has been seminal: the number of studies indexed with the search term "optogenetics" in PubMed has exploded from 10 in 2009 to 338 in 2013, and the approach was named "Method of the Year" by the journal *Nature* in 2010.

In the context of neurogenethics, I limit my discussion of optogenetics to those studies designed to manipulate memory because this topic area is of particular interest to the neuroethics community, as discussed further next. Optogenetics is particularly valuable for the study of memory encoding and retrieval because memories appear to be encoded "sparsely," that is, by sets of neurons that may only represent a small subset of all the neurons that comprise a given brain region (Gdalyahu et al., 2012), and no other technology is currently capable of targeting only these neurons.

The erasure of memory is discussed in Chapter 4 of this volume by Sargin, Yan, and Josselyn. The approach presented by these authors utilized genetically modified mice and targeted viral manipulations that were designed to either increase or diminish the expression of candidate genes believed to play a role in memory. The optogenetic approach allows for manipulation of neuronal activation at even greater temporal and spatial resolution.

For example, Huff and colleagues (Huff, Miller, Deisseroth, Moorman, & LaLumiere, 2013) used optogenetics in an inhibitory avoidance task in rats to modulate memory consolidation. Prior to training, rats received virus microinjections into the basolateral amygdala of either the excitatory cation channel channelrhodopsin-2 [ChR2(E123A)] paired with a promoter targeted at glutamatergic neurons or the inhibitory outward proton pump archaerhodopsin-3 from *Halorubrum sodomense* strain TP009 (ArchT). Rats were then trained in one session on the inhibitory avoidance task. They were placed into the lit space of a box that also contained a dark compartment. Upon entering the dark compartment, the rats received a foot shock. Immediately after this training experience, rats received optical BLA exposure to stimulate (ChR2(E123A)) or inhibit (archaerhodopsin-3) neural consolidation. Retention testing occurred 48 hours later, when rats were placed into the lit area of the box and their latency to enter the dark compartment was measured, with longer latencies interpreted to signal better memory retention. The study showed that posttraining optical stimulation of glutamatergic BLA neurons enhanced memory retention, whereas posttraining inhibition of the BLA impaired memory retention.

This study illustrates the potential for optogenetics to enhance or diminish memory. This technology could translate to humans with memory problems, such as patients with posttraumatic stress disorder (PTSD) who may desire a blunting of their memorized trauma. The study by Huff and colleagues was based on a memory intervention immediately after the to-be-encoded experience, so it remains to be seen if older memories could similarly be manipulated. Later in this section, I discuss the feasibility of optogenetics in humans and the ethical considerations regarding memory manipulations.

Liu, Ramirez, and colleagues used transgenic mice to virally transfect opsins into only those individual neurons that had been active during the encoding of an episodic memory (Liu et al., 2012; Ramirez et al., 2013). A technical tour-de-force, this approach allowed them to ask whether direct optical activation of those neurons (and *only* those neurons) that were active during an earlier learning episode would be sufficient to reactivate that memory (as measured by a behavioral response evoked by that memory). In the first study, Liu, Ramirez, and colleagues (Liu et al., 2012) exposed mice to foot shock in one environment, context A. Concurrently, neurons in the dentate gyrus of the hippocampus that were activated during this learning experience (and *only* these neurons) were labeled with ChR2. Animals were later introduced into a novel environment, context B. Although this environment had never been associated with foot shock, optical reactivation of the ChR2-labeled neurons in context B produced freezing behavior, an index of fear. Thus, activation of these neurons was sufficient to activate the memory. Several control conditions support this conclusion: ChR2 activation in context B did not produce freezing behavior if the neurons had been labeled during exposure to yet another environment, context C, in which they were not exposed to foot shock. This ruled out the possibility that mere ChR2 reactivation by itself would generate a freezing response. Another control condition was activation of other-labeled neurons, which also failed to produce freezing behavior, thus demonstrating the specificity of neurons required to be reactivated.

In a follow-up study, Ramirez, Liu, and colleagues (Ramirez et al., 2013) took the experiment one step further. They first labeled neurons in the dentate gyrus or CA1 region of the hippocampus of mice that were activated during exposure to context A in the absence of a foot shock. Thus, these labeled neurons should encode a "safe" environment. In a second step, these mice were then exposed to context B, in which they did receive foot shock. Critically, neurons that had encoded the memory of ("safe") context A were reactivated by CHR2 at the same time that animals received foot shock in context B. In the final step of the experiment, when mice were reintroduced into context A, they exhibited freezing behavior. Thus, although these animals had never experienced foot shock in this environment, their behavior suggests that the experience of foot shock in B had become associated with the reactivated memory of A, so that the fearful experience generated in B was now also associated with A. In effect, these researchers had produced a false memory of foot shock experienced in A.

Put together, these two studies demonstrate that reactivation of those neurons that were active during the encoding of an episodic memory can reinstate that memory. Furthermore, optical reactivation of a memory can be associated with the formation of a novel memory, thus modifying an existing one or creating a false memory.

How likely is it that optogenetics will be used on humans? There is a high expectation that this technology will produce translational applications, for instance in epilepsy (Bentley, Chestek, Stacey, & Patil, 2013), heart disease (Entcheva, 2013), addiction (Britt & Bonci, 2013), and psychiatric and neuropsychiatric conditions (Huang, Tang, & Jiang, 2013; Tourino, Eban-Rothschild, & de Lecea, 2013). Yet preclinical studies using nonhuman primates have been slow to catch up with other animal models (Ruiz et al., 2013), although the efforts are ongoing (Gerits & Vanduffel, 2013; Han, 2012).

Han (2012) discussed several factors that have posed technical challenges to primate optogenetic applications. One is that the most precise targeting of neurons, such as those individual neurons that are activated during the encoding of an episodic memory, requires not only viral delivery of the opsin molecules, but also transgenic manipulation of the host organism. The availability of transgenic mice, fruit flies, and worms has allowed for rapid advances in applications of optogenetics in these organisms, but few transgenic primates exist. As discussed in the previous section on gene therapy, there are some AAVs that are specific enough to reach certain cell types, and these efforts are likely to continue to be refined. Developing delivery approaches that can target neurons with great specificity without the need for transgenic manipulation is therefore critical for advancing preclinical studies and obviously for later human applications.

Another factor that poses a technical challenge to primate optogenetic applications discussed by Han (2012) is the degree to which optical control can effectively be exerted over opsin-labeled cells. There are several variables that affect the effectiveness of optical control. One of them is tissue damage to the brain stemming from the insertion of optical fibers that need to deliver the light to the target region or from heat generated by the light source. Han suggest the use of high-density optical fiber arrays and minimally powered lasers to activate neurons. These solutions, of course, do not mitigate the invasive nature of placing these devices into the brain, which limits their eventual human applications.

A novel approach is the development of noninvasive optogenetics, as demonstrated by Tsien's group (Lin, Knutsen, Muller, Kleinfeld, & Tsien, 2013). These investigators engineered a novel channelrhodopsin (ChR) called red-activatable ChR (ReaChR) that is optimally excited by orange-to-red light. Light at this wavelength is less scattered by tissue and less absorbed by blood than at other wavelengths and can therefore penetrate the intact skull. Shining red light through the intact skull of awake mice, Lin and colleagues could activate ReaChR expressing cells in the vibrissa motor cortex to drive vibrissa movements. Directing light through the external auditory canal, they could even activate ReaChR expressing cells in the facial motor nucleus in the brainstem to activate vibrissa movements. Such noninvasive transcranial optogenetic technology may expand the range of possible future applications in humans and may complement noninvasive methods for gene therapy directed at the brain discussed in the previous section.

B. *Ethics Considerations*

The previous paragraphs suggest that technical obstacles in the application of optogenetics to preclinical studies of nonhuman primates are successfully being addressed, thus opening the door for eventual human applications. I discussed the ethics implications of viral vectors targeting the human brain in the previous section. The discussion of the ethics of nonhuman primate research is well beyond the scope of this chapter. Here, I focus the ethics discussion on optogenetics applications related to the manipulation of memory, the feasibility of which was demonstrated by Huff and colleagues (Huff et al., 2013), Liu, Ramirez, and colleagues (Liu et al., 2012), Ramirez, Liu and colleagues (Ramirez et al., 2013), and others.

The unintended consequences of memory manipulation have been fictionalized in the movie "Eternal Sunshine of the Spotless Mind," in which the protagonist wants to rid himself of the memory of his girlfriend following a painful break-up but regrets this decision as the memory removal procedure unfolds. The movie comments on the value of painful experiences and on the significance of memories in defining one's self and identity. These very same points have been raised, with very differing perspectives, by the President's Council on Bioethics (2010) and Adam Kolber (Kolber, 2010).

The President's Council commented on the ethics of blunting memory, as could be the case when patients suffering from PTSD are offered treatment options that may reduce the emotional impact of traumatic memories. One concern was that the emotional blunting of episodic memories may be equivalent to "abandoning or compromising our own truthful identities" (p. 91). Without emotions to accompany a memory, individuals may be too comfortable to be moved by cruelty, have diminished capacity to appreciate the moral significance of crimes, or view psychic pain as unnecessary. Furthermore, the avoidance of psychic pain may also dull the experience of positive experiences and promote an "unmindful happiness" (p. 91). A second concern was related to the collective duty to remember historical atrocities truthfully. In the eyes of the Council, this duty applies to society in general, including those who may have experienced atrocities first-hand. They view biotechnical means to ease psychic pain as a false form of society's solidarity with victims. As the Council put it: "it exempts us from the duty to suffer-with (literally, to feel *com*-passion for) those who remember; it does not demand that we preserve the truth of their memories; it attempts instead to make the problem go away, and with it the truth of the experience in question" (p. 93). The Council then moved from a duty to remember to linking memory explicitly to moral responsibility. Their concern was that people could not be held accountable, that there would be no justice and no opportunity for forgiveness. A final concern was that painful memories serve a higher purpose, such that "[t]hese experiences become part of who we are, part of our own life as truthfully lived" (p. 95).

Kolber responded directly to the concerns raised by the President's Council. With respect to the concern that blunted memory compromises one's truthful identity, Kolber countered that, even if unaltered by biomedical means, memories are not "verbatim transcripts" of lived experiences but rather a product of "synthesis and reconstruction of our past" (p. 101). Indeed, he argued, "such selective reconstruction of our lives seems to be at the heart of the creation of a coherent life story" (p. 101). The availability of biomedical means to dampen traumatic memories may even strengthen one's identity because "[b]y preventing traumatic memories from consuming us, memory dampeners may allow us to pursue our own life projects, rather than those dictated by bad luck or past mistakes" (p. 101). In response to the Council's concern of a collective duty to remember historically significant events, Kolber stressed individual choice over legal restrictions on memory dampening. He noted that "memory-dampening drugs, by giving us the opportunity to consciously choose to keep a memory intact, may actually facilitate our identification with it. On the other hand, if an individual retains a homage memory simply because he has no choice..., the holding of the homage memory loses much of its significance" (p. 109). He acknowledged that there may be legal obligations to hold on to a particular memory; for example, in eyewitness testimony, but he does not support broad restrictions on memory dampening on the basis of this scenario. Kolber rejected the Council's assertion that "without memory,...the notion of moral responsibility would unravel" (p. 93). He pointed out that we do, in fact, hold individuals responsible for their actions if they fail to remember, such as forgetting to file their taxes. He again stressed personal choice, that the opportunity to deliberately dampen one's memory (or not) may increase one's responsibility to choose carefully. The Council's view that painful memories serve a higher purpose to become part of who we are was answered by Kolber's observation that "many experiences are simply tragic and terrifying, offering virtually no opportunity for redemption or transformation" (p. 98). Taking this argument one step further, Kolber suggested that memory dampening may actually promote life transformations that otherwise might not be possible or at least reduce the time needed for an individual to recover from a traumatic experience.

The debate on the ethics of memory dampening rests on untested assumptions about what people would do or not do if biomedical memory dampening interventions were available. Do people want their painful memories to be removed? Are there some classes of painful memories that people would be more likely to have removed than others? Are such choices based on abstract notions or personal life experiences? Are they defined by political, philosophical, religious, or other social determinants? These are empirical questions to be explored in future research. Anecdotally, the vast majority of my undergraduate neuroethics students discussing these two articles articulate the view that they would choose to hold on to painful memories, and some of their responses have gravitas because they are born out of personal traumatic experiences. It is therefore possible that people's free choices regarding memory dampening may actually assuage many of the Council's concerns.

IV. Imaging Genetics

A. Empirical Studies

I suggest that the neurogenethics of imaging genetics studies will become an increasingly "hot" topic as these studies become more precise and comprehensive in capturing gene-brain-behavior associations. I see three drivers that make imaging genetics studies more precise and comprehensive.

One driver is the transition from "small-scale" to "Big Science." Whereas the first fMRI-based imaging genetics study was based on a sample size of 28 participants (Hariri et al., 2002), current imaging genetics studies tend to have sample sizes much larger than that. The largest effort, by far, is the formation of the Enhancing NeuroImaging Genetics through Meta-Analysis (ENIGMA) Consortium, which combines data from 70 participating institutions and has access to data from close to 25,000 scan subjects (Thompson et al., 2014).

The second driver is the transition from imaging studies focused on main effects of single candidate gene polymorphisms to imaging studies of gene–gene or G×E interactions, epigenetic gene regulatory mechanisms such as gene methylation, and genome-wide data. For example, we were the first to use fMRI to discover an additive effect for two serotonergic gene polymorphisms (5-HTT-linked polymorphic region, 5-HTTLPR, and tryptophan hydroxylase-2, TPH2, rs4570625; Canli, Congdon, Constable, & Lesch, 2008) and the first to show a G×E interaction between 5-HTTLPR genotype and life stress in the brain (Canli et al., 2006). Later studies have documented similar G×E interactions for 5-HTTLPR in the amygdala in some cases (Alexander et al., 2012; Hermann et al., 2012), but not in others (Walsh et al., 2012; although they did report a G×E interaction in the lingual gyrus), suggesting that sample size limitations or other variables may obscure the true relations between 5-HTTLPR genotype and environmental factors shaping neural circuits and function.

One possible variable affecting G×E interactions could be gene methylation, an epigenetic mechanism that can regulate gene expression. Earlier work in rats demonstrated that early rearing adversity can differentially affect glucocorticoid receptor gene methylation in the hippocampus (Weaver et al., 2004). Differential glucocorticoid receptor methylation has also been discovered in postmortem hippocampal tissue as a function of history of childhood abuse (McGowan et al., 2009), and childhood trauma was found to differentially affect 5-HTT methylation in the periphery (Vijayendran, Beach, Plume, Brody, & Philibert, 2012). Ursini and colleagues (Ursini et al., 2011) combined imaging with epigenetic analysis of the catechol-O-methyltransferase (COMT) Val(158) polymorphism, which was obtained by blood draw. They found that homozygous Val allele carriers showed a negative correlation between Val allele methylation and life stress history and a corresponding prefrontal activity during a working memory task. This study suggests that environmental experiences such as lifetime stress may impact gene methylation and thus gene expression to moderate brain function.

Whereas the imaging genetics studies highlighted so far have focused on single genes or small sets of genes, future studies will likely increasingly take advantage of the availability of genome-wide data. Indeed, the very large datasets available to the ENIGMA Consortium now allow for analysis of brain imaging data in the context of genome-wide genetics results. For example, the ENIGMA Consortium used structural MRI and genome-wide single nucleotide polymorphism (SNP) data from a discovery sample of 7,795 individuals and multiple replication samples totaling 13,356 individuals to identify SNPs associated with hippocampal volume (Stein et al., 2012).

The third driver to make imaging genetics studies more precise and comprehensive is advances in analysis techniques. For example, whereas earlier fMRI studies have focused primarily on activations in regions of interest, such as amygdala, nonlinear machine-learning algorithms have been developed that take into account distributed spatial activation patterns. This approach has been used in a study of lie detection with 88 percent accuracy (Davatzikos et al., 2005), although a number of critical challenges remain in the implementation of brain imaging technology to lie detection in the real world (Farah, Hutchinson, Phelps, & Wagner, 2014).

Other analysis methods have been developed to investigate the structural and functional connectivity of the brain, the so-called "connectome" (Biswal et al., 2010; Craddock et al., 2013; Sporns, Tononi, & Kotter, 2005; Toga, Clark, Thompson, Shattuck, & Van Horn, 2012). The earliest examples of connectivity analyses in the context of imaging genetics come from Pezawas and colleagues (Pezawas et al., 2005) who reported on amygdala-anterior cingulate connectivity as a function of 5-HTTLPR genotype and from our own analyses of functional connectivity between amygdala and hippocampus to other brain regions as a function of a G×E interaction between 5-HTTLPR genotype and life stress (Canli et al., 2006). The *Human Connectome Project* (HCP; http://www.humanconnectomeproject.org/) is another example of the "Big Science" discussed earlier, in which connectomic datasets are aggregated from multiple partners and made available for analysis. Thompson and colleagues recently discussed advances in the analysis of structural and functional networks moderated by genetic factors, "connectome genetics" (Thompson, Ge, Glahn, Jahanshad, & Nichols, 2013).

B. Ethics Considerations

These drivers of increasingly precise and comprehensive imaging genetics studies have the potential to produce much better biomarkers in psychiatry. For example, connectome genetics studies have begun to identify differences in brain connectivity between carriers and noncarriers of risk alleles for Alzheimer disease (Braskie et al., 2011), schizophrenia (Braskie et al., 2012), and autism (Dennis et al., 2011). Illes and colleagues discussed the ethics of predicting Alzheimer disease from neuroimaging data (Illes, Rosen, Greicius, & Racine, 2007). For example, they noted the possibility of unintended social, economic, and ethical consequences, such as stigmatization or a vastly increased need for counseling, and they emphasized the need for validation studies that establish sensitivity and specificity of results that can be meaningfully applied to the individual patient or that have been established in longitudinal studies to demonstrate predictive validity. Presumably, these concerns become even more salient as the latest research based on connectome genetics—and soon connectome genomics (taking into account genome-wide data)—addresses ever-larger sets of medical conditions.

Tairyan and Illes discussed the ethics of imaging genetics more broadly (Tairyan & Illes, 2009). They distinguished between discriminative and cumulative ethics power of combining genetics with neuroimaging, with discriminative power representing the capacity for enhanced disease differentiation, and incidental findings and cumulative power representing the ability to better understand the brain/mind and the associated ethics challenges. With respect to the latter, the authors expressed concern that the cumulative power of imaging genetics may amplify stigma for individuals with brain diseases, increase the threat of violation of confidentiality of ever more complex datasets, and multiply the implications of commercial applications. More recently, Illes and colleagues have proposed that researchers ought to subscribe to (1) "ethical transparency" by publishing the ethics methods they employed to address study-related ethics concerns and to (2) "ethical reproducibility" by learning from ethics practices of other investigators (Eijkholt, Anderson, & Illes, 2012).

One area of intense neuroethics interest has been the ethics of fMRI to lie detection in courtroom settings or in the service of national security (Canli, Brandon, Casebeer, Crowley, Du Rousseau, et al., 2007*a*; Canli, Brandon, Casebeer, Crowley, Durousseau, et al., 2007*b*; Farah et al., 2014; Haynes, 2008; Sip, Roepstorff, McGregor, & Frith, 2008). Furthermore, studies on the genetic basis of aggression and G×E interactions (Caspi et al., 2002) already have had some impact in courtroom settings (discussed in the chapter by Morse in this volume). The combination of these two approaches produces a kind of forensic neurogenetics. Indeed, Meyer-Lindenberg and colleagues have discussed the potential of this approach but have argued against a restricted biological focus and in favor of a broader integration of neurogenetic data with insights from other fields such as psychology, sociology, criminology, and others (Dressing, Sartorius, & Meyer-Lindenberg, 2008).

A preventative framework has been adopted by those with a developmental focus (Anderson & Kiehl, 2014; Blair, 2006; Salekin & Frick, 2005; Viding et al., 2013). For example, children with high levels of callous-unemotional traits that are characterized by genetic and neurocognitive markers are at risk for developing antisocial behavior and psychopathy, but this trajectory is moderated by environmental variables (Viding & McCrory, 2012). Current work is beginning to develop effective interventions that are tailored around these cognitive and neural biomarkers (Viding & McCrory, 2012).

The ethics implications of this work were addressed by Jones and Cauffman who studied judicial perceptions and recommendations in response to a hypothetical case (Jones & Cauffman, 2008). Their study shed some light on the question of whether the stigma of psychopathy in youth affected sentencing recommendations. They discovered that, although the label of psychopathy was associated with more restrictive placement recommendations, this effect was moderated by the perceived dangerousness of the youth rather than by the label itself.

Hinshaw discussed the predictive validity and ethics of one laboratory-based test of covert antisocial behavior in children, in which children were given the opportunity to steal and deface property (Hinshaw, 2005). On the one hand, this paradigm had a high level of predictive validity; on the other hand, Hinshaw discussed the ethics of tempting children to engage in these types of covert antisocial behaviors, informed consent, and debriefing procedures.

The neurogenethics of imaging genetics is likely a fruitful area of future research as the combination of genomic and neural biomarkers produces more refined measures of biological identity than either technology alone. This enhanced power has implications for vulnerable populations such as patients or children and may require modifications to handling of datasets, informed consent procedures, and ethics considerations of stigma or behavior prediction.

V. Therapygenetics

A. *Empirical Studies*

Eley and colleagues coined the term "therapygenetics" to describe a new field concerned with understanding individual differences in response to clinical psychological interventions as a function of genetic variation (Eley et al., 2012). In their study, they stratified a sample of anxiety-disordered children ($N = 359$) who were treated with cognitive behavior therapy (CBT) by 5-HTTLPR genotype. They discovered that children who were homozygous carriers of the short (S) allele were more likely to benefit from CBT than the other pediatric patients carrying one or two copies of the long (L) allele: of these SS genotype patients, 78 percent were free of diagnosed anxiety disorder following CBT compared to 58 percent of the SL and LL genotype patients. However, the opposite result was reported in one small-scale study ($N = 45$) of adult PTSD patients, in which homozygous short-allele carriers showed poorer treatment outcome following CBT (Bryant et al., 2010). Indeed, of nine studies conducted on psychological treatment outcomes for patients with anxiety or depression-related disorders as a function of 5-HTTLPR genotype, six reported negative results (Lester & Eley, 2013).

Other therapygenetics studies of anxiety- or depression-disordered patients have begun to investigate other candidate gene polymorphisms, such as the serotonin transporter intron 2 VNTR polymorphism (STin2 VNTR); 5-hydroxytryptamine (serotonin) receptor 2_A gene (HTR2A: rs7997012); tryptophan hydroxylase 2 gene (TPH2: G-703T); monoamine oxidase-A variable number tandem repeat (MAO_A VNTR); catechol-O-methyltransferase gene (COMT: val158met); nerve growth factor (NGF: rs6330); brain-derived neurotrophic factor (BDNF: val66met, rs7934165, rs 1519480, rs11030104); glutamate receptor, ionotropic N-methyl D-asparate 2B gene (GRIN2B: rs1019385); and glutamate receptor, ionotropic kainite 4 gene (GRIK4, rs1954787; reviewed by Lester & Eley, 2013). To date, this literature has produced mixed results, which Lester and Eley attribute to small sample sizes, differences in age cohorts, differences in study designs that may or may not include pharmacotherapy, and different forms of psychological interventions (e.g., Internet-based, individual, or group-based).

The authors note that "[a]t present, genetic predictors do not have sufficient predictive power to warrant their use as a clinical marker" (p. 11). They address the potential utility of neuroimaging methods when they state "[c]ombining genetic information with clinical and demographic predictors... or perhaps with neuroimaging biomarkers... within a single algorithm may also prove to be a particularly powerful approach" (p. 11). Indeed, an earlier imaging study of patients with major depressive disorder reported brain activation patterns that distinguished treatment responders from nonresponders 8 months later (Canli et al., 2005).

The potential utility of imaging genetics studies predicting treatment outcomes has been highlighted (Di Giorgio, Sambataro, & Bertolino, 2009) but has not yet produced a large literature. One example of this approach is a study by Domschke and colleagues of several gene polymorphisms of the neuropeptide Y gene (*NPY*) in anxious-depressed patients (Domschke et al., 2010). The less active allele of one of these polymorphisms (rs16147, 399C allele) was associated with increased amygdala activation to threatening facial expressions, as well as with poor antidepressant treatment response. Another example is a study by Baune and colleagues who studied three polymorphisms of the interleukin 1β (*IL1B*) gene (Baune et al., 2010) in depressed patients. They found that the G allele of two polymorphisms (rs16944 and rs1143643) was associated with reduced amygdala and anterior cingulate cortex response to emotional stimuli, as well as with nonremission after 6 weeks of antidepressant treatment. The fact that poor treatment outcome is associated with greater amygdala activation in the Domschke study but reduced activation in the Baune study may reflect the engagement of different genetic systems, study designs, or patient characteristics and thus highlights the need for more studies using this approach.

B. *Ethics Considerations*

An ethics analysis of personalized medicine in the domain of psychopathology and pharmacogenomics was provided by Parker and Satkoske who discussed disparities in access to depression treatment among African-American and Caucasian elderly adults (Parker & Satkoske, 2012). Although their analysis is conducted with genetic variation in depression drug responses in mind, it is a useful guide to the ethics of therapygenetics as well. For example, given lower participation rates of minorities in depression and in genetics research, the authors expressed concern that—if genetic variation does predict treatment response—existing disparities may further become amplified. Another example concerns differential frequency of gene variants across ethnicities. A meta-analysis reported that the majority of gene variants associated with drug response had significantly different frequencies among African-American and Caucasian patients and that more favorable drug responses were more often observed in Caucasians (Tate & Goldstein, 2004). Parker and Satkoske commented that the association of African-American ethnicity with a genotype that is viewed as nonresponsive to treatment or difficult to treat may add another stigmatizing label to that minority. Their greatest concern, however, was that those gene variants that are more common among African Americans may not be further studied in treatment research because of difficulties in recruitment or because of commercial motivations that find no incentive in developing treatments for less common genotypes. Their recommendations for reducing disparities apply as much for pharmacogenomics studies as they would for therapygenetics studies. These recommendations included a call to address barriers to research participation, a mindfulness of other contributors to ill health (e.g., sociocultural or G×E interactions), a public policy geared toward reducing genotype-based disparities, and a responsibility to educate and set realistic expectations with regard to personalized medicine.

VI. Direct-to-Consumer Genomics

A. *Expert Versus User Perspectives*

Direct-to-consumer (DTC) genome testing began in 2007 with newly formed, privately held companies offering personal genome scanning without medical referral. These commercial developments unfolded despite skepticism about the clinical utility and validity of such scans (Hunter, Khoury, & Drazen, 2008; Janssens et al., 2008; Offit, 2008). Indeed, scientists and medical practitioners viewed the genomic scan consumer as likely lacking expertise to interpret the purchased genomic information and possibly overreacting to perceived diagnostic results when there are none (Cho, 2009; Hunter et al., 2008; McGuire, Diaz, Wang, & Hilsenbeck, 2009).

Yet the question of consumer motivations is an empirical one. Seventeen studies on the public's perspective on DTC genomic testing were systematically reviewed by Goldsmith and colleagues (Goldsmith, Jackson, O'Connor, & Skirton, 2012). Overall, these studies reported a low level of awareness that such tests exist and little evidence of use of genomic testing services. Respondents expressed concerns about privacy of medical data and reliability of the performed tests and stated a preference for genomic testing to be performed by, or the results be discussed in consultation with, a medical professional. However, the majority of reviewed studies (15 out of 17) were based on hypothetical scenarios rather than actual user experiences with DTC genomic testing services. Of the two that were based on user experiences, one analyzed online consumers' personal stories that were posted online either on DTC company websites or non-company websites (Su, Howard, & Borry, 2011). This analysis showed that actual users were primarily motivated by the desire to learn about their health or genealogy and by a general interest in genetics. Interestingly, some users also expressed the intention to donate their samples for future research; I discuss this topic in detail later in this section.

The second study to analyze actual consumer experiences used a qualitative, interview-based approach based on the responses of 23 early adopters of personal genome scanning services offered by 23andMe, deCODEme, and Navigenics (McGowan, Fishman, & Lambrix, 2010). These consumers can be characterized as "early adopters" because they had their genomes scanned between November 2007 and March 2009, within the first 18 months that such services were available.

McGowan and colleagues found that these consumers were motivated by two reasons to have their genomes scanned: one was to gain health information and the other was to learn about their individual risk factors. About half of these individuals described themselves as working in a field that was either directly or indirectly related to DTC genomics. Neither of these findings should be surprising—consumers' motivations matched the claims of the advertised product, and their professional backgrounds primed them to adopt the available commercial product sooner than other consumers further removed from the world of biotechnology and bioinformatics.

More surprising than consumers' motivations and background was their assessment of the utility of the genomic data they obtained. For example, most respondents acknowledged that the clinical utility of their results was currently very limited because they may not be medically actionable. They cited as reasons for limited medical actionability both the current limits of the technology and clinicians' lack of awareness of the technology. Some respondents speculated that limited clinical utility may lie in the application to pharmacogenomics (discussed in the previous section), such as predicting an individual's drug response based on specific gene variants. On the other hand, respondents did express optimism that such data would become more useful in the future.

Another element of consumers' assessment was in regard to the predictive power of the data. Participants reported that the predictive power of DTC services was low and would remain low until the human genome was better understood. Some respondents commented that the low predictive power of genomic data could be enhanced when integrated into a larger set of personalized data, such as one's family history, environmental exposure, and lifestyle.

Finally, consumers assessed the value of genomic testing in terms of the reliability and validity of genomic research. Although consumers trusted the reliability of the genome scanning technology itself, they were aware of the limitations of genome-wide association studies (GWAS) in the context of the current state of the field. Furthermore, they articulated an expectation that, in the future, the integration of GWAS results with clinical and other phenotypic data, and a better understanding of G×E interactions would hold potential for a better understanding of disease processes.

Given their assessment of the limited utility, predictive power, and reliability and validity of personal genomics, it is perhaps not surprising that 19 out of 23 respondents reported no change in health-related behaviors after receiving their results. Taken together, the behavior of this small cohort of early adopters does not fit the schema of DTC genomics customers portrayed by experts and medical practitioners. These consumers articulated a nuanced understanding of the promise and limitations of personal genomics and did not engage in any kind of overreaction to their test results. However, this conclusion comes with two caveats: first, these early adopters may not represent the "average" personal genomics consumer, and, second, the sample size of this study was very small and analysis of the data based on qualitative interpretation of interview responses.

These limitations are addressed by Kaufman and colleagues, who conducted an online survey of 1,048 DTC genomics customers of 23andMe, deCODEme, and Navigenics who had obtained their test results between December 2009 and March 2010 (Kaufman, Bollinger, Dvoskin, & Scott, 2012). A large proportion of this cohort did, in fact, alter their health behavior after testing: 43 percent of respondents had sought additional information about a condition that was tested and, for many, this included seeking medical advice because 28 percent discussed their test results with a health care professional. For 9 percent, these consultations led to additional lab testing. Thus, these consumers changed their health behavior by integrating their personal genome results from a commercial enterprise with medical consultation and follow-up testing. Furthermore, when asked about their attitude about regulation of DTC genomic testing companies, 66 percent wanted the tests per se to be available without governmental oversight but also wanted either governmental (73 percent) or nongovernmental (84 percent) monitoring of the scientific claims made by these companies (Bollinger, Green, & Kaufman, 2013).

Taken together, the early adopters characterized by McGowan and colleagues (McGowan et al., 2010) and the larger cohort of consumers surveyed by Kaufman and colleagues (Kaufman et al., 2012) come across as empowered, self-determinant, and sophisticated consumers of personal genomic information. These consumers demand unfettered access to their biological data but also a regulatory structure that ensures scientific fidelity of claims made. The regulation of such services has since become a major focal point of conflict between DTC companies—23andMe in particular—and the Food and Drug Administration (FDA), and will be discussed toward the end of the next section.

B. Disruptive DTC Genomics: The Social and Ethical Implications of Web-Based and Participant-Driven Research

Defining the mindset of those who access DTC services in terms of consumption is insufficient. Such a characterization leaves out what is probably the most innovative and "disruptive" (in the sense of disrupting the status quo) aspect of DTC genomics, at least as practiced by 23andMe: the emergence of web-based, participant-driven genomics research studies.

In 2009, 23andMe launched a new initiative "23andWe" to begin leveraging collected genomic data and survey responses to advance the science of genomics. The company began recruiting from among its customers and potential customers to join a "do-it-yourself revolution." By June 2011, the company claimed that 76 percent of its more than 100,000 customers had volunteered in its research surveys (23andMe, 2011), and, by the Spring of 2014, that number has risen to 500,000 (23andMe, 2014), consistent with another study of user attitudes that reported a high interest in research participation (Su et al., 2011). Indeed, based on this effort, several peer-reviewed GWAS studies have now been published, one of which reported a newly discovered GWAS association with Parkinson disease (Do et al., 2011), another that replicated known gene associations with 22 physical traits and also reported novel trait associations (Eriksson et al., 2010), and a third that demonstrated replication of more than 180 known genetic associations (Tung et al., 2011).

This is a significant innovation in research genomics. The status quo of the research enterprise is based on academics recruiting subjects for basic science studies or biotechnology companies recruiting subjects for clinical trials. In both cases, research subjects take a somewhat passive role and are usually not given any access to the data. The participants in 23andWe studies, on the other hand, are also paying customers of the company, and their research participation is based on an opt-in process during the registration process. Tutton and Prainsack (2011) characterized the essence of the 23andMe research participant as the "entrepreneurial self" who "views research as an enterprise which they can direct just as they do with their individual projects of being 'enterprising' healthy selves" (p. 1090). This view of the self is fully intertwined with the online platform on which much of the information flow is transacted: the completion of health surveys that are part of the research protocol is web-based, participants have online access to their own data, and they form an online community with access to blogs, news, and consumer stories. Tutton and Prainsack (2011) refer to this process as "the convergence of predictive genetic testing and the practices of interactive, 'social' media of Web 2.0" (p. 1090).

The social and ethical implications of online social networks and DTC genomics were discussed by Lee and Crawley (2009*a*; 2009*b*). They commented that the practice of pooling DTC genomic datasets for research purposes still holds the potential for breach of security and confidentiality, and they called for studies of consumers' understanding of such possibilities. Their concern was based on the work by Homer and colleagues (Homer et al., 2008) who demonstrated that individuals can be identified even in de-identified datasets containing 1,000 cases or more. Lee and Crawley focused particularly on how the emergence of online social networks organized around genomic identities may alter social and ethical considerations. For example, 23andMe consumers are given online tools to share their genomic results with others. Social network analysis could be used, they argue, to study novel social and ethical implications of such processes. For example, it could be used to study how information flows from consumers to health care providers or whether informed consent procedures—currently designed for the individual consumer—may have to be updated to consider secondary parties that may be affected by the sharing of genomic information.

The foregoing discussion illuminates the mindset and attitudes of the genomics consumer. This mindset can fit into the conceptualization of "biological citizenship," by which individuals think of themselves in biological terms (Rose & Novas, 2005). A neurogenethics inquiry would go one step further and ask questions about the ELSI of gene-brain associations. For example, traditional GWAS efforts to discover genes associated with psychopathology or personality traits have been largely unsuccessful. Does the new research paradigm of web-based, participant-driven research improve the odds of discovery? Are research priorities, study design, or data collection and interpretation altered when research participants view themselves as empowered co-investigators rather than passive study subjects? Another area of neurogenethics inquiry could be the genomics of mental health and online/offline behavior. For example, how does knowledge of one's genetic vulnerability for, say, Alzheimer disease affect a DTC consumer's online social network? How does information flow across this network? Does such an online social network improve or impede the communication of medical information? Does the online world impact the offline world, for example, in terms of social support, stress reduction, or behavioral changes? A third area of inquiry could be global mental health. For example, the preceding discussion of DTC genomics is based on Western biomedical conceptualizations of the self and mental health. What are the ELSI of this approach in other cultures, particularly when online information can easily flow across borders? Does it alter culturally determined understanding of mental health? What is the impact on societies when large numbers of individuals begin to share and compare their genomic information, their *biological identity*, across continents and ethnicities? These are exciting and important questions to ask because they help us think about how DTC genomics may impact our understanding of who we are.

C. *The ELSI of Regulation*

Since the early days of DTC genome services just discussed, the company landscape and marketing of such services has changed. Navigenics has been acquired by Life Technologies and has exited the DTC space, and deCODEme has also discontinued its services. Since November 2013, the services offered by 23andMe have been severely restricted by the FDA, which issued a directive in the Fall of 2013 to stop offering new consumers access to health-related genetic tests. In a warning letter from the FDA to 23andMe, the FDA stated "you are marketing the 23andMe Saliva Collection Kit and Personal Genome Service (PGS) without marketing clearance or approval in violation of the Federal Food, Drug and Cosmetic Act (the FD&C Act). This product is a device within the meaning of section 201(h) of the FD&C Act, 21 U.S.C. 321(h), because it is intended for use in the diagnosis of disease or other conditions or in the cure, mitigation, treatment, or prevention of disease, or is intended to affect the structure or function of the body" (FDA, 2013). The FDA expressed concern about the public health consequences that may stem from inaccurate results. In response, 23andMe has stopped providing access to health-related results but continues to provide access to ancestry-related genetic information and raw data, and the company has vowed to continue research and expansion of its online communities (Wojcicki, 2013).

Green and Farahany commented on the FDA's regulatory claims (Green & Farahany, 2014). They agree with the FDA that 23andMe needs to be more transparent about the accuracy of its genotyping chips and risk estimation algorithms. However, they also point out the hazards of overregulation. They begin their argument with the observation that the FDA's concerns are based on "the precautionary principle," which is "acting on the basis of speculations of potential harm rather than reported harm" (p. 286). The source of potential harm is the concern that consumers may respond to health-related results inappropriately. Yet, as Green and Farahany point out (and as has been discussed earlier in this chapter), there is no empirical evidence that this is the case. The ethical implication of their argument brings the authors to "contend that the FDA's precautionary approach may pose a greater threat to consumer health than the harms that it seeks to prevent" (p. 286).

A legal perspective provided by Green and Farahany focuses on indirect medical information. The FDA claims regulation over 23andMe's interpretation of health-related information (which is why the company is still free to disperse the raw data or ancestry-related information). But, as Green and Farahany point out, this type of information is only indirectly linked to disease. They conclude that "[t]he FDA cannot reasonably regulate all such indirect information as medical devices. Moreover, as the court cases of *Sorrell v. IMS Health* (2011) and *United States v. Caronia* (2012) demonstrate, doing so could put the FDA regulations in greater tension with the First Amendment of the US Constitution, which protects the rights of individuals to receive information, and of 'commercial speech' " (p. 286).

Green and Farahany conclude their comment with the social implications of regulating DTC genomics services. They expressed concern about the FDA's potentially widening scope of regulating not only these services, but also other kinds of products that are based on giving consumers health-related information. The nature of their concern is that, in the absence of any evidence of harm, preventing consumers access to such products would be a social harm. In their words, "[s]uch consumer products could democratize health care by enabling individuals to make choices that maximize their own health. They follow the historical trend of patient empowerment that brought informed-consent laws, access to medical records and now direct access to electronic personal health data" (p. 287).

VII. Conclusion and Future Directions

In this chapter, I introduced the idea that the confluence of genetic and neural research approaches promoted the next stage in the evolution of biologically based ethics, which I call "neurogenethics." I mapped out a problem space that takes into account different spatial levels of analysis, psychological functions, and ethics considerations. I then illustrated how this problem space can be populated with specific areas of neurogenethics inquiry in the domains of gene therapy, optogenetics, imaging genetics, therapygenetics, and DTC genomics. Given the pace of technological innovation and presidential commitment to the BRAIN initiative (Insel, Landis, & Collins, 2013), I expect that many more questions remain to be asked, and even more empirical work lies ahead.

References

23andMe. (2011). *23andMe identifies two novel genetic associations with Parkinson's disease and finds a substantial genetic component for the disease.* Retrieved March 21, 2014, from http://www.parkinson.org/About-Us/Press-Room/NPF-In-The-News/2011/June/23andMe-Identifies-Two-Novel-Genetic-Associations-23andMe. (2014). *The future of genetics in people's lives.* Retrieved March 21, 2014, from http://blog.23andme.com/news/the-future-of-genetics-in-peoples-lives/Akil, O., Seal, R. P., Burke, K., Wang, C., Alemi, A., During, M.,... Lustig, L. R. (2012). Restoration of hearing in the VGLUT3 knockout mouse using virally mediated gene therapy. *Neuron, 75*(2), 283–293. doi: S0896-6273(12)00485-0 [pii] 10.1016/j.neuron.2012.05.019

Alexander, N., Klucken, T., Koppe, G., Osinsky, R., Walter, B., Vaitl, D.,... Hennig, J. (2012). Interaction of the serotonin transporter-linked polymorphic region and environmental adversity: Increased amygdala-hypothalamus connectivity as a potential mechanism linking neural and endocrine hyperreactivity. *Biological Psychiatry, 72*(1), 49–56. doi: 10.1016/j.biopsych.2012.01.030

Anderson, N. E., & Kiehl, K. A. (2014). Psychopathy: Developmental perspectives and their implications for treatment. *Restorative Neurology and Neuroscience, 32*(1), 103–117. doi: 10.3233/RNN-139001

Appelbaum, P. S., & Grisso, T. (1988). Assessing patients' capacities to consent to treatment. *New England Journal of Medicine, 319*(25), 1635–1638. doi: 10.1056/NEJM198812223192504

Appelbaum, P. S., Roth, L. H., & Lidz, C. (1982). The therapeutic misconception: Informed consent in psychiatric research. *International Journal of Law and Psychiatry, 5*(3–4), 319–329.

Bartus, R. T., Baumann, T. L., Brown, L., Kruegel, B. R., Ostrove, J. M., & Herzog, C. D. (2013). Advancing neurotrophic factors as treatments for age-related neurodegenerative diseases: Developing and demonstrating "clinical proof-of-concept" for AAV-neurturin (CERE-120) in Parkinson's disease. *Neurobiological Aging, 34*(1), 35–61. doi: 10.1016/j.neurobiolaging.2012.07.018

Baune, B. T., Dannlowski, U., Domschke, K., Janssen, D. G., Jordan, M. A., Ohrmann, P.,... Suslow, T. (2010). The interleukin 1 beta (IL1B) gene is associated with failure to achieve remission and impaired emotion processing in major depression. *Biological Psychiatry, 67*(6), 543–549. doi: 10.1016/j.biopsych.2009.11.004

Benjamin, J., Li, L., Patterson, C., Greenberg, B. D., Murphy, D. L., & Hamer, D. H. (1996). Population and familial association between the D4 dopamine receptor gene and measures of Novelty Seeking. *Nature Genetics, 12*, 81–84.

Bentley, J. N., Chestek, C., Stacey, W. C., & Patil, P. G. (2013). Optogenetics in epilepsy. *Neurosurgical Focus, 34*(6), E4. doi: 10.3171/2013.3.FOCUS1364

Bernstein, J. G., & Boyden, E. S. (2011). Optogenetic tools for analyzing the neural circuits of behavior. *Trends in Cognitive Sciences, 15*(12), 592–600. doi: 10.1016/j.tics.2011.10.003

Bevan, A. K., Duque, S., Foust, K. D., Morales, P. R., Braun, L., Schmelzer, L.,... Kaspar, B. K. (2011). Systemic gene delivery in large species for targeting spinal cord, brain, and peripheral tissues for pediatric disorders. *Molecular Therapeutics, 19*(11), 1971–1980. doi: 10.1038/mt.2011.157

Biswal, B. B., Mennes, M., Zuo, X.-N., Gohel, S., Kelly, C., Smith, S. M.,... Milham, M. P. (2010). Toward discovery science of human brain function. *Proceedings of the National Academy of Sciences of the United States of America, 107*(10), 4734–4739.

Blair, R. J. (2006). The emergence of psychopathy: Implications for the neuropsychological approach to developmental disorders. *Cognition, 101*(2), 414–442. doi: 10.1016/j.cognition.2006.04.005

Bollinger, J. M., Green, R. C., & Kaufman, D. (2013). Attitudes about regulation among direct-to-consumer genetic testing customers. *Genetic Testing and Molecular Biomarkers, 17*(5), 424–428. doi: 10.1089/gtmb.2012.0453

Bondy, B. (2011). Genetics in psychiatry: Are the promises met? *World Journal of Biological Psychiatry, 12*(2), 81–88. doi: 10.3109/15622975.2010.546428

Boutin, S., Monteilhet, V., Veron, P., Leborgne, C., Benveniste, O., Montus, M. F., & Masurier, C. (2010). Prevalence of serum IgG and neutralizing factors against adeno-associated virus (AAV) types 1, 2, 5, 6, 8, and 9 in the healthy population: Implications for gene therapy using AAV vectors. *Human Gene Therapy, 21*(6), 704–712. doi: 10.1089/hum.2009.182

Bowers, W. J., Breakefield, X. O., & Sena-Esteves, M. (2011). Genetic therapy for the nervous system. *Human Molecular Genetics, 20*(R1), R28–R41. doi: 10.1093/hmg/ddr110

Boyden, E. S., Zhang, F., Bamberg, E., Nagel, G., & Deisseroth, K. (2005). Millisecond-timescale, genetically targeted optical control of neural activity. *Nature Neuroscience, 8*(9), 1263–1268. doi: 10.1038/nn1525

Braskie, M. N., Jahanshad, N., Stein, J. L., Barysheva, M., Johnson, K., McMahon, K. L.,... Thompson, P. M. (2012). Relationship of a variant in the NTRK1 gene to white matter microstructure in young adults. *Journal of Neuroscience, 32*(17), 5964–5972.

Braskie, M. N., Jahanshad, N., Stein, J. L., Barysheva, M., McMahon, K. L., de Zubicaray, G. I.,... Thompson, P. M. (2011). Common Alzheimer's disease risk variant within the CLU gene affects white matter microstructure in young adults. *Journal of Neuroscience, 31*(18), 6764–6770.

Britt, J. P., & Bonci, A. (2013). Optogenetic interrogations of the neural circuits underlying addiction. *Current Opinion in Neurobiology, 23*(4), 539–545. doi: 10.1016/j.conb.2013.01.010

Bryant, R. A., Felmingham, K. L., Falconer, E. M., Pe Benito, L., Dobson-Stone, C., Pierce, K. D., & Schofield, P. R. (2010). Preliminary evidence of the short allele of the serotonin transporter gene predicting poor response to cognitive behavior therapy in posttraumatic stress disorder. *Biological Psychiatry, 67*(12), 1217–1219. doi: 10.1016/j.biopsych.2010.03.016

Canli, T. (2005). When genes and brains unite: Ethical implications of genomic neuroimaging. In J. Illes (Ed.), *Neuroethics: Defining the Issues in Theory, Practice and Policy* (pp. 169–184). New York: Oxford University Press.

Canli, T., Brandon, S., Casebeer, W., Crowley, P. J., Du Rousseau, D., Greely, H. T., & Pascual-Leone, A. (2007*a*). Neuroethics and national security. *American Journal of Bioethics, 7*(5), 3–13.

Canli, T., Brandon, S., Casebeer, W., Crowley, P. J., Durousseau, D., Greely, H. T., & Pascual-Leones, A. (2007*b*). Response to open peer commentaries on "Neuroethics and national security." *American Journal of Bioethics, 7*(5), W1–3.

Canli, T., Congdon, E., Constable, T. R., & Lesch, K. P. (2008). Additive effects of serotonin transporter and tryptophan hydroxylase-2 gene variation on neural correlates of affective processing. *Biological Psychology, 79*(1), 118–125.

Canli, T., Cooney, R. E., Goldin, P., Shah, M., Sivers, H., Thomason, M. E., . . . Gotlib, I. H. (2005). Amygdala reactivity to emotional faces predicts improvement in major depression. *Neuroreport, 16*(12), 1267–1270.

Canli, T., Qiu, M., Omura, K., Congdon, E., Haas, B. W., Amin, Z., . . . Lesch, K. P. (2006). Neural correlates of epigenesis. *Proceedings of the National Academy of Sciences of the United States of America, 103*(43), 16033–16038.

Caspi, A., McClay, J., Moffitt, T. E., Mill, J., Martin, J., Craig, I. W., . . . Poulton, R. (2002). Role of genotype in the cycle of violence in maltreated children. *Science, 297*(5582), 851–854.

Chaudhury, D., Walsh, J. J., Friedman, A. K., Juarez, B., Ku, S. M., Koo, J. W., . . . Han, M. H. (2013). Rapid regulation of depression-related behaviours by control of midbrain dopamine neurons. *Nature, 493*(7433), 532–536. doi: Nature11713 [pii] 10.1038/nature11713

Cho, M. K. (2009). Translating genomics into the clinic: Moving to the post-Mendelian world. *Genome Medicine, 1*(1), 7. doi: 10.1186/gm7

Chtarto, A., Bockstael, O., Tshibangu, T., Dewitte, O., Levivier, M., & Tenenbaum, L. (2013). A next step in adeno-associated virus-mediated gene therapy for neurological diseases: Regulation and targeting. *British Journal of Clinical Pharmacology, 76*(2), 217–232. doi: 10.1111/bcp.12065

Craddock, R. C., Jbabdi, S., Yan, C.-G., Vogelstein, J. T., Castellanos, F. X., Di Martino, A., . . . Milham, M. P. (2013). Imaging human connectomes at the macroscale. *Nature Methods, 10*(6), 524–539.

Davatzikos, C., Ruparel, K., Fan, Y., Shen, D. G., Acharyya, M., Loughead, J. W., . . . Langleben, D. D. (2005). Classifying spatial patterns of brain activity with machine learning methods: Application to lie detection. *Neuroimage, 28*(3), 663–668.

Dennis, E. L., Jahanshad, N., Rudie, J. D., Brown, J. A., Johnson, K., McMahon, K. L., . . . Thompson, P. M. (2011). Altered structural brain connectivity in healthy carriers of the autism risk gene, CNTNAP2. *Brain Connectivity, 1*(6), 447–459.

Di Domenico, M., Ricciardi, C., Martone, T., Mazzarella, N., Cassandro, C., Chiarella, G., . . . Cassandro, E. (2011). Towards gene therapy for deafness. *Journal of Cell Physiology, 226*(10), 2494–2499. doi: 10.1002/jcp.22617

Di Giorgio, A., Sambataro, F., & Bertolino, A. (2009). Functional imaging as a tool to investigate the relationship between genetic variation and response to treatment with antipsychotics. *Current Pharmaceutical Design, 15*(22), 2560–2572.

Do, C. B., Tung, J. Y., Dorfman, E., Kiefer, A. K., Drabant, E. M., Francke, U., . . . Eriksson, N. (2011). Web-based genome-wide association study identifies two novel loci and a substantial genetic component for Parkinson's disease. *PLoS Genetics, 7*(6), e1002141. doi: 10.1371/journal.pgen.1002141

Domschke, K., & Dannlowski, U. (2010). Imaging genetics of anxiety disorders. *Neuroimage, 53*(3), 822 831. doi: 10.1016/j.neuroimage.2009.11.042

Domschke, K., Dannlowski, U., Hohoff, C., Ohrmann, P., Bauer, J., Kugel, H., . . . Baune, B. T. (2010). Neuropeptide Y (NPY) gene: Impact on emotional processing and treatment response in anxious depression. *European Neuropsychopharmacology, 20*(5), 301–309. doi: 10.1016/j.euroneuro.2009.09.006

Dressing, H., Sartorius, A., & Meyer-Lindenberg, A. (2008). Implications of fMRI and genetics for the law and the routine practice of forensic psychiatry. *Neurocase, 14*(1), 7–14. doi: 10.1080/13554790801992800

Duque, S., Joussemet, B., Riviere, C., Marais, T., Dubreil, L., Douar, A. M., . . . Barkats, M. (2009). Intravenous administration of self-complementary AAV9 enables transgene delivery to adult motor neurons. *Molecular Therapy, 17*(7), 1187–1196. doi: 10.1038/mt.2009.71

Durston, S. (2010). Imaging genetics in ADHD. *Neuroimage, 53*(3), 832–838. doi: 10.1016/j.neuroimage.2010.02.071

Eberling, J. L., Jagust, W. J., Christine, C. W., Starr, P., Larson, P., Bankiewicz, K. S., & Aminoff, M. J. (2008). Results from a phase I safety trial of hAADC gene therapy for Parkinson disease. *Neurology, 70*(21), 1980–1983.

Ebstein, R. P., Novick, O., Umansky, R., Priel, B., Osher, Y., Blaine, D., . . . Belmaker, R. H. (1996). Dopamine D4 receptor (D4DR) exon III polymorphism associated with the human personality trait of Novelty Seeking. *Nature Genetics, 12*, 78–80.

Eijkholt, M., Anderson, J. A., & Illes, J. (2012). Picturing neuroscience research through a human rights lens: Imaging first-episode schizophrenic treatment-naive individuals. *International Journal of Law and Psychiatry, 35*(2), 146–152. doi: 10.1016/j.ijlp.2011.12.003

Eley, T. C., Hudson, J. L., Creswell, C., Tropeano, M., Lester, K. J., Cooper, P., . . . Collier, D. A. (2012). Therapygenetics: The 5HTTLPR and response to psychological therapy. *Molecular Psychiatry, 17*(3), 236–237. doi: 10.1038/mp.2011.132

Entcheva, E. (2013). Cardiac optogenetics. *American Journal of Physiology. Heart and Circulation Physiology, 304*(9), H1179–1191. doi: 10.1152/ajpheart.00432.2012

Eriksson, N., Macpherson, J. M., Tung, J. Y., Hon, L. S., Naughton, B., Saxonov, S., . . . Mountain, J. (2010). Web-based, participant-driven studies yield novel genetic associations for common traits. *PLoS Genetics, 6*(6), e1000993. doi: 10.1371/journal.pgen.1000993

Farah, M. J., Hutchinson, J. B., Phelps, E. A., & Wagner, A. D. (2014). Functional MRI-based lie detection: Scientific and societal challenges. *Nature Reviews Neuroscience, 15*(2), 123–131. doi: 10.1038/nrn3665

Faraone, S. V., Doyle, A. E., Mick, E., & Biederman, J. (2001). Meta-analysis of the association between the 7-repeat allele of the dopamine D(4) receptor gene and attention deficit hyperactivity disorder. *American Journal of Psychiatry, 158*(7), 1052–1057.

FDA. (2013). Warning letter from 11/22/2013. Retrieved April 1, 2014, from http://www.fda.gov/ICECI/EnforcementActions/WarningLetters/2013/ucm376296.htm

Ferri, J., Izzi, S., & Canli, T. (2014). Telomere length is associated with stress-related activation of the ACC and hippocampus. Paper presented at the Annual Eastern Psychological Association meeting, Boston, MA.

Foust, K. D., Wang, X., McGovern, V. L., Braun, L., Bevan, A. K., Haidet, A. M.,... Kaspar, B. K. (2010). Rescue of the spinal muscular atrophy phenotype in a mouse model by early postnatal delivery of SMN. *Nature Biotechnology, 28*(3), 271–274. doi: 10.1038/nbt.1610

Gdalyahu, A., Tring, E., Polack, P. O., Gruver, R., Golshani, P., Fanselow, M. S.,... Trachtenberg, J. T. (2012). Associative fear learning enhances sparse network coding in primary sensory cortex. *Neuron, 75*(1), 121–132. doi: 10.1016/j.neuron.2012.04.035

Gerits, A., & Vanduffel, W. (2013). Optogenetics in primates: A shining future? *Trends in Genetics, 29*(7), 403–411. doi: 10.1016/j.tig.2013.03.004

Gizer, I. R., Ficks, C., & Waldman, I. D. (2009). Candidate gene studies of ADHD: A meta-analytic review. *Human Genetics, 126*(1), 51–90. doi: 10.1007/s00439-009-0694-x

Goldsmith, L., Jackson, L., O'Connor, A., & Skirton, H. (2012). Direct-to-consumer genomic testing: Systematic review of the literature on user perspectives. *European Journal of Human Genetics, 20*(8), 811–816. doi: 10.1038/ejhg.2012.18

Gray, S. J., Matagne, V., Bachaboina, L., Yadav, S., Ojeda, S. R., & Samulski, R. J. (2011). Preclinical differences of intravascular AAV9 delivery to neurons and glia: A comparative study of adult mice and nonhuman primates. *Molecular Therapy, 19*(6), 1058–1069. doi: 10.1038/mt.2011.72

Green, R. C., & Farahany, N. A. (2014). Regulation: The FDA is overcautious on consumer genomics. *Nature, 505*(7483), 286–287.

Han, X. (2012). Optogenetics in the nonhuman primate. *Progress in Brain Research, 196*, 215–233. doi: 10.1016/B978-0-444-59426-6.00011-2

Hardy, J., & Singleton, A. (2009). Genomewide association studies and human disease. *New England Journal of Medicine, 360*, 1759–1768.

Hariri, A. R., Mattay, V. S., Tessitore, A., Kolachana, B., Fera, F., Goldman, D.,... Weinberger, D. R. (2002). Serotonin transporter genetic variation and the response of the human amygdala. *Science, 297*(5580), 400–403.

Haynes, J. D. (2008). Detecting deception from neuroimaging signals—a data-driven perspective. *Trends in Cognitive Sciences, 12*(4), 126–127; author reply 127–128. doi: 10.1016/j.tics.2008.01.003

Hermann, A., Kupper, Y., Schmitz, A., Walter, B., Vaitl, D., Hennig, J.,... Tabbert, K. (2012). Functional gene polymorphisms in the serotonin system and traumatic life events modulate the neural basis of fear acquisition and extinction. *PLoS One, 7*(9), e44352. doi: 10.1371/journal.pone.0044352

High, K. A., & Aubourg, P. (2011). rAAV human trial experience. *Methods in Molecular Biology, 807*, 429–457. doi: 10.1007/978-1-61779-370-7_18

Hinshaw, S. P. (2005). Objective assessment of covert antisocial behavior: Predictive validity and ethical considerations. *Ethics and Behavior, 15*(3), 259–269. doi: 10.1207/s15327019eb1503_6

Homer, N., Szelinger, S., Redman, M., Duggan, D., Tembe, W., Muehling, J.,... Craig, D. W. (2008). Resolving individuals contributing trace amounts of DNA to highly complex mixtures using high-density SNP genotyping microarrays. *PLoS Genetics, 4*(8), e1000167. doi: 10.1371/journal.pgen.1000167

Huang, F., Tang, B., & Jiang, H. (2013). Optogenetic investigation of neuropsychiatric diseases. *International Journal of Neuroscience, 123*(1), 7–16. doi: 10.3109/00207454.2012.728651

Huff, M. L., Miller, R. L., Deisseroth, K., Moorman, D. E., & LaLumiere, R. T. (2013). Posttraining optogenetic manipulations of basolateral amygdala activity modulate consolidation of inhibitory avoidance memory in rats. *Proceedings of the National Academy of Sciences of the USA, 110*(9), 3597–3602. doi: 10.1073/pnas.1219593110

Hunter, D. J., Khoury, M. J., & Drazen, J. M. (2008). Letting the genome out of the bottle—will we get our wish? *New England Journal of Medicine, 358*(2), 105–107. doi: 10.1056/NEJMp0708162

Iida, A., Takino, N., Miyauchi, H., Shimazaki, K., & Muramatsu, S. (2013). Systemic delivery of tyrosine-mutant AAV vectors results in robust transduction of neurons in adult mice. *Biomed Research International.* doi: 10.1155/2013/974819

Illes, J., & Racine, E. (2005). Imaging or imagining? A neuroethics challenge informed by genetics. *American Journal of Bioethics, 5*(2), 5–18.

Illes, J., Rosen, A., Greicius, M., & Racine, E. (2007). Prospects for prediction: ethics analysis of neuroimaging in Alzheimer's disease. *Annals of the New York Academy of Sciences*, 1097, 278–95. doi: 1097/1/278 [pii]10.1196/annals.1379.030

Insel, T. R., Landis, S. C., & Collins, F. S. (2013). Research priorities. The NIH BRAIN Initiative. *Science, 340*(6133), 687–688. doi: 10.1126/science.1239276

Izzi, S., Ferri, J., & Canli, T. (2014). Resting limbic activation is associated with shorter telomere length. Paper presented at the Association for Psychological Science, San Francisco.

Jansen, L. A., Appelbaum, P. S., Klein, W. M., Weinstein, N. D., Cook, W., Fogel, J. S., & Sulmasy, D. P. (2011). Unrealistic optimism in early-phase oncology trials. *Institutional Review Board, 33*(1), 1–8.

Janssens, A. C., Gwinn, M., Bradley, L. A., Oostra, B. A., van Duijn, C. M., & Khoury, M. J. (2008). A critical appraisal of the scientific basis of commercial genomic profiles used to assess health risks and personalize health interventions. *American Journal of Human Genetics, 82*(3), 593–599. Jones, S., & Cauffman, E. (2008). Juvenile psychopathy and judicial decision making: An empirical analysis of an ethical dilemma. *Behavioral Sciences and the Law, 26*(2), 151–165. doi: 10.1002/bsl.792Kaplitt, M. G., Feigin, A., Tang, C., Fitzsimons, H. L., Mattis, P., Lawlor, P. A.,... During, M. J. (2007). Safety and tolerability of gene therapy with an adeno-associated virus (AAV) borne GAD gene for Parkinson's disease: An open label, phase I trial. *Lancet, 369*(9579), 2097–2105. doi: 10.1016/S0140-6736(07)60982-9

Kaufman, D. J., Bollinger, J. M., Dvoskin, R. L., & Scott, J. A. (2012). Risky business: Risk perception and the use of medical services among customers of DTC personal genetic testing. *Journal of Genetic Counseling, 21*(3), 413–422. doi: 10.1007/s10897-012-9483-0

Kebir, O., Tabbane, K., Sengupta, S., & Joober, R. (2009). Candidate genes and neuropsychological phenotypes in children with ADHD: Review of association studies. *Journal of Psychiatry and Neuroscience, 34*(2), 88–101.

Kheirbek, M. A., Drew, L. J., Burghardt, N. S., Costantini, D. O., Tannenholz, L., Ahmari, S. E., . . . Hen, R. (2013). Differential control of learning and anxiety along the dorsoventral axis of the dentate gyrus. *Neuron, 77*(5), 955–968. doi: S0896-6273(13)00046-9 [pii] 10.1016/j.neuron.2012.12.038

Kim, S. Y., de Vries, R., Wilson, R., Parnami, S., Frank, S., Kieburtz, K., & Holloway, R. G. (2013). Research participants' "irrational" expectations: Common or commonly mismeasured? *Institutional Review Board, 35*(1), 1–9.

Klein, R. L., Hamby, M. E., Gong, Y., Hirko, A. C., Wang, S., Hughes, J. A., . . . Meyer, E. M. (2002). Dose and promoter effects of adeno-associated viral vector for green fluorescent protein expression in the rat brain. *Experimental Neurology, 176*(1), 66–74.

Kolber, A. J. (2010). Ethical implications of memory dampening. In M. J. Farah (Ed.), *Neuroethics: An introduction with readings* (97–118). Cambridge, MA: MIT Press.

Kral, A., & O'Donoghue, G. M. (2010). Profound deafness in childhood. *New England Journal of Medicine, 363*(15), 1438–1450. doi: 10.1056/NEJMra0911225

Kramer, R. H., Mourot, A., & Adesnik, H. (2013). Optogenetic pharmacology for control of native neuronal signaling proteins. *Nature Neuroscience, 16*(7), 816–823. doi: 10.1038/nn.3424

Lee, S. S., & Crawley, L. (2009*a*). Research 2.0: Social networking and direct-to-consumer (DTC) genomics. *American Journal of Bioethics, 9*(6–7), 35–44. doi: 10.1080/15265160902874452

Lee, S. S., & Crawley, L. (2009*b*). Response to open peer commentaries on "Research 2.0: Social networking and direct-to-consumer personal genomics." *American Journal of Bioethics, 9*(6–7), W1–W3. doi: 10.1080/15265160902967009

Lesch, K. P., Bengel, D., Heils, A., Sabol, S. Z., Greenberg, B. D., Petri, S., . . . Murphy, D. L. (1996). Association of anxiety-related traits with a polymorphism in the serotonin transporter gene regulatory region. *Science, 274*(5292), 1527–1531.

Lester, K. J., & Eley, T. C. (2013). Therapygenetics: Using genetic markers to predict response to psychological treatment for mood and anxiety disorders. *Biology of Mood and Anxiety Disorders, 3*(1), 4. doi: 10.1186/2045-5380-3-4

LeWitt, P. A., Rezai, A. R., Leehey, M. A., Ojemann, S. G., Flaherty, A. W., Eskandar, E. N., . . . Feigin, A. (2011). AAV2-GAD gene therapy for advanced Parkinson's disease: A double-blind, sham-surgery controlled, randomised trial. *Lancet Neurology, 10*(4), 309–319. doi: 10.1016/S1474-4422(11)70039-4

Lin, J. Y., Knutsen, P. M., Muller, A., Kleinfeld, D., & Tsien, R. Y. (2013). ReaChR: A red-shifted variant of channelrhodopsin enables deep transcranial optogenetic excitation. *Nature Neuroscience, 16*(10), 1499–1508. doi: 10.1038/nn.3502

Liu, X., Ramirez, S., Pang, P. T., Puryear, C. B., Govindarajan, A., Deisseroth, K., & Tonegawa, S. (2012). Optogenetic stimulation of a hippocampal engram activates fear memory recall. *Nature, 484*(7394), 381–385. doi: Nature11028 [pii] 10.1038/nature11028

Lowenstein, P. R. (2008). Clinical trials in gene therapy: Ethics of informed consent and the future of experimental medicine. *Current Opinion in Molecular Therapy, 10*(5), 428–430.

Manolio, T. A., Brooks, L. D., & Collins, F. S. (2008). A HapMap harvest of insights into the genetics of common disease. *Journal of Clinical Investigation, 118*(5), 1590–1605.

Marks, W. J., Jr., Bartus, R. T., Siffert, J., Davis, C. S., Lozano, A., Boulis, N., . . . Olanow, C. W. (2010). Gene delivery of AAV2-neurturin for Parkinson's disease: A double-blind, randomised, controlled trial. *Lancet Neurology, 9*(12), 1164–1172. doi: 10.1016/S1474-4422(10)70254-4

Marks, W. J., Jr., Ostrem, J. L., Verhagen, L., Starr, P. A., Larson, P. S., Bakay, R. A., . . . Bartus, R. T. (2008). Safety and tolerability of intraputaminal delivery of CERE-120 (adeno-associated virus serotype 2-neurturin) to patients with idiopathic Parkinson's disease: An open-label, phase I trial. *Lancet Neurology, 7*(5), 400–408. doi: 10.1016/S1474-4422(08)70065-6

Mattay, V. S., Goldberg, T. E., Sambataro, F., & Weinberger, D. R. (2008). Neurobiology of cognitive aging: Insights from imaging genetics. *Biological Psychology, 79*(1), 9–22. doi: 10.1016/j.biopsycho.2008.03.015

McCown, T. J. (2011). Adeno-Associated Virus (AAV) vectors in the CNS. *Current Gene Therapy, 11*(3), 181–188.

McGowan, M. L., Fishman, J. R., & Lambrix, M. A. (2010). Personal genomics and individual identities: Motivations and moral imperatives of early users. *New Genetics and Society, 29*(3), 261–290. doi: 10.1080/14636778.2010.507485

McGowan, P. O., Sasaki, A., D'Alessio, A. C., Dymov, S., Labonte, B., Szyf, M., . . . Meaney, M. J. (2009). Epigenetic regulation of the glucocorticoid receptor in human brain associates with childhood abuse. *Nature Neuroscience, 12*(3), 342–348.

McGuire, A. L., Diaz, C. M., Wang, T., & Hilsenbeck, S. G. (2009). Social networkers' attitudes toward direct-to-consumer personal genome testing. *American Journal of Bioethics, 9*(6–7), 3–10. doi: 10.1080/15265160902928209

McWhirter, R. E. (2012). The history of bioethics: Implications for current debates in health research. *Perspectives in Biology and Medicine, 55*(3), 329–338. doi: S152987951230001X [pii] 10.1353/pbm.2012.0025

Meyer-Lindenberg, A. (2010). Imaging genetics of schizophrenia. *Dialogues in Clinical Neuroscience, 12*(4), 449–456.

Meyer-Lindenberg, A. (2012). The future of fMRI and genetics research. *Neuroimage, 62*(2), 1286–1292. doi: 10.1016/j.neuroimage.2011.10.063

Mittermeyer, G., Christine, C. W., Rosenbluth, K. H., Baker, S. L., Starr, P., Larson, P., . . . Bankiewicz, K. S. (2012). Long-term evaluation of a phase 1 study of AADC gene therapy for Parkinson's disease. *Human Gene Therapy, 23*(4), 377–381. doi: 10.1089/hum.2011.220

Offit, K. (2008). Genomic profiles for disease risk: Predictive or premature? *Journal of the American Medical Association, 299*(11), 1353–1355. doi: 10.1001/jama.299.11.1353

Packer, A. M., Roska, B., & Hausser, M. (2013). Targeting neurons and photons for optogenetics. *Nature Neuroscience, 16*(7), 805–815. doi: 10.1038/nn.3427

Parker, L. S., & Satkoske, V. B. (2012). Ethical dimensions of disparities in depression research and treatment in the pharmacogenomic era. *Journal of Law Medicine and Ethics, 40*(4), 886–903. doi: 10.1111/j.1748-720X.2012.00718.x

Pezawas, L., Meyer-Lindenberg, A., Drabant, E. M., Verchinski, B. A., Munoz, K. E., Kolachana, B. S., . . . Weinberger, D. R. (2005). 5-HTTLPR polymorphism impacts human cingulate-amygdala interactions: A genetic susceptibility mechanism for depression. *Nature Neuroscience, 8*(6), 828–834.

Potkin, S. G., Turner, J. A., Fallon, J. A., Lakatos, A., Keator, D. B., Guffanti, G., & Macciardi, F. (2009). Gene discovery through imaging genetics: Identification of two novel genes associated with schizophrenia. *Molecular Psychiatry, 14*(4), 416–428. doi: 10.1038/mp.2008.127

President's Council on Bioethics. (2010). Memory blunting: Ethical analysis. In M. J. Farah (Ed.), *Neuroethics: An introduction with readings* (pp. 88–96). Cambridge, MA: MIT Press.

Ramirez, S., Liu, X., Lin, P. A., Suh, J., Pignatelli, M., Redondo, R. L., . . . Tonegawa, S. (2013). Creating a false memory in the hippocampus. *Science, 341*(6144), 387–391. doi: 341/6144/387 [pii] 10.1126/science.1239073

Rose, N., & Novas, C. (2005). Biological citizenship. In A. Ong & S. Collier (Eds.), *Global assemblages: Technology, policy and ethics as anthropological problems.* Oxford, UK: Blackwell.

Roskies, A. L. (2007). Neuroethics beyond genethics. Despite the overlap between the ethics of neuroscience and genetics, there are important areas where the two diverge. *European Molecular Biology Organization Reports, 8*(Spec. No.), S52–S56. doi: 10.1038/sj.embor.7401009

Ruel, J., Emery, S., Nouvian, R., Bersot, T., Amilhon, B., Van Rybroek, J. M., . . . Puel, J. L. (2008). Impairment of SLC17A8 encoding vesicular glutamate transporter-3, VGLUT3, underlies nonsyndromic deafness DFNA25 and inner hair cell dysfunction in null mice. *American Journal of Human Genetics, 83*(2), 278–292. doi: 10.1016/j.ajhg.2008.07.008

Ruiz, O., Lustig, B. R., Nassi, J. J., Cetin, A., Reynolds, J. H., Albright, T. D., . . . Roe, A. W. (2013). Optogenetics through windows on the brain in the nonhuman primate. *Journal of Neurophysiology, 110*(6), 1455–1467. doi: 10.1152/jn.00153.2013

Salekin, R. T., & Frick, P. J. (2005). Psychopathy in children and adolescents: The need for a developmental perspective. *Journal of Abnormal Child Psychology, 33*(4), 403–409.

Samaranch, L., Salegio, E. A., San Sebastian, W., Kells, A. P., Foust, K. D., Bringas, J. R., . . . Bankiewicz, K. S. (2012). Adeno-associated virus serotype 9 transduction in the central nervous system of nonhuman primates. *Human Gene Therapy, 23*(4), 382–389. doi: 10.1089/hum.2011.200

Scharinger, C., Rabl, U., Sitte, H. H., & Pezawas, L. (2010). Imaging genetics of mood disorders. *Neuroimage, 53*(3), 810–821. doi: 10.1016/j.neuroimage.2010.02.019

Seal, R. P., Akil, O., Yi, E., Weber, C. M., Grant, L., Yoo, J., . . . Edwards, R. H. (2008). Sensorineural deafness and seizures in mice lacking vesicular glutamate transporter 3. *Neuron, 57*(2), 263–275. doi: 10.1016/j.neuron.2007.11.032

Sip, K. E., Roepstorff, A., McGregor, W., & Frith, C. D. (2008). Detecting deception: The scope and limits. *Trends in Cognitive Sciences, 12*(2), 48–53. doi: 10.1016/j.tics.2007.11.008

Sporns, O., Tononi, G., & Kotter, R. (2005). The human connectome: A structural description of the human brain. *PLoS Computational Biology, 1*(4), e42.

Stein, J. L., Medland, S. E., Vasquez, A. A., Hibar, D. P., Senstad, R. E., Winkler, A. M., . . . Schmidt, R. (2012). Identification of common variants associated with human hippocampal and intracranial volumes. *Nature Genetics, 44*(5), 552–561.

Su, Y., Howard, H. C., & Borry, P. (2011). Users' motivations to purchase direct-to-consumer genome-wide testing: An exploratory study of personal stories. *Journal of Community Genetics, 2*(3), 135–146. doi: 10.1007/s12687-011-0048-y

Taft, R. J., Pheasant, M., & Mattick, J. S. (2007). The relationship between non-protein-coding DNA and eukaryotic complexity. *BioEssays: News and Reviews in Molecular, Cellular and Developmental Biology, 29*(3), 288–299.

Tairyan, K., & Illes, J. (2009). Imaging genetics and the power of combined technologies: A perspective from neuroethics. *Neuroscience, 164*, 7–15.

Tate, S. K., & Goldstein, D. B. (2004). Will tomorrow's medicines work for everyone? *Nature Genetics, 36*(11 Suppl.), S34–S42. doi: 10.1038/ng1437

Tenenbaum, L., Lehtonen, E., & Monahan, P. E. (2003). Evaluation of risks related to the use of adeno-associated virus-based vectors. *Current Gene Therapy, 3*(6), 545–565.

Thompson, P. M., Ge, T., Glahn, D. C., Jahanshad, N., & Nichols, T. E. (2013). Genetics of the connectome. *Neuroimage, 80*, 475–488. doi: S1053-8119(13)00507-7 [pii] 10.1016/j.neuroimage.2013.05.013

Thompson, P. M., Stein, J. L., Medland, S. E., Hibar, D. P., Vasquez, A. A., Renteria, M. E., . . . Drevets, W. (2014). The ENIGMA Consortium: Large-scale collaborative analyses of neuroimaging and genetic data. *Brain Imaging and Behavior.* doi: 10.1007/s11682-013-9269-5

Toga, A. W., Clark, K. A., Thompson, P. M., Shattuck, D. W., & Van Horn, J. D. (2012). Mapping the human connectome. *Neurosurgery, 71*(1), 1–5. doi: 10.1227/NEU.0b013e318258e9ff 00006123-201207000-00001 [pii]

Tourino, C., Eban-Rothschild, A., & de Lecea, L. (2013). Optogenetics in psychiatric diseases. *Current Opinion in Neurobiology, 23*(3), 430–435. doi: 10.1016/j.conb.2013.03.007

Tung, J. Y., Do, C. B., Hinds, D. A., Kiefer, A. K., Macpherson, J. M., Chowdry, A. B., . . . Eriksson, N. (2011). Efficient replication of over 180 genetic associations with self-reported medical data. *PLoS One, 6*(8), e23473. doi: 10.1371/journal.pone.0023473

Tutton, R., & Prainsack, B. (2011). Enterprising or altruistic selves? Making up research subjects in genetics research. *Sociology of Health and Illness, 33*(7), 1081–1095. doi: 10.1111/j.1467-9566.2011.01348.x

Tye, K. M., & Deisseroth, K. (2012). Optogenetic investigation of neural circuits underlying brain disease in animal models. *Nature Reviews Neuroscience, 13*(4), 251–266. doi: Nrn3171 [pii] 10.1038/nrn3171

Ursini, G., Bollati, V., Fazio, L., Porcelli, A., Iacovelli, L., Catalani, A., . . . Bertolino, A. (2011). Stress-related methylation of the catechol-O-methyltransferase Val 158

allele predicts human prefrontal cognition and activity. *Journal of Neuroscience, 31*(18), 6692–6698. doi: 10.1523/JNEUROSCI.6631-10.2011
Viding, E., & McCrory, E. J. (2012). Genetic and neurocognitive contributions to the development of psychopathy. *Developmental Psychopathology, 24*(3), 969–983. doi: 10.1017/S095457941200048X
Viding, E., Price, T. S., Jaffee, S. R., Trzaskowski, M., Davis, O. S., Meaburn, E. L.,... Plomin, R. (2013). Genetics of callous-unemotional behavior in children. *PLoS One, 8*(7), e65789. doi: 10.1371/journal.pone.0065789
Vijayendran, M., Beach, S. R., Plume, J. M., Brody, G. H., & Philibert, R. A. (2012). Effects of genotype and child abuse on DNA methylation and gene expression at the serotonin transporter. *Frontiers in Psychiatry, 3*, 55. doi: 10.3389/fpsyt.2012.00055
Walsh, N. D., Dalgleish, T., Dunn, V. J., Abbott, R., St. Clair, M. C., Owens, M.,... Goodyer, I. M. (2012). 5-HTTLPR-environment interplay and its effects on neural reactivity in adolescents. *Neuroimage, 63*(3), 1670–1680. doi: 10.1016/j.neuroimage.2012.07.067
Warden, M. R., Selimbeyoglu, A., Mirzabekov, J. J., Lo, M., Thompson, K. R., Kim, S. Y.,... Deisseroth, K. (2012). A prefrontal cortex-brainstem neuronal projection that controls response to behavioural challenge. *Nature, 492*(7429), 428–432. doi: Nature11617 [pii] 10.1038/nature11617
Weaver, I. C., Cervoni, N., Champagne, F. A., D'Alessio, A. C., Sharma, S., Seckl, J. R.,... Meaney, M. J. (2004). Epigenetic programming by maternal behavior. *Nature Neuroscience, 7*(8), 847–854.
Weinberg, M. S., Samulski, R. J., & McCown, T. J. (2013). Adeno-associated virus (AAV) gene therapy for neurological disease. *Neuropharmacology, 69*, 82–88. doi: 10.1016/j.neuropharm.2012.03.004
Williams, S. C., & Deisseroth, K. (2013). Optogenetics. *Proceedings of the National Academy of Sciences of the USA, 110*(41), 16287. doi: 10.1073/pnas.1317033110
Witten, I. B., Lin, S. C., Brodsky, M., Prakash, R., Diester, I., Anikeeva, P.,... Deisseroth, K. (2010). Cholinergic interneurons control local circuit activity and cocaine conditioning. *Science, 330*(6011), 1677–1681. doi: 330/6011/1677 [pii] 10.1126/science.1193771
Wojcicki, A. (2013). *23andMe provides an update regarding FDA's review.* Retrieved April 1, 2014, from http://blog.23andme.com/news/23andme-provides-an-update-regarding-fdas- review/
Wu, J., Xiao, H., Sun, H., Zou, L., & Zhu, L. Q. (2012). Role of dopamine receptors in ADHD: A systematic meta-analysis. *Molecular Neurobiology, 45*(3), 605–620. doi: 10.1007/s12035-012-8278-5
Zhang, F., Aravanis, A. M., Adamantidis, A., de Lecea, L., & Deisseroth, K. (2007). Circuit-breakers: Optical technologies for probing neural signals and systems. *Nature Reviews Neuroscience, 8*(8), 577–581. doi: Nrn2192 [pii] 10.1038/nrn2192
Zink, C. F., & Meyer-Lindenberg, A. (2012). Human neuroimaging of oxytocin and vasopressin in social cognition. *Hormones and Behavior, 61*(3), 400–409. doi: 10.1016/j.yhbeh.2012.01.016

INDEX